Handbook of Experimental Pharmacology

Volume 114

The Pharmacology of Alcohol Abuse

Contributors

R.F. Anton, H.C. Becker, B.J. Berg, M.J. Bohn, K.E. Bremner
D.A. Ciraulo, R.K. Fuller, J. Gelernter, K.A. Grant, D. Hersh
G.A. Higgins, P.L. Hoffman, C.M. Knapp, H.R. Kranzler
A. Dzung Lê, S.W. Leslie, R.Z. Litten, J. Littleton, P. Manu
A.T. McLellan, A.K. Mehta, J. Moring, C. Müller, C.A. Naranjo
M.K. Romach, H. Rommelspacher, H.H. Samson, B.F. Sands
E.M. Sellers, W.J. Shoemaker, M.K. Ticku, D.M. Tomkins
J.R. Volpicelli, N.T. Watson, T.M. Worner

Editor

Henry R. Kranzler

Springer-Verlag
Berlin Heidelberg New York London Paris
Tokyo Hong Kong Barcelona Budapest

HENRY R. KRANZLER, M.D.
Associate Professor and Director
Division of Addictive Disorders
Department of Psychiatry, MC2103
University of Connecticut School of Medicine
263 Farmington Avenue
Farmington, CT 06030
USA

With 7 Figures and 8 Tables

ISBN-13: 978-3-642-78437-8 e-ISBN-13: 978-3-642-78435-4
DOI: 10.1007/978-3-642-78435-4

Library of Congress Cataloging-in-Publication Data. The pharmacology of alcohol abuse / contributors, R.F. Anton . . . [et al.]; editor, Henry R. Kranzler. p. cm. – (Handbook of experimental pharmacology, v. 114) Includes bibliographical references and index. ISBN 0-387-57125-6 1. Alcohol–Physiological effect. 2. Alcoholism–Physiological aspects. I. Anton, Raymond F. II. Kranzler, Henry R., 1950– . III. Series. [DNLM: 1. Alcohol, Ethyl–pharmacology. 2. Alcoholism–metabolism. 3. Alcoholism–drug therapy. W1 HA51L v. 114 1995 / QV 84 P536 1995] QP905.H3 vol. 114 [QP801.A3] 615'.1 s – dc20 [615'.7828] 94-28586

Softcover reprint of the hardcover 1st edition 1995

Typesetting: Best-set Typesetter Ltd., Hong Kong

SPIN: 10077716 27/3130/SPS – 5 4 3 2 1 0 – Printed on acid-free paper

Dedicated to
my wife Leah, and our children
Elissa, Seth, and Evan with love

Preface

In recent years alcohol abuse has received increased international attention. Such attention is justified by the enormous negative impact that this disorder has on health, as well as on economic and social well-being. Governmental and other sources of research support have promoted substantial investigation into the negative consequences of alcohol, examining both animal models and human subjects. Consequently, there has been a virtual explosion of preclinical and clinical data, much of it focused on alcohol's pharmacologic effects. However, there have been few efforts to integrate this wealth of new information into a single, manageable volume.

This volume provides an up-to-date, in-depth treatment of the pharmacology of alcohol, particularly as it relates to alcohol abuse. The overriding theme of the book is the interplay between the preclinical and clinical domains. Increasingly, these areas of investigation have served to inform one another, a trend that can be expected to grow with time. The topics covered include the effects of alcohol on biological systems and the impact of medications on those effects. In addition, recent insights obtained from molecular biological investigations are discussed in terms of their relevance for understanding the effects of alcohol at the cellular level and the implications of these for the development of medications. Consideration is also given to important methodological issues that influence the evaluation of medications in the context of clinical trials.

This integration of basic and clinical science is intended to be useful to both investigators and clinicians alike. Scientific and administrative efforts to streamline the transition from the bench to the clinic can be expected to speed development and to reward those who have broad familiarity with the process. The focus of this volume is on the integration and utility of research findings obtained using diverse approaches and how these findings may influence both clinical practice and subsequent investigation.

Farmington, CT, USA HENRY R. KRANZLER

Acknowledgement. This work was supported in part by grants K20-AA00143 (to Dr. Kranzler) and P50-AA03510 from the National Institute on Alcohol Abuse and Alcoholism, Rockville, MD, USA.

List of Contributors

ANTON, R.F., Center for Drug and Alcohol Programs (CDAP), Department of Psychiatry and Behavioral Sciences, Medical University of South Carolina, 171 Ashley Avenue, Charleston, SC 29425-0742, USA

BECKER, H.C., Center for Drug and Alcohol Programs (CDAP), Department of Psychiatry and Behavioral Sciences, Medical University of South Carolina, 171 Ashley Avenue, Charleston, SC 29425-0742, USA

BERG, B.J., Department of Psychiatry, University of Pennsylvania, Treatment Research Center, 3900 Chestnut Street, Philadelphia, PA 19104, USA

BOHN, M.J., Department of Psychiatry, B6/210 Clinical Science Center, University of Wisconsin-Madison Medical School, 600 Highland Avenue, Madison, WI 53792-2475, USA

BREMNER, K.E., Addiction Research Foundation, University of Toronto, 33 Russell Street, Toronto, Ontario, M5S 2S1, Canada

CIRAULO, D.A., Department of Veteran's Affairs Outpatient Clinic, Tufts University School of Medicine, 251 Causeway Street, Boston, MA 02114, USA

FULLER, R.K., Division of Clinical and Prevention Research, National Institute on Alcohol Abuse and Alcoholism, Room 14C-10, Parklawn Building, 5600 Fishers Lane, Rockville, MD 20857, USA

GELERNTER, J., Psychiatry 116A2, West Haven, Department of Veteran's Affairs, Medical Center, 950 Campbell Avenue, West Haven, CT 06516, USA

GRANT, K.A., Department of Physiology and Pharmacology and Comparative Medicine, Wake Forest University, Bowman Gray School of Medicine, Medical Center Boulevard, Winston-Salem NC 27157, USA

HERSH, D., Alcohol Research Center, Department of Psychiatry, MC2103, University of Connecticut School of Medicine, Farmington, CT 06030, USA

Higgins, G.A., Glaxo Unit of Behavioural Psychopharmacology, Division of Biosciences, University of Hertfordshire, Hatfield, Herts., AL10 4AB, Great Britain

Hoffman, P.L., Department of Pharmacology, University of Colorado Health Sciences Center, Campus Box C 236, 4200 East Ninth Avenue, Denver, CO 80262 USA

Knapp, C.M., Department of Veteran's Affairs Outpatient Clinic, Tufts University School of Medicine, 251 Causeway Street, Boston, MA 02114, USA

Kranzler, H.R., Alcohol Research Center, Department of Psychiatry, MC2103, University of Connecticut School of Medicine, 263 Farmington Avenue, Farmington, CT 06030-2103, USA

Lê, A. Dzung, Pre-Clinical Treatment Research Unit, Clinical Research and Treatment Institute, Addiction Research Foundation, 33 Russell Street, Toronto, Ontario, M5S 2S1, Canada

Leslie, S.W., Division of Pharmacology, College of Pharmacy and Institute of Neurosciences, University of Texas at Austin, Austin, TX 78712, USA

Litten, R.Z., Division of Clinical and Prevention Research, National Institute on Alcohol Abuse and Alcoholism, 6000 Executive Blvd., Rockville, MD 20892, USA

Littleton, J., Division of Biomedical Science, Kings College, Manresa Road, Chelsea, London, SW3 6LX, Great Britain Present address: Tobacco and Health Research Institute, University of Kentucky, Cooper and University Drives, Lexington, KY 40546-0236, USA

Manu, P., Medical Services, Long Island Medical Center, The Long Island Campus for the Albert Einstein College of Medicine, Hillside Hospital, Glen Oaks, NY 11004, USA

McLellan, A.T., Department of Psychiatry, Philadelphia VAMC and University of Pennsylvania School of Medicine, Philadelphia, PA 19104, USA

Mehta, A.K., Department of Pharmacology, University of Texas Health Science Center, 7703 Floyd Curl Drive, San Antonio, TX 78284-7764, USA

Moring, J., Alcohol Research Center, Department of Psychiatry, MC-1410, University of Connecticut Health Center, 263 Farmington Avenue, Farmington, CT 06030, USA

Müller, C., Institute of Clinical Chemistry and Biochemistry, Free University of Berlin, Augustenburger Platz 1, D-13353 Berlin, Germany

NARANJO, C.A., Psychopharmacology Research Program, Sunnybrook Health Science Centre (Room E240), Departments of Pharmacology, Psychiatry and Medicine, University of Toronto, 2075 Bayview Avenue, Toronto, Ontario, M4N 3M5, Canada

ROMACH, M.K., Clinical Research and Treatment Institute, Addiction Research Foundation and Department of Psychiatry, University of Toronto, 33 Russell Street, Toronto, Ontario, M5S 2S1, Canada

ROMMELSPACHER, H., Department of Neuropsychopharmacology, Psychiatric Hospital, Free University of Berlin, Ulmenallee 30, D-14050 Berlin, Germany

SAMSON, H.H., Department of Physiology and Pharmacology, Bowman Gray School of Medicine, Medical Center Boulevard, Winston-Salem NC 27157-1083, USA

SANDS, B.F., Substance Abuse Treatment Programs, Department of Veteran's Affairs Outpatient Clinic, 251 Causeway Street, Boston, MA 02114, USA

SELLERS, E.M., Clinical Research and Treatment Institute, Addiction Research Foundation and Departments of Pharmacology, Medicine and Psychiatry, University of Toronto, 33 Russell Street, Toronto, Ontario, M5S 2S1, Canada

SHOEMAKER, W.J., Alcohol Research Center, Department of Psychiatry, MC-1410, University of Connecticut Health Center, 263 Farmington Avenue, Farmington, CT 06030, USA

TICKU, M.K., Department of Pharmacology, Division of Molecular Pharmacology, The University of Texas, Health Science Center at San Antonio, 7703 Floyd Curl Drive, San Antonio, TX 78284-7764, USA

TOMKINS, D.M., Clinical Research and Treatment Institute, Addiction Research Foundation, University of Toronto, 33 Russell Street, Toronto, Ontario, M5S 2S1, Canada

VOLPICELLI, J.R., Department of Psychiatry, Treatment Research Center, University of Pennsylvania School of Medicine, 3900 Chestnut Street/6178, Philadelphia, PA 19104, USA

WATSON, N.T., Department of Psychiatry, University of Pennsylvania, Treatment Research Center, 3900 Chestnut Street, Philadelphia, PA 19104, USA

WORNER, T.M., State University of New York, Health Science Center, 322 East 50th Street, Brooklyn, NY 10022, USA

Contents

CHAPTER 9

Animal Models of the Alcohol Addiction Process
K.A. Grant ... 185

CHAPTER 10

Ethanol and Neurohormonal Regulation
J. Littleton ... 231

CHAPTER 11

Clinical Application of Findings from Animal Research on Alcohol Self-Administration and Dependence

CHAPTER 15

Drugs Attenuating Alcohol Consumption in Humans Through Effects on Various Neurotransmitter Systems

CHAPTER 16

Pharmacology of Gastrointestinal Comorbidity in Alcoholics

CHAPTER 17

Drugs for the Treatment of Psychiatric Comorbidity in Alcoholics: Recent Developments

CHAPTER 19

Interaction of Alcohol with Therapeutic Drugs and Drugs of Abuse

CHAPTER 1

The Pharmacology of Alcohol Abuse: An Introduction

H.R. KRANZLER

Alcohol consumption is virtually universal in the world today and has been so throughout history. Its production and psychoactive effects appear to have been identified serendipitously in all cultures in which it is used, with the exception of those in the Pacific Islands and most of North America, where it was introduced by Europeans (MARSHALL 1979). Alcohol has often been viewed as a balm and medicinal. However, some have seen it as an evil intoxicant responsible for many (if not most) of the ills of modern society. People do not readily conceive of beverage alcohol as a drug. Nonetheless, its complex pharmacologic effects, including a panoply of psychopharmacologic effects, have led people throughout the world to surround alcoholic beverages with a variety of rules and regulations governing their use (MAKELA et al. 1981; MARSHALL 1979). Despite these efforts at control, excessive drinking, with its attendant adverse effects, is widespread.

Recent efforts, including the development of standardized diagnostic interviews designed or adapted for cross-national application (ROBINS et al. 1981, 1988; WING et al. 1990), have made it possible to estimate the prevalence of alcoholism in different cultural contexts (HELZER and CANINO 1992). Lifetime prevalence rates have been shown to vary considerably within and between countries in North America, Europe, and Asia (HELZER and CANINO 1992), an effect that is not adequately explained by demographic differences in the countries studied (HELZER et al. 1990). Lifetime prevalence rates as low as 0.45% and as high as 23% have been reported in community studies that employed a structured diagnostic interview to diagnose alcoholism (HELZER and CANINO 1992). In addition to these overall differences in lifetime prevalence rates, the prevalence of alcoholism has been shown to vary considerably as a function of both sex and age (HELZER and CANINO 1992). Though differences in the conceptualization of alcoholism and in the willingness of individuals to endorse symptoms in response to questioning are important sources of variability in these estimates, there appear to be substantive differences in alcoholism among different cultural groups.

As evident in the title of this volume, the focus of the chapters that follow is on *the pharmacology of alcohol abuse*. Consequently, no effort will be made in subsequent chapters to discuss the sociocultural dimensions of alcohol abuse. However, since there is a substantial and clearly discernible interaction between pharmacologic effects and learning, the nature of this interaction and the psychological and psychiatric consequences of alcohol abuse are discussed in some detail in a number of the chapters that follow.

The term "alcohol abuse" is both widely and variously employed, attesting both to difficulty in defining the term and the importance of the phenomena to which it makes reference. For the purposes of this volume, the term "alcohol abuse" is used interchangeably with the term "alcoholism," which refers to the adverse effects of chronic alcohol consumption. Similarly, "ethanol" and "alcohol" will be used interchangeably.

A review of the pharmacology of alcohol abuse is an ambitious goal, hence it is important to acknowledge the limitations inherent in such an effort. While some of the chapters in this volume review the literature in the areas that they cover more extensively than do others, the focus is generally on *recent developments*. There has not been the expectation that the contributors to this volume provide comprehensive coverage of the pharmacology of ethanol or of alcohol abuse. Suffice it to say, however, that investigation in this area has burgeoned in recent years, so that the volume covers an enormous research literature. For illustrative purposes, Table 1 provides a partial listing of medications that have been used to treat alcoholism, some of which have been used for the treatment of alcohol withdrawal, but the majority of which have been used for the purpose of rehabilitation (i.e., to treat comorbid psychopathology or to reduce alcohol consumption directly). While at the time of this writing there are no widely accepted medications for use in alcoholism rehabilitation, recent trials have shown a number of medications to have substantial promise.

As may be evident in reviewing the volume's table of contents, the topics covered were chosen with an eye to providing equal attention to both preclinical and clinical considerations. In most of the chapters the focus is on one or the other of these domains, though in the majority there is an effort to integrate preclinical and clinical findings. The focus in a number of the predominantly preclinical chapters is on the effects of ethanol at the cellular

Table 1. Medications used in the treatment of alcoholism: a partial listing

abecarnil	cyproheptadine	lysergide (LSD)
acetophenazine	diazepam	mephenoxalone
alprazolam	dipropyltryptamine	mesoridazine
amitriptyline	disulfiram	metronidazole
amoxapine	doxepin	nalmephene
apomorphine	enalapril	naltrexone
atenolol	fenfluramine	nialamide
bromocriptine	fluoxetine	nitrefazole
buspirone	fluvoxamine	perphenazine
carbamazepine	homotaurine	phenaglycodol
carbimide	hydroxyzine	propranolol
chlordiazepoxide	imipramine	thioridazine
clonidine	lithium	thiothixene
clormethiazole	lofexidine	tiapride
		tybamate

or subcellular level, though usually as these relate to organismic (i.e., behavioral) events. Similarly, those chapters that focus predominantly on the behavioral effects of ethanol seek to relate these effects to ethanol's actions on specific neurotransmitter systems. Developments in understanding the molecular genetic mechanisms that underlie many of ethanol's effects on specific neurotransmitter systems occupy a limited role in a number of these chapters as well.

The chapters that follow begin with a review by Moring and Shoemaker (Chap. 2) of the effects of ethanol on neuronal membranes and membrane-bound receptor systems. These authors conclude that the bulk membrane disorder model of ethanol's mechanism of action does not provide an adequate explanation for many of the effects of acute and chronic ethanol administration. They point out that direct effects of ethanol on specific lipid and protein domains of neuronal membranes have also been documented. These actions, as well as others heretofore unidentified, appear to underlie the physiological and behavioral effects of ethanol. The authors point out that a limited number of mechanisms of ethanol's actions may explain these diverse actions.

In the next chapter, Leslie (Chap. 3) reviews the effects of ethanol on voltage-dependent calcium channel function. He begins by reviewing the physiologic roles of calcium channels, but then focuses specifically on L type calcium channels, the type of channel for which ethanol effects are best documented. This channel type, of which there appear to be subtypes that vary in sensitivity to ethanol's effects, has been implicated in the acute and chronic effects of ethanol. He points out that the evidence is particularly strong for the involvement of L type calcium channels in tolerance to and dependence on ethanol. The chapter concludes by relating these effects of ethanol to the effects of barbiturates and benzodiazepines on neuronal calcium channel function, which together provide evidence for calcium mediation of sedative-hypnotic effects.

Subsequent chapters review the effects of ethanol on a variety of neurotransmitter systems, including excitatory amino acids, GABA, catecholamines, serotonin and opioids. Hoffman, in Chap. 4 on excitatory amino acid function, begins with a discussion of the sites of ethanol's actions and the effects of ethanol on receptors for glutamate, the predominant CNS excitatory amino acid. She then focuses specifically on the effects of ethanol on the NMDA receptor, which appears to have one or more specific sites where ethanol is active. This activity is hypothesized to involve the interaction of the coagonists glutamate and glycine. Acutely, ethanol inhibits NMDA receptor function, while chronically it produces "up regulation" of this system. These neuronal effects appear to contribute to the cognitive effects of ethanol, as well as to the development of tolerance to ethanol and withdrawal-related effects (such as seizures).

The chapter by Ticku and Mehta (Chap. 5) reviews the effects of ethanol on GABA-mediated neurotransmission. These authors begin by highlighting the similarity of effects produced by ethanol and sedative-hypnotics, which suggests that a common modulatory system may underlie some of the pharmacologic effects of these diverse agents. GABA neurotransmission (particularly that involving $GABA_A$ receptors) is a likely mechanism for some of these effects, given that both acute and chronic ethanol administration produce a number of effects on this neurotransmitter system. Furthermore, at concentrations associated with its behavioral effects, ethanol potentiates GABA-induced chloride flux, suggesting that some of these effects of ethanol are GABA-mediated. Next there follows a discussion of the ability of inverse GABA agonists (e.g., Ro15-4513) to reverse some of the pharmacologic effects of ethanol. Finally, these authors discuss recent molecular biological aspects of investigation of the effects of ethanol on GABA neurotransmission.

Samson and Hoffman (Chap. 6) review the preclinical literature on the involvement of CNS catecholamines in ethanol self-administration, tolerance, and dependence. These authors point out that acutely ethanol causes dopamine release, particularly in mesolimbic pathways. Behavioral studies suggest that this effect is important in ethanol-induced reinforcement. Following chronic ethanol administration, dopaminergic neurons may also be involved in the development of alcohol withdrawal symptoms. In contrast, while norepinephrine appears not to play a role in ethanol self-administration or reinforcement, this neurotransmitter system appears to modulate the development of behavioral tolerance to ethanol's effects.

The chapter by Higgins and colleagues (Chap. 7) reviews the substantial literature linking changes in serotonergic function with ethanol self-administration, tolerance, and dependence. They also discuss the effects of such changes on other consummatory behavior, particularly as these effects serve to elucidate ethanol-relevant effects. Given the diversity of serotonergic receptors and their widespread distribution in the CNS, the chapter begins with an overview of the anatomy and receptor pharmacology of this system. The authors then discuss the effects of increasing serotonergic function, which in general produces decreases in alcohol consumption and food and palatable fluid intake. A review of the effects of reduced serotonergic function reveals a sparser literature and somewhat less consistent increases in ethanol self-administration and feeding behavior. The authors use these data to examine the hypothesis that the effects of changes in serotonergic tone are mediated by a satiety mechanism, which they argue does not adequately address the empirical findings. They then offer alternative hypotheses, including one in which serotonergic projections serve to oppose forebrain dopamine systems that underlie reward-related behavior. With respect to the involvement of serotonin in the development of tolerance to ethanol, the literature is quite limited. In general, however, decreases in serotonergic function impede the development of tolerance, while increases enhance tolerance development.

Volpicelli and colleagues (Chap. 8) review the literature on opioid mediation of alcohol self-administration. This literature shows that the administration of ethanol enhances opioid receptor activity, particularly in genetically susceptible organisms. Furthermore, low doses of opioid agonists increase ethanol consumption, as does opioid withdrawal. In contrast, moderate-to-high doses of opioids decrease ethanol self-administration, as do opioid antagonists. The authors interpret these preclinical and clinical findings in terms of an opioid compensation model, in which deficiencies in opioidergic activity that derive from genetic or environmental influences are compensated for by alcohol consumption. They acknowledge the limitations that exist in this model and discuss an alternative model proposed by REID and colleagues (1991) that invokes the notion of a surfeit of opioid receptor activity as an explanation for alcohol consumption.

The chapter by Grant (Chap. 9) then summarizes the diversity and utility of a number of animal models of alcoholism, which are organized in terms of stages of change. In contrast to earlier models that were dominated by tolerance and dependence as the basic elements of alcohol dependence, alcohol-seeking behavior is the primary focus of current animal models. For the purposes of this review, Grant operationally defines ethanol self-administration and ethanol-conditioned preferences as the behaviors representative of alcohol-seeking behavior. The four stages that she focuses on are the initiation of drinking, the transition to abuse and dependence (or maintenance of excessive ethanol seeking), the induction of remission, and relapse to excessive alcohol-seeking behavior. Grant argues that animal models are particularly well suited to the study of the initiation of alcohol consumption, given the ethical constraints on experimental ethanol administration to ethanol-naive humans. In the transition to abuse or dependence the focus of animal models is on the apparent qualitative shift in the reinforcing effect of the ethanol stimulus that enables it to exert such potent control over behavior. To address the processes of remission from and relapse to excessive alcohol seeking, Grant turns to the clinical literature to supplement the limited availability of animal models. Throughout the chapter the author endeavors to combine pharmacologic explanations for behavior with psychosocial ones, developing a typology of animal models that highlights the substantial gains that have accrued from them in our understanding of alcohol-related behaviors and the gaps in that understanding that remain to be filled.

The chapter by Littleton (Chap. 10) represents a particularly skillful review of the endocrine effects of ethanol in both animals and humans. This chapter helps make the transition from a preclinical focus to the clinical domain. The chapter begins with a discussion of the effects of ethanol on neurohormonal regulation. The three major neuroendocrine axes (hypothalamic-pituitary-gonadal, hypothalamic-pituitary-adrenal, and hypothalamic-pituitary-thyroid) are then examined in turn, with an emphasis on the effects of ethanol on the function of these systems. These are followed by a brief discussion of ethanol's effects on other neurohormonal systems, most notably that involving growth hormone. Subsequent sections are concerned with the role of neurohormonal systems on alcohol consumption. In this effort, specific emphasis is put on opioid peptides, the hypothalamic-pituitary-adrenal axis, and appetitive systems (under which rubric are included effects of serotonin, renin/angiotensin, and vasopressin, which are linked theoretically and to some degree empirically through their effects on food and fluid intake). With this chapter the author provides a very clear explication of the many, varied (and sometimes conflicting) interactions between ethanol and neurohormones.

The subsequent chapter by Romach and Tomkins (Chap. 11) relates a variety of animal models specifically to the clinical setting. The focus of this chapter dovetails nicely with the earlier chapter by Grant. It begins with a focus on alcohol consumption, which includes an overview of clinical aspects of alcoholism, followed by reviews of animal models and preclinical studies and a somewhat lengthier treatment of clinical studies. Subsequent sections focus on alcohol withdrawal, for which both the preclinical and clinical literatures are examined, and psychiatric comorbidity in alcoholics, concerning which there is a sparse animal literature and a somewhat more substantial clinical literature. Both of these sections briefly cover topics that are dealt with more comprehensively in subsequent chapters. The authors conclude their chapter by emphasizing how preclinical investigation has served increasingly to drive developments in the pharmacotherapy of alcoholism. They highlight means by which the transfer of knowledge from the preclinical to the clinical domain may be enhanced, as well as some of the areas in which clinical trials methodology may be refined; topics that are expanded upon further in the concluding chapter of this volume.

Gelernter, in his chapter (12) on the genetics of alcoholism, carefully reviews the clinical literature that implicates genetics in the etiology of alcoholism. He then goes on to dicuss laboratory (molecular) studies of the nature of the genetic influence on risk for the disorder. Among the clinical studies that he reviews are those based on adoption or involving twins, the effects of family history on treatment outcome in alcoholics, and trait markers of familial risk. The molecular studies are subdivided into linkage studies and association studies. Gelernter also distinguishes between genetic effects on alcoholism risk that are not expressed in brain and those that involve CNS function. He pays particular attention to the controversy surrounding the association between alcoholism and alleles of the gene for the dopamine (D2) receptor. While he concludes that the evidence supporting a genetic contribution to alcoholism is very strong, Gelernter acknowledges that much work remains to be done to identify the molecular mechanisms that may mediate major genetic effects. In this conclusion, the author also suggests how improved knowledge of genetics may lead to new developments in the pharmacotherapy of alcoholism.

Anton and Becker (Chap. 13) provide a comprehensive review of the preclinical and clinical literatures on alcohol withdrawal. They begin with a description of the clinical syndrome, followed by discussion of a variety of related issues, including predictors of severity of the alcohol withdrawal syndrome, protracted alcohol withdrawal, and the relationship between psychiatric pathology and alcohol withdrawal. The authors go on to review the neuropharmacology of alcohol withdrawal, and in this area the chapter overlaps somewhat with previous chapters on GABA, excitatory amino acids, voltage-dependent calcium channel function, and catecholamines. They make the transition to clinical management by discussing the kindling hypothesis of alcohol withdrawal. The final and largest part of the chapter is devoted to a review of treatments for the alcohol withdrawal syndrome.

In the next chapter (14), Fuller and Litten review the literature on the use of aversive agents for the treatment of alcoholism. Given that disulfiram has been in clinical use for more than 40 years, far longer than any other aversive compound, the authors focus their review primarily on this agent. They discuss the pharmacokinetics of the drug, its mechanism of action, clinical use, studies of its efficacy, its toxicity, and issues related to enhancement of compliance with the drug. Based on a large, placebo-controlled, multicenter trial of the medication, the authors conclude that disulfiram may have potential utility, particularly in a subgroup of alcoholics for whom the drug may result in decreased drinking. That study also revealed better drinking outcomes in patients who were compliant with their medication (either disulfiram or placebo). They go on to review a number of studies in which compliance with disulfiram was enhanced through specific interventions, which further highlights the potential utility of this medication for relapse prevention in some alcoholics. The authors also review the limited literature on the use of calcium carbamide, an aversive drug with comparatively rapid onset and brief duration of action. Given this pharmacologic profile, this drug has been used with some success in conjunction with psychosocial treatment. Though aversive agents have received less research attention in recent years, new findings suggest that these drugs, in combination with specific interventions to enhance compliance and efficacy, may have an important role to play in the prevention of relapse in alcoholics.

The chapter by Naranjo and Bremner (Chap. 15) provides a summary of the literature on medications that attenuate alcohol consumption in humans through effects on specific neurotransmitter systems. The preclinical basis for this chapter was provided in the earlier chapters by Higgins and colleagues, Samson and Hoffman, Volpicelli and colleagues, and Ticku and Mehta. Naranjo's considerable contributions to the literature on the effects of serotonin reuptake inhibitors on alcohol consumption in heavy drinkers enable him to review these effects clearly and succinctly. In subsequent sections, the authors review the effects of dopaminergic, opioidergic, and GABAergic medications on drinking in alcoholics. Given the relative paucity of clinical research with these medications, these sections are much briefer than the section on serotonergic medications. However, considerable research interest is now being focused in the United States on opioid antagonists and in Europe on the GABA agonist calcium acetyl-homotaurinate (homotaurine). In the not-too-distant future considerably more data from clinical trials will be available to help evaluate the relative efficacy of these various medications. The authors discuss the recent development of new experimental paradigms to screen the clinical effects and to investigate the mechanisms of action of medications to reduce alcohol consumption. They also describe some recent developments in clinical trials methodology, which are an essential ingredient in the quest for efficacious pharmacotherapies for alcoholism. This topic is discussed in some detail by Kranzler and colleagues in the final chapter of this volume.

Worner and Manu limit their chapter (16) to recent developments in the treatment of gastrointestinal comorbidity in alcoholics. They chose to limit their focus to gastrointestinal comorbidity insofar as their literature search revealed no significant recent pharmacologic advances in the treatment of cardiovascular, neurologic, renal, hematologic, endocrinologic, or metabolic complications of alcoholism. The alcohol-related gastrointestinal disorders that they discuss are hepatic disease (i.e., steatosis, hepatitis, and cirrhosis), acute pancreatitis, and chronic gastritis. Among patients with steatosis, treatment continues to depend on reduced alcohol consumption, rather than any pharmacologic intervention. Among patients with alcoholic hepatitis complicated by encephalopathy, corticosteroids appear to reduce short-term mortality. Among cirrhotics with varices who are at increased risk of bleeding, beta-blockers are effective in preventing the first episode of bleeding. The management of pancreatitis continues to depend on standard symptomatic therapy. In contrast, preliminary results in alcoholics with gastritis complicated by *Helicobacter pylori* infection suggest that a triple antibiotic regimen effectively reduces symptomatology. The authors conclude that despite a number of recent investigations, the pharmacologic treatment of gastrointestinal complications in alcoholics remains limited.

The chapter by Bohn and Hersh (Chap. 17) discusses the assessment and treatment in alcoholics of major depression, anxiety disorders, bipolar affective disorder, and schizophrenia. In each of these areas the authors emphasize the complexity in diagnosing comorbid disorders, since the acute and/or chronic effects of ethanol can minic signs and symptoms associated with these disorders in nonalcoholics. These issues are directly relevant to the validity of results obtained in clinical trials of medications for treatment of comorbid psychiatric disorders in alcoholics. The authors conclude each section with a discussion of pharmacologic treatments, of which there are a growing number that hold promise as specific remedies for comorbid psychiatric disorders. There is a striking paucity, however, of studies that have carefully examined the utility of widely used psychotropic medications for treatment of psychiatric comorbidity in alcoholics. Further study of both cyclic antidepressants and selective serotonin reuptake inhibitors for the treatment of major depression in alcoholics is warranted. In the case of comorbid anxiety disorders, the circumstances under which benzodiazepines are contraindicated remain to be identified empirically. Other anxiolytics, such as buspirone, have shown some promise for generalized anxiety, but further work is needed to guide clinicians. There are no published controlled data on the treatment of panic disorder in alcoholics. The same lack of systematic investigation exists with respect to the treatment of both bipolar disorder and schizophrenia in alcoholics. The authors of this chapter conclude with a view to how future efforts in this area may best be directed.

The chapter by Rommelspacher and Müller (Chap. 18) on clinical markers of alcohol abuse begins with a discussion of the definition of the term "marker." They differentiate between state and trait markers and between intoxication and residual markers. These dichotomies have theoretical significance for the etiology and pathophysiology of alcoholism, as well as practical relevance for the assessment of clinical outcomes in alcoholics. The authors go on to examine the utility of ethanol levels and levels of acetaldehyde, acetaldehyde adducts, and acetate as clinical markers. They also review the literature on monoamine oxidase and certain second messenger systems (including guanine nucleotide binding proteins and adenylylcyclase). This is followed by a discussion of more traditional clinical markers: namely, serum enzymes such as γ-glutamyltransferase and aspartate aminotransferase. The authors then discuss the literature on tetrahydroisoquinolines and β-carbolines, immunologic measures, and serum trace elements (including thiamine). They conclude with a detailed, but concise discussion of the status of carbohydrate-deficient transferrin as a marker of chronic heavy drinking. Though much work has been done to identify clinical markers of alcohol abuse, substantial limitations exist in this literature and considerable work remains to be done in this area. Further investigation in the area of clinical markers is likely to help elucidate the pathophysiology of alcohol abuse and to enhance the validity of treatment trials for the disorder.

The chapter by Ciraulo and colleagues provides a comprehensive review of the interactions of alcohol with both medications ad drugs of abuse. These authors begin with a general discussion of absorption and bioavailability and how these are affected by the interaction of ethanol and drugs. This is followed by consideration of the effects on distribution and metabolism of the interaction between ethanol and drugs. A general discussion is then provided concerning the interaction of ethanol and drugs. This is followed by consideration of the effects on distribution and metabolism of the interaction between ethanol and drugs. A general discussion is then provided concerning pharmacodynamic interactions of ethanol and drugs. The authors then discuss the interaction of ethanol with a wide variety of specific drugs and medications, including acetaminophen, anticonvulsants, antidepressants, anxiolytics, sedative/hypnotics, calcium channel antagonists, antihistamines, antipsychotics, hypoglycemics, and opiates and other drugs of abuse. Given the large number of interactions between ethanol and drugs, and the widespread nature of alcohol consumption, the authors conclude with an expression of concern that these interactions are of great clinical importance.

As is evident in the diversity and depth of the topics covered in this volume, much is known concerning the pharmacology of alcohol abuse.

Developments in preclinical science have begun to elucidate the neuropharmacologic basis of ethanol's effects and increasingly are being applied clinically, with resultant improvements in treatment. However, much remains to be learned. Though the effects of alcohol have been demonstrated in a variety of neurochemical systems, the molecular basis for many of alcohol's effects and the nature of the interactions among these systems remain largely unknown. Similarly, though few would deny a genetic contribution to alcohol abuse, the pathophysiology and the molecular basis for that contribution are as yet undetermined. Uncertainties and lack of understanding are equally prevalent in the clinical domain. The diagnosis of alcohol abuse is based as much on a consensus of opinion, as on scientific knowledge. Though new treatments, both pharmacologic and psychosocial, hold considerable promise for reducing the distress and social disruption attendant upon alcohol abuse, that promise remains incompletely fulfilled. The final chapter of this volume, by Kranzler and colleagues (Chap. 20), discusses issues relevant to medications development, including patient-treatment matching and clinical trials methodology, specifically as it is applied to the treatment of alcoholism.

Much work has been done and the chapters that follow detail recent pharmacologic developments and provide a basis for subsequent investigation. Though considerable work remains to be done, the confluence of a variety of technological and empirical developments suggests that the decade to come may be a watershed period in the pharmacology of alcohol abuse.

References

Helzer JE, Canino GJ, Yeh EK, Bland RC, Lee CK, Hwu HG, Newman S (1990) Alcoholism – North American and Asia. Arch Gen Psychiatry 47:313–319

Helzer JE, Canino GJ (1992) Comparative analysis of alcoholism in ten cultural regions. In: Helzer JE, Canino GJ (eds) Alcoholism in North America, Europe, and Asia. Oxford University Press, New York, pp 289–308

Makela K, Room R, Single E, Sulkunen P, Walsh B (eds) (1981) Alcohol, society, and the state, vol 1. Addiction Research Foundation, Toronto

Marshall M (1979) Introduction. In: Marshall M (ed) Beliefs, behaviors, and alcoholic beverages: A cross-cultural survey. University of Michigan Press, Ann Arbor, pp 1–11

Reid LD, Delconte JD, Nichols ML, Bilsky EJ, Hubbell CL (1991) Tests of opioid deficiency hypotheses of alcoholism. Alcohol 8:247–257

Robins LN, Helzer JE, Croughan H, Ratcliff KS (1981) National Institute to Mental Health Diagnostic Interview Schedule: Its history, characteristics, and validity. Arch Gen Psychiatry 38:381–389

Robins LN, Wing J, Wittchen HU, Helzer JE, Babor TF, Burke J, Farmer A, Jablenski A, Pickens R, Regier DA, Sartorius N, Towle LH (1988) The Composite International Diagnostic Interview: An epidemiologic instrument suitable for use in conjunction with different diagnostic systems and in different cultures. Arch Gen Psychiatry 45:1069–1077

Wing JK, Babor T, Brugha T, Burke J, Cooper JE, Giel R, Jablenski A, Regier D, Sartorius N (1990) SCAN: Schedules for Clinical Assessment in Neuropsychiatry. Arch Gen Psychiatry 47:589–593

CHAPTER 2

Alcohol-Induced Changes in Neuronal Membranes

J. MORING and W.J. SHOEMAKER

A. Introduction

A large number of effects of acute and chronic ethanol exposure on the lipid and protein components of cell membranes have been documented. These effects vary in magnitude and importance. Whether the primary site of action of ethanol is lipid or protein in nature is still unknown. The purpose of this chapter is twofold: first, to review and evaluate the evidence concerning the site of ethanol action; and second, to review the membrane effects of ethanol and to assess their relative importance in producing the characteristic physiological and behavioral effects of ethanol consumption.

Despite extensive study, the primary physiological site (or sites) of action of ethanol is still not known with certainty. Ethanol has anesthetic effects and so, on the basis of mode of action, it has been classed with general anesthetics. Because many substances with general anesthetic properties, including ethanol, are simple small molecules that are structurally unrelated to one another (for example, halothane, butanol, and xenon), a nonspecific mode of action has been considered more probable than action at a specific site. No specific, saturable binding site for ethanol has yet been discovered. Several possible sites of ethanol action on the cell membrane are illustrated in Fig. 1. Ethanol may act at these sites by: (1) disordering the bulk membrane, (2) changing membrane lipid composition, (3) altering local lipid domains, (4) altering membrane protein function, (5) affecting signal transduction proteins and their related systems, (6) changing ion channel function, and (7) regulating receptor subunit expression. Chronic exposure to ethanol leads to tolerance, dependence, and, after withdrawal, a typical set of withdrawal symptoms. We shall discuss membrane effects of chronic ethanol exposure as well as acute effects. At this time, which of the sites of acute ethanol action are responsible for the chronic effects is also not known.

B. Historical Overview

I. The Meyer-Overton Hypothesis

Early evidence suggested that the lipids of the cell plasma membrane were the most likely site of action for ethanol. In 1899, H. Meyer proposed that

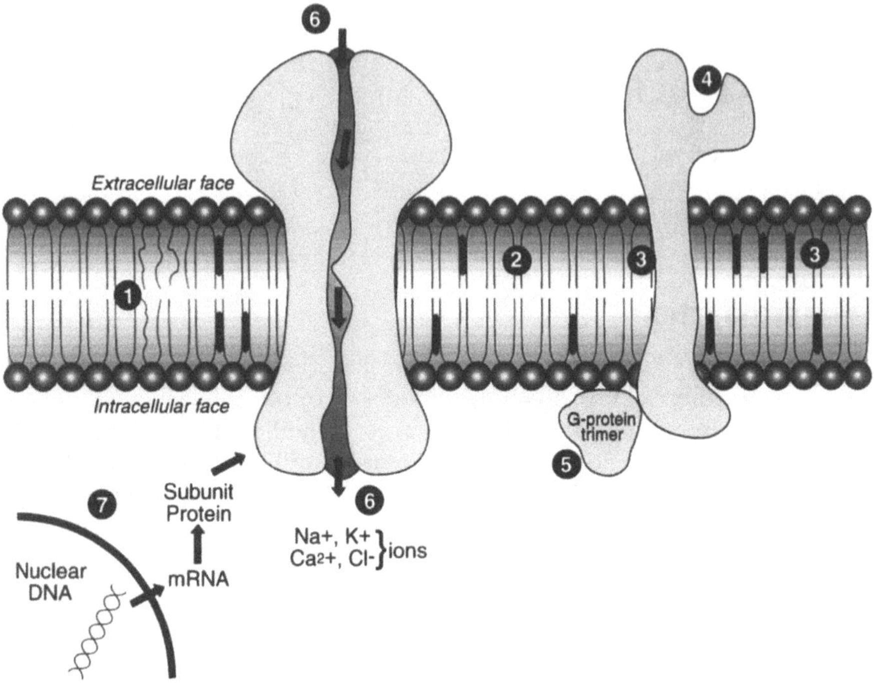

Fig. 1. Sites of ethanol action on the plasma membrane. The numbers in the figure correspond to the physical locations in the membrane or elsewhere in the cell at which ethanol might exert various effects on the membrane: ①, disordering the bulk membrane; ②, changing membrane lipid composition; ③, altering lipid domains, such as transbilayer distribution of cholesterol (moleules depicted as ovals) or the annular lipids surrounding membrane proteins; ④, altering membrane protein function; ⑤, affecting G proteins and their related systems; ⑥, changing ion channel function; ⑦, regulating receptor subunit expression. Some of these effects are acute, e.g., causing membrane disorder; some are chronic, e.g., altering membrane lipid composition, and some can be both acute and chronic, e.g., altering the function of membrane proteins. Acute and chronic effects may occur at the same site, but the mechanisms may be different. Conversely, acute or chronic effects at different sites may occur by similar mechanisms

the anesthetic potency of a substance was correlated with its oil/water partition coefficient. The partition coefficient is a measure of the lipid solubility of the substance. Meyer and his coworkers measured the olive oil/water partition coefficients of several small molecules. His results showed that higher partition coefficients, and thus higher lipid solubility, correlated with greater anesthetic potency (H. MEYER 1899, 1901). In a homologous series of alcohols, according to this principle, the concentration required to produce equal anesthetic effect would decrease with increasing lipid solubility, and thus decrease with increasing chain length. Overton independently and concurrently reached the same conclusion (OVERTON 1896, 1901).

This correlation of the lipid solubility of a substance with its anesthetic potency is now known as the Meyer-Overton rule of anesthesia. In 1937, Meyer's son, K.H. Meyer, suggested that anesthesia ensues when the anesthetic compound has reached a *critical concentration* in the cellular lipids, which would be the same for all anesthetics (K.H. MEYER 1937; LIPNICK 1989). FERGUSON (1939) refined this idea using the concept of thermodynamic activity. According to Ferguson, the activity of a substance rather than its concentration in a bulk solvent should be used as an index of potency. The thermodynamic activity, α_{50}, can be written:

$$\alpha_{50} = P_{50}/P_0$$

for volatile anesthetics, where P_0 is the vapor pressure of the anesthetic at physiological temperature and P_{50} is the partial pressure at which half of the subjects are anesthetized (MILLER 1985). According to this equation, a decrease in α_{50} corresponds to an increase in potency. Using the thermodynamic activity as an index avoids the difficult problem of measuring directly the concentration of the anesthetic substance in the tissues. The phase distribution of the substance in the tissues in vivo is usually not precisely known. Because *at equilibrium* the activities of a substance in all phases in which it is dissolved are equal, the activity in any phase is necessarily the activity at the site of action, whatever that site may be. The activities of a series of alcohols in equipotent concentrations were found to increase over a very narrow range with chain length, and to increase sharply after decanol (FERGUSON 1939). Miller found that activities for a series of alcohols were approximately equal from ethanol through octanol, but a sharp increase began with decanol and continued through tetradecanol. Tetradecanol is not anesthetic (MILLER et al. 1989). Thus these activities were consistent with the "anesthetic cutoff" effect, which is the diminished ability of longer chain alcohols (>12 carbons) to produce anesthesia. Nevertheless, because the activities of some homologous series of anesthetics, such as the alkanols, are so nearly equal, the Meyer-Overton rule appears to remain a better predictor of anesthetic potency.

In 1954, Mullins suggested that the volume of anesthetic in the membrane, rather than its concentration, correlated better with anesthetic potency. This was called the *critical volume* hypothesis. A related hypothesis was based on the ability of ethanol and other anesthetic substances to expand both natural and model membranes (SEEMAN and ROTH 1972). The expansion of biological membranes was thought to be much greater than the volume of the molecules that entered the membrane (SEEMAN 1972). That anesthesia could be reversed by increased atmospheric pressure provided support for the idea that expansion of the neuronal membrane was the basis of anesthetic action (TRUDELL et al. 1973a,b). Pressure reversal of the inhibition of bacterial luminescence by some agents that produce narcosis, including urethane and ethanol, was reported in 1942 (JOHNSON et al.). Pressure reversal of the narcotization of tadpoles was reported in 1950

(JOHNSON and FLAGLER). The model proposed to explain these phenomena was a denaturation of proteins by the narcotic agent through expansion of the protein, with the expansion being reversed by pressure (JOHNSON et al. 1942). In 1971, after experiments using nitrogen anesthesia on newts and mice, LEVER et al. (1971) proposed that anesthetics expand the dimensions of lipids in cell membranes to a critical level, although they did not rule out a protein site. In 1978 a hypothesis was advanced that included the possibility of more than one type of site for expansion (HALSEY et al. 1978). Shortly thereafter, the expansion of the membranes was shown not to be anomalously large, but to correspond to approximately the volume of the molecules in the membrane (FRANKS and LIEB 1981). At about the same time, ALKANA and MALCOLM (1980b, 1981) showed that low levels of hyperbaric pressure antagonized ethanol narcosis in mice, demonstrating in the living animal that increased pressure may reverse the intoxicating effects of ethanol.

II. Membrane Disorder

Disordering, rather than expansion, of membranes was considered another possible mechanism of anesthetic action. This process is usually referred to as "fluidization" of the membrane. Strictly, fluidity is defined as the inverse of viscosity; however, as the term is used in alcohol studies, it is more vague and dimensionless (UEDA 1991). We prefer to call the membrane phenomenon caused by ethanol and anesthetics "membrane disorder." Disordering of erythrocyte membranes by the anesthetic benzyl alcohol was demonstrated in 1968 by METCALFE et al. (1968) using nuclear magnetic resonance (NMR). In 1970, the same effect was demonstrated using electron paramagnetic resonance (EPR) (HUBBELL et al. 1970). Meanwhile, the picture of membrane structure had been evolving, with several models in existence by the 1960s. The fluid mosaic model proposed in 1972 by Singer and Nicholson, in which transmembrane and surface proteins are dispersed through a phospholipid bilayer, has become generally accepted.

Ethanol is different from other general anesthetics in that it is chronically consumed, and chronic consumption results in characteristic physiological effects. The source of these effects at the molecular level was also a subject of intense study. In 1975, Hill and Bangham proposed that one or more changes in membrane lipid composition were the basis for dependence on depressant drugs and the associated withdrawal syndrome. Their hypothesis was based on the membrane disordering effects of the added drug molecules and the possibility that changes in lipid composition could return the affected membrane to normal function in the presence of the perturbing agent. The investigators noted that the behavior of these "general depressants" would be similar to that of general anesthetics. The hypothesis was tested in 1977 by Chin and Goldstein, who added ethanol to membranes in vitro and measured its disordering effect using EPR. They discovered that membranes from normal mice were disordered ("fluidized") in a dose-dependent manner by the addition of physiologically relevant concentrations of ethanol. However, membranes from animals that had consumed ethanol for 8 days were less disordered by ethanol than the membranes from control mice. CHIN and GOLDSTEIN (1977) proposed that this resistance to the disordering effect of ethanol, called *membrane tolerance*, was an adaptation of the membrane that enabled it to maintain its normal function in the presence of ethanol. Again using EPR, this laboratory also showed that the membrane disordering abilities of a series of short-chain alcohols correlated with their membrane/buffer partition coefficients (LYON et al. 1981).

These developments were the basis for the hypothesis that ethanol acts nonspecifically on the lipids of the cell membrane to produce its characteristic physiological and behavioral effects. Since then, many effects of chronic ethanol consumption on membrane lipids have been demonstrated. However, the changes produced in membrane lipids by chronic ethanol administration were generally small. Physiologically relevant concentrations of ethanol in vitro ($\leq 100\,\mathrm{m}M$) sometimes produced rather small membrane-disordering effects. Whether effects on the bulk membrane lipid alone were sufficient to produce the characteristic effects of ethanol was questioned. Consequently, investigations of the effects of acute and chronic ethanol exposure on lipid domains and also on membrane proteins and changes in their expression have increased.

C. Membrane Lipid Effects

I. Disordering of Membranes by Acute Ethanol

In 1970, Hubbell et al. using EPR, investigated the effects of a variety of small molecules and ions on erythrocyte membrane order. Included among the substances tested were neutral anesthetics (including benzyl alcohol), which decreased membrane order, and cholesterol and Ca^{2+}, both of which increased order. In 1977, Chin and Goldstein determined that ethanol disordered various types of membranes from mice. The membranes from different sources had different intrinsic order: mitochondria < synaptosome ≤ erythrocyte < myelin. Synaptosomal and erythrocyte membranes were disordered approximately equally by addition of ethanol in vitro.

In studies of long-term ethanol treatment in animals, intrinsic membrane order was unchanged compared to controls. However, addition of ethanol in vitro to membranes from the exposed animals caused less membrane disorder than addition of ethanol to membranes from controls. The concentrations of ethanol used ranged from 22 to $347\,\mathrm{m}M$; concentrations of $100\,\mathrm{m}M$ (460 mg/dl) or less are most physiologically relevant to ordinary alcohol consumption and to ethanol's anesthetic effects. Even in membranes from ethanol-naive mice, the changes in membrane order due to $87\,\mathrm{m}M$

ethanol were rather small. A study by LYON et al. (1981) compared the hypnotic potencies in mice (determined from the loss of righting reflex) of various aliphatic alcohols (ethanol through octanol) and their abilities to disorder membranes from mouse brain. The study showed that the membrane disordering ability of the alcohols correlated with membrane solubility, as measured by the membrane/buffer partition coefficient. The intramembrane volumes of the alcohols and their intramembrane concentrations correlated equally well with membrane disordering potential, providing no evidence favoring either the critical concentration or critical volume hypothesis of action. The hypnotic potencies of the series of alcohols peaked at hexanol, with potencies for heptanol and octanol much lower than would be predicted from their chain lengths. As we shall see later on, this is a lower "anesthetic cutoff" than that seen in some other studies.

Disordering of erythrocyte and synaptosomal membranes by ethanol in vitro was found to be greater in lines of mice that were bred to be genetically sensitive to ethanol ("long-sleep") than in relatively insensitive ("short-sleep") mice (GOLDSTEIN et al. 1982). The criterion used for ethanol sensitivity was the duration of loss of righting reflex, also called "sleep time." Baseline membrane order in the two groups of mice was the same. Even in a genetically heterogeneous population of mice, those experimentally determined to be more sensitive to ethanol had membranes that were more easily disordered by ethanol. These results were taken to be evidence that the primary mechanism of ethanol's action might indeed be disordering of the bulk membrane lipids. However, recent studies conducted by AVDULOV et al. (1994) using Pyr-C_3-Pyr monomer-excimer fluorescence, as well as diphenylhexatriene (DPH) fluorescence polarization, showed that 100 mM ethanol disordered bulk and annular lipids in both HAS (high alcohol sensitivity) and LAS (low alcohol sensitivity) rats. The membrane sensitivity to ethanol was not significantly different between the two lines (AVDULOV et al. 1994). The effects in this study of ethanol on proteins will be discussed in a later section. Despite the correlation of ethanol's disordering of membrane lipid with ethanol sensitivity in mice, this recent study and other results cast doubt on the validity of the bulk membrane disorder hypothesis.

Selective disordering by ethanol of the hydrocarbon core of synaptic membranes from mice has been demonstrated using fluorescent probes, with myelin being less sensitive than synaptic membranes to ethanol disordering (HARRIS and SCHROEDER 1981). Low concentrations of ethanol (10–20 mM) disordered only the membrane interior; much higher concentrations (333 mM) were required to disorder the membrane surface. This would appear to indicate that ethanol acts deep within the membrane. However, UEDA (1991) asserts that the site of action of anesthetics is at the interface between the hydrophobic and aqueous regions of the cell membrane, because that is where anesthetics bind to the membrane. These two observations are not necessarily inconsistent. Ethanol appears to interact primarily with the membrane surface. Neutron diffraction has been used to show that ethanol preferentially binds to the hydrophilic headgroup region of the membrane lipid bilayer in rabbit skeletal muscle sarcoplasmic reticulum (HERBETTE et al. 1985). No ethanol detectable by this method was located in the more hydrophobic acyl chain region of the bilayer. Partition coefficients of ethanol into the surface and the interior of the membrane of dipalmitoylphosphatidylcholine liposomes have been determined by NMR spectroscopy (KREISHMAN et al. 1985). Different interactions between ethanol and the membrane in these two areas result in an ordering of the membrane surface and disordering of the membrane interior, though surface partition coefficients are approximately three times greater than those for the interior (HITZEMANN et al. 1986). A suggested mechanism for the ordering of the membrane surface by ethanol is hydrogen bonding of ethanol molecules to the P = O groups of the phospholipid headgroups at the lipid-water interface (CHIOU et al. 1991). This would break down the hydrogen-bonded water matrix that supports the bilayer, disordering the membrane interior while at the same time ordering the headgroup region. According to this mechanism, the most effective anesthetic agents would have both hydrophobic and hydrophilic properties; that is, they would be amphiphilic. The location of ethanol at the surface of the membrane is similar to that of many amphiphilic drugs. Some drugs with low partition coefficients do not penetrate the membrane farther than the glycerol backbone region (MORING et al. 1990); thus the action of these and other substances that have relatively low membrane/buffer partition coefficients could be affected directly by the acute administration of ethanol.

Abraham and coworkers have attempted to ascertain mathematically the relative importance of hydrogen bonding in determining the potency of general anesthetics, including the alkanols (ABRAHAM et al. 1991). Using a multiple linear regression method, they quantified the relative importance of five physical properties of the anesthetics. The aqueous potency of the anesthetics for general anesthesia in animals was found to increase with the size of the molecule and decrease with its ability to accept hydrogen bonds. The other properties tested were less important. Compared with the other 46 substances included, ethanol has a relatively small volume and high hydrogen bond accepting ability, which is in agreement with its weak anesthetic potency. According to these results, the anesthetic site of action must be a relatively good hydrogen bond donor, but a relatively poor hydrogen bond acceptor (ABRAHAM et al. 1991).

Thus ethanol added in vitro appears to affect membrane lipids by preferentially disordering the hydrocarbon core, although it interacts primarily with the hydrophilic headgroup region. Its molecular size and hydrogen bonding characteristics are consistent with weak anesthetic properties.

II. Partitioning of Ethanol into Membranes

Partition coefficients of ethanol into natural and model membranes have been determined using several methods. Some of these partition coefficients are shown in Table 1. The octanol/buffer partition coefficients are given for

Table 1. Partition coefficients of ethanol into natural and model membranes

System	Partition coefficient	Comments	Reference
Octanol/water partition coefficient ÷ 5	0.096		MCCREERY and HUNT 1978
Octanol/buffer	0.6		HERBETTE et al. 1986
Sarcoplasmic reticulum	3		HERBETTE et al. 1986
Control rat liver mitochondrial membranes	3.60	Centrifugation	ROTTENBERG et al. 1981
Liver mitochondrial membranes from chronic ethanol rats	1.17	Centrifugation	ROTTENBERG et al. 1981
Control rat brain synaptosomes	1.00	Centrifugation	ROTTENBERG et al. 1981
Synaptosomes from chronic ethanol rats	0.33	Centrifugation	ROTTENBERG et al. 1981
Dipalmitoylphosphatidylcholine (DPPC) membrane surface	0.1630	NMR, 0.350 *M* ethanol	KREISHMAN et al. 1985
DPPC membrane interior	0.0590	NMR, 0.350 *M* ethanol	KREISHMAN et al. 1985
Erythrocyte membranes from human controls	1.18	Centrifugation	STIBLER et al. 1991
Erythrocyte membranes from human alcoholics	0.48	Centrifugation	STIBLER et al. 1991
Phosphatibylcholine (PC) (16:0)	4.3	mol/mol	ROWE 1985
Phosphatidylethanolamine (PE) (16:0)	5.7	mol/mol	ROWE 1985
PE (12:0)	3.0	mol/mol	ROWE 1985

comparison. Three things are immediately clear: (1) the partition coefficient of ethanol is low, (2) the values depend on the type of membrane, and (3) chronic ethanol treatment of the organism reduces the partition coefficient substantially. For comparison, the partition coefficients of MK-801, a noncompetitive inhibitor of the NMDA receptor, into various kinds of natural and model membranes have values of about 200 to a few thousand (Moring, unpublished data); those of the 1,4-dihydropyridine calcium channel agonists and antagonists have values from a few thousand to more than 100000 (HERBETTE et al. 1986). The membrane/buffer partition coefficients of ethanol listed in Table 1 agree well, given that low partition coefficients are technically difficult to determine and allowing for the use of different membrane systems and different methods by the various investigators. The octanol/buffer partition coefficient is obviously not an adequate substitute for the experimentally determined membrane/buffer partition coefficient, although it happens to be closer to the membrane/buffer value in the case of ethanol than for drugs such as 1,4-dihydropyridine calcium channel blockers or beta adrenergic agonists (HERBETTE et al. 1986). The method of calculating the membrane/buffer partition coefficient by dividing the octanol/water partition coefficient by five yields only a crude estimate. Dipalmitoylphosphatidylcholine (DPPC) is a saturated phospholipid; a DPPC model membrane is likely to be intrinsically more ordered and thus more resistant to ethanol partitioning than a phosphatidylcholine with unsaturated acyl chains at the same temperature. The acyl chain length and the nature of the headgroup in model membranes are also important. Phosphatidylcholine (PC) and phosphatidylethanolamine (PE) are the major phospholipid constituents of many types of mammalian membranes. The PC and PE headgroups are electrically neutral at physiological pH; some other headgroup types are charged. PE unmixed with other phospholipids tends to form hexagonal phase instead of bilayers. The lowered partition coefficient of ethanol in natural membranes after chronic ethanol treatment is probably a partial cause of membrane tolerance (Rottenberg et al. 1981). In the rat study cited in Table 1 (Rottenberg et al. 1981), partition coefficients for halothane and pentobarbital, as well as for ethanol, were lowered after chronic ethanol treatment, suggesting the development of cross-tolerance.

III. Pressure Reversal of Acute Effects of Ethanol

As noted above, low levels of hyperbaric pressure [≤12 atmospheres absolute (ATA)] can antagonize the effects of ethanol in mice. The effect is independent of strain or sex (Alkana and Malcolm 1982). Pressure can also exacerbate symptoms of withdrawal from ethanol in mice after chronic ethanol exposure (Alkana et al. 1985b). It appears that pressure is acting as a direct antagonist of ethanol rather than working indirectly, for example, by increasing elimination of ethanol, decreasing body temperature, or changing the distribution of ethanol in the brain (Malcolm and Alkana 1982). Hyperbaric pressure (heliox, ≤12 ATA) decreases sleep time induced by a given amount of ethanol and also increases the brain ethanol concentration at wake-up, compared to controls at the same temperature (Malcolm and Alkana 1982). Behavioral effects of ethanol can also be antagonized by pressure (Alkana et al. 1991). Mice intubated with 2 g/kg ethanol showed a decrease in various aggressive behaviors toward intruders. At 12 ATA heliox (20% oxygen-80% helium), the behavioral inhibition due to ethanol disappeared entirely for all behaviors measured. Although at 8 ATA heliox the partial pressure of oxygen is 0.96 ATA, or more than four times the partial pressure of oxygen in the atmosphere, the pressure effects are not due to the increased amount of oxygen present (Alkana and Malcolm 1980a). The antagonism by pressure of ethanol's effects seems to be no different from pharmacological antagonism of an agonist by an antagonist, in that it is dependent on ethanol concentration and on the amount of pressure applied (Alkana et al. 1985b). These results are consistent with a membrane site of antagonism for ethanol, but do not discriminate between protein or a lipid site.

Pressure has been shown to reverse anesthesia caused by several drugs, including halothane, diazepam, and nitrous oxide, as well as ethanol (Halsey and Wardley-Smith 1975), although the pressures employed in these experiments were much higher than those used in the previously described experiments (68–75 ATA for reversal of ethanol narcosis). As yet the mechanism for pressure reversal of anesthesia is unknown. Mechanisms that have been suggested are exclusion of the molecules of anesthetic from the membrane, or reversal of expansion of the membrane caused by the anesthetic (Miller et al. 1973). If the critical volume hypothesis of anesthetic action is correct, then the relation between pressure and anesthetic effect should not only be linear, but nearly the same for all anesthetics. This has not been found to be the case (Halsey et al. 1978); thus the critical volume hypothesis is probably invalid, and anesthetics probably work at sites other than the bulk membrane lipid. The effect of hyperbaric pressure alone on animals is to increase their activity (Smith et al. 1984; Cattel 1936). Thus it is also possible that anesthetics and hyperbaric pressure exert their effects at different sites, and their observed interaction is entirely indirect.

IV. Membrane Lipid Composition Changes Due to Chronic Ethanol Exposure

1. Phospholipids

Chronic ethanol consumption has been found by several investigators to cause alterations in lipid composition of the bulk membrane. Sun et al. (1984) found that 3 weeks of daily ethanol intubation in rats produced increases in certain acidic phospholipids of brain membranes: phosphatidylserine (PS), phosphatidylinositol (PI), and phosphatidic acid (PA). Guinea pigs exhibited a 50% increase only in PS after ethanol treatment using a liquid diet (Sun and Sun 1983). According to Gustavsson (1990), the magnitude of such changes depends on whether the method of ethanol administration is continuous (e.g., an ethanol-containing liquid diet) or intermittent (e.g., intubation or injection). Intermittent administration produces greater phospholipid alterations. That intermittent administration is likely to produce a higher peak blood ethanol concentration may be the reason for the difference. Ulrichsen and coworkers found no increase in acidic phospholipids after exposure of rats to repeated episodes of intoxication and withdrawal (Ulrichsen et al. 1991). However, they did find a decrease in PI in the animals that had suffered withdrawal seizures. In platelets from human alcoholics, Alling and colleagues found no differences from control in amounts of PI or PS (Alling et al. 1986). Most of the changes reported in total amounts of major phospholipid groups are small. The inconsistency of the results concerning changes in various phospholipid types indicates that alterations in the amounts of major phospholipid groups due to chronic ethanol exposure are highly dependent upon the experimental system and conditions.

The novel acidic phospholipid phosphatidylethanol (PEth) has been reported in several organs, including brain, from rats treated acutely and chronically with ethanol (ALLING et al. 1983, 1984; KOBAYASHI and KANFER 1987). PEth appears shortly after ethanol exposure begins and is degraded after it ends, although the degradation is slow enough that PEth might accumulate in chronically ethanol-exposed tissues (GUSTAVSSON et al. 1991). Its production in the membrane from phosphatidylcholine is catalyzed by phospholipase D. No major membrane effects have yet been reported that can be ascribed to the presence of PEth; however, it has been shown to increase levels of inositol 1,4,5-triphosphate in cultured cells (LUNDQVIST et al. 1993). This may in turn increase levels of intracellular calcium and thus calcium-dependent intracellular systems. Addition of PEth in vitro to model membranes has been shown to increase membrane fluidity (OMODEO-SALE et al. 1991); PEth is a potent promoter of hexagonal phase formation in membrane lipids and thus can destabilize lipid bilayer structures (LEE et al. 1993). PEth also can activate a type of protein kinase C found exclusively in the central nervous system (ASAYOKA 1989). The amount of PEth that can be produced in lymphocytes may be a potentially useful marker. Phospholipase D was activated by a phorbol ester in the presence of ethanol in lymphocytes from males who had both a family history of alcoholism and alcohol dependence. Lymphocytes from most of the alcoholic subjects were found to produce more PEth under these conditions than lymphocytes from controls, though there was some overlap between the populations (MUELLER et al. 1988). Thus a higher potential for PEth production in lymphocytes may be a marker for the presence or risk of alcoholism.

2. Cholesterol

Addition of cholesterol to membrane preparations increases the order of phospholipid bilayers and biological membranes (HUBBELL et al. 1970). Thus membrane cholesterol might be expected to increase as a result of chronic ethanol consumption. Studies of the cholesterol content of membranes after chronic ethanol exposure suffer from the same inconsistency as do those focusing on alterations of major phospholipid groups. Some investigators have reported increases in membrane cholesterol content (ALLING et al. 1982; CHIN et al. 1978; SMITH and GERHART 1982); some have reported decreases (ALLING et al. 1984; HARRIS et al. 1984); and some report no change (WING et al. 1982; WOOD et al. 1990). CHIN et al. (1978) found increased cholesterol in both erythrocyte and brain membranes from mice exposed to ethanol vapors chronically. An intriguing finding in this regard is the identification of an endogenous, 84-amino acid peptide with benzodiazepine-like activity that stimulates cholesterol delivery to the inner mitochondrial membrane (BESMAN et al. 1989). Found in the adrenal medulla, this des-(gly-ile)-endozepine is very similar to a peptide isolated from brain that is believed to be the endogenous benzodiazepine (GUIDOTTI et al. 1983). Since ethanol mimics benzodiazepine action pharmacologically, this finding provides a possible mechanism for ethanol's actions on cholesterol levels in the inner and outer monolayers of the membrane bilayer.

3. Acyl Chain Composition

Because saturated fatty acyl chains pack more closely in the membrane than do unsaturated chains, thus conferring greater intrinsic lipid order, a decrease in the ratio of unsaturated to saturated fatty acyl chains in membrane phospholipids is another likely mechanism for the "membrane tolerance" observed by CHIN and GOLDSTEIN (1977). The effects of ethanol exposure on polyunsaturated acyl chain composition in various tissues have been recently reviewed (SALEM and WARD 1993). Littleton and colleagues found small increases in saturated acyl chains in brain of ethanol-dependent mice, as well as a decrease in polyunsaturated chains (LITTLETON and JOHN 1977; LITTLETON et al. 1979). SUN and SUN (1983) found a small increase in some saturated fatty acids in PC in synaptic plasma membranes from guinea pig brain after chronic ethanol exposure, but some mono-unsaturated fatty acids were decreased and amounts of polyunsaturated fatty acids were unchanged. In PE some mono-unsaturated acyl chains decreased, but some polyunsaturated acyl chains increased. In various tissues from rats, the content of arachidonate, which has four sites of unsaturation, was decreased after 14 days of ethanol exposure (SALEM and KARANIAN 1988). The decrease in heart and brain was less than in liver, aorta, erythrocytes, and platelets. WING et al. (1982) studied mouse erythrocyte membranes, finding decreased polyunsaturated fatty acids and increased saturated fatty acids, which would tend to increase membrane order. However, some mono-unsaturated acyl chains also decreased. Other investigators have found no changes in acyl chain composition (CREWS et al. 1983; SMITH and GERHART 1982). The alterations found in fatty acid composition of bulk membrane phospholipids in brain were small, the largest being about 20% and most being much smaller.

When fatty acid composition of particular phospholipid classes is determined, the changes can be much larger. In rat liver, arachidonate decreased by 64%, 53%, 49%, and 16% in phosphatidylserine (PS), phosphatidylinositol (PI), PC, and PE, respectively, after 14 days of ethanol inhalation. The total lipid extract from liver, however, showed only a 20% decline in arachidonate (SALEM and KARANIAN 1988). The investigators speculate that this loss may be due to stimulation of a phospholipase A_2 that is specific for polyunsaturated fatty acids such as arachidonate. Enzymatic oxygenation of these polyunsaturated lipids to biologically active compounds such as prostaglandins and leukotrienes could be the basis for some physiological effects of acute and chronic ethanol consumption (SALEM and KARANIAN 1988).

These substances modulate a large variety of physiological processes, including, for example, calcium homeostasis and inflammation. These investigators propose that acute ethanol exposure leads to increased freeing of certain polyunsaturated fatty acids from the membrane lipids. This excess fatty acid would be oxidized to prostaglandins and leukotrienes. Chronic ethanol exposure would have the opposite effect, leading to a deficiency of these compounds.

Though changes in bulk lipid composition due to chronic alcohol consumption are small and variable, TARASCHI et al. (1986) have found that certain hepatic microsomal membrane lipids from ethanol-tolerant animals could confer membrane tolerance on lipids from normal animals. When PI from tolerant rats replaced PI in lipids from normal rats, the reconstituted lipid membranes became resistant to disordering by ethanol (TARASCHI et al. 1986). These investigators also showed that rats that had been withdrawn from ethanol and lost membrane tolerance reacquired membrane tolerance in less than half the time required for ethanol-naive rats to acquire it (TARASCHI et al. 1990). The rapidity of reacquisition of tolerance suggested that considerable time is required for the enzymes that modify PI to revert to their pre-ethanol state. More recently, Rottenberg and coworkers have shown that not only PI, but also other major phospholipid classes are able to confer membrane tolerance (ROTTENBERG et al. 1992).

4. Do Lipid Composition Changes Cause Tolerance or Change Membrane Function?

Whether phospholipid changes such as these were the primary cause of tolerance and other chronic ethanol effects in vivo was the subject of much discussion in the early 1980s. The hypothesis that changes in the phospholipid composition of mitochondrial membranes were ultimately responsible for the abnormalities in mitochondrial function due to chronic ethanol consumption was proposed in 1981 (ROTTENBERG et al. 1981; WARING et al. 1981). GORDON et al. (1982) found that chronic ethanol-related depression of mitochondrial function (as reflected by oxygen consumption) was not correlated with the degree of membrane order. They also found that membrane order in chronically ethanol-fed rats was different from that in controls fed rat chow ad libitum, but not from that in pair-fed controls.

Other investigators proposed that a learning component was essential, at least in part, for the production of behavioral tolerance. Rats did not develop tolerance to the effects of ethanol on a task (walking a treadmill in a straight line to avoid footshock) unless they were given practice on the task while intoxicated (WENGER et al. 1981). However, many other experiments had shown that learning was not required for the development of tolerance (TABAKOFF et al. 1984). TABAKOFF et al. (1984) used the terms "environment-dependent" and "environment-independent" to describe types of tolerance requiring and not requiring learning, respectively, and mentioned that the latter type requires a higher level of ethanol exposure. Whether environment-dependent and environment-independent tolerance have the same origin and mechanism is not known. These findings underscore the importance, when discussing "tolerance," of defining exactly what is meant by the term in the context in which it is being used. "Membrane tolerance" appears to be a different phenomenon from "behavioral tolerance," though they may be causally related.

Further information on the mitochondrial response to ethanol has been gleaned from studies on the formation of fatty acid esters of ethanol in liver (LANGE 1991). Because the enzymes responsible for this effect also occur in brain, apparently primarily in gray matter, this mechanism appears to be active in neurons in human brain as well (BORA and LANGE 1993). Fatty acid ethyl ester synthase, which may be a membrane-associated protein, catalyzes the nonoxidative esterification of ethanol with free fatty acids. The esters migrate to mitochondrial membranes, where they are de-esterified to produce toxic free fatty acids. These fatty acids, which impair mitochondrial function, may also be the immediate cause of ethanol-related brain damage. The esters themselves disorder membranes approximately ten times more potently than does ethanol (HUNGUND et al. 1988). The fatty acid ethyl esters also appear to inhibit protein synthesis (LAPOSATA and LANGE 1986), so this pathway could indirectly mediate many effects of ethanol.

In summary, the effects of chronic ethanol exposure on the bulk membrane lipid are in general moderate, and variable among experimental paradigms. Thus, adaptation by means of alterations in the composition of the bulk membrane lipid is unlikely to be the primary cause of *physiological tolerance* to and dependence on ethanol, although it may be a major cause of *membrane tolerance*.

V. Effects of Ethanol on Membrane Lipid Domains

Ethanol exerts a more pronounced effect on certain membrane lipid domains than it does on the bulk membrane. We are using a broad definition of *domain* that includes not only three-dimensional contiguous blocks of the membrane, but also membrane components such as the inner and outer membrane monolayers, the lipids immediately surrounding membrane proteins, and various membrane lipid classes. Because there may be highly specific actions of ethanol on some parts of the lipid bilayer, the effects of ethanol on a lipid domain may actually be much more profound than a measurement of properties of the bulk membrane lipid would indicate.

1. Lipid Classes

Ethanol may act on certain types of lipids in the membrane while leaving others relatively unaffected. The effects of ethanol on fatty acid composition of various phospholipid classes have been discussed in the previous section.

Ethanol can also modulate physical properties of membrane lipids. Rowe studied the effect of alcohols on the gel-to-liquid crystalline phase transition of phosphatidylcholine (PC) and phosphatidylethanolamine (PE), following the transitions by measuring the change in absorbance at 400 nm (ROWE 1985). This transition involves melting from the more ordered gel state to the more fluid liquid crystalline state. The transitions for PC were not thermodynamically reversible at high concentrations of alcohol, while the transitions for PE were reversible at all concentrations. A thermodynamically reversible process is one in which the final state of the substance depends only on the final conditions of temperature and pressure, and not on the path used to arrive at the final state. In the case of PC, an interdigitated gel phase in which the acyl chains from the inner and outer monolayers of the bilayer fully interpenetrated formed at high ($\geq$130 m*M*) concentrations of ethanol (SIMON and MCINTOSH 1984; ROWE 1985). Because of the formation of this phase, the transition was not thermodynamically reversible (ROWE 1985). This effect did not occur in PE. Though these ethanol concentrations are too high to be physiologically relevant, these experiments are evidence that ethanol can differentially affect specific classes of lipids.

2. Transbilayer Lipid Distribution

Cholesterol is a sterol that is asymmetrically distributed between the inner and outer monolayers of cell membranes; the extent and direction of the asymmetry varies with cell type (WOOD and SCHROEDER 1988). It is a major constituent of neuronal membranes; synaptoneurosome preparations from rat cerebral cortex have a cholesterol/phospholipid mole ratio of ~0.6 (MORING et al. 1990). The mole ratio in rat synaptic plasma membranes from whole brain is 0.82 (CREWS et al. 1991). In experiments that measured the transbilayer distribution of sterol in synaptic plasma membranes from mouse brain, chronic ethanol caused a shift toward a more symmetrical transbilayer sterol distribution, although it did not produce a change in the total amount of membrane cholesterol (WOOD et al. 1990). This shift was measured using dehydroergosterol, a sterol that both behaves similarly to cholesterol in the membrane and fluoresces. By measuring fluorescence from both monolayers and from the outer monolayer alone, the investigators could determine the transbilayer distribution of dehydroergosterol. The mechanism of the sterol shift is not yet understood, but it is known that membrane cholesterol exists in different pools, for example, exchangeable and nonexchangeable cholesterol (WOOD et al. 1991b). Ethanol may change the transbilayer distribution of cholesterol by modifying these pools (WOOD et al. 1991b). To investigate effects of chronic ethanol on the exchangeable cholesterol pool, radiolabeled cholesterol was exchanged between phospholipid/cholesterol vesicles and synaptosomes from chronically ethanol-treated and pair-fed control mice. The rate constant for the exchange was slower and the time required for half of the exchange was longer in the synaptosomes from ethanol-treated mice, though the size of the exchangeable pool (~50%) was not altered (WOOD et al. 1993). Both the presence of cholesterol and an optimal membrane fluidity are known to be required for proper functioning of the acetylcholine receptor (MCNAMEE and FONG 1988). Possibly, an altered transbilayer distribution of cholesterol or kinetics of cholesterol exchange could affect the function of cholesterol-sensitive receptors, even without a change in total membrane cholesterol content. Wood and colleagues have found that chronic ethanol consumption also alters the transbilayer distribution of PC in erythrocyte membranes from miniature swine without affecting the total PC content (WOOD et al. 1991a). PC is preferentially distributed in the outer monolayer of the membrane, and the effect of ethanol is to increase that asymmetry. The increase in PC in the outer monolayer may cause the increased erythrocyte size observed in alcoholic patients. No change in PE or PS was detected in these experiments, but the sphingomyelin content of the membrane was slightly reduced.

These changes in lipid and sterol distribution may change the characteristics of the membrane by altering either the charge distribution or the degree of order of the membrane monolayers, or both. Ethanol acts more strongly on the more intrinsically disordered monolayer of the membrane than on the more rigid one (WOOD et al. 1991b). In mouse synaptic plasma membranes, the outer monolayer is the more disordered. The outer monolayer of these membranes was disordered by 25 m*M* ethanol in vitro, while the inner monolayer was unaffected (SCHROEDER et al. 1988).

3. Annular Lipids

Another membrane domain that could be differentially affected by either acute or chronic ethanol is the lipids immediately surrounding membrane proteins. These are known as the annular lipids or boundary lipids. Unfortunately, measurements of the composition and properties of annular lipids in situ in biological membranes are difficult. Although measurements of disordering of annular lipids by ethanol have been made (AVDULOV et al. 1994), most information about annular lipids has been gained by reconstituting purified membrane proteins in model membranes of various compositions and measuring protein function in these varied environments (LEE 1988). A certain range of membrane fluidity is necessary for proper function of the acetylcholine receptor (MCNAMEE and FONG 1988), but in general a precise degree of membrane disorder seems not to be required for proper function of most proteins (LEE 1988). However, the annular lipids may not have the same degree of disorder as the bulk membrane. The annular lipids have been postulated to be in the gel, or crystalline, state to help the protein keep its proper working conformation (LEE 1976). If the annular lipids were to enter the more disordered, liquid-crystalline state, the conformation of the protein might change, and thus its ability to function might be altered (LEE 1976).

Other evidence shows that the annular lipids may be more disordered than the bulk membrane. Membrane proteins can be cross-linked by dinitrodifluorobenzene to the aminophospholipids immediately surrounding them. Analysis of the cross-linked lipids has shown that some polyunsaturated lipids are preferentially associated with membrane proteins (Abood et al. 1977). Acute ethanol exposure might cause these lipids to be redistributed more randomly through the membrane, yielding an increase in disorder (Salem and Karanian 1988).

4. Lateral Membrane Domains

Treistman and colleagues provided additional evidence that the selective action of ethanol is on lipid domains rather than on the bulk lipid (Treistman et al. 1987). These investigators examined the effects of alcohols and temperature on lipid mobility in neuronal membranes from *Aplysia*. Using the technique of fluorescence recovery after photobleaching to measure diffusion of lipids labeled with fluorescent probes, they found that the two fluorescence probes they used diffused at different rates in the membrane and were differently affected by alcohols. Because the two probes report on different sets of lipid domains in the membrane, these experiments demonstrated that some membrane domains are affected more strongly than others. Further, the probes were not affected equally by temperature changes. These results show that measurement of the viscosity of only the bulk membrane could seriously underestimate the effects of ethanol on the viscosity of a particular membrane lipid domain.

No differences were found in lipid diffusion in various regions of the *Aplysia* neuronal membrane (axon, axon hillock, and cell body), and experiments to compare diffusion in each of these regions in the presence of ethanol were not conducted (Treistman et al. 1987). Fluorescence techniques have been used to demonstrate the existence of a diffusion barrier between axonal and somatodendritic domains of hippocampal neurons (Kobayashi et al. 1992). The barrier extends across both monolayers of the membrane. The two domains separated by the barrier have unlike compositions and could possibly be differentially affected by ethanol.

VI. Ethanol-Induced Hypothermia

Ethanol causes a lowering of body temperature in a variety of species, including mice and rats. The extent of hypothermia caused by ethanol is dependent on the species and the genotype (Tabakoff et al. 1980; Crabbe 1983). Body temperature during intoxication appears to influence the extent of some behavioral and physiological effects of ethanol (Pohorecky and Rizek 1981; Finn et al. 1986). The direction of this effect of hypothermia depends on the species and on the parameter being measured (Pohorecky and Rizek 1981; Finn et al. 1990). Hypothermia contributes to an increase in the duration of loss of righting reflex in some strains of mice (Finn et al. 1990), while it appears to decrease ethanol-induced depression of gross motor activity in rats (Pohorecky and Rizek 1981). During ethanol intoxication, the body temperature of the animal can be manipulated by changing the environmental temperature (Grieve and Littleton 1979; Pohorecky and Rizek 1981; Alkana et al. 1985a). Length of "sleep time" after a dose of 4.2 mg/kg ethanol varies among mouse strains. For LS (long sleep, ethanol-sensitive) mice sleep time is nearly 300 min, while SS (short sleep, ethanol-resistant) mice "slept" for less than 20 min. Body temperatures of the seven strains of mice tested dropped by as much as 5°C at 22°C, with the average drop being about 3°C. When hypothermia due to ethanol administration was offset by raising the temperature of the environment from 22°C to 34°C, the differences between the ethanol-sensitive and ethanol-resistant strains were much smaller (Finn et al. 1990). The mechanism of ethanol-induced hypothermia remains undetermined and may or may not be membrane-related. However, ethanol-induced hypothermia could affect the function of membrane proteins either directly or by perturbing surrounding lipids, or perhaps both. Especially for in vivo measurements of the behavioral effects of ethanol, differences in body temperature between ethanol-treated and control animals are an important experimental factor.

VII. Lipid Effects on Proteins

Effects of acute and chronic ethanol exposure on membrane lipids have been well documented. Indeed, such effects were a major focus of alcohol research in the 1970s and early 1980s. Whether membrane lipid effects are the primary cause of the characteristic effects of ethanol, however, remains to be decided. It is possible that some lipid alterations might lead to changes in the function of enzymes or other proteins. Na^+/K^+-ATPase, adenylyl cyclase, and some glutamate receptors have been shown to be activated by phosphatidylserine (PS) in vitro (Floreani et al. 1981; Floreani and Carpenedo 1987; Foster et al. 1982). The increase in Na^+/K^+-ATPase activity (53%) after chronic ethanol administration may be a secondary effect of an increase in PS (Sun et al. 1984; Sun and Sun 1983). In some tissues, however, adenylyl cyclase activity is stimulated by ethanol added in vitro (Saito et al. 1987), but reduced after chronic ethanol exposure (Hoffman and Tabakoff 1990). This would be inconsistent with activation of adenylyl cyclase by increased PS, so some other mechanism must also be at work. In addition, effects of ethanol on annular lipids could lead to changes in membrane protein function. Of course, these effects may be due to direct action of ethanol on enzymes themselves, or on other proteins that regulate enzymes. These possibilities will be discussed in the next section.

D. Membrane Protein Effects

In recent years, the pendulum has swung away from lipid research toward investigations of the effects of ethanol on membrane proteins. Either lipids or proteins could provide the relatively nonpolar site required for ethanol action in the membrane, and the Meyer-Overton rule might apply both to the binding of anesthetics of proteins and to lipids (FRANKS and LIEB 1990). Three types of direct anesthetic action on proteins are possible: (1) competition of the anesthetic for the specific binding site of a ligand, (2) action on a hydrophobic portion of the protein that is not accessible to membrane phospholipids, and (3) action at the protein-lipid interface (AKESON and DEAMER 1991). The first mode of action could alter the function of the protein without necessarily changing its conformation. The second type could cause an allosteric alteration of the agonist site. The third type could alter the relationship of the protein to membrane lipids that are essential for its function.

I. Direct Effects of Acute Ethanol on Proteins

Though some proteins have been shown to be acutely affected by ethanol and other anesthetics, many are not. One that appears not to be affected, at least at low ethanol concentrations, is Na^+/K^+-ATPase. In erythrocytes from humans consuming moderate amounts of ethanol this enzyme was not inhibited (PUDDEY et al. 1986), although high concentrations of ethanol inhibited it in vitro in erythrocytes and brain tissue (ISRAEL et al. 1965; ISRAEL 1970; NHAMBURO et al. 1987). A recent study has shown a direct effect of acutely administered ethanol on protein distribution in synaptic plasma membranes (SPM) from rats (AVDULOV et al. 1994). At physiologically relevant concentrations (50 and 100 mM), ethanol induced what was interpreted to be protein clustering in SPM from HAS (high alcohol sensitivity) but not LAS (low alcohol sensitivity) rats, suggesting a genetic foundation for increased ethanol sensitivity based on membrane proteins rather than lipids.

1. Firefly Luciferase

An enzyme that *is* affected directly by ethanol in vitro is firefly luciferase. Franks and Lieb showed that anesthetics at surgical concentrations, and also alcohols, inhibit the firefly enzyme luciferase (FRANKS and LIEB 1984). This enzyme has been used extensively as a model system for the study of the action of anesthetics on proteins. The anesthetics bind to the enzyme and compete with the substrate, luciferin, for its binding site; thus this is an example of the first type of possible anesthetic action. Because the luciferase was purified and no lipids were present, this was a case of direct action of anesthetics on a protein to inhibit its function. These investigators propose an amphiphilic protein pocket site of anesthetic action. The same laboratory investigated the effects of pressure on inhibition of luciferase by anesthetics (Moss et al. 1991). No pressure reversal was found, though a variety of anesthetics were used. The investigators concluded that either luciferase inhibition is an inadequate or incomplete model of the anesthetic site, or that pressure and anesthetics act at different sites.

2. The $GABA_A$ Receptor

Ethanol action on the $GABA_A$ (γ-aminobutyric acid)$_A$ receptor complex, however, does not appear to occur at the GABA binding site. The $GABA_A$ receptor complex consists of several types of peptide subunits (α, β, γ and δ) that form a ligand-gated Cl^- channel (OLSEN and TOBIN 1990); it binds GABA (the endogenous agonist), muscimol, benzodiazepines, steroid anesthetics, and antagonists such as the β-carbolines and picrotoxin (KNAPP et al. 1990; IM et al. 1990). This receptor complex is the major target of GABA in the central nervous system. Potencies of a series of alkanols for inhibition of binding of a convulsant drug, TBPS (*t*-butylbicyclophosphorothionate), to the $GABA_A$ receptor complex do not correlate well with potencies for disordering the membrane, which suggests a protein site rather than a bulk lipid site of alcohol action for this inhibition (HUIDOBRO-TORO et al. 1987). However, the anesthetic potencies of the series of alkanols correlated well with their potential for increasing Cl^- flux through the $GABA_A$ receptor-associated chloride channel (HUIDOBRO-TORO et al. 1987). This result could support either a protein or lipid site of action.

Ethanol (20–100 mM) has been shown to potentiate GABA-dependent Cl^- flux through the $GABA_A$ receptor-associated ion channel in rat cerebral cortex (SUZDAK et al. 1986), and also to potentiate muscimol-stimulated Cl^- flux in some regions of mouse brain (ALLAN and HARRIS 1986, 1987). Although Cl^- flux was enhanced by ethanol, the binding of GABA was unchanged (HUIDOBRO-TORO et al. 1987). According to these results, it is unlikely that the mechanism of ethanol's effects involves direct action at the GABA binding site (AKESON and DEAMER 1991). However, other investigators have not observed the enhancement of GABA-dependent Cl^- flux by ethanol at physiologically relevant concentrations (10–40 mM) in the absence of muscimol stimulation (MIHIC et al. 1991). A high concentration of ethanol (600 mM) was required to potentiate GABA-mediated Cl^- flux in cerebrocortical and cerebellar microsacs from mice and rats (MIHIC et al. 1991). At low ($\leq$100 mM) concentrations of ethanol, binding parameters of two benzodiazepines, the $GABA_A$ receptor agonist flunitrazepam and the antagonist flumazenil (RO15-1788), were unchanged (GREENBERG et al. 1984). Only at high concentrations (250–1000 mM) was the affinity of the binding site for flumazenil decreased, while the affinity for flunitrazepam was still unchanged (QUINLAN and FIRESTONE 1992). This argues against the direct action of ethanol at the benzodiazepine binding site at physiologically relevant concentrations.

Optical isomers of some anesthetics affect ion channel currents differently. In certain anesthetic-sensitive neurons from the pond snail *Lymnea stagnalis*, isoflurane activates a potassium current (Franks and Lieb 1988). This current persists as long as the anesthetic is present. Of the two stereoisomers, (+)-isoflurane was found to be about twice as effective as (−)-isoflurane at current activation (Franks and Lieb 1991). (+)-Isoflurane is also more effective at inhibiting K^+ currents elicited by application of acetylcholine to acetylcholine-sensitive neurons. However, both stereoisomers produced equal disorder in lipid bilayers. These results constitute evidence that the anesthetic is affecting the protein directly and is not working through a disordering of the bulk membrane. Disruption of the annular lipids, however, is not ruled out by these findings.

II. Protein Model of the Anesthetic Cutoff Effect

The anesthetic cutoff effect can be accounted for by either the lipid or protein site model. Miller and colleagues found that although tadpoles took up into their tissues both dodecanol (an anesthetic) and tetradecanol (a nonanesthetic) equally well, tetradecanol did not increase lipid disorder, while dodecanol did (Miller et al. 1989). Alkanols with chains longer than dodecanol's increased membrane order and were not anesthetic. This model could explain the anesthetic cutoff effect, if anesthesia is based on the disordering of membrane lipids. The anesthetic cutoff could also be due to inability of longer-chain alcohols to enter the membrane. The anesthetic cutoff in animals occurs at about tridecanol (Alifimoff et al. 1989). Franks and Lieb measured partition coefficients of alcohols from decanol through pentadecanol, using a method based on inhibition by the alcohols of emission of light by luciferase (Franks and Lieb 1986). The partition coefficients increased with chain length throughout the series. It was proposed that the lack of anesthetic activity of the longer chain alcohols was due, not to inability to enter the membrane, but rather to inability to interact with the protein. This model could explain the anesthetic cutoff effect, if anesthesia results from a direct interaction of the anesthetic with protein. Recent experiments on isolated frog root ganglion neurons using the patch clamp technique showed that alkanols with a molecular volume greater than 46 ml/mol (1-butanol, 1-pentanol) were unable to attenuate an ATP-activated membrane ion current. Alkanols with molecular volumes of 43 ml/mol or less (ethanol, 1-propanol) inhibited the current in a concentration-dependent manner, with potency correlating with higher lipid solubility (Li et al. 1993). The investigators attributed the ineffectiveness of the higher molecular volume alkanols to their inability to interact with the channel protein.

For the direct interaction of alcohols with luciferase, the size of the active site has been estimated. The anesthetic binding site of firefly luciferase can bind two molecules of hexanol but only one molecule of *n*-alcohols larger than heptanol (Franks and Lieb 1984). The cutoff for inhibition of luciferase by alcohols is after hexadecanol (FRANKS and LIEB 1985). Measurements of the concentration of various alkanols upon loss of righting reflex in tadpoles showed that the anesthetic cutoff in vivo in this species was at tridecanol (ALIFIMOFF et al. 1989). This has been interpreted to mean that the size of the anesthetic binding site in tadpoles is smaller than that in the luciferase model (ALIFIMOFF et al. 1989; FRANKS and LIEB 1985). It must be kept in mind that inhibition of luciferase is different from anesthesia. Although luciferase is affected by anesthetics in a physiologically reasonable way, it may be an inadequate model of the anesthetic site (ALIFIMOFF et al. 1989). Even if that is so, other evidence supports the protein site theory. A_2C (2-[2-Methoxyethoxy]-ethyl 8-[cis-2-*n*-octylcyclopropyl]-octanoate), a large, highly lipid soluble molecule, perturbs membrane lipids but does not produce anesthesia (BUCK et al. 1989). Given the molecule's large size, this result is consistent with the protein pocket model, but it is inconsistent with the lipid disorder model of anesthesia.

III. Effects of Ethanol on Calcium Channels

Many other effects of ethanol on membrane proteins have been documented, and some of the most interesting are related to calcium. Ethanol added in vitro blocks calcium entry into cells (MESSING et al. 1986; SKATTEBØL and RABIN 1987; DANIELL and LESLIE 1986). However, only $^{45}Ca^{2+}$ uptake that occurs on depolarization of the membrane is thought to be affected (HARRIS and HOOD 1980; LESLIE et al. 1990). Two phases of Ca^{2+} influx take place: the fast phase occurs within 3 s of depolarization; the slow phase occurs some time later (LESLIE et al. 1983a). Lower, physiologically relevant concentrations of ethanol preferentially inhibit the fast phase; higher concentrations inhibit the slow phase (DANIELL and LESLIE 1986; LESLIE et al. 1983b). Not all modes of calcium influx are affected. Dihydropyridine-type calcium channel blockers, which are specific for the L-type (voltage-dependent) calcium channel, were found not to alter $^{45}Ca^{2+}$ uptake into rat brain synaptosomes (DANIELL et al. 1983). This could mean either that influx through these channels is an insignificant fraction of the total $^{45}Ca^{2+}$ influx, or that the dihydropyridine binding site and the ion channel are not coupled in brain. The NMDA receptor-associated ion channel, on the other hand, was inhibited by acutely administered ethanol; [^{3}H]-MK-801 binding sites, thought to be located on the inner surface of the NMDA-gated channel, were upregulated following chronic ethanol exposure (GRANT et al. 1990; SHOEMAKER et al. 1992). Ethanol effects on voltage-dependent calcium channels are discussed by Leslie in Chap. 3 of this volume; ethanol effects on NMDA receptor function are discussed by Hoffman in Chap. 4.

Because $^{45}Ca^{2+}$ influx is inhibited by the acute administration of ethanol, chronic ethanol consumption might be expected to result in adaptive effects on calcium channels. One of these effects, which appears to be of considerable importance, is the upregulation of the dihydropyridine binding site. This site is associated with a voltage-regulated (L-type) calcium channel (DOLIN et al. 1986). The mechanism of the upregulation is unknown. If extra ion channels are present, they may be at least partly responsible for the production of withdrawal seizures. In 1983, Lynch and Littleton presented evidence that tolerance to the inhibition by ethanol of dopamine release was related to increased sensitivity of the presynaptic terminal to calcium (LYNCH and LITTLETON 1983). Nitrendipine, a dihydropyridine-type calcium channel blocker, prevented the withdrawal syndrome in mice when given together with ethanol during chronic ethanol exposure (WHITTINGTON and LITTLE 1988). A single dose of nitrendipine given before an anesthetic dose of ethanol increased the ethanol's anesthetic potency (DOLIN and LITTLE 1991) and nitrendipine given after withdrawal abolished spontaneous seizures (LITTLE et al. 1986). This evidence suggested that dihydropyridine-sensitive calcium channels were involved in the production of tolerance to and dependence upon ethanol.

L-type calcium channels, however, constitute only a small fraction of the calcium channels in brain; thus their upregulation might be secondary to some other effect. Brennan, Littleton, and colleagues have suggested that inhibition of Ca^{2+} influx through a different kind of Ca^{2+} channel may be involved in ethanol upregulation of L-type channels (LITTLETON et al. 1991; BRENNAN and LITTLETON 1990). This idea is based on the following results: (1) BAY K 8644, a dihydropyridine calcium channel agonist, increased the breakdown of [^{3}H]inositol phospholipids caused by K^+-induced depolarization in bovine adrenal chromaffin cells, (2) growth of the cells in the presence of pertussis toxin, which catalyzes the ADP ribosylation of certain G proteins, produced upregulation of L-type channels similar to that caused by ethanol, and (3) pretreatment of the cells with pertussis toxin prevented depolarization-induced breakdown of inositol phospholipids (BRENNAN and LITTLETON 1990). These effects suggested that an ion channel associated with a guanine nucleotide binding-protein (G protein) might be involved, as well as the inositol lipid-protein kinase C second messenger system. Other groups have obtained results consistent with this interpretation. Dihydropyridines alone had no effect on $^{45}Ca^{2+}$ uptake in rat brain, although BAY K 8644 increased neurotransmitter release (DOLIN et al. 1986). Chronic ethanol treatment increased the rate of breakdown of inositol lipids in rat brain, as measured by depolarization-induced accumulation of [^{3}H]-inositol phosphates (DOLIN et al. 1986). In cultured PC12 cells, a cell line of neural origin, inhibitors of protein kinase C blocked the upregulation in L-type channels due to chronic ethanol (MESSING et al. 1991). Protein kinase C is activated by the cascade of events mediated by GTP-binding proteins via 1,2-diacylglycerol; however, 1,2-diacylglycerol was not increased. Chronic treatment with nitrendipine did not change the effect of ethanol on rotarod performance of rats (DOLIN and LITTLE 1989), though chronic nifedipine treatment has been reported to decrease the number of dihydropyridine receptors in mice (PANZA et al. 1985). Chronic exposure of cultured bovine adrenal chromaffin cells to anesthetic concentrations of ethanol, alprazolam, or buspirone caused an upregulation of dihydropyridine receptors despite the fact that these substances do not work by the same mechanism (BRENNAN and LITTLETON 1991). Ethanol and alprazolam, a benzodiazepine, are supposed to have their primary effects on the $GABA_A$-receptor-associated chloride channel complex; buspirone, on the 5-HT_{1A} receptor. The investigators proposed that the upregulation of dihydropyridine binding sites might be a response to *any* chronic inhibition of the excitability of the cells by anesthetics.

IV. Effects of Ethanol on Intracellular Calcium

Ethanol administered acutely affects the *intracellular* free calcium concentration, $[Ca^{2+}]_i$, as well as Ca^{2+} influx. Davidson and colleagues found that $[Ca^{2+}]_i$ was increased in synaptosomes from rat forebrain by 50–500 m*M* ethanol in vitro, although Ca^{2+} influx was decreased (DAVIDSON et al. 1988). They proposed that the increase in $[Ca^{2+}]_i$ resulted from release of Ca^{2+} from intracellular sources, such as Ca^{2+} bound to mitochondria or intracellular buffers. A later study indicated that the endoplasmic reticulum was the source of the additional intracellular free Ca^{2+}, rather than altered Na^+-Ca^{2+} exchange or influx through Ca^{2+} channels (DAVIDSON et al. 1990). Anesthetic concentrations of ethanol (350–700 m*M*) in vitro were found to increase resting $[Ca^{2+}]_i$ in synaptosomes from mouse brain; this effect may be the actual cause of the observed increase in the release of some neurotransmitters at high ethanol concentrations (DANIELL et al. 1987). $[Ca^{2+}]_i$ was not altered by ethanol in hepatocytes. These results indicated that alterations in $[Ca^{2+}]_i$ were not due to increased influx alone. The external calcium concentration does not affect the amount of ethanol-induced $[Ca^{2+}]_i$ increase (DANIELL et al. 1987); this is evidence for release of calcium from inside the cell. Several possible mechanisms exist by which ethanol can alter $[Ca^{2+}]_i$ acutely, and through which the organism might compensate for chronic ethanol consumption. Because calcium is involved in intracellular signal transduction, its regulation is critical for maintaining intracellular homeostasis and proper responsiveness to neurotransmitters. Several systems located in the plasma membrane, including Ca^{2+} channels, could affect $[Ca^{2+}]_i$. Ethanol's effects on calcium and intracellular signaling have recently been reviewed (GANDHI and ROSS 1989).

One possible mechanism for the effect of ethanol on $[Ca^{2+}]_i$ is through a perturbation, direct or indirect, in the activity of Ca^{2+}/Mg^{2+}-ATPase. This enzyme is involved in regulation of $[Ca^{2+}]_i$ by sequestration of Ca^{2+} in vesicles of smooth endoplasmic reticulum located at the inner surface of the plasma membrane (GARRETT and ROSS 1983). Acute administration of ethanol was shown to inhibit Ca^{2+}/Mg^{2+}-ATPase activity and locomotor activity in mice (ROSS et al. 1979). In mouse synaptic membranes taken at the time of loss of righting reflex, ethanol inhibited Ca^{2+}-stimulated ATP hydrolysis and ATP-dependent Ca^{2+} uptake (GARRETT and ROSS 1983). Ca^{2+}/Mg^{2+}-ATPase activity returned to normal at recovery of the righting reflex, but ATP-dependent Ca^{2+} uptake was still inhibited. Thus, ethanol may uncouple the enzyme from Ca^{2+} uptake.

V. Effects of Ethanol on G-Protein-Related Systems

Another possible mechanism for calcium release from intracellular stores is activation of certain signal-transducing guanine nucleotide-binding proteins (G proteins) that eventually release calcium through a cascade of events. G proteins are involved in the transduction of signals from several hormones and other factors from outside the cell to the cytoplasm. They are heterotrimers composed of α-, β- and γ-subunits associated with a membrane-bound receptor. The ADP ribosylation of some of them can be catalyzed by cholera or pertussis toxins. Although the details of the mechanism of G protein action are still uncertain, the following version gives the general idea. In the resting state, the receptor-G protein complex binds GDP. When the receptor is activated, the α-subunit binds GTP and is dissociated from the complex. This leads to the generation of second messengers, initiating a cascade of events that eventually results in a cellular response. Meanwhile, the agonist dissociates from the receptor and the receptor returns to its resting state. The activated G protein α-subunit can cause several events before it is inactivated. This property makes the G protein a signal amplifier as well as a signal transducer. Inactivation takes place through hydrolysis of GTP to GDP by the α-subunit's own intrinsic GTPase activity. The α-subunit then reassociates with the $\beta\gamma$-subunits. (The β- and γ-subunits are found associated with each other in vivo.) Because of signal amplification, a small effect of ethanol on a G protein could have more pronounced effects further along the cascade of events. The α-subunit can also be phosphorylated (SAGI-EISENBERG 1989); this event is probably a regulatory mechanism. Systems coupled to G proteins include the adenylyl cyclase system, calcium and potassium channels, and polyphosphoinositide metabolism (FREISSMUTH et al. 1989).

At least 17 different G protein α-subunits are known (STERNWEIS and SMRCKA 1993), and more may be discovered. G_s, which has four different α-subunits, stimulates the adenylyl cyclase system, while G_i inhibits the same system (FREISSMUTH et al. 1989). G_i has three known α-subunits: $\alpha_{i(1)}$, $\alpha_{i(2)}$ and $\alpha_{i(3)}$. G_o, the most abundant G protein in brain, regulates K^+ channels (MANJI 1992). Its α-subunit has two known isoforms, which are splice variants (TANG et al. 1992). Some α-subunits (α_q, α_{11}, α_{14}, α_{16}) of G proteins of the G_q family (G_q, G_{11}, G_{14}, $G_{15/16}$) have been shown to activate phosphatidylinositol breakdown in several systems (RHEE and CHOI 1992; LEE et al. 1992; WU et al. 1992; SMRCKA and STERNWEIS 1993). This breakdown, due to activation of phospholipase C (PLC), produces 1,2-diacylglycerol (DAG) and inositol triphosphate (IP_3) and eventually leads to an increase in $[Ca^{2+}]_i$ and the activation of protein kinase C. G_i and/or G_o has been shown to activate PLC in hematopoietic cells and adipocytes (KIKUCHI et al. 1993; MOLSKI et al. 1985; VOLPI et al. 1983; MORENO et al. 1983).

G_s and G_i have been investigated extensively with regard to ethanol effects; G_o, G_{olf}, G_t, and the G_q family have not. The scarcity of research on ethanol effects on the G_q family is probably due to its relatively recent discovery. However, some results have been reported in these systems. WILLIAMS et al. (1993) found that in NG108-15 cells 200 mM ethanol was able to increase expression of α_o. SIMONSSON et al. (1991) demonstrated that in NG108-15 cells the cascade of events involving phosphatidylinositol hydrolysis is affected by chronic ethanol at the level of the G protein, but the G protein(s) involved were not identified. Recently, 10–200 mM ethanol has been shown to decrease expression of the α-subunits of G_q and G_{11}, which are highly homologous, in NG108-15 cells (WILLIAMS and KELLY 1993). The investigators suggest that this decrease may be responsible for the decrease in receptor-mediated phosphoinositide hydrolysis due to chronic ethanol exposure in this cell line.

The $\beta\gamma$-subunits from the various G protein are less structurally diverse than the α-subunits, such that $\beta\gamma$-subunits freed by activation of G_i, for example, could associate with and thus inactivate α-subunits of G_s. This mechanism has been proposed as an alternative pathway for the inhibition of the adenylate cyclase system by G_i. Because experiments measuring direct action of G_i on the adenylyl cyclase system show less activity than would be expected, an alternative pathway is likely to exist (LEVITZKI 1990). Recently, however, evidence for direct inhibition of adenylyl cyclase by α_i has appeared (TAUSSIG et al. 1993). Free $\beta\gamma$-subunits have been shown to activate various isoforms of PLC (BLANK et al. 1992; PARK et al. 1993).

1. Acute Effects on Protein Kinase C

As noted above, acutely administered ethanol increases $[Ca^{2+}]_i$. One possible avenue for this increase is G-protein-mediated release of calcium via inositol 1,4,5-triphosphate (IP_3) from intracellular stores. However, evidence exists that the intracellular calcium reservoirs from which ethanol and IP_3 release calcium may not be the same (DANIELL and HARRIS 1989; MACHU et al. 1989), indicating that another mechanism may be involved. Protein kinase C (PKC) is also activated through a G-protein-mediated cascade of events (via 1,2-diacylglycerol) and this membrane-bound enzyme appears to be inhibited in rat brain homogenates by acute administration of ethanol and other anesthetics (FREISSMUTH et al. 1989). These studies were conducted by assaying activated PKC, so that the point in the cascade at which ethanol exerted an inhibitory effect was not identified. Therefore, the effect may be directly on the enzyme rather than on the G protein. In cultured PC12 cells, PKC δ and PKC ε were increased by chronic ethanol exposure (MESSING et al. 1991). Further, 1,2-diacylglycerol was not increased, indicating that the upregulation did not include the G protein portion of the cascade.

2. The Adenylyl Cyclase System

The adenylyl cyclase system has been most extensively studied. Adenylyl cyclase produces cyclic AMP (cAMP), which eventually activates protein kinase A. Ethanol has been reported to increase the amount of cAMP generated by adenylyl cyclase, thus increasing the amount of signal sent to the cell interior. Ethanol effects on the adenylyl cyclase system have been reviewed (HOFFMAN and TABAKOFF 1990). Two possible sites of ethanol action on adenylyl cyclase have been suggested: the catalytic subunit of the enzyme and either the activation of the regulatory subunit or the coupling of the subunits (RABIN and MOLINOFF 1983). The activation of adenylyl cyclase by ethanol appeared to require GTP-binding proteins (LUTHIN and TABAKOFF 1984). Saito and colleagues subsequently found that a high concentration of ethanol (500 m*M*), in the absence of guanine nucleotides, produced a much greater increase in adenylyl cyclase activity in mouse cortical membrane preparations than in mouse striatal membranes (SAITO et al. 1987). The investigators reasoned that ethanol was working primarily through G proteins in the striatum, but in the cortex it was able to affect adenylyl cyclase directly, as well as through G-protein interactions. Results also varied between cell cultures. In two subclones of PC12 cells ethanol increased adenylyl cyclase activity in membrane preparations. In whole cells, however, there was enhanced cAMP accumulation in response to ethanol in one subclone, while in the other cAMP accumulation was inhibited (RABE et al. 1990). Thus extrapolation from membrane preparations to whole cells or organisms may be unwarranted. Apparently, the effects of ethanol on adenylyl cyclase are highly dependent on the experimental system and conditions.

3. Chronic Effects

Because the acute effect of ethanol is to increase the activity of adenylyl cyclase, chronic ethanol consumption would be expected to decrease hormone-stimulated adenylyl cyclase activity. Most of the experiments on chronic ethanol effects on the G proteins related to adenylyl cyclase activity have been carried out using cultured cell lines of neural origin; fewer animal studies have been conducted. In cell culture experiments, results depend on the cell type. In NG108-15 cells the amount of α_s was reduced and the amount of α_i was unchanged after long-term ethanol exposure (CHARNESS et al. 1988). In N18TG2 cells neither G protein was altered, and in N1E-115 cells α_i was increased and α_s was decreased (CHARNESS et al. 1988). In rat hepatocytes in primary culture, 100 m*M* ethanol for 48 h resulted in a 35% decrease in α_i, but no change in α_s. The observed increase in cAMP production in these cells was thus a result of attenuated inhibition by α_i (NAGY and DESILVA 1992). In mice, 7 days of ethanol exposure reduced anterior pituitary membrane α_s levels in the LS line but not in the SS line (SS, short sleep, relatively ethanol-insensitive; and LS, long-sleep, relatively ethanol-sensitive). Levels of α_i were unchanged in both lines (WAND and LEVINE 1991). The variation in these results indicates that ethanol effects on expression of G proteins are highly variable among experimental systems.

Cholera and pertussis toxins catalyze the covalent transfer of the ADP ribosyl moiety of nicotinamide adenine dinucleotide (NAD) to α_s and α_i, respectively. In the case of α_s, the activated α-subunit is ADP ribosylated; in the case of α_i, the inactive α-subunit associated with $\beta\gamma$ is labeled. Thus ADP ribosylation measures the amount of activated α_s or the amount of inactive α_i. In cerebellum and pons of SS and LS mice, expression of $\alpha_{i(1)}$ and $\alpha_{i(2)}$ was markedly increased after 7 days of ethanol exposure, but expression of α_s was unchanged (WAND et al. 1993). The same study showed that ethanol treatment of SS mice caused a twofold increase in ADP ribosylation of $\alpha_{i(1)}$, $\alpha_{i(2)}$, and perhaps α_o (combined) in cerebellar membranes. ADP ribosylation of α_s was unchanged. Results were similar in LS mice. The increases in expression and ADP ribosylation of $\alpha_{i(1)}$ and $\alpha_{i(2)}$ were associated with inhibition of adenylyl cyclase activity. The inhibition was reversed by pretreatment of the membranes with pertussis toxin, evidence that α_i was probably responsible for the effect. A smaller increase in ADP ribosylation of α_i has been observed in rat cerebral cortex after 8 weeks of a 5% ethanol liquid diet (MORING and VOLPI 1992).

Studies of human tissues have also shown effects of ethanol on second messenger systems. DIAMOND et al. (1987) measured cAMP levels in lymphocytes from alcoholic patients. Both basal and receptor-stimulated levels of cAMP were reduced by 75%, while stimulation of cAMP production by ethanol was reduced by 76%. Ethanol stimulation of cAMP production appeared to be mediated by adenosine (DIAMOND et al. 1991; NAGY et al. 1989), as shown by experiments in cell culture. In NG108-15 cells, acute exposure to ethanol produced a 60% increase in cAMP production, although ethanol did not directly activate adenylyl cyclase. However, extracellular adenosine was increased. When the extracellular adenosine was enzymically degraded, ethanol no longer stimulated cAMP production (NAGY et al. 1989).

Some G-protein effects may persist through withdrawal. In lymphocytes from abstinent alcoholics (approximately 20 days without drinking), levels of $\alpha_{i(2)}$ protein were increased threefold and levels of mRNA for $\alpha_{i(2)}$ were increased 2.9 times over controls (WALTMAN et al. 1993). mRNA for α_s was increased 2.7-fold, although α_s protein expression was unchanged. Lymphocytes from actively drinking alcoholics were similar to those from controls except that α_s mRNA was increased 1.8 times. The reduction of adenylyl cyclase activity found during abstinence was attributed to enhanced expression of $\alpha_{i(2)}$ (WALTMAN et al. 1993).

In NG108-15 cells, not only are cAMP production and α_s protein decreased as a result of chronic ethanol exposure, but mRNA for α_s is commensurately reduced (MOCHLY-ROSEN et al. 1988). The reduction in cAMP output is due to heterologous desensitization of receptors, such that cAMP production stimulated by either prostaglandin E1 or adenosine is decreased. Thus several types of receptors coupled to α_s may be affected by chronic ethanol exposure. Adenosine has also been implicated in desensitization of adenylate cyclase by chronic ethanol (NAGY et al. 1989). Evidence against this role of adenosine has been presented by WILLIAMS et al. (1993). A recent intriguing finding is that in NG108-15 cells ethanol specifically induces several genes, one of which encodes a product with extensive homology to phosducin (MILES et al. 1993). This phosducin-like protein binds to $\beta\gamma$-subunits and thus may be responsible for some of ethanol's effects on G-protein-mediated signal transduction systems.

Effects of chronic ethanol on G proteins appear to be both tissue specific and species specific. In embryonic chick myocytes cultured with ethanol and in myocardium from ethanol-fed rats, no effect on G proteins was found when adenylyl cyclase activity and ADP ribosylation were measured (BLUMENTHAL et al. 1991). A small decrease in the quantity of α_i was observed in myocytes that had been exposed to 100 m*M* ethanol for 7 days. In hepatocyte membranes of ethanol-fed rats, however, adenylyl cyclase activity stimulated by glucagon or forskolin was increased but fluoride-stimulated activity was not increased. In that study, the investigators proposed that an increased number of glucagon receptors might be responsible for the observed effects.

Clearly, the effects of acute and chronic ethanol administration on G-protein-mediated systems are highly variable among the various cell culture and animal systems studied. Furthermore, determining which part(s) of a cascade of events is (are) affected is by ethanol is difficult. If the endpoint, such as adenylyl cyclase activity, is measured, the effect may be directly at the endpoint or at any of several steps before. For example, the affected step may be the activation of the α-subunit of G_s, expression of the α-subunit of G_i, alteration of the conformation of the G-protein-coupled receptor, or, in the case of chronic ethanol exposure, production of specific G-protein mRNAs. Nevertheless, because of their ubiquity and sensitivity, G-protein-mediated systems are likely to be significant mediators of ethanol effects.

VI. Effects of Chronic Ethanol on Receptor Subunit Expression

It is puzzling that despite that fact that ethanol does not bind to any neurotransmitter receptor site as an agonist or antagonist, chronic exposure results in upregulation of both NMDA receptor binding sites (GRANT et al. 1990; SHOEMAKER et al. 1992) and dihydropyridine binding sites (DOLIN et al. 1986). In the case of the $GABA_A$ receptor, changes in binding site concentration or affinity have been less consistently observed (see Chap. 5, this volume), but changes in receptor function following chronic exposure have suggested that the subunit makeup of the $GABA_A$ receptor has been altered (MORROW et al. 1988). A plausible mechanism of ethanol action on the affected proteins would involve a change in the level of mRNA. Recent reports have demonstrated an effect of chronic ethanol exposure on mRNA levels for $GABA_A$ receptor subunit species. Although this effect of ethanol results in changes in a nucleic acid rather than in a lipid or protein molecule, the mRNAs affected specify membrane receptor proteins. It may be that many of the chronic effects of ethanol on either protein or lipid are mediated by gene activation of mRNA species, but there is not yet sufficient evidence to warrant such a conclusion.

1. The $GABA_A$ Receptor

Each of the $GABA_A$ receptor subunit classes, except δ, has two or more subtypes; six subtypes of the α-subunit are known, four of β, and two of γ (VICINI 1991, SIEGHART et al. 1992). The distribution of subunit classes and subtypes varies among brain regions. The properties of the $GABA_A$ receptor complex depend on which subunits and which variants of each subunit are present; α; β, and γ are all required for agonist and antagonist responses that mimic the in vivo situation (SIEGHART et al. 1992). For example, the presence of the γ_2-subunit confers high benzodiazepine sensitivity on the complex (LÜDDENS and WISDEN 1991); the presence of the γ2L (alternatively spliced) form is required for ethanol sensitivity (WAFFORD et al. 1991). The production of mRNA for various subtypes of the α-subunit has been investigated in regard to chronic ethanol exposure. MONTPIED et al. (1991), using a 14-day inhalation exposure in adult rats, demonstrated reductions in the $GABA_A$ subunits α_1 and α_2, but not in α_3, from cerebral cortex. The blood ethanol concentration averaged 207 ± 21 mg/dl ($\sim$45 mM) for exposed animals. A concern in studies of mRNA levels is that many mRNAs are affected similarly so that normalization for the amount applied to the electrophoretic gel can be problematic. Montpied and coworkers stripped the blots of $GABA_A$ receptor α_1-subunit cRNA probe and rehybridized with β-actin riboprobe, followed by a second densitometric measurement. This procedure assured proper normalization and indicated a 49% reduction in the 4.8-kb species of α_1-subunit and a 39% decrease in the 4.4-kb species. MHATRE and TICKU (1992) used a rat model of intragastric intubation for 6 days. They then made Northern blots of $GABA_A$ subunit mRNAs, which showed decreases in α_1 and α_2 mRNA levels, but not in α_3. These investigators also reported a decrease in subunit α_5 mRNA and an increase in α_6. Interestingly, the mRNAs for α_2 and α_3 were measured after withdrawal and returned to baseline levels by 36 h postwithdrawal. Buck, Harris, and coworkers (BUCK et al. 1991) measured $GABA_A$ receptor subunit mRNA levels in seizure-prone (WSP) and seizure-resistant (WSR) mice after 7 days of ethanol. In the WSP mice, α_1 levels decreased after the 7-day exposure, whereas α_6 levels decreased in the WSR.

2. Origin of mRNA Effects

These recent reports add a new dimension to studies of ethanol effects on membranes. If a major effect of ethanol is to activate or inactivate genes coding for membrane proteins, e.g., the $GABA_A$ receptor subunits, the NMDA receptor subunits, or the L-type Ca^{2+} channel, this effect could have profound implications for the responsiveness of the organism. This indirect effect of ethanol could occur in addition to the direct effects of ethanol on the lipid and protein domains of the membrane. Since the effects on mRNA expression have been documented only following chronic ethanol exposure, it is possible that the changes in mRNA level are in response to other acute effects of ethanol, such as changes in $[Ca^{2+}]_i$. Induction of Hsc70 gene transcription by 50–200 m*M* ethanol has been reported in NG108-15 neuroblastoma × glioma cells (MILES et al. 1991). Hsc70 is a heat-shock protein that is involved in protein trafficking. Other heat-shock proteins were not upregulated by these concentrations of ethanol; thus this was not a generalized stress protein response. Ethanol appeared to act specifically on the Hsc70 promoter. Because of the role of Hsc70 in protein trafficking its regulation by ethanol could indirectly affect many membrane proteins. It is well known that protein synthesis in the brain is depressed by both chronic and acute ethanol administration, as measured by studies of the incorporation of radiolabeled amino acids into protein (TEWARI and SYTINSKI 1985). Two-dimensional gel electrophoresis experiments showed that the membrane protein profile in rat cerebral cortex changed substantially after 70 days of chronic ethanol exposure, with some proteins being lost entirely and some being relatively unaffected (BABU et al. 1990). Total cortical protein was also decreased. These effects could possibly be due to gene regulation [or indirectly to some other effect, such as formation of fatty acid ethyl esters (BORA and LANGE 1993)]. Studies of effects of ethanol on genes coding for membrane proteins or enzymes controlling membrane constituents are quite recent. Studies of the reversibility of such changes and the long-term consequences of repeated gene activation or inhibition are needed to determine the physiological significance of ethanol's action at the gene level.

E. Conclusions

The effects of ethanol on various membrane lipids, NMDA receptors, $GABA_A$ receptors, G proteins, the adenylyl cyclase system, and other membrane components may explain some of the behavioral and physiological effects of ethanol. The question of which effects are important physiologically and which are relatively minor remains open, as does the possibility that some major effects of ethanol remain to be discovered. Ethanol's lipid disordering effects correlate well with its acute physiological effects, but so far appear not to be a major cause of those effects. The data in Table 1 indicate a significant decrease in the membrane/buffer partition coefficient for ethanol after chronic ethanol exposure. This may explain the phenomenon of membrane tolerance, but it cannot explain differences in genetic susceptibility to ethanol effects among strains of mice and rats. The reversal of ethanol intoxication by hyperbaric pressure is generally consistent with a bulk membrane disorder mechanism of action, as well as with a membrane lipid domain or protein mechanism. Lipid alterations in response to chronic ethanol exposure are small, and there is no evidence that ethanol-related lipid alterations affect membrane protein function in vivo. Whether ethanol affects membrane protein function directly or through changes in membrane disorder is unresolved. Conflicting evidence exists concerning whether the extent of membrane disorder influences protein function (Squier et al. 1988; East et al. 1984). Data suggest that membrane lipid order is not particularly important in determining the activity of membrane enzymes such as Ca^{2+}/Mg^{2+}-ATPase (Lee 1991). Whereas, the function of receptor-activated sodium channels is correlated with membrane lipid order, the function of calcium channels is not (Harris and Bruno 1985). The bulk membrane disorder model also suffers from the fact that a rise in temperature of less than 1°C produces an increase in membrane disorder similar to that produced by physiological concentrations of ethanol or anesthetics but does not produce intoxication or anesthesia (Pang et al. 1980; Franks and Lieb 1982). Furthermore, membrane disorder on a scale comparable to that associated with ethanol or anesthetic action can be caused by A_2C, an agent that does not produce anesthesia (Buck et al. 1989). At present, evidence indicates that the site of any significant lipid effects of ethanol is not the bulk lipid of the membrane, but instead one or more membrane domains that are ethanol-sensitive. Chronic ethanol effects on lipid domains, such as alteration of lipid distribution between the inner and outer monolayers of membranes, have been documented. Whether they are causally related to physiological problems due to chronic ethanol has yet to be established.

On the other hand, there is evidence for direct action of ethanol on proteins, for example, on firefly luciferase. However, this enzyme is not directly involved in anesthesia and thus is an imperfect model for the effects of ethanol and other anesthetics. The selective action of anesthetic stereoisomers on K^+ channels is better evidence for direct anesthetic action on proteins (Franks and Lieb 1991). Acute effects of ethanol on the ion-gating functions of the $GABA_A$ and NMDA receptor complexes are well documented; whether these effects are exerted directly on the protein or mediated by lipid changes has not been determined. Adaptations to chronic ethanol exposure occur in proteins as well as in lipids, usually in the form of up- or downregulation of the protein. The dihydropyridine receptor-associated calcium channel, NMDA and $GABA_A$ receptor subunits, and certain G-protein α-subunits are examples of such proteins. Alteration of mRNA levels for certain membrane proteins has been found to result from chronic ethanol exposure. The mechanism of ethanol's action, either directly at the nuclear gene site or indirectly, in producing this alteration has not been determined; neither have the long-term consequences of alterations in mRNA concentrations been determined.

Several effects of ethanol, occurring at both lipid and protein sites, have been discussed. These actions include (1) a bulk lipid disordering effect, (2) a membrane compositional change, (3) an effect on local lipid domains, (4) a direct acute effect on transmembrane proteins, (5) effects on signal transduction systems, (6) effects on ion channels, and (7) effects on receptor subunit expression (see Fig. 1). The catalog of documented effects of acute and chronic ethanol exposure continues to grow, and it would be beneficial to ascertain their relative importance in causing the drug's characteristic physiological and behavioral effects. Several apparently diverse sites of action exist. An elucidation of the mechanisms that underlie the effects exerted at these sites would help to guide the direction of future research.

References

Abraham MH, Lieb WR, Franks NP (1991) Role of hydrogen bonding in general anesthesia. J Pharm Sci 80:719–724

Abood LG, Salem N Jr, MacNeil M, Bloom L, Abood ME (1977) Enhancement of opiate binding by various molecular forms of phosphatidylserine and inhibition by other unsaturated lipids. Biochim Biophys Acta 468:51–62

Akeson M, Deamer DW (1991) Anesthetics and membranes: a critical review. In: Aloia RC, Curtain CC, Gordon LM (eds) Drug and anesthetic effects on membrane structure and function. Wiley Liss, New York, pp 71–89

Alifimoff JK, Firestone LL, Miller KW (1989) Anaesthetic potencies of primary alkanols: implications for the molecular dimensions of the anaesthetic site. Br J Pharmacol 96:9–16

Alkana RL, Malcolm RD (1980a) Antagonism of ethanol narcosis in mice by hyperbaric pressures of 4–8 atmospheres. Alcohol Clin Exp Res 4:350–353

Alkana RL, Malcolm RD (1980b) The effects of low level hyperbaric treatment on acute ethanol intoxication. Adv Exp Med Biol 126:499–507

Alkana RL, Malcolm RD (1981) Low-level hyperbaric ethanol antagonism in mice. Dose and pressure response. Pharmacology 22:199–208

Alkana RL, Malcolm RD (1982) Hyperbaric ethanol antagonism in mice: studies on oxygen, nitrogen, strain and sex. Psychopharmacology 77:11–16

Alkana RL, Boone DC, Finn DA (1985a) Temperature dependence of ethanol depression: linear models in male and female mice. Pharmacol Biochem Behav 23:309–316

Alkana RL, Finn DA, Galleisky GG, Syapin PJ, Malcolm RD (1985b) Ethanol withdrawal in mice precipitated and exacerbated by hyperbaric exposure. Science 229:772–774

Alkana RL, DeBold JF, Finn DA, Babbini M, Syapin PJ (1991) Ethanol-induced depression of aggression in mice antagonized by hyperbaric exposure. Pharmacol Biochem Behav 38:639–644

Allan AM, Harris RA (1986) Gamma-aminobutyric acid and alcohol actions: neurochemical studies of long sleep and short sleep mice. Life Sci 39:2005–2015

Allan AM, Harris RA (1987) Acute and chronic ethanol treatments alter GABA receptor-operated chloride channels. Pharmacol Biochem Behav 27:665–670

Alling C, Liljequist S, Engel J (1982) The effect of chronic ethanol administration on lipids and fatty acids in subcellular fractions of rat brain. Med Biol 60:149–154

Alling C, Gustavsson L, Änggård E (1983) An abnormal phospholipid in rat organs after ethanol treatment. FEBS Lett 152:24–28

Alling C, Becker W, Jones AW, Änggård E (1984) Effects of chronic ethanol treatment on lipid composition and prostaglandins in rats fed essential fatty acid deficient diets. Alcohol Clin Exp Res 8:238–242

Alling C, Gustavsson L, Månsson J-E, Benthin G, Änggård E (1984) Phosphatidylethanol formation in rat organs after ethanol treatment. Biochim Biophys Acta 793:119–122

Alling C, Jonsson G, Gustavsson L, Jensen L, Simonsson P (1986) Anionic glycerophospholipids in platelets from alcoholics. Drug Alcohol Depend 16:309–320

Asayoka Y (1989) Distinct effects of phosphatidylethanol on three types of rat brain protein kinase C. Kobe J Med Sci 35:229–237

Avdulov NA, Wood WG, Harris RA (1994) Effects of ethanol on structural parameters of rat brain membranes: relationship to genetic differences in ethanol sensitivity. Alcohol Clin Exp Res 18:53–59

Babu PP, Nagaraju N, Vemuri MC (1990) Differences in the plasma membrane proteins of chronic alcoholic rat brain. Membr Biochem 9:227–237

Baum F (1901) Zur Theorie der Alkoholnarkose. Der Einfluss wechselnder Temperatur auf Wirkungsstärke und Theilungscoëfficient der Narcotica. Naunyn Schmiedebergs Arch Exp Pathol Pharmakol 46:338–346

Besman MJ, Yanagibashi K, Lee TD, Kawamura M, Hall P, Shively JE (1989) Identification of des-(gly-ile)-endozepine as an effector of corticotropin-dependent adrenal steroidogenesis: stimulation of cholesterol delivery is mediated by the peripheral benzodiazepine receptor. Proc Natl Acad Sci USA 86:4897–4901

Blank JL, Brattain KA, Exton JH (1992) Activation of cytosolic phosphoinositide phospholipase C by G-protein $\beta\gamma$ subunits. J Biol Chem 267:23069–23075

Blumenthal RS, Flinn IW, Proske O, Jackson DG, Tena RG, Mitchell MC, Feldman AM (1991) Effects of chronic ethanol exposure on cardiac receptor-adenylyl cyclase coupling: studies in cultured embryonic chick myocytes and ethanol fed rats. Alcohol Clin Exp Res 15:1077–1083

Bora PS, Lange LG (1993) Molecular mechanism of ethanol metabolism by human brain to fatty acid ethyl esters. Alcohol Clin Exp Res 17:28–30

Brennan CH, Littleton JM (1990) Second messenger systems involved in genetic regulation of Ca^{2+} channels in adrenal chromaffin cells. Neuropharmacology 29:689–693

Brennan CH, Littleton JM (1991) Chronic exposure to anxiolytic drugs, working by different mechanisms causes up-regulation of dihydropyridine binding sites on cultured bovine adrenal chromaffin cells. Neuropharmacology 30:199–203

Buck KJ, Allan AM, Harris RA (1989) Fluidization of brain membranes by A_2C does not produce anesthesia and does not augment muscimol-stimulated $^{36}Cl^-$-flux. Eur J Pharmacol 160:359–367

Buck KJ, Hahner L, Sikela J, Harris R (1991) Chronic ethanol treatment alters brain levels of γ-aminobutyric $acid_A$ receptor subunit mRNAs: relationship to genetic differences in ethanol withdrawal seizure severity. J Neurochem 57:1452–1455

Cattel MK (1936) The physiological effects of pressure. Biol Rev 11:441–476

Charness ME, Querimet LA, Henteleff M (1988) Ethanol differentially regulates G proteins in neural cells. Biochem Biophys Res Commun 155:138–143

Chin JH, Goldstein DB (1977) Drug tolerance in biomembranes: a spin label study of the effects of ethanol. Science 196:684–685

Chin JH, Parsons LM, Goldstein DB (1978) Increased cholesterol content of erythrocyte and brain membranes in ethanol-tolerant mice. Biochim Biophys Acta 513:358–363

Chiou J-S, Kuo C-C, Lin SH, Kamaya H, Ueda I (1991) Interfacial dehydration by alcohols: hydrogen bonding of alcohols to phospholipids. Alcohol 8:143–150
Crabbe JC (1983) Sensitivity of ethanol in inbred mice: genotypic correlations among several behavioral responses. Behav Neurosci 97:280–289
Crews FT, Camacho A, Phillips I, Tjeenk Willink EC, Calderini G, Hirata FL, Axelrod J, McGivney A, Siraganian R (1991) Effects of membrane fluidity on mast cell and nerve cell function. In: Horrocks LA, Kanfer JN, Porcellati G (eds) Phospholipids in the nervous system, vol 1: metabolism. Raven, New York, pp 237–245
Crews FT, Majchrowicz ER, Meeks R (1983) Changes in cortical synaptosomal plasma membrane fluidity and composition in ethanol-dependent rats. Psychopharm 81:208–213
Daniell LC, Leslie SW (1986) Inhibition of fast phase calcium uptake and endogenous norepinephrine release in rat brain region synaptosomes by ethanol. Brain Res 377:18–28
Daniell LC, Harris RA (1989) Ethanol and inositol 1,4,5-triphosphate release calcium from separate stores of brain microsomes. J Pharmacol Exp Ther 250:875–881
Daniell LC, Barr EM, Leslie SW (1983) $^{45}Ca^{2+}$ uptake into rat whole brain synaptosomes unaltered by dihydropyridine calcium antagonists. J Neurochem 41:1455–1459
Daniell LC, Brass EP, Harris RA (1987) Effect of ethanol on intracellular ionized calcium concentrations in synaptosomes and hepatocytes. Mol Pharmacol 32: 831–837
Davidson M, Wilce P, Shanley B (1988) Ethanol increases synaptosomal free calcium concentration. Neurosci Lett 90:165–169
Davidson M, Wilce P, Shanley B (1990) Ethanol and synaptosomal calcium homeostasis. Biochem Pharmacol 39:1283–1288
Diamond I, Wrubel B, Estrin W, Gordon A (1987) Basal and adenosine receptor-stimulated levels of cAMP are reduced in lymphocytes from alcoholic patients. Proc Natl Acad Sci USA 84:1413–1416
Diamond I, Nagy L, Mochly-Rosen D, Gordon A (1991) The role of adenosine and adenosine transport in ethanol-induced cellular tolerance and dependence. Possible biologic and genetic markers of alcoholism. Ann NY Acad Sci 625: 473–487
Dolin SJ, Little HJ (1989) Are changes in neuronal calcium channels involved in ethanol tolerance? J Pharmacol Exp Ther 250:985–991
Dolin SJ, Little HJ (1991) Augmentation by calcium channel antagonists of general anaesthetic potency in mice. Br J Pharmacol 88:909–914
Dolin SJ, Little HJ, Hudspith M, Pagonis C, Littleton JM (1986) Increased dihydropyridine calcium channels in rat brain may underlie ethanol physical dependence. Neuropharmacology 26:270–275
East JM, Jones WT, Simmonds AC, Lee AG (1984) Membrane fluidity is not an important physiological regulator of the (Ca^{2+}-Mg^{2+})-dependent ATPase of sarcoplasmic reticulum. J Biol Chem 259:8070–8071
Ferguson J (1939) The use of chemical potentials as indices of toxicity. Proc R Soc B 127:387–404
Finn DA, Bejanian M, Jones BL, McGivern FR, Syapin PJ, Crabbe JC, Alkana RL (1990) Body temperature differentially affects ethanol sensitivity in both inbred strains and selected lines of mice. J Pharmacol Exp Ther 253:1229–1235
Finn DA, Boone DC, Alkana RL (1986) Temperature dependence of ethanol depression in rats. Psychopharmacology 90:185–189
Floreani M, Bonetti AC, Carpenedo F (1981) Increase of Na^+/K^+ ATPase activity in intact rat brain synaptosomes after their interaction with phosphatidylserine vesicles. Biochem Biophys Res Commun 101:1337–1344
Floreani M, Carpenedo F (1987) Phosphatidylserine vesicles increase rat brain synaptosomal adenylyl cyclase activity. Biochem Biophys Res Commun 145:631–636

Foster AC, Fagg GE, Harris EW, Cotman CW (1982) Regulation of glutamate receptors: possible role of phosphatidylserine. Brain Res 242:374–377
Franks NP, Lieb WR (1981) Is membrane expansion relevant to anesthesia? Nature 292:248–251
Franks NP, Lieb WR (1982) Molecular mechanisms of general anesthesia. Nature 300:487–493
Franks NP, Lieb WR (1984) Do general anesthetics act by competitive binding to specific receptors? Nature 310:599–601
Franks NP, Lieb WR (1985) Mapping of general anesthetic target sites provides a molecular basis for cutoff effects. Nature 316:349–351
Franks NP, Lieb WR (1986) Partitioning of long-chain alcohols into lipid bilayers: implications for mechanisms of general anesthesia. Proc Natl Acad Sci USA 83:5116–5120
Franks NP, Lieb WR (1988) Volatile anesthetics activate a novel neuronal K^+ current. Nature 333:662–664
Franks NP, Lieb WR (1990) Mechanisms of general anesthesia. Environ Health Perspect 87:199–205
Franks NP, Lieb WR (1991) Stereospecific effects of inhalational general anesthetic optical isomers on nerve ion channels. Science 254:427–430
Freissmuth P, Casey J, Gilman AG (1989) G proteins control diverse pathways of transmembrane signaling. FASEB J 3:2125–2131
Gandhi CR, Ross DH (1989) Influence of ethanol on calcium, inositol phospholipids and intracellular signalling mechanisms. Experientia 45:407–413
Garrett KM, Ross DH (1983) Effects of in vivo ethanol administration of Ca^{2+}/Mg^{2+} ATPase and ATP-dependent Ca^{2+} uptake activity in synaptosomal membranes. Neurochem Res 8:1013–1028
Goldstein DB, Chin JH, Lyon RC (1982) Ethanol disordering of spin-labeled mouse brain membranes: correlation with genetically determined ethanol sensitivity of mice. Proc Natl Acad Sci USA 799:4231–4233
Gordon ER, Rochman J, Arai M, Lieber CS (1982) Lack of correlation between hepatic mitochondrial membrane structure and functions in ethanol-fed rats. Science 216:1319–1321
Grant KA, Valverius P, Hudspith M, Tabakoff B (1990) Ethanol withdrawal seizures and the NMDA receptor complex. Eur J Pharmacol 176:289–296
Greenberg DA, Cooper EC, Gordon A, Diamond I (1984) Ethanol and the gamma-aminobutyric acid-benzodiazepine receptor complex. J Neurochem 42:1062–1068
Grieve SJ, Littleton JM (1979) Ambient temperature and the development of functional tolerance to ethanol by mice. J Pharm Pharmacol 31:707–708
Guidotti A, Forchetti CM, Corda MG, Konkel D, Bennett CD, Costa C (1983) Isolation, characterization, and purification to homogeneity of an endogenous polypeptide with agonistic action on benzodiazepine receptors. Proc Natl Acad Sci USA 80:3531–3535
Gustavsson L (1990) Brain lipid changes after ethanol exposure. Up J Med Sci Suppl 48:245–266
Gustavsson L, Moehren G, Hoek JB (1991) Phosphatidylethanol formation in rat hepatocytes. Ann NY Acad Sci 625:438–440
Halsey MJ, Wardley-Smith B (1975) Pressure reversal of narcosis produced by anesthetics, narcotics and tranquilizers. Nature 257:811–813
Halsey MJ, Wardley-Smith B, Green CJ (1978) Pressure reversal of general anaesthesia – a multi-site expansion hypothesis. Br J Anaesth 50:1091–1097
Harris RA, Bruno P (1985) Membrane disordering by anesthetic drugs: relationship to synaptosomal sodium and calcium fluxes. J Neurochem 44:1274–1281
Harris RA, Hood WF (1980) Inhibition of synaptosomal calcium uptake by ethanol. J Pharmacol Exp Ther 213:562–568

Harris RA, Schroeder F (1981) Ethanol and the physical properties of brain membranes. Fluorescence studies. Mol Pharmacol 20:128–137
Harris RA, Baxter DM, Mitchell MA, Hitzemann RJ (1984) Physical properties and lipid composition of brain membranes from ethanol tolerant-dependent mice. Mol Pharmacol 25:401–409
Herbette L, Napolitano CA, Messineo FC, Katz AM (1985) Interaction of amphiphilic molecules with biological membranes. A model for nonspecific and specific drug effects with membranes. Adv Myocardiol 5:333–346
Herbette LG, Chester DW, Rhodes DG (1986) Structural analysis of drug molecules in biological membranes. Biophys J 49:91–94
Hill MW, Bangham AD (1975) General depressant drug dependency: a biophysical hypothesis. Adv Exp Med Biol 59:1–9
Hitzemann RJ, Schueler HE, Graham-Brittain C, Kreishman GP (1986) Ethanol-induced changes in neuronal membrane order. Biochim Biophys Acta 859:189–197
Hoffman PL, Tabakoff B (1990) Ethanol and guanine nucleotide binding proteins: a selective interaction. FASEB J 4:2612–2622
Hubbell WL, Metcalfe JC, Metcalfe SM, McConnell HM (1970) The interaction of small molecules with spin-labelled erythrocyte membranes. Biochim Biophys Acta 219:415–427
Huidobro-Toro JP, Bleck V, Allan AM, Harris RA (1987) Neurochemical actions of anesthetic drugs on the γ-aminobutyric acid receptor-chloride channel complex. J Pharmacol Exp Ther 242:963–969
Hungund BL, Goldstein DB, Villegas F, Cooper TB (1988) Formation of fatty acid ethyl esters during chronic ethanol treatment in mice. Biochem Pharmacol 37:3001–3004
Im WB, Blakeman DP, Davis JP, Ayer DE (1990) Studies on the mechanism of interactions between anesthetic steroids and γ-aminobutyric $acid_A$ receptors. Mol Pharmacol 37:429–434
Israel Y (1970) Cellular effects of alcohol: a review. Q J Stud Alcohol 31:293–316
Israel Y, Kalant H, Laufer I (1965) Effect of ethanol on electrolyte transport and electrogenesis in animal tissues. J Cell Comp Physiol 65:127–132
Johnson FH, Brown ES, Marsland DA (1942) Pressure reversal of the action of certain narcotics. J Cell Comp Physiol 20:269–276
Johnson FH, Flagler EA (1950) Hydrostatic pressure reversal of narcosis in tadpoles. Science 112:91–92
Kikuchi A, Kozawa O, Kaibuchi K, Katada T, Ui M, Takai Y (1993) Direct evidence for involvement of a guanine nucleotide-binding protein chemotactic peptide-stimulated formation of inositol biphosphate and trisphosphate in differentiated human leukemic (HL-60) cells. J Biol Chem 261:11558–11562
Knapp RJ, Malatynska E, Yamamura HI (1990) From binding studies to the molecular biology of GABA receptors. Neurochem Res 15:105–112
Kobayashi M, Kanfer JN (1987) Phosphatidylethanol formation via transphosphatidylation by rat brain synaptosomal phospholipase D. J Neurochem 48:1597–1603
Kobayashi T, Storrie B, Simons K, Dotti CG (1992) A functional barrier to movement of lipids in polarized neurons. Nature 359:647–650
Kreishman GP, Graham-Brittain C, Hitzemann RJ (1985) Determination of ethanol partition coefficients to the interior and the surface of dipalmityl-phosphatidylcholine liposomes using deuterium nuclear magnetic resonance spectroscopy. Biochem Biophys Res Commun 130:301–301
Lange LG (1991) Mechanism of fatty acid ethyl ester formation and biological significance. Ann NY Acad Sci 625:802–805
Laposata E, Lange LG (1986) Presence of non-oxidative ethanol metabolism in human organs commonly damaged by ethanol abuse. Science 231:497–499

Lee AG (1976) Model for action of local anesthetics. Nature 262:545–548
Lee AG (1988) Annular lipids and the activity of the calcium-dependent ATPase. In: Aloia RA, Curtain CC, Gordon LM (eds) Lipid domains and the relationship to membrane function. Alan R. Liss, Inc., New York, pp 111–139
Lee AG (1991) Lipids and their effects on membrane proteins: evidence against a role for fluidity. Prog Lipid Res 30:323–348
Lee CH, Park D, Wu D, Rhee SG, Simon MI (1992) Members of the G_q subunit gene family activate phospholipase C β isozymes. J Biol Chem 2267:16044–16047
Lee Y-C, Taraschi TF, James N (1993) Support for the shape concept of lipid structure based on a headgroup volume approach. Biophys J 65:1429–1432
Leslie SW, Barr E, Chandler LJ (1983a) Comparison of voltage-dependent $^{45}Ca^{2+}$ uptake rates by synaptosomes isolated from rat brain regions. J Neurochem 41:1602–1605
Leslie SW, Barr E, Chandler J, Farrar RP (1983b) Inhibition of fast- and slow-phase depolarization dependent synaptosomal calcium uptake by ethanol. J Pharmacol Exp Ther 225:571–575
Leslie SW, Brown LM , Dildy JE, Sims JS (1990) Ethanol and neuronal calcium channels. Alcohol 7:233–236
Lever MJ, Miller KW, Paton WD, Smith EB (1971) Pressure reversal of anesthesia. Nature 231:368–371
Levitzki A (1990) Dual control of adenylate cyclase. In: Houslay MD, Milligan G (eds) G-proteins as mediators of cellular signaling processes. John Wiley and Sons, New York, pp 1–14
Li C, Peoples RW, Weight FF (1993) Alcohols inhibit ATP-activated ion current by a direct interaction with the channel protein. Society for Neuroscience Abstracts 19:283
Lipnick RL (1989) Hans Horst Meyer and the lipoid theory of narcosis. Trends Pharmacol Sci 10:265–269
Little HJ, Dolin LH, Halsey MJ (1986) Calcium channel antagonists decrease the ethanol withdrawal syndrome. Life Sci 39:2059–2065
Littleton JM, John G (1977) Synaptosomal membrane lipids of mice during continuous exposure to ethanol. J Pharm Pharmacol 29:579–580
Littleton JM, Brennan C, Bouchenafa O (1991) The role of calcium flux in the central nervous system actions of ethanol. Ann NY Acad Sci 625:388–394
Littleton JM, John GR, Grieve SJ (1979) Alterations in phospholipid composition in ethanol tolerance and dependence. Alcohol Clin Exp Res 3:50–56
Lüddens H, Wisden W (1991) Function and pharmacology of multiple $GABA_A$ receptor subunits. Trends Pharmacol Sci 12:49–51
Lundqvist C, Rodriguez FD, Simonsson P, Alling C, Gustavsson L (1993) Phosphatidylethanol affects inositol 1,4,5-triphosphate levels in NG108-15 neuroblastoma × glioma hybrid cells. J Neurochem 60:738–744
Luthin GR, Tabakoff B (1984) Activation of adenylyl cyclase by alcohols requires the nucleotide-binding protein. J Pharmacol Exp Ther 288:579–587
Lynch MA, Littleton JM (1983) Possible association of alcohol tolerance with increased synaptic Ca^{2+} sensitivity. Nature 303:175–176
Lyon RC, McComb JA, Schreurs J, Goldstein DB (1981) A relationship between alcohol intoxication and the disordering of brain membranes by a series of short-chain alcohols. J Pharmacol Exp Ther 218:669–675
Machu T, Woodward JJ, Leslie SW (1989) Ethanol and inositol 1,4,5-trisphosphate mobilize calcium from rat brain microsomes. Alcohol 6:431–436
Malcolm R, Alkana RL (1982) Hyperbaric ethanol antagonism: role of temperature, blood and brain ethanol concentrations. Pharmacol Biochem Behav 16:341–346
Manji HK (1992) G-protein: implications for psychiatry. Am J Psychiatry 149:749–760
McCreery MJ, Hunt WA (1978) Physico-chemical correlates of alcohol intoxication. Neuropharmacology 17:451–461

McNamee MG, Fong TM (1988) Effects of membrane lipids and fluidity on acetylcholine receptor function. In: Aloia RC, Curtain CC, Gordon LM (eds) Lipid domains and the relationship to membrane function. Liss, New York, pp 43–62
Messing RO, Carpenter CI, Diamond I, Greenberg DA (1986) Ethanol regulates calcium channels in clonal neural cells. Proc Natl Acad Sci USA 83:6213–6215
Messing RO, Petersen PJ, Henrich CJ (1991) Chronic ethanol exposure increases levels of protein kinase C δ and ε and protein kinase C-mediated phosphorylation in cultured neural cells. J Biol Chem 266:23428–23432
Metcalfe JC, Seeman P, Burgen ASV (1968) The proton relaxation of benzyl alcohol in erythrocyte membranes. Mol Pharmacol 4:87–95
Meyer H (1899) Zur Theorie der Alkoholnarkose. Welche Eigenschaft der Anästhetica bedingt ihre narkotische Wirkung? Naunyn Schmiedebergs Arch Exp Pathol Pharmakol 42:109–118
Meyer H (1901) Zur Theorie der Alkoholnarkose. 3. Mittheilung: Der Einfluss wechselnder Temperatur auf Wirkungsstärke und Theilungscoefficient der Narcotica. Naunyn Schmiedebergs Arch Exp Pathol Pharmakol 46:338–346 (see also corrections, ibid. p 431)
Meyer KH (1937) Contributions to the theory of narcosis. Trans Faraday Soc 33:1062–1068
Mhatre MC, Ticku MK (1992) Chronic ethanol administration alters γ-aminobutyric acid$_A$ receptor gene expression. Mol Pharmacol 42:415–422
Mihic SJ, Wu PH, Kalant H (1991) Potentiation of γ-aminobutyric acid-mediated chloride flux by pentobarbital and diazepam but not ethanol. J Neurochem 58:745–751
Miles MF, Barhite S, Sganga M, Elliott M (1993) Phosducin-like protein: an ethanol-responsive potential modulator of guanine nucleotide-binding protein function. Proc Natl Acad Sci USA 90:10831–10835
Miles MF, Diaz JE, DeGuzman VS (1991) Mechanisms of neuronal adaptation to ethanol. J Biol Chem 266:2409–2414
Miller KW (1985) The nature of the site of general anesthesia. Int Rev Neurobiol 27:1–61
Miller KW, Paton DM, Smith EB (1973) The pressure reversal of general anesthesia and the critical volume hypothesis. Mol Pharmacol 9:131–143
Miller KW, Firestone LL, Alifimoff JK, Streicher P (1989) Nonanesthetic alcohols dissolve in synaptic membranes without perturbing their lipids. Proc Natl Acad Sci USA 86:1084–1087
Mochly-Rosen D, Chang F-H, Cheever L, Kim M, Diamond I, Gordon AS (1988) Chronic ethanol causes heterologous desensitization of receptors by reducing α_s messenger RNA. Nature 333:848–850
Molski TFP, Naccache PH, Marsh ML, Kermode J, Becker EL, Sha'afi RI (1985) Pertussis toxin inhibits the rise in the intracellular concentration of free calcium that is induced by chemotactic factors in rabbit neutrophils: possible role of the "G proteins" in stimulus-response coupling. Biochem Biophys Res Commun 126:1174–1181
Montpied P, Morrow A, Karanian J, Ginns EI, Martin BM, Paul SM (1991) Prolonged ethanol inhalation decreases γ-aminobutyric acid$_A$ receptor α subunit mRNAs in the rat cerebral cortex. Mol Pharmacol 39:157–163
Moreno FJ, Mills I, Gracia-Sainz JA, Fain JN (1983) Effects of pertussis toxin treatment on the metabolism of rat adipocytes. J Biol Chem 25:10938–10943
Moring J, Volpi M (1992) ADP-ribosylation of G-proteins in rat cerebral cortex: effects of acute and chronic ethanol exposure. Alcohol Clin Exp Res 16: 392
Moring J, Shoemaker WJ, Skita V, Mason RP, Hayden HC, Salomon RM, Herbette LG (1990) Rat cerebral cortical synaptoneurosomal membranes. Structure and interactions with imidazobenzodiazepine and 1,4-dihydropyridine calcium channel drugs. Biophys J 58:513–531

Morrow AL, Suzdak PD, Karanian JW, Paul SM (1988) Chronic ethanol administration alters γ-aminobutyric acid, pentobarbital and ethanol-mediated $^{36}Cl^-$ uptake in cerebral cortical synaptoneurosomes. J Pharmacol Exp Ther 246:158–164
Moss GWJ, Lieb WR, Franks NP (1991) Anesthetic inhibition of firefly luciferase, a protein model for general anesthesia, does not exhibit pressure reversal. Biophys J 60:1309–1314
Mueller GC, Fleming MF, LeMahieu MA, Lybrand CS, Barry KJ (1988) Synthesis of phosphatidylethanol – a potential marker for adult males at risk for alcoholism. Proc Natl Acad Sci USA 85:9778–9782
Mullins LJ (1954) Some physical mechanisms in narcosis. Chem Rev 54:289–323
Nagy LE, DeSilva SEF (1992) Ethanol increases receptor-dependent cyclic AMP production in cultured hepatocytes by decreasing G_i-mediated inhibition. Biochem J 286:681–686
Nagy LE, Diamond I, Collier K, Lopez L, Ullman B, Gordon AS (1989) Adenosine is required for ethanol-induced heterologous desensitization. Mol Pharmacol 36:744–748
Nhamburo PT, Salafsky BP, Tabakoff B, Hoffman PL (1987) Effects of ethanol on ouabain inhibition of mouse brain (Na^+,K^+)ATPase activity. Biochem Pharmacol 36:2027–2033
Olsen RW, Tobin AJ (1990) Molecular biology of $GABA_A$ receptors. FASEB J 4:1469–1480
Omodeo-Sale F, Lindi C, Palestini P, Masserini M (1991) Role of phosphatidylethanol in membranes. Effects on membrane fluidity, tolerance to ethanol, and activity of membrane-bound enzymes. Biochem 30:2477–2482
Overton E (1896) Über die osmotischen Eigenschaften der Zelle in ihrer Bedeutung für die Toxicologie und Pharmakologie. Z Phys Chem 22:189–209
Overton E (1901) Studien über die Narkose, zugleich ein Beitrag zur allgemeinen Physiologie. Gustav Fischer, Jena
Pang K-YY, Braswell LM, Chang L, Sommer TJ, Miller KW (1980) The perturbation of lipid bilayers by general anesthetics: a quantitative test of the disordered lipid hypothesis. Mol Pharmacol 18:84–90
Panza G, Grebb JA, Sanna E, Wright AG Jr, Handbauer I (1985) Evidence for down-regulation of ^3H-nitrendipine recognition sites in mouse brain after long-term treatment with nifedipine or verapamil. Neuropharmacology 24:1113–1117
Park D, Jhon D-K, Lee C-W, Lee K-H, Rhee SG (1993) Activation of phospholipase C isozymes by G protein $\beta\gamma$ subunits. J Biol Chem 268:4573–4576
Pohorecky LA, Rizek AE (1981) Biochemical and behavioral effects of acute ethanol in rats at different environmental temperatures. Psychopharmacol 72: 205–209
Puddey IB, Beilin LJ, Vandongen R (1986) Lack of effect of acute alcohol ingestion of erythrocyte Na^+,K^+-ATPase activity or passive sodium uptake in vivo in man. J Stud Alcohol 47:489–494
Quinlan JJ, Firestone LL (1992) Ligand-dependent effects of ethanol and diethylether at brain benzodiazepine receptors. Pharmacol Biochem Behav 42:787–790
Rabe CS, Giri PR, Hoffman PL, Tabakoff B (1990) Effect of ethanol on cyclic AMP levels in intact PC12 cells. Biochem Pharmacol 40:565–571
Rabin RA, Molinoff PB (1983) Multiple sites of action of ethanol on adenylyl cyclase. J Pharmacol Exp Ther 227:551–556
Rhee SG, Choi KD (1992) Regulation of inositol phospholipid-specific phospholipase C isozymes. J Biol Chem 267:12393–12396
Ross DH, Mutchler TM, Grady MM (1979) Calcium and glycoprotein metabolism as correlates for ethanol preference and sensitivity. Alcohol Clin Exp Res 3:64–69
Rottenberg H, Bittman R, Li J-L (1992) Resistance to ethanol disordering of membranes from ethanol-fed rats is conferred by all phospholipid classes. Biochim Biophys Acts 1123:282–290

Rottenberg H, Waring A, Rubin E (1981) Tolerance and cross-tolerance in chronic alcoholics: reduced membrane binding of ethanol and other drugs. Science 213:583–585

Rowe ES (1985) Thermodynamic reversibility of phase transitions. Specific effects of alcohols on phosphatidylcholines. Biochim Biophys Acta 813:321–330

Sagi-Eisenberg R (1989) GTP-binding proteins as possible targets for protein kinase C action. Trend Biochem Sci 14:355–357

Saito T, Luthin GR, Lee JM, Hoffman PL, Tabakoff B (1987) Differential effects of ethanol on the striatal and cortical adenylyl cyclase system. Japan J Pharmacol 43:133–141

Salem N Jr, Karanian JW (1988) Polyunsaturated fatty acids and ethanol. Adv Alcohol Subst Abuse 7:183–197

Salem N Jr, Ward G (1993) The effects of ethanol on polyunsaturated fatty acid composition. In: Alling C, Sun G (eds) Alcohol, cell membranes and signal transduction in the brain. Plenum Press, New York, pp 33–46

Schroeder F, Morrison WJ, Gorka C, Wood WG (1988) Transbilayer effects of ethanol on fluidity of brain membrane leaflets. Biochim Biophys Acta 946:85–94

Seeman P (1972) The membrane actions of anesthetics and tranquilizers. Pharmacol Rev 24:583–655

Seeman P, Roth S (1972) General anesthetics expand cell membranes at surgical concentrations. Biochim Biophys Acta 255:171–177

Shoemaker WJ, Moring J, Ganley L, Shaw J, Xu J, Seale E (1992) Chronic ethanol effects on MK-801 binding: age and brain regional differences. Alcohol Clin Exper Res 16:365

Sieghart W, Fuchs K, Zezula J, Buchstaller A, Zimprich F, Lassman H (1992) Biochemical, immunological, and pharmacological characterization of $GABA_A$-benzodiazepine receptor subtypes. In: Biggio G, Concas A, Costa E (eds) GABAergic synaptic transmission: molecular, pharmacological, and clinical aspects. Raven, New York, pp 155–162 (Advances in biochemical pharmacology, vol 4)

Simon SA, McIntosh TJ (1984) Interdigitated hydrocarbon chain packing causes the biphasic transition behavior in lipid/alcohol suspensions. Biochim Biophys Acta 773:169–172

Simonsson P, Rodriguez FS, Loman N, Alling C (1991) G proteins coupled to phospholipase C: molecular targets of long-term ethanol exposure. J Neurochem 56:2018–2026

Singer SJ, Nicholson GL (1972) The fluid mosaic model of the structure of cell membranes. Science 175:720–731

Skattebøl A, Rabin RA (1987) Effects of ethanol on $^{45}Ca^{2+}$ uptake in synaptosomes and in PC12 cells. Biochem Pharmacol 36:2227–2229

Smith EB, Bowser-Riley F, Daniels S, Dunbar IT, Harrison CB, Paton WDM (1984) Species variation and the mechanism of pressure-anaesthetic interactions. Nature 311:56–57

Smith TL, Gerhart MJ (1982) Alterations in brain lipid composition of mice made physically dependent to ethanol. Life Sci 31:1419–1425

Smrcka AV, Sternweis PC (1993) Regulation of purified subtypes of phosphatidylinositol-specific phospholipase C β by G protein α and $\beta\gamma$ subunits. J Biol Chem 268:9667–9674

Squier TC, Bigelow DJ, Thomas DD (1988) Lipid fluidity directly modulates the overall protein rotational mobility of the Ca-ATPase in sarcoplasmic reticulum. J Biol Chem 263:9178–9186

Sternweis PC, Smrcka AV (1993) G proteins in signal transduction: the regulation of phospholipase C. In: Marsh J, Good J (eds) The GTPase superfamily. John Wiley & Sons, New York, pp 96–111

Stibler H, Beaugé F, Leguicher A, Borg S (1991) Biophysical and biochemical alterations in erythrocyte membranes from chronic alcoholics. Scand J Clin Lab Invest 51:309–319

Sun GY, Sun AY (1983) Chronic ethanol administration induced an increase in phosphatidylserine in guinea pig synaptic plasma membranes. Biochem Biophys Res Commun 113:262–268

Sun Gy, Huang H-M, Lee D-Z, Sun AY (1984) Increased acidic phospholipids in rat brain membranes after chronic administration. Life Sci 35:2127–2133

Suzdak PD, Schwartz RD, Skolnick P, Paul SM (1986) Ethanol stimulates gamma-aminobutyric acid receptor-mediated chloride transport in rat brain synaptoneurosomes. Proc Natl Acad Sci USA 83:4071–4075

Tabakoff B, Melchior C, Hoffman P (1984) Factors in ethanol tolerance. Science 224:523–524

Tabakoff B, Ritzmann RF, Raju TS, Deitrich RA (1980) Characterization of acute and chronic tolerance in mice selected for inherent differences in sensitivity to ethanol. Alcohol Clin Exp Res 4:70–73

Tang W-J, Iñiguez-Lluhi JA, Mumby S, Gilman AG (1992) Regulation of mammalian adenylyl cyclases by G-protein α and $\beta\gamma$ subunits. Cold Spring Harbor Symposia on Quantitative Biology 52:135–144

Taraschi TF, Ellingson JS, Wu A, Zimmerman R, Rubin E (1986) Phosphatidylinositol from ethanol-fed rats confers membrane tolerance to ethanol. Proc Natl Acad Sci USA 83:9398–9402

Taraschi TF, Ellingson JS, Wu-Sun A, Zimmerman R, Rubin E (1990) Rats withdrawn from ethanol rapidly re-acquire membrane tolerance after resumption of ethanol feeding. Biochim Biophys Acta 1021:51–55

Taussig R, Inguiñez-Lluhi JA, Gilman AG (1993) Inhibition of adenylyl cyclase by Gi α. Science 261:218–221

Tewari S, Sytinski IA (1985) Alcohol. In: Lajtha A (ed) Alterations of metabolites in the nervous system. Plenum, New York, pp 219–261, (Handbook of neurochemistry, 2nd edn, vol 9)

Treistman SN, Moynihan MM, Wolf DF (1987) Influence of alcohol, temperature, and region on the mobility of lipids in neuronal membrane. Biochim Biophys Acta 898:109–120

Trudell JR, Hubbell WL, Cohen EN (1973a) Pressure reversal of inhalation anesthetic-induced disorder in spin-labeled phospholipid vesicles. Biochim Biophys Acta 291:328–334

Trudell JR, Hubbell WL, Cohen EN, Kendig JJ (1973b) Pressure reversal of anesthesia. Anesthesiology 38:207–211

Ueda I (1991) Interfacial effects of anesthetics on membrane fluidity. In: Aloia RC, Curtin CC, Gordon LM (eds) Drug and anesthetic effects on membrane structure and function. Wiley Liss, New York, pp 15–33

Ulrichsen J, Gustavsson L, Alling C, Clemmesen L, Hemmingsen R (1991) Acidic phospholipids in synaptosomal plasma membranes during repeated episodes of physical ethanol dependence in the rat. Alcohol Alcohol 26:323–328

Vicini S (1991) Pharmacologic significance of the structural heterogeneity of the $GABA_A$ receptor-chloride ion channel complex. Neuropsychopharmacology 4:9–15

Volpi M, Naccache PH, Molski TFP, Shefcyk J, Huang C-K, Marsh ML, Munoz J, Becker EL, Sha'afi RI (1985) Pertussis toxin inhibits fMet-Leu-Phe- but not phorbol ester-stimulated changes in rabbit neutrophils: role of G proteins in excitation response coupling. Proc Natl Acad Sci USA 82:2708–2712

Wafford KA, Burnett DM, Leidenheimer NH, Burt DR, Wang JB, Kofuji P, Dunwiddie TV, Harris RA, Sikela JM (1991) Ethanol sensitivity of the $GABA_A$ receptor expressed in Xenopus oocytes requires 8 amino acids contained in the γ2L subunit. Neuron 7:27–33

Waltman C, Levine MA, McCaul ME, Svikis DS, Wand GS (1993) Enhanced expression of the inhibitory protein Gi2α and decreased activity of adenylyl cyclase in lymphocytes of abstinent alcoholics. Alcohol Clin Exp Res 17:315–320

Wand GS, Levine MA (1991) Hormonal tolerance to ethanol is associated with decreased expression of the GTP-binding protein, $G_s\alpha$, and adenylyl cyclase activity in ethanol-treated LS mice. Alcohol Clin Exp Res 15:705–710

Wand GS, Diehl AM, Levine MA, Wolfgang D, Samy S (1993) Chronic ethanol treatment increases expression of inhibitory G-proteins and reduces adenylyl-cyclase activity in the central nervous system of two lines of ethanol-sensitive mice. J Biol Chem 268:2595–2601

Waring AJ, Rottenberg H, Ohnishi T, Rubin E (1981) Membranes and phospholipids of liver mitochondria from chronic alcoholic rats are resistant to membrane disordering by alcohol. Proc Natl Acad Sci USA 78:2582–2586

Wenger JR, Tiffany TM, Bombardier C, Nicholls K, Woods S (1981) Ethanol tolerance in the rat is learned. Science 213:575–577

Williams RJ, Kelly E (1993) Chronic ethanol reduces immunologically detectable $G_{q\alpha/11\alpha}$ in NG108-15 cells. J Neurochem 61:1163–1166

Williams RJ, Veale MA, Horne P, Kelly E (1993) Ethanol differentially regulates guanine nucleotide-binding protein α subunit expression in NG108-15 cells independently of extracellular adenosine. Mol Pharmacol 43:158–166

Whittington MA, Little HJ (1988) Nitrendipine prevents the ethanol withdrawal syndrome, when administered chronically with ethanol prior to withdrawal. Br J Pharmacol 92:385P

Wing DR, Harvey DJ, Hughes J, Dunbar PG, McPherson KA, Paton WDM (1982) Effects of chronic ethanol administration on the composition of membrane lipids in the mouse. Biochem Pharmacol 31:3431–3439

Wood WG, Schroeder F (1988) Membrane effects of ethanol: bulk lipid versus lipid domains. Life Sci. 43:467–475

Wood WG, Schroeder F, Hogy L, Rao AM, Nemecz G (1990) Asymmetric distribution of a fluorescent sterol in synaptic plasma membranes: effects of chronic ethanol consumption. Biochim Biophys Acts 1025:243–264

Wood WG, Gorka C, Johnson JA, Sun GY, Sun AY, Schroeder F (1991a) Chronic ethanol consumption alters transbilayer distribution of phosphatidylcholine in erythrocytes of Sinclair (S-1) miniature swine. Alcohol 8:395–399

Wood WG, Schroeder F, Murali Rao A (1991b) Significance of ethanol-induced changes in membrane lipid domains. Alcohol Alcohol [Suppl 1]:221–225

Wood WG, Rao AM, Igbavboa U, Semotuk M (1993) Cholesterol exchange and lateral cholesterol pools in synaptosomal membranes of pair-fed control and chronic ethanol-treated mice. Alcohol Clin Exp Res 17:345–350

Wu D, Lee CH, Rhee SG, Simon MI (1992) Activation of phospholipase C by the α subunits of the G_q and G_{11} proteins in transfected Cos-7 cells. J Biol Chem 267:1811–1817

CHAPTER 3

Effects of Ethanol on Voltage-Dependent Calcium Channel Function

S.W. LESLIE

A. Voltage-Dependent Calcium Channels

I. Introduction

Before discussing the effects of ethanol on voltage-dependent calcium channels it is important to consider the physiological roles of calcium channels in the central nervous system. Indeed, it is only in recent years that investigators have begun to recognize that numerous types of calcium channels exist, that their cellular and brain regional localizations may differ, and that activation of these channel types may initiate different cellular responses. It is important to state at the outset, however, that this area of research is still at an early stage and many important questions remain to be answered and several controversies need to be resolved. The details of these questions and controversies are beyond the scope of this review. However, an attempt will be made to highlight some of the key questions in order to allow for a timely perspective on those effects of ethanol that may be linked with alterations in calcium channel function.

Calcium channels in the central nervous system can be separated into high voltage activated (HVA) and low voltage activated (LVA) types (SHER et al. 1991). At least three distinct HVA calcium channels exist and these are characterized predominantly in terms of their sensitivity to specific pharmacological agonists and antagonists, but also by differences in electrophysiological activation and inactivation parameters and magnitudes of channel conductances. HVA channels characterized thus far include L, N, and P type channels (for reviews see MILLER 1987; TSIEN et al. 1991; SHER et al. 1991). L (long-lasting) and N (neither L nor T) channels were first identified by NOWYCKY et al. (1985). This report also identified an LVA calcium channel that was named T (transient) type for its activation at low voltage and rapid inactivation leading to the transient nature of the channel response. P type channels have been identified only recently (TSIEN et al. 1991; SHER et al. 1991).

II. L Type Channels

Of the calcium channels identified thus far, L type channels have been studied most extensively. These channels are blocked selectively by clinically used dihydropyridine, phenylalkylamine, and benzothiazepine calcium channel blockers (e.g., nifedipine, verapamil, and diltiazem, respectively), the binding sites for which are pharmacologically distinct but reside on the α_1-subunit of the L type calcium channel complex (TSIEN et al. 1991). L type calcium channels are known to exist in many vascular and nonvascular tissues (GODFRAIND et al. 1986). It is noteworthy that L type calcium channel blockers have potent pharmacological effects in the cardiovascular system but produce very few clinically noticeable side effects that involve central nervous system function. This is striking since the brain possesses high concentrations of dihydropyridine receptors. Receptor binding studies show that binding affinities for dihydropyridine antagonists are essentially the same, with K_d values below 1 nM, in brain, cardiovascular, and other tissues (GODFRAIND et al. 1986). Many dihydropyridines are lipid soluble and, therefore, cross the blood-brain barrier. Nimodipine, in particular, is an example of a dihydropyridine calcium channel blocker that crosses the blood-brain barrier and its usefulness is derived from its ability to concentrate in the brain (SCRIABINE et al. 1989). As is the case for other dihydropyridine calcium channel blockers, nimodipine produces few behavioral side effects. However, it is a drug that has cerebrovasodilatory and consequent antiischemic effects in the brain. Studies have shown that nimodipine, and possibly other dihydropyridine calcium channel blockers, may be effective in the treatment of cerebral ischemia and may have benefit in treating cognitive deficiencies associated with aging (SCRIABINE et al. 1989). These benefits of nimodipine are attributable, to some extent, to its selectivity for cerebrovasculature; however, these is also significant evidence that nimodipine and other dihydropyridines have direct effects on neurons (MILLER 1987; SCRIABINE et al. 1989).

The pharmacological profile of calcium channel blocker actions in the brain is consistent with what is known about the physiological role of L type calcium channels on neurons. L type calcium channels appear to exist largely on dendrites and cell bodies in the brain (MILLER 1987; AHLIJANIAN et al. 1990). Activation of these channels may mediate long-lasting increases in cytosolic calcium that may be linked with second messenger functions of calcium in the brain (AHLIJANIAN et al. 1990). Overactivation of L type calcium channels may result in neuronal toxicity associated with excessive calcium entry. This role of L type calcium channels may help explain the potential therapeutic effect of dihydropyridine calcium channel blockers described above (SCRIABINE et al. 1989).

Some controversy exists as to the possible involvement of L type calcium channels in neurotransmitter release from nerve terminals. Early biochemical work on the question of the effects of L type calcium channel blockers on neurotransmitter release showed that these drugs were ineffective in blocking potassium-induced calcium entry into and neurotransmitter release from synaptosomes (DANIELL et al. 1983; SUSZKIW et al. 1986; see also MILLER 1987). Interestingly, similar studies conducted on adrenal medullary, PC-12, and neuroblastoma cells showed a highly potent inhibition of voltage-dependent calcium entry and neurotransmitter release by dihydropyridine calcium channel blockers. However, exposure of PC-12 cells to nerve growth factor caused these cells to differentiate such that the potassium-induced release of catecholamines was predominantly insensitive to inhibition by dihydropyridine antagonists or stimulation by the dihydropyridine agonist BAY K 8644 (KONGSAMUT and MILLER 1986). Thus, these studies suggest that calcium entry into and neurotransmitter release from brain nerve terminals may not be sensitive to inhibition by L type calcium channel blockers. However, adrenal medullary and neuroblastoma cells may possess L type calcium channels which are involved in the control of neurotransmitter release. More recent evidence implicates L-type calcium channels in neurotransmitter release from synaptosomes and brain slices. These studies show that exposure of synaptosomes and brain slices to BAY K 8644 in the presence of partial depolarization with potassium results in BAY K 8644-stimulated calcium uptake and neurotransmitter release that is sensitive to inhibition by dihydropyridine calcium channel blockers (MIDDLEMISS and SPEDDING 1985; WOODWARD and LESLIE 1986; E.J. WHITE and BRADFORD 1986; WOODWARD et al. 1988a; GANDHI and JONES 1990). The dihydropyridine-sensitive component of calcium entry and neurotransmitter release shown in these studies was quite small. Thus, calcium-dependent neurotransmitter release appears to be predominantly dihydropyridine insensitive.

As stated above, research on L type calcium channels is still at a very early stage of understanding. Recent work has characterized four distinct rat brain cDNAs for the α_1-subunit of calcium channels (SNUTCH et al. 1990). These appear to be derived from distinct genes or gene families. The significance of these findings is not yet clear, although one might speculate that they implicate the α_1-subunit of calcium channels in the expression of distinct calcium channel types in brain. Furthermore, recent evidence indicates that molecular diversity exists within the L type class of calcium channels (PEREZ-REYES et al. 1990). Why these different forms of L type channels exist and what functional differences might occur in response to their expression is not clear. Finally, dihydropyridine-sensitive calcium channels may be present on astrocyte membranes (HERTZ et al. 1989). Potassium-stimulated calcium uptake into astrocytes was inhibited by very low concentrations of nimodipine (IC_{50} of approximately 3 nM). How the presence of L type calcium channels on astrocyte membranes might physiologically affect glial/neuronal interactions is not known, but is certainly a question that is worth pursuing.

The conclusions that can be drawn from the above discussion are as follows: (1) High concentrations of dihydropyridine receptor sites located on L type calcium channels exist in the brain. (2) The cellular location of dihydropyridine binding appears to be predominantly on dendrites and neuronal cell bodies. Activation of these channels may be linked with long-lasting calcium entry associated with calcium-dependent second messenger function. Excessive activation of these channels may be associated with neurotoxicity. Thus, alcohol, or any other substance that may serve to modify the function of L type channels, would be expected to produce symptomatology consistent with these physiological functions. (3) To a limited extent, L type calcium channels may also be present on presynaptic nerve terminals. They may have a small role in the activation and/or control of calcium-dependent neurotransmitter release. L type channels may have a more significant involvement in the contol of peptide release from a variety of exocrine tissues and in brain (GODFRAIND et al. 1986; WANG et al. 1991a,b).

III. N, P, and T Type Channels

While L type channels appear to be localized predominantly on dendrites and neuronal cell bodies, N type channels exist in high concentrations on nerve terminals (SHER et al. 1991). N type channels are not inhibited by dihydropyridine calcium channel blockers, but are blocked by the GVIA fraction of the *Conus geographus* sea snail toxin, ω-conotoxin (OLIVERA et al. 1985; MILLER 1987; TSIEN et al. 1991). Autoradiographic studies with radiolabeled dihydropyridines and ω-conotoxin indicate that L and N type channels have similarities and differences in brain regional distributions (SHER et al. 1991). Taken together, these findings suggest that the functional roles of L and N type channels may be different. Indeed, ω-conotoxin is a potent inhibitor of neurotransmitter release (HIRNING et al. 1988; WOODWARD et al. 1988b). This observation, in combination with the recognition that N type channels are localized on nerve terminals, suggests the involvement of N type calcium channels in calcium-dependent neurotransmitter release.

Both N and L type calcium channels are HVA channels. They differ somewhat in their electrophysiological properties (TSIEN et al. 1991). The characterization of neuronal N type channels has relied, to a great extent, on the use of ω-conotoxin as a specific pharmacological antagonist (SHER and CLEMENTI 1991). Indeed, ω-conotoxin has been a highly valuable tool and studies with this toxin have led to the general conclusion that ω-conotoxin-sensitive calcium channels are linked with neurotransmitter release at presynaptic terminals (SHER and CLEMENTI 1991). There are, however, some reasons for doubt concerning this conclusion. For example, while ω-conotoxin-binding sites are widely distributed in the brain, some brain areas have few or no binding sites (SHER and CLEMENTI 1991). In these brain regions some other voltage-dependent calcium channel type(s) must be linked with neurotransmitter release. Secondly, ω-conotoxin is highly potent in blocking potassium-stimulated calcium entry into synaptosomes isolated from frog and chick brain, but calcium entry into rat brain synaptosomes is unaffected even by high concentrations of ω-conotoxin (SUSZKIW et al. 1986; WOODWARD et al. 1988b; LUNDY et al. 1991). WOODWARD et al. (1988b) reported that there was no apparent correlation between the ability of ω-conotoxin to block potassium-stimulated endogenous dopamine release from striatal synaptosomes isolated from rat brain and its ability to inhibit calcium entry into the same preparation. Concentrations of ω-conotoxin as low as 10 n*M* significantly inhibited dopamine release, but 1 m*M* ω-conotoxin was required to produce a small inhibition of calcium entry into the same synaptosomal preparations. The reason for the apparent disassociation between ω-conotoxin inhibition of neurotransmitter release and its lack of inhibition of calcium entry is not known. A possible explanation is that while N-type calcium channels associated with neurotransmitter release may represent a small percentage (20%–25% at best) of the total calcium channels present on nerve terminals, they are positioned at key "hot spots" in close proximity to neurotransmitter release sites.

As stated above, ω-conotoxin is more potent in antagonizing potassium-stimulated calcium uptake in synaptosomes isolated from frog and chick brain than in synaptosomes isolated from mammalian brain. The reason for the apparent species differences is not yet clear, although recent evidence suggests that multiple N type channels may exist (Williams et al. 1992). It may be that evolutionary changes have taken place such that different forms of N type calcium channels exist throughout the phylogenetic scale. It is known that animals lower on the evolutionary scale are considerably more sensitive to ω-conotoxin than higher mammals (Olivera et al. 1985). For example, intraperitoneal injection of GVIA ω-conotoxin into fish, the natural prey of the *Conus geographus* snail, resulted in paralysis and death. Intraperitoneal injection of ω-conotoxin into mice, however, produced no visible effects. This may be due to the presence of subtypes of N type channels in the two species of animals.

Recent work has shown that another type of HVA calcium channel that exists in the brain is insensitive to both dihydropyridines and ω-conotoxin. This channel is referred to as the P type channel (Tsien et al. 1991; Sher et al. 1991). ω-Agatoxin IVA (ω-aga-IVA), a component of the venom from the funnel web spider, *Agelenopsis aperta*, is a specific inhibitor of P type calcium channels. In cerebellar Purkinje cells, approximately 85%–90% of the high threshold calcium current is not blocked by dihydropyridine calcium channel blockers or ω-conotoxin, but is blocked completely by ω-aga-IVA. However, only 20% of the calcium current was blocked by the ω-aga-IVA toxin in hippocampal CA1 neurons. Sympathetic neurons and some hippocampal CA3 neurons have no ω-aga-IVA sensitive currents (Mintz et al. 1992a). Thus, P type channels may comprise a higher percentage of calcium channels in certain brain regions than in others. ω-Aga-IVA blocks 70%–80% of synaptosomal calcium entry (Mintz et al. 1992b) and inhibits neuromuscular transmission in insects by blocking presynaptic function (Adams et al. 1990). P type channels are also present on nerve terminals of squid axon (Sher et al. 1991). These findings suggest that P type channels may be highly concentrated on presynaptic terminals and may be involved in neurotransmitter release. P type channels are not restricted to presynaptic terminals, however, since they are also known to exist on dendrites of cerebellar Purkinje cells (SHER et al. 1991).

T type calcium channels activate in response to low voltage and are characterized as LVA channels (TSIEN et al. 1991; SHER et al. 1991). Unlike the HVA channels described above, T type channels inactivate rapidly (NOWYCKY et al. 1985). These channels have not received as much attention as HVA channels and are not as well characterized, owing largely to the fact that specific pharmacological tools are not yet available to study these channels. Octanol and amiloride have been shown to inhibit more potently T type currents than they do currents of HVA channels (TSIEN et al. 1991); but these drugs are not selective inhibitors of T type channels.

1. Conclusions

The conclusions that can be drawn from studies of N, P, and T channels are as follows: (1) N type calcium channels have a wide distribution in the brain and are found predominantly on neurons where they are localized on presynaptic nerve terminals. ω-Conotoxin is a selective inhibitor of most N type channels. L and N channels share many electrophysiological similarities and it is sometimes difficult to differentiate these channels electrophysiologically. Different subtypes of N type calcium channels have been recently identified. The role of these channel subtypes is not yet clear. (2) P type calcium channels were discovered only recently. While a great deal of information has been obtained about these channels in a short period of time, much is yet to be learned. While P type channels are found throughout the brain, they are particularly concentrated in cerebellar Purkinje cells and are prominent on presynaptic nerve terminals. P type channels can be clearly differentiated from L and N type HVA channels. P type channels are inhibited specifically by ω-agatoxin IVA, a component of the venom of the funnel web spider, *Agelenopsis aperta*. (3) Unlike L, N, and P type channels T type channels are opened by small depolarizations and are thus classified as LVA channels. These channels are found in both excitable and nonexcitable tissues and may be involved in pacemaker type activity. Specific pharmacological probes for these channels have not yet been identified.

B. Effects of Ethanol and Other Sedative-Hypnotic Drugs on Voltage-Dependent Calcium Channels

I. Ethanol Effects on Ion Channels

Many biochemical and electrophysiological studies have shown that pharmacologically relevant ethanol concentrations inhibit voltage-dependent calcium channels (LESLIE 1987; LESLIE et al. 1990). Some reports indicate that relatively high concentrations of ethanol are required to inhibit voltage-dependent calcium channels while other ion channels, e.g., GABA-activated chloride channels and *N*-methyl-*d*-aspartate (NMDA)-activated ion channels, are much more sensitive (HARRIS and ALLEN 1989; LESLIE et al. 1990). A review of the literature, as described below, reveals that the modification of ion channel function, including voltage-dependent calcium channels, GABA-activated chloride channels, and NMDA-activated ion channels, is probably quite complex and may depend upon receptor subtypes in each of the ion channel categories, some of which may be highly sensitive to ethanol while others are not.

NMDA receptors, which are generally regarded as being highly sensitive to inhibition by ethanol, may exist in various isoforms in the brain (MONYER et al. 1992). LIMA-LANDMAN and ALBUQUERQUE (1989) and SIMON et al. (1991) reported that the NMDA response in some neurons is sensitive to inhibition by ethanol while in others it is not. The difference in ethanol sensitivity of various neuronal NMDA responses may be linked to the presence of different subtypes of NMDA receptors which differ in sensitivity to ethanol.

A similar picture may exist for the sensitivity of GABA-activated chloride conductance to ethanol. Some laboratories report that GABA-activated chloride conductance is highly sensitive to activation by ethanol while other reports indicate no ethanol effect (G. WHITE et al. 1990). Furthermore, brain regional and genetic differences in ethanol sensitivity of GABA-activated chloride channels are known to exist (HARRIS and ALLAN 1989; PROCTOR et al. 1992). Subtypes of GABA-activated chloride channels exist in the brain. Recent evidence suggests that ethanol sensitivity is found in GABA receptors which possess a γ2L-subunit of the receptor complex, while receptors with a γ2S-subunit may not be ethanol sensitive (WAFFORD et al. 1991). Other recent findings suggest that differences in post-translational processing of the $GABA_A$ receptor may be responsible for the differences in genetic sensitivity to ethanol (ZAHNISER et al. 1992). Thus, sensitivity of ion channels, including voltage-dependent calcium channels, may depend upon molecular properties of the channel subtype studied. This will be discussed in detail below.

Another point to consider concerning the physiological implications of ethanol on ion channels, including voltage-dependent calcium channels, is the cellular role of the ion channel and the extent to which it is involved in a physiological amplification process. For example, ion channels that are only slightly inhibited by ethanol but exist at the leading edge of a cellular cascade system may be just as involved in ethanol's effects as ion channels that are potently inhibited by ethanol but at the end of an amplification system. For example, KATZ and MILEDI (1967) showed that depolarization of presynaptic terminals of the giant squid did not result in a postsynaptic response until a critical threshold of depolarization was reached (approximately 25–40 mV). After this threshold was reached, further depolarization of presynaptic terminals resulted in an exponential increase in the postsynaptic response. A slight modification of presynaptic excitability within this dynamic range would produce a robust postsynaptic change. Thus, if alcohol were to produce a small percentage change of presynaptic voltage-dependent calcium channel function involved in neurotransmitter release, for example, the result may be reflected as an exponential change postsynaptically. Indeed, ethanol inhibits potassium-stimulated calcium uptake into isolated presynaptic nerve terminals (synaptosomes) (HARRIS and HOOD 1980; LESLIE et al. 1983). Concentrations of ethanol as low as 25 m*M* produced a small (approximately 8%), but statistically significant, inhibition of the fast component of potassium-stimulated calcium uptake into cerebrocortical synaptosomes (LESLIE et al. 1983). HARRIS and HOOD (1980) reported that 45 m*M* ethanol produced a small, but significant, inhibition of potassium-stimulated calcium uptake into whole mouse brain synaptosomes. Thus, relatively low concentrations of ethanol (25–50 m*M*) may produce small decreases in voltage-dependent calcium uptake in some cases. Nevertheless, this small presynaptic inhibition may be quite significant physiologically since, as described above, in the intact brain it may be amplified exponentially resulting in large decreases in postsynaptic responses (KATZ and MILEDI 1967).

II. Ethanol Effects on Different Types of Calcium Channel

Voltage-dependent calcium channels on some nerve terminals may be highly sensitive to ethanol. GRUOL (1982) reported that low concentrations of ethanol (10–40 m*M*) reduced synaptic activity in cultured spinal neurons in a manner consistent with inhibition of voltage-dependent calcium channels. WANG et al. (1991b) found that, in patch-clamped nerve terminals isolated from rat neurohypophysis, both fast inactivating (dihydropyridine-insensitive) and long-lasting calcium currents (dihydropyridine-sensitive) were significantly inhibited by concentrations of ethanol as low as 10 m*M*. These effects were selective for calcium channels since transient potassium currents were not altered by ethanol concentrations as high as 100 m*M* in this preparation. The magnitude of inhibition by ethanol was found to be quite large. Ethanol, 10 m*M*, inhibited the long-lasting currents by approximately 30%. These currents, which were likely derived from L type channels, were more sensitive to ethanol than the fast-inactivating currents, which were identified as N_t and are known to be dihydropyridine insensitive.

Potassium-stimulated release of arginine vasopressin (AVP) from these nerve terminals was also studied. Ethanol, 10 m*M*, inhibited AVP release by approximately 50%. This inhibitory effect was eliminated in permeabilized terminal preparations, indicating that the mechanism for inhibition of AVP release occurred prior to the entry of calcium into the terminals. Thus, the potent inhibition of AVP release by ethanol was probably derived mechanistically from the inhibition of voltage-dependent calcium channels.

A later study using intact neurohypophysis confirmed the potent effect of ethanol on AVP release (WANG et al. 1991a). Whole cell patch-clamp studies confirmed that ethanol has a potent inhibitory effect on voltage-dependent calcium channels in this preparation. Ethanol, 25 m*M*, produced more than a 30% inhibition of calcium currents, but 10 m*M* ethanol did not appear to have a statistically significant effect on calcium channels in this study (WANG et al. 1991a).

TWOMBLY et al. (1990) also examined the effects of ethanol on transient (T type) and long-lasting calcium currents (L type) in N1E-115 neuroblastoma and NG108-15 neuroblastoma × glioma cells. Their findings agreed with those of WANG et al. (1991a,b) in that both types of calcium currents were found to be inhibited by ethanol. However, these calcium currents were much less sensitive to inhibition by ethanol than those of neurohypophyseal terminals. TWOMBLEY et al. (1990) found that transient calcium currents were inhibited slightly but significantly by 30 m*M* ethanol. Long-lasting currents were inhibited significantly by alcohol concentrations in the range of 100–300 m*M* but not by 30 m*M*. Ethanol, 300 m*M*, inhibited both types of calcium currents by approximately 40%. These findings suggest that in N1E-115 and NG108-15 cells T type channels are more sensitive to ethanol than L type channels. This differs from the results of WANG et al. (1991a,b), who found that both fast-inactivating and long-lasting calcium currents were highly sensitive to ethanol but that long-lasting, L type channels were more sensitive than fast-inactivating channels. The reasons for these discrepancies are not known. A possible explanation may reside in the presence of distinct isoforms of calcium channel types (SNUTCH et al. 1990; PEREZ-REYES et al. 1990) on neurohypophyseal, N1E-115, and NG108-15 cells. These isoforms may vary in sensitivity to ethanol.

In addition to the electrophysiological studies described above, recent biochemical and behavioral studies have focused on the potential involvement of L type channels in the neuronal actions of ethanol. Much of this work focuses on the involvement of these channels in the chronic effects of ethanol. This aspect of ethanol/L type calcium channel interaction is discussed later in this review. Ethanol concentrations as low as 50 m*M* significantly inhibited potassium-stimulated $^{45}Ca^{2+}$ uptake into PC-12 cells (MESSING et al. 1986), which is known to occur through L type channels. The IC_{50} for inhibition of $^{45}Ca^{2+}$ uptake into PC-12 cells was 211 m*M*. SKATTEBOL and RABIN (1987) also reported a similar IC_{50} (238 m*M*) for ethanol inhibition of $^{45}Ca^{2+}$ uptake into PC-12 cells. These investigators also reported a similar IC_{50} (220 m*M*) for ethanol inhibition of the fast component of $^{45}Ca^{2+}$ uptake into rat cerebrocortical synaptosomes. In agreement with a previous report (LESLIE et al. 1983), SKATTEBOL and RABIN (1987) found that the fast component of $^{45}Ca^{2+}$ uptake by cerebrocortical synaptosomes was much more sensitive to inhibition by ethanol than was the slow component of $^{45}Ca^{2+}$ uptake. This observation is significant since fast-phase $^{45}Ca^{2+}$ uptake is thought to occur exclusively through voltage-dependent calcium channels while slow-phase $^{45}Ca^{2+}$ uptake involves other cellular processes such as Na^{+}/Ca^{2+} exchange (NACHSHEN and BLAUSTEIN 1980). The inhibition of dihydropyridine-sensitive calcium channels by ethanol probably does not involve a competitive interaction with the dihydropyridine recognition site, however, since concentrations of ethanol much larger than those required to inhibit the channels are necessary to displace ^{3}H-nitrendipine binding (GREENBERG and COOPER 1984; HARRIS et al. 1985).

Dihydropyridine calcium channel blockers potentiate the sedative properties of ethanol. ISAACSON et al. (1985) found that nimodipine, 5 mg/kg, potentiated both the motor incoordination and hypothermia produced by ethanol in mice. DOLIN and LITTLE (1986) reported that a high dose of nitrendipine (100 mg/kg) increased the anesthetic potencies of both ethanol and pentobarbital. Similar findings were reported for the calcium channel blockers verapamil and flunarizine (DOLIN and LITTLE 1986).

Other electrophysiological studies also indicate that ethanol may have actions on neurons that are mediated through voltage-dependent calcium channels. Calcium currents in *Aplysia* neurons are inhibited by concentrations of ethanol as low as 50 m*M* (CAMACHO-NASI and TREISTMAN 1986). Sodium and potassium conductances were found to be much less sensitive to ethanol in the *Aplysia* preparation. ESKURI and POZOS (1987) reported that ethanol reduced calcium currents in dorsal root ganglion cells with an IC_{50} of 148 m*M* at 30°C, 73 m*M* at 37°C, and 44 m*M* at 43°C. These investigators reported that ethanol did not alter the amplitude or rate of rise of calcium action potentials and did not alter resting potential of the dorsal root ganglion cells. Thus, taken together with the increasing potency of ethanol with increasing temperature, the results suggest that ethanol may modify the lipid environment around calcium channels or exert direct effects on the channel protein. This suggestion is consistent with the findings of WANG et al. (1991a,b) and TWOMBLY et al. (1990), who showed that calcium channel inhibition by ethanol occurred without changing the voltage dependence of activation or inactivation. The suggestion that a lipid microdomain around calcium channels may be important for the effects of ethanol is also supported by studies showing that the potency of a series of alcohols in blocking voltage-dependent calcium channels is correlated closely with their membrane/buffer partition coefficient (STOKES and HARRIS 1982).

III. Brain Regional Differences in the Effects of Ethanol

Ethanol's ability to inhibit voltage-dependent calcium channels may vary from one brain region to another. STOKES and HARRIS (1982) found that inhibition of slow-phase potassium-stimulated calcium uptake by ethanol was greater in synaptosomes isolated from cerebellum and striatum compared to cerebral cortex and brain stem. Fast-phase calcium uptake, however, was inhibited by 25 m*M* ethanol in cerebrocortical synaptosomes but 80 m*M* ethanol inhibited only the slow-phase component in cerebellum, brain stem, and midbrain synaptosomes (LESLIE et al. 1983). Ethanol, 200 mM, was needed to inhibit the fast component of calcium entry into and endogenous neurotransmitter release from striatal synaptosomes (LESLIE et al. 1986b). In a study examining the effects of ethanol on fast-phase calcium uptake and endogenous norepinephrine release from brain regions, ethanol was found to be most potent in inhibiting potassium-stimulated calcium entry into hypothalamic synaptosomes and least potent in blocking calcium uptake into brain stem synaptosomes (DANIELL and LESLIE 1986). The reasons for these brain regional differences in ethanol sensitivity are not known, but, again, point to the possibility that molecular differences may exist in calcium channels or in microenvironment sites among calcium channels in different brain regions.

IV. Chronic Ethanol Effects on Calcium Channels

Regional differences also appear to exist in the responses of calcium channels to chronic ethanol administration. Chronic administration of ethanol results in adaptation of calcium channels in cerebrocortical synaptosomes such that ethanol added in vitro has a reduced ability to inhibit calcium entry (LESLIE et al. 1983). A similar adaptive response in synaptosomal calcium uptake after chronic ethanol exposure was reported by HARRIS and HOOD (1980). Chronic ethanol exposure was also shown to result in adaptation to the inhibitory effects of ethanol in hypothalamic synaptosomes, but no adaptation was observed in brain stem or cerebellar synaptosomes (DANIELL and LESLIE 1986). Ethanol did not produce an adaptive response in synaptosomes isolated from striatum after chronic ethanol administration. To the contrary, chronic ethanol treatment resulted in an apparent uncoupling of calcium entry from dopamine (LESLIE et al. 1986b). These studies showed that chronic ethanol exposure did not alter potassium-stimulated calcium entry into striatal synaptosomes but markedly reduced the release of endogenous dopamine from the same synaptosomal preparations. Thus, responses to chronic ethanol administration may result in adaptation of certain calcium channels involved in the development of functional tolerance (LESLIE et al. 1983) and no adaptation in others (DANIELL and LESLIE 1986). Other calcium channels may not adapt to the presence of chronic ethanol but may, in fact, deteriorate in function (LESLIE et al. 1986b).

Evidence implicates dihydropyridine-sensitive, L type calcium channels in the chronic effects of ethanol. Chronic exposure of PC-12 cells to ethanol results in an increased uptake of $^{45}Ca^{2+}$ (MESSING et al. 1986; GREENBERG et al. 1987) that is correlated with a chronic ethanol-induced increase in the number of ^{3}H-nitrendipine-binding sites (MESSING et al. 1986). Chronic administration of ethanol to rats has also been shown to increase dihydropyridine binding in brain homogenates (DOLIN and LITTLE 1989). Furthermore, chronic exposure of cultured bovine adrenal chronmaffin cells to 200 mM ethanol for 4 days resulted in a significant increase in dihydropyridine binding (BRENNAN et al. 1989; HARPER et al. 1989). However, dihydropyridine binding in human autopsy brain tissue isolated less than 36 h postmortem did not show increased binding in alcoholics compared to nonalcoholics (KRIL et al. 1989). MARKS et al. (1989) also reported that dihydropyridine binding was not increased in cerebrocortical autopsy tissue isolated from alcoholics. Many of the patients in this study were hospitalized for weeks to months prior to their death and were not exposed to alcohol during hospitalization. MESSING et al. (1986) showed that chronic alcohol-induced increases in dihydropyridine binding in PC-12 cells returned to control values within 16 h after alcohol removal. Thus, a possible explanation for the lack of increased dihydropyridine binding in experiments on alcoholic human brain is that dihydropyridine binding may have returned to nonalcoholic levels if individuals had been without alcohol for more than 16 h prior to their death.

L type calcium channels may be involved in the development of tolerance to ethanol and in the ethanol withdrawal syndrome. Concurrent administration of chronic dihydropyridine calcium channel blockers together with chronic ethanol prevented the development of ethanol tolerance (WU et al. 1987; DOLIN and LITTLE 1989). Chronic nitrendipine treatment also prevented the increase in dihydropyridine binding caused by chronic ethanol exposure (WHITTINGTON et al. 1991). Thus, dihydropyridine calcium channel blockers may prevent adaptive increases in L type calcium channels that occur during chronic ethanol exposure. Furthermore, dihydropyridine calcium channel blockers given chronically with ethanol also prevented the withdrawal syndrome (WHITTINGTON et al. 1991). Other studies indicate that calcium channel blockers are effective in decreasing the severity of the withdrawal syndrome after chronic ethanol treatment (LITTLE et al. 1986; LITTLETON et al. 1990). Thus, changes in dihydropyridine-sensitive calcium channels may be linked with ethanol withdrawal symptomatology.

V. Calcium Channel Blockers and Ethanol Preference

Recent studies have shown that treatment of rats with nifedipine (ENGEL et al. 1988) and verapamil (REZVANI and JANOWSKY 1990) caused a significant decrease in alcohol preference. This finding was recently extended in a study examining ethanol consumption in monkeys (REZVANI et al. 1991). Verapamil, 10 mg/kg s.c. for 2 days, significantly decreased ethanol but not water intake in adult macaque monkeys. Ethanol intake by the monkeys increased to pre-verapamil levels in a test conducted 4 days after verapamil treatment. This effect on ethanol consumption was not seen with diltiazem, 10 mg/kg s.c. for 6 days. Diltiazem did not alter either ethanol or water consumption.

VI. Conclusions

The conclusions that can be drawn concerning ethanol effects on voltage-dependent calcium channels are as follows: (1) Electrophysiological and biochemical studies indicate that some neuronal calcium channels are highly sensitive to inhibition by ethanol while others are not. Some calcium channels are inhibited by ethanol concentrations as low as 10–25 m*M*, while others are unaffected by much higher concentrations. (2) There are brain regional differences in ethanol sensitivity of voltage-dependent calcium channels. The reasons for these observations are not yet known, but may result from differences in ethanol sensitivity of calcium channel subtypes. (3) L type calcium channels have been implicated in the acute and chronic effects of ethanol. Ethanol, 10 m*M*, has been reported to inhibit L type channel activity in neurohypophyseal nerve terminals (Wang et al. 1991b). L-type channels in neuroblastoma (Twombly et al. 1990) and PC-12 cells (Messing et al. 1986) appear to be less sensitive to ethanol inhibition. Recent evidence indicates that various subtypes of L type calcium channels may exist. It may be that ethanol has more potent effects on some of these channel subtypes than on others. The answer to this possibility awaits investigation. (4) Adaptive changes in calcium channels have been implicated in both ethanol tolerance and dependence. Evidence is particularly strong for the involvement of L type channels in tolerance and dependence. (5) Much less is known about the effects of ethanol on other types of calcium channels. Ethanol may have some effects on T type channels (Wang et al. 1991a,b; Twombly et al. 1990). Recent evidence indicates that intraventricular injection of ω-conotoxin prolongs ethanol sleep-times (Brown et al. 1993). Very little information is available concerning the effects of ethanol on N and P channels.

C. Effects of Barbiturates and Benzodiazepines on Calcium Channels

Blaustein and Ector (1975) first reported that barbiturates inhibit potassium-stimulated calcium uptake into synaptosomes isolated from whole brain. These findings were confirmed and extended in a study on the acute and chronic effects of barbiturates on potassium-stimulated calcium uptake into synaptosomes isolated from rat whole brain tissue (Leslie et al. 1980) and various brain regions (Elrod and Leslie 1980). Sedative-hypnotic concentrations of pentobarbital inhibited calcium uptake. However, considerable differences were found in the magnitude of inhibition of calcium uptake in the various brain regions studied (Elrod and Leslie 1980). Potassium-stimulated calcium uptake was inhibited by more than 70% in synaptosomes isolated from brain stem and cerebellum. Cerebrocortical synaptosomes were inhibited by approximately 40%, while calcium uptake into midbrain, striatal, and hypothalamic synaptosomes was not significantly inhibited by pentobarbital. At the time these studies were conducted, it was still not known that multiple calcium channel types existed in the brain. Thus, the reasons for the variability in sensitivity of these brain regions to pentobarbital were not clear. GROSS and MACDONALD (1987) showed, in electrophysiological studies on mouse dorsal root ganglion (DRG) cells in culture, that barbiturates do not alter T type currents but do inhibit both L and N currents. Both dihydropyridines and barbiturates inhibited L type channels in DRG cells, but only barbiturates inhibited N type channels. This led these investigators to conclude that the potent action of barbiturates on the central nervous system may be derived, at least in part, from their inhibitory effects on N type channels (GROSS and MACDONALD 1987). The effects of barbiturates on P type channels have not yet been determined.

Several lines of evidence suggest that inhibition of voltage-dependent calcium channels may be closely linked with the sedative-hypnotic properties of drugs. ELROD and LESLIE (1980) reported that chronic administration of barbiturates resulted in an adaptive response to inhibition of potassium-stimulated calcium uptake into synaptosomes in a manner that correlated closely with the development of functional tolerance to their sedative effects. Electrophysiological studies showed that the duration of calcium-dependent action potentials in cultured spinal neurons and dorsal root ganglion cells was reduced by sedative-hypnotic and anesthetic concentrations of barbiturates (HEYER and MACDONALD 1980; WERZ and MACDONALD 1985; GROSS and MACDONALD 1987).

Benzodiazepines, in micromolar concentrations, also inhibit voltage-dependent calcium channels (LESLIE et al. 1980a; TAFT and DELORENZO 1984). This action of benzodiazepines is independent of their effects at nanomolar concentrations on γ-aminobutyric acid receptors (CHERUBINI and NORTH 1985). As was the case with barbiturates, chronic administration of chlordiazepoxide resulted in the development of tolerance to its sedative properties; this tolerance development was correlated with tolerance to the ability of chlordiazepoxide to inhibit synaptosomal calcium uptake. The ability of a series of nine benzodiazepines to inhibit potassium-stimulated calcium uptake into synaptosomes was correlated significantly with behavioral measures of sedative-hypnotic activity. However, no correlation was observed between measures of anticonvulsant activity and calcium uptake (LESLIE et al. 1986a). Other findings suggest, however, that a link may exist between the anticonvulsant actions of benzodiazepines and their ability to inhibit voltage-dependent calcium channels (FERRENDELLI and DANIELS-MCQUEEN 1982; RAMPE et al. 1986).

Studies on the interactions of micromolar concentrations of benzodiazepines with the various calcium channels types are limited. Taft and DeLorenzo (1984) reported that the dihydropyridine calcium channel antagonist nitrendipine inhibited photoaffinity labeling of ^{3}H-flunitrazepam to micromolar benzodiazepine-binding sites. RAMPE et al. (1986) found that diazepam, 60 mM, blocked ^{3}H-nitrendipine binding in a competitive manner. Thus, benzodiazepines may interact with L type calcium channels on neuronal membranes.

I. Conclusions

(1) Electrophysiological and biochemical studies indicate that inhibition of neuronal calcium channels by barbiturates and micromolar concentrations of benzodiazepines is correlated closely with the sedative-hypnotic effects of these drugs. (2) Chronic administration of these drugs results in adaptive responses that are consistent with the involvement of calcium channels in functional tolerance development. (3) Taken together with the literature on the effects of ethanol on voltage-dependent calcium channels described above, the findings of barbiturate and benzodiazepine inhibition of neuronal calcium channels provide strong evidence for the possible involvement of voltage-dependent calcium channels in the acute and chronic actions of sedative-hypnotic drugs.

References

Adams ME, Bindokas VP, Hasegawa L, Venema VJ (1990) ω-Agatoxins: novel calcium channel antagonists of two subtypes from funnel web spider (Agelenopsis aperta) venom. J Biol Chem 265:861–867
Ahlijanian MK, Westenbroek RE, Catterall WA (1990) Subunit structure and localization of dihydropyridine-sensitive calcium channels in mammalian brain, spinal cord, and retina. Neuron 4:819–832
Blaustein MP, Ector AC (1975) Barbiturate inhibition of calcium uptake by depolarized nerve terminals in vitro. Mol Pharmacol 11:369–378
Brennan CH, Lewis A, Littleton JM (1989) Membrane receptors, involved in up-regulation of calcium channels in bovine adrenal chromaffin cells, chronically exposed to ethanol. Neuropharmacology 28:1303–1307
Brown LM, Sims JS, Randall P, Wilcox RE, Leslie SW (1993) ω-Conotoxin increases sleep times following ethanol injection. Alcohol 10:159–162
Camacho-Nasi P, Treistman SN (1986) Ethanol effects on voltage-dependent membrane conductance: comparative sensitivity of channel populations in Aplysia neurons. Cell Mol Neurobiol 6:263–279
Cherubini E, North RA (1985) Benzodiazepines both enhance γ-aminobutyrate responses and decrease calcium action potentials in guinea-pig myenteric neurons. Neuroscience 14:309–315
Daniell LC, Leslie SW (1986) Inhibition of fast phase calcium uptake and endogenous norepinephrine release in rat brain region synaptosomes by ethanol. Brain Res 377:18–28
Daniell LC, Barr EM, Leslie SW (1983) $^{45}Ca^{2+}$ uptake into rat whole brain synaptosomes unaltered by dihydropyridine calcium antagonists. J Neurochem 41:1455–1459
Dildy JE, Leslie SW (1989) Ethanol inhibits NMDA-induced increases in free intracellular Ca^{2+} in dissociated brain cells. Brain Res 499:383–387
Dolin SJ, Little HJ (1986) Augmentation by calcium channel antagonists of general anaesthetic potency in mice. Br J Pharmacol 88:909–914

Dolin SJ, Little HJ (1989) Are changes in neuronal calcium channels involved in ethanol tolerance? J Pharmacol Exp Ther 250:985–991
Elrod SV, Leslie SW (1980) Acute and chronic effects of barbiturates on depolarization-induced calcium influx into synaptosomes from rat brain regions. J Pharmacol Exp Ther 212:131–136
Engel JA, Fahlke C, Hulthe P, Hard E, Johannessen K, Snape B, Svensson L (1988) Biochemical and behavioral evidence for an interaction between ethanol and calcium channel antagonist. J Neural Transm 74:181–193
Eskuri SA, Pozos RS (1987) The effect of ethanol and temperature on calcium-dependent sensory neuron action potentials. Alcohol Drug Res 7:153–162
Ferrendelli JA, Daniels-McQueen S (1982) Comparative actions of phenytoin and other anticonvulsant drugs on potassium- and veratridine-stimulated calcium uptake in synaptosomes. J Pharmacol Exp Ther 220:29–34
Gandhi VC, Jones DJ (1990) Modulation of [^{3}H]serotonin release by dihydropyridines in spinal cord synaptosomes. Eur J Pharmacol 187:271–280
Godfraind T, Miller R, Wibo M (1986) Calcium antagonism and calcium entry blockade. Pharmacol Rev 38:321–416
Greenberg DA, Cooper EC (1984) Effect of ethanol on [^{3}H]nitrendipine binding to calcium channels in brain membranes. Alcohol Clin Exp Res 8:568–571
Greenberg DA, Carpenter CL, Messing RO (1987) Ethanol-induced component of $^{45}Ca^{2+}$ uptake in PC12 cells is sensitive to Ca^{2+} channel modulating drugs. Brain Res 410:143–146
Gross RA, Macdonald RL (1987) Barbiturates and nifedipine have different and selective effects on calcium currents of mouse DRG neurons in culture: a possible basis for differing clinical actions. Neurology 38:443–451
Gruol DL (1982) Ethanol alters synaptic activity in cultured spinal cord neurons. Brain Res 243:25–33
Harper JC, Brennan CH, Littleton JM (1989) Genetic up-regulation of calcium channels in a cellular model of ethanol dependence. Neuropharmacology 28:1299–1302
Harris RA, Allan AM (1989) Alcohol intoxication: ion channels and genetics. FASEB J 3:1689–1695
Harris RA, Hood WF (1980) Inhibition of synaptosomal calcium uptake by ethanol. J Pharmacol Exp Ther 213:562–567
Harris RA, Jones SB, Bruno P, Byland DB (1985) Effects of dihydropyridine derivatives and anticonvulsant drugs on [^{3}H]nitrendipine binding and calcium and sodium fluxes in brain. Biochem Pharmacol 34:2187–2191
Hertz L, Bender AS, Woodbury DM, White HS (1989) Potassium-stimulated calcium uptake in astrocytes and its potent inhibition by nimodipine. J Neurosci Res 22:209–215
Heyer EJ, Macdonald RL (1980) Barbiturate reduction of calcium-dependent action potentials: correlation with anesthetic action. Brain Res 236:157–171
Hirning LD, Fox AP, McCleskey EW, Olivera BM, Thayer SA, Miller RM, Tsien RW (1988) Dominant role of N-type Ca^{2+} channels in evoked release of norepinephrine from sympathetic neurons. Science 239:57–60
Isaacson RL, Molina JC, Draski LJ, Johnston JE (1985) Nimodipine's interactions with other drugs: I. Ethanol. Life Sci 36:2195–2199
Katz B, Miledi R (1967) A study of synaptic transmission in the absence of nerve impulses. J Physiol (Lond) 192:407–436
Kongsamut S, Miller RJ (1986) Nerve growth factor modulates the drug sensitivity of neurotransmitter release from PC-12 cells. Proc Natl Acad Sci USA 83: 2243–2247
Kril JJ, Gundlach AL, Dodd PR, Johnston GAR, Harper CG (1989) Cortical dihydropyridine binding sites are unaltered in human alcoholic brain. Ann Neurol 26:395–397

Leslie SW (1987) Calcium channels: interactions with ethanol and other sedative-hypnotic drugs. In: Galanter M (ed) Recent developments in alcoholism, vol 5. Plenum, New York, p 289

Leslie SW, Friedman MB, Coleman RR (1980a) Effects of chlordiazepoxide on depolarization-induced calcium influx into synaptosomes. Biochem Pharmacol 29:2439–2443

Leslie SW, Friedman MB, Wilcox RE, Elrod SV (1980b) Acute and chronic effects of barbiturates on depolarization-induced calcium influx into rat synaptosomes. Brain Res 185:409–417

Leslie SW, Barr E, Chandler J, Farrar RP (1983) Inhibition of fast- and slow-phase depolarization-dependent synaptosomal calcium uptake by ethanol. J Pharmacol Exp Ther 225:571–575

Leslie SW, Chandler LJ, Chweh AY, Swinyard EA (1986a) Correlation of the hypnotic potency of benzodiazepines with inhibition of voltage-dependent calcium uptake into mouse brain synaptosomes. Eur J Pharmacol 126:129–134

Leslie SW, Woodward JJ, Wilcox RE, Farrar RP (1986b) Chronic ethanol treatment uncouples striatal calcium entry and endogenous dopamine release. Brain Res 368:174–177

Leslie SW, Brown LM, Dildy JE, Sims JS (1990) Ethanol and neuronal calcium channels. Alcohol 7:233–236

Lima-Landman MTR, Albuquerque EX (1989) Ethanol potentiates and blocks NMDA-activated single-channel currents in rat hippocampal pyramidal cells. FEBS Lett 247:61–67

Little HJ, Dolin SJ, Halsey MJ (1986) Calcium channel antagonists decrease the ethanol withdrawal syndrome. Life Sci 39:2059–2065

Littleton JM, Little HJ, Whittington MA (1990) Effects of dihydropyridine calcium channel antagonists in ethanol withdrawal; doses required, stereospecificity and actions of Bay K 8644. Psychopharmacology 100:387–392

Lundy PM, Frew R, Fuller TW, Hamilton MG (1991) Pharmacological evidence for an ω-conotoxin, dihydropyridine-insensitive neuronal Ca^{2+} channel. Eur J Pharmacol Mol Pharmacol Sect 206:61–68

Marks SS, Watson DL, Carpenter CL, Messing RO, Greenberg DA (1989) Calcium channel antagonist receptors in cerebral cortex from alcoholic patients. Brain Res 478:196–198

Messing RO, Carpenter CL, Diamond I, Greenberg DA (1986) Ethanol regulates calcium channels in clonal neural lines. Proc Natl Acad Sci USA 83:6213–6215

Middlemiss DN, Spedding M (1985) A functional correlate for the dihydropyridine binding site in rat brain. Nature 314:94–96

Miller RJ (1987) Multiple calcium channels and neuronal function. Science 235: 46–52

Mintz IM, Adams ME, Bean BP (1992a) P-type calcium channels in rat central and peripheral neurons. Neuron 9:85–95

Mintz IM, Venema VJ, Swiderek K, Lee T, Bean BP, Adams ME (1992b) P-type calcium channels blocked by the spider toxin ω-Aga-IVA. Nature 355:827–829

Monyer H, Sprengel R, Schoepfer R, Herb A, Higuchi M, Lomeli H, Burnashev N, Sakmann B, Seeburg PH (1992) Heteromeric NMDA receptors: molecular and functional distinctions of subtypes. Science 256:1217–1221

Nachshen DA, Blaustein MP (1980) Some properties of potassium-stimulated calcium influx in presynaptic nerve endings. J Gen Physiol 76:709–728

Nowycky MC, Fox AP, Tsien RW (1985) Three types of neuronal calcium channel with different calcium agonist sensitivity. Nature 316:440–443

Olivera BM, Gray WR, Zeikus R, McIntosh JM, Varga J, Rivier J, de Santos V, Cruz LJ (1985) Peptide neurotoxins from fish-hunting cone snails. Science 230:1338–1343

Perez-Reyes E, Wei X, Castellano A, Birnbaumer L (1990) Molecular diversity of L-type calcium channels. J Biol Chem 265:20430–20436

Proctor WR, Allan AM, Dunwiddie TV (1992) Brain region-dependent sensitivity of $GABA_A$ receptor-mediated responses to modulation by ethanol. Alcohol Clin Exp Res 16:480–489

Rampe D, Ferrante J, Triggle D (1986) The actions of diazepam and diphenylhydantoin on fast and slow Ca^{2+} uptake processes in guinea pig cerebral cortex synaptosomes. Can J Physiol Pharmacol 65:538–543

Rezvani AH, Janowsky DS (1990) Decreased alcohol consumption by verapamil in alcohol preferring rats. Prog Neuropsychopharmacol Biol Psychiatry 14:623–631

Rezvani AH, Grady DR, Janowsky DS (1991) Effect of calcium-channel blockers on alcohol consumption in alcohol-drinking monkeys. Alcohol and Alcoholism 26:161–167

Scriabine A, Schuurman T, Traber J (1989) Pharmacological basis for the use of nimodipine in central nervous system disorders. FASEB J 3:1799–1809

Sher E, Clementi F (1991) ω-Conotoxin-sensitive voltage-operated calcium channels in vertebrate cells. Neuroscience 42:301–307

Sher E, Biancardi E, Passfaro M, Clementi F (1991) Physiopathology of neuronal voltage-operated calcium channels. FASEB J 5:2677–2683

Simon PE, Criswell HE, Johnson KB, Hicks RE, Breese GB (1991) Ethanol inhibits NMDA-evoked electrophysiological activity in vivo. J Pharmacol Exp Ther 257:225–231

Skattebol A, Rabin RA (1987) Effects of ethanol on $^{45}Ca^{2+}$ uptake in synaptosomes and in PC12 cells. Biochem Pharmacol 36:2227–2229

Snutch TP, Leonard JP, Gilbert MM, Lester HA, Davidson N (1990) Rat brain expresses a heterogeneous family of calcium channels. Proc Natl Acad Sci USA 87:3391–3395

Stokes JA, Harris RA (1982) Alcohols and synaptosomal calcium transport. Mol Pharmacol 22:99–104

Suszkiw JB, O'leary ME, Murawsky MM, Wang T (1986) Presynaptic calcium channels in rat cortical synaptosomes: fast-kinetics of phasic calcium influx, channel inactivation, and relationship to nitrendipine receptors. J Neurosci 6:1349–1357

Taft WC, DeLorenzo RJ (1984) Micromolar-affinity benzodiazepine receptors regulate voltage-sensitive calcium channels in nerve terminal preparations. Proc Natl Acad Sci USA 81:3118–3122

Tsien RW, Ellinor PT, Horne WA (1991) Molecular diversity of voltage-dependent Ca^{2+} channels. Trends Pharmacol Sci 12:349–354

Twombly DA, Herman MD, Kye CH, Narahashi T (1990) Ethanol Effects on two types of voltage-activated calcium channels. J Pharmacol Exp Ther 254: 1029–1037

Wafford KA, Burnett DM, Leidenheimer NJ, Burt DR, Wang JB, Kofuji P, Dunwiddie TV, Harris RA, Sikela JM (1991) Ethanol sensitivity of the $GABA_A$ receptor expressed in Xenopus oocytes requires 8 amino acids contained in the g2L subunit. Neuron 7:27–33

Wang X, Dayanithi G, Lemos JR, Nordmann JJ, Treistman SN (1991a) Calcium currents and peptide release from neurohypophysial terminals are inhibited by ethanol. J Pharmacol Exp Ther 259:705–711

Wang X, Lemos JR, Dayanithi G, Nordmann JJ, Treistman SN (1991b) Ethanol reduces vasopressin release by inhibiting calcium currents in nerve terminals. Brain Res 551:338–341

Werz MA, Macdonald RL (1985) Barbiturates decrease voltage-dependent calcium conductance of mouse neurons in dissociated cell culture. Mol Pharmacol 28: 269–277

White EJ, Bradford HF (1986) Enhancement of depolarization-induced synaptosomal calcium uptake and neurotransmitter release by BAY K8644. Biochem Pharmacol 35:2193–2197

White G, Lovinger DM, Weight FF (1990) Ethanol inhibits NMDA-activated current but does not alter GABA-activated current in an isolated adult mammalian neuron. Brain Res 507:332–336

Whittington MA, Dolin SJ, Patch TL, Siarey RJ, Butterworth AR, Little HJ (1991) Chronic dihydropyridine treatment can reverse the behavioral consequences of and prevent adaptations to, chronic ethanol treatment. Br J Pharmacol 103: 1669–1676

Williams ME, Brust PF, Feldman DH, Patthi S, Simerson S, Maroufi A, McCue AF, Velicelebi G, Ellis SB, Harpold MM (1992) Structure and functional expression of an ω-conotoxin-sensitive human N-type calcium channel. Science 257: 389–395

Woodward JJ, Leslie SW (1986) Bay K 8644 stimulation of calcium entry and endogenous dopamine release in rat striatal synaptosomes antagonized by nimodipine. Brain Res 370:397–400

Woodward JJ, Cook ME, Leslie SW (1988a) Characterization of dihydropyridine-sensitive calcium channels in brain synaptosomes. Proc Natl Acad Sci USA 85:7389–7393

Woodward JJ, Rezazadeh SM, Leslie SW (1988b) Differential sensitivity of synaptosomal calcium entry and endogenous dopamine release to ω-conotoxin. Brain Res 475:141–145

Wu PH, Pham T, Naranjo CA (1987) Nifedipine delays the acquisition of tolerance. Eur J Pharmacol 139:233–236

Zahniser NR, Buck KJ, Curella P, McQuilkin SJ, Wilson-Shaw D, Miller CL, Klein RL, Heidenreich KA, Keir WK, Sikela JM, Harris RA (1992) $GABA_A$ receptor function and regional analysis of subunit mRNAs in long-sleep and short-sleep mouse brain. Mol Brain Res 14:196–206

CHAPTER 4

Effects of Alcohol on Excitatory Amino Acid Receptor Function

P.L. HOFFMAN

A. Introduction

I. Site of Action of Ethanol: Protein Versus Lipid

For several years, the major hypothesis to explain the behavioral effects of ethanol has been the "membrane hypothesis" (HUNT 1985). This postulate is based on work demonstrating that ethanol, which is an amphipathic substance, can penetrate into and perturb the structure of ("fluidize") cell membrane lipids (CHIN and GOLDSTEIN 1977; HARRIS and SCHROEDER 1981). While there is evidence to suggest that bulk lipid perturbation may be involved in the high-dose, hypnotic, or anesthetic effects of ethanol (GOLDSTEIN et al. 1982), this nonspecific interaction of ethanol with biological systems does not well explain low-dose effects of ethanol such as intoxication, ataxia, or reinforcement. Recent research has demonstrated that the function of certain neurochemical systems, notably those which consist of multiple membrane-associated protein subunits (e.g., the $GABA_A$ receptor, receptor-coupled adenylate cyclase), is very sensitive to modulation by low concentrations of ethanol. These systems have been designated "receptive elements" for ethanol (TABAKOFF and HOFFMAN 1987). Knowledge that the function of particular neurochemical systems is sensitive to modification by ethanol allows investigators to develop hypotheses regarding the neurochemical basis for certain behavioral responses to low doses of ethanol, based on the involvement of the neurochemical systems in the particular behavior (e.g., ethanol potentiation of $GABA_A$ receptor-coupled ion flux might be expected to contribute to the anxiolytic effect of ethanol).

While ethanol could perturb the proteins in the "receptive elements" via an action on the membrane lipids which surround them, the fact that these proteins are sensitive to very low ethanol concentrations, which would have little effect on bulk membrane lipid properties (CHIN and GOLDSTEIN 1977; HARRIS and SCHROEDER 1981), suggests that ethanol may interact directly with hydrophobic sites on the proteins themselves or with protein-protein interactions. Furthermore, there is now evidence that, in the case of the $GABA_A$ receptor, a specific amino acid sequence in one receptor subunit confers sensitivity to ethanol, perhaps based on its phosphorylation state (WAFFORD et al. 1991).

The initial reports that ethanol is a potent and selective inhibitor of the function of the *N*-methyl-D-aspartate (NMDA) subtype of glutamate receptor (HOFFMAN et al. 1989; LOVINGER et al. 1989) have led, during the past few years, to intense investigation of the interactions of ethanol with excitatory amino acid receptors. One reason for this interest is the sensitivity of the system to ethanol: substantial inhibition is observed using in vitro concentrations of ethanol that are well within the range believed to be associated with reinforcement, intoxication, ataxia, and cognitive impairment in vivo (HOFFMAN et al. 1989; LOVINGER et al. 1989). Furthermore, the NMDA receptor has been implicated in a number of phenomena that are affected acutely and/or chronically by ethanol, including learning and memory (long-term potentiation), CNS development and plasticity, and epileptiform seizure activity (COLLINGRIDGE and LESTER 1989). Together, these data suggest that inhibition of excitatory amino acid action in the brain could underlie certain specific pharmacological effects of ethanol, e.g., ethanol-induced cognitive dysfunction, developmental deficits induced by ethanol (fetal alcohol syndrome), and, after chronic ethanol exposure, ethanol withdrawal seizures.

B. Ethanol and Excitatory Amino Acid Receptors

The purpose of this chapter is to provide an up-to-date and integrated review of the literature regarding ethanol and glutamate receptors and to evaluate the importance of these receptors in particular pharmacological responses to ethanol.

I. Characteristics of Glutamate Receptors

Glutamate is believed to be the major excitatory neurotransmitter in the CNS, and it interacts with at least three subtypes of receptor. On the basis of interactions with specific agonists and antagonists, these were originally designated the NMDA, kainate, and quisqualate (or AMPA) receptors (COLLINGRIDGE and LESTER 1989). With the advent of molecular biological investigations, this receptor classification has become (temporarily) more complex. For example, at least six different non-NMDA receptor subunits (GluR) have been cloned (DINGLEDINE 1991). The first was GluR 1 (HOLLMANN et al. 1989), and a family of four related subunits, GluR 1-4 (DINGLEDINE 1991) [or GluR A-D (KEINÄNEN et al. 1990)], has now been identified. When expressed in oocytes or mammalian cells, these proteins form functional receptors that are activated by quisqualate, α-amino-3-hydroxy-5-methylisoxazole-4-propionate (AMPA), and kainate (DINGLEDINE 1991; HEINEMANN et al. 1991). These findings supported the proposal that kainate and quisqualate receptors might not be distinct entities. However, more recently, two more GluR subunits have been cloned (GluR 5 and GluR 6) that have 80% sequence identity with each other and less than

40% identity with GluR 1-4 (DINGLEDINE 1991). When expressed, GluR 6 showed high affinity for kainate and no response to AMPA, suggesting that it could be considered to be a kainate receptor (DINGLEDINE 1991). In addition, a protein called KA-1 has been described, which has high affinity for kainate and low affinity for AMPA, and is related structurally to GluR 5 and 6 (DINGLEDINE 1991). Based on these findings, as well as ligand binding, in situ hybridization, and immunohistochemical analyses (DINGLEDINE 1991; HEINEMANN et al. 1991; EGEBJERG et al. 1991; YOUNG and FAGG 1990; MONAGHAN and COTMAN 1982), there appear to be distinct kainate and AMPA receptors in brain, as originally proposed on the basis of pharmacological results. In addition, two kainate-binding proteins from chick and frog have been characterized, but no functional activity was detected when the frog protein was expressed in oocytes (GREGOR et al. 1989; WADA et al. 1989).

Although many GluR subunits have been cloned, there is as yet little information regarding the nature of subunit combinations that actually form native non-NMDA glutamate receptors. These in situ combinations are probably crucial for defining receptor function. For example, it has been demonstrated that activation by kainate or AMPA of receptors formed from GluR 1, GluR 3, or GluR 1 plus GluR 3, expressed in oocytes, resulted in a calcium current; however, permeability to calcium was not seen in cells expressing GluR 2 or GluR 2 plus GluR 3 (HOLLMAN et al. 1991). It is likely that the proportions and assembly of GluR subunits in various brain areas will significantly affect the function, including the pharmacological responses, of the receptors, similar to the situation with the $GABA_A$ receptor (VICINI 1991).

The NMDA receptor has also recently been cloned. In one study, a single protein from rat was identified (NR 1) (MORIYOSHI et al. 1991). When expressed in oocytes, this protein demonstrated many of the pharmacological properties of NMDA receptors (see below). A nearly identical protein has also been cloned from mouse (ζ 1) (YAMAZAKI et al. 1992). The distribution in brain of mRNA coding for these proteins was widespread, and not entirely consistent with previous ligand binding and autoradiographic studies of the NMDA receptor (BOWERY et al. 1988; MONAGHAN et al. 1988). However, recently, other NMDA receptor subunits from mouse and rat have been cloned (NR 2A, B, and C [MONYER et al. 1992]; ε 1, 2, and 3 [MEGURO et al. 1992; KUTSUWADA et al. 1992]), which have more specific localizations in brain. These subunits, when expressed alone in oocytes, do not respond to NMDA. When they are expressed together with NR 1 or ζ 1 respectively, however, the responses to NMDA are larger than those obtained with homomeric NR 1 or ζ 1 (MONYER et al. 1992; MEGURO et al. 1992; KUTSUWADA et al. 1992). There has also been another report of the cloning of a glutamate-binding protein, which was suggested to represent one subunit of a multisubunit complex with the characteristics of the NMDA receptor (KUMAR et al. 1991). This complex has been reported to consist of several distinct proteins that contain binding sites for NMDA agonists, antagonists, and other modulators of receptor function (WANG et al. 1992). Thus, the NMDA receptor, like other receptor-coupled ion channels (VICINI 1991; RAFTERY et al. 1980), appears to consist of a number of subunits which can exist in various combinations, resulting in differences in pharmacological and physiological properties. Variability in the subunit composition of receptors in different brain areas may contribute to the heterogeneity of NMDA receptor properties that has previously been reported (MONAGHAN et al. 1988; YONEDA and OGITA 1991; MONAGHAN 1991).

The pharmacology of NMDA receptor function has been investigated in detail because of the availability of selective agonists and antagonists (COLLINGRIDGE and LESTER 1989). This receptor is coupled to an ion channel which, when activated, is permeable to calcium as well as monovalent cations. Activation of the NMDA receptor by glutamate or NMDA is voltage dependent, i.e., the response increases as the cell is depolarized (COLLINGRIDGE and LESTER 1989). This voltage dependence is mediated by Mg^{2+}, which binds to a site within the ion channel and blocks channel function. Mg^{2+} is released from the channel upon cellular depolarization, thus relieving the channel blockade (NOWAK et al. 1984). In addition to a Mg^{2+} site, there is also a binding site within the channel for phencyclidine (PCP), a dissociative anesthetic. This site can also be occupied by ketamine or dizocilpine (MK-801), and all of these drugs are uncompetitive blockers of NMDA receptor function (COLLINGRIDGE and LESTER 1989; ANIS et al. 1983; WONG et al. 1986). Other modulators of NMDA receptor function include glycine, which is *required* for the action of glutamate at the NMDA receptor (i.e., glutamate and glycine are co-agonists) (KLECKNER and DINGLEDINE 1988), and which binds to a strychnine-insensitive site on the receptor complex. Zn^{2+} has an inhibitory effect on NMDA receptor function (PETERS et al. 1987), while polyamines appear to enhance the response to glutamate (RANSOM and STEC 1988; SACAAN and JOHNSON 1989); however, the physiological role of these latter two modulators is not clear.

The unique properties of the NMDA receptor are important for its physiological functions. For example, the release of glutamate in the synapse is expected to activate ionotropic kainate and AMPA receptors, leading to rapid cellular depolarization. Once the neuron is depolarized, glutamate can also activate the NMDA receptor. Thus, activation of NMDA receptors depends on both pre- and postsynaptic activity; simultaneous pre- and postsynaptic activity were also postulated to be necessary for synaptic strengthening (i.e., LTP or learning) to occur (HEBB 1949). The response to glutamate at the NMDA receptor is slow, allowing for summation of responses and entry into the cell of large amounts of calcium (COTMAN et al. 1989). This calcium influx is important for LTP and synaptic plasticity, but can also lead to excitotoxicity and neuronal death, as well as seizures, if there is excess NMDA receptor activity (COTMAN et al. 1989).

II. Ethanol and NMDA Receptor Function: Acute Effects

The acute inhibitory effect of ethanol on NMDA receptor function was first well documented in biochemical studies in primary cultures of cerebellar granule cells, in which ethanol was shown to be a potent inhibitor of NMDA-stimulated calcium influx and cyclic GMP generation (HOFFMAN et al. 1989), and in brain slice preparations, where ethanol inhibited NMDA-stimulated neurotransmitter release (GÖTHERT and FINK 1989). The inhibitory effect of ethanol was confirmed in electrophysiological (whole cell patch clamp) studies of dissociated embryonic hippocampal neurons in culture (LOVINGER et al. 1989; LIMA-LANDMAN and ALBUQUERQUE 1989). In each case, the effect of ethanol was observed at low concentrations (5–20 mM), while much higher ethanol concentrations were needed to inhibit the response to kainate (HOFFMAN et al. 1989; LOVINGER et al. 1989) or quisqualate (LOVINGER et al. 1989). These initial investigations were followed by a number of studies which have, overall, been remarkably consistent in demonstrating a potent and selective effect of ethanol on NMDA receptor function. For example, ethanol inhibited NMDA-induced currents in dorsal root ganglion neurons of adult rats, and inhibited NMDA receptor-mediated population excitatory postsynaptic potentials, as well as depolarization of pyramidal cells, in hippocampal slices of adult rats (WHITE et al. 1990a; LOVINGER et al. 1990). Intracellular recordings from hippocampal cells of adult rats also demonstrated that ethanol affected the threshold for cellular activation via an inhibitory effect on NMDA-induced conductance (YUEN et al. 1991). While these studies were carried out in vitro, it was also shown that ethanol, when administered systemically to rats, inhibited the activation of certain medial septal neurons by iontophoretically applied NMDA (SIMSON et al. 1991) and inhibited the excitation of locus coeruleus neurons by NMDA as well as glutamate or quisqualate (ENGBERG and HAJÓS 1992).

Biochemical studies have also yielded consistent results. Ethanol selectively inhibited NMDA-stimulated Ca^{2+} influx in primary cultures of cerebellar granule cells and in dissociated cells from whole brain of neonatal rats (measured either as $^{45}Ca^{2+}$ uptake or with the use of fura-2 to measure intracellular $[Ca^{2+}]$) (HOFFMAN et al. 1989; DILDY and LESLIE 1989). Furthermore, several investigations using brain slices or synaptosomal preparations from various brain areas have shown that ethanol inhibits NMDA-stimulated neurotransmitter release (FINK and GÖTHERT 1990; ÖTHERT and FINK 1991; GONZALES and WOODWARD 1990; WOODWARD and GÖONZALES 1990). The results indicate that ethanol can inhibit the response to agonists both at pre- and postsynaptic NMDA receptors in neuronal tissue (GÖTHERT and FINK 1991; PITTALUGA and RAITERI 1990). Interestingly, a recent report indicated that ethanol (30–100 mM) also inhibits glutamate-induced transient contractions of the guinea pig ileum longitudinal muscle myenteric plexus, which are mediated by an NMDA-type glutamate receptor (FRYE 1991).

1. Mechanism of Action of Ethanol

The mechanism by which ethanol inhibits NMDA receptor function has been investigated in several studies. The inhibition does not appear to be "competitive" in nature, i.e., in most instances, ethanol inhibition is not overcome by increasing the concentration of NMDA (Rabe and Tabakoff 1990; Dildy-Mayfield and Leslie 1991). However, in a recent investigation, it was reported that preincubation of rat cerebral cortical slices with NMDA resulted in a reduced ability of NMDA to stimulate norepinephrine release ("desensitization") (Fink and Göthert 1991). Both ethanol (320 m*M*) and the competitive antagonist AP5 (2-amino-5-phosphopentanoic acid) prevented this desensitization, while dizocilpine did not. The conclusion in this study was that ethanol, like AP5, might be acting at the NMDA recognition site, and not at the dizocilpine site, to inhibit "desensitization." However, it was recognized that the potency of ethanol to inhibit desensitization was at least threefold less than its potency (acutely) to inhibit NMDA-induced norepinephrine release (Fink and Göthert 1991).

In studies of NMDA-stimulated increases in cyclic GMP production in cerebellar granule cells, there was no interaction between the effects of ethanol and PCP, suggesting that ethanol does not act at the PCP-binding site (Hoffman et al. 1989). However, in a later study, measuring NMDA-stimulated increases in intracellular calcium in dissociated brain cells from neonatal rats, an interaction between ethanol and dizocilpine was observed (i.e., in the presence of high concentrations [100–400 n*M*] of dizocilpine, ethanol no longer inhibited the NMDA response; conversely, the IC_{50} for dizocilpine was increased from 134 n*M* to 240 n*M* in the presence of 100 m*M* ethanol) (Dildy-Mayfield and Leslie 1991). It is possible that, under some conditions, there may be an interaction of ethanol with the PCP site on the NMDA receptor. However, as pointed out by the authors, at high concentrations, dizocilpine is known to interact with sites other than the NMDA receptor-coupled ion channel (Ramoa et al. 1990).

There does not appear to be an interaction of ethanol and Mg^{2+} inhibition of NMDA responses. In studies of $^{45}Ca^{2+}$ influx in cerebellar granule cells, and intracellular calcium in dissociated brain cells, inhibition by ethanol and Mg^{2+} were additive (Rabe and Tabakoff 1990; Dildy-Mayfield and Leslie 1991). On the other hand, in a study of NMDA receptor-mediated population synaptic potentials in area CA1 of the hippocampus of the adult rat, ethanol was a much more potent inhibitor in the presence of 1 m*M* Mg^{2+} than in its absence (Martin et al. 1991a). However, in experiments in which NMDA-induced depolarizations in CA1 were measured, and in which ethanol and Mg^{2+} were covaried, it was concluded that the two acted by distinct mechanisms to inhibit NMDA responses (Morrisett et al. 1991).

There are more substantial data to indicate that ethanol may affect the interaction of the co-agonists, glutamate and glycine. In cerebellar granule cells and in dissociated brain cells, glycine reversed the inhibitory effect of ethanol on NMDA-stimulated increases in calcium influx (Rabe and Tabakoff 1990; Dildy-Mayfield and Leslie 1991). Similarly, in striatal slices, addition of glycine attenuated the ability of ethanol to inhibit NMDA-stimulated dopamine release (Woodward and Gonzales 1990). However, reversal of ethanol inhibition by glycine was not observed for NMDA-stimulated norepinephrine release in cerebral cortical slices (Gonzales and Woodward 1990). It is possible that this brain regional difference could be associated with the reported brain regional heterogeneity of NMDA receptors, which includes differences in sensitivity to glycine (Yoneda and Ogita 1991).

In contrast to the biochemical studies, there is less evidence from electrophysiological investigations (Peoples and Weight 1992) that ethanol inhibition of NMDA receptor function is affected by glycine. One difference is that lower concentrations of glycine ($<1\,\mu M$) are usually used in the electrophysiological studies, while higher concentrations of glycine ($10-100\mu M$) were necessary to reverse ethanol inhibition in the biochemical studies (Rabe and Tabakoff 1990; Dildy-Mayfield and Leslie 1991). In one investigation of depolarization of rat hippocampal CA1 pyramidal cells, $100\,\mu M$ glycine did not alter the inhibitory effect of ethanol (Martin et al. 1991a). However, at the concentration used in this study ($170\,\mathrm{m}M$), ethanol inhibits responses at receptors other than the NMDA receptor (e.g., the kainate receptor).

We have recently found that, although ethanol does not alter equilibrium binding of glycine in mouse brain tissue, there is an inhibitory effect of ethanol on the nonequilibrium binding (i.e., on the association rate) of dizocilpine, and this inhibitory effect is reversed by glycine (Snell et al. 1993). These analyses were carried out using well-washed brain membranes and in the presence of glutamate and a specific glycine antagonist, 5,7-dichlorokynurenic acid, to eliminate any influence of endogenous glycine. The association rate for dizocilpine binding has been suggested to reflect the kinetics of NMDA receptor-coupled channel opening (Javitt and Zukin 1989; Bonhaus and McNamara 1988; Sircar and Zukin 1991). NMDA receptor agonists, and glycine in the presence of these agonists, enhance the rate of association, thought to reflect dizocipline binding to the open channel (Bonhaus and McNamara 1988; Sircar and Zukin 1991). The inhibition of dizocilpine association by ethanol may therefore indicate decreased probability of channel opening, which can be reversed by glycine (in the presence of glutamate) in an algebraically additive manner (Bonhaus and McNamara 1988; Sircar and Zukin 1991).

2. Effects of Anesthetics and Sedative Hypnotics

The effects of alcohols other than ethanol, as well as of some inhalational anesthetics, on glutamate receptor function, have also been investigated.

Both NMDA-induced currents, in whole cell patch clamp experiments using hippocampal cells, and NMDA-stimulated norepinephrine release in rat cerebral cortical slices, were inhibited by a series of short-chain alcohols, and in both cases potency of inhibition increased with carbon chain length (LOVINGER et al. 1989; GONZALES et al. 1991). These data suggest that the degree of inhibition of NMDA-induced responses is related to the hydrophobicity of the alcohols. While such a relationship has often been cited to support the hypothesis that ethanol and other alcohols exert their effects through partition into and perturbation of cell membrane lipids (HUNT 1985), it has also been demonstrated that this relationship may be associated with an interaction of the alcohols with hydrophobic areas of *proteins* (FRANKS and LIEB 1984).

In addition to alcohols, ethers can inhibit NMDA responses. Electrophysiological investigations of hippocampal neurons revealed that diethyl ether (80 m*M*), like ethanol, had a selective inhibitory effect on NMDA-induced currents, in that kainate- and quisqualate-induced currents were less affected (WEIGHT et al. 1991). The effect of enflurane on ^{3}H-MK-801 binding to rat cerebral cortical membranes has also been studied. Enflurane inhibited glutamate stimulation of dizocilpine binding (reportedly under equilibrium conditions), but apparently not by directly interfering with glutamate binding (MARTIN et al. 1991b). Interestingly, the effect of enflurane was reversed by the addition of glycine, similar to the results with ethanol described above. These findings suggest that enflurane, as well as ethanol, may disrupt the interaction of the co-agonists at the NMDA receptor, although there was no confirmation that the effect of enflurane on dizocilpine binding was in fact associated with reduced function of the NMDA receptor. The concentration of enflurane used was reported to be consistent with that which produces anesthesia in vivo (MARTIN et al. 1991b). While inhibition of NMDA receptor function could, in this case, be associated with anesthesia, it should be emphasized that the inhibition by ethanol occurs at concentrations that are well below those necessary in vivo to produce an anesthetic effect (e.g., WALLGREN and BARRY 1970).

Other sedative hypnotics, i.e., barbiturates, appear to have different effects from ethanol on glutamate receptor function. In electrophysiological studies of hippocampal neurons, pentobarbital inhibited the responses to kainate and quisqualate (IC_{50}, approx. 50 μM), but had no effect on the response to NMDA (WEIGHT et al. 1991; MILJKOVIC and MACDONALD 1986). In cerebellar granule cells in culture, pentobarbital (100 μM, a concentration producing sedative/hypnotic effects similar to 50 m*M* ethanol) was more effective at inhibiting kainate-induced calcium influx than NMDA-induced influx (TABAKOFF et al. 1991). In this study, neither diazepam nor flurazepam had any effect on the response to kainate or NMDA, consistent with the postulate that ethanol and barbiturates do not exert their effects on glutamate receptor function indirectly, via potentiation of GABA effects at the $GABA_A$ receptor. Similarly, the ability of ethanol to inhibit NMDA-stimulated calcium influx in cerebellar granule cells was not altered in the presence of the $GABA_A$ receptor-coupled channel blocker, picrotoxin (HOFFMAN et al. 1989).

The findings to date indicate that ethanol is a selective and potent inhibitor of the response to agonists at the NMDA receptor, and is less effective at other subtypes of glutamate receptor. In a recent study, in which rat hippocampal mRNA was expressed in *Xenopus* oocytes, it was observed that ethanol (50 and 100 m*M*) produced similar inhibition of electrophysiological responses to NMDA and a low concentration (12.5 *μM*) of kainate (DILDY-MAYFIELD and HARRIS 1992). It should be noted, however, that these investigations of the NMDA receptor were performed in the presence of 10 μ*M* added glycine, which could significantly reverse ethanol-induced inhibition of NMDA responses (RABE and TABAKOFF 1990).

The inhibitory effect of ethanol on NMDA receptor function appears to involve a hydrophobic reaction site, and, as mentioned, can be reversed by high concentrations of glycine acting at the strychnine-insensitive glycine-binding site on the NMDA receptor (LOVINGER et al. 1989; RABE and TABAKOFF 1990; DILDY-MAYFIELD and LESLIE 1991; GONZALES et al. 1991). However, it is not yet clear if ethanol specifically interferes with the co-agonist interaction, or if other compounds which, like glycine, increase the probability of channel opening, would also reverse ethanol's inhibitory effect [however, the effect of ethanol is *not* overcome by increasing concentrations of NMDA (RABE and TABAKOFF 1990; DILDY-MAYFIELD and LESLIE 1991)]. It may be pertinent to point out that, although dizocilpine association, as well as biochemical studies of calcium influx and neurotransmitter release, are believed to reflect the physiological consequences of NMDA receptor-coupled channel activation, the time scales for measurement of biochemical and electrophysiological events are quite different. This difference may influence the findings with respect to reversal of ethanol's effect by glycine. Nevertheless, both techniques provide comparable and compatible results regarding the selective effect of ethanol on NMDA receptor function. The different selectivity of barbiturates argues against the possibility that inhibition of NMDA receptor function is solely responsible for the sedative/hypnotic or anesthetic effects of the drugs, and suggests that this inhibition may instead contribute to specific pharmacological effects of ethanol, such as particular aspects of intoxication or memory impairment.

3. Ethanol and the NMDA Receptor In Vivo

Since glycine can reverse the effect of ethanol on the NMDA receptor, and glycine concentrations in brain and extracellular fluid are reported to be in the high micromolar range (TOSSMAN et al. 1986), it may be questioned whether ethanol would inhibit responses to NMDA in vivo. As mentioned above, ethanol, administered systemically to rats, inhibited the response of some (but not all) medial septal neurons, as well as locus coeruleus neurons, to iontophoresed NMDA (Simson et al. 1991; Engberg and Hajós 1992). However, although this study demonstrates an in vivo action of ethanol, the NMDA was exogenously applied, which complicates the analysis. Several studies have suggested that the glycine site on the NMDA receptor is not saturated in vivo. For example, intrahippocampal administration of D-serine (an agonist at the strychnine-insensitive glycine site) increased the magnitude of long-term potentiation (Thiels et al. 1992). Furthermore, electrophysiological responses to NMDA can be potentiated by very small (nanomolar) increases in glycine concentration, so that even if the glycine concentration is close to saturating, increases in extracellular glycine might still be effective (Johnson and Ascher 1987). Finally, administration of glycine to animals has been shown to modulate susceptibility to NMDA-induced seizures (Singh et al. 1990). Taken together with the sensitivity of NMDA receptor function to ethanol in vitro – i.e., inhibition is observed at concentrations found in brains of humans who have consumed modest amounts of ethanol – these results support the hypothesis that ethanol in vivo would inhibit the activity of the NMDA receptor, leading to behavioral consequences.

In order to assess this possibility experimentally, studies of the discriminative stimulus properties of ethanol have been performed. These studies are designed to determine whether interoceptive cues produced by ethanol and specific NMDA receptor antagonists are perceived as being similar by animals. White Carneau pigeons or CD-1 mice were trained to discriminate ethanol and water by standard drug discrimination techniques (i.e., animals are trained to respond on a particular lever for food reinforcement after being given either ethanol or water) (Grant et al. 1991). They were then given various specific NMDA receptor antagonists (phencyclidine, dizocilpine, or ketamine) and allowed to respond on either of the levers. In both mice and pigeons, phencyclidine or ketamine administration resulted in more than 80% responding on the lever associated with ethanol administration; the same was true when the pigeons were tested with dizocilpine (Grant et al. 1991). Thus, the animals "perceived" the effects of NMDA receptor antagonists as being similar to those of ethanol in vivo.

Other behavioral/physiological results of ethanol-induced inhibition of NMDA receptor function may be observed in studies of the protective effect of ethanol against neurotoxicity and convulsions. As mentioned earlier, the relatively slow nature of the neuronal response to NMDA allows for summation of neuronal responses and entry of large amounts of calcium into the cell. This calcium influx can produce neurotoxicity (Choi 1988), possibly through the recently recognized mechanism of nitric oxide synthesis (Dawson et al. 1991), as well as seizures. In rats, ethanol (2 g/kg) provided protection against the convulsant activity of NMDA, and was less effective against kainate (Kulkarni et al. 1990), in accord with the biochemical and electrophysiological experiments showing selective ethanol inhibition of NMDA responses. Although ethanol itself is often thought to be a neurotoxin, its toxic effects are usually seen after prolonged chronic consumption or exposure. The acute effect of ethanol on NMDA receptor function might also be expected to reduce glutamate-induced neurotoxicity, and recent studies have demonstrated that ethanol does protect rat embryonic brain cells in culture from NMDA-induced cytotoxicity (TAKADERA et al. 1990; LUSTIG et al. 1992; CHANDLER et al. 1991). This finding could have important implications with respect to glutamate-induced toxicity that may be associated with hypoxia and stroke (CHANDLER et al. 1991).

A few other studies have investigated the interactions of ethanol and NMDA receptor agonists and/or antagonists on various behavioral responses. NMDA decreased the response to ethanol both of mice bred selectively for sensitivity ["long-sleep" (LS) mice] or resistance ["short-sleep" (SS) mice] to the hypnotic effect of ethanol, while NMDA antagonists increased sensitivity in both lines (WILSON et al. 1990). However, there was some evidence of differential sensitivity, with SS mice being more affected by dizocilpine than LS mice (WILSON et al. 1990). These results were suggested by the authors to be consistent with the possibility that antagonism of NMDA receptor function by ethanol plays a role in ethanol-induced hypnosis (loss of righting reflex). Similarly, both dizocilpine and a competitive NMDA receptor antagonist were found to potentiate the sedative (muscle relaxant) effect of ethanol, and thus to inhibit ethanol-induced locomotor stimulation, in NMRI mice (LILJEQUIST 1991a). Ethanol was also found to act in a manner similar to specific NMDA receptor antagonists in a model in which locomotor activation was measured in monoamine-depleted mice (CARLSSON and ENGBERG 1992). Thus, in mice pretreated with reserpine, either ethanol, or other NMDA antagonists, in combination with clonidine, resulted in locomotor stimulation. On the other hand, ethanol and dizocilpine had differential effects on locomotor activity in rats (ROBLEDO et al. 1991). This finding may reflect the fact that ethanol (and possibly "specific" NMDA receptor antagonists) also affects neuronal systems other than the NMDA receptor (see TABAKOFF and HOFFMAN 1987), and behavioral responses to ethanol can reflect the composite of these neuronal activities.

4. Ethanol and the NMDA Receptor in Development

There is substantial evidence to implicate NMDA receptor function in neuronal development, particularly in activity-dependent synaptic modification (see COLLINGRIDGE and LESTER 1989). For example, administration of an NMDA receptor antagonist blocked the ability of the visual cortex in the newborn cat to respond to a period of monocular deprivation, which normally alters the sensitivity of neurons to visual stimulation (KLEINSCHMIDT et al. 1987). These types of experiments support the hypothesis that exposure to ethanol during gestation could interfere with NMDA receptor-dependent establishment of normal neuronal connections. It is possible that such interference may contribute to certain of the characteristics of the fetal alcohol syndrome (FAS), a pattern of congenital malformations seen in children of alcoholic mothers (Clarren and Smith 1978). One consequence of FAS is mental retardation (Abel 1980). In animal models of FAS as well, various deficits of learning and memory, and also hippocampal abnormalities, have been observed in offspring of animals given ethanol during gestation (Abel 1980).

There has, as yet, been little investigation of the possible importance of ethanol-induced NMDA receptor inhibition in producing the symptoms of FAS. However, there have been investigations of the effect of prenatal alcohol exposure on the function of the NMDA receptor. For example, hippocampal slices were obtained from the adult offspring of ethanol-treated rats (ethanol in a liquid diet was administered throughout gestation), and the response of extracellular population field potentials to NMDA was found to be significantly reduced, compared to controls (Morrisett et al. 1989). The reduction was reported to result from an enhanced sensitivity to Mg^{2+} (Morrisett et al. 1989). In addition, NMDA-sensitive glutamate binding was found to be reduced in hippocampus of rats that had been exposed chronically to ethanol during the last third of gestation (Savage et al. 1991). In other studies of adult rats whose dams were fed an ethanol-containing diet during gestation, somewhat similar results were found. The stimulation of hippocampal polyphosphoinositide metabolism by agonists acting at the quisqualate receptor, and the inhibition of agonist-stimulated polyphosphoinositide metabolism by NMDA or kainate, were all reduced in tissue of the animals that had been exposed to ethanol in utero (Noble and Ritchie 1989). Thus, in the latter study, there was a generalized reduction in response to agonists acting at several subtypes of glutamate receptor, while in the electrophysiological study there was a selective reduction in the NMDA response, with little or no change in response to agonists acting at the non-NMDA glutamate receptors (Morrisett et al. 1989). These authors also found no change in ligand binding to *ionotropic* kainate or AMPA receptors located in the apical dendritic field regions of the hippocampus in the offspring of rats given ethanol throughout gestation (Martin et al. 1992). Reduced functioning of the NMDA receptor could contribute to deficits in neuronal development and/or to behavioral and intellectual impairment in the offspring of the ethanol-treated mothers.

III. Ethanol and NMDA Receptor Function: Chronic Effects

While ethanol treatment during development appears to result in a reduced response to NMDA, in the adult animal there is evidence for an "upregulation" of NMDA receptor function after chronic ethanol ingestion. This change may underlie symptoms of ethanol withdrawal, specifically, ethanol withdrawal seizures. In adult mice that were fed ethanol chronically in a liquid diet, and were tolerant to and physically dependent on ethanol, there was an increased number of dizocilpine-binding sites in hippocampus, as measured in membrane-binding studies, and in several other brain areas, determined by autoradiography (Grant et al. 1990; Gulya et al. 1991). More recent work has also shown an increase in NMDA-specific glutamate binding in hippocampus of ethanol-fed mice, but no change in strychnine-insensitive glycine binding or binding of the competitive NMDA receptor antagonist, CGS-19755 (Snell et al. 1993). An increase in glutamate binding was also observed in brains of ethanol-exposed rats in an earlier study (Michaelis et al. 1978), and, more recently, in a postmortem study, glutamate binding was found to be higher in hippocampal synaptic membranes of alcoholics than controls (Michaelis et al. 1990).

The increase in ligand binding to the NMDA recognition site, and in dizocilpine binding, are consistent with an increased number of NMDA receptors in brains of ethanol-fed mice; however, an increase in glycine and competitive antagonist binding might also have been expected to occur. As mentioned earlier, it has been suggested that the NMDA receptor consists of a multisubunit complex, in which distinct protein subunits bind glutamate, glycine, and antagonists (Wang et al. 1992). This postulate is supported by the demonstration that, during development, ligand binding to each site appears at different times (Mcdonald et al. 1990). The recent cloning and characterization of the function of the NR 1, NR 2A, B, and C, ζ 1 and ε 1, 2, and 3 subunits, described above, also supports the hypothesis that the NMDA receptor is a heteromeric protein complex. Changes in the composition of NMDA receptor subunits caused by chronic ethanol ingestion, similar to those that appear to occur in the $GABA_A$ receptor in brains of ethanol-treated mice and rats (Buck et al. 1991; Montpied et al. 1991; Morrow et al. 1992), may account for the observed changes in ligand-binding characteristics.

Some caveats must be mentioned with respect to the ligand-binding studies described above. Although quantitative autoradiography revealed an increase in dizocilpine binding in several brain areas, including cerebral cortex, of ethanol-fed mice, the increase in cerebral cortical binding of dizocilpine in these mice, as measured in a membrane-binding assay, was not statistically significant (Snell et al. 1993). It seems possible that additional factors, present in the more intact brain sections used for autoradiography, may contribute to the ethanol-induced increase in dizocilpine binding in some brain areas. These factors may be removed during the preparation of well-washed brain membranes, and, thus, only the more robust changes, such as those in hippocampus, are clearly detected in membrane-binding analysis. Further evidence for the influence of differing conditions on ligand binding to the NMDA receptor comes from a recent report that ethanol-dependent rats exhibited a *decrease* in NMDA-specific glutamate binding measured by autoradiography (Cummins et al. 1990), which contrasts with the other studies in which glutamate binding was increased when assayed in membrane preparations (Grant et al. 1990; Michaelis et al. 1978).

Although changes in ligand binding suggest a change in the quantity of NMDA receptors in the ethanol-treated animals, it is important to determine whether this change is also reflected in receptor *function*. In one study of ethanol-fed rats, no change in NMDA-stimulated neurotransmitter release from brain slices was observed (Brown et al. 1991). However, the process of agonist-stimulated neurotransmitter release involves many complex steps between the initial agonist interaction with the receptor and the final effect. This process could "buffer" a change in receptor number. In addition, this system measures primarily the function of presynaptic NMDA receptors (Göthert and Fink 1991), which may be affected differently from postsynaptic receptors during chronic ethanol exposure. To address the question of ethanol-induced changes in NMDA receptor function more directly, we have investigated the response to NMDA and other glutamate receptor agonists in primary cultures of cerebellar granule cells, by measuring the agonist-induced increase in intracellular calcium with the fluorescent dye fura-2. Following exposure of the cells to 100 m*M* ethanol for 2 or more days, the response to NMDA (and glycine) was significantly increased, as compared to control cells, with no change in the EC_{50} value for either NMDA or glycine (Iorio et al. 1992). A similar increase ws observed when cells were exposed to 20 m*M* ethanol for 3 or more days. This change appeared to reflect an increase in the number of receptors, with little or no change in receptor properties, since inhibition of the NMDA response by competitive and noncompetitive antagonists was not altered by chronic ethanol exposure. In addition, acute inhibition of the NMDA response by ethanol, determined as percent change from the response to NMDA alone, was not affected by chronic ethanol exposure (Iorio et al. 1992).

The chronic effect of ethanol on NMDA receptor function in the cerebellar granule cells was relatively selective, since there was no change in the depolarization (30 m*M* KCl)-dependent increase in intracellular calcium in the ethanol-treated cells (Iorio et al. 1992). The response to KCl was inhibited by nifedipine, indicating the involvement of L-type voltage-sensitive calcium channels (VSCC). These VSCC have been reported to be increased in brain after chronic ethanol treatment of animals, and to be increased in PC 12 cells exposed chronically to ethanol (Dolin et al. 1987; Messing et al. 1986). This latter increase was postulated to be mediated by protein kinase C (Messing et al. 1990). The apparent lack of change of L-type VSCC in ethanol-exposed cerebellar granule cells supports the hypothesis that chronic effects of ethanol on VSCC depend on complex intracellular mechanisms that may vary in different cell types (Messing et al. 1990).

1. Role of NMDA Receptors in Ethanol Withdrawal (Physical Dependence)

The finding of increased function of NMDA receptors in cerebellar granule cells exposed chronically to ethanol is compatible with the binding studies in ethanol-fed animals discussed earlier, and suggests that chronic ethanol exposure results in an upregulation of NMDA receptors that could be involved in the generation or expression of ethanol withdrawal seizures. For example, the time course of appearance and disappearance of the changes in dizocilpine binding in hippocampus of ethanol-fed mice paralleled the time course for withdrawal seizures in these mice (GULYA et al. 1991). Furthermore, several studies have shown that competitive and noncompetitive antagonists at the NMDA receptor attenuate ethanol withdrawal seizures in mice and rats (GRANT et al. 1990; MORRISETT et al. 1990; LILJEQUIST 1991b). Similarly, NMDA, at a dose that had little or no effect in control animals, exacerbated withdrawal seizures in ethanol-dependent mice (GRANT et al. 1990).

The role of NMDA receptors in ethanol withdrawal has also been examined using mice that have been selectively bred to be prone (WSP) or resistant (WSR) to ethanol withdrawal seizures. Such selected lines of animals are, ideally, genetically invariant for the genes that contribute to the selected trait, while all unrelated gene frequencies remain variable (PHILLIPS et al. 1989). Therefore, under identical environmental conditions, a biochemical difference between the lines can be presumed to play a role in the selected trait. A greater number of dizocilpine-binding sites was found in the hippocampus of the WSP mice than the WSR mice. In addition, glycine was a more effective enhancer of dizocilpine binding in brain tissue of WSP mice (VALVERIUS et al. 1990). After chronic ethanol ingestion, the number of hippocampal dizocilpine-binding sites was increased in both lines of mice, such that the total number of sites remained significantly higher in the WSP mice (in the ethanol-treated WSR mice, the number of dizocilpine-binding sites was similar to that in ethanol-naive WSP mice) (VALVERIUS et al. 1990). The finding of a difference in NMDA receptors in these selected lines of mice provides strong support for a role of NMDA receptors in ethanol withdrawal seizures.

While an increased number of NMDA receptors may contribute to symptoms of ethanol withdrawal, a question remains as to the biochemical consequences of NMDA receptor activation (in addition to increased intracellular Ca^{2+} levels) that may be involved. For example, activation of NMDA receptors has been linked to stimulation of protein dephosphorylation (HALPAIN et al. 1990) and of tyrosine hydroxylase activity (ARIAS-MONTANO et al. 1992), among other effects. A recent report that is of interest in this context is the finding that concentrations of extracellular dopamine, measured by microdialysis, were increased in the ventral striatum of ethanol-withdrawn rats, and that dizocilpine treatment reversed the fall in dopamine levels that occurred over time after withdrawal (ROSSETTI et al. 1991). These results would suggest that increased NMDA receptor function may contribute to changes in dopamine metabolism after ethanol withdrawal, and the molecular mechanism of such changes needs further elucidation.

Current evidence suggests that, in contrast to NMDA receptors, other glutamate receptors are *not* altered by chronic ethanol exposure. In mice fed ethanol chronically, there was no significant change in hippocampal or cerebral cortical kainate binding, although there was a tendency for kainate binding to decrease (SNELL et al. 1993). Similarly, in cerebellar granule cells exposed chronically (3 days) to 100 m*M* ethanol in vitro, there was no change in response to kainate, as determined by increases in intracellular calcium measured with fura-2 (K.R. IORIO, B. TABAKOFF and P.L. HOFFMAN, unpublished observation). However, different results were found after chronic treatment with another sedative hypnotic, barbital. When rats were fed a diet containing barbital for 8 weeks, and then injected intracerebroventricularly with glutamate receptor agonists, the response of cerebellar cyclic GMP levels to kainate, but not NMDA or quisqualate, was found to be enhanced (MCCASLIN and MORGAN 1989). This finding bears on the issue of whether changes in glutamate receptors observed after chronic drug treatment represent adaptive responses to the initial effects of the drugs. Thus, ethanol, which is a more selective inhibitor of NMDA responses, appears to produce an upregulation of NMDA receptors after chronic treatment, while barbiturates, which, acutely, more selectively inhibit kainate receptor function (TABAKOFF et al. 1991; MORGAN et al. 1991), produce enhanced kainate receptor responses after chronic exposure. On the other hand, more recent results have indicated that, in mice fed barbiturates chronically, dizocilpine binding is increased, while kainate binding is *decreased*, in the cerebral cortex (SHORT and TABAKOFF 1993).

Whether the changes in glutamate receptor characteristics, observed after chronic ethanol or barbiturate treatment in cells or certain brain areas, are in fact adaptive in nature can be further addressed by investigating receptor upregulation after chronic exposure of animals or cells to specific receptor antagonists. However, the available data with regard to NMDA receptor function are somewhat equivocal. While chronic phencyclidine infusion in rats appeared to produce an increase in NMDA receptors (phencyclidine binding) in brain (MASSEY and WESSINGER 1990), chronic in vivo dizocilpine administration had less consistent effects on ligand binding, and resulted in a *decreased* electrophysiological response to NMDA (BEART and LODGE 1990; MANNALLACK et al. 1989). Therefore, the postulate that chronic ethanol-induced changes in NMDA receptor function represent a neuroadaptive response to the acute effect of ethanol needs further investigation.

While the data discussed earlier provide convincing evidence that NMDA receptor systems are involved in ethanol withdrawal seizures/physical dependence, it should be clearly stated that the NMDA receptor system is not the only neurochemical system involved in ethanol withdrawal. There is good evidence, for example, that downregulation of $GABA_A$ receptor function, as well as upregulation of VSCC, may also contribute to the symptoms of ethanol withdrawal (physical dependence on ethanol) (see HOFFMAN and TABAKOFF 1991; HOFFMAN et al. 1992). The precise role of each of these systems, and of their interactions, in ethanol withdrawal symptomatology is under active investigation (Hoffman and Tabakoff 1991; Hoffman et al. 1992).

2. Role of NMDA Receptors in Ethanol Tolerance

It has been argued that ethanol tolerance and physical dependence do not arise from a single unitary change in CNS function. For example, treatment of mice with the neurotoxin 6-hydroxydopamine blocked the development of functional ethanol tolerance, but did not affect physical dependence (withdrawal symptoms) (Tabakoff and Ritzmann 1977). In WSP and WSR mice, which differ significantly in the severity of ethanol withdrawal seizures, there was no difference in the degree of chronic ethanol tolerance developed (Crabbe and Kosobud 1986). Thus, while the NMDA receptor seems to play a key role in physical dependence on ethanol (ethanol withdrawal seizures), its role in ethanol tolerance is not as clear.

The simplest definition of ethanol tolerance in an acquired resistance to the effect of ethanol in an organism that has previously been exposed to ethanol. However, ethanol tolerance is a rather complex phenomenon (Tabakoff et al. 1982). For example, tolerance may be classified as functional or metabolic. In the first case, resistance to the effect of ethanol is a result of a change in neuronal structure or function; in the second, there is a change in the metabolism, distribution, or excretion of ethanol such that the individual is exposed to a lower concentration of ethanol after a given dose (Tabakoff et al. 1982). Tolerance may also be classified as acute or chronic. Acute tolerance is that which occurs during exposure of an individual to a single dose of ethanol (within-session tolerance), while chronic tolerance refers to the tolerance that is seen after more than one exposure to, or after longer-term ingestion of, ethanol (Tabakoff et al. 1982). Some studies have investigated a form of chronic tolerance designated "rapid" tolerance to ethanol, in which tolerance is measured after a single dose of ethanol has been administered to and metabolized by the animal (e.g., at 24 h after the first dose) (Khanna et al. 1991).

In addition to these factors, environmental cues can play a role in ethanol tolerance, and, in certain situations, conditioning or learning may be crucial for tolerance development. One form of tolerance has been termed "behaviorally-augmented," and occurs when an animal or human performs a task under the influence of ethanol. In this case, tolerance develops more rapidly than in an individual who has the same amount of practice in the task, but has performed the task *prior* to being given ethanol (see Tabakoff et al. 1982). Conditional, or "learned" tolerance, also called "environment-dependent" tolerance, occurs when an individual pairs cues from the environment with drug administration, and generates a compensatory response, which reduces the effect of ethanol, upon being exposed to those cues (Melchior and Tabakoff 1981). For example, one effect of ethanol in mice and rats is hypothermia. If an animal is given ethanol repeatedly in a particular environment, it develops tolerance to the hypothermic effect of ethanol in that environment. However, if the animal is placed in a different environment, tolerance is not observed (MELCHIOR and TABAKOFF 1981). An explanation for this phenomenon can be seen if the animal is given placebo in the ethanol-associated environment. In this case, one can observe a compensatory *hyper*thermia, which would counteract the effect of ethanol and therefore lead to tolerance. The compensatory response is not observed if placebo is administered in the environment not associated with ethanol treatment. This type of "environment-dependent" tolerance has a significant learned component. If the NMDA receptor is involved in ethanol tolerance, it is likely to play a role in functional, rather than metabolic, tolerance. Since the NMDA receptor has been implicated in learning and memory processes (see COLLINGRIDGE and LESTER 1989), it might also be expected to influence environment-dependent, conditional tolerance to a greater degree than environment-independent tolerance.

In a preliminary electrophysiological study, there was no change in the acute inhibitory effect of ethanol on NMDA responses in hippocampal slices from rats exposed chronically to ethanol, or in cultured hippocampal cells exposed chronically to ethanol (WHITE et al. 1990b). Similarly, BROWN et al. (1991) found that ethanol inhibited NMDA-stimulated neurotransmitter efflux equally in brain slices from control rats and those treated chronically with ethanol. In cerebellar granule cells, the acute inhibitory effect of ethanol on NMDA receptor function was also unchanged after chronic exposure of the cells to ethanol, when percentage inhibition was measured. These data suggested that the change ("upregulation") of NMDA receptor function did not contribute to "tolerance" to the effect of ethanol. However, because the baseline response to NMDA was increased in the ethanol-treated cerebellar granule cells, the inhibitory effect of ethanol was, in fact, *reduced* in these cells, if one uses the overall level of intracellular calcium as a measure. In effect, while *each* NMDA receptor remained sensitive to the acute effect of ethanol (percentage inhibition), the overall system, with a greater number of NMDA receptors (IORIO et al. 1992), may be viewed as being "tolerant" to ethanol. Even if one takes this point of view, however, the question remains as to whether or how tolerance to ethanol at the cellular level may translate into tolerance at the behavioral level.

We have previously used the construct of intrinsic and extrinsic neuronal systems to discuss ethanol tolerance (see TABAKOFF and HOFFMAN 1989). Intrinsic systems are those that encode tolerance to particular effects of ethanol within themselves, presumably as a result of changes in synaptic efficacy. Extrinsic systems are those that modulate the development, expression, or dissipation of tolerance, but do not encode tolerance in themselves. The data available to date are most compatible with the postulate that NMDA receptors represent an extrinsic neuronal system that modulates functional ethanol tolerance.

Khanna and his colleagues have examined the effect of dizocilpine and ketamine (which also interacts with the phencyclidine site within the NMDA receptor-coupled ion channel) on the development of rapid tolerance to ethanol. In their studies, rats were given a single dose of ethanol, and the ataxic (measured on a tilting plane) and hypothermic effects of ethanol were tested. The animals were injected with dizocilpine, ketamine, or vehicle prior to the ethanol injection. Twenty-four hours later, to assess tolerance, the identical procedure was carried out, except that no dizocilpine or ketamine was administered (KHANNA et al. 1991, 1992a). It was found that the active, but not the inactive, isomer of dizocilpine, as well as ketamine, prevented the development of rapid tolerance to ethanol, and this effect was not due to changes in ethanol metabolism. The blockade of tolerance was observed even if animals were tested 5 days after the single injection of ethanol (i.e., those receiving vehicle prior to the first ethanol treatment were still tolerant, and those receiving dizocilpine were not) (KHANNA et al. 1992a). Interestingly, when dizocilpine was given *following* the initial ethanol injection and testing for ataxia, it did *not* prevent the development of tolerance.

These data were suggested to be compatible with the putative role of the NMDA receptor in learning and memory, since it has been shown that learning can play a role in the rapid ethanol tolerance that was measured in these studies (BITRÁN and KALANT 1991). It was suggested that dizocilpine, when given before intoxicated practice, blocked the acquisition of the learned tolerance; when dizocilpine was given *after* practice, consolidation had already occurred, and dizocilpine was ineffective in attenuating tolerance (KHANNA et al. 1991, 1992a). Although these observations are certainly of interest, and may provide important information regarding the mechanisms underlying ethanol tolerance, some questions remain. For example, since ethanol, dizocilpine, and ketamine all acutely inhibit NMDA receptor function, the mechanism by which NMDA receptor antagonists attenuate ethanol tolerance development is not immediately obvious. However, ethanol clearly has effects other than those at the NMDA receptor, and other neuronal systems are involved in tolerance development (see TABAKOFF and HOFFMAN 1989); the interactions among these neuronal systems and NMDA receptor-mediated events need further elucidation. To address some of these questions, it was necessary to compare the effects of NMDA antagonists on the development and/or expression of environment-independent tolerance (see TABAKOFF et al. 1982) to more definitively evaluate the importance of the learning component of ethanol tolerance with respect to the role of the NMDA receptor. It is of interest that a recent report by Khanna and his colleagues (KHANNA et al. 1992b) indicated that ketamine, which blocked the development of "rapid" tolerance to ethanol, also blocked the development of chronic ethanol tolerance (10 days of ethanol injections), although it did not affect acute tolerance that occurs as a single dose of ethanol is metabolized. On the other hand we have found that dizocilpine does not block tolerance that developed in mice after chronic (6 day) ingestion of ethanol in a liquid diet (environment-independent tolerance), although it is an effective blocker of environment-dependent ethanol tolerance in mice (SZABÓ, et al. 1994).

3. Role of NMDA Receptors in Opiate Tolerance and Dependence

Supporting evidence for NMDA receptor involvement in the neuroadaptive mechanisms of tolerance comes from studies of opiate tolerance. In the rat, the NMDA receptor antagonists kynurenic acid and dizocilpine both interfered with the development of tolerance to the analgesic effect of morphine (MAREK et al. 1991a; TRUJILLO and AKIL 1991). This effect was observed when the antagonists were administered daily prior to morphine injections, and whether tolerance to morphine-induced analgesia was measured on a hot plate or tail-flick apparatus (MAREK et al. 1991a; TRUJILLO and AKIL 1991). The effect of dizocilpine was dose dependent, and was observed under conditions where dizocilpine had no acute effect on morphine analgesia (TRUJILLO and AKIL 1991). In contrast to the results described above with ethanol, dizocilpine was also reported to attenuate morphine tolerance when injected daily *after* morphine (MAREK et al. 1991b).

Tolerance and physical dependence on opiates, in contrast to ethanol, have often been considered to reflect a single underlying process, although that view is now being questioned (WÜSTER et al. 1985). It is of interest, however, that dizocilpine treatment appeared to block development of opiate dependence, and/or certain symptoms of opiate withdrawal, as well as tolerance. In rats that received chronic daily injections of dizocilpine and morphine, naloxone-induced withdrawal symptoms (escape jumps) were significantly reduced, in comparison to those in animals that received chronic vehicle and morphine injections (TRUJILLO and AKIL 1991). In this instance, dizocilpine did not simply affect the symptoms of withdrawal, since a single treatment with dizocilpine (0.1 mg/kg) prior to naloxone, in rats that had received chronic vehicle and morphine injections, did *not* attenuate withdrawal symptoms (TRUJILLO and AKIL 1991). Furthermore, when rats received chronic dizocilpine and morphine treatments, they did not display withdrawal signs even if they were given vehicle (rather than dizocilpine) prior to naloxone injection (TRUJILLO and AKIL 1991). Therefore, in this case, chronic dizocilpine treatment appeared to interfere with the development of opiate dependence.

Higher doses of NMDA receptor antagonists have also been reported to attenuate withdrawal *symptoms* in opiate-dependent animals, similar to the results in ethanol-dependent mice and rats (GRANT et al. 1990; MORRISETT et al. 1990; LILJEQUIST 1991b). When opiate dependence was achieved by implantation of a morphine pellet in rats, treatment with dizocilpine (0.3 or 0.5 mg/kg) 15 min before naltrexone attenuated many symptoms of withdrawal (although not jumping behavior). Treatment with a competitive

NMDA antagonist blocked many of the same withdrawal symptoms as dizocilpine (RASMUSSEN et al. 1991). Dizocilpine (0.02 or 0.1 mg/kg) was also reported to attenuate "motivational" aspects of opiate withdrawal in rats, as measured using a place conditioning (aversion) technique (HIGGINS et al. 1992). Furthermore, dizocilpine (0.3 or 1 mg/kg) reduced withdrawal symptoms in opiate-dependent mice and guinea pigs (TANGANELLI et al. 1991), and, in guinea pigs, dizocilpine reversed the withdrawal-induced increase in cerebral cortical acetylcholine outflow (TANGANELLI et al. 1991). In contrast, in rats, dizocilpine treatment that suppressed behavioral withdrawal signs did not reduce withdrawal-induced increases in activity of locus coeruleus neurons or norepinephrine turnover in hippocampus, cerebral cortex, or hypothalamus (RASMUSSEN et al. 1991). To accurately interpret these results, it is necessary to elucidate the contribution of noradrenergic and cholinergic systems to the symptoms of opiate withdrawal that were evaluated in these studies. It is also necessary to determine whether the withdrawal symptoms themselves may affect neurotransmitter turnover and neuronal activity.

C. Summary: Ethanol and the NMDA Receptor

Overall, the data reviewed indicated that ethanol is a potent and selective inhibitor of the function of the NMDA receptor. There appears to be a specific site (or sites) of action for ethanol at this receptor, and the most substantial evidence at this time supports the hypothesis that ethanol may influence the interaction of the co-agonists glutamate and glycine to affect the kinetics of activation of at least some forms of the NMDA receptor-coupled ion channel. Chronic exposure to ethanol, in vivo (in adult animals) or in vitro, produces a selective "upregulation" of NMDA receptor function that may mediate certain symptoms of ethanol withdrawal (i.e., withdrawal seizures). The role of the NMDA receptor in synaptic plasticity, learning and memory, has also led to the investigation of its importance for ethanol tolerance, another neuroadaptive process. The current data support the postulate that NMDA receptor function does contribute to certain forms of ethanol tolerance (i.e., environment-dependent, or conditional tolerance), as well as to opiate tolerance. It seems possible that the NMDA receptor system may represent an "extrinsic system" that modulates the development, expression, or dissipation of tolerance (or memory), but does not encode tolerance/memory within its pathways. If so, the NMDA receptor could be a contributor to ethanol and/or opiate tolerance, while other neuronal systems ("intrinsic systems") would be responsible for mediating the specific behavioral/pharmacological responses that become tolerant to the drugs (e.g., ataxia, hypothermia). The sensitivity of NMDA receptor function to ethanol, and the (possibly) adaptive changes in this receptor after chronic ethanol exposure, indicate its crucial role in mediating certain acute and chronic pharmacological effects of ethanol.

References

Abel EL (1980) The fetal alcohol syndrome: behavioral teratology. Psychol Bull 87:29–50
Anis NA, Berry SC, Burton NR, Lodge D (1983) The dissociative anaesthetics, ketamine and phencyclidine, selectively reduce excitation of central mammalian neurones by N-methyl-D-aspartate. Br J Pharmacol 79:565–575
Arias-Montano JA, Martinez-Fong D, Aceves J (1992) Glutamate stimulation of tyrosine hydroxylase is mediated by NMDA receptors in the rat striatum. Brain Res 569:317–322
Beart PM, Lodge D (1990) Chronic administration of MK-801 and the NMDA receptor: further evidence for reduced sensitivity of the primary acceptor site from studies with the cortical wedge preparation. J Pharm Pharmacol 42:354–355
Bitrán M, Kalant H (1991) Learning factor in rapid tolerance to ethanol-induced motor impairment. Pharmacol Biochem Behav 39: 917–922
Bonhaus DW, McNamara JO (1988) N-Methyl-D-aspartate receptor regulation of uncompetitive antagonist binding in rat brain membranes: kinetic analysis. Mol Pharmacol 34:250–255
Bowery NG, Wong EHF, Hudson AL (1988) Quantitative autoradiography of [^{3}H]MK-801 binding sites in mammalian brain. Br J Pharmacol 93:944–954
Brown LM, Leslie SW, Gonzales RA (1991) The effects of chronic ethanol exposure on N-methyl-D-aspartate-stimulated overflow of [^{3}H]catecholamines from rat brain. Brain Res 547:289–294
Buck KJ, Hahner L, Sikela J, Harris RA (1991) Chronic ethanol treatment alters brain levels of γ-aminobutyric acid$_A$ receptor subunit mRNAs: relationship to genetic differences in ethanol withdrawal seizure severity. J Neurochem 57: 1452–1455
Carlsson ML, Engberg G (1992) Ethanol behaves as an NMDA antagonist with respect to locomotor stimulation in monoamine-depleted mice. J Neural Transm Gen Sect 87:155–160
Chandler LJ, Summers C, Crews FT (1991) Ethanol inhibits NMDA-stimulated excitoxicity. Alcohol Clin Exp Ther 15:323
Chin JH, Goldstein DB (1977) Effects of low concentrations of ethanol on the fluidity of spin-labeled erythrocyte and brain membranes. Mol Pharmacol 13:435–441
Choi DW (1988) Calcium-mediated neurotoxicity: relationship to specific channel types and role in ischemic damage. Trends Neurosci 11:465–469
Clarren SK, Smith DW (1978) The fetal alcohol syndrome. N Engl J Med 298:1063–1067
Collingridge GL, Lester RAJ (1989) Excitatory amino acid receptors in the vertebrate central nervous system. Pharmacol Rev 40: 143–210
Cotman CW, Bridges RJ, Taube JS, Clark AS, Geddes JW, Monaghan DT (1989) The role of the NMDA receptor in central nervous system plasticity and pathology. J NIH Res 1:65–74
Crabbe JC, Kosobud A (1986) Sensitivity and tolerance to ethanol in mice bred to be genetically prone or resistant to ethanol withdrawal seizures. J Pharmacol Exp Ther 239:327–333
Cummins JT, Sack M, von Hungen K (1990) The effect of chronic ethanol on glutamate binding in human and rat brain. Life Sci 47:877–882
Dawson VL, Dawson TM, London ED, Bredt DS, Snyder SH (1991) Nitric oxide mediates glutamate neurotoxicity in primary cortical cultures. Proc Natl Acad Sci USA 88:6368–6371
Dildy JE, Leslie SW (1989) Ethanol inhibits NMDA-induced increases in free intracellular Ca^{2+} in dissociated brain cells. Brain Res 499:383–387

Dildy-Mayfield JE, Harris RA (1992) Acute and chronic ethanol exposure alters the function of hippocampal kainate receptors expressed in *Xenopus* oocytes. J Neurochem 58:1569–1572

Dildy-Mayfield JE, Leslie SW (1991) Mechanism of inhibition of N-methyl-D-aspartate-stimulated increases in free intracellular Ca^{2+} concentration by ethanol. J Neurochem 56:1536–1543

Dingledine R (1991) New wave of non-NMDA excitatory amino acid receptors. Trends Pharmacol Sci 12:360–362

Dolin S, Little H, Hudspith M, Pagonis C, Littleton J (1987) Increased dihydropyridine-sensitive calcium channels in rat brain may underlie ethanol physical dependence. Neuropharmacology 26:275–279

Egebjerg J, Bettler B, Hermans-Borgmeyer I, Heinemann S (1991) Cloning of a cDNA for a glutamate receptor subunit activated by kainate but not AMPA. Nature 351:745–748

Engberg G, Hajós M (1992) Ethanol attenuates the response of locus coeruleus neurons to excitatory amino acid agonists in vivo. Naunyn Schmiedeberg's Arch Pharmacol 345:222–226

Fink K, Göthert M (1990) Inhibition of N-methyl-D-aspartate-induced noradrenaline release by alcohols is related to their hydrophobicity. Eur J Pharmacol 191:225–229

Fink K, Göthert M (1991) Ethanol inhibits the N-methyl-D-aspartate (NMDA)-induced attenuation of the NMDA-evoked noradrenaline release in the rat brain cortex: interaction with NMDA-induced desensitization. Naunyn Schmiedeberg's Arch Pharmacol 344:167–173

Franks NP, Lieb WR (1984) Do general anesthetics act by competitive binding to specific receptors? Nature 310:599–601

Frye GD (1991) Interaction of ethanol and L-glutamate in the guinea pig ileum myenteric plexus. Eur J Pharmacol 192:1–7

Goldstein DB, Chin JH, Lyon RC (1982) Ethanol disordering of spin-labeled mouse brain membranes – correlation with genetically-determined ethanol sensitivity of mice. Proc Natl Acad Sci USA 79:4231–4233

Gonzales RA, Woodward JJ (1990) Ethanol inhibits N-methyl-D-aspartate-stimulated [^{3}H]norepinephrine release from rat cortical slices. J Pharmacol Exp Ther 253:1138–1144

Gonzales RA, Westbrook SL, Bridges LT (1991) Alcohol-induced inhibition of N-methyl-D-aspartate-evoked release of [^{3}H]norepinephrine from brain is related to lipophilicity. Neuropharmacology 30:441–446

Göthert M, Fink K (1989) Inhibition of N-methyl-D-aspartate (NMDA)- and L-glutamate-induced noradrenaline and acetylcholine release in the rat brain by ethanol. Naunyn Schmiedeberg's Arch Pharmacol 340:516–521

Göthert M, Fink K (1991) Stimulation of noradrenaline release in the cerebral cortex via presynaptic N-methyl-D-aspartate (NMDA) receptors and their pharmacological characterization. J Neural Transm [Suppl 34]:121–127

Grant KA, Valverius P, Hudspith M, Tabakoff B (1990) Ethanol withdrawal seizures and the NMDA receptor complex. Eur J Pharmacol 176:289–296

Grant KA, Knisely JS, Tabakoff B, Barrett JE, Balster RL (1991) Ethanol-like discriminative stimulus effects of non-competitive N-methyl-D-aspartate antagonists. Behav Pharmacol 2:87–95

Gregor P, Mano I, Maoz I, McKeown M, Teichberg VI (1989) Molecular structure of the chick cerebellar kainate-binding subunit of a putative glutamate receptor. Nature 342:689–692

Gulya K, Grant KA, Valverius P, Hoffman PL, Tabakoff B (1991) Brain regional specificity and time course of changes in the NMDA receptor-ionophore complex during ethanol withdrawal. Brain Res 547:129–134

Halpain S, Girault J-A, Greengard P (1990) Activation of NMDA receptors induces dephosphorylation of DARPP-32 in rat striatal slices. Nature 343:369–372

Harris RA, Schroeder F (1981) Ethanol and the physical properties of brain membranes: fluorescence studies. Mol Pharmacol 20:128–137
Hebb DO (1949) The organization of behavior. Wiley, New York
Heinemann S, Bettler B, Boulter J, Deneris E, Gasic G, Hartley M, Hollmann M, Hughes TE, O'Shea-Greenfield A, Rogers S (1991) The glutamate receptors: genes, structure and expression. In: Costa E, Joh TH (eds) Neurotransmitter regulation of gene transcription. vol 7. Thieme, New York, p 143
Higgins GA, Nguyen P, Sellers EM (1992) The NMDA antagonist dizocilpine (MK-801) attenuates motivational as well as somatic aspects of naloxone precipitated opioid withdrawal. Life Sci 50:PL167–PL172
Hoffman PL, Tabakoff B (1991) The contribution of voltage-gated and NMDA receptor-gated calcium channels to ethanol withdrawal seizures. In: Kalant H, Khanna JM, Israel Y (eds) Advances in biomedical alcohol research. Pergamon, New York, p 171
Hoffman PL, Rabe CS, Moses F, Tabakoff B (1989) N-Methyl-D-aspartate receptors and ethanol: inhibition of calcium flux and cyclic GMP production. J Neurochem 52:1937–1940
Hoffman PL, Grant KA, Snell LD, Reinlib L, Iorio KR (1992) NMDA receptors: role in ethanol withdrawal seizures. Ann NY Acad Sci 654:52–60
Hollmann M, O'Shea-Greenfield A, Rogers SW, Heinemann S (1989) Cloning by functional expression of a member of the glutamate receptor family. Nature 342:643–648
Hollmann M, Hartley M, Heinemann S (1991) Ca^{2+} permeability of KA-AMPA-gated glutamate receptor channels depends on subunit composition. Science 252:851–853
Hunt WA (1985) Alcohol and biological membranes. Guilford, New York
Iorio KR, Reinlib L, Tabakoff B, Hoffman PL (1992) Chronic exposure of cerebellar granule cells to ethanol results in increased NMDA receptor function. Mol Pharmacol 41:1142–1148
Javitt DC, Zukin SR (1989) Biexponential kinetics of [^{3}H]MK-801 binding: evidence for access to closed and open N-methyl-D-aspartate receptor channels. Mol Pharmacol 35:387–393
Johnson JW, Ascher P (1987) Glycine potentiates the NMDA response in cultured mouse brain neurones. Nature 325:529–531
Keinänen K, Wisden W, Sommer B, Werner P, Herb A, Verdoon TA, Sakmann B, Seeburg PH (1990) A family of AMPA-selective glutamate receptors. Science 249:556–560
Khanna JM, Wu PH, Weiner J, Kalant H (1991) NMDA antagonist inhibits rapid tolerance to ethanol. Brain Res Bull 26:643–645
Khanna JM, Kalant H, Shah G, Chau A (1992a) Effect of (+)MK-801 and ketamine on rapid tolerance to ethanol. Brain Res Bull 28:311–314
Khanna JM, Kalant H, Weiner J, Chau A, Shah G (1992b) Ketamine retards chronic but not acute tolerance to ethanol. Pharmacol Biochem Behav 42:347–350
Kleckner NW, Dingledine R (1988) Requirement for glycine in activation of NMDA receptors expressed in *Xenopus* oocytes. Science 241:835–837
Kleinschmidt A, Bear MF, Singer W (1987) Blockade of NMDA receptors disrupts experience-dependent plasticity of kitten striate cortex. Science 238:355–358
Kulkarni SK, Mehta AK, Ticku MK (1990) Comparison of anticonvulsant effect of ethanol against NMDA-, kainic acid- and picrotoxin-induced convulsions in rats. Life Sci 46:481–487
Kumar KN, Tilakaratne N, Johnson PS, Allen AE, Michaelis EK (1991) Cloning of cDNA for the glutamate-binding subunit of an NMDA receptor complex. Nature 354:70–73
Kutsuwada T, Kashiwabuchi N, Mori H, Sakimura K, Kushiya E, Araki K, Meguro H, Masaki H, Kumanishi T, Arakawa M, Mishina M (1992) Molecular diversity of the NMDA receptor channel. Nature 358:36–41

Liljequist S (1991a) NMDA receptor antagonists inhibit ethanol-produced locomotor stimulation in NMRI mice. Alcohol 8:309–312

Liljequist S (1991b) The competitive NMDA receptor antagonist, CGP 39551, inhibits ethanol withdrawal seizures. Eur J Pharmacol 192:197–198

Lima-Landman MT, Albuquerque EX (1989) Ethanol potentiates and blocks NMDA-activated single-channel currents in rat hippocampal pyramidal cells. FEBS Lett 247:61–67

Lovinger DM, White G, Weight FF (1989) Ethanol inhibits NMDA-activated ion current in hippocampal neurons. Science 243:1721–1724

Lovinger DM, White G, Weight FF (1990) NMDA receptor-mediated synaptic excitation selectively inhibited by ethanol in hippocampal slice from adult rat. J Neurosci 10:1372–1379

Lustig HS, Chan J, Greenberg DA (1992) Ethanol inhibits excitotoxicity in cerebral cortical cultures. Neurosci Lett 135:259–261

Mannallack DT, Lodge D, Beart PM (1989) Subchronic administration of MK-801 in the rat decreases cortical binding of ^{3}H-D-AP5, suggesting down-regulation of the cortical N-methyl-D-aspartate receptors. Neuroscience 30:87–94

Marek P, Ben-Eliyahu S, Gold M, Liebeskind JC (1991a) Excitatory amino acid antagonists (kynurenic acid and MK-801) attenuate the development of morphine tolerance in the rat. Brain Res 547:77–81

Marek P, Ben-Eliyahu S, Vaccarino AL, Liebeskind JC (1991b) Delayed application of MK-801 attenuates development of morphine tolerance in rats. Brain Res 558:163–165

Martin D, Morrisett RA, Bian X-P, Wilson WA, Swartzwelder HS (1991a) Ethanol inhibition of NMDA mediated depolarizations is increased in the presence of Mg^{2+}. Brain Res 546:227–234

Martin DC, Abraham JE, Plagenhoef M, Aronstam RS (1991b) Volatile anesthetics and NMDA receptors. Enflurane inhibition of glutamate-stimulated [^{3}H]MK-801 binding and reversal by glycine. Neurosci Lett 132:73–76

Martin D, Savage DD, Swartzwelder HS (1992) Effects of prenatal ethanol exposure on hippocampal ionotropic-quisqualate and kainate receptors. Alcohol Clin Exp Res 16:816–821

Massey BW, Wessinger WD (1990) Changes in phencyclidine (PCP) receptor binding following cessation of chronic PCP administration. Pharmacologist 32:192

McCaslin PP, Morgan WW (1989) Increased response of cerebellar cGMP to kainate but not NMDA or quisqualate following barbital withdrawal from dependent rats. Eur J Pharmacol 173:127–132

McDonald JW, Johnston MV, Young AB (1990) Differential ontogenic development of three receptors comprising the NMDA receptor/channel complex in the rat hippocampus. Exp Neurol 110:237–247

Meguro H, Mori H, Araki K, Kushiya E, Kutsuwada T, Yamazaki M, Kumanishi T, Arakawa M, Sakimura K, Mishina M (1992) Functional characterization of a heteromeric NMDA receptor channel expressed from cloned cDNAs. Nature 357:70–74

Melchior CL, Tabakoff B (1981) Modification of environmentally cued tolerance to ethanol in mice. J Pharmacol Exp Ther 219:175–180

Messing RO, Carpenter CL, Greenberg DA (1986) Ethanol regulates calcium channels in clonal neural cells. Proc Natl Acad Sci USA 83:6213–6215

Messing RO, Sneade AB, Savidge B (1990) Protein kinase C participates in upregulation of dihydropyridine-sensitive calcium channels by ethanol. J Neurochem 55:1383–1389

Michaelis EK, Mulvaney MJ, Freed WJ (1978) Effects of acute and chronic ethanol intake on synaptosomal glutamate binding activity. Biochem Pharmacol 27: 1685–1691

Michaelis EK, Freed WJ, Galton N, Foye J, Michaelis ML, Phillips I, Kleinman JE (1990) Glutamate receptor changes in brain synaptic membranes from human alcoholics. Neurochem Res 15:1055–1063

Miljkovic A, MacDonald JF (1986) Voltage-dependent block of excitatory amino acid currents by pentobarbital. Brain Res 376:396–399
Monaghan DT (1991) Differential stimulation of [^{3}H]MK-801 binding to subpopulations of NMDA receptors. Neurosci Lett 122:21–24
Monaghan DT, Cotman CW (1982) The distribution of [^{3}H]kainic acid binding sites in rat CNS as determined by autoradiography. Brain Res 252:91–100
Monaghan DT, Overman HJ, Nguyen L, Watkins J, Cotman CW (1988) Two classes of N-methyl-D-aspartate recognition site: differential distribution and differential regulation by glycine. Proc Natl Acad Sci USA 85:9836–9840
Montpied P, Morrow AL, Karanian JW, Ginns EI, Martin BM, Paul SM (1991) Prolonged ethanol inhalation decreases γ-aminobutyric acid$_A$ receptor α subunit mRNAs in the rat cerebral cortex. Mol Pharmacol 39:157–163
Monyer H, Sprengel R, Schoepfer R, Herb A, Higuchi J, Lomeli H, Burnashev N, Sakmann B, Seeburg PH (1992) Heteromeric NMDA receptors: molecular and functional distinction of subtypes. Science 256:1217–1221
Morgan WW, Bermudez J, Chang X (1991) The relative potency of pentobarbital in suppressing the kainic acid- or the N-methyl-D-aspartic acid-induced enhancement of cGMP in cerebellar cells. Eur J Pharmacol 204:335–338
Moriyoshi K, Masu M, Ishii T, Shigemoto R, Mizuno N, Nakaniski S (1991) Molecular cloning and characterization of the rat NMDA receptor. Nature 354:31–37
Morrisett RA, Martin D, Wilson WA, Savage DD, Swartzwelder HS (1989) Prenatal exposure to ethanol decreases the sensitivity of the adult rat hippocampus to N-methyl-D-aspartate. Alcohol 6:415–420
Morrisett RA, Rezvani AH, Overstreet D, Janowsky DS, Wilson WA, Swartzwelder HA (1990) MK-801 potently inhibits alcohol withdrawal seizures in rats. Eur J Pharmacol 176:103–105
Morrisett RA, Martin D, Oetting TA, Lewis DV, Wilson WA, Swartzwelder HS (1991) Ethanol and magnesium ions inhibit N-methyl-D-aspartate-mediated synaptic potentials in an interactive manner. Neuropharmacology 30:1173–1178
Morrow AL, Herbert JS, Montpied P (1992) Differential effects of chronic ethanol administration of GABA$_A$ receptor α1 and α6 subunit mRNA levels in rat cerebellum. Mol Cell Neurosci 3:251–258
Noble EP, Ritchie T (1989) Prenatal ethanol exposure reduces the effects of excitatory amino acids in the rat hippocampus. Life Sci 45:803–810
Nowak L, Bregestovski P, Ascher P, Herbet A, Prochiantz Z (1984) Magnesium gates glutamate-activated channels in mouse central neurones. Nature 307:462–465
Peoples RW, Weight FF (1992) Ethanol inhibition of N-methyl-D-aspartate-activated ion current in rat hippocampal neurons is not competitive with glycine. Brain Res 571:342–344
Peters S, Koh J, Choi DW (1987) Zinc selectively blocks the action of N-methyl-D-aspartate on cortical neurons. Science 236:589–593
Phillips TJ, Feller DJ, Crabbe JC (1989) Selective mouse lines, alcohol, and behavior. Experientia 45:805–827
Pittaluga A, Raiteri M (1990) Release-enhancing glycine-dependent presynaptic NMDA receptors exist on noradrenergic terminals of hippocampus. Eur J Pharmacol 191:231–234
Rabe CS, Tabakoff B (1990) Glycine site directed agonists reverse ethanol's actions at the NMDA receptor. Mol Pharmacol 38:753–757
Raftery MA, Hunkapiller MW, Strader DC, Hood LE (1980) Acetylcholine receptor: complex of homologous subunits. Science 208:1454–1457
Ramoa AS, Alkondon M, Aracava Y, Irons J, Lunt GG, Deshpande SS, Wonnacott S, Aronstam RS, Albuquerque EX (1990) The anticonvulsant MK-801 interacts with peripheral and central nicotinic acetylcholine receptor ion channels. J Pharmacol Exp Ther 254:71–82

Ransom RW, Stec NL (1988) Cooperative modulation of [^{3}H]-MK-801 binding to the N-methyl-D-aspartate receptor-ion channel complex by L-glutamate, glycine and polyamines. J Neurochem 51:830–836

Rasmussen K, Fuller RW, Stockton ME, Perry KW, Swinford RM, Ornstein PL (1991) NMDA receptor antagonists suppress behaviors but not norepinephrine turnover or locus coeruleus unit activity induced by opiate withdrawal. Eur J Pharmacol 197:9–16

Robledo P, Kaneko W, Ehlers CL (1991) Combined effects of ethanol and MK-801 on locomotor activity in the rat. Pharmacol Biochem Behav 39:513–516

Rossetti ZL, Melis F, Carboni S, Gessa GL (1991) Marked decrease of extraneuronal dopamine after alcohol withdrawal in rats: reversal by MK-801. Eur J Pharmacol 200:371–372

Sacaan AI, Johnson KM (1989) Spermine enhances binding to the glycine site associated with the N-methyl-D-aspartate receptor complex. Mol Pharmacol 36:836–839

Savage DD, Queen SA, Sanchez CF, Paxton LL, Mahoney JC, Goodlett CR, West JR (1991) Prenatal ethanol exposure during the last third of gestation in rat reduces hippocampal NMDA agonist binding site density in 45-day-old offspring. Alcohol 9:37–41

Short KR, Tabakoff B (1993) Chronic barbiturate treatment increases NMDA receptors but decreases kainate receptors in mouse brain. Eur J Pharmacol 230:11–14

Simson PE, Criswell HE, Johnson KB, Hicks RE, Breese GR (1991) Ethanol inhibits NMDA-evoked electrophysiological activity in vivo. J Pharmacol Exp Ther 257:225–231

Singh L, Oles RJ, Tricklebank MD (1990) Modulation of seizure susceptibility in the mouse by the strychnine-insensitive glycine recognition site of the NMDA receptor/ion channel complex. Br J Pharmacol 99:285–288

Sircar R, Zukin SR (1991) Kinetic mechanisms of glycine requirement for N-methyl-D-aspartate channel activation. Brain Res 556:280–284

Snell LD, Tabakoff B, Hoffman PL (1993) Radioligand binding to the N-methyl-D-aspartate receptor/ionophore complex: alterations by ethanol in vitro and by chronic in vivo ethanol ingestion. Brain Res 602:91–98

Szabó G, Tabakoff B, Hoffman PL (1994) The NMDA receptor antagonist dizocilpine differentially affects environment-dependent and environment-independent ethanol tolerance. Psychopharmacol 113:511–517

Tabakoff B, Hoffman PL (1987) Biochemical pharmacology of alcohol. In: Meltzer HY (ed) Psychopharmacology – the third generation of progress. Raven, New York, p 1521

Tabakoff B, Hoffman PL (1989) Adaptive responses to ethanol in the central nervous system. In: Goedde HW, Agarwal DP (eds) Alcoholism: biomedical and genetic aspects. Pergamon, New York, p 99

Tabakoff B, Ritzmann RF (1977) The effects of 6-hydroxydopamine on tolerance to and dependence on ethanol. J Pharmacol Exp Ther 203:319–331

Tabakoff B, Melchior CL, Hoffman PL (1982) Commentary on ethanol tolerance. Alcohol Clin Exp Res 6:252–259

Tabakoff B, Rabe CS, Hoffman PL (1991) Selective effects of sedative/hypnotic drugs on excitatory amino acid receptors in brain. Ann NY Acad Sci 625:488–495

Takadera T, Suzuki R, Mohri T (1990) Protection by ethanol of cortical neurons from N-methyl-D-aspartate-induced neurotoxicity is associated with blocking calcium influx. Brain Res 537:109–114

Tanganelli S, Antonelli T, Morari M, Bianchi C, Beani L (1991) Glutamate antagonists prevent morphine withdrawal in mice and guinea pigs. Neurosci Lett 122:270–272

Thiels E, Weisz DJ, Berger TW (1992) In vivo modulation of N-methyl-D-aspartate receptor-dependent long-term potentiation by the glycine modulatory site. Neuroscience 46:501–509
Tossman U, Jonsson G, Ungerstedt U (1986) Regional distribution and extracellular level of amino acids in rat central nervous system. Acta Physiol Scand 127:533–545
Trujillo KA, Akil H (1991) Inhibition of morphine tolerance and dependence by the NMDA receptor antagonist MK-801. Science 251:85–87
Valverius P, Crabbe JC, Hoffman PL, Tabakoff B (1990) NMDA receptors in mice bred to be prone or resistant to ethanol withdrawal seizures. Eur J Pharmacol 184:185–189
Vicini S (1991) Pharmacologic significance of the structural heterogeneity of the $GABA_A$ receptor-central nervous system actions. Pharmacol Rev 41:489–537
Wada K, Dechesne CJ, Shimasaki S, King RG, Kusano K, Buonanno A, Hampson DR, Banner C, Wenthold RJ, Nakatani Y (1989) Sequence and expression of a frog brain complementary DNA encoding a kainate-binding protein. Nature 342:684–689
Wafford KA, Burnett DM, Leidenheimer NJ, Burt DR, Wang JB, Kofuji P, Dunwiddie TV, Harris RA, Sikela JM (1991) Ethanol sensitivity of the $GABA_A$ receptor expressed in *Xenopus* oocytes requires eight amino acids contained in the γ_{2L} subunit of the receptor complex. Neuron 7:27–33
Wallgren H, Barry H III (1970) Actions of alcohol, vol 1. Elsevier, New York
Wang H, Kumar KN, Michaelis EK (1992) Isolation of glutamate-binding proteins from rat and bovine brain synaptic membranes and immunochemical and immunocytochemical characterization. Neuroscience 46:793–806
Weight FF, Lovinger DM, White G, Peoples RW (1991) Alcohol and anesthetic actions on excitatory amino acid-activated ion channels. Ann NY Acad Sci 625:97–107
White GD, Lovinger DM, Grant KA (1990b) Ethanol (EtOH) inhibition of NMDA-activated ion current is not altered after chronic exposure of rats or neurons in culture. Alcohol Clin Exp Res 14:352
White G, Lovinger DM, Weight FF (1990a) Ethanol inhibits NMDA-activated current but does not affect GABA-activated current in an isolated adult mammalian neuron. Brain Res 507:332–336
Wilson WR, Bosy TZ, Ruth JA (1990) NMDA agonists and antagonists alter the hypnotic response to ethanol in LS and SS mice. Alcohol 7:389–395
Wong EHF, Kemp JA, Preistley T, Knight AR, Woodruff GN, Iversen LL (1986) The anticonvulsant MK-801 is a potent N-methyl-D-aspartate antagonist. Proc Natl Acad Sci USA 83:7104–7108
Woodward JJ, Gonzales RA (1990) Ethanol inhibition of N-methyl-D-aspartate stimulated endogenous dopamine release from rat striatal slices: reversal by glycine. J Neurochem 54:712–715
Wüster M, Schulz R, Herz A (1985) Opioid tolerance and dependence: re-evaluating the unitary hypothesis. Trend Pharmacol Sci 6:64–67
Yamazaki M, Mori H, Araki K, Mori K, Mishina J (1992) Cloning, expression and modulation of a mouse NMDA receptor subunit. FEBS Lett 300:39–45
Yoneda Y, Ogita K (1991) Heterogeneity of the N-methyl-D-aspartate receptor ionophore complex in rat brain, as revealed by ligand binding techniques. J Pharmacol Exp Ther 259:86–96
Young AB, Fagg GE (1990) Excitatory amino acid receptors in the brain-membrane binding and receptor autoradiographic approaches – EAA pharmacology. Trends Pharmacol Sci 11:126–133
Yuen GLF, Patil M, Durand D (1991) Effects of ethanol on the excitability of hippocampal granule neurons. Brain Res 563:315–320

CHAPTER 5

Effects of Alcohol on GABA-Mediated Neurotransmission

M.K. Ticku and A.K. Mehta

A. Introduction

Ethanol is the oldest and most widely consumed and abused drug by our society. The exact mechanism of its action, including its ability to produce tolerance and withdrawal following chronic administration, is not yet known. Ethanol has been shown to affect a variety of neurotransmitters in the CNS. Recent evidence supports the notion that ethanol may produce many of its effects by modulating ligand-gated ion channels (mediated by receptors for GABA, NMDA, 5-HT_3) in the brain. The pharmacological profile of ethanol is very similar to that of benzodiazepines and barbiturates. These classes of drugs are anxiolytics, muscle relaxants, and anticonvulsants, and their chronic use results in tolerance (Belleville and Frazer 1957). There also exists a cross-tolerance between ethanol and barbiturates, and ethanol and benzodiazepines. Furthermore, benzodiazepines and barbiturates are effective in the management of ethanol withdrawal syndromes (Goldstein 1973). These observations suggest that a common modulatory system may be involved at least in some of the pharmacological effects of these drugs. Since the GABAergic system is involved in the pharmacological actions of benzodiazepines and barbiturates (Ticku and Rastogi 1980; Olsen 1982; Maksay and Ticku 1985; Biggio et al. 1992), many investigations have focused on the effects of ethanol on GABAergic pathways.

There are two types of GABA receptors ($GABA_A$ and $GABA_B$), which differ in terms of their selectivity for agonists/antagonists and coupling mechanisms. Recent molecular biological research has demonstrated the heterogeneity of $GABA_A$ receptors (Wisden et al. 1988; Pritchett et al. 1989). So far, six α-, four β-, three γ-, and one δ-subunits have been cloned.

It appears that a minimum combination of α-, β- and γ_2-subunits is needed for generating functional $GABA_A$ responses in expression systems. However, the exact stoichiometry (pentameric or tetrameric) of the subunits needed to generate functional $GABA_A$ receptors that may resemble the in vivo situation is yet to be established. Furthermore, the distribution of α-subunits exhibits heterogeneity in terms of brain regional distribution (Wisden et al. 1988). Thus, the cerebellum has a great abundance of the α_1-subunits, whereas the cerebral cortex, hippocampus, and spinal cord have a mixture of α_3-, α_2-, and α_1-subunits, and these α-subunits have different affinities for the agonist GABA. The γ_2-subunit appears to be crucial for demonstrating benzodiazepine potentiation, and the γ_2L-subunit for ethanol potentiation of GABA responses (see below). Currently, the effects of acute and chronic ethanol on changes in the levels of different subunits of the GABA receptor are being investigated in various laboratories. In this chapter, an attempt has been made to review the behavioral, binding, functional, and gene expression studies of ethanol as they relate to GABAergic transmission.

B. Behavioral Studies

GABA agonists, and substances that raise endogenous GABA levels, such as aminooxyacetic acid (AOAA), increased the sedative, hypnotic, ataxic, and anticonvulsant effects of ethanol, whereas GABA antagonists had opposite effects (Hakkinen and Kulonen 1976; Liljequist and Engel 1982; Martz et al. 1983). These data suggest that potentiation of GABA transmission was an important factor in the mechanism of these properties of ethanol. The GABAergic system has also been implicated in ethanol withdrawal-induced audiogenic seizures (Guerrero-Figueroa et al. 1970; Cooper et al. 1979; Goldstein 1979; Aaronson et al. 1982; Frye et al. 1983a,b). The observation that a reduced GABA concentration in CSF of seizure-susceptible alcoholics during acute ethanol withdrawal supports the above fact (Greenblatt and Greenblatt 1972; Kramp and Rafaelson 1978; Thompson 1978; Goldman et al. 1981). Furthermore, chronic alcoholic males were reported to have significantly lower plasma levels of GABA than control subjects (Coffman and Petty 1985). In contrast to these observations, it has been reported recently that there are no differences in CSF levels of GABA between abstinent alcoholic patients and normal control subjects, and there was no significant difference in GABA levels between alcoholic patients with histories of withdrawal seizures and those without such a history (Roy et al. 1990). Intranigral administration of muscimol, a $GABA_A$ agonist, produced a suppression of audiogenic convulsions following ethanol withdrawal, thereby indicating the importance of the GABAergic system in the substantia nigra in the mediation of withdrawal seizures (Gonzales and Hettinger 1984). Recently, it has been suggested that the inferior colliculus is important in the GABAmimetic suppression of audiogenic seizures in rats during ethanol withdrawal (Frye et al. 1983b). Ethanol has been reported to have an anticonvulsant effect against chemoconvulsants such as bicuculline, picrotoxin, pentylenetetrazole, and strychnine (Liljequist and Engel 1982; Rastogi and Ticku 1986; Mehta and Ticku 1989b; Kulkarni et al. 1990). Anticonvulsant properties have also been reported for smaller doses of ethanol in combination with ineffective doses of facilitators of GABAergic transmission against seizures and mortality induced by chemoconvulsants which act via the GABAergic system (Rastogi and Ticku 1986).

The GABAergic system has also been implicated in the ataxic, narcotic, and hypothermic effects of ethanol. GABAmimetics lengthened ethanol narcosis while picrotoxin shortened ethanol narcosis (MARTZ et al. 1983). Both AOAA and GABA antagonists such as bicuculline and picrotoxin enhanced ethanol hypothermia (LILJEQUIST and ENGEL 1982; DAR and WOOLES 1985). Muscimol decreased the locomotor stimulant actions of low doses of ethanol at a dose that did not elicit any effect when given alone, while picrotoxin had the opposite effect (LILJEQUIST and ENGEL 1982).

Chronic ethanol treatment has been reported to decrease the behavioral effects of GABA agonists (TABERNER and UNWIN 1981; MARTZ et al. 1983), suggesting the occurrence of either downregulation of GABA receptors or alteration in coupling mechanisms. GABA receptor agonists and agents that increase GABA concentrations, such as AOAA and sodium valproate, are reported to decrease the ethanol withdrawal syndrome (GOLDSTEIN 1973, 1979; COOPER et al. 1979; FRYE et al. 1983a; GONZALES and HETTINGER 1984). While benzodiazepines are the drugs of choice in treating the ethanol withdrawal syndrome, these drugs merely decrease withdrawal symptomatology without preventing ethanol dependence (NUTT et al. 1989).

C. Binding Studies

Since ethanol shares several of its behavioral effects with that of barbiturates and benzodiazepines, there has been speculation that the GABA/benzodiazepine (BZ) receptor-coupled chloride channel is the site of action of ethanol in brain (COTT et al. 1976; HUNT 1983; TICKU et al. 1983; LILJEQUIST and ENGEL 1984). In contrast to this speculation, the addition of ethanol to brain membrane homogenates in vitro had no effect either on [^{3}H]diazepam binding to BZ receptors or [^{3}H]muscimol binding to GABA receptors (GREENBERG et al. 1984). However, ethanol has been reported to increase [^{3}H]diazepam binding to solubilized benzodiazepine receptors (DAVIS and TICKU 1981) and to decrease the binding of [^{35}S]*t*-butylbicyclophosphorothionate (TBPS) to a site closely associated with the Cl^- channel in rat and mouse brain homogenates (SQUIRES et al. 1983; RAMANJANEYULU and TICKU 1984; LILJEQUIST et al. 1986). Since ethanol inhibited [^{35}S]TBPS binding at lethal concentration, it is unlikely that this mechanism is responsible for the pharmacological effects of ethanol. Acute ethanol administration produced an increase or no change in the binding capacity of the low-affinity GABA receptor sites (TICKU 1980; TICKU and BURCH 1980; UNWIN and TABERNER 1980). In contrast to this finding, there is also a report that acute ethanol treatment caused a decrease in specific [^{3}H]GABA binding in cerebellum with no change in other brain areas (REGGIANI et al. 1980).

Chronic administration of ethanol has been reported to alter GABA receptor sensitivity depending upon the method of ethanol administration and animal species studied. TICKU (1980), using rats, reported that treatment with 11–17 g/kg per day of ethanol for 21 days, a schedule sufficient to cause withdrawal symptoms, decreased the affinity of the low-affinity GABA-binding site. Chronic ethanol administration (10% v/v) for 14 days has been reported to produce a decrease in B_{max} of the low-affinity GABA receptor sites in mice (TICKU and BURCH 1980). However, rats chronically treated with ethanol (7%) in a liquid diet for 21 days did not exhibit any changes in the binding constants of the high- or low-affinity GABA receptor sites (TICKU 1980). It is also reported that chronic ethanol administration does not alter either K_d or B_{max} value for [^{3}H]flunitrazepam and [^{35}S]TBPS binding, nor did it modify the modulatory effect of GABA on [^{3}H]flunitrazepam or [^{35}S]TBPS binding (RASTOGI et al. 1986).

A recent report indicated that chronic ethanol treatment (50 m*M* for 3 days) of intact cultured mice spinal cord neurons did not alter binding constants for [^{3}H]flunitrazepam or [^{3}H]Ro15-1788 binding (MHATRE and TICKU 1989). Chronic ethanol administration selectively increased Ro15-4513 binding in cerebral cortex and cerebellum but not in striatum and hippocampus (MHATRE et al. 1988). However, another study reported no change in the binding of a β-carboline inverse agonist in cortex, cerebellum, hippocampus, or olfactory bulb after chronic ethanol treatment (TAMBORSKY and MARANGOS 1986). TRAN et al. (1981) demonstrated an increase in [^{3}H]muscimol binding in human alcoholics. Chronic ethanol treatment is reported either to decrease (FREUND 1980; KOCHMAN et al. 1981; VOLICER and BIAGIONI 1982) or leave unchanged (KAROBATH et al. 1980; HEMMINGSEN et al. 1982; SCHOEMAKER et al. 1983) binding at the benzodiazepine receptor. It has been suggested that a longer duration of chronic ethanol treatment is required to cause changes in benzodiazepine receptor binding than is needed to produce tolerance and dependence on ethanol (FREUND 1980; ROTTENBERG 1985).

D. Functional Studies

I. Electrophysiological Studies

There are reports based on in vitro and in vivo electrophysiological studies that ethanol either potentiates or fails to alter GABAergic transmission. Thus ethanol potentiates the $GABA_A$ receptor-mediated responses in amphibian spinal cord neurons (DAVIDOFF 1973), cultured spinal cord neurons (CELENTANO et al. 1988), cerebral cortical neurons (NESTOROS 1980; AGUAYO 1990), hippocampal neurons (GAGE and ROBERTSON 1985; AGUAYO 1990), lateral septum (GIVENS and BREESE 1990), and rat dorsal root ganglion neurons (NISHIO and NARAHASHI 1990). However, negative results have also been reported in many regions (e.g., GAGE and ROBERTSON 1985; SIGGINS et al. 1987; HARRISON et al. 1987; WHITE et al. 1990). Studies of the effects of ethanol on GABAergic transmission in hippocampal neurons have been mostly negative or inconsistent. Thus, ethanol has been reported to enhance GABA-mediated inhibition of electrically evoked responses in CA_1 and CA_3 regions (TAKADA et al. 1989); however, it did not alter the spontaneous inhibitory responses in the CA_1 region (GAGE and ROBERTSON 1985). Recently, ethanol was shown to potentiate the GABA responses in medial but not lateral septum suggesting not only regional but site-selective action (GIVENS and BREESE 1990). Furthermore, a recent report has demonstrated that ethanol potentiates the GABA inhibition of Purkinje cells in the presence of β-adrenergic agonists (LIN et al. 1991). This is a unique observation suggesting possible involvement of phosphorylation mediated by protein kinase A in the ethanol modulation of GABAergic transmission. However, it contrasts with the conventional notion that phosphorylation is principally involved in receptor desensitization phenomena.

II. Chloride Flux

The biochemical interaction of ethanol with GABA or GABA receptor agonists (i.e., to increase chloride flux) has been more consistent than the electrophysiological changes. It has been reported that GABA produces a concentration-dependent influx of ^{36}Cl in cultured spinal cord neurons. This effect of GABA on ^{36}Cl influx is additive with glycine, specific for $GABA_A$ agonists, and blocked by bicuculline and picrotoxin (LEHOULLIER and TICKU 1989; MEHTA and TICKU 1989a). Ethanol (5–100 mM) potentiated the effect of GABA on ^{36}Cl influx in spinal cord neurons (TICKU et al. 1986; MEHTA and TICKU 1988) and cortical neurons (A.K. MEHTA and M.K. TICKU, unpublished observation). The potentiating effect of ethanol on GABA-induced ^{36}Cl influx was blocked by GABA receptor antagonists and inverse agonists of the BZ receptor sites (MEHTA and TICKU 1988). Ethanol at a concentration of $\geq$50 mM had a direct effect in the absence of GABA in spinal cord cultured neurons. The direct effect of ethanol on ^{36}Cl influx was also blocked by $GABA_A$ receptor antagonists and inverse agonists of BZ receptors (MEHTA and TICKU 1988). However, $GABA_B$ receptors do not appear to be involved in the action of ethanol on ^{36}Cl influx in spinal cord neurons (MEHTA and TICKU 1990). ALLAN and HARRIS (1987a) demonstrated that ethanol potentiates the action of muscimol on chloride uptake into membrane vesicles. SUZDAK et al. (1986a) showed that ethanol (10 mM) potentiates the effect of GABA on ^{36}Cl uptake in synaptoneurosomes. They also found a direct effect of ethanol, in the absence of GABA agonists, on ^{36}Cl uptake in synaptoneurosomes. The ED_{50} for this direct effect was between 25 and 35 mM (SUZDAK et al. 1986a). Ethanol is reported to potentiate the effect of muscimol on ^{36}Cl uptake in cortical and cerebellar membranes prepared from ethanol-sensitive mice whereas it failed to elicit the same effect in ethanol-resistant mice (ALLAN et al. 1988). This suggests that genetic differences in ethanol hypnosis may be related to differences in the sensitivity of GABA-operated chloride channels to ethanol.

There are also some functional studies of the GABAergic system following chronic administration of ethanol. It has been reported that chronic ethanol treatment abolishes the potentiation of muscimol-stimulated chloride uptake by ethanol into microsacs (ALLAN and HARRIS 1987a). Chronic inhalation of ethanol results in a tolerance to the action of ethanol on chloride flux (ALLAN and HARRIS 1987b; MORROW et al. 1988). DURAND and CARLEN (1984) reported a decreased amplitude and duration of IPSPs 3 weeks following withdrawal from an ethanol liquid diet. The potentiation of the inhibitory effect of GABA on the simultaneous discharge of pyramidal cells by ethanol at concentrations above 70 mM, seen in control preparations, was lost after chronic treatment with ethanol (TAKADA et al. 1989). In addition, Ro15-4513 has been reported to bind to diazepam-insensitive sites (57 kDa) in cerebellar granule cells and olfactory tubercle with the same affinity (~4 nM), as shown for diazepam-sensitive sites (SIEGHART et al. 1987). Recent studies indicate that diazepam-insensitive Ro15-4513-binding sites may also be present in other regions including cortex and hippocampus (TURNER et al. 1991). This is true despite the fact that the α_6-subunit, which encodes diazepam-insensitive binding sites, is not found in these regions. Interestingly, our studies have shown that chronic ethanol treatment increases the binding of [^{3}H]Ro15-4513 in rat cerebral cortex and cerebellum (MHATRE et al. 1988).

E. Ro15-4513: Ethanol Antagonist

Ro15-4513 is an azido analogue of the classical benzodiazepine receptor antagonist Ro15-1788 which has the ability to bind to BZ receptors (MOHLER et al. 1984; SIEGHART et al. 1987). It acts as a partial inverse agonist.

There has been great interest recently in the effects of BZ receptor inverse agonists, such as Ro15-4513 and FG 7142, on the actions of ethanol. These drugs are reported to decrease some of the behavioral actions of ethanol, including ataxia and anesthesia (BONETTI et al. 1985; SUZDAK et al. 1986b; LISTER 1987). The prior administration of Ro15-4513 protected the animals against intoxication and mortality due to massive doses (5–15 g/kg, po) of ethanol (BONETTI et al. 1985; POLC 1985; SUZDAK et al. 1986b; FADDA et al. 1987). The protective effect of Ro15-4513 was not related to its effects on blood ethanol concentrations (BONETTI et al. 1985; FADDA et al. 1987). However, a recent study has failed to observe such a reversal of ethanol-induced narcosis in CD-1 mice, whether given before or after ethanol administration (HATCH and JERNIGAN 1988). Ro15-4513 also failed to reverse ethanol-induced hypothermia in mice (HOFFMAN et al. 1987). Pretreatment of rats with Ro15-4513 is reported to block both the anticonflict activity of low doses of ethanol (SUZDAK et al. 1986b) and reductions in the number of head dips caused by ethanol in mice in the holeboard apparatus (LISTER 1987). Ro15-4513 resulted in a dose-dependent suppression of oral ethanol self-administration in an operant situation in non-food or fluid-deprived rats (SAMSON et al. 1987). Ro15-4513 is also reported to reverse the anticonvulsant actions of ethanol against bicuculline as well as picrotoxin-induced chemoconvulsions in rats (KULKARNI and TICKU 1989). BONETTI et al. (1989) demonstrated that the interactions between Ro15-4513 and ethanol were prevented by Ro15-1788 (flumazenil), a BZ receptor antagonist.

Biochemical studies have also shown the effectiveness of Ro15-4513 in reversing ethanol's effects. Ro15-4513 reversed the effect of ethanol on chloride flux in synaptoneurosomes (SUZDAK et al. 1986b). These investigators reported that this effect was peculiar to Ro15-4513 but MEHTA and TICKU (1988) reported that both FG 7142 and Ro15-4513 had this effect. Additionally, Ro15-1788 was able to reverse the antagonist effects of Ro15-4513 on ethanol potentiation of GABA-gated ^{36}Cl flux in spinal cord neurons (MEHTA and TICKU 1988). These results suggest that most of the effects of Ro15-4513 are mediated via the diazepam-sensitive benzodiazepine receptors and not via the diazepam-insensitive sites, which are encoded by the α_6-subunit and localized in cerebellar granule cells (see below). There are also reports that FG 7142 is more effective than Ro15-4513 in blocking the action of ethanol (R.A. HARRIS et al. 1988; PALMER et al. 1988). The electrophysiological effects of Ro15-4513, recorded in vivo using chronically implanted electrodes, were very similar to those of other inverse agonists (MARROSU et al. 1989). Ro15-4513 decreased the effect of ethanol on the EEG and on single unit activity in the pars reticulata (MARROSU et al. 1989).

The amino acid residue involved in the binding of benzodiazepine agonists and Ro15-4513 at the receptor site is not known. In 1983 investigators in our laboratory demonstrated that a crucial "histidine" residue was involved in the binding of benzodiazepine agonists at the GABA receptor complex (BURCH et al. 1983; MAKSAY and TICKU 1984). Furthermore, irreversible modification of histidine residues abolished the binding in cerebellum (type I, α_1-subunit abundant), cortex, and hippocampus (type II, α_2-, α_3-subunit abundant). This indicated that the histidine residue was crucial in both type I and type II benzodiazepine receptors. These findings have recently been confirmed by a single point mutation study, which showed that replacement of the histidine residue with arginine in the α_1-subunit caused this subunit to exhibit the pharmacology of the α_6-subunit (WIELAND et al. 1992).

It has become apparent that α_4- and α_6-subunits lack a crucial histidine at the 100 position, which is present in most other α-subunits at an equivalent position (101). In contrast α_6- and α_4-subunits contain an arginine residue at this position. This histidine-arginine switch is responsible for the binding of Ro15-4513 to the diazepam-insensitive sites, which involves the α_6-subunit in the cerebellum. This is consistent with binding studies which have indicated that [^{3}H]Ro15-4513, besides binding to central BZ receptors, also binds to an additional protein (P57 or 56 kDa) in the cerebellum (SIEGHART et al. 1987). This suggests a unique binding site for this ligand. Further support is provided by cloning studies which have demonstrated that the α_6-subunit, in combination with β- and γ_2-subunits, binds [^{3}H]Ro15-4513 to diazepam-insensitive sites in cerebellum (PRITCHETT et al. 1989). The α_6-subunit is unique to the granule cells in the cerebellum. Recently, however, diazepam-insensitive sites have also been observed in the cortex, a region that lacks the α_6-subunit (TURNER et al. 1991). Thus, it is feasible that [^{3}H]Ro15-4513 binds to diazepam-in-sensitive sites in cortex (probably the α_4-subunit), which also has a histidine-arginine switch (WIELAND et al. 1992) and shares substantial homology with the α_6-subunit. Since Ro15-4513 reverses many of the behavioral effects of ethanol, as well as ethanol potentiation of GABA-induced ^{36}Cl-influx in regions other than cerebellum, these studies suggest that the α_6-subunit alone is not responsible for blocking the major biochemical and behavioral effects of Ro15-4513 in the CNS. However, the ability of Ro15-4513, but not FG 7142, to reverse ethanol-induced ataxia could be mediated by the α_6-subunit receptor subtype.

F. Molecular Biological Studies

Individual humans and rodents differ markedly in their behavioral sensitivity to ethanol (WILSON and PLOMIN 1986). There are reports based on behavioral and biochemical studies that long-sleep (LS) and short-sleep (SS) mice have similar ethanol metabolism and pharmacokinetics but differ in the response of neuronal GABA receptors to ethanol (ALLAN and HARRIS 1986). This suggests that the genetic differences seen in response to ethanol in these mouse strains may be the result of modifications of the $GABA_A$ receptor-chloride channel complex. This complex is composed of multiple subunits, many of which have been cloned, sequenced, and expressed in *Xenopus* oocytes or transfected cells (HOUAMED et al. 1984; SCHOFIELD et al. 1987; LEVITAN et al. 1988; PRITCHETT et al. 1989). It has been reported that while ethanol facilitated the GABA responses in oocytes injected with mRNA from LS mice and antagonized responses in oocytes injected with mRNA from SS mice, it inhibited NMDA responses equally in the two lines (WAFFORD et al. 1990). It thus seems that genes coding for the $GABA_A$ receptor or associated proteins may be critical determinants of various differences to ethanol sensitivity among individuals and species.

Molecular cloning experiments indicate that the $GABA_A$ receptor is a hetero-oligomer, and that the drug sensitivity of the complex depends upon its subunit composition (OLSEN and TOBIN 1990). Studies of the expression of mRNAs coding for different subunits have revealed that the γ2-subunit appears to be particularly important for enhancement of GABA action by benzodiazepines (PRITCHETT et al. 1989; SIGEL et al. 1990). It has recently been reported that there exists an alternative spliced form of γ2, which contains 24 additional nucleotides (WHITING et al. 1990; KOFUJI et al. 1991). This spliced form is termed γ2L, and the unspliced variant, previously termed γ2, is now called γ2S (WHITING et al. 1990). The γ2L-subunit differs from γ2S by the presence of eight amino acids, but more importantly it also contains a serine residue that can be phosphorylated by protein kinase C. The β-subunit can also be phosphorylated by protein kinase C (BROWNING et al. 1990). Activation of protein kinase C has also been shown to inhibit GABA responses in *Xenopus* oocytes and microsacs (SIGEL and BAUR 1988; SIGEL et al. 1990; LEIDENHEIMER et al. 1992). It has also been reported recently that ethanol enhancement of the action of GABA was prevented only by antisense oligonucleotides to γ2L. Expression of either the $\alpha 1\beta 1\gamma 2$S- or the $\alpha 1\beta 1\gamma 2$L-subunit mRNA combination in oocytes resulted in GABA responses that were enhanced by diazepam or pentobarbital, but only the combination containing the γ2L-subunit was affected by ethanol (WAFFORD et al. 1991). These observations indicate that enhancement of GABA responses by ethanol displays a subunit specificity, such that the alternatively spliced γ2L-subunit is required for this action of ethanol.

These results, however, do not define the primary site of ethanol action, but suggest that ethanol may enhance GABA receptor function by altering activity of a protein kinase or protein phosphatase. It is also worth noting that hippocampus, a region where many studies have shown an absence of effects of ethanol, appears to lack γ2L-subunit (WHITING et al. 1990).

More recent studies, however, do not support the relationship between the presence of specific subunits and sensitivity to ethanol's effects. ZAHNISER et al. (1992) demonstrated that LS and SS mice do not differ in terms of α_1-, γ_2L-, and γ_2S-subunits in various brain regions, including hippocampus. Their findings indicate that lack of ethanol modulation of GABA responses in SS mice and in some regions of LS mice cannot be due to the absence of the γ_2L-subunit. Thus the notion that the γ_2L-subunit requirement is crucial for demonstrating ethanol modulation of GABAergic transmission may not hold. Region-specific ethanol enhancement of GABA responses has also been reported in medial, but not lateral, septum (GIVENS and BREESE 1990). A crucial role for β_2-subunit in ethanol effects on GABA receptors is suggested by a recent preliminary report indicating that this subunit was present in medial septum (where ethanol enhances) and not in lateral septum (where ethanol does not enhance; BREESE et al. 1992). This study is of interest and needs to be pursued further. However, β_2-subunit is localized in hippocampus (GAMBARANA et al. 1991), a region where many negative results have been reported in electrophysiological experiments. Furthermore, the absolute requirement of β_2-subunit is in question, since $\alpha_1\beta_1\gamma_2$ expression in oocytes has previously been shown to demonstrate ethanol potentiation (WAFFORD et al. 1990).

Thus, the involvement of particular subunits in ethanol modulation of GABAergic transmission may be more complicated than is currently understood. This complex intracellular signaling system includes translational and post-translational mechanisms that are affected by ethanol and which result in the augmentation of function of a subtype of $GABA_A$ receptor containing specific subunits. This system may also be altered by endogenous neuromodulators that regulate phosphorylation of these subunits.

G. Chronic Ethanol Treatment and GABA Receptor Gene Expression

The $GABA_A$ receptor has been implicated not only in the actions of ethanol but also in the development of ethanol tolerance and physical dependence. Biochemical studies have demonstrated tolerance to the ethanol enhancement of GABA-mediated ^{36}Cl flux (ALLAN and HARRIS 1987b; MORROW et al. 1988; M. Mhatre and M.K. Ticku, unpublished observation). The molecular basis of this subsensitivity or adaptive responsiveness is not clear. Additionally, chronic ethanol treatment increases the sensitivity of benzodiazepine inverse agonists (LISTER and KARANIAN 1987; MEHTA and TICKU 1989b).

We have recently reported that chronic ethanol administration produced ~40% reduction in α_1-(4.2 and 3.8 kb), α_2-(6 and 3 kb), and α_5-(2.8 kb) subunit mRNA levels in the cerebral cortex (TICKU and MHATRE 1991; MHATRE and TICKU 1992). Additionally, we have observed that chronic ethanol treatment decreased the α_1-subunit and increased the α_6-subunit mRNAs in the cerebellum. MONTPIED et al. (1991) have also reported that chronic ethanol treatment decreased α_1- and α_2-subunit mRNA in the cerebral cortex. An increase in the α_6-subunit mRNA levels in the cerebellum is consistent with an increase in the binding and photolabeling of [^{3}H]Ro15-4513 to the 56-kDa band in the cerebellum following chronic ethanol treatment (MHATRE and TICKU 1992). We were unable to detect the α_6-subunit in the cerebral cortex following chronic ethanol treatment. Finally, using polyclonal antibodies, we have confirmed that chronic ethanol treatment decreased the α_1-subunit (51 kDa), α_2-subunit (53 kDa), and α_3-subunit (59 kDa) polypeptides in cerebral cortex (MHATRE et al. 1993). These studies are of interest and suggest that chronic ethanol treatment alters $GABA_A$ receptor gene expression, which may have implications for ethanol-induced tolerance and withdrawal. Further examination of ethanol-induced effects on gene expression may provide greater insight into the neuroadaptive changes produced by chronic ethanol exposure.

H. Conclusions

We have attempted in this review to evaluate the evidence related to the effects of ethanol on GABAergic transmission. It has long been thought that the pharmacological actions of ethanol on GABAergic transmission are of great importance. It is quite evident from the literature cited above that ethanol, administered either acutely or chronically, has many effects on the GABAergic system. There are reports that the potentiation of GABA-induced chloride flux by ethanol in vitro occurs at concentrations of ethanol that are associated with the behavioral effects of ethanol, thereby suggesting that some of the behavioral effects of ethanol are mediated through the GABAergic system. Early reports relating selective antagonism of the actions of ethanol by Ro15-4513 created great interest. It is now clear that Ro15-4513 is not a unique compound in reversing the pharmacological effects of ethanol, since other inverse agonists of BZ receptors also possess the same properties. Currently, research on ethanol is focused mainly on molecular biological aspects. Preliminary evidence indicates that chronic ethanol treatment alters $GABA_A$ receptor gene expression. Thus the possibility exists that altered $GABA_A$ receptor gene expression may constitute a molecular basis for ethanol-induced tolerance and withdrawal.

Acknowledgments. We thank Mrs. Sadie Phillips for excellent secretarial help. This work was supported by NIAAA grant (NIAAA AA04090).

References

Aaronson LM, Hinman DJ, Okamoto M (1982) Effects of diazepam on ethanol withdrawal. J Pharmacol Exp Ther 221:319–325

Aguayo LG (1990) Ethanol potentiates the $GABA_A$-activated $C1^-$ current in mouse hippocampal and cortical neurons. Eur J Pharmacol 187:127–130

Allan AM, Harris RA (1986) Gamma-aminobutyric acid and alcohol actions: neurochemical studies of long sleep and short sleep mice. Life Sci 39:2005–2015

Allan AM, Harris RA (1987a) Acute and chronic ethanol treatments alter GABA receptor-operated chloride channels. Pharmacol Biochem Behav 27:665–670

Allan AM, Harris RA (1987b) Involvement of neuronal chloride channels in ethanol intoxication, tolerance and dependence. Recent Dev Alcohol 5:313–325

Allan AM, Spuhler KP, Harris RA (1988) γ-Aminobutyric acid-activated chloride channels: relationship to genetic differences in ethanol sensitivity. J Pharmacol Exp Ther 244:866–870

Belleville RE, Frazer HF (1957) Tolerance to some effects of barbiturates. J Pharmacol Exp Ther 120:469–474

Biggio G, Concas A, Costa E (1992) GABAergic synaptic transmission: molecular, pharmacological and clinical aspects. Adv Biochem Psychopharmacol 47: Raven, New York

Bonetti EP, Burkard WP, Gabl M, Mohler H (1985) The partial inverse benzodiazepine agonist Ro15-4513 antagonises acute ethanol effects in mice and rats. Br J Pharmacol 86:463P

Bonetti EP, Burkard WP, Gabl M, Hinkler W, Lorez HP, Martin JR, Moehler H, Osterrieder W, Schaffner R (1989) Ro15-4513: partial inverse agonism at the benzodiazepine receptor and interaction with ethanol. Pharmacol Biochem Behav 31:733–749

Breese GR, Keir WJ, Simson PE, Criswell HE, Duncan GE, Morrow AL (1992) Localization of the β_2-subunit of the $GABA_A$ receptor complex to sites where ethanol enhances GABA. Alcohol Clin Exp Res 16:1356

Browning MD, Bureau M, Dudek EM, Olsen RW (1990) Protein kinase C and cAMP-dependent protein kinase phosphorylates the β subunit of the purified γ-aminobutyric acid A receptor. Proc Natl Acad Sci USA 87:1315–1318

Burch TP, Thyagarajan R, Ticku MK (1983) Group selective modification of the benzodiazepine γ-aminobutyric acid receptor ionophore complex reveals that low-affinity γ-aminobutyric acid receptor stimulate benzodiazepine binding. Mol Pharmacol 23:52–59

Carlen PL, Gurevich N, Durand D (1982) Ethanol in low doses augments calcium-mediated mechanisms measured intracellularly in hippocampal neurones. Science 215:306–309

Celentano JJ, Gibbs TT, Farb DH (1988) Ethanol potentiates GABA- and glycine-induced chloride currents in chick spinal cord neurones. Brain Res 455:377–380

Coffman JA, Petty F (1985) Plasma GABA levels in chronic alcoholics. Am J Psychiatry 142:1204–1205

Cooper BR, Virk K, Ferris RM, White HL (1979) Antagonism of the enhanced susceptibility to audiogenic seizures during alcohol withdrawal in the rat by gamma-aminobutyric acid (GABA) and "GABA-mimetic" agents. J Pharmacol Exp Ther 209:396–403

Cott J, Carlsson A, Engel J, Lindquist M (1976) Suppression of ethanol-induced locomotor stimulation by GABA-like drugs. Naunyn Schmiedeberg's Arch Pharmacol 295:203–209

Dar MS, Wooles WR (1985) GABA mediation of the central effects of acute and chronic ethanol in mice. Pharmacol Biochem Behav 22:77–84

Davidoff RA (1973) Alcohol and presynaptic inhibition in an isolated spinal cord preparation. Arch Neurol 28:60–63

Davis WC, Ticku MK (1981) Ethanol enhances [^{3}H]diazepam binding at the benzodiazepine-γ-aminobutyric acid receptor ionophore complex. Mol Pharmacol 20:287–294

Durand D, Carlen PL (1984) Decreased neuronal inhibition in vitro after long-term administration of ethanol. Science 224:1359–1361

Fadda F, Mosca E, Colombo G, Gessa GL (1987) Protection against ethanol mortality in rats by imidazobenzodiazepine Ro15-4513. Eur J Pharmacol 136:265–266

Freund G (1980) Benzodiazepine receptor loss in brains of mice after chronic alcohol consumption. Life Sci 27:987–992

Frye GD, McCown TJ, Breese GR (1983a) Differential sensitivity of ethanol withdrawal signs in the rat to γ-aminobutyric acid (GABA) mimetics: blockade of audiogenic seizures but not forelimb tremors. J Pharmacol Exp Ther 226: 720–725

Frye GD, McCown TJ, Breese GR (1983b) Characterization of susceptibility to audiogenic seizures in ethanol-dependent rats after microinjection of γ-aminobutyric acid (GABA) agonists into the inferior colliculus, substantia nigra or medial septum. J Pharmacol Exp Ther 227:663–670

Gage PW, Robertson B (1985) Prolongation of inhibitory postsynaptic currents by pentobarbitone, halothane and ketamine in CA1 pyramidal cells in rat hippocampus. Br J Pharmacol 85:675–681

Gambarana C, Beattlie CE, Rodriguez ZR, Siegel RE (1991) Region-specific expression of messenger RNAs encoding $GABA_A$ receptor subunits in the developing rat brain. Neuroscience 45:423–432

Givens BS, Breese GR (1990) Site-specific enhancement of gamma-aminobutyric acid-mediated inhibition of neural activity by ethanol in the rat medial septal areas. J Pharmacol Exp Ther 254:528–538

Goldman GD, Volicer L, Gold BI, Roth RH (1981) Cerebrospinal fluid GABA and cyclic nucleotides in alcoholics with and without seizures. Alcohol Clin Exp Res 5:431–434

Goldstein DB (1973) Alcohol withdrawal reactions in mice: effects of drugs that modify neurotransmission. J Pharmacol Exp Ther 186:1–9

Goldstein DB (1979) Sodium bromide and sodium valproate: effective suppressants of ethanol withdrawal reactions in mice. J Pharmacol Exp Ther 208:223–227

Gonzales LP, Hettinger MK (1984) Intranigral muscimol suppresses ethanol withdrawal seizures. Brain Res 298:163–166
Greenberg DA, Cooper EC, Gordon A, Diamond I (1984) Ethanol and the γ-aminobutyric acid-benzodiazepine receptor complex. J Neurochem 42: 1062–1068
Greenblatt DJ, Greenblatt M (1972) Which drug for alcohol withdrawal? J Clin Pharmacol 12:429–431
Guerrero-Figueroa R, Merrill M, Rye D (1970) Electrographic and behavioral effects of diazepam during alcohol withdrawal stage in cat. J Pharmacol 9: 143–150
Hakkinen HM, Kulonen E (1976) Ethanol intoxication and gamma-aminobutyric acid. J Neurochem 27:631–633
Harris DP, Sinclair JG (1984) Ethanol-GABA interactions at the rat Purkinje cell. Gen Pharmacol 15:449–454
Harris RA, Allan AM, Daniell LC, Nixon C (1988) Antagonism of ethanol and pentobarbital actions by benzodiazepine inverse agonists: neurochemical studies. J Pharmacol Exp Ther 247:1012–1017
Harrison NL, Majewska MD, Harrington JW, Barker JL (1987) Structure-activity relationships for steroid interaction with the gamma-aminobutyric acid A receptor complex. J Pharmacol Exp Ther 241:346–353
Hatch RC, Jernigan AD (1988) Effect of intravenously administered putative and potential antagonists of ethanol on sleep time in ethanol narcotized mice. Life Sci 42:11–19
Hemmingsen R, Braestrup C, Nielsen M, Barry DI (1982) The benzodiazepine-GABA receptor complex during severe ethanol intoxication and withdrawal in the rat. Acta Psychiatr Scand 65:120–126
Hoffman PL, Tabakoff B, Szabo G, Suzdak PD, Paul SM (1987) Effect of an imidazobenzodiazepine, Ro15-4513, on the incoordination and hypothermia produced by ethanol and pentobarbital. Life Sci 41:611–619
Houamed KM, Bilbe G, Smart TG, Constanti A, Brown DA, Barnard EA, Richards BM (1984) Expression of functional GABA, glycine and glutamate receptors in Xenopus oocytes injected with rat brain mRNA. Nature 310:318–321
Hunt WA (1983) The effect of ethanol on GABAergic transmission. Neurosci Biobehav Rev 7:87–95
Karobath M, Rogers J, Bloom FE (1980) Benzodiazepine receptors remain unchanged after chronic ethanol treatment. Neuropharmacology 19:125–128
Kochman RL, Hirsch JD, Clay GA (1981) Changes in [^{3}H]-diazepam receptor binding after subacute ethanol administration. Res Commun Subst Abuse 2: 135–144
Kofuji P, Wang JB, Moss SJ, Huganir RL, Burt DR (1991) Generation of two forms of the γ-aminobutyric acid receptor$_A$ γ_2-subunit in mice by alternative splicing. J Neurochem 56:713–715
Kramp P, Rafaelson OJ (1978) Delirium tremens: a double-blind comparison of diazepam and barbital treatment. Acta Psychiatr Scand 58:174–190
Kulkarni SK, Ticku MK (1989) Ro15-4513 but not FG-7142 reverses anticonvulsant effects of ethanol against bicuculline- and picrotoxin-induced convulsions in rats. Pharmacol Biochem Behav 32:233–240
Kulkarni SK, Mehta AK, Ticku MK (1990) Comparison of anticonvulsant effect of ethanol against NMDA-, kainic acid- and picrotoxin-induced convulsions in rats. Life Sci 46:481–487
Lehoullier PF, Ticku MK (1989) The pharmacological properties of GABA receptor-coupled chloride channels using ^{36}Cl-influx in cultured spinal cord neurons. Brain Res 487:205–214
Leidenheimer N, McQuilkin SJ, Hanner LD, Whiting P, Harris RA (1992) Activation of protein kinase C selectively inhibits the γ-aminobutyric acid$_A$ receptor: role of desensitization. Mol Pharmacol 41:1116–1123

Levitan ES, Schofield PR, Burt DR, Rhee LM, Wisden W, Kohler M, Fujita N, Rodriguez HF, Stephenson A, Darlison MG, Barnard EA, Seeburg PH (1988) Structural and functional basis for $GABA_A$ receptor heterogeneity. Nature 335:76–79
Liljequist S, Engel J (1982) Effects of GABAergic agonists and antagonists on various ethanol-induced behavioral changes. Psychopharmacology 78:71–75
Liljequist S, Engel J (1984) The effects of GABA and benzodiazepine receptor antagonists on the anticonflict actions of diazepam or ethanol. Pharmacol Biochem Behav 21:521–525
Liljequist S, Culp S, Tabakoff B (1986) Effect of ethanol on the binding of [^{35}S]t-butylbicyclophosphorothionate to mouse brain membranes. Life Sci 38: 1931–1939
Lin AM-Y, Freund RK, Palmer MR (1991) Ethanol potentiation of GABA-induced electrophysiological responses in cerebellum requirement of catecholamine modulation. Neurosci Lett 122:154–158
Lister RG (1987) The benzodiazepine receptor inverse agonists FG-7142 and Ro15-4513 both reverse some of the behavioral effects of ethanol in a holeboard test. Life Sci 41:1481–1489
Lister RG, Karanian JW (1987) Ro15-4513 induces seizures in DBA/2 mice undergoing alcohol withdrawal. Alcohol 4:409–411
Maksay G, Ticku MK (1984) Characterization of GABA-benzodiazepine receptor complex by protection against inactivation by group-specific reagents. J Neurochem 43:261–268
Maksay G, Ticku MK (1985) Dissociation of [^{35}S]t-butylbicyclophosphorothionate binding differentiates convulsant and depressant drugs that modulate GABAergic transmission. J Neurochem 44:480–486
Mancillas JR, Siggins GR, Bloom FE (1986) Systemic ethanol: selective enhancement of responses to acetylcholine and somatostatin in hippocampus. Science 231:161–163
Marrosu F, Carcangiu G, Passino N, Aramo S, Mereu G (1989) Antagonism of ethanol effects by Ro15-4513: an electrophysiological analysis. Synapse 3: 117–128
Martz A, Deitrich RA, Harris RA (1983) Behavioral evidence for the involvement of γ-aminobutyric acid in the actions of ethanol. Eur J Pharmacol 89:53–62
Mehta AK, Ticku MK (1988) Ethanol potentiation of GABAergic transmission in cultured spinal cord neurons involves gamma-aminobutyric $acid_A$-gated chloride channels. J Pharmacol Exp Ther 246:558–564
Mehta AK, Ticku MK (1989a) Benzodiazepine and beta-carboline interactions with $GABA_A$ receptor-gated chloride channels in mammalian cultured spinal cord neurons. J Pharmacol Exp Ther 249:418–425
Mehta AK, Ticku MK (1989b) Chronic ethanol treatment alters the behavioral effects of Ro15-4513, a partially negative ligand for benzodiazepine binding sites. Brain Res 489:93–100
Mehta AK, Ticku MK (1990) Are $GABA_B$ receptors involved in the pharmacological effects of ethanol? Eur J Pharmacol 182:473–480
Mhatre M, Ticku MK (1989) Chronic ethanol treatment selectively increases the binding of inverse agonists for benzodiazepine binding sites in cultured spinal cord neurons. J Pharmacol Exp Ther 251:164–168
Mhatre M, Ticku MK (1992) Chronic ethanol administration alters $GABA_A$ receptor gene expression. Mol Pharmacol 42:415–422
Mhatre M, Mehta AK, Ticku MK (1988) Chronic ethanol administration increases the binding of the benzodiazepine inverse agonist and alcohol antagonist [^{3}H]Ro15-4513 in rat brain. Eur J Pharmacol 153:141–145
Mhatre MC, Pena G, Sieghart W, Ticku MK (1993) Antibodies specific for $GABA_A$ receptor α subunits reveal that chronic alcohol treatment downregulates α-subunit expression in rat brain regions. J Neurochem 61:1620–1625

Mohler H, Sieghart W, Richards JG, Hunkeler W (1984) Photo-affinity labeling of benzodiazepine receptors with a partial inverse agonist. Eur J Pharmacol 102: 191–192
Montpied P, Morrow AL, Karanian JW, Ginns EI, Martin BM, Paul SM (1991) Prolonged ethanol inhalation decreases γ-aaminobutyric acid$_A$ receptor α-subunit in mRNAs in the rat cerebral cortex. Mol Pharmacol 39:157–163
Morrow AL, Suzdak PD, Karanian JW, Paul SM (1988) Chronic ethanol administration alters GABA, pentobarbital and ethanol mediated ^{36}Cl – uptake in cerebral cortical synaptoneurosomes. J Pharmacol Exp Ther 246:158–164
Nestoros JN (1980) Ethanol specifically potentiates GABA mediated neurotransmission in feline cerebral cortex. Science 209:708–710
Nishio M, Narahashi T (1990) Ethanol enhancement of GABA-activated chloride current in rat dorsal root ganglion neurons. Brain Res 518:283–286
Nutt DJ, Adinoff B, Linnoila M (1989) Benzodiazepines in the treatment of alcoholism. Recent Dev Alcohol 7:283–313
Olsen RW (1982) Drug interactions at the GABA receptor ionophore complex. Ann Rev Pharmacol Toxicol 22:245–277
Olsen RW, Tobin AJ (1990) Molecular biology of GABA$_A$ receptors. FASEB J 4:1469–1480
Palmer MR, van Horne CG, Harlan JT, Moore EA (1988) Antagonism of ethanol effects on cerebellar purkinje neurones by the benzodiazepine inverse agonists Ro15-4513 and FG 7142. J Pharmacol Exp Ther 247:1018–1024
Polc P (1985) Interactions of partial inverse benzodiazepine agonists Ro15-4513 and FG 7142 with ethanol in rats and cats. Br J Pharmacol 86:465P
Pritchett DB, Sontheimer H, Shivers BD, Ymer S, Kettenmann H, Schofield PR, Seeburg PH (1989) Importance of a novel GABA$_A$ receptor subunit for benzodiazepine pharmacology. Nature 338:582–585
Ramanjaneyulu R, Ticku MK (1984) Binding characteristics and interaction of depressant drugs with [^{35}S]t-butylbicyclophosphorothionate, a ligand that binds to the picrotoxinin site. J Neurochem 42:221–229
Rastogi SK, Ticku MK (1986) Anticonvulsant profile of drugs which facilitate GABAergic transmission on convulsions mediated by a GABAergic mechanism. Neuropharmacology 25:175–185
Rastogi SK, Thyagarajan R, Clothier J, Ticku MK (1986) Effect of chronic treatment of ethanol on benzodiazepine and picrotoxin sites on the GABA receptor complex in regions of the brain of the rat. Neuropharmacology 25:1179–1184
Reggiani A, Barbaccia MK, Spano PF, Trabucchi M (1980) Acute and chronic ethanol treatment on specific [^{3}H]GABA binding in different rat brain areas. Psychopharmacology 67:261–264
Rottenberg H (1985) Alcohol modulation of benzodiazepine receptors. Alcohol 2:203–207
Roy A, DeJong J, Ferraro T, Adinoff B, Ravitz B, Linnoila M (1990) CSF γ-aminobutyric acid in alcoholics and control subjects. Am J Psychiatry 147: 1294–1296
Samson HH, Tolliver GA, Pfeffer AO, Sadeghi KG, Mills FG (1987) Oral ethanol reinforcement in the rat: effect of the partial inverse benzodiazepine agonist Ro15-4513. Pharmacol Biochem Behav 27:517–519
Schoemaker H, Smith TL, Yamamura HI (1983) Effect of chronic ethanol consumption on central and peripheral type benzodiazepine binding sites in the mouse brain. Brain Res 258:347–350
Schofield PR, Darlison MG, Fujita N, Burt DR, Stephenson FA, Rodriguez H, Rhee LM, Ramachandran J, Reale V, Glencorse TA, Seeburg PH, Barnard EA (1987) Sequence and functional expression of the GABA$_A$ receptor shows a ligand-gated receptor super-family. Nature 328:221–227
Sieghart W, Eichinger A, Richards JG, Mohler H (1987) Photo-affinity labeling of benzodiazepine receptor proteins with the partial inverse agonists [^{3}H]Ro15-4513: a biochemical and autoradiographic study. J Neurochem 48:46–52

Sigel E, Baur R (1988) Activation of protein kinase C differentially modulates neuronal Na^+, Ca^{++} and γ-aminobutyric acid type A channel. Proc Natl Acad Sci USA 85:6192–6196
Sigel E, Baur R, Trube G, Mohler H, Malherbe P (1990) The effect of subunit composition of rat brain $GABA_A$ receptors on channel function. Neuron 5: 703–711
Siggins GR, Pittman QJ, French ED (1987) Effects of ethanol on CA1 and CA3 pyramidal cells in the hippocampal slice preparation: an intracellular study. Brain Res 414:22–34
Squires RF, Casida JE, Richardson M, Saederup E (1983) [^{35}S]t-Butylbicyclophosphorothionate binds with high affinity to brain sites coupled to the γ-aminobutyric $acid_A$ and ion recognition sites. Mol Pharmacol 23:326–336
Suzdak PD, Schwartz RD, Skolnick P, Paul SM (1986a) Ethanol stimulates gamma-aminobutyric acid receptor-mediated chloride transport in rat brain synaptoneurosomes. Proc Natl Acad Sci USA 83:4071–4075
Suzdak PD, Glowa JR, Crawley JN, Schwartz RD, Skolnick P, Paul SM (1986b) A selective imidazobenzodiazepine antagonist of ethanol in the rat. Science 234: 1243–1247
Taberner PV, Unwin JW (1981) Behavioral effects of muscimol, amphetamine and chlorpromazine on ethanol tolerant mice. Br J Pharmacol 74:276P
Takada R, Saito K, Matsura H, Inoki R (1989) Effect of ethanol on hippocampal receptors in the rat brain. Alcohol 6:115–119
Tamborsky E, Marangos PJ (1986) Brain benzodiazepine binding sites in ethanol dependent and withdrawal states. Life Sci 38:465–472
Thompson WL (1978) Management of alcohol withdrawal syndromes. Arch Intern Med 138:278–283
Ticku MK (1980) The effects of acute and chronic ethanol administration and its withdrawal on γ-aminobutyric acid receptor binding in rat brain. Br J Pharmacol 70:403–410
Ticku MK, Burch T (1980) Alterations in GABA receptor sensitivity following acute and chronic ethanol treatment. J Neurochem 34:417–423
Ticku MK, Mhatre M (1991) Chronic ethanol administration induces changes in $GABA_A$ receptor gene expression. Neurosci Abstr 17:263(#109.7)
Ticku MK, Rastogi SK (1980) Barbiturate-sensitive sites in the benzodiazepine-GABA receptor-ionophore complex. In: Roth SH, Miller WK (eds) Molecular and cellular mechanisms of anesthetics. Plenum, New York, pp 179–188
Ticku MK, Burch TP, Davis WC (1983) The interaction of ethanol with the benzodiazepine-GABA receptor-ionophore complex. Pharmacol Biochem Behav 18: 15–18
Ticku MK, Lowrimore P, Lehoullier P (1986) Ethanol enhances GABA-induced ^{36}Cl-influx in primary spinal cord cultured neurons. Brain Res Bull 17:123–126
Tran VT, Snyder SH, Major LF, Hawley RJ (1981) GABA receptors are increased in brains of alcoholics. Ann Neurol 9:289–292
Turner DM, Sapp DW, Olsen RW (1991) The benzodiazepine/alcohol antagonist Ro15-4513: binding to a $GABA_A$ receptor subtype that is insensitive to diazepam. J Pharmacol Exp Ther 257:1236–1242
Unwin JW, Taberner PV (1980) Sex and strain differences in GABA receptor binding after chronic ethanol drinking in mice. Neuropharmacology 19: 1257–1259
Volicer L, Biagioni TM (1982) Effect of ethanol administration and withdrawal on benzodiazepine receptor binding in the rat brain. Neuropharmacology 21: 283–286
Wafford KA, Burnett DM, Dunwiddie TV, Harris RA (1990) Genetic differences in the ethanol sensitivity of $GABA_A$ receptors expressed in Xenopus oocytes. Science 249:291–293

Wafford KA, Burnett DM, Leidenheimer NJ, Burt DR, Wang JB, Kofuji P, Dunwiddie TV, Harris RA, Sikela JM (1991) Ethanol sensitivity of the $GABA_A$ receptor expressed in Xenopus oocytes requires 8 amino acids contained in the γ2L subunit. Neuron 7:27–33

White G, Lovinger DM, Weight FF (1990) Ethanol inhibits NMDA-activated current but does not alter GABA-activated current in an isolated adult mammalian neuron. Brain Res 507:332–336

Whiting P, McKernan RM, Iversen LL (1990) Another mechanism for creating diversity in γ-aminobutyrate type A receptors: RNA splicing directs expression of two forms of γ2 subunit, one of which contains a protein kinase C phosphorylation site. Proc Natl Acad Sci USA 87:9966–9970

Wieland HA, Luddens H, Seeburg PH (1992) A single histidine in $GABA_A$ receptors is essential for benzodiazepine agonist binding. J Biol Chem 267:1426–1429

Wiesner JB, Henriksen SJ, Bloom FE (1987) Ethanol enhances recurrent inhibition in the dentate gyrus of the hippocampus. Neurosci Lett 79:169–173

Wilson JR, Plomin R (1986) Individual differences in sensitivity and tolerance to alcohol. Soc Biol 32:162–184

Wisden W, Morris BJ, Darlison MG, Hunt SP, Barnard EA (1988) Distinct $GABA_A$ receptor α subunit mRNAs show differential patterns of expression in bovine brain. Neuron 1:937–947

Zahniser NR, Buck KJ, Curella P, McQuilkin SJ, Wilson-Shaw D, Miller CL, Klein RL, Heidenreich KA, Keir WJ, Sikela JM, Harris RA (1992) $GABA_A$ receptor function and regional analysis of subunit mRNAs in long-sleep and short-sleep mouse brain. Mol Brain Res 14:196–206

CHAPTER 6

Involvement of CNS Catecholamines in Alcohol Self-Administration, Tolerance and Dependence: Preclinical Studies*

H.H. SAMSON and P.L. HOFFMAN

A. Introduction

The role of the CNS catecholamines dopamine (DA) and norepinephrine (NE) has been studied for many years in relation to ethanol's actions. This is partially because these two neurotransmitters have been intensively studied in the field of neuroscience. A complete review of all of the studies carried out over the last 25 years is beyond the scope and limitations of this chapter. For an overview of earlier work, the reader is referred to prior reviews (HOFFMAN and TABAKOFF 1985; HUNT 1990; POHORECKY and BRICK 1988; TABAKOFF and HOFFMAN 1991).

B. Acute Effects of Investigator-Administered Ethanol: Potential for Catecholamine Involvement in Ethanol Reinforcement

I. Norepinephrine

A variety of studies over the years have indicated that acute administration of ethanol has a biphasic effect upon NE release and metabolism (for a review see HOFFMAN and TABAKOFF 1985; and POHORECKY and BRICK 1988). Increases in brain NE turnover occur after administration of a low dose of ethanol to animals, as well as at early time points after administration of higher doses, when brain ethanol levels are rising. However, at later times after administration of high doses of ethanol, there is a decrease in NE turnover (BACOPOULOS et al. 1985; HUNT and NAJCHROWICZ 1974; see also HOFFMAN and TABAKOFF 1985). Although the rate of NE turnover is affected by ethanol, steady-state endogenous levels of NE are apparently not altered (HELLEVUO and KIIANMAA 1988; MURPHY et al. 1988). On the other hand, in keeping with the increased turnover rate, a recent study found an increase in frontal cortical MHPG-SO_4, and NE metabolite, after acute ethanol administration (SHIRO et al. 1988). The effect of ethanol on NE turnover

* Some portions of this chapter have appeared previously in the referenced reviews.

depends on the brain area under investigation, as the reported increase in NE turnover measured after a low dose of ethanol (2 g/kg) was confined primarily to brain stem (BACOPOULOS et al. 1985).

The biochemical effects of ethanol are reflected in activity of NE cell bodies in the locus coeruleus (LC). At low concentrations (1–10 mM), ethanol increased the spontaneous firing frequency of some neurons, although it inhibited others (SHEFNER and TABAKOFF 1985). However, concentrations of ethanol greater than 30 mM generally inhibited the firing of LC neurons (SHEFNER and TABAKOFF 1985).

Ethanol also affects the signal transduction system that mediates NE effects at β-adrenergic receptors. Activation of these receptors leads to the stimulation of adenylate cyclase (AC) activity, and production of cyclic AMP. The β-adrenergic receptors are coupled to adenylate cyclase through the stimulatory guanine nucleotide binding protein G_s (see STRADER et al. 1989). In cerebral cortical tissue, ethanol, added in vitro, increases guanine nucleotide- and isoproterenol-activated AC activity, and increases the rate of activation of AC by the guanine nucleotide analog GppNHp (SAITO et al. 1985), suggesting that G_s may play a role in the action of ethanol. Thus, whether ethanol affects NE turnover or neuronal activity per se in cortical areas, it may well affect synaptic transmission at NE synapses by potentiating β-adrenergic receptor-mediated events.

There is some evidence that the ability of ethanol to affect NE systems may contribute to the intoxicating and/or sedative effects of ethanol. Norepinephrine acts not only at β-adrenergic receptors, but also at α_1-receptors, which are coupled to polyphosphoinositide turnover and increases in intracellular calcium, and at presynaptic α_2-receptors, which regulate NE release. It has been reported that certain α_1-agonists can antagonize the sedative/hypnotic effects of ethanol in mice (MENON and KODAMA 1985a,b). However, a selective α_2-antagonist, atipamezole, which would be expected to increase NE release through blockade of the presynaptic autoreceptor, was also reported to attenuate the sedative/hypnotic and "intoxicating" effects of ethanol (LISTER et al. 1988). This may not be surprising, since sedation is a common side effect of α_2-agonists such as clonidine. There has also been a report that inhibitors of phenylethanolamine N-methyltransferase antagonized ethanol (and, to a lesser extent, pentobarbital)-induced "intoxication" in rats, suggesting a role for central epinephrine synthesis in the sedation and intoxication produced by those drugs (MEFFORD et al. 1990). Epinephrine has also been shown to have sedative effects, possibly mediated by α_2-receptors.

On the other hand, the effects of investigator-administered ethanol on NE turnover, except for potential actions in cerebral cortical areas, are mainly observed in brain areas not associated with reinforcement. This would suggest that there is most likely only a minor role for NE in self-administered ethanol reinforcement processes.

II. Dopamine

Ethanol's ability to alter brain dopaminergic activity has been extensively studied (for recent reviews see MODELL et al. 1990; SAMSON 1992). The effects of investigator-administered ethanol on DA have been observed in a variety of studies using physiologically relevant ethanol doses. In experiments using acute doses of ethanol, ranging from 1 g/kg to 3 g/kg, increased DA release and/or turnover, as evaluated by increased levels of the DA metabolites DOPAC and HVA, has been a common finding for most DA-rich brain areas (LUCCHI et al. 1983a; RUSSELL et al. 1988; FADDA et al. 1990). However, it should be noted that the effects of ethanol on brain levels of dopamine metabolites have been found to differ, depending on the strain of animal being tested (KIIANMAA and TABAKOFF 1983), suggesting a pharmacogenetic component to the effect of ethanol on DA metabolism that may influence individual responses to ethanol-induced reinforcement.

In addition to the biochemical studies, electrophysiological analysis of DA cell bodies in the ventral tegmental area (VTA) of the rat has shown an increase in firing after i.v. administration of low doses of ethanol (0.125–0.5 g/kg) (FADDA and GESSA 1985; GESSA et al. 1985). While increases were found for DA cell bodies in the substantia nigra region (SN), there was a marked shift to the right of the ethanol dose-effect curve for that area compared to the VTA, suggesting that brain DA areas believed to be critical for reinforcement in general (KOOB 1992) are selectively more sensitive to ethanol. BRODIE et al. (1990) have demonstrated that ethanol, at concentrations as low as 20 m*M* infused into the bathing medium of a VTA slice preparation, results in increased firing of DA cells. Similar results were recently reported for both in vivo and in vitro VTA DA cell activity (VERBANACK et al. 1990). Taken together, the data suggest that ethanol can stimulate increased activity in DA cell bodies in a brain area believed to be important for reinforcement, and this increased activity could result in the increases in DA and DA metabolites observed following investigator-administered ethanol.

The apparent preferential level of stimulation of VTA over SN cells at low ethanol doses appears to be supported by studies of DA release in the terminal fields of these respective cells. IMPERATO and DI CHIARA (1986), using brain microdialysis, demonstrated that low doses of ethanol elevated DA release in nucleus accumbens to a much greater extent than in caudate. It is important to note that this differential stimulation occurs at ethanol doses which result in rats orally self-administering ethanol in a reinforcement paradigm (GRANT and SAMSON 1985; SAMSON 1986).

It has recently been reported that the administration of ethanol directly into the n. accumbens (YOSHIMOTO et al. 1992), or both n. accumbens and caudate, results in the release of DA, measured using a microdialysis technique (WOZNIAK et al. 1991). Thus, these results suggest that ethanol may increase DA release both as a result of activation of DA cell bodies, either indirectly or directly, and by a direct action on DA terminals in the n. accumbens and caudate nucleus. This latter possibility is also supported by the finding that concentrations of ethanol as low as 20 m*M* released DA from an in vitro n. accumbens slice preparation (RUSSELL et al. 1988). On the other hand, recent data obtained using microinjection of ethanol in the n. accumbens and voltammetry have failed to indicate any measurable DA release when pharmacological levels of ethanol are used (SAMSON and HODGE 1993).

C. Oral Ethanol Self-Administration in Nonoperant Situations

There are a variety of methods which can be used to examine oral ethanol self-administration. These include: (1) limited fluid access paradigms in which ethanol and sometimes another fluid (usually water) are available for a short period each day (LINSEMAN 1990; FADDA et al. 1989); (2) continuous fluid access paradigms where ethanol and another fluid source (again, usually water) are available at all times (DAOUST et al. 1986; PFEFFER and SAMSON 1986); and (3) studies using operant procedures in which response-contingent ethanol presentation is examined (SAMSON et al. 1990).

I. Limited Access Ethanol Drinking Situations

1. Norepinephrine

To our knowledge, there have been no studies using this paradigm exploring the actions of NE.

2. Dopamine

The systemic administration of both specific and nonspecific DA receptor agonists and antagonists has yielded mixed results in limited access ethanol drinking situations. In animals in which an alternate day ethanol exposure procedure was used, doses of agonists and antagonists which altered ethanol consumption also decreased water consumption (LINSEMAN 1990). Using rats who were first initiated to drink ethanol in the home cage using a sucrose-substitution procedure, and not restricted in terms of their daily water access, 30-min limited access ethanol consumption was substantially decreased by both amphetamine and pimozide (PFEFFER and SAMSON 1986).

In animals selectively bred for ethanol preference (sP) or avoidance (sNP), orally ingested ethanol in a 2-h limited access procedure increased DA metabolite levels in the caudate nucleus, medial prefrontal cortex, and olfactory tubercle, and reduced DA levels compared to water-drinking controls (FADDA et al. 1989). The baseline DA levels were higher in the preferring sP rats compared to the nonpreferring sNP animals, but it is important to note that even in the nonpreferring animals ethanol consumption resulted in decreased DA levels compared to water controls.

These studies suggest that DA activity is involved during oral ethanol consumption in this paradigm, but the exact role for DA cannot be assessed, as both antagonists and agonists decreased consumption. In a recent report, microinjection of the D2 antagonist sulpiride into the n. accumbens increased ethanol consumption in a 1-h limited access situation in the alcohol-preferring (P) rat (LEVY et al. 1991). The D1 antagonist SCH 23390 had no effect upon ethanol drinking. These results could suggest that antagonism of the D2 receptor in the n. accumbens can increase ethanol consumption in the limited access situation. However, opposite results using D2 antagonists microinjected into n. accumbens in operant self-administration situations (see below) make the interpretation of these data premature.

II. Continuous Access Ethanol Drinking Situations

1. Norepinephrine

In their initial work, AMIT et al. (1977) found that the dopamine-β-hydroxylase inhibitor FLA-57 decreased ethanol consumption during 24 h unlimited access with water available (BROWN et al. 1977). It was found that in these animals whole brain NE levels were significantly decreased, while there was no change or a slight increase in brain DA levels. In the alcohol-preferring P rat, the mixed NE and 5-HT uptake blocker desipramine decreased ethanol intake (MURPHY et al. 1985). As well, in this later study, the α-adrenergic antagonist phentolamine and the β-adrenergic antagonist propranolol had no significant effect on ethanol intake. These data are difficult to reconcile, as both decreases in NE levels resulting from FLA-57 and increased NE levels resulting from desipramine reduced ethanol consumption. Given the potential of 5-HT involvement with desipramine, the difference between the studies could be somewhat explained, but the failure of either the α- or β-adrenergic antagonist to decrease drinking suggests a minimal role for NE in oral ethanol consumption in this paradigm. Using α_2-agonists and antagonists, KORPI (1990) also failed to find any significant alterations in ethanol consumption in the AA (alcohol-preferring) rats. These data support the above conclusion that NE plays a minimal role in oral self-administration in continuous access situations.

2. Dopamine

With a single daily injection of the D2 antagonist pimozide, no effect on 24-h ethanol intakes were observed when water was also present (BROWN et al. 1982). While these results were replicated by PFEFFER and SAMSON (1986), they found ethanol drinking was reduced by pimozide during the first several hours of the 24-h period. It appeared that as the drug was metabolized rebound ethanol drinking occurred, such that total daily ethanol consumption was unaltered.

Using the mixed catecholamine reuptake blocker nomifensine, and rats selected from a heterogeneous stock which demonstrated a high ethanol preference, DAOUST et al. (1986) failed to find any change in ethanol consumption in a 24-H two-bottle ethanol-water choice situation after drug administration. No pattern of intake was examined in this study. Using the selectively bred alcohol-preferring (P) rat, both alteration of brain DA levels and alteration of ethanol consumption by DA agents (D2 agonist, bromocriptine; DA mixed agonist, amphetamine; DA uptake inhibitor, GBR 12090) have been reported in the 24-h two-bottle ethanol-water drinking situations (MCBRIDE et al. 1990). While these studies have indicated a mixed role for DA in the continuous access drinking situation, it is clear that the DA system does play some role in ethanol consumption under certain conditions. The need to examine pattern of intake over the 24-h period is especially important in these studies, as rebound drinking may distort initial drug effects.

III. Operant Paradigms of Oral Ethanol Self-Administration

1. Norepinephrine

As with the limited access paradigm, we have failed to find any studies specifically examining the role of NE in this paradigm of ethanol self-administration.

2. Dopamine

Following systemic administration of DA agonists, antagonists, and reuptake blockers, ethanol-reinforced responding has been found to decrease (SAMSON et al. 1990; WEISS et al. 1990). However, the nature of the decrease of ethanol self-administration is different for DA agonists and antagonists. The DA agonists amphetamine and apomorphine decreased total ethanol-reinforced responding as a result of an overall disruption of the response pattern (PFEFFER and SAMSON 1985, 1988). The DA antagonists haloperidol and pimozide decreased total responding by terminating what appears as a normal response pattern soon after it begins, with little or no resumption during the 30-min session. The reuptake blocker, bupropion, resulted in a response pattern more like that seen with the DA agonists, but later session responding was also suppressed (SAMSON et al. 1990).

The effects of microinjection of DA agonists and antagonists into the n. accumbens on ethanol-reinforced behavior have recently been reported (RASSNICK et al. 1992; SAMSON et al. 1992). Microinjection of the indirect DA agonist *d*-amphetamine, at doses of 4.0–20.0 μg/brain, were found to increase total session ethanol-reinforced responding at the higher doses (SAMSON et al. 1992; HODGE et al. 1992). While there was an increase in total session responding, there was a reduction of the initial high rate of responding typically observed with ethanol reinforcement. A similar effect on response pattern was observed with 4 μg/brain bilateral microinjections of the specific D2 agonist quinpirole (LY 171555). These rate changes are somewhat similar to those observed when either *d*-amphetamine or apomorphine are administered systemically. At this time it is difficult to explain this increase in responding resultant from DA activation. In studies using i.v. cocaine self-administration as the reinforcer, n. accumbens microinjection of amphetamine decreased the self-administration behavior (see KOOB 1992 for review). The finding of increased ethanol self-administration after microinjection of amphetamine suggests that processes different from those involved in cocaine-mediated reinforcement underlie the relationship between DA release and ethanol reinforcement.

Bilateral microinjection into the n. accumbens of the D2 antagonist raclopride (SAMSON et al. 1993) or the DA antagonist fluphenazine (RASSNICK et al. 1992) have resulted in similar effects on responding to those observed following systemic injections of the D2 antagonists haloperidol and pimozide (PFEFFER and SAMSON 1985, 1988). Total session responses are reduced with no major effects upon the initial response pattern.

The data obtained after microinjection of DA agonists and antagonists into the n. accumbens suggest that the mesolimbic DA pathway is involved in the processing of ethanol reinforcement. As well, a recent report using microdialysis of the n. accumbens has demonstrated an increase in extracellular DA in rats self-administering ethanol (WEISS et al. 1992). This supports the hypothesis that ethanol reinforcement shares this neural substrate with a variety of other drugs and natural reinforcers (KOOB 1992). However, it remains unclear how ethanol accesses this system or the nature of activation of this system in maintaining ethanol-reinforced behavior.

D. Chronic Effects of Ethanol on Noradrenergic and Dopaminergic Activity

I. Norepinephrine

After chronic ethanol ingestion by animals, the activity of NE neurons in brain appears to be increased, even in the presence of high blood ethanol levels (HUNT and MAJCHROWICZ 1974; KAROUM et al. 1976; POHORECKY 1974; see also HOFFMAN and TABAKOFF 1985). Similarly, intoxicated human alcoholics were reported to have significantly elevated CSF levels of the NE metabolite MHPG (BORG et al. 1981). These data indicate a developed resistance to the effects of an acute, high dose of ethanol on brain NE neurons (i.e., tolerance develops to the inhibitory effect of high doses of ethanol on NE turnover). In animals, NE metabolism remained elevated throughout ethanol withdrawal, and was still elevated following the disappearance of overt withdrawal symptoms (POHORECKY 1974). Thus, the increased NE turnover does not appear to reflect the stress of withdrawal, or to be a determinant of the signs of ethanol withdrawal.

Chronic ethanol ingestion also results in altered interactions of NE with β-adrenergic receptors. In early studies, a decreased stimulation of adenylate cyclase (AC) activity by NE was found in cerebral cortex of rats treated chronically with ethanol, and a decrease in antagonist binding to β-adrenergic receptors was also reported (BANERJEE et al. 1978; FRENCH et al. 1975; ISRAEL et al. 1972). More recent work demonstrated decreased stimulation of AC activity by guanine nucleotides and isoproterenol in cerebral cortical tissue of C57BL/6 mice fed ethanol in a liquid diet (SAITO et al. 1987). In addition, ligand binding analysis indicated a loss of high-affinity agonist (isoproterenol) binding in cerebral cortex of the ethanol-fed mice, with no change in the total number of receptors (VALVERIUS et al. 1987). Since high-affinity agonist binding may reflect the formation of the ternary complex of ligand, receptor and G_s (BIRNBAUMER 1990), or of ligand, receptor, and G_s-AC complex, the results suggested that chronic ethanol exposure resulted in uncoupling of the receptor from G_s and AC (SAITO et al. 1987). A similar reduction in high-affinity isoproterenol binding was observed in postmortem brain tissue of alcoholics who had measurable levels of blood or urine alcohol at the time of death (VALVERIUS et al. 1989a). Support for the postulate that chronic ethanol ingestion results in uncoupling of the β-adrenergic receptor-G protein-AC system in brain also comes from studies indicating a reduction of high-affinity forskolin binding [believed to reflect the AC and/or G_s-AC complex (YAMASHITA et al. 1986)] in several brain areas of ethanol-fed mice (VALVERIUS et al. 1989b). The most obvious explanation for these results would be that chronic ethanol ingestion results in a change in the quantity or function of G_s. Alpha-subunits of several G proteins, as well as β-subunits, have been quantitated by Western analysis in various brain areas of ethanol-fed mice, using antibodies as described by SPIEGEL et al. (1987). However, chronic ethanol ingestion resulted in no change in the quantity of any form of $G_{s\alpha}$ or $G_{i\alpha}$, or in G-protein β-subunits, in cortex, hippocampus, or cerebellum. More recent research suggests that, in brain, the catalytic subunit of AC may be affected by chronic ethanol (TABAKOFF et al. 1994).

In certain cultured cell lines, chronic ethanol exposure has been reported to alter the quantity of $G_{s\alpha}$ or other G protein subunits (MOCHLY-ROSEN et al. 1988; CHARNESS et al. 1988), and it has been suggested that these changes mediate reduced stimulation of AC activity by agonists (GORDON et al. 1986; MOCHLY-ROSEN et al. 1988; RICHELSON et al. 1986). Similarly, reduced levels of G_s were reported in the anterior pituitary of LS mice (selectively bred for sensitivity to the hypnotic effect of ethanol) given chronic ethanol injections (WAND and LEVINE 1991). The data at present seem to indicate that, while chronic ethanol exposure results in decreased stimulation of AC activity by β-adrenergic and other agonists in a number of neuronal systems, the molecular mechanism by which this change occurs may vary among cell types and/or brain regions.

II. Dopamine

Chronic ingestion of ethanol by animals results in reduced effectiveness of ethanol to stimulate dopamine (DA) release (i.e., tolerance develops to the DA-releasing effect of ethanol) (BARBACCIA et al. 1982; KIIANMAA and TABAKOFF 1983). This tolerance seems to be fully developed after 7 days of ethanol injections or of oral ingestion of ethanol in mice and rats (see HOFFMAN and TABAKOFF 1985). In addition, during ethanol withdrawal, striatal dopamine turnover has been reported to be decreased in rats and mice (see HOFFMAN and TABAKOFF 1985; EISENHOFER et al. 1990; HUNT and MAJCHROWICZ 1974; LUCCHI et al. 1983b; TABAKOFF and HOFFMAN 1978). In human alcoholics undergoing withdrawal, CSF levels of the dopamine metabolite homovanillic acid (HVA) were reduced, also suggesting a decrease in DA turnover, in comparison to alcoholics not showing withdrawal symptoms (MAJOR et al. 1977).

In animals, alterations in dopamine receptor sensitivity have also been reported after chronic ethanol exposure, including subsensitivity to the behavioral and biochemical effects of drugs acting at dopamine receptors (see HOFFMAN and TABAKOFF 1984; RABIN et al. 1980; TABAKOFF and HOFFMAN 1978; TABAKOFF et al. 1978). Furthermore, dopamine-stimulated adenylate cyclase activity was found to be reduced in striatal tissue (but not mesolimbic tissue) of mice and rats exposed chronically to ethanol (LUCCHI et al. 1983b; SAFFEY et al. 1988; TABAKOFF and HOFFMAN 1979). In the rats, the ability of GTP to modify apomorphine binding to striatal dopamine receptors was also disrupted (LUCCHI et al. 1983b). These changes in response to dopamine were reversed rapidly after ethanol withdrawal, which may account for the lack of observation of a change in some studies (RABIN et al. 1987).

Overall, the data suggest a decrease in dopaminergic activity, and a decrease in responsiveness of DA neurons to ethanol, in the brains of animals and humans after chronic ethanol exposure and/or ethanol withdrawal. It is possible that this decreased dopaminergic activity contributes to certain ethanol withdrawal symptoms. Since striatal tissue has been used for most of the animal investigations, and given the differences in regulation and response to ethanol of the mesolimbic and striatal dopaminergic systems (BANNON and ROTH 1983; BUSTOS et al. 1981), it is difficult to determine whether the decrease in dopaminergic activity would alter ethanol reinforcement. However, the development of tolerance to the effect of ethanol on DA release suggests that reinforcement would be reduced in chronically ethanol treated animals unless higher doses of ethanol were ingested (see

TABAKOFF and HOFFMAN 1988). It might also be postulated that, under conditions where the reinforcing effect of ethanol was reduced, the aversive properties of ethanol would become more prominent, tending to decrease ethanol intake (see TABAKOFF and HOFFMAN 1988).

E. Role of Catecholamines in Ethanol Tolerance and Physical Dependence

I. Tolerance

Tolerance to ethanol is most simply defined as an acquired resistance to various effects of the drug. However, there are a number of forms of tolerance (see TABAKOFF et al. 1982). Tolerance can be classified as either functional or metabolic, reflecting, respectively, a cellular change that produces resistance to ethanol, or a change in the metabolism, distribution, or excretion of the drug such that, after a given dose, the organism is exposed to a lower level of ethanol. Tolerance can occur during a single exposure to ethanol (acute tolerance) or after multiple exposures (chronic tolerance). Tolerance can be influenced by the environment, in that an individual can "learn" to associate ethanol administration or ingestion with a particular environment, and only display tolerance within the environment in which ethanol is "expected" (conditional or environment-dependent tolerance).

Noradrenergic systems have been demonstrated to play an important role in chronic, functional tolerance. A construct used to analyze the neurochemistry of tolerance is that of intrinsic and extrinsic systems. Intrinsic systems are those that encode tolerance to specific effects of ethanol within themselves, while extrinsic systems are those that modify the development, expression, or dissipation of tolerance (HOFFMAN and TABAKOFF 1984; SQUIRE and DAVIS 1981). Noradrenergic pathways in the brain may represent an extrinsic system that affects ethanol tolerance. In mice, partial destruction of catecholaminergic systems with 6-hydroxydopamine (6-OHDA) blocked the development of tolerance to the hypnotic and hypothermic effects of ethanol, without affecting ethanol withdrawal symptoms (TABAKOFF and RITZMANN 1977). In these studies, when mice were treated with desmethylimipramine prior to 6-OHDA, noradrenergic neurons were protected, and tolerance developed normally (TABAKOFF and RITZMANN 1977). More recent results suggested that the interaction of NE with β-adrenergic receptors coupled to adenylate cyclase may be important for the development of tolerance in mice. The blockade of tolerance development produced by 6-OHDA treatment was found to be overcome by daily intracerebroventricular treatment with forskolin, an activator of adenylate cyclase (SZABO et al. 1988).

The tolerance discussed above is chronic, environment-independent tolerance (MELCHIOR and TABAKOFF 1985); however, 6-OHDA treatment was also found to slow tolerance development in a paradigm (multiple injections) that results in environment-dependent, or conditional tolerance in mice (MELCHIOR and TABAKOFF 1981). Thus, noradrenergic systems seem to play an important role in the development of several forms of ethanol tolerance. On the other hand, depletion of catecholamines after tolerance had developed did not affect the *expression* of tolerance (TABAKOFF and RITZMANN 1977).

Depletion of NE did not block the development of tolerance to the hypnotic effect of ethanol in rats, in contrast to mice (LÊ et al. 1981; WOOD and LAVERTY 1979). However, development of tolerance to the hypothermic effect of ethanol in rats was blocked after administration of the NE neurotoxin DSP-4 (TRZASKOWSKA et al. 1986). Interestingly, although NE depletion in the mouse was sufficient to block the development of tolerance to the hypnotic effect of ethanol, combined depletion of both noradrenergic and serotonergic systems was necessary in order to completely block the development of tolerance to this effect of ethanol in the rat (LÊ et al. 1981).

In the neurotoxin studies just described, neurotransmitters were often depleted prior to the animals' ingestion of ethanol. Thus, tolerance did not develop in the lesioned animals, even though they were exposed to the same levels of ethanol as the intact animals. These results emphasize a key characteristic of ethanol tolerance, i.e., that the presence of ethanol in the brain is a necessary, but not sufficient, factor for the development of chronic functional ethanol tolerance. The normal activity of certain neuronal systems, including the noradrenergic system, is also required for development of tolerance to many effects of ethanol.

A key factor (extrinsic system) in the *dissipation* of tolerance is the neuropeptide vasopressin. Administration of this peptide results in the maintenance of functional tolerance to ethanol in animals that have acquired such tolerance, even in the absence of further ethanol intake (HOFFMAN et al. 1978). In mice, intact noradrenergic systems were necessary in order for vasopressin to maintain ethanol tolerance (HOFFMAN et al. 1983), suggesting that an interaction between two extrinsic systems in brain (vasopressin and norepinephrine) influences the loss of tolerance.

There is little evidence for a role of dopaminergic systems in the development of ethanol tolerance, even though these systems do appear to become "tolerant" to the acute effect of ethanol on dopamine release, as discussed above.

II. Physical Dependence

There is little evidence that noradrenergic systems play a key role in physical dependence on ethanol, although NE turnover is increased during withdrawal. However, as mentioned earlier, it is possible that decreased dopaminergic activity after chronic ethanol ingestion may contribute to certain specific signs of ethanol withdrawal (i.e., dopaminergic systems may play a role in physical dependence on ethanol). For example, extrapyramidal symptoms have been reported to be associated with alcohol withdrawal (SHEN 1984). In an early study, GOLDSTEIN (1972) reported that chlorpromazine (a dopamine receptor antagonist) increased the severity of alcohol withdrawal seizures in mice. More recently, BORG and WEINHOLDT (1982) found that bromocriptine, a dopamine agonist, reduced anxiety, restlessness, depression, thremor, sweating, and nausea in alcoholics undergoing withdrawal. Similarly, LEPOLA et al. (1984) reported that dopamine agonists, as well as neuroleptics that increase dopamine turnover, were effective in the treatment of alcohol withdrawal in humans. Although other neurotransmitter systems are undoubtedly involved in the development of physical dependence on ethanol and in ethanol withdrawal symptomatology (see TABAKOFF and HOFFMAN 1992), the available data suggest that ethanol-induced alterations in dopaminergic pathways may contribute to certain specific signs of alcohol withdrawal.

F. Summary

Acutely, ethanol affects the activity of the noradrenergic and dopaminergic systems in brain. The effect of ethanol on norepinephrine turnover is biphasic, with low doses increasing turnover, and higher doses decreasing turnover. The acute effect of ethanol on dopaminergic function appears to be an increase in dopamine release, both as a result of activation of dopaminergic neurons, and possibly an effect on dopaminergic terminals. In general, the results indicate that the mesolimbic dopaminergic pathways are most sensitive to ethanol. Studies of ethanol self-administration and operant responding are consistent with a role of dopamine in ethanol-induced reinforcement, but the exact mechanism of this involvement is not clear. There is little evidence that noradrenergic systems are involved in ethanol self-administration or reinforcement.

Chronically, ethanol produces an increase in noradrenergic activity, and a decreased sensitivity of the β-adrenergic receptor-adenylate cyclase system to stimulation by agonists. The noradrenergic system appears to be an "extrinsic" system that modulates the development of ethanol tolerance: in mice and rats, depletion of brain norepinephrine can block the development of tolerance to certain effects of ethanol, without affecting ethanol withdrawal. Chronic ethanol ingestion results in a decreased activity of brain dopaminergic neurons, and a subsensitivity of dopamine receptors to stimulation by agonists. These effects may be involved in symptoms of ethanol withdrawal, some of which have been shown to be alleviated by administration of dopaminergic agonists. Although dopamine systems in chronically ethanol treated animals become "tolerant" to the acute effects of ethanol on dopamine release, there is little evidence for a role of dopaminergic neurons in the development of behavioral tolerance per se. However, the resistance of the dopaminergic neurons to the acute effect of ethanol (i.e., dopamine release) could reduce the reinforcing effect of ethanol.

References

Amit Z, Brown W, Levitan DE, Ogren S-O (1977) Noradrenergic mediation of the positive reinforcing properties of ethanol. I. Suppression of ethanol consumption in laboratory rats following dopamine-beta-hydroxylase inhibition. Arch Int Pharmacodyn Ther 230:65–75

Bacopoulos NG, Bhatanger RK, van Orden LS III (1985) The effects of subhypnotic doses of ethanol on regional catecholamine turnover. J Pharmacol Exp Ther 204:1–10

Banerjee SP, Sharma VK, Khanna JM (1978) Alterations in β-adrenergic receptor binding during ethanol withdrawal. Nature 276:407–409

Bannon MJ, Roth RH (1983) Pharmacology of mesocortical dopamine neurons. Pharmacol Rev 35:53–68

Barbaccia ML, Bosio A, Lucchi L, Spano PF, Trabucchi M (1982) Neuronal mechanisms regulating ethanol effects on the dopaminergic system. J Neural Transm 53:169–177

Birnbaumer L (1990) G proteins in signal transduction. Annu Rev Pharmacol Toxicol 30:675–705

Borg V, Weinholdt T (1982) Bromocriptine in the treatment of the alcohol withdrawal syndrome. Acta Psychiatr Scand 65:101–111

Borg S, Kvande H, Sedvall G (1981) Central norepinephrine metabolism during alcohol intoxication in addicts and healthy volunteers. Science 213:1135–1137

Brodie MS, Shefner SA, Dunwiddie TV (1990) Ethanol increases the firing rate of dopamine neurons of the rat ventral tegmental area in vitro. Brain Res 508:65–69

Brown ZW, Amit Z, Levitan DE, Ogren O-S, Sutherland ES (1977) Noradrenergic mediation of the positive reinforcing properties of ethanol. II. Extinction of ethanol drinking behavior in laboratory rats by inhibition of dopamine-beta-hydroxylase. Implications for treatment procedures in human alcoholics. Arch Int Pharmacodyn Ther 230:76–82

Brown ZW, Gill K, Abitobol M, Amit Z (1982) Lack of effect of dopamine receptor blockade on voluntary ethanol consumption. Behav Neural Biol 36:291–294

Bustos G, Liberona JL, Gysling K (1981) Regulation of transmitter synthesis and release in mesolimbic dopaminergic nerve terminals. Effect of ethanol. Biochem Pharmacol 30:2157–2164

Charness ME, Querimit LA, Henteleff M (1988) Ethanol differentially regulates G Proteins in neural cells. Biochem Biophys Res Commun 155:138–143

Daoust M, Moore N, Saligaut C, Lhuintre JP, Chretien P, Boismare F (1986) Striatal dopamine does not appear involved in the voluntary intake of ethanol by rats. Alcohol 3:15–17

Eisenhofer G, Szabó G, Hoffman PL (1990) Opposite changes in turnover of noradrenaline and dopamine in the CNS of ethanol-dependent mice. Neuropharmacology 29:37–45

Fadda F, Gessa GL (1985) Role of dopamine in the CNS effect of ethanol. In: Parvez S, Burov Y, Parvez, Burns E (eds) Progress in alcohol research, vol 1. VNU, Utrecht, pp 147–162

Fadda F, Mosca E, Colombo G, Gessa GL (1989) Effects of spontaneous ingestion of ethanol on brain ethanol dopamine metabolism. Life Sci 44:281–287

Fadda F, Mosca E, Colombo G, Gessa GL (1990) Alcohol-preferring rats: genetic sensitivity to alcohol-induced stimulation of dopamine metabolism. Physiol Behav 47:727–729

French SW, Palmer DS, Narod ME, Reid PE, Ramey CW (1975) Noradrenergic sensitivity of the cerebral cortex after chronic ethanol ingestion and withdrawal. J Pharmacol Exp Ther 194:319–326

Gessa GL, Muntoni F, Collu M, Vargui L, Mereu G (1985) Low doses of ethanol activate dopaminergic neurons in the ventral tegmental area. Brain Res 348: 201–203

Goldstein DB (1972) An animal model for testing effects of drugs on alcohol withdrawal reactions. J Pharmacol Exp Ther 183:14–22

Gordon AS, Collier K, Diamond I (1986) Ethanol regulation of adenosine receptor-stimulated cAMP levels in a clonal neural cell line: an in vitro model of cellular tolerance to ethanol. Proc Natl Acad Sci USA 83:2105–2108

Grant KA, Samson HH (1985) Oral self-administration of ethanol in free-feeding rats. Alcohol 2:317–322

Hellevuo K, Kiianmaa K (1988) Effects of ethanol, barbital, and lorazepam on brain monoamines in rats lines selectively outbred for differential sensitivity to ethanol. Pharmacol Biochem Behav 29:183–188

Hodge CW, Samson HH, Haraguchi M (1992) Microinjections of dopamine agonists in n. accumbens increase ethanol reinforced responding. Pharmacol Biochem Behav 43:249–254

Hoffman PL, Tabakoff B (1984) Neurohypophyseal peptides maintain tolerance to the incoordinating effects of ethanol. Pharmacol Biochem Behav 21:539–543

Hoffman PL, Tabakoff B (1985) Ethanol's action on brain biochemistry. In: Tarter RE, van Thiel DH (eds) Alcohol and the brain: chronic effects. Plenum, New York, pp 19–63

Hoffman PL, Ritzmann RF, Walter R, Tabakoff B (1978) Arginine vasopressin maintains ethanol tolerance. Nature 276:614–616

Hoffman PL, Melchior CL, Tabakoff B (1983) Vasopressin maintenance of ethanol tolerance requires intact brain noradrenergic systems. Life Sci 32:1065–1071

Hunt WA (1990) Biochemical bases for the reinforcing effects of ethanol. In: Cox WM (ed) Why people drink: parameters of alcohol as a reinforcer. Gardner, New York, pp 51–91

Hunt WA, Majchrowicz E (1974) Alterations in the turnover of brain norepinephrine and dopamine in the alcohol-dependent rat. J Neurochem 23:549–552

Imperato A, Di Chiara G (1986) Preferential stimulation of dopamine release in the nucleus accumbens of freely moving rats by ethanol. J Pharmacol Exp Ther 239:219–228

Israel MA, Kimura H, Kuriyama K (1972) Changes in activity and hormonal sensitivity of brain adenylyl cyclase following chronic ethanol administration. Experientia 28:1322–1323

Karoum F, Wyatt RJ, Majchrowicz E (1976) Brain concentrations of biogenic amine metabolites in acutely-treated and ethanol-dependent rats. Br J Pharmacol 56:403–411

Kiianmaa K, Tabakoff B (1983) Neurochemical correlates to tolerance and strain differences in the neurochemical effects of ethanol. Pharmacol Biochem Behav 18 [Suppl 1]:383–388

Koob GF (1992) Drugs of abuse: anatomy, pharmacology and the function of the reward pathways. TIPS 13:177–184

Korpi ER (1990) Effects of alpha2-adrenergic drugs on the alcohol consumption of alcohol-preferring rats. Pharmacol Toxicol 66:283–286

Lê AD, Khanna JM, Kalant H, LeBlanc AE (1981) Effect of modification of brain serotonin (5-HT), norepinephrine (NE) and dopamine (DA) on ethanol tolerance. Psychopharmacology (Berl) 75:231–235

Lepola U, Kokko S, Nuutila J, Gordin A (1984) Tiapride and chlordiazepoxide in acute alcohol withdrawal. A controlled clinical trial. Int J Clin Pharmacol Res 6:321–326

Levy AD, Murphy JM, McBride WJ, Lumeng L, Li T-K (1991) Microinjections of sulpiride into the nucleus accumbens increases ethanol drinking in alcohol-preferring (P) rats. Alcohol Alcohol [Suppl 1]:417–420

Linseman MA (1990) Effects of dopaminergic agents on alcohol consumption by rats in a limited access paradigm. Psychopharmacology (Berl) 100:195–200

Lister RG, Durcan MJ, Nutt DJ, Linnoila M (1988) Attenuation of ethanol intoxication by alpha-2 adrenoceptor antagonists. Life Sci 44:111–119

Lucchi L, Lupini M, Govoni S, Covelli V, Spano PF, Trabucchi M (1983a) Ethanol and dopaminergic systems. Pharmacol Biochem Behav 18 [Suppl 1]:379–382

Lucchi L, Covelli V, Anthopoulou H, Spano PF, Trabucchi M (1983b) Effect of chronic ethanol treatment on adenylate cyclase activity in rat striatum. Neruosci Lett 40:187–192

Major LF, Ballenger JC, Goodwin FK, Brown GL (1977) Cerebrospinal fluid homovanillic acid in male alcoholics: effects of disulfiram. Biol Psychiatry 12:635–642

McBride WJ, Murphy JM, Lumeng L, Li T-K (1990) Serotonin, dopamine and GABA involvement in ethanol drinking of selectively bred rats. Alcohol 7:191–205

Mefford IN, Lister RG, Ota M, Linnoila M (1990) Antagonism of ethanol intoxication in rats by inhibitors of phenylethanolamine N-methlytransferase. Alcohol Clin Exp Res 14:53–57

Melchior CL, Tabakoff B (1981) Modification of environmentally cued tolerance to ethanol in mice. J Pharmacol Exp Ther 219:175–180

Melchior CL, Tabakoff B (1985) Features of environment-dependent tolerance to ethanol. Psychopharmacology (Berl) 87:94–100

Menon MK, Kodama CK (1985a) Antagonism of the hypnotic effect of ethanol in mice by alpha-1 adrenoceptor agonist. Neuropharmacology 24:927–930

Menon MK, Kodama CK (1985b) Further studies on the ethanol antagonism exhibited by 2(2-chloro-5-trifluoromethyl phenylimino) imidazolidine (St 587). Life Sci 37:2091–2098

Mochly-Rosen D, Chang F-H, Cheever L, Kim M, Diamond I, Gordon AS (1988) Chronic ethanol causes heterologous desensitization of receptors by reducing α_s messenger RNA. Nature 333:848–850

Modell JG, Mountz JM, Beresford TP (1990) Basal ganglia/limibic striatal and thalamocortical involvement in craving and loss of control in alcoholism. J Neuropsychiatry 2:123–144

Murphy JM, Waller MB, Gatto GJ, McBride WJ, Lumeng L, Li T-K (1985) Monoamine uptake inhibitors attenuate ethanol intake in alcohol-preferring rats. Alcohol 2:349–352

Murphy JM, McBride WJ, Gatto GJ, Lumeng L, Li T-K (1988) Effects of acute ethanol administration on monoamine and metabolic content in forebrain regions of ethanol tolerant and nontolerant alcohol-preferring (P) rats. Pharmacol Biochem Behav 29:169–174

Pfeffer AO, Samson HH (1985) Oral ethanol reinforcement: interactive effects of amphetamine, pimozide and food-restriction. Alcohol Drug Res 6:37–48

Pfeffer AO, Samson HH (1986) Effects of pimozide on homecage ethanol drinking in the rat: dependence on drinking session length. Drug Alcohol Depend 17:47–55

Pfeffer AO, Samson HH (1988) Haloperidol and apomorphine effects on ethanol reinforcement in free-feeding rats. Pharmacol Biochem Behav 29:343–350

Pohorecky LA (1974) Effects of ethanol on central and peripheral noradrenergic neurons. J Pharmacol Exp Ther 189:380–391

Pohorecky LA, Brick J (1988) Pharmacology of ethanol. Pharmacol Ther 36: 335–427

Rabin RA, Wolfe BB, Dibner MD, Zahniser NR, Melchior C, Molinoff PB (1980) Effects of ethanol administration and withdrawal on neurotransmitter receptor systems in C57 mice. J Pharmacol Exp Ther 213:491–496

Rabin RA, Baker RC, Deitrich RA (1987) Effects of chronic ethanol exposure on adenylate cyclase activities in the rat. Pharmacol Biochem Behav 26:693–697

Rassnick S, Pulvirenti L, Koob GF (1992) Oral ethanol self-administration in rats is reduced by the administration of dopamine and glutamate receptor antagonists into the nucleus accumbens. Psychopharmacology (Berl) 109:92–98

Richelson E, Stenstrom S, Forray C, Enloe L, Pfenning M (1986) Effects of chronic exposure to ethanol on the prostaglandin E_1 receptor-mediated response and binding in murine neuroblastoma clone (N1E-115). J Pharmacol Exp Ther 239:687–692

Russell VA, Lamm MCL, Taljaard (1988) Effect of ethanol on [3H]dopamine release in rat nucleus accumbens and striatal slices. Neurochem Res 13:487–492

Saffey K, Gillman MA, Cantrill RC (1988) Chronic in vivo ethanol administration alters the sensitivity of adenylate cyclase coupling in homogenates of rat brain. Neurosci Lett 84:317–322

Saito T, Lee JM, Tabakoff B (1985) Ethanol's effects on cortical adenylate cyclase activity. J Neurochem 44:1037–1044

Saito T, Lee JM, Hoffman PL, Tabakoff B (1987) Effects of chronic ethanol treatment on the β-adrenergic receptor-coupled adenylate cyclase system of mouse cerebral cortex. J Neurochem 48:1817–1822

Samson HH (1986) Initiation of ethanol reinforcement using a sucrose-substitution procedure in food- and water-sated rats. Alcohol Clin Exp Res 10:436–442

Samson HH (1992) The function of brain dopamine in ethanol reinforcement. In: Watson RA (ed) Alcohol and neurobiology: receptors, membranes, and channels. CRC, Boca Raton, pp 91–107

Samson HH, Hodge CW (1993) The role of the mesoaccumbens dopamine system in ethanol reinforcement: studies using the techniques of microinjection and voltammetry. Adv Biomed Alcohol Res Alcohol Alcohol [Suppl 2]:469–474

Samson HH, Tolliver GA, Schwarz-Stevens K (1990) Oral ethanol self-administration: a behavioral pharmacological approach to CNS control mechanisms. Alcohol 7:187–191

Samson HH, Tolliver GA, Haraguchi M, Hodge CW (1992) Alcohol self-administration: role of mesolimbic dopamine. Ann NY Acad Sci 654:242–253

Samson HH, Hodge CW, Tolliver GA, Haraguchi M (1993) Effects of dopamine agonists and antagonists on ethanol reinforced behavior: the involvement of the nucleus accumbens. Brain Res Bull 30:133–141

Shefner SA, Tabakoff B (1985) Basal firing rate of rat locus coeruleus neurons affects sensitivity to ethanol. Alcohol 2:239–243

Shen WW (1984) Extrapyramidal symptoms associated with alcohol withdrawal. Biol Psychiatry 19:1037–1043

Shiro I, Tsuda A, Ida Y, Tsujimaru S, Satoh H, Oguchi M, Tanaka M, Inanaga K (1988) Effect of acute ethanol administration on noradrenaline metabolism in brain regions of stressed and nonstressed rats. Pharmacol Biochem Behav 30:769–773

Spiegel AM, Carter A, Brann M, Collins R, Goldsmith P, Simonds W, Vinitsky R, Eide B, Rossiter K, Weinstein L, Woodard C (1987) Signal transduction by guanine nucleotide-binding proteins. Recent Prog Horm Res 44:337–375

Squire LR, Davis HP (1981) The pharmacology of memory: a neurobiological perspective. Annu Rev Pharmacol Toxicol 21:323–256

Strader CD, Sigal IS, Dixon RA (1989) Structural basis of β-adrenergic receptor function. FASEB J 3:1825–1832

Szabó G, Hoffman PL, Tabakoff B (1988) Forskolin promotes the development of ethanol tolerance in 6-hydroxydopamine-treated mice. Life Sci 42:615–621

Tabakoff B, Hoffman PL (1978) Alterations in receptors controlling dopamine synthesis after chronic ethanol ingestion. J Neurochem 31:1223–1229
Tabakoff B, Hoffman PL (1979) Development of functional dependence on ethanol in dopaminergic systems. J Pharmacol Exp Ther 208:216–222
Tabakoff B, Hoffman PL (1988) A neurobiological theory of alcoholism. In: Chaudron CD, Wilkinson DA (eds) Theories on alcoholism. Addiction Research Foundation, Toronto, pp 29–72
Tabakoff B, Hoffman PL (1991) Neurochemical effects of alcohol. In: Frances RJ, Miller SI (eds) Clinical textbook of addictive disorders. Guidford, New York, pp 501–525
Tabakoff B, Hoffman PL (1992) Alcohol: neurobiology. In: Lowinson JH, Ruiz P, Millman RB (eds) Substance abuse: a comprehensive textbook, 2nd edn. Williams and Wilkins, Baltimore, pp 152–185
Tabakoff B, Ritzmann RF (1977) The effects of 6-hydroxydopamine on tolerance to and dependence on ethanol. J Pharmacol Exp Ther 203:319–332
Tabakoff B, Hoffman PL, Ritzmann RF (1978) Dopamine receptor function after chronic ingestion of ethanol. Life Sci 23:643–648
Tabakoff B, Melchior CL, Hoffman PL (1982) Commentary on ethanol tolerance. Alcohol Clin Exp Res 6:252–259
Tabakoff B, Whelan JP, Ovchinnikova L, Nhamburo P, Yoshimura M, Hoffman PL (1994) Quantitative changes in G protein do not mediate ethanol-induced down-regulation of adenylyl cyclase in mouse cerebral cortex. Alcohol Clin Exp Res (in press)
Trzaskowska E, Pucilowski O, Dyr W, Kostowski W, Hauptmann M (1986) Suppression of ethanol tolerance and dependence in rats treated with DSP-4, a noradrenergic neurotoxin. Drug Alohnol Depend 18:349–353
Valverius P, Hoffman PL, Tabakoff B (1987) Effects of ethanol on mouse cerebral cortical β-adrenergic receptors. Mol Pharmacol 32:217–222
Valverius P, Borg S, Valverius MR, Hoffman PL, Tabakoff B (1989a) β-$_{\mathrm{A}}$drenergic receptor binding brain of alcoholics. Exp Neurol 105:280–286
Valverius P, Hoffman PL, Tabakoff B (1989b) Brain forskolin binding in mice dependent on and tolerant to ethanol. Brain Res 503:38–43
Verbanack P, Seutin V, Dresse A, Scuvee J, Massotte L, Giesbers I, Kornreich C (1990) Electrophysiological effects of ethanol on monoaminergic neurons: an in vivo and in vitro study. Alcohol Clin Exp Res 14:728–735
Wand GS, Levine MA (1991) Hormonal tolerance to ethanol is associated with decreased expression of the GTP-binding protein, $G_{s\alpha}$, and adenylyl cyclase activity in ethanol-treated LS mice. Alcoholism Clin Exp Res 15:705–710
Weiss F, Mitchiner M, Bloom FE, Koob GF (1990) Free-choice responding for ethanol versus water in alcohol preferring (P) and unselected Wistar rats is differentially modified by naloxone, bromocriptine, and methylsergide. Psychopharmacology (Berl) 101:178–186
Weiss F, Hurd YL, Ungerstedt U, Markou A, Plotsky PM, Koob GF (1992) Neurochemical correlates of cocaine and ethanol self-administration. Ann NY Acad Sci 654:220–241
Wood JM, Laverty R (1979) Effect of depletion of brain catecholamines on ethanol tolerance and dependence. Eur J Pharmacol 58:285–293
Wozniak KM, Pert A, Mele A, Linnoila M (1991) Focal application of alcohols elevates extracellular dopamine in rat brain: a microdialysis study. Brain Res 540:31–40
Yamashita A, Kurokawa T, Higashi K, Dan'ura T, Ishibashi S (1986) Forskolin stabilizes a functionally coupled state between activated guanine nucleotide-binding stimulatory regulatory protein, N_s, and catalytic protein of adenylate cyclase in rat erythrocytes. Biochem Biophys Res Commun 137:190–194
Yoshimoto K, McBride WJ, Lumeng L, Li T-K (1992) Alcohol stimulates the release of dopamine and serotonin in the nucleus accumbens. Alcohol 9:17–22

CHAPTER 7

5-HT Mediation of Alcohol Self-Administration, Tolerance and Dependence: Pre-Clinical Studies

G.A. Higgins, A.D. Lê, and E.M. Sellers

A. Introduction

An association between the neurotransmitter 5-hydroxytryptamine (5-HT, serotonin) and ethanol drinking was probably first indicated during the late 1960s by Myers and Veale (1968). These workers demonstrated that treating rats with the 5-HT synthesis inhibitor, parachlorophenylalanine (pCPA), produced a profound suppression in subsequent alcohol preference – an effect that persisted even 1 month after cessation of pCPA treatment. While this dramatic finding has since been explained by the potent aversive properties of pCPA (Parker and Radow 1976; see Zabik 1989), the work of Myers and Veale served to stimulate research, both at the clinical and preclinical levels, examining how 5-HT systems may modify ethanol drinking behaviour.

Today, the one principal finding that most researchers engaged in this area would agree upon is that increasing 5-HT neurotransmission by pharmacological means (e.g., by stimulating 5-HT release, blocking 5-HT re-uptake or by direct 5-HT agonists) will produce reductions in alcohol self-administration and preference. This observation has been made in rodents, primates, and humans.

In humans, alcohol abuse is a behaviour maintained by a variety of factors, not least the euphorigenic or other subjective effects of alcohol and adaptive changes subsequent to its repeated usage such as tolerance and dependence. Consequently, preclinical research in these areas will be discussed. However, it must be noted that apart from certain genetically selected alcohol-preferring strains (e.g., alcohol P-rats, Waller et al. 1982), animals rarely self-administer alcohol in sufficient quantities to show physical dependence. Consequently in the majority of studies described, the effect of manipulations to the 5-HT system on alcohol drinking is in animals that are not physically dependent on this substance.

B. Overview of the 5-HT System: Morphology and Receptors

5-Hydroxytryptamine is distributed throughout the body, although the majority is contained within the GI tract and blood platelets. Indeed, less than 1% of the body's endogenous 5-HT is found within the CNS. Despite this preponderance of peripheral 5-HT, most research is directed towards the involvement of central 5-HT systems and alcohol drinking. While this approach is clearly valid, as evidenced for example by the fact that the injection of 5-HT into the cerebral ventricles of rats reduces voluntary ethanol consumption (HILL 1974), such a distribution profile makes it premature to exclude totally a peripheral 5-HT component in alcohol drinking. Indeed peripheral 5-HT systems have been implicated in the regulation of feeding behaviour in rats (FLETCHER and BURTON 1984; MONTGOMERY and BURTON 1986).

In terms of the distribution of 5-HT-containing neurons within the CNS of mammalian species, the majority of the ascending 5-HT fibres, which comprise part of the median forebrain bundle, originate from the dorsal and median raphe (DRN, MRN; see TORK 1985 for further reading). These midbrain raphe nuclei are aligned along the midline, dorsal to the fourth ventricle. Several brain regions such as the hypothalamus, cortex and nucleus accumbens appear to be innervated by both nuclei, although in some cases a delineation can be made between the projections of the DRN and MRN. For instance, the septohippocampal region is innervated primarily by the MRN although a minor projection from the DRN to the ventral hippocampus has been described. Serotonergic inputs to the substantia nigra, striatum and amygdala appear to derive almost exclusively from the DRN. Other raphe nuclei exist (e.g., raphe pontis, magnus and obscuras) and it is from these sites that 5-HT inputs to hindbrain and spinal regions primarily originate. It seems that the raphe system shows little variation across species (see TORK 1985 for further reading).

In recent years it has become abundantly clear that there are a variety of pharmacologically distinct subtypes of 5-HT receptor. These are currently designated 5-$HT_{1A\text{-}D}$, 5-HT_2, 5-HT_3 and 5-HT_4 (see MIDDLEMISS and TRICKLEBANK 1992; PEROUTKA 1993 for recent reviews) although other subtypes almost certainly exist. Most, if not all, of these subtypes are localized both centrally and in peripheral tissues, and each has a unique functional molecular and/or neuroanatomical profile (see PEROUTKA 1993 for a recent review). For instance, the 5-HT_{1A} receptor serves as the inhibitory autoreceptor found on the soma of cells within the DRN and MRN. Activation of this subtype results in an inhibition of 5-HT cell firing and a consequent reduction of 5-HT release from terminals localized within the many regions innervated by these nuclei (SHARP et al. 1989). In contrast, the 5-HT_3 receptor does not appear to be localized to raphe cell bodies but instead is found in low densities within limbic areas (see KILPATRICK et al. 1990; BARNES et al. 1992 for recent reviews). Although at these sites, the functional role of the 5-HT_3 receptor is unclear, it does seem that this may be a mechanism through which 5-HT can regulate other neurotransmitter (e.g., cholinergic, dopaminergic) systems (see reviews by KILPATRICK et al. 1990; BARNES et al. 1992).

5-Hydroxytryptamine-containing neurons are distributed extensively throughout the brain and, perhaps not surprisingly, manipulations of this system have been shown to influence a wide range of behaviour (SOUBRIE 1986; WILKINSON and DOURISH 1991; DEAKIN and GRAEFF 1991). Given the pharmacological complexity of the 5-HT system, there has been an extensive development of subtype-specific drugs designed to study the role of discrete aspects of the 5-HT system in determining animal behaviour (see WILKINSON and DOURISH 1991). The following account reviews the use of such agents, as well as non-selective 5-HT drugs, in studies that elucidate the role of this neurotransmitter in alcohol self-administration.

C. 5-HT and Alcohol Self-Administration

I. Increasing 5-HT Function

1. Indirect 5-HT Agonists and Alcohol Self-Administration

Compounds that prevent the re-uptake of released 5-HT (e.g., fluoxetine, zimeldine, sertraline), those that release endogenous 5-HT (e.g., fenfluramine or its active isomer dexfenfluramine), and 5-HT precursors [e.g., 5-hydroxytryptophan (5-HTP)] have been the most extensively studied classes of drugs with respect to alcohol self-administration. It is a consistent observation that these drugs reduce alcohol intake, and, when water is offered as an alternative, reduce alcohol preference) AMIT et al. 1984; MURPHY et al. 1985, 1988; HARAGUCHI et al. 1990; MCBRIDE et al. 1990; HIGGINS et al. 1992a,b; LYNESS and SMITH 1992; ROWLAND and MORIAN 1992; see Fig. 1). The generality of these findings extends from limited to continual access schedules and via oral, intragastric and intravenous routes of self-administration. This latter finding negates the possibility that such drugs reduce ethanol drinking simply by an oropharyngeal factor such as taste (i.e., making the ethanol solution less palatable) or by inducing conditioned taste aversions (see GILL et al. 1986; ZABIK 1989). Furthermore, these changes are seen in heterogeneous animal strains showing moderate ethanol intake/preference and in genetically selected, high-preferring animal strains (MURPHY et al. 1985, 1988). Therefore, indirect 5-HT agonists reduce ethanol self-administration across a wide range of test situations.

An alternative means of increasing endogenous 5-HT function is to block the enzyme primarily responsible for its metabolic degradation, monoamine oxidase A (MAO-A). A number of MAO-A inhibitors have recently become available, including meclobemide and brofaromine (HAEFELY et al. 1992) and we have recently investigated the effects of each drug on alcohol drinking in a continual access model. Briefly, both meclobemide and brofaromine reduced alcohol intake (and preference) in a heterogeneous Wistar rat population, although the effects of brofaromine seemed more robust

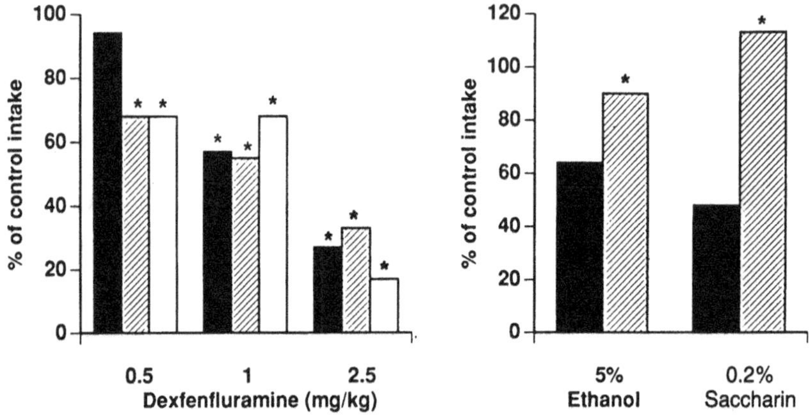

Fig. 1. A Effect of dexfenfluramine on 5% ethanol intake (■), 0.2% saccharin intake (▨), and wet mash intake (□). Data are expressed as a percentage of intake compared to vehicle pretreatment. Ethanol study: 12h consumption in vehicle-treated rats = 21.8 ± 2.5 ml (2.8 g/kg). Saccharin study: 2h intake in vehicle-treated rats = 12.0 ± 1.2 ml. Wet mash study: 1h intake in vehicle-treated rats = 7.7 ± 0.7 g. Percentages were calculated from these means. **B** Effect of metergoline (1 mg/kg) pretreatment against dexfenfluramine (1 mg/kg)-induced suppression of 12h ethanol intake and 2h saccharin intake. ■, vehicle/dexfenfluramine; ▨, metergoline/dexfenfluramine. Data are expressed as a percentage of intake compared to vehicle/vehicle pretreatment. Ethanol study: 12h consumption in vehicle/vehicle group = 32.6 ± 3.4 (2.9 g/kg). Saccharin study: 2h intake in vehicle/vehicle-treated group = 12.5 ± 1.0 ml. Percentages were calculated from these means

(Tomkins et al., unpublished data). Further studies are currently in progress to elucidate more clearly the nature of this response.

Theoretically, such drugs increase synaptic levels of 5-HT, resulting in activation of 5-HT receptors, which ultimately leads in some way to reduced alcohol drinking. However, attempts to identify the 5-HT receptor subtypes involved (see Sect. B) have been largely unsuccessful. MURPHY et al. (1985) reported marked reductions in ethanol intake in alcohol P-rats following fluoxetine pretreatment, an effect that was resistant to blockade by the non-selective $5\text{-HT}_{1/2}$ antagonist methysergide and the 5-HT_2 antagonist LY53857. These results essentially confirmed those of Amit and coworkers, who found that the ethanol suppressant effect of zimeldine was unaffected by metergoline (a $5\text{-HT}_{1/2}$ receptor antagonist) or by prior 5-HT depletion with parachloroamphetamine (PCA) (ROCKMAN et al. 1982; GILL et al. 1985). However, it is also worth noting that the anorectic effect of fluoxetine (see Sect. C.I.3) is unaffected by metergoline and/or other 5-HT receptor antagonists (WONG et al. 1988; GRIGNASCHI and SAMANIN 1992). This has since prompted some speculation as to whether the effects of drugs such as fluoxetine and zimeldine are mediated via neurotransmitter systems other than 5-HT (GILL and AMIT 1989), although if this is the case no obvious candidates have emerged.

It was because of this inconsistency that we recently decided to perform a pharmacological analysis of the ethanol suppressant effects of dexfenfluramine (HIGGINS et al. 1992b). 5-HT antagonists, notably metergoline, have consistently been shown to attenuate other pharmacological effects of dexfenfluramine (or fenfluramine) (see BLUNDELL 1986 and references therein) and we found that metergoline pretreatment also blocked the reductions in alcohol drinking produced by this agent (HIGGINS et al. 1992b; see Fig. 1). Further studies showed that the $5\text{-HT}_{1C/2}$ receptor antagonist ritanserin, but not the 5-HT_3 antagonist ondansetron or the peripheral 5-HT antagonist xylamidine, could also block suppression in ethanol intake produced by dexfenfluramine (HIGGINS et al. 1992b). Together these findings suggest that centrally located $5\text{-HT}_{1C/2}$ receptors may mediate the dexfenfluramine reduction of ethanol drinking.

Of final note, ZABIK et al. (1985) reported that xylamidine blocked a 5-HTP-induced conditioned taste aversion to ethanol, suggesting the involvement of peripheral 5-HT mechanisms (see also ERVIN et al. 1984). Therefore it seems that indirect 5-HT agonists may reduce ethanol drinking by at least two mechanisms, a peripheral effect to induce a conditioned taste aversion and a central effect to reduce ethanol preference. However, by taking appropriate precautions (i.e., timing drug administration relative to ethanol drinking and prior experience of ethanol solutions before drug treatment), the involvement of conditioned aversions can be minimized and thus, in most ethanol self-administration studies, the involvement of this component is negligible.

2. 5-HT Receptor Agonists and Alcohol Self-Administration

Currently, the most widely studied 5-HT agonists are those that specifically interact with the 5-HT_{1A} receptor (e.g., 8-OH DPAT, buspirone, gepirone) and suppression of alcohol drinking has been reported with each drug in rats (SVENSSON et al. 1989; MCBRIDE et al. 1990; KNAPP et al. 1992; KOSTOWSKI and DYR 1992) and monkeys (COLLINS and MYERS 1987). While this may at first sight seem consistent with the hypothesis that increasing 5-HT function reduces alcohol preference/intake, it should be remembered that the 5-HT_{1A} receptor functions as the cell body inhibitory autoreceptor (see Sect. B). Preferential activation at this site seems to occur following low systemic doses of 5-HT_{1A} agonists due either to greater accessibility and/or responsiveness of the receptor to drug. Thus low doses of 8-OH DPAT actually *reduce* central 5-HT function (DOURISH et al. 1986; SHARP et al. 1989; WILKINSON and DOURISH 1991), which might be predicted to increase alcohol intake/preference (see Sect. C.II.1,2). We have recently found that in a limited access schedule 8-OH DPAT (30–60 μg/kg s.c.) stimulates ethanol drinking, while at higher doses (125 μg/kg and above) reductions in this measure are found (TOMKINS et al. 1993; see Fig. 2). This suppression was accompanied by components of the 5-HT behavioural syndrome, which, like

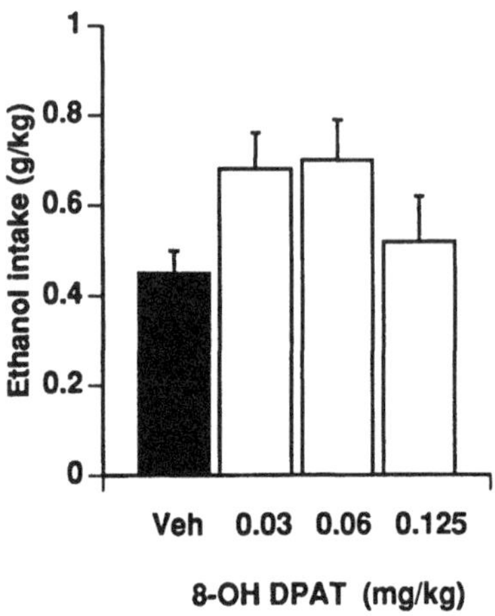

Fig. 2. Effect of 8-OH DPAT on the consumption of a 12% ethanol solution during a 40-min limited access model. The rats received each treatment s.c. according to a Latin square design. A significant increase in ethanol consumption was recorded at the 0.03- and 0.06-mg/kg doses. Data are modified from TOMKINS et al. (1993) and reproduced with permission

the reduction in ethanol preference, may be a consequence of activation of postsynaptic 5-HT_{1A} receptors (TRICKLEBANK et al. 1984; SVENSSON et al. 1989; TOMKINS et al. 1993). Therefore, 5-HT_{1A} agonists may both enhance and reduce alcohol preference, depending on the doses and experimental paradigm used, and this could account for the biphasic effects reported for 8-OH DPAT in some studies (MURPHY et al. 1987a; HIGGINS et al. 1992b). Because the reductions in alcohol drinking are often accompanied by motoric changes that may be incompatible with this consummatory response, caution should be exercised in interpreting the relevance of this suppression.

Unlike the 5-HT_{1A} receptor, truly subtype-selective agonists for other 5-HT receptors are not currently available (see VAN WIJNGAARDEN et al. 1990). Thus quipazine binds to 5-HT_1, 5-HT_2 and 5-HT_3 receptors, as well as to the 5-HT uptake site. DOI and MK212 show reasonable selectivity for both 5-$HT_{1C/2}$ receptors, and TFMPP and m-chlorophenylpiperazine (mCPP) are agonists at 5-$HT_{1B/1C}$ receptors and also bind with moderate affinity to other subtypes (HOYER 1988; VAN WIJNGAARDEN et al. 1990). Each of these drugs has been reported to reduce alcohol drinking/preference in rats (LAWRIN et al. 1986; ZABIK et al. 1985; MURPHY et al. 1987a; MCBRIDE et al. 1990; HIGGINS et al. 1992b; BUCZEK et al. 1993), but as yet no attempt has been made to test empirically which receptors and neuroanatomical mechanisms underlie these responses. It is noteworthy, however, that central 5-$HT_{1C/2}$ receptors may mediate the suppression in alcohol drinking produced by dexfenfluramine (see Sect. C.I.1), and agonists showing affinity for these sites (e.g., mCPP, TFMPP) also produce similar effects. In common with the 5-HT_{1A} receptor agonists, each of these drugs may produce motoric changes in animals and thus may influence ethanol drinking in a non-specific way. However, in most cases distinctions can be drawn and reductions in ethanol drinking are not due simply to a generalized behavioural disruption.

3. Treatments that Enhance 5-HT Function and Other Consummatory Behaviours

In addition to reducing ethanol drinking, elevations in 5-HT function have been shown to reduce other consummatory behaviours. Again, the drugs primarily used to investigate this have been 5-HT releasers and re-uptake blockers. It is a consistent observation that these drugs reduce food and palatable fluid intake (BLUNDELL 1986; BLUNDELL and LAWTON 1990) and at least one mechanism may involve an enhancement of satiety. Thus following ingestion of a meal, rats normally display a characteristic behavioural pattern of activity followed by grooming and terminating in resting behaviour (ANTIN et al. 1975). This behavioural sequence is taken as an index of satiety (i.e., the postprandial satiety sequence). Preclinical studies indicate that both fluoxetine and fenfluramine, without affecting the initiation of feeding, produce an early termination of eating behaviour, resulting in less food eaten followed by an earlier onset of the complete satiety sequence (BURTON et al. 1981; BLUNDELL 1986; CLIFTON et al. 1989; however, see MONTGOMERY and WILLNER 1988). Although these results are consistent with a satiety hypothesis, recent studies suggest that other factors may be involved in the suppressant effects of 5-HT drugs on consummatory behaviour. For instance, rats chronically implanted with an open gastric fistula (i.e., sham feeding rats) will ingest copious amounts of a palatable fluid. These animals show a pronounced satiety deficit and, instead, the ingestion appears to be under oropharyngeal control (WEINGARTEN and WATSON 1982). Both fluoxetine and dexfenfluramine potently suppress sucrose sham feeding in the gastric-fistulated rat, a finding that seems incompatible with the hypothesis that these drugs enhace satiety (NEILL and COOPER 1988). Further studies also support the contention that 5-HT agonists (direct and indirect) may suppress feeding where palatability serves as the primary impetus for this behaviour. Thus in non-food-deprived rats, dexfenfluramine and fluoxetine have been reported to reduce sucrose and saccharin consumption, respectively (BORSINI et al. 1985; LEANDER 1987) and recently we found that dexfenfluramine reduces saccharin intake at doses similar to those that lower ethanol preference (see Fig. 1; HIGGINS et al. 1993a). GILL et al. (1985, 1988a,b) have reported equivalent findings with respect to the 5-HT uptake blockers zimeldine and sertraline. Microstructural studies designed to compare, in a manner analogous to feeding research, how 5-HT drugs reduce alcohol drinking have yielded conflicting data. GILL et al. (1988b) reported that sertraline reduced ethanol intake in rats by affecting bout size, not number – which may seem consistent with the satiety hypothesis. However, in a recent study we found that dexfenfluramine, sertraline, and mCPP produced the opposite effect (i.e., reduced the number of bouts of alcohol drinking but did not affect their size) (HIGGINS et al. 1992b; see Table 1). Also, we have found that both sertraline and fluoxetine delay the initiation of ethanol drinking (LAWRIN 1988; HIGGINS et al. 1992a,b). Together these findings argue against a satiety explanation and suggest that enhancement of 5-HT function suppresses alcohol self-administration and/or preference by an alternative mechanism, which is perhaps related to the effect on palatability-induced ingestion.

We have recently conducted a study designed to examine this question. Non-food-deprived rats were trained to consume saccharin solutions of varying concentration, which promoted varying degrees of intake and thus provided a definable index of palatability. Following dexfenfluramine, saccharin intake was markedly reduced, particularly at the most favoured saccharin concentration (0.2%), although in percentage terms the magnitude of this suppression was similar for each concentration (HIGGINS et al. 1993a). These results contrast slightly with those of LEANDER (1987) and GILL and AMIT (1989), who reported that the most marked suppressions of saccharin intake occurred at the preferred concentrations following fluoxetine and zimeldine treatment, suggesting that at certain doses a selective effect on palatability-induced intake may be achieved. Therefore it is clear that drugs that enhance 5-HT function will potently suppress feeding initiated by

Table 1. Effect of dexfenfluramine (1 mg/kg) on drinking behaviour continuously recorded during the 12 h dark period using the drinkometer system. The criterion for a "drink" was one or more consecutive minutes during which the fluid level changed by 0.2 ml or more. Food consumption and weight gain were measured daily. Note that dexfenfluramine reduced the amount of 5% ethanol consumed by decreasing the *number* of drinks and not their size. (Data modified from HIGGINS et al. 1992b)

	Vehicle	Dexfenfluramine 1 mg/kg
Total ethanol (ml)	25.8 ± 3.2	16.5 ± 2.8*
Total ethanol (g/kg)	2.4 ± 0.3	1.4 ± 0.3*
Number of drinks (*N*)	32.3 ± 4.5	20.9 ± 2.7*
Mean drink size (ml)	0.9 ± 0.1	0.8 ± 0.1
Total water (ml)	25.4 ± 3.0	29.7 ± 3.1
Number of drinks (*N*)	34.1 ± 3.7	35.4 ± 3.7
Mean drink size (ml)	0.8 ± 0.1	0.9 ± 0.1
Food consumption (g)	32.5 ± 0.9	30.6 ± 0.8
Weight gain (g)	4.5 ± 0.5	4.3 ± 0.5

* $P < 0.01$ vs. vehicle controls.

palatability. Furthermore, under some circumstances, these agents may show selectivity for substances of higher incentive value to the animal.

4. Treatments that Enhance 5-HT Function and Other Drug-Reinforced Behaviour

Of the limited studies conducted to date, in addition to alcohol, the self-administration of a variety of other drugs is reduced by 5-HT receptor activation. Dietary or acute pretreatment with the 5-HT precursor L-tryptophan reduces cocaine self-administration (CARROLL et al. 1990a; MCGREGOR et al. 1993). Similarly, 5-HT uptake inhibitors have also been reported to suppress the intravenous self-administration of cocaine (CARROLL et al. 1990b; RICHARDSON and ROBERTS 1991) and amphetamine (LECCESE and LYNESS 1984), as well as the oral self-administration of morphine (RONNBACK et al. 1984). Also, dexfenfluramine reduces intravenous heroin self-administration in non-dependent rats (HIGGINS et al. 1993b,c). Although a central site of action for these effects seems most likely, the only direct evidence for this is derived from the work of LECCESSE and LYNESS (1984), who reported that 5,7-DHT lesions to the median forebrain bundle abolished the suppressant effects of both fluoxetine and of L-tryptophan, on amphetamine self-administration.

II. Reducing 5-HT Function

1. 5-HT Receptor Antagonists and Alcohol Self-Administration

In contrast to the many studies conducted to date that have examined the effects of 5-HT agonists on alcohol intake, relatively few have specifically studied the effects of 5-HT antagonists on this behaviour.

Generally, non-selective $5\text{-HT}_{1/2}$ receptor antagonists such as metergoline and methysergide do not appear to influence markedly alcohol intake or preference (ROCKMAN et al. 1982; MURPHY et al. 1985; WEISS et al. 1990; HIGGINS et al. 1992b), although WEISS et al. (1990) did observe a mild enhancement following methysergide pretreatment, particularly in genetically heterogeneous Wistar rats consuming only moderate amounts of alcohol. Further studies using lower doses of methysergide and/or other $5\text{-HT}_{1/2}$ antagonists in relatively low-alcohol-consuming animals (e.g., following a partial satïation) may prove to be of interest, for there may be parallels to this in the feeding literature (see Sect. C.II.3).

An increase in alcohol self-administration following 5-HT receptor blockade is what might be predicted from the substantial evidence that enhancement of 5-HT function reduces this behaviour. Therefore, the recent reports describing decreased alcohol intake and preference following 5-HT_2 and 5-HT_3 receptor antagonist pretreatment seem paradoxical. There is some evidence (see Sect. B) that the 5-HT_3 receptor may represent a mechanism through which the 5-HT system modulates other neurotransmitters, such as dopamine (BLANDINA et al. 1989; JIANG et al. 1990; WOZNIAK et al. 1990; CHEN et al. 1991). Therefore antagonism of 5-HT_3 receptors may not reduce the functioning of 5-HT systems per se, but rather may modify the way in which 5-HT interacts with other neurotransmitter systems. The suppression in alcohol drinking produced by 5-HT_3 antagonists (e.g., MDL72222 and ondansetron) may therefore be secondary to changes in other neurotransmitters (e.g., dopamine) (see Chap. 6, this volume).

Over recent years, a number of reports have appeared describing reductions in alcohol intake and preference following pretreatment with the 5-HT_3 antagonists ondansetron (OAKLEY et al. 1988; HIGGINS et al. 1992b; MEERT 1993; see Fig. 3), MDL72222 (FADDA et al. 1991), zacopride (KNAPP and POHORECKY 1992), and ICS205-930 (HODGE et al. 1993). While most of these studies utilized rats, this observation has also been extended to primates (OAKLEY et al. 1988) and humans (SELLERS et al. 1991; JOHNSON et al. 1993). In our studies using rats in a continual access model, the magnitude of this suppression was small (approximately 17%), not clearly related to dose, and also had a delayed onset with peak effects seen approximately 5 h after ondansetron injection (HIGGINS et al. 1992b). However, other studies, notably FADDA et al. (1991), have reported greater reductions in alcohol preference and clear dose-related effects following MDL72222 pretreatment. Interestingly, these effects were not maximal until the 3rd day of treatment (FADDA et al. 1991) and this may well suggest that treatment regimen is a critical factor in determining the effects of 5-HT_3 receptor antagonists on alcohol drinking. It has been hypothesized that, in a

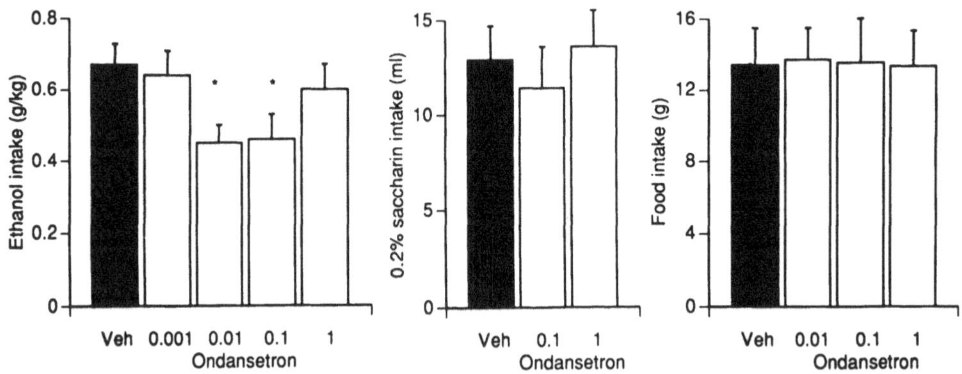

Fig. 3A–C. Effect of the 5-HT_3 receptor antagonist ondansetron on: **A** 12% ethanol intake in a 1-h limited access schedule; **B** 0.2% saccharin intake in a 2-h limited access schedule; and **C** 1 h wet mash intake. In each study, all animals received each treatment according to a balanced design and in all cases the animals were neither food nor water deprived. Only in the ethanol study was a significant suppression of intake seen following ondansetron (0.01–0.1 mg/kg) pretreatment. Saccharin and wet mash data are modified from HIGGINS et al. (1992b, 1993a). The ethanol data were kindly donated by Dr. Denise Tomkins (unpublished observations)

manner similar to DA receptor blockade, 5-HT_3 receptor antagonists may diminish the reinforcing properties of alcohol, leading to a gradual reduction, or early extinction, of alcohol drinking (HIGGINS et al. 1992b; KNAPP and POHORECKY 1992). Perhaps related to these observations are the findings that 5-HT_3 antagonists may block an ethanol discriminative cue (GRANT and BARRETT 1991; GRANT 1992). Interestingly, in humans ondansetron has been reported to reduce the subjective desire for alcohol (SELLERS et al. 1991; JOHNSON et al. 1993). However, further research needs to be conducted in both the preclinical and clinical areas, not least to gain a clearer insight into how 5-HT_3 receptors may modulate forebrain DA systems. A further point in this issue of 5-HT_3 antagonists reducing alcohol intake is the question of what drug dose is required to produce this effect.

In the studies of KNAPP et al. (1992) and HODGE et al. (1993), the doses of zacopride and ICS205-930, respectively, were high in comparison to doses considered to be behaviourally active in other 5-HT_3 models. Because both of these drugs have reasonable 5-HT_4 receptor affinity, it is possible that the observed effects reported on alcohol consumption in these studies may be attributable to an interaction with this receptor subtype.

Regarding selective 5-$HT_{1C/2}$ antagonists, some recent work, principally by MEERT (1993), has shown modest decreases in alcohol intake/preference following ritanserin pretreatment. These effects have been reported in moderate- and high-preferring rats. This result, however, was not confirmed in two recent studies, again using a heterogeneous rat strain in a continual access model (HIGGINS et al. 1992b; MYERS and LANKFORD 1993), and further studies are now awaited.

2. 5-HT Lesions and Alcohol Self-Administration

Two main approaches have been employed in this area, the use of 5-HT synthesis inhibitors (e.g., pCPA, PCA) and 5-HT cell neurotoxins (e.g., 5,6- and 5,7-dihydroxytryptamine). Both have yielded conflicting data. Thus MYERS and VEALE (1968) reported that pCPA markedly reduced alcohol preference in rats, an effect that persisted long after discontinuation of this treatment. This suppression has since been attributed primarily to a conditioned taste aversion to alcohol produced by pCPA (PARKER and RADOW 1976; see ZABIK 1989 for further detail), for when pCPA is not administered in close proximity to the time that alcohol consumption would normally occur (i.e., minimal association between pCPA and alcohol), an increase in preference has been reported (GELLER 1973; HILL and GOLDSTEIN 1974). The use of neurotoxin lesions to destroy central 5-HT-containing pathways has also been shown to increase alcohol intake/preference in some (Ho et al. 1974; MELCHIOR and MYERS 1976; RICHARDSON and NOVAKOVSKI 1978), but not in all, studies (KIIANMAA 1976; MELCHIOR and MYERS 1976). Unlike pCPA, non-specific aversive effects of the neurotoxin treatment would seem unlikely, although differences in animal strain, neurotoxin and baseline alcohol preference may explain anomalous findings (see MELCHIOR and MYERS 1976; ZABIK 1989).

3. Treatments that Reduce 5-HT Function and Other Consummatory Behaviours

Results from these studies, like those described for alcohol self-administration, are inconsistent and may again be dependent on certain procedural variables. Thus, while in some studies central 5-HT neurotoxin lesions and pCPA pretreatment have been reported to increase food intake (see BLUNDELL 1986 and references therein), on balance no effect of the manipulation is, perhaps, the most common finding. In food-deprived rats, 5-HT antagonists (including metergoline, ritanserin and ondansetron) have little, if any, effect on subsequent intake, except that at high drug doses a suppression of intake, probably related to a non-specific behavioural disruption, has been observed (FLETCHER 1988 and references therein; HIGGINS et al. 1992b). However, in pre-fed satiated rats, clear increases in food consumption are seen following pretreatment with a range of 5-HT_1 and perhaps 5-HT_2, but not 5-HT_3, receptor antagonists (DOURISH et al. 1989; FLETCHER 1988; HIGGINS et al. 1992b). Also, in a task involving food-deprived rats trained to run repeatedly down an alleyway to obtain food reinforcement, metergoline increased performance during the latter stages of this test when controls were partially satiated (NEILL et al. 1990). Therefore, low baseline responding, presumably reflecting satiation and/or a low consummatory drive, seems to be an important factor underlying the observation of increased feeding following pharmacological blockade of 5-HT receptors. This is also true for hyperphagia seen following low systemic injections or direct intra-raphe (DRN and MRN) infusions of 5-HT_{1A} agonists such as 8-OH DPAT (DOURISH et al. 1986; FLETCHER and DAVIES 1990).

4. Treatments that Reduce 5-HT Function and Other Drug-Reinforced Behaviour

Depletion of 5-HT within the CNS has been reported to increase both cocaine (LOH and ROBERTS 1990) and amphetamine (LECCESE and LYNESS 1984) self-administration. The study of LOH and ROBERTS (1990) was particularly interesting because cocaine was made available under a progressive ratio schedule of reinforcement. 5,7-DHT infusions into either the median forebrain bundle or amygdala of rats significantly increased the "breakpoint" before cocaine self-administration extinguished. These findings indicate that depletion of forebrain 5-HT increased the incentive value of cocaine to the animal (LOH and ROBERTS 1990). As yet, similar studies have not been conducted with respect to other self-administered drugs, including ethanol, but this clearly seems a worthwhile line of research (see Sect. C.III with respect to 5-HT neuroanatomy of the alcohol P-rat).

The results of investigations using 5-HT antagonists are, like much research in this area, inconsistent. Lyness and coworkers (LYNESS and MOORE 1983; LECCESSE and LYNESS 1984) reported increases in amphetamine self-administration following metergoline pretreatment, and decreases with methysergide and cyproheptidine. Limitations in dosage and the range of antagonists used make interpretation of these findings difficult and a non-specific behavioural disruption (e.g., sedation) may well have contributed to suppressant effects (see also FLETCHER 1988). Recently, LACOSTA and ROBERTS (1993) reported that the non-selective $5\text{-HT}_{1/2}$ antagonist methysergide and the 5-HT_2 antagonist ketanserin failed to influence the break-point of cocaine reinforcement available under a progressive ratio schedule. Of final note, despite initial enthusiasm (CARBONI et al. 1989), it seems that 5-HT_3 receptor antagonists do not influence the intravenous self-administration of cocaine (PELTIER and SCHENK 1991; LANE et al. 1992; LACOSTA and ROBERTS 1993), nicotine (Corrigall and Coen, unpublished findings) or heroin (HIGGINS et al. 1993b). However, these drugs do appear to antagonize place preference conditioning to opiates and nicotine (CARBONI et al. 1989; HIGGINS et al. 1992c) and the hyperactivity response to morphine (PEI et al. 1993). This might imply that certain aspects of drug-induced reinforcement are dependent on 5-HT_3 receptor function, although their clinical relevance will require critical evaluation.

III. Biochemical Factors

Changes in 5-HT neurochemistry following acute and chronic ethanol challenge are largely inconsistent (POHORECKY et al. 1978; GOTHONI and AHTEE 1980; HUNT and MAJCHROWICZ 1983; MORINAN 1987). However, recently a number of genetically selected, high-alcohol-preferring rodent strains have been reported to show a deficiency of 5-HT function relative to their non-preferring counterparts. Most of this work has focused on the alcohol P-rat (MURPHY et al. 1982, 1987b; WONG et al. 1990), although similar changes have been reported for the HAD (high alcohol drinking) and fawn-hooded (REZVANI et al. 1990; HULIHAN-GIBLIN et al. 1992) rat strains, and also certain mouse strains (YOSHIMOTO and KOMURA 1987). This is not a universal finding, however, for the high-preferring AA rats show the converse (i.e., greater 5-HT activity) compared to their non-preferring ANA counterparts (AHTEE and ERIKSSON 1972).

Despite this exception, these results may imply that, at least in some animal strains, a causal relationship exists between low central 5-HT function and preference for alcohol. Studies with the alcohol P-rat are particularly interesting. Aside from the neurochemical changes that are found in a number of forebrain areas, including the nucleus accumbens, hippocampus and cortex (MURPHY et al. 1982, 1987b), immunohistochemical studies have revealed clear reductions in the number of 5-HT-containing fibres in each area (ZHOU et al. 1991). Alcohol P-rats satisfy most of the criteria for an animal model of alcoholism, not least of which is that they may self-administer such amounts as to show tolerance and physical dependence to alcohol (WALLER et al. 1982). Based on the foregoing account of diminished 5-HT function increasing certain reinforced behaviours (see Sects. C.II.2, C.II.4), it would be of clear interest to see if this excessive behaviour of P-rats extends to particular foods/palatable fluids and other self-administered drugs. In this regard, recent research has shown that P-rats show a greater preference for saccharin by comparison to NP-rats (SINCLAIR et al. 1992). Procedures such as the progressive ratio schedule of reinforcement may provide a means to determine how motivated P and NP rats are to self-administer a variety of drugs, including cocaine (LOH and ROBERTS 1990). HYYTIA and SINCLAIR (1993) have reported that AA rats may also orally consume the opioid ethoketazocine to a greater degree than do ANA rats. Thus, it does seem that at least some rat strains selected for ethanol preference are likely to show a higher preference for other drug reinforcers.

Of final note, electrophysiological studies by LOVINGER (1991) suggest a direct interaction between ethanol and the 5-HT_3 receptor ionophore. Using an NCB-20 neuroblastoma cell line, LOVINGER (1991) reported that, at pharmacologically relevant concentrations in vivo, ethanol potentiated a 5-HT-mediated ion current at the 5-HT_3 receptor. This may explain the apparent selectivity of 5-HT_3 receptor antagonists for alcohol relative to other self-administered drugs or foods (see Sects. C.II.1, C.II.3, C.II.4).

IV. Summary

Throughout this section we have tried to emphasize that, apart from reducing alcohol intake/preference, increasing 5-HT function also reduces a variety of other consummatory and drug-reinforced behaviours. Conversely, reductions in the functioning of this neurotransmitter system may enhance these same behaviours, particularly under conditions of low baseline responding, such as partial satiation. Attempts to define the processes that underline these effects are inconclusive. Enhancement of satiety has been implicated as a mechanism through which elevated 5-HT function may suppress food (BLUNDELL 1986) and alcohol (GILL et al. 1988b) intake. However, our own studies on alcohol drinking (HIGGINS et al. 1992b) are not entirely compatible with this hypothesis and in the feeding literature there are further inconsistencies to a satiety hypothesis (see Sect. C.I.3). An alternative view has been proposed by GRUPP and colleagues (1991). 5-HT is a releaser of renin, which in turn activates the renin-angiotensin system. The biologically active product of this pathway is angiotensin II (AII), which, in addition to stimulating thirst and vasopressin/aldosterone release, also reduces alcohol preference (see GRUPP et al. 1991, for review). Grupp and coworkers suggest that drugs such as fluoxetine reduce alcohol intake indirectly by stimulating endogenous All release. Interestingly, pretreatment with the angiotensin-converting enzyme (ACE) inhibitor enalapril (which blocks All synthesis) attenuates the suppressant effect of fluoxetine on alcohol self-administration (GRUPP et al. 1991).

However, in our view there are perhaps two interrelated hypotheses that provide the most intriguing explanation for the diversity of effects produced by 5-HT drugs. Recent reviews by SOUBRIE (1986) and DEAKIN and GRAEFF (1991) suggest that ascending 5-HT projections may serve to oppose forebrain dopamine systems important in mediating "approach" or reward-related behaviour. Thus, activation of this system by 5-HT agonists will suppress dopaminergic function and consequently reduce motivated behaviour. Within this scheme can be incorporated the broad influence of 5-HT manipulations on behaviours maintained by a variety of positive reinforcers, including alcohol (see Table 2). This account may also explain why palatability-induced feeding is particularly sensitive to 5-HT agonist pretreatment, for DA antagonists are also highly effective in reducing palatability-induced ingestion (VACCARINO et al. 1989). Furthermore, SOUBRIE (1986) has proposed that central 5-HT systems may comprise a behavioural inhibition system, serving to control ongoing behaviour particularly under situations of punishment/non-reward or delay to reward. Functional deficiencies in this system may result in behaviours indicative of lack of control or over-responding towards environmental reinforcers including, but not limited to, alcohol (SELLERS et al. 1992 and references therein). Evidence to support this hypothesis is provided by the observation that 5-HT uptake inhibitors such as fluoxetine and clomipramine may be efficacious in restoring some behavioural control to such individuals. With respect to the present chapter, it is intriguing to note that the rat model that most clearly satisfies the definition as an animal model of alcoholism, viz. the alcohol P rat, is typified by a deficiency in forebrain 5-HT content and reduced 5-HT neuronal content (ZHOU et al. 1991). It would be of interest to assess the behaviour of P and NP rats under situations of delay to reward (SOUBRIE 1986; WOGAR et al. 1993), or presentation of reward under a differential reinforcement of low rate (DRL) schedule (WOGAR et al. 1993).

Ultimately the suppressant effects of 5-HT drugs on alcohol self-administration may be shown to depend on any combination of the above factors, or by alternative, as yet undefined, mechanisms. The majority of drugs belonging to this class produce fairly modest reductions in alcohol drinking in humans (SELLERS et al. 1992). There is a clear need to enhance these effects, which can only come by a greater appreciation of the mechanisms by which 5-HT controls animal behaviour. Ultimately this may also assist in the use of such drugs in the most appropriate population of alcohol abusers and also in conjunction with the most appropriate form of behavioural therapy.

A final comment should be made concerning the observation that 5-HT_3 antagonists reduce alcohol self-administration. It is presently uncertain as to how these drugs modify ethanol drinking in relation to the foregoing account that a functional enhancement of the serotonin system will produce

Table 2. Effect of manipulations to the 5-HT system on alcohol self-administration: comparison with feeding and other drug-reinforced behaviour

	Alcohol self-administration	Feeding behavior	Other drug-reinforced behavior			Intracranial self-stimulation
			Psychostimulant	Opioid	Nicotine	
Increasing 5-HT function						
Precursor loading (e.g., 5-HTP, tryptophan)	↓	↓	↓	NAD	NAD	↓
5-HT releasers (e.g., dexfenfluramine)	↓	↓	NAD	↓	NAD	↓
5-HT agonists						
5-HT_{1A} (e.g., 8-OH DPAT)	↑ (low doses) ↓ (high doses)	↑ (low doses) ↓ (high doses)	↓ (high doses)	NAD	NAD	↑ (low doses) ↓ (high doses)
Mixed 5-$HT_{1B/1C}$ (e.g., TFMPP, mCPP) and 5-$HT_{1C/2}$ (e.g., DOI, MK212)	↓	↓	NAD	NAD	NAD	NAD
5-HT uptake blockers (e.g., fluoxetine)	↓	↓	↓	NAD	NAD	↓
MAO type A inhibitors (e.g., bromfaromine)	↓	↓	NAD	NAD	NAD	NAD
Decreasing 5-HT function						
Synthesis inhibition (e.g., pCPA)	↓/O/↑	O/↑	NAD	NAD	NAD	↓/O/↑
Central 5-HT lesions (e.g., 5,6-DHT, 5,7-DHT)	O/↑	O/↑	↑	↑	NAD	O
5-$HT_{1/2}$ antagonists (e.g., metergoline)	O	O/↑	↑	NAD	NAD	NAD
5-HT_3 antagonists (e.g., ondansetron)	↓	O	O	O	O	O

↓, decrease in behaviour; ↑, increase in behaviour; O, no effect on behaviour; NAD, no available data. Where more than one symbol is used this suggests that the result is dependent upon the paradigm or experimental protocol used. The most obvious example of this may be seen in studies using pCPA, which has marked aversive properties, although with appropriate caution this feature may be minimized.

Observations on ICSS were summarized from KATZ and CARROLL (1977), MONTGOMERY et al. (1991, 1993), POSCHEL and NINTEMAN (1971), HOEBEL et al. (1988), and MCCLELLAND et al. (1989). All other references may be taken from appropriate portions of the text.

the same behaviour. One fundamental difference between 5-HT_3 antagonists and indirect/direct 5-HT agonists is that, although each suppress alcohol self-administration, only the latter appear to suppress other drug- and food-reinforced behaviours. This suggests a specific interaction between alcohol and 5-HT_3 receptors, and the work of LOVINGER (1991) (see Sect. C.III) would seem relevant in this respect. However, as yet there has been little research to elaborate on this initial observation. Also, the work of GRANT (1992) and colleagues (GRANT and BARRETT 1991), suggesting that the discriminative stimulus properties of ethanol may be modified by 5-HT_3 antagonist pre-treatment, seems worthy of further investigation. It may have relevance to recent clinical studies conducted in normals (non-alcohol abusers) (JOHNSON et al. 1993) and moderate alcohol abusers (SELLERS et al. 1991), demonstrating that 5-HT_3 antagonists may affect some of the subjective effects of alcohol. Each of these studies requires replication and further characterization in order to determine whether 5-HT_3 receptor antagonists may represent effective therapies for the treatment of alcohol abuse.

D. 5-HT, Alcohol Tolerance, and Physical Dependence

I. General Aspects of Alcohol Tolerance

1. Definition and Classification

Tolerance and physical dependence are features associated with chronic intake of alcohol in humans and animals. Acquired tolerance to ethanol is defined as a reduction in the effects produced by a given dose of ethanol as a consequence of repeated exposure to the drug. Acquired tolerance is distinct from innate or initial tolerance, which refers to the variation in central nervous sensitivity to ethanol among individuals. Acquired tolerance can result from both an increase in the rate of metabolic elimination (dispositional tolerance) and from a decrease in sensitivity of the target tissues, particularly the central nervous system (KALANT et al. 1971). Based on a temporal framework, acquired tolerance can be classified into two different forms: acute and chronic tolerance. Acute tolerance refers to a change in drug response that can be observed within a single exposure, for example, a reduced effect of ethanol on the descending portion of the blood ethanol curve relative to the effect of the same blood ethanol level on the ascending portion of the curve. In contrast, chronic tolerance refers to changes in drug effect after repeated administration, usually over a number of days or weeks (KALANT et al. 1971).

Physical dependence is defined by the occurrence of an ethanol withdrawal reaction upon the cessation of ethanol treatment, which is usually characterized by a hyperexcitability of the CNS. While there are a variety of sensitive tests of ethanol effects for assessing tolerance, the assessment of physical dependence often involves relatively crude measures such as precipitation of motor seizure. Tolerance has been investigated much more extensively than physical dependence. There is considerable debate as to whether tolerance and physical dependence are simply two manifestations of the same adaptive process (see KALANT 1988).

2. Features of Ethanol Tolerance

It is beyond the scope of this chapter to examine the various factors that have been shown to affect ethanol tolerance. Detailed information can be obtained from various chapters in GOUDIE and EMMETT-OGELSBY (1989). In addition, tolerance is discussed in Chaps. 6, 8, and 11, this volume. Basically, research over the last several decades has shown that ethanol tolerance is a complex phenomenon. Tolerance is influenced by a number of factors ranging from pharmacological and behavioural to the response system employed. Moreover, these factors interact to regulate the development and manifestation of tolerance. For example, the development and manifestation of tolerance can be influenced by opportunity for learning during intoxicated practice or by the presence or absence of external cues previously associated with ethanol administration. The influences of intoxicated-practice or conditioning on tolerance, however, are dependent on the treatment doses employed. Generally, intoxicated practice and conditioning can be more dominant factors in the expression of tolerance at lower rather than higher doses of ethanol (KALANT 1988; LÊ 1990).

II. Role of 5-HT in Ethanol Tolerance/Dependence

The involvement of learning in tolerance and the implication of 5-HT in learning have stimulated much research into the role of 5-HT in the development of tolerance to ethanol. Evidence for the involvement of brain 5-HT in chronic tolerance to ethanol was first reported by FRANKEL and coworkers (1975). These investigators demonstrated that depletion of brain 5-HT by administration of pCPA retarded the rate of development of tolerance to the motor impairment produced by ethanol in rats. Subsequent research by these workers and by other investigators has shown that depletion of brain 5-HT by administration of pCPA or by intraventricular (i.c.v.) injection of 5,7-DHT retarded the development of tolerance to motor impairment, hypothermia, and hypnotic effects of ethanol in mice and rats (LÊ et al. 1980; KHANNA et al. 1980; MELCHIOR and TABAKOFF 1981). Raising brain 5-HT levels by daily administration L-tryptophan, on the other hand, accelerates tolerance development (LÊ et al. 1979). In all these studies, manipulation of brain 5-HT was found to affect mainly the rate at which tolerance is acquired rather the final level of tolerance attained. The failure by WOOD (1980) to observe any effects of 5-HT depletion on ethanol tolerance might be due to the fact that tolerance was assessed at a single time point, before which maximum tolerance had already been established.

The mesolimbic 5-HT pathway, or more precisely 5-HT in septal and hippocampal areas, plays a crucial role in regulating tolerance development. Electrolytic lesions of the median raphe, but not the dorsal raphe, retarded tolerance development in a manner similar to that induced by pCPA or 5,7-DHT (LÊ et al. 1981a). Furthermore, selective depletion of septal and hippocampal 5-HT through microinjection of 5,7-DHT into the cingulum bundle and fimbria-fornix also retarded tolerance development (SPEISKY and KALANT 1985). The 5-HT system does not act independently, but interacts with other systems to regulate tolerance development. For example, while depletion of brain NE alone did not affect tolerance development in the rat, combined depletion of both 5-HT and NE in brain inhibits or blocks tolerance development (LÊ et al. 1981b). Similarly, the mesolimbic 5-HT system has been shown to interact with vasopressin in maintaining ethanol tolerance (LÊ et al. 1982; SPEISKY and KALANT 1985).

Beside chronic tolerance, depletion of 5-HT has been shown to affect acute tolerance to ethanol (CAMPANELLI et al. 1988), as well as the development of cross-tolerance between ethanol and pentobarbital (KHANNA et al. 1987). In addition, manipulation of brain 5-HT has also been shown to affect the development of tolerance to morphine (Ho et al. 1972, 1975; TILSON and RESCH 1974) and various barbiturates (KHANNA et al. 1980; LYNESS and MYCEK 1980; KHANNA et al. 1987). It is likely that 5-HT affects tolerance development through its action on learning processes.

In contrast to studies that investigated the role of 5-HT on ethanol drinking, little work has been carried out to examine the involvement of specific types of 5-HT receptors in ethanol tolerance. Recent work by WU et al. (1993) has indicated that 5-HT_2 and 5-HT_3 receptors may be involved in the regulation of ethanol tolerance. These workers reported that the effects of 5-HT depletion induced by i.c.v. administration of 5,7-DHT, on vasopressin-maintained tolerance, can be reversed by i.c.v. administration of the 5-HT receptor agonist methyl-5-HT or the 5-HT_3 receptor agonist 2-methyl-5-HT.

III. Differences in 5-HT Regulation of Ethanol Drinking and Ethanol Tolerance

The ingestion of alcohol or other drugs of abuse produces a variety of effects, some of which are rewarding, while others are aversive. Tolerance has been shown to occur to a variety of the aversive effects of alcohol. For these reasons, it is believed that tolerance to ethanol's aversive effects may enhance drinking behaviour (KALANT 1989). As the 5-HT system is involved in both ethanol drinking and in tolerance development, the question arises as to whether both ethanol drinking and tolerance are modulated by the same 5-HT pathway or receptor subtypes.

The available data indicate that different 5-HT pathways and receptor types are involved in the regulation of ethanol intake and ethanol tolerance.

Raising 5-HT levels by L-tryptophan administration facilitates the development of tolerance to ethanol or to morphine (HO et al. 1975; LÊ et al. 1979), while a variety of manipulations that increase brain 5-HT reduce ethanol intake. The 5-HT_2 receptor appears to be involved in the regulation of tolerance (WU et al. 1993), but not necessarily in ethanol intake (MYERS and LANKFORD 1993). Stimulation of 5-HT pathways that originate from the dorsal or median raphe by microinjection of 8-OH-DPAT into these areas has been shown to enhance ethanol drinking (TOMKINS et al. 1993). On the other hand, the median but not the dorsal raphe, or only septal and hippocampal 5-HT, have been shown to be involved in ethanol tolerance (LÊ et al. 1981b; SPEISKY and KALANT 1985). Together these studies suggest that the 5-HT system regulates ethanol tolerance through its modulation of learning processes via its interaction or regulation of septal and hippocampal activities. On the other hand, the regulation of ethanol intake by 5-HT may operate through its modulation of the rewarding effects of ethanol via its interaction with the mesolimbic DA pathway.

IV. 5-HT and Ethanol Dependence

There are only two studies that examined the effects of 5-HT depletion on the development of physical dependence to ethanol. Basically, both of these studies revealed that depletion of brain 5-HT prior to ethanol exposure did not modify ethanol withdrawal reactions in the rat (FRYE and ELLIS 1977; WOOD 1980). However, alterations in brain 5-HT function during ethanol withdrawal have been shown. For example, a decrease in 5-HT activity during ethanol withdrawal has been demonstrated (TABAKOFF et al. 1977). In addition, enhanced seizure activity has been induced by administration of the non-selective 5-HT antagonist methysergide during ethanol withdrawal (BLUM et al. 1976). These studies suggest that reduced 5-HT activity may be a secondary consequence of ethanol withdrawal.

E. Effects of Chronic Ethanol Administration on 5-HT Receptors

I. Intoxication

Chronic ethanol administration increases the levels of the 5-HT metabolite 5-hydroxyindole-acetic acid (5-HIAA) in rat striatum (HUNT and MAJCHROWICZ 1983; HOLMAN and SNAPE 1985; KHATIB et al. 1988), indicating that chronic ethanol modulates serotonin neurotransmission, perhaps by increasing central turnover. The effects of chronic administration on receptor subtypes are conflicting. Using the non-selective receptor label of [^{3}H]5-HT (PEROUTKA 1988), 5-HT_1 binding was increased in the striatum and brainstem but decreased in the hippocampus in rats treated chronically with ethanol (MULLER et al. 1980). In contrast, HUNT and DALTON (1981) found no differences in binding in these regions. With more specific and selective ligands, BUCKHOLTZ et al. (1989) found no alterations in the density of 5-HT_{1A}- and 5-HT_2-binding sites in mice consuming a liquid ethanol diet for 7 days. In rats made physically dependent on ethanol by administration of 12 g/kg ethanol intragastrically in divided doses for 4 days, 8-OH-DPAT binding (B_{max}) was decreased by 25% during intoxication and 17% during withdrawal in the hippocampus. The binding of labelled ketanserin in cortex, cyanopindolol in striatum and hypothalamus, and 8-OH-DPAT in the cortex was not affected during chronic intoxication (ULRICHSEN 1991). These data indicate regional specificity in the changes within the serotonergic system.

F. Interaction of 5-HT and Opioids

Considerable data exist indicating an important role for 5-HT in the development of tolerance to analgesia and the various behavioural actions of opiates (e.g., CONTRERAS et al. 1980; FUCHS and COPER 1980; CHENEY and GOLDSTEIN 1971; Ho et al. 1975) and physical dependence after opiate administration. These effects are generally robust and suggest that the endogenous opioid system itself plays an important role in the development of tolerance. Chronic treatment with dexfenfluramine has been reported to increase enkephalin levels. Interestingly, tolerance develops quickly to the effects of dexfenfluramine on food and heroin self-administration (HIGGINS et al., unpublished observation), but not to ethanol in the limited access paradigm (Tomkins et al., unpublished observation). Changes in opioid activity have been suggested to be a mechanism for the development of the dexfenfluramine effect on feeding. This dissociation of feeding and heroin self-administration on one hand and ethanol self-administration on the other suggests that the interaction of these two neurotransmitter systems may show some reinforcer selectivity.

G. General Summary

Increasing 5-HT reduces alcohol intake/preference but also reduces a variety of other consummatory and drug-reinforced behaviours. Conversely, reductions in the functioning of this neurotransmitter system may enhance these same behaviours, particularly under conditions of low baseline responding. Attempts to define the processes that underlie these effects are currently inconclusive.

Ascending 5-HT projections may serve to oppose forebrain dopamine systems important in mediating "approach" or reward-related behaviour. Thus, activation of this system by 5-HT agonists will suppress dopaminergic function and consequently reduce motivated behaviour. Within this scheme can be incorporated the broad influence of 5-HT manipulations on behaviours maintained by a variety of positive reinforcers, including alcohol (see Table 2). Functional deficiencies in this system may conceivably result in behaviours indicative of lack of control or over-responding towards environmental reinforcers including, but not limited to, alcohol.

The serotonergic receptor subtypes play different roles in the regulation of consummatory behaviours. The 5-HT_{1A} presynaptic receptors increase ethanol consumption, whereas antagonists at the 5-HT_3 receptor decrease self-administration. The suppressant effects of dexfenfluramine on ethanol consumption can be completely reversed by metergoline but are only partially reversible by the 5-$HT_{2A/2C}$ antagonist ritanserin.

The complex neurochemical events underlying the phenomenon of tolerance and the intimate interplay of the behavioural task and the pattern of tolerance development has severely limited the amount of critical research concerning the role of 5-HT in tolerance. In general, depletion of 5-HT delays or decreases the development of tolerance in learned behaviours. Conversely, generally enhanced function accelerates tolerance.

The 5-HT system does not appear to play an important role in the development of physical dependence on ethanol and consequent acute withdrawal.

Central microinjection into the raphe nuclei, nucleus accumbens and VTA of specific and selective 5-HT, opioid and dopaminergic agents in animals engaged in ethanol self-administration and its related behaviours should be a fruitful research strategy to elucidate the mechanism of regulation of ethanol self-administration and the development of tolerance to ethanol's effects.

Acknowledgements. First and foremost we would like to thank Cathy Van Der Giessen for her help in preparing this manuscript and Dr. Denise Tomkins for the data used in Fig. 3. We also gratefully acknowledge the skilled technical assistance of Xiaoping Zhang, Jeff Tighe, Peter Nguyen, Yvona Buczek and Narges Joharchi in some of the research described. Finally, we would like to thank Drs. Paul Fletcher, Denise Tomkins, Bill Corrigall and Constantine Poulos for their valuable comments on particular aspects of the research reviewed in this article.

References

Ahtee L, Eriksson K (1972) 5-Hydroxytryptamine and 5-hydroxyindolylacetic acid content in brain of rat strains selected for their alcohol intake. Physiol Behav 8:123–126

Amit Z, Sutherland EA, Gill K, Ogren SO (1984) Zimeldine: a review of its effects on ethanol consumption. Neurosci Biobehav Rev 8:35–54

Antin J, Gibbs J, Holt J, Young RC, Smith GP (1975) Cholecystokinin elicits the complete behavioural sequence of satiety in rats. J Comp Physiol Psychol 89:784–790

Barnes JN, Barnes NM, Cooper SJ (1992) Behavioural pharmacology of 5-HT_3 receptor ligands. Neurosci Biobehav Rev 16:107–113

Blandina P, Goldfarb J, Craddock-Royal B, Green JP (1989) Release of endogenous dopamine by stimulation of 5-hydroxytryptamine$_3$ receptors in rat striatum. J Pharmacol Exp Ther 251:803–809
Blum K, Wallace JE, Schwertner HA, Eubanks JD (1976) Enhancement of ethanol-induced withdrawal convulsions by blockade of 5-hydroxytryptamine receptors. J Pharm Pharmacol 28:832–835
Blundell JE (1986) Serotonin manipulations and the structure of feeding behaviour. Appetite 7:39–56
Blundell JE, Lawton CL (1990) Serotonin receptor sub-types and the organisation of feeding behaviour: experimental models. In: Paoletti R (ed) Serotonin: from cell biology to pharmacology and therapeutics. Kluwer Academic, Norwell, Massachusetts, pp 213–219
Borsini F, Bendotti C, Samanin R (1985) Salbutamol, d-amphetamine and d-fenfluramine reduce sucrose intake in freely fed rats by acting on different neurochemical mechanisms. Int J Obes 9:277–283
Buckholtz NS, Zhou D, Tabakoff B (1989) Ethanol does not affect serotonin receptor binding in rodent brain. Alcohol 6:277–280
Buczek Y, Tomkins DM, Higgins GA, Sellers EM (1993) Separate effects of m-chlorophenylpiperazine (mCPP) on anxiety, pharmacologic conditioning, and ethanol self-administration. Abstracts of College on problems of drug dependence, 55th annual scientific meeting, Toronto, Canada
Burton MJ, Cooper SJ, Popplewell DA (1981) The effect of fenfluramine on the microstructure of feeding and drinking in the rat. Br J Pharmacol 72:621–633
Campanelli C, Lê AD, Khanna JM, Kalant H (1988) Effects of raphe lesions on the development of acute tolerance to ethanol and pentobarbital. Psychopharmacology 96:194–221
Carboni E, Acquas E, Leone P, Di Chiara G (1989) 5-HT$_3$ receptor antagonists block morphine- and nicotine- but not amphetamine-induced reward. Psychopharmacology 97:175–178
Carroll ME, Lac ST, Asencio M, Kragh R (1990a) Intravenous cocaine self-administration in rats is reduced by dietary L-tryptophan. Psychopharmacology 100:293–300
Carroll ME, Lac ST, Asencio M, Kragh R (1990b) Fluoxetine reduces intravenous cocaine self-administration in rats. Pharmacol Biochem Behav 35:237–244
Chen J, van Praag HM, Gardner E (1991) Activation of 5-HT$_3$ receptor by 1-phenylbiguanide increases dopamine release in the rat nucleus accumbens. Brain Res 543:354–357
Cheney DL, Goldstein A (1971) The effect of p-chlorophenylalanine on opiate-induced running, analgesia, tolerance and physical dependence in mice. J Pharmacol Exp Ther 177(1):309–315
Clifton PG, Barnfield AMC, Philcox L (1989) A behavioural profile of fluoxetine-induced anorexia. Psychopharmacology 97:89–95
Collins DM, Myers RD (1987) Buspirone attenuates volitional alcohol intake in the chronically drinking monkey. Alcohol 4:49–56
Contreras E, Tamayo L, Quijada L, Munoz M (1980) Decrease of tolerance and dependence on morphine by drugs affecting brain serotonin. Cell Mol Biol 26:389–393
Deakin JFW, Graeff FG (1991) 5-HT and mechanisms of defence. J Psychopharmacol 5:305–315
Dourish CT, Hutson PH, Kennett GA, Curzon G (1986) 8-OH DPAT-induced hyperphagia: its neural basis and possible therapeutic relevance. Appetite 7:127–140
Dourish CT, Clark ML, Fletcher A, Iversen SD (1989) Evidence that blockade of post-synaptic 5-HT$_1$ receptors elicits feeding in satiated rats. Psychopharmacology 97:54–58

Ervin GN, Carter RB, Webster EL, Moore SI, Cooper BR (1984) Evidence that taste aversion learning induced by 1-5-hydroxytryptophan is mediated peripherally. Pharmacol Biochem Behav 20:799–802
Fadda F, Garau B, Marcheli F, Colombo G, Gessa GL (1991) MDL72222, a selective 5-HT_3 receptor antagonist, suppresses voluntary ethanol consumption in alcohol-preferring rats. Alcohol Alcohol 26:107–110
Fletcher PJ (1988) Increased food intake in satiated rats induced by the 5-HT antagonists methyersgide, metergoline and ritanserin. Psychopharmacology 96:237–242
Fletcher PJ, Burton MJ (1984) Effects of manipulation of peripheral serotonin on feeding and drinking in the rat. Pharmacol Biochem Behav 20:835–840
Fletcher PJ, Davies M (1990) The involvement of 5-hydroxytryptaminergic and dopaminergic mechanisms in the eating induced by buspirone, gepirone and ipsapirone. Br J Pharmacol 99:519–525
Frankel D, Khanna JM, Le Blanc AE, Kalant H (1975) Effect of p-chlorophenylalanine on the acquisition of tolerance to ethanol and pentobarbital. Psychopharmacologia (Berl) 44:247–252
Frye GD, Ellis FW (1977) Effects of 6-hydroxydopamine or 5,7-dihydroxytryptamine on the development of physical dependence on ethanol. Drug Alcohol Depend 2:349–359
Fuchs V, Coper H (1980) The influence of p-chlorophenylalanine on different morphine effects. Drug Alcohol Depend 5:367–377
Geller I (1973) Effects of para-chlorophenylalanine and 5-hydroxytryptophan on alcohol intake in the rat. Pharmacol Biochem Behav 1:361–365
Gill K, Amit Z (1989) Serotonin uptake blockers and voluntary alcohol consumption. A review of recent studies. In: Galanter M (ed) Recent developments in alcoholism, vol 7. Plenum, New York, pp 225–248
Gill K, Amit Z, Ogren SO (1985) The effects of zimeldine on voluntary ethanol consumption: studies on the mechanism of action. Alcohol 2:343–347
Gill K, Shatz, K, Amit Z, Ogren SO (1986) Conditioned taste aversion to ethanol induced by zimeldine. Pharmacol Biochem Behav 24:463–468
Gill K, Amit Z, Ogren SO (1988a) Treatment with sertraline, a new serotonin uptake inhibitor, reduces voluntary ethanol consumption in rats. Alcohol 5:349–354
Gill K, Fillon Y, Amit Z (1988b) A further examination of the effects of sertraline on voluntary ethanol consumption. Alcohol 5:355–358
Gothoni P, Ahtee L (1980) Chronic ethanol administration decreases 5-HT and increases 5-HIAA concentration in rats. Acta Pharmacol Toxicol 46:113–120
Goudie AJ, Emmett-Oglesby MW (1989) Psychoactive drugs: tolerance and sensitization. Humana, Clinton, New Jersey
Grant KA (1992) The 5-HT_3 antagonist MDL72222 selectively blocks the discriminative stimulus effects of ethanol in rats. In: Harris LS (ed) Problems of drug dependence 1991: 53rd annual scientific meeting of the committee on problems of drug dependence. US Government Printing Office, Washington DC, p 211 (NIDA research monograph 119)
Grant KA, Barrett JE (1991) Blockade of the discriminative stimulus effects of ethanol with 5-HT_3 receptor antagonists. Psychopharmacology 104:451–456
Grignaschi G, Samanin R (1992) Role of serotonin and catecholamines in brain in the feeding suppressant effect of fluoxetine. Neuropharmacology 31:445–449
Grupp LA, Perlanski E, Stewart RB (1991) Regulation of alcohol consumption by the renin-angiotensin system: a review of recent findings and a possible mechanism of action. Neurosci Biobehav Rev 15:264–275
Haefely W, Burkard WP, Cesura AM, Kettler R, Lorez HP, Martin JR, Richards JG, Scherschlicht R, Da Prada M (1992) Biochemistry and pharmacology of moclobemide, a prototype RIMA. Psychopharmacology 106:S6–S14
Haraguchi M, Samson HH, Tolliver GA (1990) Reduction in oral ethanol self-administration in the rat by the 5-HT uptake blocker fluoxetine. Pharmacol Biochem Behav 35:259–262

Higgins GA, Lawrin MO, Sellers EM (1992a) Serotonin and alcohol consumption. In: Naranjo CA, Sellers EM (eds) Novel pharmacological interventions for alcoholism. Springer, Berlin Verlag, New York, pp 83–91

Higgins GA, Tomkins DM, Fletcher PJ, Sellers EM (1992b) Effects of drugs influencing 5-HT function on ethanol drinking and feeding behaviour in rats: studies using a drinkometer system. Neurosci Biobehav Res 16:535–552

Higgins GA, Joharchi N, Nguyen P, Sellers EM (1992c) Effect of the 5-HT_3 receptor antagonists. MDL72222 and ondansetron on morphine place conditioning. Psychopharmacology 106:315–320

Higgins GA, Tomkins DM, Poulos CX, Sellers EM (1993a) Effect of dexfenfluramine on saccharin drinking: behavioural and pharmacological studies. Pharmacol Biochem Behav 47:307–315

Higgins GA, Wang Y, Corrigall WA, Sellers EM (1993b) Influence of 5-HT_3 receptor antagonists and the indirect 5-HT agonist, dexfenfluramine, on heron self-administration in rats. Psychopharmacology 114:611–619

Higgins GA, Wang Y, Sellers EM (1993c) Preliminary findings with the indirect 5-HT agonist dexfenfluramine on heroin discrimination and self-administration in rats. Pharmacol Biochem Behav 45:963–966

Hill SY (1974) Intraventricular injection of 5-hydroxytryptamine and alcohol consumption in rats. Biol Psychiatry 8:151–158

Hill SY, Goldstein R (1974) Effect of parachlorophenylalanine and stress on alcohol consumption by rats. Q J Stud Alcohol 35:34–41

Ho AKS, Tsai CS, Chen RCA, Begleiter H, Kissin B (1974) Experimental studies on alcoholism. 1. Increased in alcohol preference by 5,6-dihydroxytryptamine and brain acetylcholine. Psychopharmacologia (Berl) 40:101–107

Ho IK, Lu SE, Stolman S, Loh HH, Way EL (1972) Influence of p-chlorophenylalanine on morphine tolerance and physical dependence and regional brain serotonin turnover in morphine tolerant-dependent mice. J Pharmacol Exp Ther 182:155–165

Ho IK, Brase DA, Loh HH, Way EL (1975) Influence of L-tryptophan on morphine analgesia, tolerance and physical dependence. J pharmacol Exp Ther 193(1): 35–43

Hodge CW, Samson HH, Lewis RS, Erickson HL (1993) Specific decreases in ethanol- but not water-reinforced responding produced by the 5-HT_3 antagonist ICS205-930. Alcohol 10(3):191–196

Hoebel BG, Hernandez L, McClelland RC, Schwartz D (1988) Dexfenfluramine and feeding reward. Clin Neuropharmacol 11 [Suppl 1]:S72–S85

Holman RB, Snape BM (1985) Effects of ethanol on 5-hydroxytryptamine release from rat corpus striatum in vivo. Alcohol 2:249–253

Hoyer D (1988) Functional correlates of 5-HT_1 recognition sites. J Recept Res 8:59–81

Hulihan-Giblin BA, Park YD, Aulakh CS, Goldman D (1992) Regional analysis of 5-HT_{1A} and 5-HT_2 receptors in the fawn-hooded rat, Neuropharmacology 11:1095–1099

Hunt WA, Dalton TK (1981) Neurotransmitter receptor binding in various brain regions in ethanol-dependent rats. Pharmacol Biochem Behav 14:733–739

Hunt WA, Majchrowicz E (1983) Studies of neurotransmitter interactions after acute and chronic ethanol administration. Pharmacol Biochem Behav 18:371–374

Hyytia P, Sinclair JD (1993) Oral etonitazene and cocaine consumption by AA, ANA and Wistar rats. Psychopharmacology 111:409–414

Jiang LH, Ashby CR, Kasser RJ, Wang RY (1990) The effect of intraventricular administration of the 5-HT_3 receptor agonist 2-methylserotonin on the release of dopamine in the nucleus accumbens: an in vivo chronocoulometric study. Brain Res 513:156–160

Johnson BA, Campling GM, Griffiths P, Cowen PJ (1993) Attenuation of some alcohol-induced mood changes and the desire to drink by 5-HT_3 receptor

blockade: a preliminary study in healthy male volunteers. Psychopharmacology 112:142–144
Kalant H (1988) Alcohol tolerance, dependence and withdrawal: an overview of current issues. Aust Drug Alcohol Rev 7:27–34
Kalant H (1989) Drug tolerance and sensitization: a pharmacological overview. In: Goudie AJ, Emmett-Oglesby MW (eds) Psychoactive drugs. Tolerance and sensitization. Human, Clifton, pp 547–577
Kalant H, Le Blanc AE, Gibbins RJ (1971) Tolerance to, and dependence on, some nonopiate psychotropic drugs. Pharmacol Rev 23(3):135–191
Katz RJ, Carroll BJ (1977) Intracranial reward after Lilly 110140 (fluoxetine HCI): evidence for an inhibitory role for serotonin. Psychopharmacology 51:189–193
Khanna JM, Kalant H, Lê AD, Mayer J, Le Blanc AE (1980) Effect of p-chlorophenylalanine on the acquisition of tolerance to the hypnotic effects of pentobarbital, barbital, and ethanol. Can J Physiol Pharmacol 58:1031–1041
Khanna JM, Campanelli C, Lê AD, Kalant H (1987) The effects of lesions in the dorsal and median raphe nuclei on the development of tolerance to pentobarbital and cross-tolerance to ethanol. Psychopharmacology 91:473–478
Khatib SA, Murphy JM, McBride WJ (1988) Biochemical evidence for activation of specific monoamine pathways by ethanol. Alcohol 5:295–299
Kiianmaa K (1976) Alcohol intake in the rat after lowering brain 5-hydroxytryptamine content by electrolytic midbrain raphe lesions, 5,6 dihydroxytryptamine or p-chlorophenylalanine. Med Biol 54:203–209
Kilpatrick GJ, Bunce KT, Tyers MB (1990) 5-HT_3 receptors. Med Res Rev 10: 441–475
Knapp DJ, Pohorecky LA (1992) Zacopride, a 5-HT_3 receptor antagonist, reduces voluntary ethanol consumption in rats. Pharmacol Biochem Behav 41(4):847–850
Knapp DJ, Benjamin D, Pohorecky LA (1992) Effects of gepirone on ethanol consumption, exploratory behavior, and motor performance in rats. Drug Dev Res 26:319–341
Kostowski W, Dyr W (1992) Effect of 5-HT_{1A} receptor agonists on ethanol preference in the rat. Alcohol 9:283–286
Lacosta S, Roberts DCS (1993) MDL72222, ketanserin, and methysergide pretreatments fail to alter breaking points on a progressive ratio schedule reinforced by intravenous cocaine. Pharmacol Biochem Behav 44:161–165
Lane JD, Pickering CL, Hooper ML, Fagan K, Tyers MB, Emmett-Oglesby MW (1992) Failure of ondansetron to block the discriminative or reinforcing stimulus effects of cocaine in the rat. Drug Alcohol Depend 30:151–162
Lawrin MO (1988) Serotonin uptake inhibitors: effect on the macro- and microstructure of ethanol drinking. Doctoral thesis, University of Toronto, Toronto, Canada
Lawrin MO, Naranjo CA, Sellers EM (1986) Identification and testing of new drugs for modulating alcohol consumption. Psychopharmacol Bull 22:1020–1025
Lê AD (1990) Factors regulating ethanol tolerance. Ann Med 22:265–268
Lê AD, Khanna JM, Kalant H, Le Blanc AE (1979) Effect of L-tryptophan on the acquisition of tolerance to ethanol-induced motor impairment and hypothermia. Psychopharmacology 61:125–129
Lê AD, Khanna JM, Kalant H, Le Blanc AE (1980) Effect of 5,7-dihydroxytryptamine on the acquisition of tolerance to ethanol. Psychopharmacology 67:143–146
Lê AD, Khanna JM, Kalant H, Le Blanc AE (1981a) The effects of lesions in the dorsal, median and magnus raphe nuclei on the development of tolerance to ethanol. J Pharmacol Exp Ther 218:525–529
Lê AD, Khanna JM, Kalant H, Le Blanc AE (1981b) Effect of modification of brain serotonin (5-HT), norepinephrine (NE) and dopamine (DA) on ethanol tolerance. Psychopharmacology 75:231–235

Lê AD, Kalant H, Khanna JM (1982) Interaction between des-glycinamide-Arg-vasopressin and serotonin on ethanol tolerance. Eur J Pharmacol 80:337–345
Leander JD (1987) Fluoxetine suppresses palatability-induced ingestion. Psychopharmacology 91:285–287
Leccese AP, Lyness WH (1984) The effects of putative 5-hydroxytryptamine receptor active agents on D-amphetamine self-administration in controls and rats with 5,7-dihydroxytryptamine median forebrain bundle lesions. Brain Res 303:153–162
Loh EA, Roberts DCS (1990) Break-points on a progressive ratio schedule reinforced by intravenous cocaine increase following depletion of forebrain serotonin. Psychopharmacology 101:262–266
Lovinger DM (1991) Ethanol potentiation of 5-HT_3 receptor-mediated ion current in NCB-20 neuroblastoma cells. Neurosci Lett 122:57–60
Lyness WH, Moore KE (1983) Increased self-administration of d-amphetamine by rats pretreated with metergoline. Pharmacol Biochem Behav 18:721–724
Lyness WH, Mycek MJ (1980) The role of cerebral serotonin in the development of tolerance to phenobarbital. Brain Res 187:443–456
Lyness WH, Smith FL (1992) Influence of dopaminergic and serotonergic neurons on intravenous ethanol self-administration in the rat. Pharmacol Biochem Behav 42:187–192
McBride WJ, Murphy JM, Lumeng L, Li T-K (1990) Serotonin, dopamine and GABA involvement in alcohol drinking of selectively bred rats. Alcohol 7:199–205
McClelland RC, Sarfaty T, Hernandez L, Hoebel BG (1989) The appetite suppressant, d-fenfluramine, decreases self-administration at a feeding site in the lateral hypothalamus. Pharmacol Biochem Behav 32:411–414
McGregor A, Lacosta S, Roberts DCS (1993) L-Tryptophan decreases the breaking point under a progressive ratio schedule of intravenous cocaine reinforcement in the rat. Pharmacol Biochem Behav 44:651–655
Meert TF (1993) Effects of various serotonergic agents on alcohol intake and alcohol preference in Wistar rats selected at two different levels of alcohol preference. Alcohol Alcoholism 28(2):157–170
Melchior CL, Myers RD (1976) Genetic differences in ethanol drinking of the rat following injection of 6-OHDA, 5,6-DHT or 5,7-DHT into the cerebral ventricles. Pharmacol Biochem Behav 5:63–72
Melchior CL, Tabakoff B (1981) Modification of environmentally-cued tolerance to ethanol in mice. J Pharmacol Exp Ther 219:175–180
Middlemiss DN, Tricklebank MD (1992) Centrally active 5-HT receptor agonists and antagonists. Neurosci Biobehav Rev 16:75–82
Montgomery AMJ, Burton MJ (1986) Effects of peripheral 5-HT on consumption of flavoured solutions. Psychopharmacology 88:262–266
Montgomery AMJ, Willner P (1988) Fenfluramine disrupts the behavioural satiety sequence in rats. Psychopharmacology 94:397–401
Montgomery AMJ, Rose IC, Herberg LJ (1991) 5-HT_{1A} agonists and dopamine: the effects of 8-OH-DPAT and buspirone on brain stimulation reward. J Neural Trans 83:139–148
Montgomery AMJ, Rose IC, Herberg LJ (1993) The effect of a 5-HT_3 receptor antagonist, ondansetron, on brain stimulation reward, and its interaction with direct and indirect stimulants of central dopaminergic transmission. J Neural Trans 91:1–11
Morinan A (1987) Reduction in striatal 5-hydroxytryptamine turnover following chronic administration of ethanol to rats. Alcohol 22:56–60
Muller S, Britton RS, Seeman P (1980) The effects of long-term ethanol on brain receptors for dopamine, acetylcholine, serotonin and noradrenaline. Eur J Pharmacol 65:31–37

Murphy JM, McBride WJ, Lumeng L, Li T-K (1982) Regional brain levels of monoamines in alcohol-preferring and non-preferring lines of rats. Pharmacol Biochem Behav 16:145–149
Murphy JM, Waller MB, Gatto GJ, McBride WJ, Lumeng L, Li T-K (1985) Monoamine uptake inhibitors attenuate ethanol intake in alcohol-preferring (P) rats. Alcohol 2:349–352
Murphy JM, McBride WJ, Lumeny L, Li T-K (1987a) Effects of serotonergic agents on ethanol intake of the high alcohol drinking (HAD) line of rats. Alcohol Clin Exp Res 11:208
Murphy JM, McBride WJ, Lumeng L, Li T-K (1987b) Contents of monoamines in forebrain regions of alcohol-preferring (P) and non-preferring (NP) lines of rats. Pharmacol Biochem Behav 26:389–392
Murphy JM, Waller MB, Gatto GJ, McBride WJ, Lumeng L, Li T-K (1988) Effects of fluoxetine on the intragastric self-administration of ethanol in the alcohol preferring P-line of rats. Alcohol 5:283–286
Myers RD, Lankford MF (1993) Failure of the 5-HT_2 receptor antagonist, ritanserin, to alter preference for alcohol in drinking rats. Pharmacol Biochem Behav 45:233–237
Myers RD, Veale WL (1968) Alcohol preference in the rat: reduction following depletion of brain serotonin. Science 160:1469–1471
Neill JC, Cooper SJ (1988) Evidence for serotonergic modulation of sucrose sham-feeding in the gastric-fistulated rat. Physiol Behav 44:453–459
Neill JC, Bendotti C, Samanin R (1990) Studies on the role of 5-HT receptors in satiation and the effect of d-fenfluramine in the runway test. Eur J Pharmacol 190:105–112
Oakley NR, Jones BJ, Tyers MB, Costall B, Domeney AM (1988) The effect of GR38032F on alcohol consumption in the marmoset. Br J Pharmacol 95:870P
Parker LF, Radow BL (1976) Effects of parachlorophenylalanine on ethanol self-selection in the rat. Pharmacol Biochem Behav 4:535–540
Pei O, Zetterstrom T, Leslie RA, Grahame-Smith DG (1993) 5-HT_3 receptor antagonists inhibit morphine-induced stimulation of mesolimbic dopamine release and function in the rat. Eur J Pharmacol 230:63–68
Peltier R, Schenk S (1991) GR38032F, a serotonin 5-HT_3 antagonist, fails to alter cocaine self-administration in rats. Pharmacol Biochem Behav 39:133–136
Peroutka SJ (1988) 5-Hydroxytryptamine receptor subtypes: molecular, biochemical and physiological characterization. Trends Neurosci 11:496–500
Peroutka SJ (1993) 5-Hydroxytryptamine receptors. J Neurochemistry 60:408–416
Pohorecky LA, Newman B, Sun J, Baile WH (1978) Acute and chronic ethanol ingestion and serotonin metabolism in the rat brain. J Pharmacol Exp Ther 224:424–432
Poschel BPH, Ninteman FW (1971) Intracranial reward and the forebrain's serotonergic mechanism: studies employing parachlorophenylalanine and parachloroamphetamine. Physiol Behav 7:39–46
Rezvani AM, Overstreet DM, Janowski DS (1990) Fawn-hooded rats with serotonin deficiency have a high preference for alcohol: a novel strain of alcohol-preferring rats. Alcohol Clin Exp Res 14:331
Richardson JS, Novakovski DM (1978) Brain monoamines and free choice ethanol consumption in rats. Drug Alcohol Depend 3:253–264
Richardson NR, Roberts DCS (1991) Fluoxetine pretreatment reduces breaking points on a progressive ratio schedule reinforced by intravenous cocaine self-administration in the rat. Life Sci 49:833–840
Rockman GE, Amit Z, Brown ZW, Bourque C, Ogren SO (1982) An investigation of the mechanisms of action of 5-hydroxytryptamine in the suppression of ethanol intake. Neuropharmacology 21:341–347
Ronnback L, Zeuchner L, Rosengren L, Wronski A, Ogren SO (1984) Decreased morphine intake by opiate addicted rats administered zimelidine, a 5-HT uptake inhibitor. Psychopharmacology 82:30–35

Rowland NE, Morian KR (1992) Effect of dexfenfluramine on alcohol intake in alcohol-preferring P-rats. Alcohol 9:559–561
Schreiber R, Opitz K, Glaser T, DeVry J (1993) Involvement of pre-synaptic 5-HT_{1A} receptors in the ethanol preference reducing effects of ipsapirone (I) and 8-OH-DPAT in the rat. Alcohol Alcohol 28:228 (abstract C2.21)
Sellers EM, Romach MK, Frecker RC, Higgins GA (1991) Efficacy of the 5-HT_3 antagonist ondansetron in addictive disorders. Biol Psychiatry 2:894–897
Sellers EM, Higgins GA Sobell MB (1992) 5-HT and alcohol abuse. TIPS 13:69–75
Sharp T, Bramwell SR, Clark D, Graham-Smith DG (1989) In vivo measurement of extracellular 5-hydroxytryptamine in hippocampus of the anaesthetised rat using microdialysis: changes in relation to 5-hydroxytryptaminergic neuronal activity. J Neurochem 53:234–240
Sinclair JD, Kampovpolevoy A, Stewart R, Li T-K (1992) Taste preferences in rat lines selected for low and high alcohol consumption. Alcohol 9(2):155–160
Soubrie P (1986) Reconciling the role of central serotonin neurons in human and animal behavior. Behav Brain Sci 9:319–364
Speisky MB, Kalant H (1985) Site of interaction of serotonin and desglycinamide-arginine-vasopressin in maintenance of ethanol tolerance. Brain Res 326:281–290
Svensson L, Engel J, Hard E (1989) Effects of the 5-HT receptor agonist 8-OH DPAT on ethanol preference in the rat. Alcohol 6:17–21
Tabakoff B, Hoffman PL, Moses F (1977) Neurochemical correlates of ethanol withdrawal: alterations in serotonergic function. J Pharm Pharmacol 29:471–476
Tilson HA, Resch RH (1974) The effects of p-chlorophenylalanine on morphine analgesia, tolerance and dependence in two strains of rats. Psychopharmacologia (Berl) 35:45–60
Tomkins DM, Higgins GA, Sellers EM (1993) Low doses of the 5-HT_{1A} agonist, 8-hydroxy-2-(di-N-propylamino) tetralin (8-OH DPAT) increase ethanol intake. Psychopharmacology 115:173–179
Tork I (1985) Raphe nuclei and serotonin containing systems. In: Paxinos G (ed) Hindbrain and spinal cord. Academic, Sydney, pp 43–78 (The rat nervous system, vol 2)
Tricklebank MD, Forler C, Fozard JR (1984) The involvement of subtypes of the 5-HT_1 receptor and of catecholaminergic systems in the behavioural response to 8-hydroxy-2-(di-N-propylamino) tetralin in the rat. Eur J Pharmacol 106:271–282
Ulrichsen J (1991) Alterations in serotonin receptor subtypes in ethanol-dependent rats. Alcohol Alcohol 26(5/6):567–573
Vaccarino FJ, Schiff BB, Glickman SE (1989) Biological view of reinforcement. In: Klein SB, Mowrer RR (eds) Contemporary learning theories. Lawrence Erlbaum, New Jersey, pp 111–141
van Wijngaarden I, Tulp MTM, Soudijn W (1990) The concept of selectivity in 5-HT receptor research. Eur J Pharmacol (Mol Biol) 188:301–312
Waller MB, McBride WJ, Lumeng L, Li T-K (1982) Induction of dependence on ethanol by free-choice drinking in alcohol-preferring rats. Pharmacol Biochem Behav 16:501–507
Weingarten HP, Watson SD (1982) Sham feeding as a procedure for assessing the influence of diet palatability on food intake. Physiol Behav 28:401–407
Weiss F, Mitchiner M, Bloom FE, Koob GF (1990) Free-choice responding for ethanol versus water in alcohol preferring (P) and unselected Wistar rats is differentially modified by naloxone, bromocriptine, and methysergide. Psychopharmacology 101:178–186
Wilkinson LO, Dourish CT (1991) Serotonin and animal behaviour. In: Peroutka SJ (ed) Serotonin receptor subtypes: basic and clinical aspects. Wiley-Liss, New York, pp 147–210
Wogar MA, Bradshaw CM, Szabadi E (1993) Effects of lesions of the ascending 5-hydroxytryptaminergic pathways on choice between delayed reinforcers. Psychopharmacology 111:239–243

Wong DT, Reid LR, Threlkeld PG (1988) Suppression of food intake in rats by fluoxetine: comparison of enantiomers and effects of serotonin antagonists. Pharmacol Biochem Behav 31:475–479

Wong DT, Threlkeld PG, Lumeng L, Li T-K (1990) Higher density of serotonin-1A rteceptors in the hippocampus and cerebral cortex of alcohol-preferring P rats. Life Sci 46:231–235

Wood JM (1980) Effect of depletion of brain 5-hydroxytryptamine by 5,7-dihydroxytryptamine on ethanol tolerance and dependence in the rat. Psychopharmacology 67:67–72

Wozniak KM, Pert A, Linnoila M (1990) Antagonism of 5-HT_3 receptors attenuates the effects of ethanol on extracellular dopamine. Eur J Pharmacol 187:287–289

Wu PH, Liu JF, Lanca AJ, Grupp LA, Kalant H (1993) Involvement of central 5-HT_2 receptors in the maintenance of tolerance to ethanol by arginine-8-vasopressin. Abstr Soc Neurosci 19(2):1381

Yoshimoto K, Komura S (1987) Re-examination of the relationship between alcohol preference and brain monoamines in inbred strains of mice including senescence-accelerated mice. Pharmacol Biochem Behav 27:317–322

Zabik JE (1989) Use of serotonin-active drugs in alcohol preference studies. In: Galanter M (ed) Recent developments in alcoholism, vol 7. Elsevier, New York, pp 211–223

Zabik JE, Binkerd K, Roach JD (1985) Serotonin and ethanol aversion in the rat. In: Naranjo CA, Sellers EM (eds) Research advances in new psychopharmacological treatments for alcoholism. Elsevier, New York, pp 87–103

Zhou FC, Bledsoe S, Lumeng L, Li T-K (1991) Immunostained serotonergic fibers are decreased in selected brain regions of alcohol-preferring rats. Alcohol 8:425–431

CHAPTER 8

Opioid Mediation of Alcohol Self-Administration: Pre-Clinical Studies

J.R. Volpicelli, B.J. Berg, and N.T. Watson

A. Introduction

Once viewed as separate and distinct addictions, alcoholism and opiate dependence may share a common neurobiological mechanism of reinforcement. In the review that follows, we discuss recent advances in the understanding of alcohol reinforcement, the addictive process, and pharmacotherapy for addiction, as well as how these developments relate to one another. The utility of this approach is demonstrated by recent data on the use of the opiate antagonist naltrexone in the treatment of alcohol dependence.

There are five lines of investigation that demonstrate interactions between alcohol drinking and opioid system activity: (a) administration of alcohol enhances opiate receptor activity; (b) the effects of alcohol on opiate receptor activity are greater in gene lines associated with heavy alcohol drinking; (c) administration of exogenous opiates influences alcohol drinking; (d) following procedures which enhance endogenous opioidergic activity, such as experience with uncontrollable shock, alcohol drinking is increased; and (e) opiate antagonists reliably decrease alcohol drinking. This paper will review each of these lines of evidence and defend the hypothesis that at least some forms of alcohol abuse result from genetic or environmentally induced deficiencies in opioid system activity.

B. Alcohol Drinking Enhances Opioid System Activity

Alcohol drinking can enhance opiate receptor activity although the mechanism for this effect is not clear. Three mechanisms have been suggested: (1) opioid peptides are formed from the condensation products of alcohol metabolites and catecholamines, (2) alcohol can directly affect the sensitivity of opiate receptors, and (3) alcohol stimulates the release of endogenous opioids.

I. Opioid Peptides Are Produced in the Metabolism of Alcohol (TIQ Alkaloid Hypothesis)

In 1970, two independent laboratories showed that alcohol drinking led to the in vivo formation of salsolinol, a biologically active alkaloid with morphine-like effects (COHEN and COLLINS 1970; DAVIS and WALSH 1970). In fact, a variety of salsolinol-like compounds, collectively known as tetrahydroisoquinoline (TIQ) alkaloids, have now been shown to be formed in vivo in brain and other tissue as a result of alcohol drinking (MYERS 1990a). Salsolinol is formed in the brain as a condensation product of acetaldehyde (the initial oxidation product of ethanol) and dopamine. Like the other TIQ alkaloids, salsolinol is an opiate receptor agonist, and has morphine-like effects that are blocked by the opiate antagonist naloxone (FERTEL et al. 1980).

Salsolinol levels have been shown to increase in the limbic system of rats following either acute or chronic consumption of alcohol, or its metabolite acetaldehyde (MATSUBARA et al. 1987; MYERS et al. 1985). COLLINS et al. (1990) found TIQ concentrations to be elevated in the hypothalamus of rats following chronic exposure (3 weeks) to alcohol, although TIQs were not elevated in other brain regions nor were they elevated at 23-week follow-up. The failure to observe increased levels of salsolinol throughout the limbic system is probably due to the elimination of dietary salsolinol in the COLLINS et al. (1990) study. Human studies have found salsolinol to be elevated both in dopamine-rich regions of the brain and in the CSF of alcoholic individuals (SJOQVIST et al. 1981, 1982). However, differences in dietary salsolinol may also have affected the human studies.

II. Alcohol Drinking Leads to Changes in Opioid Receptor Sensitivity

A second mechanism by which alcohol can affect the central opioid system is by directly affecting the sensitivity of opiate receptors. In general, alcohol acts as a CNS depressant which diminishes the firing rates of CNS neurons. However, this simple description of alcohol's effects obscures both the complex interactions of alcohol with various neuronal components and the overall complexity of the CNS.

Since alcohol is known to disrupt the lipid matrix of neuronal cell membranes, changes in membrane conformation may be one of the mechanisms by which alcohol affects receptors (including opiate receptors) located on these membranes. Ethanol is a simple aliphatic compound which, despite its short chain length, is lipid soluble. Aliphatic alcohols such as ethanol act to increase the fluidity of cell membranes, causing changes in membrane permeability that can disrupt the function of membrane-bound proteins, including membrane-bound receptors. Because all biomembranes are affected, alcohol can be expected to have effects on multiple neurotransmitter (or neuromodulator) receptors and hence can affect multiple neuronal systems. In fact, alcohol has been shown to affect opiate, GABA, glutamate-NMDA, dopamine, serotonin, and norepinephrine neurotransmitter systems (US Department of Health 1990).

HOFFMAN ET AL. (1982; TABAKOFF and HOFFMAN 1987) demonstrated that alcohol affects different receptor subtypes preferentially, based upon the sensitivity of a particular membrane-bound receptor complex to changes in its lipid environment. In this way, alcohol can exert selective effects on specific receptors and neuronal systems even though it is not a ligand for any specific receptor. For example, alcohol was shown to inhibit binding of opiates to δ-receptors far more than to μ-receptors (HILLER et al. 1981).

Alcohol's effects on opiate receptor sensitivity are dose dependent. With respect to μ-receptors, physiologically obtainable acute doses of alcohol (50 mM) enhance μ-receptor binding in vitro (TABAKOFF and HOFFMAN 1983). However, concentrations greater than 50 mM decrease μ-receptor affinity (TABAKOFF and HOFFMAN 1983). The increased affinity of μ-receptors following low, acute doses of alcohol may contribute to alcohol's initial reinforcing effect. With respect to δ-receptors, alcohol produces a simple dose-response inhibition of binding to the receptors (CHARNESS et al. 1983; TABAKOFF and HOFFMAN 1983). Alcohol's effects on receptor sensitivity are also dependent on the duration of alcohol administration. Chronic alcohol exposure was shown to decrease μ opiate receptor function in living mice (HOFFMAN et al. 1982). Thus, long-term alcohol use may cause a decrease in the sensitivity of receptors for both exogenous opiate agonists and possibly the brain's own endogenous peptides.

III. Alcohol Drinking Leads to the Release of Endogenous Opioids

Alcohol drinking may stimulate the endogenous opioid system by triggering the release of endogenous opioids such as β-endorphins and enkephalins. Rats given free access to a 10% alcohol solution for 8 days had a 150% increase in plasma β-endorphin levels and a 52% increase in whole brain β-endorphin levels (ESKELSON et al. 1980). Acute alcohol administration (2.5 mg/kg) increased β-endorphin levels in specific hypothalamic areas, although β-endorphin levels were reduced relative to controls following chronic alcohol treatment (SCHULTZ et al. 1980). In order to fully understand these findings, however, one must realize that a β-endorphin level is the net result of three processes: endorphin production, release, and catabolism. Changes in any of these processes will affect endorphin levels and could influence the response obtained from drinking alcohol.

C. Genetic Susceptibility to Alcohol Abuse

Studies using animal strains selectively bred to self-administer alcohol show significant differences in baseline endogenous opioid activity (CRABBE et al. 1981; FROEHLICH and LI 1990; GIANOULAKIS et al. 1992; GIANOULAKIS and GUPTA 1986). These differences in endorphin activity, presumed to be genetically based, may mediate the increased drinking seen in these strains. Of particular interest are findings that alcohol-preferring strains show increased endorphin response to alcohol. For example, β-endorphin levels increase in the hypothalami of alcohol-preferring mice after in vitro administration of alcohol. This β-endorphin response to alcohol is greater in alcohol-preferring mice (C57BL/6J) than nonprefering mice (DBA/2) (DE WAELE et al. 1992). This effect was also demonstrated in whole brain extracts of alcohol-preferring mice (BLUM and BRIGGS 1988).

Preclinical human studies also support a link between opioid systems and genetic susceptibility to alcohol abuse. GIANOULAKIS et al. (1990) showed that baseline levels of plasma β-endorphin were significantly lower in a group of abstinent alcoholics and in nonalcoholic individuals who had a strong positive family history for alcoholism, in comparison to nonalcoholic individuals who had a strong negative family history for alcoholism. GIANOULAKIS et al. (1990) also demonstrated that subjects with a positive family history had a marked increase (nearly twofold) in their plasma β-endorphin levels after drinking alcohol, while the family-history-negative subjects actually exhibited a slight decrease in β-endorphin levels. Equally intriguing was the finding that plasma β-endorphin levels in family-history-positive subjects rose after drinking to approach those of family-history-negative subjects.

These findings are provocative in that they suggest that alcohol may be consumed to correct or "self-medicate" an abnormally low level of endogenous opioids. The animal studies in particular provide strong support for opioid involvement in alcohol preference. Since alcohol-preferring strains are selected purely on the basis of a behavioral preference for alcohol, the strain differences in baseline endorphin activity and response to alcohol suggest that endorphins are partially responsible for increased alcohol preference. Although the findings of GIANOULAKIS et al. (1990) need to be replicated in future research and the relevance of peripheral levels of β-endorphin to central levels is uncertain, the human data provide additional support for the hypothesis that alcohol consumption enhances opioid system activity.

The hypothesis that alcoholics may drink to correct a deficiency in opioid activity is further supported by the finding that detoxifying alcoholics have low levels of central endorphins. GENAZZANI et al. (1982) found significantly depressed CSF levels of β-endorphin in detoxifying alcoholics (3–5 days after cessation of drinking) compared to normal volunteers.

D. Modulating Opiate Receptor Activity Influences Alcohol Drinking

Alcohol self-selection is influenced by the administration of exogenous opioids. In general, moderate-to-high doses of morphine suppress alcohol consumption, while withdrawal from opioids is associated with increased alcohol drinking (HO et al. 1976; SINCLAIR 1974; VOLPICELLI et al. 1991). However, low doses of morphine can enhance alcohol drinking, particularly in organisms that are not initially alcohol preferring (Reid and Hunter 1984).

I. Low Doses of Opioids Increase Alcohol Drinking

Reid and associates have extensively tested the small-dose morphine effect in rats. They find that the injections of low doses of morphine (<2.5 mg/kg) act to increase alcohol drinking (Hubbell et al. 1987; Reid et al. 1991; Reid and Hunter 1984). Reid and Hunter (1984) utilized an unusual procedure designed to maximize alcohol drinking. Typically, animals were fluid deprived for 22 h and then given a time-limited opportunity to drink unsweetened water or a sucrose-sweetened solution of alcohol. After 3 weeks of adaption to this procedure, morphine was injected into the rats 30 min prior to access to water and alcohol. Small doses of morphine (less than 2.5 mg/kg) increased absolute levels of alcohol drinking and decreased water consumption. Total fluid consumption remained relatively constant.

This small-dose morphine effect was consistent over a variety of procedural manipulations. For example, the small-dose morphine effect was found to increase alcohol intake over a dose range of 1.0–2.5 mg/kg (Hubbell et al. 1986). Sweetened alcohol was not essential since low doses of morphine also increased selection of unsweetened alcohol, though to a much lesser degree (Hubbell et al. 1986, 1987; Wild et al. 1988). The effect was consistent in both male and female animals (Czirr et al. 1987), and regardless of whether testing was conducted during the light or dark phase of the diurnal cycle (Hubbell et al. 1986). Low doses of other opioids such as methadone (Mudar et al. 1986) and fentanyl (Czirr et al. 1987) have also been reported to increase alcohol drinking.

It is interesting that effects similar to low-dose morphine administration have also been obtained with intracerebroventricular (ICV) infusions of TIQ alkaloids (Myers and Melchior 1977a; Myers and Privette 1989). Both small, acute doses and chronic infusions of TIQ alkaloids evoked excessive alcohol drinking in alcohol nonpreferring rats when a choice of alcohol or water was given (Clow et al. 1983; Myers et al. 1982; Myers and Oblinger 1977). The significance of this increased alcohol drinking was confirmed by the fact that rats drank pharmacologically significant blood alcohol levels.

The brain areas responsible for the low-dose opioid effect have been mapped in vivo using tetrahydropapaveroline (THP), another type of TIQ. When THP is microinjected into certain brain areas, a low-dose enhancement of alcohol drinking can be observed (Duncan and Fernando 1991; Myers 1990b; Myers and Privette 1989; Privette and Myers 1989). These mapping studies reveal a distinct "circuitry" of neuronal structures which seem to mediate abnormal intake of alcohol. This "circuit" extends from the ventral tegmental area of the midbrain rostrally to structures within the limbic forebrain and medial prefrontal cortex. The hypothesized anatomical "circuit" encompasses both dopaminergic and enkephalinergic pathways and has a striking pattern of concordance with the mesolimbic system. The mesolimbic system is the hypothesized site of the drug reward system, i.e., the medial forebrain bundle and associated loci.

II. Moderate to High Doses of Opioids Suppress Alcohol Drinking

Several studies have shown that moderate to high doses of opioids decrease alcohol drinking. For example, in a study by VOLPICELLI et al. (1991), rats given free access to food, water, and 5% alcohol were randomly assigned to receive subcutaneous injections of saline or one of two different dose levels of morphine sulfate. In comparison to the saline-treated rats, rats treated with high doses of morphine (10 mg/kg) daily for 8 days had a 60% reduction in alcohol self-administration during the last 4 days of morphine administration. Rats treated with lower doses (2.5 mg/kg) had only a 20% reduction in alcohol consumption. These effects were relatively specific, in that the doses of morphine used by VOLPICELLI et al. (1991) had only modest, nonsignificant effects in reducing both total fluid consumption and body weight relative to the saline-injected rats.

A similar effect has been observed following a single injection of morphine (60 mg/kg) in rats with 20 weeks experience with alcohol drinking. Following the morphine injection, alcohol drinking was suppressed while water consumption increased (SINCLAIR et al. 1973). In rats with either 1 day or 32 days previous experience with drinking, a single 30-mg/kg morphine injection suppressed alcohol drinking for 8 days (SINCLAIR 1974). The effect of morphine injections was specific to alcohol drinking since water consumption was not affected.

Ho et al. (1976) used morphine injections of 10, 30, and 60 mg/kg and derived a similar dose-response curve that also demonstrated an inverse relationship between alcohol preference and opiate receptor activity (Ho et al. 1976). A single subcutaneous injection of morphine produced a dose-dependent decrease in alcohol drinking in both rats and mice, with no effect on water consumption. In a separate experiment on the chronic effects of morphine, rats received daily morphine injections on an incremental dose regimen (from 10 mg/kg up to 200 mg/kg twice daily) for 2 weeks. When morphine injections were stopped, alcohol consumption was found to be increased above baseline levels and relative to consumption levels in saline-treated control rats.

Administration of endogenous opioids can also suppress alcohol drinking. Administration of the endogenous opioid met-enkephalin has been shown to reduce alcohol intake. For instance, ICV injections of 200 μg met-enkephalin reduced consumption of a 5% alcohol solution by rats (Ho and ROSSI 1982). However, the use of enkephalinase inhibitors to artificially increase endogenous opioid levels produced conflicting results. While BLUM et al. (1987) found that the enkephalinase inhibitor *d*-phenylalanine reduced alcohol intake in mice, Froehlich and colleagues (FROEHLICH and LI 1990;

FROEHLICH et al. 1991) found that a different enkephalinase inhibitor (thiorphan) increased alcohol consumption in rats.

III. Alcohol Drinking Increases During Opiate Withdrawal

During the postinjection or opiate withdrawal phase, when opioid system activity can be considered to be deficient, rats show increased alcohol preference relative to saline-treated controls (HO et al. 1976; VOLPICELLI et al. 1991). VOLPICELLI et al. (1991) found that morphine-treated rats showed an increase in alcohol drinking when morphine injections were discontinued. For example, rats in the low-dose morphine group drank nearly twice as much alcohol as the saline-injected subjects during the postinjection phase.

Similarly, when rats were retested 1–6 months following intraventricular infusions of TIQs, it was found that the preference for alcohol induced by TIQ alkaloids persisted. Because TIQ alkaloids are rapidly degraded in the brain, such a long-term effect on alcohol preference suggests that TIQs can produce a lasting effect on neuronal processes (DUNCAN and DEITRICH 1980; HUTTUNEN and MYERS 1987; MYERS and MELCHIOR 1977a,b; MYERS and PRIVETTE 1989).

Results from human studies that have considered the effects of methadone administration on alcohol drinking are consistent with the animal data (SIEGEL 1986). Opiate addicts, maintained on a regular, stable dose of the opiate agonist methadone, generally show a decrease in their levels of alcohol consumption. However, as predicted, opiate abusers typically increase their alcohol use during opiate withdrawal or when a steady, stable opiate supply is not available.

E. Uncontrollable Stress and Alcohol Abuse

If an important motivation for alcohol drinking was its tension-reducing properties, it would likely follow that stress would result in increased alcohol consumption. However, animal studies have generally failed to demonstrate a consistent effect of stress on alcohol drinking (see CAPPELL and HERMAN 1972).

Exposure to uncontrollable stress increases endogenous opiate receptor activity. For example, in animal studies, intermittent electric footshock produces a period of opioid-mediated, stress-induced analgesia, which is reversible by opioid antagonists (CHESHER and CHAN 1977; GRAU 1984; HYSON et al. 1982). The fact that this transient analgesia is blocked by opioid antagonists suggest that the analgesia is the result of stress-induced release of endogenous opioids such as β-endorphin. Given the ability of exogenous opiate administration to affect alcohol drinking, we would expect perturbations in endogenous opioid activity to also affect alcohol drinking.

I. Post-Stress Alcohol Drinking

The lack of agreement concerning the effects of stress on alcohol drinking can be attributed in part to the need to distinguish between drinking that occurs during stress and that which occurs after stress. Studies of alcohol drinking *during* stress show inconsistent effects, while studies of drinking *following* stress typically show a large increase in alcohol intake. For example, VOLPICELLI et al. (1982) found that rats exposed to 4 days of inescapable shock did not show an increase in alcohol consumption until after the shock was discontinued.

VOLPICELLI et al. (1990) found that the increase in alcohol preference following stress was independent of initial preference for alcohol. However, alcohol preference during uncontrollable stress was found to depend upon initial alcohol preference. During the shock phase, rats with high initial alcohol preference decreased alcohol preference, while alcohol nonpreferring rats increased preference for alcohol.

Post-stress drinking has also been observed in animals exposed to stressors other than electric footshock. For example, rats given a choice between water and a sweetened 10% alcohol solution were found to have markedly increased alcohol preference in the 3 weeks following isolation stress or immobilization stress, but not during periods of stress (NASH and MAICKEL 1985).

A study of the temporal patterns of drinking in college students shows that the post-stress drinking research applies to humans as well (ORCUTT and HARVEY 1991). Consideration of the times when humans are most likely to drink – at the end of the work day or work week, on vacation and during times of leisure, and after a crisis or important life event – supports the notion that individuals increase alcohol drinking upon relief from a stress and not during the actual stress. This "weekends are for Michelob" effect seems to have been noted by people who are not typically thought of as scientific researchers.

F. Opiate Antagonists Attenuate Alcohol Drinking

It has been observed that some of the behavioral effects of alcohol could be blocked by opiate antagonists. For example, naloxone reduced the increase in motor activity elicited by alcohol (MIDDAUGH et al. 1978). Opiate antagonists also antagonize other pharmacological effects of alcohol. Naloxone increased the LD_{50} of alcohol approximately twofold (Ho and Ho 1979). These observations led to the notion that opiate antagonists might attenuate alcohol preference.

Studies of the effects of opiate antagonists on alcohol drinking in animals have generally shown that opiate antagonists reliably decrease alcohol preference across a range of conditions (ALTSHULER et al. 1980; HUBBELL et al. 1986; MARFAING-JALLET et al. 1983; MYERS et al. 1986;

SAMSON and DOYLE 1985; VOLPICELLI et al. 1986). Monkeys bar pressing for intravenous alcohol decreased their bar press responses by 50% when they were administered naltrexone (20 mg/kg) for 15 days. Initially, naltrexone injections led to an increase in responding for alcohol but over the subsequent 10 days alcohol drinking gradually decreased relative to baseline (ALTSHULER et al. 1980).

In a similar experiment using rodents bar pressing for either water or a 5% alcohol solution, naloxone (20 mg/kg) also decreased responding for alcohol (SAMSON and DOYLE 1985). This effect was observed at the highest dose of naloxone employed (20 mg/kg), but not at lower doses. Since high doses of opiate antagonists can reduce many appetitive behaviors (not just alcohol drinking), it is important to control for the nonspecific appetite-reducing effects of opiate antagonists. No change in water responding was found.

The effect of opioid antagonists on alcohol drinking has been shown in less traditional paradigms as well. A metabolite of alcohol, acetaldehyde, can also reinforce bar pressing in rats. This effect was also blocked by the administration of an opioid antagonist. After a period of stable responding for acetaldehyde, three closely spaced (30 min apart) injections of naloxone at 10 mg/kg decreased bar pressing for i.v. injections of acetaldehyde. Smaller doses of naloxone did not have this effect (MYERS et al. 1984).

Alcohol drinking initiated by administration of small doses of opioids is also reduced by administration of opiate antagonists. For example, the increased alcohol preference induced by administration of TIQ alkaloids can also be suppressed by opiate antagonists (CRITCHER et al. 1983; MYERS and CRITCHER 1982). This is further evidence that the increase in alcohol drinking induced by TIQ alkaloids is mediated by a mechanism involving opiate receptors.

Opiate antagonists also block the increase in alcohol drinking observed following uncontrollable stress. While post-shock alcohol preference increased (as predicted) in placebo-treated rats, naltrexone-treated rats did not show an increase in alcohot consumption following exposure to inescapable shock (VOLPICELLI et al. 1986).

The animal data previously described have led to studies on the effect of naltrexone on alcohol drinking by alcohol-dependent humans. If alcohol use leads to increased opioid system activity and increased opioid activity reinforces alcohol drinking, then opiate antagonists such as naltrexone should reduce alcohol drinking. Two recent studies have shown that naltrexone is effective in reducing alcohol relapse in detoxified alcohol-dependent subjects. These double-blind, placebo-controlled studies compared naltrexone 50 mg daily to placebo for 12 weeks as an adjunct to outpatient alcohol rehabilitation (O'MALLEY et al. 1992; VOLPICELLI et al. 1992).

VOLPICELLI et al. (1992), using a sample of 70 detoxified alcoholics, found that daily administration of 50 mg naltrexone significantly reduced the number of days on which alcohol was consumed. Craving for alcohol, as measured by self-report, was significantly lower in natrexone-treated subjects than placebo controls. Relapse rates were also significantly lower for the naltrexone group. Finally, among subjects that had a "slip" from abstinence, nearly all placebo-treated subjects continued drinking, such that they met criteria for relapse, while only half of the naltrexone-treated samplers relapsed. O'MALLEY et al. (1992) reported similar findings in a different sample of 97 alcoholics. Naltrexone-treated subjects had significantly fewer drinking days than placebo subjects. Naltrexone subjects also had significantly lower relapse rates, and a lower total number of drinks during the study. At termination, naltrexone subjects had significantly lower clinician-rated Addiction Severity Index (ASI) scores. The reduction in alcohol use and relapse rates found in alcoholics treated with the opioid antagonist naltrexone can only be understood in the context of an opioid system-based theory of alcohol reinforcement.

I. Why Do Organisms Abuse Alcohol?

If alcohol is reinforcing because of its effects on the opioid system activity, how might this mechanism explain excessive alcohol drinking? One theory suggested by several researchers (GIANOULAKIS et al. 1990; TRACHTENBERG and BLUM 1987; VOLPICELLI 1987) is that alcohol drinking is motivated by deficiencies in opiate receptor activity. This opioid compensation hypothesis proposes that deficiencies in opiate receptor activity can be the result of genetic deficiencies in opioidergic activity, or environmentally induced deficiencies following excessive opiate receptor activity. Such relative deficiencies might be produced by prolonged exposure to exogenous opioids, or by the transient elevation in endogenous opioids resulting from experience with uncontrollable stress. Alcohol drinking may temporarily compensate for these deficiencies, but chronic drinking would be expected to lead to compensatory changes. Finally, the withdrawal of alcohol itself might also lead to a deficiency in endogenous opioids, thereby setting the occasion for resumed abusive drinking (see VOLPICELLI 1987).

It would appear that the opioid compensation hypothesis cannot account for the finding that low doses of opioids increase alcohol drinking. REID et al. (1991) have argued that the propensity to drink alcohol increases when a "surfeit" of opiate receptor activity exists. The low-dose morphine effect can account for the difficulty alcoholics have in trying to stop once drinking is begun. If alcohol drinking enhances opiate receptor activity and enhanced opiate receptor activity increases the motivation to drink alcohol, then a vicious cycle is created when organisms drink alcohol.

An analogy to eating behavior illustrates that it is possible for both models to be correct. Just as eating can be motivated by hunger (food deficiency), alcohol drinking may be motivated by opiate receptor deficiencies. Yet, eating can also be stimulated by a small appetizer and, similarly, a small dose of morphine can "prime" alcohol drinking. The biochemical basis of this priming effect may involve other neurotransmitters. For example, the serotonin reuptake inhibitor fluoxetine was found to block the enhanced alcohol drinking elicited by low doses of morphine (HUBBELL et al. 1991).

Opioid mechanisms may mediate reinforcement of alcohol drinking, but alcohol's sedative effects seem due to interactions with the benzodiazepine receptor complex (the GABAergic system). A recent meta-analysis has supported the hypothesis that subjects with a positive family history for alcoholism show decreased sensitivity to the intoxication produced by alcohol (POLLOCK 1992). SCHUCKIT (1985) found that subjects with a positive family history for alcoholism show decreased body sway after alcohol drinking. He proposed that tolerance to the alcohol's intoxicating effects may be a genetic marker for alcoholism. In other words, alcohol may be addicting to persons who are less sedated by alcohol, which suggests that they are relatively insensitive to alcohol's effects on GABA. While GABAergic mechanisms for alcohol abuse have considerable experimental support (see the chapter by Ticku and Mehta, this volume), the research reviewed above shows that the opioid system must considered as well in a comprehensive view of the etiology of alcoholism.

A synthesis of the GABA and opioid system approaches to alcoholism might help to explain how the process of addiction develops. One theoretical possibility is that risk for alcoholism is the sum of sensitivity to the opioid-mediated reinforcing effects and tolerance to the GABA-mediated sedative effect. Thus, persons who are less sedated by alcohol may still be at relatively little risk if their sensitivity to alcohol's rewarding effects is low. Similarly, high sensitivity to both GABA and opioid effects may result in low risk because a drinker might fall asleep or lose consciousness before a substantial quantity of alcohol was consumed. At greatest risk are those patients who are not sedated, but are highly reinforced by alcohol.

Other neurotransmitter systems involved in the etiology of alcohol abuse are serotonin and dopamine. Several converging lines of evidence show the importance of serotonin in alcohol use: alcohol-preferring strains of rat show abnormally low serotonin levels, alcohol may stimulate serotonin release, and serotonin reuptake inhibitors depress alcohol drinking in humans (LITTEN and ALLEN 1993; see also the chapter by Sellers, this volume). The parallels between the serotonin evidence and the opioid system data reviewed above suggest that these two systems may work additively or synergistically to reinforce and maintain alcohol drinking.

Catecholamine mechanisms for alcohol drinking have also received some experimental support. Although rats bred for alcohol preference show depressed dopamine levels in some brain areas, and alcohol administration stimulates dopamine release from the nucleus accumbens, dopamine agonists and antagonists have generally failed to produce effects specific to alcohol drinking (LITTEN and ALLEN 1993; see also the chapter by Samson, this volume). Collectively, these data show that future research on alcoholism must continue to closely examine the action of alcohol at multiple receptor systems. Future research may reveal the particular neuronal system imbalances and appropriate pharmacotherapies for different alcoholism subtypes.

G. Summary

Taken in their entirety, the data presented above provide a compelling argument for the importance of alcohol-opioid interactions in alcoholism. Alcohol drinking can produce an increase in opiate receptor activity. This pharmacological effect is specific to opioids, and is localized to brain regions thought to be involved in the reinforcing effects of drugs. As predicted, alcohol exerts more pronounced pharmacological effects on opiate receptor activity in organisms with high alcohol preference.

Opiate agonists also have specific pharmacological effects on alcohol preference. Low doses of opioids increase alcohol preference even in alcohol nonpreferring animals, while moderate to high doses of opioids decrease alcohol preference. The effect is specific for alcohol preference, i.e., it occurs at doses that do not significantly affect food or water preference. Upon withdrawal of exogenous opioids following chronic administration, organisms show a prolonged preference for alcohol.

Uncontrollable stress, which is known to affect the activity of the endogenous opioid system, also influences alcohol drinking. Similar to the effect of small doses of exogenous opioids, mild stress increases alcohol preference in rats with low alcohol preference, but decreases preference in alcohol-preferring rats. Regardless of initial alcohol preference, organisms show a reliable increase in alcohol preference upon relief from stress.

Opiate antagonists reliably decrease excessive alcohol drinking, regardless of whether the excessive drinking is the result of genetic factors, follows exposure to exogenous opioids, or is related to perturbations in endogenous opioids. This effect is relatively specific, in that doses of opiate antagonists that reduce alcohol preference do not affect consumption of other appetitive reinforcers such as food or water. Though opiate agonists and antagonists have opposite effects on alcohol consumption, they show comparable specificity in these effects.

While we have focused our attention in this chapter almost exclusively on alcohol-opioid interactions, clearly alcohol has a variety of other pharmacological effects. Though it is simplest to study a single neurotransmitter system in isolation, it is quite likely that alcohol's reinforcing effects are mediated by several neurotransmitters. Nevertheless, the work presented above demonstrates that the opioid system is important in mediating the reinforcing effects of alcohol drinking.

Acknowledgment. Preparation of this article was supported in part by National Institute on Drug Abuse research center grant DA05186, National Institute on Alcohol Abuse and Alcoholism grant AA07517, and the Penn Veterans Affairs Addiction Research Center, Philadelphia, PA.

References

Altshuler HL, Phillips PE, Feinhandler DA (1980) Alterations of ethanol self-administration by naltrexone. Life Sci 26:679–688

Blum K, Briggs AH (1988) Opioid peptides and genotypic responses to ethanol. Biogenic Amines 5:527–533

Cappell H, Herman CP (1972) Alcohol and tension reduction: a review. Q J Stud Alcohol 33:33–64

Charness ME, Gordon AS, Diamond I (1983) Ethanol modulation of opiate receptors in cultured neural cells. Science 222:1246–1248

Chesher GB, Chan B (1977) Footshock induced analgesia in mice: its reversal by naloxone and cross tolerance with morphine. Life Sci 21:1569–1574

Clow A, Stolerman IP, Murray RM, Sandler M (1983) Ethanol preference in rats: increased consumption after intraventricular administration of tetrahydropapaveroline. Neuropharmacology 22:563–565

Cohen G, Collins MA (1970) Alkaloids from catecholamines in adrenal tissue: possible role in alcoholism. Science 167:1749–1751

Collins MA, Ung-Chhun N, Cheng BY, Pronger D (1990) Brain and plasma tetrahydroisoquinolines in rats: effects of chronic ethanol intake and diet. J Neurochem 55:1507–1514

Crabbe JC, Allen RG, Gaudette D, Young ER, Kosobud A, Stack J (1981) Strain differences in pituitary beta-endorphin and ACTH content in inbred mice. Brain Res 219:219–223

Critcher EC, Lin CI, Patel J, Myers RD (1983) Attenuation of alcohol drinking in tetrahydroisoquinoline-treated rats by morphine and naltrexone. Pharmacol Biochem Behav 18:225–229

Czirr SA, Hubbell CL, Milano WC, Frank JM, Reid LD (1987) Selected opioids modify intake of sweetened ethanol solution among female rats. Alcohol 4:157–160

Davis VE, Walsh MD (1970) Alcohol, amines and alkaloids: a possible basis for alcohol addiction. Science 167:1005–1007

De Waele JP, Papachristou DN, Gianoulakis C (1992) The alcohol-preferring C57BL/6 mice present an enhanced sensitivity of the hypothalamic b-endorphin system to ethanol than the alcohol-avoiding DBA/2 mice. *J Pharmacol Exp Ther* 261:778–791

Duncan C, Deitrich RA (1980) A critical evaluation of tetrahydroisoquinoline induced ethanol preference in rats. Pharmacol Biochem Behav 13:265–281

Duncan C, Fernando PW (1991) Effects of tetrahydropapaveroline in the nucleus accumbens and ventral tegmental area on ethanol preference in the rat. Alcohol 8:87–90

Eskelson CD, Hameroff SR, Kanel JS (1980) Ethanol increases serum beta-endorphin levels in rats. Anesth Analg 59:537–538

Fertel RH, Greenwald JE, Schwar R, Wong L, Bianchine J (1980) Opiate receptor binding and analgesic effects of the tetrahydroisoquinolines salsolinol and tetrahydropapaveroline. Res Commun Chem Pathol Pharmacol 27:3–16

Froehlich JC, Li TK (1990) Enkephalinergic involvement in voluntary drinking of alcohol. In: Reid LE (ed) Opioids, bulimia, and alcohol abuse and alcoholism. Springer, Berlin Heidelberg New York, pp 217–228

Froehlich JC, Zweifel M, Harts J, Lumeng L, Li T-K (1991) Importance of delta opioid receptors in maintaining high alcohol drinking. Psychopharmacology (Berl) 103:467–472

Genazzani AR, Nappi G, Facchinetti F, Mazzella GL, Parrini D, Sinforiani E, Petraglia, F, Savoldi F (1982) Central deficiency of beta-endorphin in alcohol addicts. *J Clin Endocrinol Metab* 55:583–586

Gianoulakis C, Gupta A (1986) Inbred strains of mice with variable sensitivity to ethanol exhibit differences in the content and processing of beta-endorphin. Life Sci 39:2315–2325

Gianoulakis C, Angelogianni P, Meaney M, Thavundayil J, Tawar V (1990) Endorphins in individuals with high and low risk for development of alcoholism. In: Reid LD (ed) Opioids, bulimia, and alcohol abuse and alcoholism. Springer, Berlin Heidelberg New York, pp 229–246

Gianoulakis C, de Waele JP, Kiianmaa K (1992) Differences in the brain and pituitary beta-endorphin system between the alcohol-preferring AA and alcohol-avoiding ANA rats. Alcohol Clin Exp Res 16:453–459

Grau JW (1984) Influence of naloxone on shock-induced freezing and analgesia. Behav Neurosci 98:278–292

Hiller JM, Angel LM, Simon EJ (1981) Multiple opiate receptors: alcohol selectively inhibits binding to delta receptors. Science 214:468–469

Ho AKS, Ho CC (1979) Toxic interactions of ethanol with other central depressants: antagonism by naloxone to narcosis and lethality. Pharmacol Biochem Behav 11:111–114

Ho AKS, Rossi N (1982) Suppression of ethanol consumption by MET-enkephalin in rats. J Pharm Pharmacol 34:118–119

Ho AKS, Chen RCA, Morrison JM (1976) Interactions of narcotics, narcotic antagonists, and ethanol during acute, chronic, and withdrawal states. Ann NY Acad Sci 281:297–310

Hoffman PL, Urwyler S, Tabakoff B (1982) Alterations in opiate receptor function after chronic ethanol exposure. J Pharmacol Exp Ther 222:182–189

Hubbell CL, Czirr SA, Hunter GA, Beaman CM, LeCann NC, Reid LD (1986) Consumption of ethanol is potentiated by morphine and attenuated by naloxone persistently across repeated daily administrations. Alcohol 3:39–54

Hubbell CL, Czirr SA, Reid LD (1987) Persistence and specificity of small doses of morphine on intake of alcoholic beverages. Alcohol 4:149–156

Hubbell CL, Marglin SH, Spitalnic SJ, Abelson ML (1991) Opioidergic, serotonergic, and dopaminergic manipulations and rats' intake of a sweetened alcohol beverage. Alcohol 8:355–367

Huttunen P, Myers RD (1987) Anatomical localization in hippocampus of tetrahydro-beta-carboline-induced alcohol drinking in the rat. Alcohol 4:181–187

Hyson RL, Ashcraft LH, Drugan RC, Grau JW, Maier SF (1982) Extent and control of shock affects naltrexone sensitivity of stress-induced analgesia and reactivity to morphine. Pharmacol Biochem Behav 17:1019–1025

Litten RZ, Allen JP (1993) Reducing desire to drink: neurobiology, pharmacologic agents, and future research directions. In: Galanter M (ed) Recent developments in alcoholism: ten years of progress. Plenum, New York

Marfaing-Jallet P, Miceli D, Le Magnen J (1983) Decrease in ethanol consumption by naloxone in naive and dependent rats. Pharmacol Biochem Behav 18: 537–539

Matsubara K, Fukushima S, Fukui Y (1987) A systematic regional study of brain salsolinol levels during and immediately following chronic ethanol ingestion in rats. Brain Res 413:336–343

Middaugh LD, Read E, Boggan WO (1978) Effects of naloxone on ethanol induced alterations of locomotor activity in C57BL/6 mice. Pharmacol Biochem Behav 11:157–160

Mudar PJ, Le Cann NC, Czirr SA, Hubbell CL, Reid LD (1986) Methadone, pentobarbitol, pimozide and ethanol-intake. Alcohol 3:303–308

Myers RD (1990a) Anatomical "circuitry" in the brain mediating alcohol drinking revealed by THP-reactive sites in the limbic system. Alcohol 7:449–459

Myers RD (1990b) Neurobiological basis of alcohol reinforcement. In: Eng C (ed) Controversy in the addiction field. Hendall/Hunt, Dubuque, pp 25–40
Myers RD, Critcher EC (1982) Naloxone alters alcohol drinking induced in the rat by tetrahydropapaveroline (THP) infused ICV. Pharmacol Biochem Behav 16:827–836
Myers RD, Melchior CL (1977a) Alcohol drinking: abnormal intake caused by tetrahydropapaveroline in the brain. Science 196:554–556
Myers RD, Melchior CL (1977b) Differential actions on voluntary alcohol intake of tetrahydroisoquinolines or a beta-carboline infused chronically in the ventricle of the rat. Pharmacol Biochem Behav 7:381–392
Myers RD, Oblinger MM (1977) Alcohol drinking in the rat induced by acute intracerebral infusion of two tetrahydroisoquinolines and a beta-carboline. Drug Alcohol Depend 2:469–483
Myers RD, Privette TH (1989) A neuroanatomical substrate for alcohol drinking: identification of tetrahydropapaveroline (THP)-reactive sites in the rat brain. Brain Res Bull 22:899–911
Myers RD, McCaleb ML, Ruwe WD (1982) Alcohol drinking induced in the monkey by tetrahydropapaveroline (THP) infused into the cerebral ventricle. Pharmacol Biochem Behav 16:827–836
Myers WD, Ng KT, Singer G (1984) Ethanol preference in rats with a prior history of acetaldehyde self-administration. Experientia 40:1008–1010
Myers WD, McKenzie L, Ng KT, Singer G, Smythe GA, Duncan MW (1985) Salsolinol and dopamine in rat medial basal hypothalamus after chronic ethanol exposure. Life Sci 36:309–314
Myers RD, Borg S, Mossberg R (1986) Antagonism by naltrexone of voluntary alcohol selection in the chronically drinking macaque monkey. Alcohol 3: 383–388
Nash JF, Maickel RP (1985) Stress-induced consumption of ethanol by rats. Life Sci 37:757–765
O'Malley SS, Jaffe AJ, Chang G, Schottenfeld RS, Meyer RE, Rounsaville B (1992) Naltrexone and coping skills therapy for alcohol dependence: a controlled study. Arch Gen Psychiatry 49:881–887
Orcutt JD, Harvey LK (1991) The temporal pattern of tension reduction: stress and alcohol use on weekdays and weekends. J Stud Alcohol 52:415–424
Pollock VE (1992) Meta-analysis of subjective sensitivity to alcohol in sons of alcoholics. Am J Psychiatry 149:1534–1538
Privette TH, Myers RD (1989) Anatomical mapping of tetrahydropapaveroline reactive sites in brain mediating suppression of alcohol drinking in the rat. Brain Res Bull 22:1039–1048
Reid LD, Hunter GA (1984) Morphine and naloxone modulate intake of ethanol. Alcohol 1:33–37
Reid LD, Delconte JD, Nichols ML, Bilsky EJ, Hubbell CL (1991) Tests of opioid deficiency hypotheses of alcoholism. Alcohol 8:247–257
Samson HH, Doyle TF (1985) Oral ethanol self-administration in the rat: effect of naloxone. Pharmacol Biochem Behav 22:91–99
Schuckit MA (1985) Ethanol-induced changes in body sway in men at high alcoholism risk. Arch Gen Psychiatry 42:375–379
Schultz R, Wister M, Duka, T, Herz A (1980) Acute and chronic ethanol treatment changes endorphin levels in brain and pituitary. Psychopharmacology (Berl) 68:221–227
Siegel S (1986) Alcohol and opiate dependence: re-evaluation of the Victorian perspective. In: Cappell HD, Glaser FB, Israel Y, Kalant H, Schmidt W, Sellers EM, Smart RG (eds) Research advances in alcohol and drug problems. Plenum, New York, pp 279–314
Sinclair JD (1974) Morphine suppresses alcohol drinking regardless of prior alcohol access duration. Pharmacol Biochemist Behav 2:409–412

Sinclair JD, Adkins J, Walker S (1973) Morphine-induced suppression of voluntary alcohol drinking in rats. Nature 246:425–427

Sjoqvist B, Borg S, Kvande H (1981) Catecholamine derived compounds in urine and cerebrospinal fluid from alcoholics during and after long-standing intoxication. Subst Alcohol Actions Misuse 2:63–72

Sjoqvist B, Eriksson A, Winblad B (1982) Salsolinol and catecholamines in human brain and their relation to alcoholism. In: Bloom F, Barchas J, Sandler M, Usdin E (eds) Beta-carbolines and tetrahydroisoquinolines. Liss, New York, pp 57–67

Tabakoff B, Hoffman PL (1983) Alcohol interactions with brain opiate receptors. Life Sci 32:197–204

Tabakoff B, Hoffman PL (1987) Biochemical pharmacology of alcohol. In: Meltzer HY (eds) Psychopharmacology: the third generation of progress. Raven, New York, pp 1521–1526

Trachtenberg MC, Blum K (1987) Alcohol and opioid peptides: neuropharmacological rationale for physical craving of alcohol. Am J Drug Alcohol Abuse 13:365–372

US Department of Health (1990) Seventh special report to the US Congress on alcohol and health. US Department of Health and Human Services, Rockville

Volpicelli JR (1987) Uncontrollable events and alcohol drinking. Br J Addict 82:381–392

Volpicelli JR, Tiven J, Kimmel SC (1982) The relationship between tension reduction and ethanol consumption in rats. Physiol Psychol 10:114–116

Volpicelli JR, Davis MA, Olgin JE (1986) Naltrexone blocks the post-shock increase of ethanol consumption. Life Sci 38:841–847

Volpicelli JR, Ulm RR, Hopson N (1990) The bidirectional effects of shock on alcohol preference in rats. Alcohol Clin Exp Res 14:913–916

Volpicelli JR, Ulm RR, Hopson N (1991) Alcohol drinking in rats during and following morphine injections. Alcohol 8:289–292

Volpicelli JR, Alterman AI, Hayashida M, O'Brien CP (1992) Naltrexone in the treatment of alcohol dependence. Arch Gen Psychiatry 49:876–880

Wild KD, Marglin SH, Reid LD (1988) Small doses of morphine enhance intake of a solution of only ethanol and water. Bull Psychon Soc 26:129–131

CHAPTER 9

Animal Models of the Alcohol Addiction Process

K.A. GRANT

A. Introduction

Historically, animal models of alcohol addiction have been reflective of the current definitions and understanding of alcohol addiction in humans. Thus, in conjunction with the increasing biomedical approach to alcoholism, the hallmarks of human alcohol dependence in the early 1960s and into the 1970s were tolerance and physical dependence. During this period, most animal models of alcoholism were designed to demonstrate both an increased ability to tolerate the effects of alcohol and the emergence of withdrawal signs following the cessation of alcohol treatment. The major use of animal models was to define the neuropharmacological adaptation to prolonged exposure to alcohol in the hopes that these adaptations would reveal the critical stages of the addiction process.

Although tolerance to and physical dependence upon alcohol are robust consequences of prolonged exposure to alcohol, their place as central phenomena in the addiction process has been supplanted by the recognition that the excessive nature of the behavior maintained by alcohol consumption is also a robust, defining characteristic. The inclusion of more behavioral criteria into the diagnosis of alcoholism was reflected in the 1978 revision of the World Health Organization definition of alcohol dependence syndrome. In this revision, the alcohol dependence syndrome included several criteria for behavioral manifestations of alcohol dependence, with alcohol-seeking as a central variable. Since the late 1970s and throughout the 1980s drug-seeking behavior, as opposed to the development of tolerance of dependence, has become the hallmark of addiction and addictive behaviors (STOLERMAN 1988, 1992). This analysis places the emphasis on the behavior directed towards acquiring and consuming alcohol and allows the incorporation of both pharmacological and environmental stimuli as modulators to this behavior. This analysis also allows the incorporation of information from other fields of behavioral analysis as well as neurobiological disciplines such as neurochemistry, neurophysiology, and molecular biology.

Within this analysis, alcohol-seeking behavior results from the reinforcing effects of alcohol. These reinforcing effects are determined by an interplay of several variables including: the pharmacological characteristics of alcohol, the environmental events that occurred concurrent with the initial and subsequent alcohol use, the genetics of the individual, and previously established behaviors not necessarily related to the effects of alcohol. To study these influences, numerous conceptual models of addictive processes have emerged that acknowledge the multidimensional nature of addiction (Tarter 1988). To organize this review, a stages-of-change analysis of the addiction process, drawn from the human psychosocial field, was chosen. In this analysis, there are roughly four major stages in the addiction process: initiation, transition and maintenance of addiction, remission, and relapse (Marlatt et al. 1988). The primary dependent variable is drug-seeking behavior, as this is the primary variable of interest in humans. Information from animal studies, focusing both within stages and on the transitions between stages, will be presented.

This review is confined to discussing the behavioral effects of alcohol related to the addiction process. The other contributions in this book are dedicated to reviewing the pharmacological basis of ethanol's behavioral effects and, consequently, will not be reviewed here. However, it should be clear that ethanol does not act at a single set of receptors. The multiple pharmacological effects of ethanol are reflected in the multiple behavioral effects, which change quantitatively and qualitatively with the dose of ethanol.

B. Alcohol-Seeking Behavior

As noted in the introduction, alcohol-seeking has become the primary behavioral variable of interest in animal models of the alcohol addiction process. For the purposes of this review, two behaviors, found under different experimental conditions, will be operationally defined as representative of alcohol-seeking. These behaviors are alcohol self-administration and alcohol-conditioned preferences. Other behavioral procedures may be used to reflect the ability of alcohol to serve as a reinforcer. For example, procedures that measure alcohol effects such as the anxiolytic, aversive, discriminative, motor, or amnestic effects each address important aspects of alcohol's ability to serve as a reinforcer. However, these behavioral outcomes are not direct measures of alcohol-seeking and will be discussed in a later section (Sect. C.IV). Likewise, the effects of alcohol on the threshold for intracranial self-stimulation (ICSS) are believed to reflect the reinforcing effects of alcohol (Wise et al. 1992). This procedure, however, is also an indirect measure of alcohol-seeking.

Self-administration refers to the process of a laboratory animal or human engaging in a behavior that results in the administration of alcohol. The most common self-administration procedure involves the presentation of alcohol following a set number of specific responses in a distinct environment. By definition, the intake of alcohol must be greater than the intake of the vehicle (water, sweetened solutions, or saline are common vehicles) for alcohol to be serving as a reinforcer. The presentation of alcohol in self-administration procedures has been accomplished by several routes, including oral, intravenous, intragastric, and intracranial delivery (see CARROLL et al. 1990). In short, self-administration procedures emphasize the consequences of behavior in the role of alcohol-seeking.

Conditioned preference refers to the process of a laboratory animal becoming attracted to the place associated with the delivery of alcohol (MUCHA et al. 1982; D.C. HOFFMAN 1989). Procedurally, the animal is given a specific dose of alcohol and placed in a distinct environment. Through the association of alcohol's effects with the specific environemntal stimuli, these stimuli serve as incentive stimuli and elicit approach behaviors. When given a choice between two environments, only one of which is associated with the effects of alcohol, the animal spends a greater proportion of time in the environment associated with alcohol. These procedures emphasize environmental stimuli as generating motivational states, reflected in approach or avoidance behaviors that are indicative of the reinforcing effects of alcohol. While self-administration and place conditioning are derived from different procedures, the process of alcohol-seeking in humans and in animal models undoubtedly incorporates aspects from both paradigms.

In order to examine the progression of alcohol-seeking behavior from initiation through dependency and into relapse, it is useful to isolate variables and divide them into biological components, consisting of genetic and pharmacological influences, and environmental components. However, the lack of evidence for any intrinsically or fundamentally reinforcing effects of alcohol, or any other psychoactive drug, must be emphasized. For example, it is possible that alcohol and other abused substances are self-administered because of their euphoric or mood-elevating effects. Indeed the majority of drugs which are self-administered by animals are reported as euphorigenic in humans (SCHUSTER et al. 1981). However, euphorigenic effects do not consistently covary with the reinforcing effects of drugs (JOHANSON and UHLENHUTH 1980, 1981). For example, humans report amphetamine as euphorigenic on self-report rating scales; however, when given an opportunity to self-administer the drug, not all subjects do so. In addition, there are reports of alcohol increasing anxiety and depression in alcoholics, even though they continue to self-administer alcohol (MELLO 1972). Therefore, since the euphoric effects of a drug, including alcohol, are not necessarily related to the self-administration of alcohol, these effects cannot be solely responsible for its reinforcing effects.

Animal models, involving several drug classes, further support the dissociation between stimuli inducing positive mood states and subsequent self-administration. For example, several studies have shown that the responding of morphine-dependent monkeys can be maintained by the administration of naloxone (GOLDBERG et al. 1971, 1978; WOODS et al. 1975) at doses which produce avoidance responding in the same (GOLDBERG et al. 1971) or other (D.A. KANDEL and SCHUSTER 1977) morphine-dependent monkeys. Furthermore, within subject designs have shown that the same doses of cocaine can either be actively avoided or self-administered (SPEALMAN 1979). Thus, any other intrinsic effect of a drug or alcohol, for example dopamine release in the nucleus accumbens, is likely only to *influence*, rather than *determine*, alcohol-seeking. Alcohol's ability to serve as a reinforcer at any given time will be due to a combination of antecedent events, current environmental contingencies, and the pharmacological basis of alcohol's stimulus effects. Clearly, drug- and alcohol-seeking behavior can be viewed as malleable and not wholly determined by intrinsic pharmacological effects.

C. Initiation

Human studies suggest that there are many influences on the initiation of alcohol consumption, including genetic load, personality traits, parental drinking practices, cultural patterns, age, peer pressure, the taste of the alcoholic beverage, and degree of sensitivity to some of alcohol's effects (see MARLATT et al. 1988; SAMSON 1987; TARTER et al. 1985). Animal models are particularly suitable for the study of influences on the initiation of alcohol consumption because of the obvious ethical constraints on human studies. However, it is a well-known fact that the typical laboratory animal does not readily consume intoxicating quantities of alcohol. This reluctance has been attributed to the taste of alcohol, the delay between the consumption of alcohol and its pharmacological effects, the volume of alcohol needed for a pharmacological effect, and the particular pharmacological effects of alcohol. These effects of alcohol make it difficult to establish alcohol-seeking behavior. However, one arguable advantage of investigating alcohol-seeking compared to seeking behavior maintained by other drugs of abuse has been the necessity to explore the role of genetic and conditioned effects in the initiation of alcohol consumption. Unlike many opiates and psychomotor stimulants, it is often not effective to simply allow access to alcohol, even by intravenous or intragastric routes of administration, in order to engender alcohol self-administration (MEISCH 1977; CARROLL et al. 1990; SAMSON 1987). Alcohol researchers have been confronted with the rather large task of determining appropriate conditions for the establishment of alcohol-seeking. Thus, the development of animal models of alcohol-seeking has required an emphasis on the interplay among the individual organism, the pharmacology of alcohol, and the environment. As a result, alcohol researchers have become keenly aware of the potential contribution of conditioned reinforcement to the development of alcohol-seeking behavior. Alcohol is not alone in this regard; nicotine, caffeine, and minor tranquilizers all have less than robust reinforcing effects in the uninitiated animal (YOUNG and HERLING 1986). Yet, all these drugs are widely used by humans. The question remains, what variables influence the initiation of alcohol use?

I. Models to Assess Environmental Effects in the Initiation of Alcohol-Seeking

The first set of variables to be considered are those related to environmental conditioning. Models that address these variables encompass conditioning in the presence as well as the absence of alcohol. Thus, some of the models address predisposing factors that may influence the initiation of alcohol-seeking, while others address conditioning that occurs during the initial encounters with alcohol.

Early environmental influences, independent of alcohol exposure, can exert powerful control over the establishment of alcohol-seeking. Animal models are just beginning to address these early influences. Human studies point to rearing environments as important predictors of later alcohol use. For example, family structure and child-rearing practices have been noted as potential influences that affect risk for problem drinking (TARTER et al. 1984). In a recent study investigating the effects of rearing conditions on subsequent alcohol consumption in rhesus monkeys, HIGLEY et al. (1991) found that monkeys reared by their mothers for the first 6 months were less likely to consume alcohol compared to monkeys taken from their mothers and reared by similar aged monkeys. The data suggest that the stressful rearing conditions contributed to the initiation of alcohol consumption and demonstrate the powerful effects on subsequent alcohol consumption of factors that exist prior to alcohol exposure.

Other early environmental influences on the ability of alcohol to serve as a reinforcer occur concurrent with the initial uses of alcohol. These are conditioned effects, wherein the effects of alcohol are associated with environmental stimuli and events. Of utmost importance in the human situation may be the effects of social interactions. An animal study of the influence of social stress on the initiation of alcohol consumption used monkeys that were all reared by their mothers for the first 6 months, then housed in groups of four (KRAEMER and MCKINNEY 1985). When the monkeys were adults, they were divided into two groups: one group consisted of animals continuously housed individually for 12 weeks and another group that for 12 weeks was intermittently separated from and then reunited with their social group for 1-week periods. Overall the intermittently separated monkeys drank more alcohol than the continuously separated monkeys. The intermittently separated monkeys also drank more alcohol when they were separated compared to when they were housed together. The authors interpreted the findings as evidence that the stress of intermittent isolation promoted alcohol consumption, which eventually transferred to the social situation (KRAEMER and MCKINNEY 1985). Social stress was also invoked as an explanation of induced drinking in subordinate male rats housed in social groups (BLANCHARD et al. 1987). Both of these studies suggest that social interactions can be important determinants of alcohol consumption in laboratory animals.

Apart from social stress, positive social interactions can also promote alcohol intake. Several studies have shown that initial drug use is associated with peer models and friendship groups (JESSOR and JESSOR 1975; JESSOR et al. 1972; D.B. KANDEL et al. 1978). One way in which ethanol consumption may be socially reinforced is by drinking in a manner that is defined by peers as appropriate and which receives approval. Drinking initially maintained by peer approval can be conceived of as being a conditioned reinforcer. An animal model designed to investigate the early conditioning of alcohol-seeking used a secondary conditioning procedure (see SAMSON 1987; GRANT and JOHANSON 1988). In a secondary conditioning procedure, one set of stimuli are paired with the presentation of a reinforcing event (the primary reinforcer). Following a number of pairings, presentation of the stimuli associated with the primary reinforcer, can, by itself, maintain responding (a secondary reinforcer). It was hypothesized that the initial stages of alcohol drinking in humans may be maintained by peer approval, and only later in the drinking history do the pharmacological effects of alcohol maintain drinking. In an animal model of this process, rats and monkeys were required to drink alcohol in order to gain access to a highly palatable sugar solution (GRANT and SAMSON 1985; GRANT and JOHANSON 1988). After a number of pairings, presentation of the sugar solution was no longer necessary to maintain the alcohol consumption. Thus, alcohol served as a reinforcer following secondary conditioning in animal models, strengthening the possibility that such conditioning can indeed play a role in the initiation of alcohol drinking in humans. The secondary conditioning procedure is not specific to alcohol, and has been successful in the initiation of chlodiazepoxide, cannabis, and phencyclidine self-administration in laboratory animals (see GRANT and SAMSON 1985).

It is possible that secondary conditioning procedures are simply serving to acclimate the animal to the taste of alcohol. However, this does not appear to be a sufficient explanation for the initiation of alcohol-seeking. For example, high alcohol intake in laboratory animals can be induced by scheduled food presentation and intermittent access to alcohol. Yet when these inducing procedures are suspended, alcohol-seeking behavior is not maintained or cannot be initiated. Rather, conditioned drinking appears to be dependent upon the conditioning environment. For example, in an early study of conditioning effects, rats drank large volumes of alcohol in order to gain access to food and to avoid foot shock (SENTER and PERSENSKY 1968). The conditioning was then discontinued and the rats were divided into two groups. One group of rats received access to alcohol in operant chambers while the second group was given access to alcohol in the home cage. The alcohol intakes of the home-cage group fell dramatically compared to the rats that had access to alcohol in the conditioning environment. Thus, an important factor for maintaining conditioned alcohol drinking appears to be the presence of the conditioning environment. These effects have been replicated and explained in terms of both operant (SAMSON 1987) and classical (KRANK 1989) conditioning frameworks.

There are several explanations for the facilitating effect of the environment on the initiation and maintenance of alcohol consumption. For example, stimuli in the environment may function as discriminative stimuli, signaling the availability of alcohol and therefore setting the occasion for drinking (BICKEL and KELLY 1988). An alternative explanation is that the environment acts to stimulate incentive motivation and elicit approach behaviors (J. STEWART et al. 1984). A third explanation invokes the hypothesis of conditioned withdrawal in anticipation of alcohol administration (HINSON and SIEGEL 1980); however, in the early stages in the initiation of alcohol-seeking, this latter explanation is probably not applicable. Finally, the role of associative factors in the development of tolerance (see next section) may also underlie the importance of the conditioning environment. Regardless of the explanation, the importance of environmental cues in initiating alcohol-seeking suggests that an avenue for early intervention may be the avoidance of places where alcohol has already been sampled and its use reinforced by peers. However, there may be only a small window of opportunity for such intervention, since the hallmark of the addiction process is the ability of alcohol to serve as a reinforcer in numerous environments. Apparently, early in the addiction process the pharmacological effects of ethanol begin to reinforce drinking, shifting the relative control over drinking from the exteroceptive cues to the interoceptive effects of alcohol (see Sects. B.II, III).

Another early environmental influence that appears to be important in the initiation of alcohol-seeking is the taste of alcohol. Evidently, the taste of alcohol is unpleasant in both uninitiated humans and laboratory animals. One way to circumvent the unpleasant taste of alcohol is to mask it with other, preferred tastes. The most common procedure use sweet solutions (e.g., soft drinks) as mixers for alcoholic solutions. In animal models of this process, animals are initially trained to respond on a lever by the presentation of a highly sweet solution. Over a period of time, usually weeks, the concentration of sucrose is gradually decreased and alcohol is added to the solution in increasing concentrations, until alcohol in water maintains responding (see SAMSON 1987). This example of acquired taste is not unique to alcohol. A striking example of acquired taste in laboratory animals is the demonstration that chimpanzees can acquire a preference for chili peppers which are initially avoided and result in lacrimation upon ingestion (ROZIN and KENNEL 1983). The suggestion that alcohol preference is acquired indicates that masking is an important process in desensitizing the animal (or human) to initially aversive effects of alcohol. The observation that repeated exposure to ethanol is required to show alcohol-seeking in place preference procedures also suggests a desensitization (REID et al. 1985). This apparent desensitization, or tolerance, to the aversive effects of alcohol is discussed at length in a later section.

An alternative explanation for the increased alcohol consumption observed when acquisition techniques are employed is that the adulterated taste of alcohol increases the likelihood that the animal will engaged in the act of drinking. In other words, increasing the baseline rates of behavior (drinking) increases the probability that when alcohol is available it also will be consumed. Evidence from both monkey and rodent studies supports the hypothesis that increases in baseline rates of responding and access to alcohol are sufficient to establish the self-administration of alcohol. For example, monkeys that initially do not self-administer alcohol through the intravenous route can be induced to do so if response rates are increased by allowing access to cocaine or methohexital (WINGER and WOODS 1973). Likewise, the technique of prandial drinking, in which food-deprived animals given food also tend to drink, is useful in establishing alcohol consumption (MEISCH 1977). Thus, techniques that establish relatively high amounts of drinking, or other behaviors consistent with the consumption of alcohol, may be sufficient to initiate alcohol-seeking.

In summary, animal models have been used to explore the effects of environmental events prior to, or concurrent with, the use of alcohol on the initiation of alcohol-seeking. To place these influences in perspective, however, it is also necessary to consider the pharmacological effects of alcohol and genetically determined vulnerability to the reinforcing effects of alcohol. For example, in each study that manipulated social interaction, rearing conditions, taste of alcohol, or secondary conditioning to establish alcohol-seeking in monkeys, there were consistent individual differences in alcohol intakes (KRAEMER and MCKINNEY 1985; GROWLEY and ANDREWS 1987; GRANT and JOHANSON 1988; HIGLEY et al. 1991). The demonstration that some individuals consistently consumed more alcohol compared to others within each induction procedure suggests that a genetically determined risk factor modulates the expression of alcohol-seeking. This apparent interaction between genetic and early environmental influences in determining subsequent alcohol-seeking appears to be a robust finding. Such interactions have been noted in studies of the etiology of alcoholism in humans (MARLATT et al. 1988; TARTER 1988) and emphasize the conclusion that neither genetic nor environmental factors operate alone to establish alcohol-seeking.

II. Models to Assess the Genetic Basis of Initiation of Alcohol-Seeking

Genetic effects can be viewed as, but are not inclusive of, predisposing or risk factors. As such, the contributing role of the genetics to alcohol abuse and alcoholism has been the focus of intensive research efforts. Human studies show that there is a genetic contribution to the expression of alcoholism (GOODWIN 1983; SCHUCKIT and GOLD 1988; CLONINGER et al. 1981). This contribution appears to be polygenetic resulting in heterogeneous typologies in families of alcoholics (CLONINGER 1987, 1988) and the possibility of more than one type of genetic predisposition to alcoholism (CLONINGER et al. 1981).

The hypothesis that alcoholism has a genetic basis has been extensively explored using animal models (see CRABBE and HARRIS 1991). Several approaches have been used to characterize the genetic effects, including inbred strains, recombinant inbred strains, selectively bred lines, and mutants (CRABBE and BELKNAP 1992). There is an enormous amount of behavioral data derived from animal studies of genetic differences in alcohol effects, which have recently been reviewed (PHILLIPS and CRABBE 1991). In general, these investigations have clearly shown that many alcohol-related phenotypes have a genetic basis. In addition, both human and animal studies suggest a large interaction of environmental influences on the development of alcohol-related problems (or traits) in at-risk individuals (TARTER 1988; PHILLIPS and CRABBE 1991). Although some animal models have directly addressed human phenotypes believed to be important in the initiation or maintenance of alcohol-seeking (e.g., anxiolytic, reinforcing, tolerance-producing, or aggressive effects of alcohol), the genetic basis of preference is the most widely studied behavioral characteristic of the alcohol addiction process. A few of the procedures and results of the animal models are reviewed here for illustrative purposes. For more in-depth reviews see GEORGE (1990), CARROLL et al. (1990), and PHILLIPS and CRABBE (1991). In addition, the most common use of genetic models has been to test hypotheses concerning the pharmacological basis for the effects of alcohol and these studies have also been reviewed recently (ALLAN and HARRIS 1992; DEITRICH et al. 1989).

The majority of animal models that have explored the influence of genetic factors in the initiation of alcohol-seeking have used both selected lines and inbred strains. Basically, selected lines are developed by breeding animals that express a particular phenotype. Thus, selective pressure is placed on sequestering the genes that control the expression of the chosen trait. Theoretically, after generations of selective breeding, lines are created that contain the genetic material necessary to express the phenotype (these genes are fixed) without altering the distribution of genes unrelated to the phenotype. These animals demonstrate a genetic basis for alcohol-related traits and serve as powerful tools to investigate the potential of a common genetic basis for related traits. For example, selected lines can be used to test the hypothesis that rats bred to drink alcohol preferentially are also less sensitive to the aversive effects of alcohol. Given some qualifications (see CRABBE and BELKNAP 1992, for the necessity of replicate and control lines), if the two traits covary in a selected line, then the possibility that the traits are manifestations of common genes is strengthened.

Animals have been selectively bred for numerous alcohol-related traits that may be important in the initiation of alcohol-seeking. These include preference for alcohol, sensitivity to the hypnotic effects of alcohol, ethanol withdrawal severity, and expression of the motor-stimulant effects of alcohol. The vast majority of studies designed to measure the initiation of alcohol-seeking have used rats selectively bred to prefer 10% alcohol (see PHILLIPS and CRABBE 1991). The most widely studied lines include the preferring (P) and non-preferring (NP), the high alcohol drinkers (HAD) and low alcohol drinkers (LAD), the alcohol-accepting (AA) and alcohol-non-accepting (ANA), and the Sardinian preferring (sP) and Sardinian non-preferring (sNP) rats. The alcohol-preferring lines can consume relatively large amounts of alcohol, and develop tolerance and signs of physical dependence (Li et al. 1987, 1988). Several studies have investigated alcohol-seeking in these selected lines using either place preference procedures or self-administration procedures. When trained to self-administer alcohol orally, alcohol intakes are greater than vehicle intakes in AA, P, and HAD rats (Murphy et al. 1989; Schwartz-Stevens et al. 1991; Sinclair 1974; George 1987). In contrast, under similar conditions, alcohol does not serve as a reinforcer to the ANA or NP rats (Ritz et al. 1989; Murphy et al. 1989). However, it is important to emphasize that preference for alcohol does not reside solely in the oral cavity. NP rats avoid the intragastric administration of ethanol (Waller et al. 1984) and are more sensitive to conditioned taste aversions induced by ethanol administration (Froehlich et al. 1988). Despite this innate low preference for alcohol, environmental manipulations can alter ethanol intakes in rats selectively bred for low ethanol preference. For example, food restriction or increased palatability of ethanol solutions increase absolute ethanol intake in NP rats and in ANA rats (Sinclair and Li 1989; Waller et al. 1982). Furthermore, NP rats can be trained to self-administer 10% ethanol with a sucrose-fading procedure (Samson et al. 1989), although the overall amount and pattern of ethanol consumption by the NP rats remains below that of preferring (P) rats (Samson et al. 1989). Thus, genetic predispositions are not entirely predictive of the ability of ethanol to serve as a reinforcer, but they can act as a buffer to environmental events that may promote alcohol-seeking.

Another tool to investigate the genetic basis of vulnerability in animal models is the characterization of inbred strains. An inbred strain refers to a group of animals that are genetically identical and are typically derived from a series of brother-sister matings. These strains are useful in testing hypotheses concerning correlative traits; however, each inbred strain should be viewed as a single data point (i.e., all their genes are identical). Therefore, it is recommended that six or more panels of strains be used to properly address possible underlying mechanisms for the expression of complex behavior (Phillips and Crabbe 1991). Unfortunately, most behavioral studies of inbred strains report correlative data in only two strains (e.g., C57BL/6 and BALB/c mice). On the other hand, one advantage of research using inbred strains is the cumulative nature of the data, since the genetic material is identical from generation to generation. Thus, investigators can continually add information to the inbred strain database, strengthening or weakening correlations.

There are several reports concerning the self-administration of alcohol and conditioned preference to alcohol in inbred strains. In examining the ability of alcohol to establish a conditioned place preference, Cunningham et al. (1991, 1992) found that DBA/2J but not C57BL/6J mice develop a conditioned place preference to alcohol. The self-administration of alcohol in inbred strains of mice and rats that differ in their preference for alcohol has also been examined (GEORGE 1990). Both strains of mice that prefer (C57BL/6J mice) and those that do not prefer alcohol (BALB/c mice) can be induced to drink alcohol with prandial drinking (see Sect. C.I); however, only C57BL/6J mice continued to self-administer alcohol in quantities greater than water. Thus, the mice from the C57BL/6J strain, who prefer alcohol over water, also self-administer alcohol. On the other hand, mice from the BALB/c strain, who prefer water over ethanol, do not selectively self-administer ethanol in the operant situation (ELMER et al. 1987; GEORGE 1990). In contrast, Lewis rats drink greater quantities of alcohol, reach greater response requirements, and show more motor activation than Fisher 344 rats. However, alcohol can serve as a reinforcer for the Fisher 344 rats. Thus it is important to emphasize that techniques that induce alcohol self-administration may circumvent initial preferences for alcohol (see CARROLL et al. 1990).

In general, both the selected lines and the inbred strains provide strong evidence that initial preferences are fairly predictive of future alcohol-seeking. In particular, the selected lines may represent an animal model of vulnerability to alcohol addiction. As such, these animal models are potentially useful for testing hypotheses concerning prevention strategies and/or in understanding the progression of alcohol addiction. However, an application of these models that has not been adequately explored is the possibility that vulnerability to the initiation of alcohol-seeking also predicts vulnerability to the reinforcing effects of other drugs of abuse. That is, do common sets of genes underlie a predisposition to self-administer opiates or stimulants as well as alcohol? Since some investigators view addiction as a general phenomenon, rather than a substance-specific process (MARLATT et al. 1988; PEELE 1985), cross-drug generalities may enhance the use of animals bred for alcohol-related traits in studies of the addiction process.

III. Models to Assess the Psychopharmacological Basis of Initiation of Alcohol-Seeking

The psychopharmacology of alcohol that contributes to its initial abuse potential is the result of several interacting processes, including the positive reinforcing effects, the punishing effects, the discriminative effects, and the motoric effects. Each of these behavioral effects can be hypothesized to positively or negatively affect the initiation of alcohol consumption. The following models are examples of procedures used to isolate and study a particular psychopharmacological effect of alcohol that is believed to influence the initiation of alcohol consumption.

The positive reinforcing effects of alcohol are, by definition, those effects that increase the probability of future alcohol consumption. The positive effects are most often demonstrated in animal models using self-administration and conditioned preference procedures. As such, these are also the procedures that reflect alcohol-seeking (reviewed above). As stated previously, the efficacy of ethanol's reinforcing effects in animal models is moderate, typically lower than the reinforcing efficacy of stimulants, opiates, and barbiturates. For example, intravenous alcohol does not appear to serve as a reinforcer in rodents, although cocaine and barbiturates are readily self-administered through the same catheter (see Carroll et al. 1990). Likewise, the ability of alcohol to condition approach behaviors in place and taste preference procedures is equivocal (Asin et al. 1985; J. Sherman et al. 1983; R.B. Stewart and Grupp 1981). Indeed, one aim of this chapter is to review the conditions that enhance the reinforcing effects of alcohol. However, in addition to the traditional measures of drug-seeking, there are other, indirect, measures of the positive reinforcing effects of alcohol. Two of these procedures are intracranial self-stimulation (ICSS) and assessment of the anxiolytic effects of alcohol. The ICSS model is most useful in assessing the reinforcing efficacy of alcohol and determining the underlying neuroanatomical and neuropharmacological pathways that mediate the reinforcing effects of alcohol. In contrast, the anxiolytic effects of ethanol provide a basis for the situational and potentially malleable ability of alcohol to serve as a reinforcer.

Intracranial self-stimulation refers to the ability of electrical stimulation in particular areas of the brain to engender and maintain responding (Olds and Milner 1954). Thus, electrical stimulation of these areas of the brain are reinforcing and may represent pivotal neural areas that underlie and mediate the process of reinforcement. Two measures are frequently utilized in the application of ICSS to drug abuse research. The first measure is the threshold, or intensity, of the electrical stimulation necessary to maintain ICSS. The second measure is the frequency of the electrical pulses necessary to maintain ICSS. Drugs of abuse typically decrease the threshold of ICSS in sensitive brain areas (Kornetsky et al. 1979; Wise et al. 1992), an effect believed to reflect a summation of the reinforcing effects of the stimulation with the reinforcing drug effects through a common pathway (Wise et al. 1992). The effects of ethanol on ICSS threshold have been mixed. Using probes located in the lateral hypothalamus, acute injections of alcohol result in both decrease (Lewis and Phelps 1987) and increases (Carlson and Lydic 1976) in current threshold that maintains ICSS. A recent study, using probes located in the nucleus accumbens, found no consistent change in threshold during an intravenous drip of ethanol (Wise et al. 1992). Thus, the positively reinforcing effects of ethanol, as measured by ICSS, appear to be weak and less efficacious in comparison to the robust effects of opiates, stimulants, and nicotine (Wise et al. 1992). In general, the findings of the ICSS studies agree with the results from the self-administration studies, in which the reinforcing effects of ethanol in uninitiated rats appear weak, at best.

Another approach to investigate the reinforcing effects of alcohol is to determine the circumstances under which a specific behavioral effect of alcohol may be beneficial to the animal and thereby increase the probability of future consumption. It has been hypothesized for many years that the ability of alcohol to reduce stress underlies its ability to serve as a reinforcer (WILLIAMS 1966; POHORECKY 1981). The most common animal models that assess the anxiolytic effects of alcohol have recently been reviewed (KOOB and BRITTON, in press). One procedure frequently used involves conflict, in which responding maintained by food or water presentation in a deprived animal is occasionally punished, usually by the delivery of shock (POHORECKY 1981). These procedures result in a suppression of behavior that can be reinstated with typical anxiolytics such as benzodiazepines. In general, ethanol increases suppressed behavior in conflict paradigms, although at doses that also disrupt other behaviors, including nonpunished responding (see KOOB and BRITTON, in press).

Two additional animal models of anxiety that do not use punishment or food deprivation are the elevated plus maze and the social interaction tests (FILE and HYDE 1978; PELLOW et al. 1985). In the elevated plus maze procedure, rats or mice are placed in a maze with a central post that elevates the maze. Radiating out from the central post are four arms that are perpendicular to each other. Two of the maze arms, opposite to one another, have sides (closed arms) and the other two arms have no sides (open arms). When placed in the maze, undrugged animals spend most of their time in the closed arms of the maze, while anxiolytics, including alcohol, increase the amount of time spent in the open arms of the maze (PELLOW and FILE 1986; LISTER 1987). The social interaction test capitalizes on the observation that social interactions decrease when animals are placed in unfamiliar or brightly illuminated environments. Typically, anxiolytics increase the social interactions of rats in these environments and alcohol has been shown to increase social interactions under these conditions (FILE 1990). However, ethanol decreases social interaction between mice (LISTER and HILAKIVI 1988); therefore the use of this particular procedure to study the anxiolytic effects of ethanol may be limited.

A final procedure that is gaining use as an animal model of anxiety is the discriminative stimulus effects of pentylenetetrazol (PTZ), a convulsant that increases the subjective reports of anxiety in humans (RODIN 1958). Subjective effects of drugs in humans are interoceptive effects that are distinct and discriminable. These discriminable effects of drugs are modeled in laboratory animals with drug discrimination procedures (see Sect. D.V). The discriminative stimulus effects of PTZ are blocked by anxiolytics such as benzodiazepines (G.T. SHERMAN and LAL 1980). The anxiolytic effects of ethanol as assessed in a PTZ discrimination are weaker than those of the benzodiazepines and barbiturates. For example, ethanol will not block a training dose of 20 mg/kg PTZ, but is effective in blocking a training dose of 10 mg/kg PTZ (EMMETT-OGLESBY et al. 1990). In addition to serving as an animal model of anxiety, PTZ discriminations have also been used as animal models of the subjective effects of drug withdrawal (LAL and EMMETT-OGLESBY 1983; EMMETT-OGLESBY et al. 1990). The application of PTZ discrimination to assess the influence of ethanol withdrawal on drug-seeking is addressed in the next section.

In summary, the anxiolytic effects of ethanol have been demonstrated using a wide variety of procedures. This research has shown that these effects of ethanol are fairly robust, although alcohol is somewhat less efficacious as an anxiolytic compared to benzodiazepines. The anxiolytic effects of ethanol occur at a low-to-moderate dosage, and are not always separable from other behavioral effects of ethanol (KOOB and BRITTON, in press).

In order for the anxiolytic effects of alcohol to promote the initiation of alcohol-seeking in the early stages of the addiction process, an animal must be able to associate these effects with the consumption of alcohol. For example, in humans alcohol attenuates stress reactions in young adults at risk for alcohol dependence (SHER and LEVENSON 1982). It is possible that the early association of alcohol's anxiolytic effects with the use of alcohol can serve as a basis for promoting future alcohol consumption. Evidence that the anxiolytic effects of alcohol can be associated with the consumption of alcohol comes from studies using human social drinkers. When these drinkers believe that they have consumed alcohol, imbibing the nonalcoholic drink results in decreased levels of anxiety (ABRAMS and WILSON 1979). Thus, humans can, apparently, associate the consumption of alcohol with its anxiolytic effects. In the animal models reviewed above, early rearing conditions and social separations are associated with stressful reactions and these conditions are sufficient to initiate and maintain alcohol consumption (HIGLEY et al. 1991; KRAEMER and MCKINNEY 1985; BLANCHARD et al. 1987). However, in contrast to social stress, other forms of stress have not clearly established ethanol-seeking. In these latter studies, the consumption of alcohol is increased only following a stressful event, but not while the stressor is present (M.A. KAPLAN and PUGLISI 1986). Clearly the interactions and associations between the nature of the stress, the anxiolytic effects of alcohol, and initiation of alcohol consumption, although documented, need further exploration.

Another, well-documented, effect that may alter the initiation of alcohol-seeking is the aversive effect of alcohol. Animal models that demonstrate ethanol's aversive effects are typically the conditioned place and taste procedures using intraperitoneal injections. However, aversive effects of ethanol are also found with intravenous ethanol (NUMAN 1981; GRUPP 1981; GRUPP and STEWART 1983). Generally, these procedures demonstrate that animals avoid environments or foods that are associated with the effects of alcohol. The dose range for the aversive effects of alcohol are 1.0–2.0 g/kg for rats (GAUVIN and HOLLOWAY 1992; DAVIES and PARKER 1990; HOLLOWAY et al. 1992; SCHECTER 1992; SCHECTER and KRIMMER 1992) and 2.0–3.0 g/kg for mice (CUNNINGHAM et al. 1991; RISINGER and CUNNINGHAM 1992).

There is evidence that sensitivity to some of the aversive effects of ethanol may be under genetic control. For example, NP rats, but not P rats, develop conditioned taste aversion associated with 1.0 g/kg ethanol (FROEHLICH et al. 1988). However, at higher doses of alcohol, P rats also show conditioned taste aversions (FROEHLICH et al. 1988). Using place preference procedures, rather than conditioned taste procedures, both the P and NP selected lines developed place aversions following 1.0 g/kg ethanol (SCHECTER 1992). Where conditioned preferences are demonstrable, repeated exposure to alcohol is necessary (see GRANT et al. 1990).

The necessity of multiple exposures to alcohol implies that tolerance to the aversive effects of alcohol may expose the positive reinforcing effects of alcohol (the general phenomenon of tolerance is discussed in Sect. D.I). However, it is worth noting that at least some strains and lines of mice do not show the initial aversive effects of alcohol when tested in place preferences paradigms. Both DBA/2J mice and mice selected for hyperthermia in response to alcohol develop conditioned place preferences associated with 2–4 g/kg ethanol (CUNNINGHAM et al. 1991, 1992; RISINGER et al. 1992b). Thus, it may be better to use rats, rather than mice, to study the potential influence of the aversive effects of ethanol on alcohol-seeking.

Finally, motor activation is another psychopharmacological effect that may influence the initial use of alcohol. The ability of psychoactive drugs to induce motor stimulation has been suggested to be a characteristic of drug-mediated reward (WISE and BOZARTH 1987). Ethanol increases activity in rats following 0.5–1.5 g/kg ethanol and in mice following 1.5–2.5 g/kg ethanol. The increase in activity is most often measured in an open field (CRABBE et al. 1987), but can also be documented in place preference (RISINGER et al. 1992a,b; CUNNINGHAM and NOBLE 1992) and operant procedures (GRANT and COLOMBO 1993). The increase in motor activity has been compared to increases in exploratory activity in humans, a trait related to novelty seeking (CLONINGER 1988). Therefore, the increase in motor activity in animals associated with low doses of ethanol may be a model for novelty-seeking or risk-taking behavior of humans. However, there are difficulties in this analogy. A recent study investigating both the stimulant and rewarding effects of ethanol in a place preference procedure reported blockade of the motor-activating effects of ethanol with a dopaminergic antagonist (see H.H. Samson and P.L. Hoffman, this volume) without blocking the expression of place preference (RISINGER et al. 1992a). Thus, motor-activating effects of ethanol are not necessary for the development of alcohol-seeking. In addition, mice that were selectively bred to have high motor activation in response to alcohol (FAST mice) do not show differences in 24 h preference for alcohol (PHILLIPS and CRABBE 1991). Therefore, the relationship between the motor-activating effects and the reinforcing effects of ethanol is unclear.

Apart from risk taking, the activating effects of ethanol are believed to be reflected in increased social interaction and conversation. Alcohol increases rates of social conversation in both alcoholics and social drinkers (STITZER et al. 1981; BABOR et al. 1983; SAMSON and FROMME 1984). Alcohol, up to four drinks at a setting, also increased speaking in isolated humans, whereas the same doses decreased rates on a nonverbal behavioral task (HIGGINS and STITZER 1988). There is evidence that the increase in speech is restricted to the rising limb of the blood alcohol curve, and therefore may correspond to the euphorigenic effects of alcohol in humans (BABOR et al. 1983; HIGGINS and STITZER 1988). Perhaps animal models investigating social behavior, as opposed to motor activity in isolation, will be more useful in determining how the stimulatory effects of alcohol influence subsequent alcohol-seeking behavior.

In summary, animal models illustrate that rearing conditions, operant and classical conditioning, desensitization, and increases in baseline consumption rates are some of the environmental manipulations sufficient to initiate alcohol-seeking. These events, concurrent with the psychopharmacological effects of alcohol and the genetic load of the animal interact to establish different patterns of intake, which in turn determine the rate and extent to which the individual progresses through the addiction process. However, a note of caution is necessary when designing and evaluating animals models to investigate the factors controlling the initiation of alcohol-seeking. Investigation of human risk factors clearly demonstrates the difficulty in delineating the specificity of the potential risk factor to the alcohol addiction process. For example, although the human experimental literature suggests a strong association between antisocial behavior and alcoholism, antisocial behavior is not specific to alcoholism (TARTER et al. 1985). Our animal models need to begin to explore the generality and specificity of identified variables that appear to control the establishment of alcohol-seeking.

There remain several areas in the initiation of alcohol-seeking that have not been adequately explored using animal models. First, there is a lack of studies investigating the possibility of protecting the individual from the development of alcohol-seeking behavior. For example, are there certain periods in the initial use of alcohol during which punishment is particularly effective in decreasing future alcohol-seeking? Due to the difficulties in establishing alcohol-seeking behavior, the possibility of investigating protective variables are understandably few. Thus, the use of animal models to study strategies for preventing the initiation of alcohol-seeking will continue to be limited by advances in our understanding of how alcohol comes to serve as a reinforcer.

Second, animal models of the initiation of alcohol-seeking have not adequately addressed the area of gender differences. There are clear human gender differences regarding the age of onset of alcoholism (NIAAA 1990). However, whether this effect is largely due to social or genetic factors is unclear. In general, animal studies have noted that female rodents drink more alcohol than males, although female rodents also drink more caffeine and morphine compared to male rodents (see CARROLL et al. 1990). Data on alcohol consumption in nonhuman female primates is sparse. One study used a secondary conditioning procedure to initiate alcohol self-administration and found that female rhesus monkeys were less likely to initiate alcohol consumption than males (GRANT and JOHANSON 1988). However, other studies have clearly shown that intravenous alcohol can serve as a reinforcer in female rhesus monkeys (MELLO et al. 1983). A greater understanding of the basis of gender differences in the initiation of alcohol-seeking behavior may help elucidate neurochemical mechanisms of alcohol reinforcement.

Finally, an early environmental influence that does not have an appropriate animal model is the phenomenon of modeling. Modeling refers to drinking in patterns and situations set by other individuals within a social context. Modeling appears to be an important determinant of the initiation and continued use of alcohol (CAUDILL and MARLATT 1975; CAUDILL and LIPSCOMB 1980). If may be possible to develop animal models to investigate the effects of modeling on alcohol consumption, particularly with the use of nonhuman primates. For example, nonhuman primates exhibit observational learning in making their nests, gathering and preparing their food, and engaging in reproductive and maternal behaviors (NAPIER and NAPIER 1985). Thus, primate models may be appropriate to investigate alcohol-seeking by individuals exposed to high-versus-low drinking cohorts during the initiation of alcohol consumption.

D. Transition to Abuse and Dependence (Maintenance)

Available evidence suggests that the ability of alcohol to serve as a reinforcer early in the initiation of alcohol consumption is qualitatively different and under different sets of controls than the ability to maintain drinking in the problem drinker. Characteristics of the transition from the initiation of alcohol-seeking to episodes of abuse and dependence have been identified (EDWARDS and GROSS 1976). These characteristics include: narrowing of one's drinking repertoire, increased salience of alcohol-seeking behavior, increased tolerance, repeated withdrawal symptoms, relief drinking to avoid alcohol withdrawal, subjective awareness of the compulsion to drink, impaired control over drinking, compulsive drinking style, and tendency to relapse. It is clear from the clinical literature that the transition from "social" to "pathologica" consumption of alcohol is not an all-or-none phenomenon. For example, not all the elements listed above need be present to diagnose abuse or dependence. However, the development of dependence does lead to a greater coherence among the elements, indicating that, during the transition to problem drinking, alcohol seeking is channeled into distinctive constellations of alcohol-directed behaviors. Therefore, the transition from low-level, low-priority drinking to high-level, high-priority drinking is pivotal to the addiction process. Animal models that address this process investigate how the alcohol stimulus increases control over behavior. Under limited access conditions, where alcohol is available for discrete, relatively short periods of time during a 24-h period, intakes are stable and even defended. For example, pretreatment with alcohol prior to the drinking session results in the animal adjusting its intake during the session (GRANT and SAMSON 1985; KAROLY et al. 1978; MEISCH 1977). However, under excessive intake patterns these pharmacological feedbacks no longer control self-administration (WINGER and WOODS 1977). Apparently, during the transition to abusive intake, a fundamental difference emerges in the control of alcohol-seeking. In short, the basic question to address becomes: what processes are responsible for the transition from controlled intakes following initiation to excessive intakes that occur later in the addiction process?

I. Role of Tolerance in the Maintenance of Excessive Alcohol-Seeking

The development of tolerance is believed to play a pivotal role in the addiction process. Tolerance is a diminished physiological or behavioral effect of a particular dose of ethanol. Tolerance is also indicated if a greater dose of ethanol is necessary to achieve an effect of ethanol that was previously present at lower doses. Acute tolerance develops within the course of a single exposure, while chronic tolerance develops over the course of multiple exposures (see B. Tabakoff and P.L. Hoffman, this volume). Under a number of circumstances, tolerance to the effects of ethanol involves associative learning. Classical conditioning has been implicated in many forms of tolerance where the response is physiological in nature. Often, the adaptations in the physiological response that impart tolerance are compensatory to the direct, acute effects of alcohol. For example, ethanol (the unconditioned stimulus) given acutely results in hypothermia (the unconditioned response). Following several injections of alcohol, animals become tolerant to this effect, in that the same dose of alcohol no longer results in hypothermia. However, if saline is administered to these animals in the same environment (the conditioned stimulus) in which alcohol was administered, hyperthermia (the conditioned response) is noted. That is, a compensatory increase in body temperature occurs that compensates for the hypothermia induced by the alcohol, with the net result being no change in temperature when alcohol is given (see GOUDIE and DEMELLWEEK 1986). Insofar as environmental cues may become associated with the drug effect and thereby elicit a physiological response, the environment in which alcohol is consumed can be an important influence on the effects of alcohol.

Tolerance due to associative processes is not restricted to classical conditioning. Using before-after designs, several investigations have shown that, for tolerance to develop, the effects of ethanol must be experienced in the context of performing the target behavior. For example, ethanol will disrupt the coordinated motor activity of locomotion on a moving belt. Tolerance develops to the disrupting effects of ethanol after several trials. However, tolerance does not develop in a second group of rats that receive the same amount of ethanol following, rather than prior to, the same number of trails on the moving belt (LeBlanc et al. 1973; Wenger et al. 1980). The before-after design in tolerance research emphasizes the consequences of drug-impaired behavior (response-stimulus) in the induction of tolerance within an operant framework as opposed to the stimulus-response emphasis of classical conditioning.

Through the use of classical and operant conditioning procedures, it is clear that the demonstration of tolerance can be dependent upon environmental variables (Corkfield-Sumner and Stolerman 1978; J. Stewart and Eikelboom 1987). These procedures use discrete and repeated administration of ethanol within a consistent environmental setting. On the other hand, there are a number of procedures in which conditioning effects are minimized (Tabakoff and Hoffman 1988). In this latter case, tolerance develops under conditions where ethanol is administered insidiously, as part of a liquid diet or by inhalation techniques, and responses to ethanol are tested after the administration of ethanol in a novel environment. Thus, tolerance can be viewed as a conditioned effect (i.e., an associative process) in some circumstances and as a nonassociative process in other circumstances (Goudie and Demellweek 1986). This distinction is important because it implies that drinking patterns can determine different forms of tolerance, more or less influenced by the prevailing environmental circumstances.

Tolerance is hypothesized to influence alcohol-seeking by altering at least three effects of alcohol. The first is the development of tolerance to the aversive effects of alcohol. Evidence for this is derived primarily from taste and place preference procedures. As reviewed in Sect. C.III, alcohol exposure results in both place and taste aversions, particularly in rats. However, there are several reports that repeated exposure to ethanol can result in both conditioned taste and conditioned place preferences (see J.E. Sherman et al. 1984; Reid et al. 1985; Holloway et al. 1992). Thus, tolerance to the aversive effects of alcohol may allow for the reinforcing effects of ethanol to maintain alcohol-seeking. The second important influence of tolerance may be under circumstances where ethanol intake is elevated. The self-limiting variable in a drinking binge may be the depressant effects of alcohol that constrain alcohol-seeking due to incapacitation. There is clear evidence of tolerance to the hypnotic effects of alcohol (Tabakoff and Hoffman 1988); however, its role in promoting alcohol-seeking is unknown. Thirdly, tolerance to the reinforcing effects of alcohol could increase alcohol-seeking. However, little empirical data exist to support or dispute this hypothesis. In one study that addressed this issue, rats were placed under an ICSS procedure and the acute and chronic effects of alcohol were investigated. Following chronic ethanol, tolerance to the rate-decreasing effects of ethanol was apparent, but there was no change in the current threshold necessary to maintain ICSS in the presence of ethanol (Magnuson and Reid 1977). As noted before, an increase in the threshold necessary to maintain ICSS can be viewed as a decrease in reinforcing magnitude, an outcome that would have supported the hypothesis of tolerance to the reinforcing effects of ethanol. Using a self-administration procedure, ethanol intake prior to and following chronic exposure to alcohol did not change, indicating chronic exposure did not alter the reinforcing effects of ethanol (SAMSON 1987). Thus, tolerance to the motor and aversive effects of ethanol are common results of chronic ethanol exposure, but tolerance to the reinforcing effects following high intakes of ethanol has yet to be demonstrated.

Animal models have been used to examine the genetic influence of tolerance to alcohol. These studies have primarily looked at correlated responses in inbred, heterogeneous, or selected lines. For example, tolerance to the ataxic effects of ethanol in mice is strain dependent and correlated to initial sensitivity (CRABBE 1983). Furthermore, preference ratios are positively correlated with acute tolerance to the motor-impairing effects of alcohol in heterogeneous stock (*HS/Ibg*) mice (ERWIN et al. 1980). Rats selectively bred to prefer alcohol also show more tolerance to alcohol and prolonged maintenance of tolerance following acute administration (WALLER et al. 1983; GATTO et al. 1987). These data can be interpreted as evidence that there is a genetic component to tolerance and that this trait may be related to both initial sensitivity and preference for alcohol. This finding is intriguing since sons of alcoholics have shown decreased sensitivity to the ataxic effects of alcohol (SCHUCKIT and GOLD 1988). However, given the comlexity of the process of tolerance development and the number of ethanol effects investigated (e.g., hypothermia, ataxia), broad generalizations concerning the genetic basis of tolerance and correlated traits are not yet possible (see also PHILLIPS and CRABBE 1992).

One animal model that is potentially useful for studying the effects of tolerance on the transition from the initiation of alcohol consumption to abusive drinking is the vasopressin deficient (diabetes insipidus, DI) rat. These rats were derived from a mutation that resulted in a lack of functional vasopressin. Vasopressin is a hormone that plays an important role in water regulation but has also been implicated in general learning and memory processes as well as the maintenance of tolerance to alcohol (P.L. HOFFMAN 1987). Althoug the DI rats can apparently acquire tolerance, they do not retain tolerance to the hypnotic or hypothermic effects of alcohol (P.L. HOFFMAN et al. 1990). Thus, these animals may provide a suitable model for exploring the influence of tolerance maintenance on increases in alcohol-seeking and on the production of physical dependence.

II. Role of Sensitization in the Maintenance of Excessive Alcohol-Seeking

An effect opposite to tolerance is sensitization. In sensitization, the effects of a drug become more pronounced with repeated exposures. Sensitization has been most extensively studied using locomotor activity induced by psychomotor stimulants (see KALIVAS et al. 1992). Typically the locomotor activity is assessed in an open field equipped with photobeams to record motion and location. With repeated injections of a stimulant the locomotor activity increases. This sensitization has been cited as evidence supporting a psychomotor stimulant theory of drug reward, which proposes that a common characteristic of drugs of abuse is their ability to stimulate motor activity (WISE and BOZARTH 1987). According to this theory, the incentive properties of the drug increase with continued use, accounting for increased use with repeated exposures. Although alcohol in low-to-moderate doses results in increased locomotor activity, evidence for sensitization to the locomotor stimulant effects of ethanol is lacking (TABAKOFF and KIIANMAA 1982). Thus, the evidence to date does not support a role for sensitization to the locomotor effects of ethanol in promoting excessive alcohol consumption.

Sensitization may occur to the cognitive and amnestic effects of alcohol in human alcoholics. However, the effects of age, socioeconomic factors, education, etc., make interpretation of these data difficult (see RIEGE 1987). The implications that sensitization to these effects of ethanol has on the promotion of alcohol consumption are also unclear. Certainly, poor judgment and memory may affect the ability to maintain a job and social contacts, which in turn may promote further consumption (see Sect. D.IV). However, this hypothesis is untested, insofar as no animal experiments have addressed sensitization to the amnestic or cognitive impairing effects of alcohol.

Finally, sensitization to the withdrawal effects of ethanol has been noted. Alcoholics that have undergone multiple detoxifications are more likely to experience seizures compared to alcoholics that have not been repeatedly detoxified (BROWN et al. 1988). In animal models, repeated cycles of ethanol dependence result in sensitization to withdrawal seizures (BECKER and HALE 1993; POHORECKY and ROBERTS 1991). These results have been explained in terms of kindling (BALLENGER and POST 1978), a progressive intensification of seizure activity following electrical stimulation of specific brain regions at levels below the seizure threshold (GODDARD et al. 1969). It is not known if sensitization occurs only to withdrawal seizures or if other manifestations of the withdrawal syndrome (e.g., the anxiogenic effects) are also intensified with repeated dependence cycles. Sensitization to the withdrawal effects of ethanol may be hypothesized to maintain excessive drinking if the individual is consuming alcohol to avoid the unpleasant effects of withdrawal (see EDWARDS 1990).

Procedures very similar to those used to demonstrate tolerance or sensitization have shown an effect of expectancies in humans. That is, under certain conditions, when humans are told they will receive alcohol and are given placebo (or told they will receive placebo and are given alcohol) the behavioral outcomes are influenced more by what they expect, rather than what they actually receive (MARLATT and ROHSENOW 1980). Thus, some of the effects of alcohol in humans may be related to conditioned expectancies and not the pharmacological effects of alcohol. Expectancies differ between heavy and light drinkers. Heavy drinkers expect more positive and fewer negative consequences from alcohol consumption compared to light drinkers (Oei and Jones 1986). For example, heavy drinkers are more likely to believe that alcohol will improve psychomotor performance, sustain attention, enhance learning or memory performance, and social interactions (see Marlatt et al. 1988). Expectancies may also influence the self-administration of alcohol. While craving is elicited in alcoholics following the consumption of an alcoholic drink, if the taste of the alcoholic beverage is masked, craving is not reported (Merry 1966). While the modeling of these complex cognitive effects may not be possible with animal models, studies of tolerance and sensitization may serve to identify circumstances that enhance the development of expectancies and may increase our knowledge of conditions necessary to extinguish expectancies once they are established.

III. Role of Dependence in the Maintenance of Excessive Alcohol-Seeking

Physical dependence on alcohol is defined primarily by the presence of characteristic symptoms that appear when the administration of relatively high doses of ethanol is abruptly terminated. These symptoms are, in many cases, opposite to the signs of acute intoxication and follow a characteristic time course of appearance after the cessation of ethanol intake (Majchrowicz 1975). Alcohol withdrawal in humans consists of autonomic nervous system hyperactivity, central nervous system excitability, and distorted sensory perception (R.F. Anton and H.C. Becker, Chap. 14 in this volume). Alcohol withdrawal seizures have received the most attention in the animal models, including a genetic selection study that has produced alcohol withdrawal seizure prone and alcohol withdrawal seizure resistant mice (Crabbe et al. 1985). However, the contribution of withdrawal seizures to alcohol-seeking is unclear.

Apart from the acute signs of physical dependence, there appears to be a protracted withdrawal syndrome that may continue for weeks or months following the beginning of abstinence (Alling et al. 1982; Begleiter and Porjesz 1979; R.F. Anton and H.C. Becker, Chap. 14 in this volume). The symptomotology of the protracted withdrawal syndrome includes tension, insomnia, anxiety, and depression. These protracted symptoms may have profound implications for craving, conditioning effects with environmental stimuli, and the role of anxiolytic effects of alcohol in alcoholic relapse. The role of conditioned withdrawal in relapse will be considered in Sect. F.I. Here, evidence is reviewed concerning the influence of physical dependence, which is a result of excessive ethanol intakes upon continued ethanol self-administration.

The paramount question asked with animal models has been whether animals, once they have become physically dependent upon ethanol, will self-administer alcohol to avoid withdrawal. In behavioral terms, the discomfort of withdrawal may act as a negative reinforcer and increase the probability that alcohol will be consumed. Inherent in this argument is that the initial determinants of excessive alcohol consumption are displaced, or at least supplemented, by the ability of alcohol to relieve or avoid withdrawal effects. In other words, the transition from initial drinking patterns to compulsive drinking patterns is accompanied by fundamental changes in how ethanol serves as a reinforcer. Unfortunately, this is one area where clinical data and laboratory data are not in complete agreement (see EDWARDS 1990). Data from controlled experiments in both humans and laboratory animals lend only weak support to the hypothesis that relief or avoidance of withdrawal maintains excessive drinking. Evidence against the "relief of withdrawal" hypothesis is derived from studies that show spontaneous declines in episodes of heavy alcohol intakes in both monkeys and humans, and subsequently the failure to avoid overt signs of physical withdrawal (MELLO and MENDELSON 1977; WINGER and WOODS 1971). Furthermore, in a handful of studies, rats or monkeys made physically dependent on ethanol and subjected to forced abstinence did not consume alcohol to avoid withdrawal (MYERS et al. 1972; SAMSON 1987; TANG et al. 1982). Finally, a recent study using rhesus monkeys found that ethanol self-administration was suppressed during the presence of withdrawal signs and returned to normal only after the withdrawal signs disappeared (WINGER 1988). These studies suggest that self-administering alcohol to alleviate the effects of acute withdrawal is not a robust effect of physical dependence on alcohol.

On the other hand, other investigations have yielded some evidence that the reinforcing effects of ethanol increase following several cycles of dependence and withdrawal (DEUTSCH and WALTON 1977; HUNTER et al. 1974; NUMAN 1981), and that rats will self-administer alcohol at the onset of withdrawal and thereby avoid withdrawal signs (NUMAN 1981). Similarly, several cycles of induced dependence and withdrawal from barbital were necessary to produce sustained barbital self-administration in rats (JANICKE and COPER 1984). Together, the animal studies that suggest physical dependence may increase the reinforcing efficacy of ethanol also demonstrate the necessity of multiple exposures to the withdrawal syndrome. Thus, the association between alcohol consumption and relief of alcohol withdrawal may be a learning process that requires conditioning on several trials. Alternatively, these studies all investigated the effects of relatively severe withdrawal effects on alcohol-seeking. It is possible that other, more subtle withdrawal signs that do not render the animal incapacitated are more efficacious at enhancing alcohol-seeking.

Anxiety is a common feature of alcohol withdrawal and has been reported to last for months following the cessation of acute withdrawal signs (DE SOTO et al. 1985; MOSSBERG et al. 1985). Relief from the anxiety associated with alcohol withdrawal has been suggested as a motivating factor in alcoholic drinking (BARRETT 1985). A recently developed animal model that studies the subjective effects associated with ethanol withdrawal is a pentylenetetrazol (PTZ) discrimination procedure. Earlier, this procedure was described as a model for investigating the anxiolytic effects of ethanol. However, PTZ discrimination also shows considerable promise for studying the anxiety associated with alcohol withdrawal (LAL et al. 1988; EMMETT-OGLESBY et al. 1990). Following termination of chronic alcohol treatment rats respond on the PTZ-appropriate lever, indicating withdrawal has effects similar to the PTZ stimulus. This state lasts from 12 h to 48 h, at which time the percentage of rats choosing the PTZ lever is 80% and 30%, respectively. The length of chronic ethanol treatment necessary to show this withdrawal effect is apparently 3 days at 12.5 g/kg per day (EMMETT-OGLESBY et al. 1990). However, in a more recent analysis of the ethanol withdrawal state, acute administration of moderate-to-large (2–4 g/kg) doses of ethanol resulted in PTZ generalization (GAUVIN et al. 1989, 1992). Thus, PTZ discrimination appears to be sensitive to ethanol withdrawal effects following acute (i.e., hangover) or chronic (i.e., withdrawal) ethanol treatment. Therefore, it may be possible to use this model to quantify and characterize the severity and time course of interoceptive stimuli associated with alcohol withdrawal. The important link to alcohol-seeking will then be made by determining if the bout of drinking following a short period of abstinence (e.g., a few hours of sleep) is reliably initiated following the onset of this subjective state of withdrawal.

IV. Role of Environmental Interactions in the Maintenance of Excessive Alcohol-Seeking Behavior

Just as the interaction between alternative reinforcers and alcohol is important in initiating alcohol-seeking, these interactions are also important in the maintenance of seeking behavior. A variety of alternative reinforcers and their schedules of availability affect alcohol consumption (VUCHINICH and TUCKER 1988; CARROLL et al. 1990; SAMSON and GRANT 1990). In general, alcohol consumption declines when alternative reinforcers of high reinforcing magnitudes are introduced. In animal models, the alternative reinforcer is often food or a highly palatable substance. For example, in a situation where two levers are available in a chamber and each lever is exclusively associated with the presentation of water or a solution of ethanol, the previously initiated rat will respond primarily on the lever associated with ethanol delivery. However, if a highly palatable solution of sucrose is available instead of water, the rat shifts its response choice, such that very few responses are made on the ethanol lever and very little ethanol is consumed (SAMSON et al. 1982). Thus, the introduction of an alternative reinforcer can decrease the reinforcing efficacy of alcohol.

Alternatively, removal of an available reinforcer can increase consumption. For example, social isolation increases alcohol consumption in monkeys (KRAEMER and MCKINNEY 1985). The effect of social isolation can be viewed as an overall restriction in the sources of reinforcement that increases the value of the remaining reinforcing stimuli. Similar effects are found if the relative cost of the alternative reinforcer is increased, thereby functionally restricting the options for reinforcement (SAMSON and GRANT 1990). For example, in the case of the rat given a choice between ethanol and sucrose, if the response requirement for sucrose is increased eightfold, a marked increase in ethanol intake occurs (SAMSON and LINDBERG 1984). This increased ethanol consumption following the added restriction on access to sucrose is greater than intakes when water is paired with alcohol and similarly restricted. The relationship between restricted access to alternative reinforcers and alcohol consumption is extremely important because it shows that the alternative reinforcer does not need to vanish, *just be relatively harder to acquire*, for ethanol consumption to increase.

A related effect of restricting access to alternative reinforcers is the induction of adjunctive, or schedule-induced, behavior (FALK 1971). In these procedures, the constraints placed on the availability of an alternative reinforcer are temporal rather than related to effort (i.e., an increased response requirement). The most common method to induce excessive drinking is to schedule the delivery of food pellets intermittently and allow free access to a solution of alcohol. The amount of ethanol consumed is an inverted U function of the interval of pellet delivery. That is, when the pellets are delivered too frequently or too infrequently, drinking is not excessive. Thus, there exists a "window," where the intermittent delivery of alternative reinforcers induces excessive alcohol consumption. Schedule-induced alcohol consumption has been reported in rats, monkeys, and humans using a variety of alternative reinforcers (see FALK 1981), and can be used both to initiate alcohol consumption and to induce intakes that result in physical dependence. Schedule-induced drinking has profound implications for the role of the environment in generating (and reducing) excessive alcohol consumption. For example, a variety of reinforcers in our environment are available under intermittent schedules. These include daily schedules such as meals and working hours, weekly schedules such as Friday and Saturday nights, and monthly schedules such as paychecks. How the intermittent delivery of reinforcers other than alcohol interacts in the life of the alcoholic is most certainly complex, but clearly open for examination.

A more direct environmental influence on alcohol consumption is the availability of alcohol. In animal models, the effects of either limited or unlimited access conditions have been investigated. Limited access refers to the procedure of allowing ethanol consumption within specified periods of time during a 24-h period. Unlimited access is a situation in which alcohol is always available, 24 h a day. Following the initiation of alcohol self-administration, unlimited access conditions can result in ethanol intakes that are variable, resembling binges (see Carroll et al. 1990). The binge pattern of intake is particularly striking when intravenous routes of administration are used. Similar patterns of intake have been noted in alcoholics given unlimited access to alcohol in a laboratory setting (Bigelow et al. 1975). In contrast, under limited access conditions, intakes are stable. Although the intakes within the limited access period are usually higher than those over a similar period in unlimited access conditions, total 24-h intakes are lower. Limited access conditions are primarily used in animal research to achieve a stable baseline prior to assessing the effects of pharmacological manipulation on the intakes of alcohol. However, the availability of alcohol is not always determined by the investigator. For example, in a social setting, low-ranking monkeys drink less than dominant monkeys. Apparently, dominant monkeys can control access to the drinking stations where alcohol solution is dispensed, therefore restricting the opportunity of the subordinate monkeys to sample the alcohol solution (Crowley and Andrews 1987).

Other environmental influences related to the maintenance of alcohol consumption include conditioned reinforcers. Conditioned reinforcers are stimuli that, through a process of association with the presentation of a primary reinforcer, function to reinforce behavior. For example, following a number of pairings where a red light consistently precedes the presentation of cocaine, presentation of the red light alone can maintain responding. Under a variety of procedures, conditioned reinforcers maintain long chains and high rates of behavior (Goldberg et al. 1979).

When associated with drug administration, the presentation of conditioned stimuli alone can maintain drug-seeking (Goldberg et al. 1979; Kelleher and Goldberg 1977). Conditioned reinforcers have been hypothesized to increase alcohol intakes in animal models. In one study, rats that had been initiated to self-administer ethanol in limited access conditions were placed in a schedule-induced drinking situation, where intakes averaged 10–13 g ethanol/kg per day. Following a period of schedule-induced drinking, the rats were placed back into the initial environment where alcohol self-administration was established. Here they continued to self-administer alcohol, even though they had just consumed large amounts of alcohol by schedule induction. Thus, the environment associated with ethanol self-administration, rather than the pharmacological effects of large alcohol intakes, controlled the amount of alcohol consumed (Roehrs and Samson 1981; Samson 1987). It is possible that the conditioned stimuli associated with the environment induce the onset of a drinking episode, independent of the amount of alcohol already consumed. An analogous situation in human drinking may be the phenomenon of "bar hopping" where excessive intakes can occur by visiting several drinking establishments in an evening. For example, entering a number of bars that each trigger a drinking episode can result in greater intakes of alcohol than if only a single bar were visited for the same amount of time.

A related phenomenon may be modeling, where the behavior of an individual is determined by the behavior of others in the setting. For example, subjects will model the intake patterns of a confederate, whether the consumption of alcohol is rapid or slow (CAUDILL and LIPSCOMB 1980; CAUDILL and Marlatt 1975). These data suggest that one of the external cues in the environment that is associated with alcohol-seeking is the behavior of the drinking group. That is, modeling can serve to modulate alcohol-seeking and consumption (MACANDREW and EDGERTON 1969). This effect has also been reported for heroin addicts in which the best predictor for heroin intake was the heroin intake by other subjects on the ward (MEYER 1988). However, as pointed out in Sect. C, animal studies on the effects of modeling have not been developed in the areas of alcohol and drug abuse.

V. Role of Conditioned Effects in the Maintenance of Excessive Alcohol-Seeking Behavior

Discriminative stimuli are defined as internal or external stimuli that signal when reinforcement is available and the appropriate behavior necessary to obtain reinforcement. Thus, discriminative stimuli are both differentiable and informational and can exert control over behavior. In short, a sensory effect, such as a light, a sound, or a dizzy feeling, can become associated with certain behavioral outcomes. For example, a green light at an intersection consistently signals to the pedestrian to cross (Boston and other New England cities excepted). By their associations with particular outcomes, these sensory effects, or stimuli, function as cues to signal the appropriate behavior necessary for reinforcement (e.g., avoidance of an automobile accident). Similarly, drugs that produce distinct subjective effects can also be associated with particular outcomes. Thus, the internal effects of psychoactive drugs, including alcohol, can function as stimuli to set the occasion for particular behaviors. The animal literature has numerous examples of increases in drug-seeking under the presence of stimuli that have been associated with the self-administration of drugs. These stimuli can be either exteroceptive cues, such as lights or specific environments (reviewed above), or interoceptive cues, such as the subjective effects produced by the drug (BICKEL et al. 1987; BICKEL and KELLY 1988).

Alcohol produces a characteristic set of interoceptive cues that are believed to be a reflection of its pharmacological activity in the brain. These interoceptive effects can become associated with self-administration behaviors in the same manner that exteroceptive cues become associated with consuming alcohol. The relative shift in the stimulus control over drinking from the exteroceptive cues to interoceptive effects has profound implications for the development of effective treatment strategies. For example, if the interoceptive, or subjective, effects of alcohol can set the occasion for a drinking episode, then the relative sphere of influence over drinking may expand to include mood states. At the same time, environmental control over drinking, such as the time of day or the setting, diminishes. The ability of the interoceptive effects of alcohol to result in alcohol-seeking behavior has been demonstrated in alcoholics living in a research ward. When the consumption of alcohol was suppressed by behavioral procedures such as time-out from social contact, the noncontingent administration of a drink of alcohol reliably increased the consumption of alcohol (Bigelow et al. 1977). Likewise abstinent alcoholics have difficulty in refraining from drinking after tasting an alcoholic beverage (Hodgson and Rankin 1976). Thus, the interoceptive effects of a fraction of the total dose normally self-administered can reinstate drug-seeking. These studies have implications not only for control of drinking following a single dose, but also for the ability of subjective effects, or moods, to reinstate suppressed drinking (see Sect. F).

Discriminative stimulus effects of alcohol are also useful in understanding the neurochemical mediation of alcohol's effects. These effects provide researchers with a method to test hypotheses concerning the ability of specific receptor agonists and antagonists to mimic or block the subjective effects of alcohol. Over the 40 years that the discriminative stimulus effects of alcohol have been studied, several neurotransmitter systems have been implicated in mediating the discriminative stimulus effects of alcohol (see Barry 1991; Colombo and Grant 1992). The cumulative data have led to the conclusion that alcohol produces a mixed cue, composed of separate input from a number of systems (Grant and Colombo 1993). Furthermore, drugs that produce stimulus effects similar to any one component of the ethanol cue can substitute for ethanol (Colombo and Grant 1992). That is, not every effect of ethanol needs to be present in order for the animal to report the stimulus as similar to alcohol. Thus, it is possible for a variety of mood effects to be similar to the effects associated with alcohol consumption. While a detailed review of the pharmacology of alcohol's discriminative stimulus effects are beyond the scope of this chapter, it is important to emphasize that mood may be a predictor of drinking and relapse (see Bickel and Kelly 1988), and understanding the neurochemical mediation of these effects can lead to important advances in the development of pharmacotherapies to treat excessive drinking.

E. Remission/Treatment

Remission, in which the use of alcohol declines to levels that are no longer problematic, has been the subject of several research efforts (see Marlatt et al. 1988). Remission of excessive drinking can be divided into either self-change (spontaneous) or treatment induced. Very few animal models have demonstrated spontaneous remission, perhaps because the levels of alcohol self-administration are relatively low, resulting in little loss of health or resources. One study demonstrated spontaneous remission in monkeys self-administering alcohol intravenously, which resulted in severe withdrawal symptoms (WOODS et al. 1971). However, the monkeys in this study also reinitiated self-administration within 3–4 days. Thus, this demonstration of decreased alcohol use may represent a specific case of remission due to the direct toxic effects of high alcohol intakes. Although this may be one reason humans discontinue alcohol use, many other reasons are also reported (LUDWIG 1985).

Remission can also be achieved through direct treatment interventions. The development of pharmacological and/or behavioral treatment methodology is an active area of research, and is discussed in several other chapters in this volume. Several processes that appear to interact with alcohol-seeking may predict the outcome of treatment strategies and are readily studied with animal models. These include the efficacy of ethanol as a reinforcer in the individual, the immediacy of the negative effects following drinking, and the differential reinforcement of abstinence.

I. Nonpharmacological Treatments to Reduce Alcohol Consumption

Reduction of alcohol intakes can be accomplished by punished procedures where drinking alcohol results in negative consequences or by explicitly reinforcing low alcohol intakes. In alcoholics, response-contingent shock (WILSON et al. 1975) or removal of a reinforcer contingent on alcohol consumption decreases alcohol intake in humans (BIGELOW et al. 1975; GRIFFITHS et al. 1978). Similarly, CROWLEY (1988) found that contract contingencies were powerful in decreasing drug use, but only as long as the contracts involved punishment (e.g., losing one's medical license) that was enforceable. In animals self-administering alcohol, response contingent shock or delay in food presentation can decrease alcohol intakes (POLING and THOMPSON 1977a,b). Taken together, the evidence from punishment procedures in controlled situations show some efficacy in the treatment of alcohol abuse. However, it is also clear that punishment procedures decrease consumption only when the contingencies are in effect. Another approach to lowering alcohol consumption is explicitly to reinforce low intakes or abstinence. This approach has successfully decreased the use of opiates (see STITZER and KIRBY 1991), and alcohol (COHEN et al. 1971) in humans. In particular, alcohol consumption of alcoholics was reduced by making an alternative reinforcer (money) contingent on low alcohol intake (COHEN et al. 1971). However, animal studies of the effectiveness of reinforcing abstinence are lacking.

II. Pharmacological Treatments to Reduce Alcohol Consumption

The development of pharmacotherapies to decrease alcohol consumption can be categorized into several approaches. One is to develop compounds that substitute for the reinforcing and discriminative stimulus effects of alcohol but lack the behavioral toxicity associated with alcohol. It is possible that serotonin reuptake blockers or specific serotonin agonists act through this mechanism (see other chapters in this volume). This strategy may be efficacious because the medication substitutes for the effects of alcohol that maintained drinking, thereby decreasing alcohol self-administration. Alternatively, substitution therapy may work due to the continued presence of similar discriminative stimulus effects, which have become entrenched in the alcoholic's life, and which may be necessary to set the occasion for all of life's activities (SCHUSTER 1986). Through either mechanism, substitution therapy may result in the dissociation of exteroceptive discriminative stimulus effects from the reinforcing effects of drinking. For example, if a drug constantly produces effects that resemble those of alcohol, but in a less efficacious manner, these internal stimuli (as well as those of alcohol) may no longer set the occasion for alcohol consumption and self-administration.

An alternative approach to reduce alcohol-seeking is to selectively antagonize the pharmacological basis of the reinforcing effects of alcohol, which should lead to extinction of alcohol self-administration. Theoretically, this approach is similar to desensitization procedures used in treating phobias. Although this appears to be sound goal, it is clear from the numerous studies cited above that excessive consumption of alcohol is not solely due to the pharmacological effects of alcohol. In addition, few animal or clinical studies have explicitly tried extinction procedures to decrease alcohol intakes. Studies that have used extinction procedures in alcoholics, even to the point of conducting extinction sessions within the drinker's favorite pub, have reported some success in decreasing the desire to drink (BLAKEY and BAKER 1980; HODGSON and RANKIN 1976; RANKIN et al. 1983). However, investigations of the efficacy of extinction procedures with cocaine addicts have reported limited success (O'BRIEN et al. 1991). It has been suggested that the extinction of stimulus control over self-administration may be more difficult than its establishment (BICKEL et al. 1988). Therefore, perhaps coping skills in conjunction with extinction procedures would enhance the outcomes (CHILDRESS et al. 1988).

Another pharmacological approach to reduce alcohol-seeking is to punish the drinking of alcohol. One medication that serves to punish alcohol consumption is the drug disulfiram. This drug results in the accumulation of acetaldehyde, a toxic by-product in the oxidation of alcohol in the liver. When alcohol is consumed following the administration of disulfiram, many aversive effects, including nausea and tachycardia result. The effectiveness of disulfiram in the treatment of alcohol abuse is reviewed in detail in Chap. 15 by R.K. Fuller and R.Z. Litten (this volume).

Finally, pharmacological treatment of the alcohol withdrawal syndrome may decrease drinking that is maintained by avoidance of withdrawal symptomotology. However, before an animal model can be applied to this hypothesis, it will be necessary to develop an animal model that clearly shows that avoidance of withdrawal is an important factor in the maintenance of excessive alcohol consumption.

F. Relapse

The clinical literature recognizes that specific situations increase the probability of relapse. These high-risk situations include specific mood states, the presence of alcohol, and exposure to environments that alcoholics frequented while drinking (MARLATT et al. 1988). All of these situations have been discussed in this chapter in terms of conditioned effects of alcohol in the initiation or maintenance of drinking. These effects will now be discussed with respect to the reinstatement of alcohol intake in abstinent individuals with a history of excessive alcohol use.

I. Role of Conditioned Withdrawal in the Relapse Process

Research with human opiate addicts suggested to WIKLER (1965) that environmental stimuli associated with opiate withdrawal could trigger relapse. WIKLER (1965) hypothesized that abstinent addicts upon returning to environments in which they experienced withdrawal would experience conditioned withdrawal symptoms. These conditioned reactions would, in turn, produce a craving for the drug in order to relieve the discomfort of withdrawal (see O'BRIEN 1991 or SCHUSTER 1986). There is some evidence that alcohol-associated cues, either exteroceptive, such as a bar, or interoceptive such as a mood, can elicit withdrawal symptoms, craving, and consumption in alcoholics (LUDWIG et al. 1974, 1977). Indeed, many other conditioned responses to alcohol have been documented, including changes in heart rate, galvanic skin response, affective state, salivation, and insulin release (R. KAPLAN et al. 1983, 1984; COONEY et al. 1984; POMERLEAU et al. 1983; DOLINSKY et al. 1987; MONTI et al. 1987; STAIGER and WHITE 1988). These conditioned responses are reported to result in craving for alcohol (POMERLEAU et al. 1983; DOLINSKY et al. 1987; MONTI et al. 1987), such that the degree of conditioning is predictive of an impaired ability to refuse alcohol (MONTI et al. 1988), increased alcohol-seeking in a choice procedure (R. KAPLAN et al. 1983), and an increase in the likelihood relapse (MONTI et al. 1988).

Interestingly, one of the conditioned reactions following the presentation of alcohol cues in alcoholics in anxiety (MEANKER 1967), a state indicative of ethanol withdrawal and hangover in both human and animal models (see Sect. D.III). Alternatively, the anxious state reported in alcoholics exposed to alcohol cues may be due to the possibility of engaging in a behavior that is not acceptable, or has negative consequences (e.g., approach-avoidance conflict) (CAPPELL 1975). It is possible that anxiety, and negative mood states in general, result in increased craving for alcohol. Evidence for this hypothesis comes from clinical studies with opiate addicts. For example, during opiate detoxification, negative mood states such as anxiety or depression have been shown to increase craving and withdrawal in addicts while positive mood states such as euphoria reduce these symptoms (CHILDRESS et al. 1989). In addition, negative interpersonal interactions have been reported to increase the likelihood of relapse to heavy drinking (MARLATT et al. 1988). It is possible that the anxious state induced by alcohol withdrawal is sufficient to trigger craving and relapse. However, anxiety associated with events other than withdrawal may also induce craving.

It should be noted that not all investigations have supported a role for conditioned withdrawal in relapse. In one study, at least one-third of opiate addicts report no withdrawal sickness when shown drug stimuli, while all of the subjects report craving in response to the stimuli (CHILDRESS et al. 1988). In addition, addicts given access to heroin reported no decrease in their high levels of craving as the heroin injections continued (MEYER 1988). One conclusion is that craving is not solely a reflection of conditioned withdrawal and events independent from conditioned withdrawal episodes can reestablish drug-seeking in abstinent individuals.

Animal studies that have successfully demonstrated conditioned withdrawal have used monkeys trained to self-administer morphine. In these procedures, a red light was associated with the administration of nalorphine, an opiate antagonist, which resulted in withdrawal signs in these monkeys. After morphine was discontinued, illumination of the red light elicited signs of opiate withdrawal for over 4 months, demonstrating a long-lasting conditioned withdrawal (GOLDBERG and SCHUSTER 1970). A subsequent study showed that stimuli paired with nalorphine to monkeys self-administering morphine not only elicited signs of opiate withdrawal, but also increased rates of morphine self-administration (GOLDBERG et al. 1969). Thus, there is evidence for both conditioned withdrawal and the association of conditioned withdrawal with the self-administration of opiates (SCHUSTER and WOODS 1968; WOODS et al. 1973). Unfortunately, comparable studies have not been conducted with alcohol. Such studies may have to await the development of an alcohol antagonist that can reliably induce signs of withdrawal. Alternatively, the use of selectively bred animals that differ in their withdrawal severity may prove valuable in the effort to determine whether conditioned withdrawal occurs.

II. Role of Conditioned Drug-Like Effects in the Relapse Process

In the opiate literature, a group of former addicts reported euphoric effects when they injected saline (LEVINE 1974; O'BRIEN et al. 1988). In an analogous situation, some addicts under a naltrexone blockade reported that heroin injections were rewarding (MEYER and MIRIN 1979). In these people, the stimuli associated with the administration of opiates results in conditioned reinforcement, although these responses can be readily extinguished

(Childress et al. 1988). Craving is also linked to conditioned stimuli and is reported to be highest when addicts perceive that heroin is available (Meyer 1988). Furthermore, environmental stimuli can reverse behavioral disruption caused by opiate withdrawal (Thompson and Schuster 1964; Tye and Iversen 1975). Collectively, this evidence indicates that drug-like conditioned responses occur under some circumstances of opiate administration. Conditioned alcohol-like responses have been reported in alcoholics who anticipate consuming alcohol (Ludwig 1985). Furthermore, there are conditioned alcohol-like neuroendocrine responses (i.e., decreased leutinizing hormone and testosterone) in response to stimuli associated with the consumption of alcohol (Dolinsky et al. 1987). These data suggest that environments associated with alcohol use can result in alcohol-like effects and may "prime" the onset of relapse.

Evidence that drug-like effects can initiate the onset of relapse comes from reinstatement paradigms. In these studies animals are first trained to self-administer a drug and then placed under extinction conditions where saline replaces the drug for a number of sessions. If a small "priming" dose of the drug that was previously self-administered is then given noncontingently to the animal immediately prior to a session under extinction conditions, drug-seeking is reinstated (J. Stewart et al. 1984). It is argued that the interoceptive cues following noncontingent drug administration act as conditioned incentive stimuli and reinstate drug-seeking (J. Stewart 1992). Apparently, exteroceptive cues can reinstate drug-seeking. For example, Schuster and Woods (1968) found that response-contingent presentation of stimuli previously associated with morphine self-administration increased responding under extinction conditions. Similarly, rats withdrawn from morphine drank more vehicle when placed in environments where morphine self-administration had been acquired than did rats placed in environments not associated with morphine self-administration (Thompson and Ostland 1965; Hinson et al. 1986). There is also evidence that the presence of exteroceptive stimuli associated with drug administration may be necessary for reinstatement of drug-seeking. For example, noncontingent amphetamine given to monkeys under extinction conditions can reinstate responding previously maintained by amphetamine only if a masking noise, present during the self-administration sessions, is also present (Stretch et al. 1971). There is also evidence to support the conditioned effects of exteroceptive cues in determining relapse, or reinstatement, in humans. Addicts that are detoxified and returned to the environment in which they used the drugs showed nearly 100% relapse (Chushman 1974), whereas Vietnam veterans, who became addicted in Southeast Asia, then detoxified and returned home to an environment not associated with drug use, remained relatively abstinent (Vietnam vets, Robins et al. 1974; O'Brien et al. 1980). Thus, contact with the drug may not be sufficient to elicit drug-seeking behavior outside an environment in which the drug was normally taken.

The reinstatement literature suggests that relapse can result from interoceptive cues (possibly including mood states) or exteroceptive cues (e.g., passing a favorite drinking establishment). An important issue in relapse is the influence of other drugs that were used in conjunction with alcohol. For example, a person that co-abused alcohol and marijuana may relapse to alcohol consumption when exposed to marijuana. Likewise, alcohol itself is a compound cue. Consequently, drugs such as barbiturates, benzodiazepines, and phencyclidine which have discriminative stimulus effects similar to alcohol (see BARRY 1991; GRANT and COLOMBO 1993) may, upon administration, reinstate alcohol-seeking. Thus, depending on the history of alcohol use, exposure to a number of compounds may be capable of triggering relapse. Animal studies in the area of alcohol conditioning and relapse are sparse and most of our information is drawn from the opiate literature or from clinical studies. Clearly, this is an area where animal studies are necessary to investigate the conditioning effects that may be specific to the effects of alcohol.

G. Summary

In this review information derived from animal studies is organized in terms of the dynamic progression of the addiction process as it is described in humans. Characteristic stages of addiction imply that the reinforcing efficacy of alcohol is altered as the individual progresses through the dependence cycle. The results of animals studies have made it possible to identify several influences on the reinforcing effects of ethanol at each stage. These influences are most readily classified as environmental, genetic, and pharmacological, with the process of conditioning representing the interaction among these elements. Although the information gathered to date is substantial, these reults must be incorporated and integrated into a framework that makes them easily applicable to the human situation. It is also important that in many cases alcohol addiction appears to be progressive. More information is needed concerning the rate of progression, which appears to show a large degree of individual variability. This analysis will require characterization of the dynamic processes that underlie progression from the initial stages of alcohol consumption, through the emergence of problematic drinking, and into the stage of severe dependence.

Acknowledgment. This review was supported, in part, by Grant AA09346 from the National Institute on Alcohol Abuse and Alcoholism.

References

Abrams DG, Wilson GT (1979) Effects of alcohol on social anxiety in women: cognitive versus physiological processes. J Abnorm Psychol 88:161–173

Allan AM, Harris RA (1992) Neurochemical studies of genetic differences in alcohol action. In: Crabbe JC, Harris RA (eds) The genetic basis of alcohol and drug actions. Plenum, New York, pp 105–152

Alling C, Balldin J, Bokstrom K, Gottfries CG, Karlsson I, Langstrom G (1982) Studies on duration of a late recovery period after chronic abuse of ethanol. Acta Psychiatr Scand 66:384–397
Asin KE, Wirtshafter D, Tabakoff B (1985) Failure to establish a conditioned place preference with ethanol in rats. Pharmacol Biochem Behav 22:169–173
Babor TE, Berglas S, Mendelson JH, Ellingboe J, Miller K (1983) Alcohol, affect and the disinhibition of verbal behavior. Psychopharmacology 80:53–60
Ballenger JC, Post RM (1978) Kindling as a model for alcohol withdrawal syndromes. Br J Psychiatry 133:1–14
Barrett RJ (1985) Behavioral approaches to individual differences in substance abuse. In: Galizio M, Maisto S (eds) Determinants of substance abuse: biological, psychological and environmental factors. Plenum, New York, pp 125–174
Barry H (1991) Distinct discriminative effects of ethanol. In: Glennon RA, Jarbe TUC, Frankenheim J (eds) Drug discrimination: applications to drug abuse research. US Government Printing Office, Washington DC, pp 131–144 (National Institute on Drug Abuse, Monograph 116)
Becker HC, Hale RL (1993) Repeated episodes of ethanol withdrawal potentiate the severity of subsequent withdrawal seizures: an animal model of alcohol withdrawal "kindling". Alcohol Clin Exp Res 17:94–98
Begleiter H, Projesz B (1979) Persistence of a "subacute withdrawal syndrome" following chronic ethanol intake. Drug Alcohol Depend 4:353–357
Bickel WK, Kelly TH (1988) The relationship of stimulus control to the treatment of substance abuse. In: Ray B (ed) Learning factors in substance abuse. US Government Printing Office, Washington DC, pp 122–140 (National Institute on Drug Abuse, Monograph 84)
Bickel WK, Mathis D, Emmett-Oglesby M, Harris C, Lal H, Gauvin D, Young A, Schindler C, Williams BA (1987) Stimulus control: continuity or discontinuity? Views from the study of interoceptive and exteroceptive stimuli. Psychol Rec 37:153–198
Bigelow GR, Griffiths R, Liebson I (1975) Experimental models for the modification of human drug self-administration: methodological developments in the study of ethanol self-administration by alcoholics. Fed Proc 34:1785–1792
Bigelow GR, Griffiths RR, Liebson IA (1977) Pharmacological influences upon ethanol self-administration. In: Gross MM (ed) Alcohol intoxication and withdrawal, vol 111B. Plenum, New York, pp 523–538
Blakey R, Baker R (1980) An exposure approach to alcohol abuse. Behav Res Ther 18:319–323
Blanchard RJ, Hori K, Tom P, Blanchard C (1987) Social structure and ethanol consumption in the laboratory rat. Pharmacol Biochem Behav 28:437–442
Brown ME, Anton RF, Malcolm R, Ballenger JC (1988) Alcohol detoxification and withdrawal seizures: clinical support for a kindling hypothesis. Biol Psychiatry 23:507–514
Cappell H (1975) An evaluation of tension models of alcohol consumption. In: Gibbins RJ, Isreal Y, Kalant H, Popham RE, Schmidt W, Smart RG (eds) Research developments in alcohol and drug problems, vol 2. Wiley, New York, pp 177–209
Cappell H, LeBlanc AE (1981) Tolerance and physical dependence. Do they play a role in self-administration? In: Israel Y, Glaser FB, Kalantt H, Popham RE, Schmidt W, Smart RG (eds) Recent advances in alcohol and drug problems, vol 6. Plenum, New York, pp 159–196
Carlson RH, Lydic R (1976) The effects of ethanol upon threshold and response rate for self-stimulation. Psychopharmacology 50:61–64
Carroll ME, Stitzer ML, Strain E, Meisch RA (1990) The behavioral pharmacology of alcohol and other drugs. In: Galanter M (ed) Recent developments in alcoholism, vol 8. Plenum, New York, pp 5–46

Caudill BD, Lipscomb TR (1980) Modeling influences on alcoholics' rates of alcohol consumption. J Appl Behav Anal 13:355–365
Caudill BD, Marlatt GA (1975) Modeling influences in social drinking: an experimental analogue. J Consult Clin Psychol 43:405–415
Childress AR, McLellan AT, Ehrman R, O'Brien C (1988) Classically conditioned responses in opioid and cocaine dependence: a role in relapse? In: Ray B (ed) Learning factors in substance abuse. US Government Printing Office, Washington DC, pp 25–43 (National Institute on Drug Abuse, Monograph 84)
Cloninger CR (1987) Neurogenetic adaptive mechanisms in alcoholism. Science 236:410–416
Cloninger CR (1988) Etiologic factors in substance abuse: an adoption study perspective. In: Pickens RW, Svikis DS (eds) Biological vulnerability to drug abuse. US Government Printing Office, Washington DC, pp 52–71 (National Institute on Drug Abuse, DHHS publication no (ADM) 88-1590)
Cloninger CR, Bohman M, Sigvardsson S (1981) Inheritance of alcohol abuse. Arch Gen Psychiatry 38:861–868
Cohen M, Liebson I, Faillace L, Speers W (1971) Alcoholism: controlled drinking and incentives for abstinence. Psychol Rep 28:575–580
Colombo G, Grant KA (1992) NMDA receptor complex antagonists have ethanol-like discriminative stimulus effects. Ann NY Acad Sci 654:192–198
Cooney NL, Baker L, Pomerleau D, Josephy B (1984) Salivation to drinking cues in alcohol abusers: toward the validation of a physiological measure of craving. Addict Behav 9:91–94
Corkfield-Sumner PK, Stolerman IP (1978) Behavioural tolerance. In: Blackman DE, Sanger DJ (eds) Contemporary research in behavioral pharmacology. Plenum, New York, pp 391–448
Crabbe JC (1983) Sensitivity to ethanol in inbred mice: genotypic correlations among several behavioral responses. Behav Neurosci 97:280–289
Crabbe JC, Belknap JK (1992) Genetic approaches to drug dependence. Trends Pharmacol Sci 13:212–219
Crabbe JC, Harris RA (1991) The genetic basis of alcohol and drug addictions. Plenum, New York
Crabbe JC, Kosobud A, Young ER, Tam BR, McSwigan JD (1985) Bidirectional selection for susceptibility to ethanol withdrawal seizures in Mus musculus. Behav Genet 15:521–536
Crabbe JC, Young ER, Cuetsch CM, Tam BR, Kosobud A (1987) Mice genetically selected for differences in open-field activity after ethanol. Pharmacol Biochem Behav 27:577–581
Crowley TJ (1988) Learning and unlearning drug abuse in the real world: clinical treatment and public policy. In: Ray B (ed) Learning factors in substance abuse. US Government Printing Office, Washington DC, pp 100–121 (National Institute on Drug Abuse, Monograph 84)
Crowley TJ, Andrews AE (1987) Alcoholic-like drinking in simian social groups. Psychopharmacology 92:196–205
Cunningham CL, Noble D (1992) Conditioned activation induced by ethanol: role in sensitization and conditioned place preference. Pharmacol Biochem Behav 43: 307–313
Cunningham CL, Hallett CL, Niehus DR, Hunter JS, Nouth L, Risinger FO (1991) Assessment of ethanol's hedonic effects in mice selectively bred for sensitivity to ethanol-induced hypothermia. Psychopharmacology 105:84–92
Cunningham CL, Niehus DR, Malott DH, Prather LK (1992) Genetic differences in the rewarding and activating effects of morphine and ethanol. Psychopharmacology 107:385–393
Davies BT, Parker LA (1990) Novel versus familiar ethanol: a comparison of aversive and rewarding properties. Alcohol 7:523–529
Deitrich RA, Dunwiddie TV, Harris RA, Erwin VG (1989) Mechanism of action of ethanol: initial central nervous system actions. Pharmacol Rev 41:491–537

De Soto CB, O'Donnell WE, Allred LJ, Lopes CE (1985) Symptomology in alcoholics at various stages of abstinence. Alcohol Clin Exp Res 9:505–512
Deutsch JA, Walton NY (1977) A rat alcoholism model in a free choice situation. Behav Biol 19:349–360
Dolinsky ZS, Morse DE, Kaplan RF, Meyer RE, Corry D, Pomerleau OF (1987) Neuroendocrine, psychophysiological and subjective reactivity to an alcohol placebo in alcoholic subjects. Alcohol Clin Exp Res 11:296–300
Edwards G (1990) Withdrawal symptoms and alcohol dependence: fruitful mysteries. Br J Addict vol 1. 85:447–461
Edwards G, Gross MM (1976) Alcohol dependence: a provisional description of a clinical syndrome. Br Med J 1:1058–1061
Elmer GI, Meisch RA, George FR (1987) Oral ethanol reinforced behavior in inbred mice. Behav Genet 17:439–451
Emmett-Oglesby MW, Mathis DA, Moon RTY, Lal H (1990) Animal models of drug withdrawal symptoms. Psychopharmacology 101:292–309
Erwin VG, McClearn GE, Kuse AR (1980) Interrelationships of alcohol consumption, actions of alcohol, and biochemical traits. Pharmacol Biochem Behav 13:297–302
Falk JL (1971) The nature and determinants of adjunctive behavior. Physiol Behav 6:577–588
Falk JL (1981) The nevironmental generation of excessive behavior. In: Mule SJ (ed) Behavior in excess: an examination of the volitional disorders. Free Press, New York, pp 313–377
File SF (1990) The use of social interaction as a method for detecting anxiolytic activity of chlordiazepoxide-like drugs. J Neurosci Methods 2:219–238
File SF, Hyde JR (1978) Can social interaction be used to measure anxiety? Br J Pharmacol 62:19–28
Froehlich JC, Harts J, Lumeng L, Li T-K (1988) Differences in response to the aversive properties of ethanol in rats selectively bred for oral ethanol preference. Pharmacol Biochem Behav 31:215–222
Gatto GJ, Murphy JM, Waller MB, McBride WJ, Lumeng L, Li T-K (1987) Persistence of tolerance to a single dose of ethanol in the selectively-bred alcohol-preferring P rat. Pharmacol Biochem Behav 28:105–110
Gauvin DV, Holloway FA (1992) Ethanol tolerance developed during intoxicated operant performance in rats prevents subsequent ethanol-induced conditioned taste aversion. Alcohol 9:167–170
Gauvin DV, Harland RD, Criado JR, Michaelis RC, Holloway FA (1989) The discriminative stimulus properties of ethanol and acute ethanol withdrawal states in rats. Drug Alcohol Depend 24:103–113
Gauvin DV, Youngblood BD, Holloway FA (1992) The discriminative stimulus properties of acute ethanol withdrawal (hangover) in rats. Alcohol Clin Exp Res 16:336–341
George FR (1987) Genetic and environmental factors in ethanol self-administration. Pharmacol Biochem Behav 27:379–384
George FR (1990) Genetic tools in the study of drug-self-administration. Alcohol Clin Exp Res 12:586–590
Goddard GV, McIntyre DC, Leech CK (1969) A permanent change in brain function resulting from daily electrical stimulation. Exp Neurol 25:295–330
Goldberg SR, Schuster CR (1970) Conditional nalorphine-induced abstinence changes: persistence in post morphine-dependent monkeys. J Exp Anal Behav 14:33–39
Goldberg SR, Woods JH, Schuster CR (1969) Conditioned increases in self-administration in rhesus monkeys. Science 166:1306–1307
Goldberg SR, Hoffmeister F, Schlichting UU, Wuttke W (1971) Aversive properties of nalorphine and naloxone in morphine-dependent monkeys. J Pharmacol Exp Ther 179:268–276

Goldberg SR, Hoffmeister F, Schlichting UU (1978) Morphine antagonists: modification of behavioral effects by morphine dependence. In: Singh JM, Miller L, Lal H (eds) Experimental pharmacology. Futura, New York, pp 31–48
Goldberg SR, Spealman RD, Kelleher RT (1979) Enhancement of drug-seeking by environmental stimuli associated with cocaine or morphine injections. Neuropharmacology 18:1015–1017
Goodwin D (1983) Alcoholism. In: Tarter R (ed) The child at psychiatric risk. Oxford University Press, New York, pp 195–213
Goudie AJ, Demellweek C (1986) Conditioning factors in drug tolerance. In: Goldberg S, Stolerman I (eds) Behavioral analysis of drug dependence. Academic, New York, pp 225–283
Grant KA, Colombo G (1993) The discriminative stimulus effect of ethanol: effects of training dose on the substitution of N-methyl-D-aspartate antagonists. J Pharmacol Exp Ther 264:1241–1247
Grant KA, Johanson CE (1988) Oral ethanol self-administration in free-feeding rhesus monkeys. Alcohol Clin Exp Res 12:780–784
Grant KA, Samson HH (1985) Induction and maintenance of ethanol self-administration without food deprivation in the rat. Psychopharmacology 86:475–479
Grant KA, Hoffman PL, Tabakoff B (1990) Neurobiological and behavioral approaches to tolerance and dependence. In: Edwards G, Lader M (eds) The nature of dependence. Oxford University Press, New York, pp 135–169
Griffiths R, Bigelow G, Liebson I (1978) Relationship of social factors to ethanol self-administration in alcoholics. In: Nathan PE, Marlatt GA, Loberg T (eds) Alcoholism: new directions in behavioral research and treatment. Plenum, New York, pp 351–379
Grupp LA (1981) Ethanol as a negative reinforcer in an active avoidance paradigm. Prog Neuropsychopharmacol 5:241–244
Grupp LA, Stewart RB (1983) Active and passive avoidance behaviour in rats produced by intravenous infusions of ethanol. Psychopharmacology 79:318–321
Higgins ST, Stitzer ML (1988) Effects of alcohol speaking in isolated humans. Psychopharmacology 95:189–194
Higley JD, Hasert MF, Suomi SJ, Linnoila M (1991) Nonhuman primate model of alcohol abuse: effect of early experience, personality, and stress on alcohol consumption. Proc Natl Acad Sci USA 88:7261–7265
Hinson RE, Siegel S (1980) The contribution of Pavlovian conditioning to ethanol tolerance and dependence. In: Rigter H, Crabbe JC (eds) Alcohol tolerance, dependence and addiction. Elsevier, Amsterdam, pp 181–199
Hinson RE, Poulos CX, Thomas W, Cappell H (1986) Pavlovian conditioning and addictive behavior: relapse to oral self-administration of morphine. Behav Neurosci 100:368–375
Hodgson RJ, Rankin HJ (1976) Modification of excessive drinking by cue exposure. Behav Res Ther 14:305–307
Hoffman DC (1989) The use of place conditioning in studying the neuropharmacology of drug reinforcement. Brain Res Bull 23:373–397
Hoffman PL (1987) Central nervous system effects of neurohypophyseal peptides. In: The peptides, vol 8. Academic, New York, pp 239–295
Hoffman PL, Ishizawa H, Giri PR, Dave JR, Grant KA, Liu L-I, Gulya K, Tabakoff B (1990) The role of arginine vasopressin in alcohol tolerance. Ann Med 22:269–274
Holloway FA, King DA, Bedingfield JB, Gauvin DV (1992) Role of context in ethanol tolerance and subsequent hedonic effects. Alcohol 9:109–116
Hunter BE, Walker DW, Riley JN (1974) Dissociation between physical dependence and volitional ethanol consumption: role of multiple withdrawal episodes. Pharmacol Biochem Behav 2:523–529

Janicke U-A, Coper H (1984) Repeated withdrawal from barbital as a drive for drug taking behavior in rats. Drug Alcohol Depend 13:43–54
Jessor R, Jessor SL (1975) Adolescent development and the onset of drinking: a longitudinal study. J Stud Alcohol 36:27–51
Jessor R, Collins MI, Jessor SL (1972) On becoming a drinker: social-psychological aspects of an adolescent transition. Ann NY Acad Sci 197:199–213
Johanson CE, Uhlenhuth EH (1980) Drug preference and mood in humans: d-amphetamine. Psychopharmacology 71:275–279
Johanson CE, Uhlenhuth EH (1981) Drug preference and mood in humans: repeated assessment of d-amphetamine. Pharmacol Biochem Behav 14:159–163
Kalivas PW, Striplin CD, Steketee JD, Klitenick MA, Duffy P (1992) Cellular mechanisms of behavioral sensitization to drugs of abuse. Ann NY Acad Sci 654:128–135
Kandel DA, Schuster CR (1977) An investigation of nalorphine and perphenazine as negative reinforcers in an escape paradigm. Pharmacol Biochem Behav 6:61–71
Kandel DB, Kessler RC, Margulies RZ (1978) Antecedents of adolescent initiation into stages of drug use: a developmental analysis. In: Kandel DB (ed) Longitudinal research on drug use. Halsted, New York, pp 73–99
Kaplan MA, Puglisi K (1986) Stress and conflict conditions leading to and maintaining voluntary alcohol consumption in rats. Pharmacol Biochem Behav 24: 271–280
Kaplan R, Meyer R, Stroebel C (1983) Alcohol dependence and responsivity to an alcohol stimulus as predictors of alcohol consumption. Br J Addict 78:259–267
Kaplan RE, Meyer RE, Virgilio LM (1984) Physiological reactivity to alcohol cues and the awareness of an alcohol effect in a double blind placebo design. Br J Addict 79:439–442
Karoly AJ, Winger GD, Ikomi F, Woods JH (1978) The reinforcing property of ethanol in the rhesus monkey. Psychopharmacology 58:19–25
Kelleher RT, Goldberg SR (1977) Fixed-interval responding under second-order schedules of food presentation or cocaine injection. J Exp Anal Behav 28: 221–231
Koob GF, Britton KT (in press) Neurobiological substrates for the anti-anxiety effects of ethanol. In: Begleiter H, Kissin J (eds) Alcohol and alcoholism. Oxford University Press, New York
Kornetsky DR, Esposito RU, McLean S, Jacobson JD (1979) Intracranial self-stimulation thresholds: a model for the hedonic effects of drugs of abuse. Arch Gen Psychiatry 36:289–292
Kraemer GW, McKinney WT (1985) Social separation increases alcohol consumption in rhesus monkeys. Psychopharmcology 86:182–189
Krank MD (1989) Environmental signals for ethanol enhance free-choice ethanol consumption. Behav Neurosci 103:365–372
Lal H, Emmett-Oglesby MW (1983) Behavioral analogs of anxiety: animal models. Neuropharmacology 22:1423–1441
Lal H, Harris CM, Benjamin D, Springfield AC, Bhadra S, Emmett-Obelsby MW (1988) Characterization of a pentylienetetrazol-like interoceptive stimulus produced by ethanol withdrawal. J Pharm Exp Ther 247:508
LeBlanc AE, Gibbins RH, Kalant H (1973) Behavioural augmentation of tolerance to ethanol in the rat. Psychopharmacology 30:117–122
Levine DG (1974) Needle freaks: compulsive self-injections by drug users. Am J Psychiatry 131:297–300
Lewis MJ, Phelps RW (1987) A multifunctional on-line brain stimulation system: investigation of alcohol and aging effects. In: Bozarth MA (ed) Methods of assessing the reinforcing properties of abused drugs. Springer, Berlin Heidelberg New York Tokyo, pp 463–478
Li T-K, Lumeng L, McBridge WJ, Murphy JM (1987) Rodent lines selected for factors affecting alcohol consumption. Alcohol Alcoholism [Suppl 1]:91–96

Li TK, Lemeng L, Froehlich JC, Murphy JM, McBride WJ (1988) Genetic influence on response to the reinforcing properties of ethanol in rats. In: Kuriyama K, Takeda A, Ishii H (eds) Biochemical and social aspects of alcohol and alcoholism. Elsevier, Amsterdam, pp 487–490

Lister RG (1987) The use of a plus-maze to measure anxiety in the mouse. Psychopharmacology 92:180–185

Lister RG, Hilakivi LA (1988) The effects of novelty, isolation light and ethanol on the social behavior of mice. Psychopharmacology 96:181–187

Ludwig AM (1985) Cognitive processes associated with "spontaneous" recovery in alcoholism. J Stud Alcohol 46:53–58

Ludwig AM, Wikler A, Stark LH (1974) The first drink: psychobiological aspects of craving. Arch Gen Psychiatry 30:529–547

Ludwig AM, Cain RB, Wikler A, Taylor RM, Bendfelt F (1977) Physiologic and situational determinants of drinking behavior. In: Gross MM (ed) Alcohol intoxication and withdrawal-IIIb: studies in alcohol dependence. Plenum, New York, pp 589–600

MacAndrew C, Edgerton RB (1969) Drunken comportment. Aldine, Chicago

Magnuson DJ, Reid LD (1977) Addictive agents and intracranial stimulation (ICS): pressing for ICS under the influence of ethanol before and after physical dependence. Bull Psychol Soc 10:364–366

Majchrowicz E (1975) Induction of physical dependence upon ethanol and the associated behavioral changes in rats. Psychopharmacology 43:245–254

Marlatt GA, Rohsenow DJ (1980) Cognitive processes in alcohol use: expectancy and the balanced placebo design. In: Mello NK (ed) Advances in substance abuse: behavioral and biological research, vol 1. JAI Press, Greenwich CT, pp 159–199

Marlatt GA, Baer J, Donovan DM, Kivlahan DR (1988) Addictive behaviors: etiology and treatment. Annu Rev Psychol 39:223–252

Meanker T (1967) Anxiety about drinking in alcoholics. J Abnorm Psychol 72:43–48

Meisch RA (1977) Ethanol self-administration: infrahuman studies. In: Thompson T, Dews PB (eds) Advances in behavioral pharmacology, vol 1. Academic, New York, pp 35–84

Mello NK (1972) Behavioral studies of alcoholism. In: Kissin B, Begleiter H (eds) The biology of alcoholism, vol 2. Physiology and behavior. Plenum, New York, pp 219–297

Mello NK, Mendelson JH (1977) Clinical aspects of alcohol dependence. In: Martin WR (ed) Drug addiction: handbook of experimental pharmacology. Springer, Berlin Heidelberg New York, pp 613–666

Mello NK, Bree MP, Mendelson JH, Ellingboe J (1983) Alcohol self-administration disrupts reproductive function in female macaque monkeys. Science 221:677–679

Merry J (1966) The 'loss of control' myth. Lancet i:1257–1258

Meyer RE, Mirin SM (1979) The heroin stimulus: Implications for a theory of addiction. Plenum, New York, pp 231–247

Meyer RE (1988) Conditioning phenomena and the problem of relapse in opioid addicts and alcoholics. In: Ray B (ed) Learning factors in substance abuse. US Government Printing Office, Washington DC, pp 161–179 (National Institute on Drug Abuse, Monograph 84)

Monti PM, Binkoff JA, Abrams DB, Zwick WR, Nirenberg TD, Liepman MR (1987) Reactivity of alcoholics and nonalcoholics to drinking cues. J Abnorm Psychol 96:122–126

Monti PM, Rohsenow DJ, Abrams DB, Binkoff JA (1988) Social learning approaches to alcohol relapse: selected illustrations and implications. In: Ray B (ed) Learning factors in substance abuse. US Government Printing Office, Washington DC, pp 141–160 (National Institute on Drug Abuse, Monograph 84)

Mossberg D, Liljeberg P, Borg S (1985) Clinical conditions in alcoholics during long-term abstinence: a descriptive, longitudinal treatment study. Alcohol 2:551–553

Mucha RF, van der Kooy D, O'Shaughnessy M, Bucenieks P (1982) Drug reinforcement studies by the use of place conditioning in rat. Brain Res 243:91–105

Murphy JM, Gatto GJ, McBride WJ, Lumeng L, Li TK (1989) Operant responding for oral ethanol in the alcohol-preferring and alcohol-nonpreferring lines of rats. Alcohol 6:127–131

Myers RD, Staltman WP, Martin GE (1972) Effects of ethanol dependence induced artificially in the rhesus monkey on the subsequent preference for ethyl alcohol. Physiol Behav 9:43

Napier JR, Napier PH (1985) The natural history of primates. MIT Press, Cambridge, pp 80–83

NIAAA (National Institute on Alcohol Abuse and Alcoholism) (1990) Seventh special report to the US Congress on Alcohol and Health. US Government Printing Office, Washington DC (PHS publication no ADM-281-88-0002)

Numan R (1981) Multiple exposures to ethanol facilitate intravenous self-administration of ethanol by rats. Pharmacol Biochem Behav 15:101–108

O'Brien CP, Nace EP, Mintz J, Meyers AL, Ream N (1980) Follow-up of Vietnam veterans. 1. Relapse to drug use after Vietnam service. Drug Alcohol Depend 5:333–340

O'Brien CP, Childress AR, McLellan AT, Ehrman R, Ternes JW (1988) Types of conditioning found in drug-dependent humans. In: Ray B (ed) Learning factors in substance abuse. National Institute on Drug Abuse Monograph 84. US Government Printing Office, Washington, DC, pp 44–63

O'Brien CP, Childress AR, McLellan AT (1991) Conditioning factors may help to understand and prevent relapse in patients who are recovering from drug dependence. In: Pickens RW, Leukefeld CG, Schuster CR (eds) Improving drug abuse treatment. US Government Printing Office, Washington DC, pp 293–312 (National Institute on Drug Abuse, Monograph 106)

Oei TPS, Jones R (1986) Alcohol-related expectancies: have they a role in the understanding and treatment of problem drinking? Adv Alcohol Subst Abuse 6:89–105

Olds J, Milner P (1954) Positive reinforcement produced by electrical stimulation of septal area and other regions of rat brain. J Comp Physiol Psychol 47:419–427

Peele S (1985) The meaning of addiction. Heath, Lexington MA

Pellow S, File SE (1986) Anxiolytic and anxiogenic drug effects on exploratory activity in an elevated plus-maze: a novel test of anxiety in the rat. Pharmacol Biochem Behav 24:525–529

Pellow S, Chopin B, Briley M (1985) Validation of the open: closed arm entries in an elevated plus maze as a measure of anxiety in the rat. J Neurosci Methods 14:149–167

Philips TJ, Crabbe JC (1991) Behavioral studies of genetic differences in alcohol action. In: Crabbe JC, Harris RA (eds) The genetic basis of alcohol and drug actions. Plenum, New York, pp 25–104

Pohorecky LA (1981) The interaction of ethanol and stress. Neurosci Biobehav Rev 5:209–229

Pohorecky LA, Roberts P (1991) Development of tolerance to and physical dependence on ethanol: daily versus repeated cycles treatment with ethanol. Alcohol Clin Exp Res 15:824–833

Poling A, Thompson T (1977a) Effects of delaying food availability contingent on ethanol-maintained lever pressing. Psychopharmacology 51:289–291

Poling A, Thompson T (1977b) Suppression of ethanol-reinforced lever pressing by delayed food availability. J Exp Behav 28:271–283

Pomerleau OF, Fertig J, Baker L, Cooney N (1983) Reactivity to alcohol cues in alcoholics and nonalcoholics: implications for a stimulus control analysis of drinking. Addict Behav 8:1–10

Rankin H, Hodgson R, Stockwell T (1983) Cue exposure and response prevention with alcoholics: a controlled trial. Behav Res Ther 21:435–446
Reid LD, Hunter GA, Beaman CM, Hubbell CL (1985) Toward understanding ethanol's capacity to be reinforcing: a conditioned place preference following injections of ethanol. Pharmacol Biochem Behav 22:483–487
Riege WH (1987) Specificity of memory deficits in alcoholism. In: Galanter M (ed) Recent developments in alcoholism, vol 5. Plenum, New York, pp 81–109
Risinger FO, Cunningham CL (1992) Genetic differences in ethanol-induced hyperglycemia and conditioned taste aversion. Life Sci 50:PL113–PL118
Risinger FO, Dickinson SD, Cunningham CL (1992a) Haloperidol reduced ethanol-induced motor activity stimulation but not conditioned place preference. Psychopharmacology 107:453–456
Risinger FO, Malott DH, Riley AL, Cunningham CL (1992b) Effect of Ro 15-4513 on ethanol-induced conditioned place preference. Pharmacol Biochem Behav 43:97–102
Ritz MC, George FR, Meisch RA (1989) Ethanol self-administration in ALKO rats: I. Effects of selection and concentration. Alcohol 6:227–233
Robins LN, Davis DH, Goodwin DW (1974) Drug use by U.S. army enlisted men in Vietnam: a follow-up upon their return home. Am J Epidemol 99:235–249
Rodin E (1958) Metrazol tolerance in a "normal" volunteer population. Electroencephalogr Clin Neurophysiol 10:433–446
Roehrs TA, Samson HH (1981) Ethanol reinforced behavior: effect of chronic ethanol overdrinking. Alcohol Clin Exp Res 5:165
Rozin P, Kennel K (1983) Acquired preferences for piquant foods by chimpanzees. Appetite J Intake Res 4:67–77
Samson HH (1987) Initiation of ethanol-maintained behavior: a comparison of animal models and their implication to human drinking. In: Thomspon T, Dews PB, Barrett JE (eds) Neurobehavioral pharmacology, vol 6. Erlbaum, Hillsdale NJ, pp 221–248
Samson HH, Fromme K (1984) Social drinking in a simulated tavern: an experimental analysis. Drug Alcohol Depend 14:141–163
Samson HH, Grant KA (1990) Some implications of animal model self-administration studies for human alcohol problems. Drug Alcohol Depend 25:141–144
Samson HH, Lindberg K (1984) Comparison of sucrose-sucrose to sucrose-ethanol concurrent responding in the rat: reinforcement schedule and fluid concentration effects. Pharmacol Biochem Behav 20:973–977
Samson HH, Roehrs TA, Tolliver GA (1982) Ethanol reinforced responding in the rat: a concurrent analysis using sucrose as the alternate choice. Pharmacol Biochem Behav 17:333–339
Samson HH, Tolliver GA, Lumeng L, Li TK (1989) Ethanol reinforcement in the alcohol nonpreferring rat: initiation using behavioral techniques without food restriction. Alcohol Clin Exp Res 13:378–385
Schecter MD (1992) Locomotor activity but not conditioned place preference is differentially affected by a moderate dose of ethanol administered to P and NP rats. Alcohol 9:185–188
Schecter MD, Krimmer EC (1992) Difference in response to the aversive properties and activity of low dose ethanol in LAS and HAS selectively bred rats. Psychopharmacology 107:564–568
Schuckit MA, Gold E (1988) A simultaneous evaluation of multiple markers of ethanol/placebo challenges in sons of alcoholics and controls. Arch Gen Psychiatry 45:211–216
Schuster CR (1986) Implications of laboratory research for the treatment of drug dependence. In: Goldberg S, Stolerman I (eds) Behavioral analysis of drug dependence. Academic, New York, pp 357–385
Schuster CR, Woods JH (1968) The conditioned reinforcing effects of stimuli associated with morphine reinforcement. Int J Addict 3:223–230

Schuster CR, Fishman MW, Johanson CE (1981) Internal stimulus control and subjective effects of drugs. In: National Institute on Drug Abuse, Research Monograph 37. US Government Printing Office, Washington DC, pp 116–179
Schwartz-Stevens K, Samson HH, Tolliver GA, Lumeng L, Li T-K (1991) The effects of ethanol initiation procedures on ethanol reinforced behavior in the alcohol-preferring rat. Alcohol Clin Exp Res 15:277–285
Senter RJ, Persensky JJ (1968) Effects of environment on ethanol consumption in rats after conditioning, Q J Stud Alcohol 29:856–862
Sher KJ, Levenson RW (1982) Risk for alcoholism and individual differences in the stress-response dampening effect of alcohol. J Abnorm Psychol 19:350–367
Sherman GT, Lal H (1980) Generalization and antagonism studies with convulsant, GABAergic and anticonvulsant drugs in rats trained to discriminate pentylenetetrazol from saline. Neuropharmacology 19:473–479
Sherman J, Hicks CF, Rice AG (1983) Preferences and aversions for stimuli paired with ethanol in hungry rats. Anim Learn Behav 11:101–106
Sherman JE, Rusiniak KW, Garcia J (1984) Alcohol-ingestive habits: the role of flavor and effect. In: Galanter M (ed) Recent developments in alcoholism, vol 2. Plenum, New York, pp 59–79
Sinclair JD (1974) Rats learning to work for alcohol. Nature 249:590–592
Sinclair JD, Li T-K (1989) Long and short alcohol deprivation: effects on AA and P alcohol-preferring rats. Alcohol 6:505–509
Spealman RD (1979) Behavior maintained by termination of a schedule of self-administration of cocaine. Science 204:1231–1233
Staiger PK, White JM (1988) Conditioned alcohol-like and alcohol-opposite responses in humans. Psychopharmacology 95:87–91
Stewart J (1992) Neurobiology of conditioning to drugs of abuse. Ann NY Acad Sci 654:335–346
Stewart J, Eikelboom R (1987) Conditioned drug effects. In: Iversen LL, Iversen SD, Synder SH (eds) New directions in behavioral pharmacology. Plenum, New York, p 57 (Handbook of Psychopharmacology, vol 19)
Stewart J, de Wit H, Eikelboom R (1984) Role of unconditioned drug effects in the self-administration of opiates and stimulants. Psychol Rev 91:251–268
Stewart RB, Grupp LA (1981) An investigation of the interaction between the reinforcing properties of food and ethanol using the place preference paradigm. Prog Neurol Psychopharmacol 5:609–613
Stitzer ML, Kirby KC (1991) Reducing illicit drug use among methadone patients. In: Pickens RW, Leukefeld CG, Schuster CR (eds) Improving drug abuse treatment. US Government Printing Office, Washington DC, pp 178–203 (National Institute on Drug Abuse, Monograph 106)
Stitzer ML, Griffiths RR, Bigelow GE, Liebson I (1981) Human social conversation: effects of ethanol, secobarbital and chlorpromazine. Pharmacol Biochem Behav 14:353–360
Stolerman I (1988) Behavioral analysis of addiction: what future does it have? Br J Addict 83:991–993
Stolerman I (1992) Durgs of abuse: behavioral principles, methods and terms. Trends Pharmacol Sci 13:170–176
Stretch R, Gerber GJ, Wood SM (1971) Factors affecting behavior maintained by response-contingent intravenous infusions of amphetamine by squirrel monkeys. Can J Physiol Pharmacol 49:581–589
Tabakoff B, Hoffman, PL (1988) Tolerance and etiology of alcoholism; hypothesis and mechanisms. Alcohol Clin Exp Res 12:184–186
Tabakoff B, Kiianmaa K (1982) Does tolerance develop to the activating, as well as the depressant, effects of ethanol? Pharmacol Biochem Behav 17:1073–1076
Tang M, Brown B, Falk JL (1982) Complete reversal of chronic ethanol polydipsia by schedule withdrawal. Pharmacol Biochem Behav 16:155–158

Tarter RE (1988) The high risk paradigm in alcohol and drug abuse research. In: Pickens RW, Svikis DS (eds) Biological vulnerability to drug abuse. Us Government Printing Office, Washington DC, pp 73–86 (DHHS publication no (ADM) 88-1590)
Tarter, RE, Hegedug A, Winsten N, Alterman A (1984) Neuropsychological, personality and familial characteristics of physically abused juvenile delinquents. J Am Acad Child Psychiatry 23:668–674
Tarter RE, Alterman A, Edwards K (1985) Vulnerability to alcoholism in men: a behavior-genetic perspective. J Stud Alcohol 46:329–356
Thompson T, Ostland W (1965) Susceptibility to readdiction as a function of the addiction and withdrawal environments. J Comp Physiol Psychol 60:388–392
Thompson TI, Schuster CR (1964) Morphine self-administration, food reinforced and avoidance behaviors in rhesus monkeys. Psychopharmacologia 5:87–94
Tye NC, Iversen SD (1975) Some behavioral signs of morphine withdrawal blocked by conditional stimuli. Nature 255:416–418
Vuchinich RE, Tucker JA (1988) Contributions from behavioral theories of choice to an analysis of alcohol abuse. J Abnorm Psychol 97:181–195
Waller MB, McBride WJ, Lumeng L, Li T-K (1982) Induction of dependence on ethanol by free-choice drinking in alcohol-preferring rats. Pharmacol Biochem Behav 16:501–507
Waller MB, McBride WJ, Lumeng L, Li T-K (1983) Initial sensitivity and acute tolerance to ethanol in the P and NP lines of rats. Pharmacol Biochem Behav 19:683–686
Waller MB, McBride WJ, Gatto GJ, Lumeng L, Li TK (1984) Intragastric self-infusion of ethanol by ethanol-preferring and non-preferring lines of rats. Science 225:78–80
Wenger JR, Berlin V, Woods SC (1980) Learned tolerance to the behaviorally disruptive effects of ethanol. Behav Neural Biol 28:418–430
Wikler A (1965) Conditioning factors in opiate addictions and relapse. In: Wilner DM, Kasebaum GG (eds) Narcotics. McGraw-Hill, New York, pp 85–100
Williams AF (1966) Social drinking, anxiety and depression. J Pers Soc Psychol 3:689–693
Wilson GT, Leaf RC, Nathan PE (1975) The aversive control of excessive alcohol consumption by chronic alcoholics in the laboratory setting. J Appl Behav Anal 8:13–26
Winger GD, Woods JH (1973) The reinforcing property of ethanol in the rhesus monkey. I. Initiation, maintenance and termination of intravenous ethanol reinforced responding. Ann NY Acad Sci 215:162–175
Winger G (1988) Effects of ethanol withdrawal on ethanol-reinforced responding in rhesus monkeys, Drug Alcohol Depend 22:235–240
Winger GD, Woods JH (1973) The reinforcing property of ethanol in the rhesus monkey. I. Initiation, maintenance and termination of intravenous ethanol-reinforced responding. Ann NY Acad Sci 215:162–175
Wise RA, Bozarth MA (1987) A psychomotor stimulant theory of addiction. Psychol Rev 94:469–492
Wise RA, Bauco P, Carlezon WA, Trojniar W (1992) Self-stimulation and drug reward mechanisms. Ann NY Acad Sci 654:192–198
Woods JH, Fumio I, Winger G (1971) The reinforcing property of ethanol In: Roach MK, McIssac WM, Creaven PJ (eds) Biological aspects of alcohol. University of Texas Press, Austin, pp 371–388
Woods JH, Down DA, Villarreal JE (1973) Changes in operant behavior during deprivation- and antagonist-induced withdrawal states. In: Goldberg L, Hoffmeister F (eds) Psychic dependence. Springer, Berlin Heidelberg New York, pp 114–123
Woods JH, Downs DA, Carney J (1975) Behavioral functions of narcotic antagonists: response-drug contingencies. Fed Proc 34:1777–1784

World Health Organization (1978) Mental disorders: glossary and guide to their classification in accordance with the ninth revision on the international classification of diseases. WHO, Generva

Young AM, Herling S (1986) Drugs as reinforcers: studies in laboratory animals. In: Goldberg SR, Stolerman IP (eds) Behavioral analysis of drug dependence. Academic, New York, pp 9–69

CHAPTER 10

Ethanol and Neurohormonal Regulation

J. LITTLETON

The neurohormonal system is responsible for the regulation of a wide variety of physiological and behavioural responses of the organism, so alterations of this system by ethanol have widespread and serious functional consequences. The excellent review by CICERO (1982) leaves no doubt whatsoever that ethanol does perturb the neurohormonal system at several different levels, including the synthesis, release and receptor-mediated responses of all the chemical messengers involved, i.e., including hypothalamic-releasing hormones, trophic hormones from the pituitary and the endocrine hormones themselves. These effects alone would be sufficient to justify several chapters on ethanol and neurohormones, but they are only half the story because the effects of ethanol on neurohormones are not just "one way" so that alterations in neurohormones (including some caused by ethanol itself) may affect ethanol *consumption* (and ethanol tolerance and dependence). This offers the possibility that understanding interactions with neurohormones may provide clues to potential pharmacological modification of ethanol consumption in man.

These two aspects of the interaction between ethanol and neurohormones will be the focus of this chapter. In the first section I will deal with the consequences of exposure to ethanol for the synthesis, release and effects of a variety of neurohormones, and in the second section I will discuss the role of neurohormones in regulating consumption of ethanol. Since this is so vast a field it has been necessary to be very selective in the material covered. This inevitably represents a personal view of the area and I apologize for any omissions of material which others may consider to be vitally important. A more extensive review of the earlier literature is provided by CICERO (1982) and I will concentrate on concepts that have developed since that time.

A. Effects of Ethanol on Neurohormones

In general terms the neurohormones are peptides that are released by neurons originating in the hypothalamus. Some of these neurons have long axons which run down the pituitary stalk ending in terminals that secrete neurohormones, for example oxytocin and vasopressin, into the general circulation in the posterior pituitary. These neurohormones act directly on peripheral target tissues, such as the breast and the kidney in the examples used above. Other hypothalamic neurons have terminals within the median eminence of the hypothalamus which secrete their specific neurohormones into the portal blood supply to the anterior pituitary. Many of these neurohormones are the "releasing factors" or "releasing hormones" which regulate the secretion of trophic hormones from the anterior pituitary. These trophic hormones then travel in the general circulation to influence the secretions of their target endocrine organs, such as the gonads, the adrenals and the thyroid. The endocrine hormones then circulate in the blood and alter the activity and metabolism of their target organs (see GREENSPAN 1991, for general references). The effect of ethanol on "neurohormonal regulation" therefore encompasses actions on all this cascade of chemical messengers, as well as actions on the transmitters that act at the suprahypothalamic level to regulate the neurons that release the neurohormones.

There are at least two major feedback loops built into this neurohormonal system. First, the levels of the hormones released from the endocrine gland may feed back onto the secretory cells in the pituitary and modify the secretion of the trophic hormone. Second, and less controversially, the levels of circulating trophic hormones feed back onto the hypothalamus and alter the neurosecretion of releasing hormones by the hypothalamic neurons (see GREENSPAN 1991).

These feedback loops have important implications for the effects of ethanol on the endocrine system because they tend to offset any perturbation caused by the drug. Thus for example an ethanol-induced reduction in the release of a trophic hormone from the pituitary would activate the feedback loop, resulting in an increase in secretion of releasing hormone from the hypothalamus, thus tending to restore the system toward normality. This has two important implications, first, many of the acute effects of ethanol on the endocrine system are not maintained (i.e., they show tolerance to ethanol) because feedback control eventually overcomes the effect of the drug and, second, the effects of ethanol that *are* maintained with chronic use are likely to be those that are either "outside" the feedback loops or which disrupt the feedback loop. In other words, effects of ethanol that are either suprahypothalamic or are due to an action of ethanol on the responses of target organs to the endocrine hormones are those most likely to have important functional consequences in the alcoholic patient. I will try to illustrate this general principle at different stages in this chapter.

Another general point relating to effects of ethanol on the neuroendocrine system is the variety of changes ascribed to the drug. Medical students used to be taught that "the great impostor" was syphilis, because this disease could mimic any other condition you cared to name. With syphilis under relative control in the West there is a case to be made for labelling alcohol-related disease as the current impostor. Nowhere is this more evident than in the field of clinical endocrinology, where almost every type of endocrine dysfunction can be caused by excessive ethanol consumption (e.g., CICERO 1982; MILLER et al. 1989; ADLER 1992). Trying to make sense of this massive and often contradictory clinical literature is not a rewarding task and so I have concentrated on a rather mechanistic approach to the subject, trying to explain what the most common effects of ethanol may be on each neurohormonal system.

There are several other complicating factors associated with the effects of ethanol that I have also tried to ignore, but which can be very important in the clinical picture. Among the most important factors are intercurrent disease and nutritional deficiency (see ADLER 1992). In particular, alcoholic liver disease can alter the circulating levels of neurohormones by affecting their catabolism and in some cases their synthesis from prohormones. Unless they seem to be of particular importance, such complications have been excluded.

Lastly there is an ongoing theoretical argument about the toxic effects of ethanol which I will also try to ignore. The argument goes that the toxic compound is not ethanol itself but its primary metabolite, acetaldehyde. Thus, cellular damage which might lead to alterations in the synthesis and release of neurohormones could be due to circulating acetaldehyde or to acetaldehyde produced from ethanol within the endocrine cells themselves (e.g., JOHNSON et al. 1987). This is an important principle, but in most cases it is not possible to make a definite pronouncement. In the cases where there does seem to be good evidence one way or the other I will mention this, but for the most part the possibility will be ignored.

The organization of this section was difficult. A logical approach would have been to divide the material into effects of ethanol on neurohormone synthesis, effects on release, effects on catabolism, etc., but this would have made the chapter almost useless as a reference work. It is now divided on the basis of the anatomical and physiological organization of the neurohormonal system.

I. Hypothalamic-Pituitary-Gonadal Axis

The hypothalamic-pituitary-gonadal axis (HPG) axis is responsible for differences in sexual development in the fetus and at puberty, and for sexual function and behaviour in the adult. The HPG axis of course differs considerably between the sexes but in this introduction I will try to stress the similarities (see GREENSPAN 1991, for general references). As in all the neurohormonal systems, the hypothalamus provides the driving force for hormone secretion and, in this case, the hypothalamus secretes gonadotropin-releasing hormones (GnRH) in a phasic manner and these cause release of follicle-stimulating hormone (FSH) and luteinizing hormone (LH) from the anterior pituitary. Although named for their functions in the female, FSH and LH are also secreted in the male and are responsible for normal sexual function in both sexes. In the male LH is the controlling factor in causing testosterone secretion from the testis and adrenal, and both FSH and LH are necessary for normal testicular spermatogenesis and spermatic maturity. In the female, FSH and LH control the development of the follicle and the corpus luteum respectively during the reproductive cycle, and both hormones are responsible for controlling the release of oestrogen from the ovaries. Also in the female, prolactin is released from the pituitary to influence breast development (with oestrogen) and initiate lactation, and chorionic gonadotropin is secreted to regulate the synthesis and release of progesterone from the placenta.

It can be seen that any perturbation in the HPG axis will cause widespread changes in reproductive function, but this of course is only part of the story because the sex hormones have important developmental effects (those on the fetus are particularly relevant here) as well as important effects on behaviour. Before discussing these consequences I will consider some of the more important effects of ethanol on the HPG axis. These effects are, almost without exception, inhibitory (see CICERO 1982; BOYDEN and PARMENTER 1983).

1. Gonadotropin-Releasing Hormone Secretion

The secretion of GnRH by the hypothalamus is under tonic inhibition by neurons releasing β-endorphin and dopamine and the secretion of prolactin also is inhibited by dopaminergic neurons. Most of the published work on the effects of ethanol on this system relates to luteinizing hormone releasing hormone (LHRH) secretion; thus ethanol inhibits the cyclic secretion of LHRH (BOYDEN and PARMENTER 1983) and blunts the secretory response of LHRH to naloxone in rats (ADAMS and CICERO 1991). Alcohol also reduces levels of LHRH in the pituitary portal blood (CHUNG et al. 1988) while increasing the levels in the hypothalamus (DEES et al. 1984). This implies that ethanol decreases the release of LHRH by an action on the hypothalamus without inhibiting synthesis of the hormone to the same extent.

These findings suggest that ethanol disrupts the secretory mechanism for LHRH either by some suprahypothalamic mechanism or at the level of neurosecretion itself. Since ethanol has many direct effects on calcium-dependent neurosecretory processes (see Chap. 3, this volume) the latter explanation might seem a strong possibility; however, the release of LHRH from the hypothalamus in vitro was not inhibited by ethanol (EMANUELE et al. 1989). This leaves a suprahypothalamic mechanism of inhibition of LHRH release by ethanol as the favoured explanation (e.g., RIVIER et al. 1992).

2. Gonadotropin Secretion

Here again most work relates to the synthesis and release of LH and again much of the evidence suggests inhibition by ethanol (CICERO 1982). Thus, in rats, ethanol reduces the levels of LH in the systemic circulation (CHUNG et al. 1988) and causes a reduction in the mRNA for LH in the pituitary (EMANUELE et al. 1991). In addition to this effect on LH transcription, other, post-transcriptional, effects are likely mechanisms for ethanol-induced inhibition of secretion (EMANUELE et al. 1991). The inhibition of synthesis appears to be a relatively specific effect on LH because the mRNA affected is that for the β-subunit of the LH polypeptide rather than the α-subunit which is shared with thyroid-stimulating hormone (TSH), FSH, and chorionic gonadotropin (CG) (EMANUELE et al. 1991, 1992).

Again the exact mechanism by which ethanol reduces the release of LH from the pituitary is unknown, but it may be direct since in some experiments it can be demonstrated in vitro (SCHADE et al. 1983). In vivo, however, the effect of ethanol appears to depend on the level of pituitary stimulation by LHRH. Thus, acutely, ethanol does not inhibit the release of gonadotropins when this is stimulated by the "normal" mechanism, e.g., by synthetic LHRH or by LHRH released by naloxone (MELLO et al. 1986a,b; MENDELSON and MELLO 1988; MENDELSON et al. 1987, 1989). In addition, alcoholic women may show reduced "resting" LH levels in plasma but they can still respond to synthetic LHRH by secretion of LH (VALUNIAKI et al. 1984), so the LHRH receptor system and the gonadotropic secretory mechanism must be relatively intact (MELLO et al. 1989). It seems likely that the most important functional disturbance in alcoholics at the hypothalamic-pituitary level in this system is at the level of LHRH secretion (or above) and that reduced levels of circulating LH, when observed, are a consequence of this (CICERO 1982).

The action of ethanol on FSH secretion may well be similar to that on LH, since when levels are measured they also tend to be low in alcohol abusers (VAN THIEL 1983). However, these effects of ethanol cannot be due to a *general* inhibition of pituitary gonadotropin secretion because prolactin secretion is increased by ethanol acutely (CICERO 1982; PHIPPS et al. 1987). In the female alcoholic the circulating levels of prolactin are often high (TEOH et al. 1992), suggesting that these effects of ethanol on prolactin secretion are maintained.

3. Gonadal Endocrine Secretion

Most of the early work on the effects of ethanol on sex hormones was related to observations of testicular atrophy and impotence in alcoholic males (VAN THIEL et al. 1974; CICERO 1982). This sex bias in observation and experimentation can be defended only on the basis that male alcoholics were probably in the majority at that time, but in truth it probably has more to do with the preponderance of male investigators rather than patients (VANICELLI and NASH 1984). Anyway, it is now certain that the effects of ethanol are equally devastating on female sexual function (see MELLO et al. 1992). Because (and only because) of the historical progression I will begin with the effects of ethanol on gonadal endocrine function in the male.

There is no doubt that ethanol acutely can directly reduce the release of testosterone from the testis in vitro (see CICERO 1982; VAN THIEL et al. 1983). These "direct" effects of ethanol on testosterone production by the testis may, however, not be so direct in one sense, since they may depend on the metabolism of ethanol to acetaldehyde (CICERO and BELL 1980). Whether or not this effect of ethanol (or acetaldehyde) is important in vivo is uncertain because testosterone release, when stimulated by gonadotropins in vivo, has been reported to be *increased* by ethanol (PHIPPS et al. 1987). There seems to be no obvious way of reconciling these different experimental results, but there is better agreement that alcoholics have reduced circulating levels of testosterone (CICERO 1982; VAN THIEL 1983). This, of course, could be due to their reduced levels of circulating gonadotropins, as described above, rather than a direct effect of ethanol on the testis (WIDENIUS et al. 1989). This question of a direct effect of ethanol on the testis versus an indirect effect via reduced gonadotropins seems no nearer being resolved, and in reality both factors probably contribute. Thus ethanol may directly inhibit the testosterone production and release by the testes, but also prevent the feedback increase in LH release by the pituitary which would normally overcome this "direct" effect (CICERO 1982).

There is an additional complicating factor in the effects of ethanol on testicular function and that is whether the metabolism of ethanol alters the *type* of androgen produced by the testis. For example, ethanol may not only reduce the production of testosterone, it may increase the production of other steroids, or "proandrogens," by the testis (or adrenal) that are subsequently metabolized in the alcohol abuser to form oestrogens (CHAO and VAN THIEL 1983). In addition there are many other potential explanations for the demasculinization and feminization that can accompany alcoholism in males, e.g., the hyperprolactinaemia may contribute to gynaecomastia (GAVALER and VAN THIEL 1983).

The situation is even less clear in the female. Levels of circulating gonadotropins are of course low, as described above, and this inevitably reduces secretion of oestrogens from the ovary (and presumably androgens from the adrenal) (MELLO et al. 1989). However, alcohol does not reduce the LHRH-induced secretion of oestradiol (MENDELSON et al. 1989) and the secretion of oestrogens in response to gonadotropins is close to normal in alcoholic women (MELLO et al. 1989). Both these results argue that any "direct" effects of ethanol on gonadal secretion of oestrogen are comparatively unimportant here. The same may, of course, not necessarily apply to progesterone secretion, which is regulated differently.

4. Gonadal Steroid Metabolism

In the introductory paragraphs I stated that I would ignore the complications introduced by concurrent alcoholic liver damage as much as possible. In this section it is impossible to ignore the liver, both because it has such profound effects on steroid metabolism and because ethanol can alter this without any other evidence of overt liver dysfunction. In general terms, in the male the effect of the metabolism of ethanol together with adaptive changes in liver enzymes and blood flow lead to increased breakdown of testosterone (see CICERO 1982) and reduced removal of proandrogens from the circulation (e.g., GORDON et al. 1979) and their increased metabolism to oestrogen in other peripheral tissues (VAN THIEL and GAVALER 1990). In the female, oestrogen levels may be low primarily because of the reduced circulating levels of gonadotropins and so there is no feminizing corollary to the effect in males; on the contrary alcoholic women show defeminization (GAVALER and VAN THIEL 1987).

Not only are the circulating levels of appropriate gonadal steroid hormones low after chronic alcohol administration, there are also changes in the numbers of steroid receptors on the target cells. Thus, in the male, androgen receptor numbers are reduced whereas oestrogen receptor numbers are increased (EAGON et al. 1981a,b), inevitably contributing to the demasculization and feminization seen. In this case there may be a corollary in the female because, in postmenopausal women, moderate alcohol intake is associated with oestrogenization and a reduction in some of the consequences of the menopause (e.g., coronary disease, strokes) (GAVALER 1990). These beneficial effects are both inhibited and outweighed by the onset of alcohol-induced liver damage in postmenopausal women (GAVALER et al. 1990).

5. Consequences of Alterations in the HPG Axis

The consequences of effects of ethanol on the neurohormonal system in *adult* males and females have been introduced above. In males the reduction in LH and FSH result in reduced sperm production and maturity as well as reduced testosterone production by the testis. This, together with increased oestrogen formation, leads to loss of body hair, hypogonadism, impotence and loss of libido (VAN THIEL and GAVALER 1990). Adult females show signs of defeminization due to low oestrogen levels, including reductions in secondary sexual characteristics and anovulatory cycles, oligomenorrhoea (or amenorrhoea) and even premature menopause (BECKER et al. 1989). Reproductive dysfunction in women is also common, with spontaneous abortion being much more frequent in alcoholic women than controls (MELLO et al. 1989; BECKER et al. 1989). When alcoholic women do go to term there are commonly consequences of ethanol on fetal development, many of which can be ascribed to effects on gonadal hormones during pregnancy (see below).

6. Hypothalamic-Pituitary-Gonadal Axis and Fetal Development

There are many ways in which ethanol-induced alterations in endocrines could contribute to abnormal fetal development, and alterations in the HPG axis are a prime candidate (see ANDERSON 1981). Maternal oestrogen is an important factor in the development and sexual differentiation of the fetus because oestrogen has (paradoxically) masculinizing effects on the fetus via oestrogen receptors in the developing brain (see ARNOLD and GORSKI 1984). Indeed several gonadal and adrenal hormones can have masculinizing effects on the fetus, at least in the rat. In this regard, circulating testosterone is converted intracellularly to oestrogen by aromatase enzymes before this steroid activates oestrogen receptors, causing masculinization. This may be particularly relevant at the time of hypothalamic differentiation (ARNOLD and GORSKI 1984) when there is a testosterone surge from the fetal endocrine system. In this respect therefore, either reduced maternal oestrogen or reduced testosterone secreted by the fetus (or reduced conversion of testosterone) can all lead to reduced masculinization during fetal development.

Based on the previous sections, hormonal oestrogen levels in pregnant alcoholic women would be expected to be reduced, and this should reduce the masculinizing effect of the hormone during the fetal development of both sexes. However, no longitudinal measurements of oestrogen levels during pregnancy have been made in alcoholic women so this possibility is unresolved. The fetal testosterone surge, occurring around parturition in the rat, is also important for masculinization and is suppressed by ethanol (MCGIVERN et al. 1988). Both these effects of ethanol would thus be expected to reduce masculinization of the fetus, but the extent to which this bears on *human* fetal development is very uncertain.

In addition to these hormonal effects, there is another important factor which modifies the masculinizing effects of maternal steroids on fetal development. This is the synthesis of α-fetoprotein, which binds oestrogen, thus preventing it from activating oestrogen receptors in the fetus. α-Fetoprotein thus protects the female fetus from the masculinizing effects of oestrogen but, since it does not bind testosterone, it allows masculinization of the male fetus to occur. Ethanol seems to reduce the formation of α-fetoprotein (HANNIGAN et al. 1992a), and this might differentially affect the female fetus, tending to allow increased masculinization. The reduction of α-fetoprotein synthesis may well play some part in the abnormalities in births to alcoholic women; indeed it has been suggested that a low α-fetoprotein level at amniocentesis is helpful in predicting fetal alcohol effects (HALMESMAKI et al. 1987).

The consequences of these effects of ethanol on the HPG axis can only be guessed at in the human, but they can be investigated in animals. They include an ethanol-induced reduction in differentiation of the sexually dimorphic areas of the brain (RUDEEN 1984) and changes in sexually determined behaviour (e.g., MCGIVERN et al. 1984; BARRON and RILEY 1985). The changes produced by ethanol are considerably easier to observe in male animals, but there is evidence for defeminization of behaviour in female offspring also (e.g., BARRON and RILEY 1985). There is also evidence that the hormonal responses of prenatally exposed animals may be disturbed after birth (UDANI et al. 1985; HAUDA et al. 1985). Sometimes there are anatomical correlates of these neurohormonal changes, for example there are long-term deficits in GnRH secretion (MORRIS et al. 1989) which may be due to altered numbers (SCOTT et al. 1990) or morphology (McGIVERN and YELLON 1992) of GnRH-secreting neurons in the hypothalamus. Alcohol-induced alterations in sexual function and behaviour after fetal exposure to the drug almost certainly owe a great deal to effects on maternal and fetal sex hormones.

7. Conclusions

There is a wealth of evidence that acute and chronic ethanol administration alters the neuroendocrine balance of sex hormones, in a way that is almost uniformly detrimental. The HPG axis may be the most sensitive neurohormonal system to perturbation by ethanol and effects of ethanol occur at all stages of the control process, but it is extremely difficult to decide which part of this system is *most* sensitive to the drug (see CICERO 1982). As indicated in the introductory paragraphs, it is probable that perturbations of the endocrine system that are outside the physiological feedback control loops are those that persist on chronic administration of ethanol, making suprahypothalamic influences and effects on endocrine metabolism and responses likely to be of particular importance in chronic alcoholics. The consequences of the disruptions caused by ethanol are to reduce sexual characteristics, either after adult or fetal exposure to ethanol. Although these effects are more *noticeable* in the male, female reproductive function is at least equally disturbed by alcohol abuse.

II. Hypothalamic-Pituitary-Adrenal Axis

The regulation of the endocrine functions of the adrenal cortex forms an important part of the neurohormonal system. The pattern of control for glucorticoid hormones is the same as for the gonadal steroids (see GREENSPAN 1991, for general references). Thus the hypothalamus secretes the corticotropic-releasing hormone (CRH) into the pituitary portal blood and this stimulates the production and release of adrenocorticotropic hormone (ACTH). ACTH in turn causes the release of glucocorticoid hormones from the adrenal cortex and these have very widespread functional effects on the organism, most importantly in the generalized stress response, and probably including regulation of the immune system (e.g., IRWIN and HAUGER 1987). The secretion of the other adrenal cortical steroids, the mineralocorticoids such as aldosterone, is regulated by renin secretion from the juxtaglomerular cells of the kidney and controls fluid and electrolyte secretion (see GREENSPAN 1991). I will deal with the renin-angiotensin system in the second section of this chapter (Sect. B) because there is evidence for its involvement in ethanol consumption.

Additionally to the steroid hormones there are two related systems which should be mentioned here because they interact strongly with the corticosteroids. First, the secretion of catecholamines from the adrenal medulla (controlled by the sympathetic nervous system) is closely involved with stress and immune responses (IRWIN et al. 1988) and both corticosteroid secretion and sympathetic nervous system activity are increased by ethanol administration (see CICERO 1982). Second, other systems in addition to aldosterone are involved in the control of fluid and electrolyte balance and are also altered by ethanol. For example, the secretion of vasopressin, the "antidiuretic hormone", by hypothalamic neurons terminating in the posterior pituitary is inhibited by ethanol although this depends on several additional factors (e.g., HELDERMAN et al. 1978) including genetic selection for voluntary ethanol consumption (see LINKOLA et al. 1974, and last section). These effects of ethanol must clearly be taken into account when considering the functional consequences of ethanol use related to its effects on the HPA axis.

1. Corticotropin-Releasing Hormone Secretion

Most evidence suggests that ethanol stimulates the synthesis and release of CRH in the hypothalamus. For example, ethanol acutely increases the hypothalamic mRNA for CRH (RIVIER et al. 1990) and increases the release of CRH from the hypothalamus in vitro (REDEI et al. 1988). In the latter experiments preparations from animals treated in vivo with ethanol maintained an increased frequency of pulsatile CRH release ex vivo compared to controls (REDEI et al. 1988). The duration of the ethanol-induced increase in CRH in vivo differed between two mouse lines genetically selected for ethanol sensitivity (WAND 1990a) but the regulatory difference did not extend to the in vitro preparation (WAND 1990a). The return of CRH secretion toward normal levels in the face of chronic ethanol exposure is presumably because of feedback control or because of ethanol tolerance (WAND and LEVINE 1991), suggesting that genetic differences in this adaptive response to ethanol could determine the consequences of chronic exposure to the drug. If this occurs in humans then individuals that are genetically unable to operate this control mechanism would suffer a much more severe and prolonged disruption of CRH secretion [and hypothalamic-pituitary-adrenal (HPA) function] during chronic exposure to ethanol (see WAND and DOBS 1991). It is probably safe to assume that ethanol produces an acute increase in the synthesis and secretion of CRH, but it is not certain how long this persists, what mechanism is responsible, or whether this effect is the primary cause of alterations in circulating corticosteroids (see below).

2. Adrenocorticotropic Hormone Secretion

Ethanol has been shown to increase the release of ACTH from pituitary preparations in vitro (REDEI et al. 1986) and it was proposed that multiple mechanisms of ethanol were responsible. One mechanism may be an enhanced synthesis of ACTH. Thus, the production of ACTH involves the synthesis of much larger precursor polypeptides including pro-opiomelanocortin (POMC), and in mice it has been shown that ethanol acutely increases the synthesis of POMC in the pituitary (WAND 1990b). However, this effect regresses after a few days and could simply be a consequence of the putative increase in CRH levels reaching the pituitary since it is genetically influenced in the same way (WAND 1990b). There is some evidence that POMC gene expression is under the control of cAMP-dependent protein kinase and this may be the level at which adaptation to the effects of ethanol are mediated (WAND and LEVINE 1991). After chronic administration of ethanol, circulating levels of ACTH are usually normal (KIKIHANA et al. 1971), but this may be because they are subject to inhibitory feedback control from elevated corticosteroid levels. Again most of the data can be explained on the basis of an initial increase in CRH secretion producing a new elevated steady-state level of endocrine secretion with subsequent feedback reduction in the intermediary hormones originating from the hypothalamus and pituitary.

3. Adrenal Steroids

Almost all the evidence from animal and human studies suggests that ethanol acutely in vivo causes an increased secretion of corticosteroids (particularly glucocorticoids) from the adrenal glands (FAZEBAS 1966; KIKIHANA et al. 1971). Many of these studies, however, include stressful procedures (handling, injection, sampling, etc.), making it very difficult to separate the physiological response (to stress) from the pharmacological response (to ethanol) (see CICERO 1982, for experimental criticisms). There is some evidence that pharmacological concentrations of ethanol can directly influence the synthesis and release of corticosteroids from the adrenal in vitro (COBB and VAN THIEL 1982), but in vivo hypophysectomy markedly reduces the secretory response to ethanol (WRIGHT 1978), suggesting that the effect is at least partly secondary to the transient increases in CRH and ACTH described above. In animal experiments, chronic exposure to ethanol produces variable results in circulating corticosteroid levels and this may depend on several factors, for example some workers do not observe elevated glucocorticoid levels until high (stressful?) concentrations of ethanol are reached (see CICERO 1982), whereas other authors (e.g., WAND 1989) suggest that the response differs due to genetic differences between animals so that some animals maintain higher circulating glucocorticoid levels during chronic ethanol exposure. In humans the results are also variable, with some alcoholic individuals showing very high levels of the adrenal endocrine hormones (e.g., REES et al. 1977) particularly when exhibiting the pseudo-Cushing's syndrome (see below). The picture is complicated by lack of information on the drinking status of the alcoholics in some of the studies, for example, alcohol withdrawal causes an increase in corticosteroid concentrations in plasma (MERRY and MARKS 1972), but this may reflect a nonspecific response to the associated stress.

4. Consequences of Alterations in the HPA Axis

The increased circulating levels of corticosteroids found in some alcohol-abusing individuals can have widespread effects which can mimic other conditions. For example, when increased circulating glucocorticoids are produced pathologically by hypersecretion of ACTH by pituitary tumours, the features of Cushing's disease are seen (see GREENSPAN 1991). These include, among others, obesity, hirsutism, weakness and diabetes, and a similar syndrome can be produced by chronic alcohol abuse (REES et al. 1977; SMALS et al. 1977). This pseudo-Cushing's syndrome occurs in only about 5% of alcoholics and, as mentioned previously, animal work suggests that this incidence could have a genetic component in addition to the causal role of ethanol (WAND 1990a,b).

Pseudo-Cushing's syndrome is relatively unique (and relatively easy to diagnose) so there is little doubt about the role of alcohol in the condition (VAN THIEL et al. 1982); however, this is by no means true for other complications which may be caused by alterations in the HPA axis. One of the potentially most important (and least well understood) roles of the HPA axis in the response to stress may be the regulation of the immune system (e.g., IRWIN and HAUGER 1987) so that any alteration in the HPA axis will disrupt the immune response. In general terms, the early phase of the immune response is activated by release of corticosteroids, whereas if these steroids remain elevated they suppress the continuation of the immune response (IRWIN and HAUGER 1987). In other words the corticosteroids form a kind of negative feedback loop for the immune system and in this the opioid peptides also play a role (see SHAVIT et al. 1984, and last section). Both systems can account for some of the detrimental effects of ethanol on immune responses.

The early phase of the immune response involves release of cytokines from macrophages which then stimulate CRH secretion in the hypothalamus, leading to ACTH release from the pituitary and steroid release from the adrenal (e.g., SAPOLSKY et al. 1987). This early response could be blunted by the presence of ethanol and may also be reduced in those alcoholics who show continually elevated ACTH and corticosteroid levels, because these suppress both early and late immune responses (RIVIER et al. 1990; IRWIN and HAUGER 1987). This has fairly obvious implications for the resistance to intercurrent infections of individuals who abuse alcohol, but it also has potentially more sinister consequences. This is because the HPA axis is involved specifically in host responses which involve natural killer (NK) cells (IRWIN et al. 1990a,b), so that high circulating levels of corticosteroids and/or ACTH suppress the activity of these cells. In addition there is extensive evidence that chronic ethanol administration suppresses NK cell activity (Abadallah et al. 1988; Blank et al. 1991; Meadows et al. 1992). Since NK cells are probably involved in resistance to tumour growth and prevention of tumour metastasis, the increased incidence of tumours in alcohol abusers (e.g., Breeden 1984) might be related to the effects of ethanol on the HPA axis (Irwin et al. 1990).

Raised levels of glucocorticoids in alcohol abusers also have extremely important implications for one of the most common and debilitating complications of alcohol abuse, the mental deterioration of alcohol-related dementia. Thus, there is considerable evidence of a causal relationship between raised corticosterone levels and the onset of deficits in normal aging and the pathological changes of Alzheimer's dementia (see Landfield and Eldridge 1991) and the same aetiology may be important in alcohol-related dementia. The reason for the relationship between corticosteroids and dementia is unknown but it is hypothesized to be because activation of glucocorticoid receptors in the hippocampus potentiates mechanisms leading to neurodegeneration. This may be either because glucocorticoids increase the neurotoxic effects of calcium entry through glutamate receptor-ion channels (Sapolsky 1990) or because they cause calcium entry through voltage-operated ion channels (Landfield and Eldridge 1991), thus increasing the neurodegeneration that causes the dementia. Since chronic ethanol administration is known to increase both glutamate receptors (Grant et al. 1990) and voltage-operated calcium channels (Dolin et al. 1987) in the CNS, and to cause neurodegeneration in the hippocampus (Riley and Walker 1978), it is not hard to see that there is a potential interaction. It remains to be seen to what extent raised corticosteroid levels associated with chronic alcohol abuse exacerbate the neuronal damage which ethanol causes directly.

5. Hypothalamic-Pituitary-Adrenal Axis and Fetal Development

Steroid hormones produced by the maternal adrenal glands have subtle effects on the development of the fetus and on postnatal development and behaviour, but the extent to which alterations in the HPA axis contribute to alterations in these in the offspring of alcoholic women is unknown (Anderson 1981). One of the most important areas of fetal development that *could* be altered is sexual differentiation. For example, as discussed in the section on the HPG axis, several steroid hormones, including androgens from the adrenal, are capable of having masculinizing effects on the fetus. Ethanol-induced alterations in adrenal steroid synthesis or release during pregnancy would inevitably alter fetal development in this way, but whether this plays a role in fetal alcohol effects is unknown.

In addition to effects on sexual differentiation it is likely that the fetus shows corollaries of the effects of ethanol on the HPA axis in adults; thus the cytokine stimulation of ACTH release is blunted after prenatal exposure to alcohol (LEE et al. 1990) and it would therefore be expected that the postnatal immune response would be suppressed. Reports that the *behavioural* reaction to stress is increased in animals prenatally exposed to ethanol (WEINBERG 1988) may be relevant to this. By and large, however, the consequences of fetal exposure to ethanol on the HPA axis are still a mystery.

6. Conclusions

The acute effect of ethanol on the HPA axis in animals and man appears to be stimulatory, probably as a result of increased secretion of CRH stimulated at the hypothalamic or suprahypothalamic level. With continued exposure to ethanol this effect is overcome so that HPA-related hormone levels and functions return towards normal levels. This may be a consequence of normal feedback control or the development of tolerance to ethanol. Whatever the reason, the loss of the effect of ethanol does not seem to be universal with the result that circulating levels of corticosteroids remain high in some individuals. When this occurs these endocrine hormones have the well-recognized effects of pseudo-Cushing's syndrome and may also have potentially much more damaging effects associated with other alcohol-related morbidity.

III. Hypothalamic-Pituitary-Thyroid Axis

The hypothalamic-pituitary-thyroid (HPT) axis shows the common pattern of regulation by secretion of thyroid-stimulating hormone releasing hormone (TRH) from the hypothalamus into the pituitary portal system, this in turn regulating the release of thyroid-stimulating hormone (TSH) from the pituitary and release of thyroid hormone from the thyroid gland (see GREENSPAN 1991). It has the added complication that a high proportion of the "thyroid hormone" (thyroxine, T_4) released from the endocrine gland is, in fact, a pro-hormone, requiring partial de-iodination by the liver to produce the effective endocrine, tri-iodothyronine (T_3). As a result liver disease can affect thyroid function (NOMURA et al. 1975) by affecting the formation of T_3. This endocrine hormone has widespread effects on almost all organs both in adult life and, particularly, during development. In most respects the effects of ethanol on this system appear to be inhibitory (see LOOSEN 1988).

1. Thyrotropin-Releasing Hormone and TSH Secretion

Ethanol does not appear to have marked effects on hypothalamic secretion of TRH either acutely or chronically (CICERO 1982) although there is rather little direct evidence (see LOOSEN 1988). The level of mRNA for the α-subunit of TRH (shared with LHRH) is probably unchanged by ethanol administration (EMANUELE et al. 1991). In contrast to the absence of effects of ethanol on the *release* of TRH there is some evidence that the *effects* of

TRH are altered by ethanol. Thus the effect of TRH on TSH secretion differs markedly in mouse lines selected for their sensitivity to ethanol (ERICKSON et al. 1991) and the TRH-induced response is blunted in alcoholics (ADINOFF et al. 1991). It is possible that the normal circadian rhythm of TSH release is lost in alcoholics as it is in depressive patients (LOOSEN 1988; HEIN and JACKSON 1990). Indeed, similarities in the findings in alcoholics and depressive patients are common to several other endocrine disturbances. The blunted effect of TRH in alcoholics appears to persist even after abstinence (ADINOFF et al. 1991), so that it is presumably caused by some semipermanent change in the anterior pituitary thyrotrophs. The reduced response of TSH release to stimulation by TRH is found postnatally in rats exposed to ethanol in utero (HANNIGAN and BELLISARIO 1990). This too suggests a semipermanent change in the hypothalamic-pituitary axis at the level of pituitary function.

2. Thyroid Hormones

There is little evidence that pharmacological concentrations of ethanol directly reduce the release of thyroid hormone from the endocrine gland, but chronic alcohol treatment does reduce the effects of TSH on the gland (LOOSEN 1988). This, together with reduced release of TSH itself, is probably sufficient to explain occasional reports of low circulating levels of thyroid hormone (see CICERO 1982). The clinical situation is probably similar in that chronic alcoholics often show small fibrosed thyroid glands (HEGEDUS et al. 1988) with presumably low thyroid hormone output. However, this is probably not the limiting factor because the output from the gland is mainly the ineffective thyroxine (T_4), and its hepatic conversion to circulating T_3 determines the functional response to alterations in activity of the HPT axis (NOMURA et al. 1975). Since this T_4 to T_3 conversion appears to be deficient in the face of alcohol-related liver dysfunction (NOMURA et al. 1975) one would expect correspondingly decreased HPT function.

As a result of these alterations in thyroid hormone production and metabolism it is likely that the thyroid status of alcohol abusers tends to deficiency (CICERO 1982; VAN THIEL and GAVALER 1990). However, this is difficult to confirm clinically because concomitant alterations in plasma binding by albumin and other proteins confuse the issue (e.g., CHOPRA et al. 1974). Even in the case of alcoholic liver disease there is confusion because the disease itself has been ascribed to an ethanol-induced *hyper*-thyroid state in hepatic tissue (ISRAEL et al. 1979). Whatever the situation in liver it currently seems likely that other tissues exhibit reduced input from the HPT axis during chronic ethanol administration. These comments also apply to levels of maternal and fetal circulating thyroid hormones during pregnancy in animal experiments (LEE and WAKABAYASHI 1986), where thyroid function appears to be significantly reduced with potentially important consequences for fetal development (see below).

3. Consequences of Alterations in the HPT Axis

It is not more common for hypothyroid states to be diagnosed in the adult alcoholic than it is in the general population (ADLER 1992) so any consequences of reduced thyroid function do not appear to be serious, at least in the adult. However, there are several features of the fetal alcohol syndrome (FAS) which *are* suggestive of an involvement of the HPT axis in producing some of the effects associated with alcohol use in pregnancy. FAS is associated with a generalized delay in growth and skeletal development (CASTELLS et al. 1981), both characteristic of early hypothyroidism, and deficient cerebellar development in FAS has also been linked with reduced serum thyroxine levels (KORNGUTH et al. 1979). There are other defects in behaviour, physiology and neuroanatomy which are similar in the two conditions and which can be reversed by thyoxine treatment (GOTTFELD and SILVERMAN 1990).

Although there seems a good case to be made for alterations in the HPT axis playing some causal role in the developmental complications of alcohol use in pregnancy (ANDERSON 1981), there is little direct experimental evidence for this. In animal experiments, chronic fetal exposure to ethanol is detrimental to HPT function, and these alterations persist into postnatal life (PORTOLES et al. 1988; HANNIGAN and BELLISARIO 1990), but there is currently no corresponding data for human infants affected with FAS (see HANNIGAN et al. 1992b). The role of the HPT axis in fetal development during exposure to ethanol remains an interesting but unproven possibility.

4. Conclusions

There seems little doubt that the HPT axis is affected by chronic abuse of alcohol and that this can have functional consequences probably in the adult and certainly in the fetus. The question that cannot yet be answered is how *important* these changes are relative to other alterations in endocrines. In adults with severe alcoholic liver disease, thyroid deficiency (or thyroid excess!) is just one of a constellation of hormonal disturbances that are relatively unimportant in the prognosis. In the fetus, however, the situation is very different, and developmental effects of thyroid hormone deficit in concert with alterations in growth hormone (see below) could be vital in some of the generalized disturbances associated with ethanol use in pregnancy.

IV. Other Neurohormonal Systems

Of the pituitary hormones which are either neurohormones themselves or which are under neurohormonal control there are a few which have not been dealt with under previous headings. These include opioid peptides, melanophore-stimulating hormone and growth hormone from the anterior pituitary, and vasopressin and oxytocin from the posterior pituitary. The opioid peptides (the endorphins) are legitimate neurohormones both in respect of their function in controlling secretion of hypothalamic-releasing hormones and in their other central and peripheral functions, and they will be considered later. Of the others I will deal only with growth hormone here because there is no evidence that alterations in oxytocin or melanophore-stimulating hormone cause serious alcohol-related defects (see CICERO 1982). Vasopressin will be considered with the endorphins in the next section because there is evidence for interactions of these with ethanol beyond the pathophysiological role of the other neurohormones.

1. Growth Hormone

In the adult, GH release is pulsatile and, as with the other anterior pituitary hormones, secretion is under hypothalamic control (EDEN 1979). The stimuli for GH release are not well understood but, at the hypothalamic level, dopamine and norepinephrine (via α_2-adrenoceptors) will cause GH release, whereas TRH can block stimulated release. Any acute effects of ethanol could therefore be due to interactions with these control mechanisms. In fact ethanol does not seem to have dramatic effects, though it may decrease the amplitude of phasic GH secretion (e.g., REDMOND 1980). The reason for this is not certain but the inhibitory effect of ethanol on GH secretion is probably at or above the level of the hypothalamus (CICERO 1982), providing a link with the likely mechanism for inhibition of gonadotropic hormone release.

The situation is similar after chronic administration of ethanol where subtle changes in the pulsatile nature of unstimulated circulating levels of GH (REDMOND 1980) are seen, such as a change from high-amplitude low-frequency release to low-amplitude high-frequency release (BADGER et al. 1993). Such changes could be important because the pulsatile pattern of GH release confers sex differences in biochemical parameters such as hepatic enzyme activity (WAXMAN et al. 1991; BADGER et al. 1993) and the change induced by alcohol represents a switch from a "male" pattern to a "female" pattern. Relatively subtle changes are also found in alcoholics where the sleep-induced rise in GH secretion is absent (OTHNER et al. 1982), and there are alterations in the responses to stimulation or inhibition of GH release by TRH and dopamine, dependent on the stage of alcohol withdrawal (ANNUNZIATO et al. 1983) and the presence of liver disease (ZANBONI et al. 1983).

As with the effects of ethanol on thyroid hormone the most compelling reason for studying growth hormone is its potential for causing the developmental deficits associated with FAS (e.g., ANDERSON 1981). Growth hormone is released during pregnancy both from the maternal and fetal pituitary and has generalized effects on fetal growth and development. The alterations that ethanol produces, both in GH release and the circulating GH levels, seem to be fairly subtle but, given our ignorance of the importance of the pulsatile nature of release of this hormone, this does not mean that they are unimportant. GH release is altered by ethanol, probably at the hypothalamic or suprahypothalamic level, and this is a good candidate for some of the alterations in growth and development caused by the drug.

B. Role of Neurohormonal Alterations in Ethanol Consumption

Worldwide, the consequences of alcohol abuse provide an overwhelming series of social, economic and medical problems, but despite this there is no consensus on treatment of any kind. As a result, any understanding of the factors which control alcohol consumption in individuals, or which lead to differences between individuals, would be extremely valuable. Not surprisingly there have been several theories about the biochemical basis of ethanol consumption and, since neurohormones are involved in other "appetitive" behaviour such as eating and drinking, it is equally unsurprising that many have focussed on hormonal alterations as the cause.

Although there is no shortage of theories there is a distinct paucity of good experimental data in this area. One of the major problems in studying ethanol consumption is that experimental animals will not usually consume ethanol voluntarily (e.g., KEANE and LEONARD 1989). Clearly this made early attempts to model ethanol consumption rather suspect! Later attempts which relied on genetically distinct strains of animals under specialized experimental conditions succeeded in achieving levels of ethanol intake comparable with those in humans, but neglected to take into account the much faster metabolism of ethanol in these species. Thus, *blood* ethanol concentrations in these experiments often did not achieve pharmacologically relevant levels, and ethanol consumption may have been for completely different reasons that those that are important in the human (KEANE and LEONARD 1989). Despite these criticisms there are now several experimental paradigms that seem to provide data relevant to human ethanol consumption and I will concentrate on some of these recent experiments in this very brief review.

In general terms neurohormones and endocrines could theoretically influence ethanol consumption in several ways and I will first try to classify these to provide a framework for what follows. Ethanol could be consumed for the following reasons:

1. *Rewarding effects*. The consumption of ethanol could release a neurohormone that had rewarding effects and this would act as the stimulus for repeated consumption. Differences between individuals might be because some had a greater response to ethanol than others. For example, ethanol causes release of opioid peptides (see below) and individual differences between this response could dictate susceptibility to alcohol abuse.

2. *Avoidance of punishing effects*. The consumption of ethanol might reduce the release of a neurohormone that had punishing effects and individual susceptibility to alcohol abuse might be because this effect of ethanol was greater in such individuals. For example, ethanol can reduce the endocrine response to stress in some individuals (see below) and this might reinforce the use of the drug for them.
3. *Appetitive effects of hormones*. Ethanol may influence the release of neurohormones that themselves provide the appetite for further ethanol consumption. For example, ethanol inhibits the release of vasopressin (see below) and this causes a diuresis and increases thirst, leading to further ethanol consumption. This is really no more than a restatement of the previous two mechanisms in a physiological form, but it provides a convenient label for the effects of ethanol on mechanisms that influence physiological drives rather than psychological states.

There are a very large number of potential interactions between hormones and ethanol consumption in each of these categories. I have selected examples of each, but this is not intended to be an exhaustive survey of the literature and I apologize if my selection neglects examples that prove to be important at a later date. In this respect an obvious omission is the role of cholecystokinin in modifying dopaminergic reward pathways but there is presently very little literature concerning potential interactions with ethanol.

I. Opioid Peptides

The opioid peptides are released from neurons in many areas of the brain, and from many peripheral nerves, as neurotransmitters or neuromodulators, but they are also released from the anterior pituitary and circulate as neurohormones. In the brain they subserve roles in pain reduction and in reward systems and there is evidence that effects of ethanol on endorphins in brain may underlie the rewarding action of the drug (Widdowson and Holman 1992). In addition the release of β-endorphin (the major circulating opioid peptide) from the pituitary is probably also pleasurable or rewarding. If ethanol caused the release of β-endorphin from this site it would be a good candidate for a rewarding effect of the drug leading to further consumption of ethanol.

In the pituitary, as elsewhere, the synthesis of the opioid peptides is from the large polypeptide precursor pro-opiomelanocortin (POMC). Stimulation of β-endorphin synthesis and release is therefore associated with an increase in the mRNA for POMC. As described previously ethanol increases the synthesis of POMC in mice with a genetic difference in the duration of this effect (Wand 1989). In rats genetically selected for alcohol "preference" or avoidance there are differences in the mRNA for POMC and in the levels of endorphins in various regions of the brain (Gianoulakis et al. 1992). In addition there are also differences in the response to ethanol administration of POMC and endorphins in a similar pair of genetically selected rat lines (LI et al. 1992). This suggests that this potentially rewarding effect of ethanol (on endorphin synthesis and release) may be the reason for ethanol preference in these lines. The argument is strengthened by the observation that voluntary ethanol consumption in the ethanol-preferring, "P", rat line is inhibited by the nonspecific opiate receptor antagonist naloxone (FROELICH et al. 1990) and by specific μ- and δ-opiate receptor antagonists (FROELICH et al. 1991).

These results have their corollary in clinical research where alcohol use is associated with an increase in β-endorphin levels in plasma following ethanol challenge (BORG et al. 1982), which may be under genetic control (GIANOULAKIS et al. 1989). In addition to providing a potential genetic explanation for alcohol abuse these results suggest that opiate receptor antagonists, such as naltrexone, may be useful in modifying alcohol consumption in man. It has long been suspected that there is some common path linking opiate and ethanol dependence (see CICERO 1982), and this now seems to be very likely. Whether the opioid peptides prove to be the most important link between ethanol consumption and neurohormones remains to be seen, but the connection at least has logic (and some data) on its side.

II. Hypothalamic-Pituitary-Adrenal Axis

The argument for considering the HPA axis as involved in voluntary ethanol consumption is that it is part of the stress response and stress is generally unrewarding. However, just as stress releases β-endorphins to *overcome* some of the associated mental and physical consequences, so it also activates the HPA axis to release CRH, ACTH and corticosteroids, and some of these hormones may also *limit* the effects of stress. Thus it is far from certain that the psychological state these hormones produce is the same as that caused by stress, and this would be important in the psychological consequences of activation of the HPA axis by any drug, including ethanol. For example, subjects' ratings of "euphoria" after ethanol correlate with the rise in plasma ACTH (LUKAS and MENDELSON 1988), suggesting that ethanol-induced activation of the HPA axis (in the absence of stress) may be rewarding in some way. Using this argument one might expect that susceptibility to alcoholism would correlate *positively* with the increase in circulating corticosteroids induced by an alcohol challenge, so that the bigger the hormonal response the greater should be the susceptibility to alcoholism.

On the other hand if alcohol did initiate the *whole* stress response (because it produced *dysphoria* for example) then this would be associated with activation of the HPA axis with a corresponding increase in CRH, ACTH and corticosteroids. Under these conditions one would predict that the increase in levels of these hormones would correlate with the unrewarding stressful effects of ethanol. Susceptibility to alcoholism would then correlate *negatively* with corticosteroid levels, the lower the levels the more susceptible the individual should be to alcoholism.

Clearly both these arguments cannot be right. Or can they? It has recently been shown that acute alcohol challenge to *women* with a family history of alcoholism caused a significantly *higher* corticosterone response than it did in women who were "family history negative" (LEX et al. 1991). This supports the first contention, that HPA activation may produce rewarding effects and the extent of this activation may therefore underlie susceptibility to alcoholism. What about the second hypothesis, that HPA activation reflects stress and that susceptibility to alcoholism should therefore correlate with a reduced level of stress hormones after alcohol? Well, it had previously been shown in a series of papers on *males* with a family history of alcoholism that these subjects had *lower* corticosteroid levels after an alcohol challenge than the "family history negative" controls (e.g., SCHUCKIT et al. 1987a).

These results do suggest that both arguments may be correct, but apply under different circumstances. However, the immediate conclusion, that there is a complete divergence in the hormonal basis for male and female alcoholism, is premature! It must be remembered that alterations in the HPA axis can *follow* alterations in psychological state as well as cause them. Subtle differences in the experimental design may have emphasized a "stress-buffering" effect of alcohol in one group (with *reduction* in corticosteroids correlating with effectiveness of alcohol) and an excitement/euphoria effect in the other (with an *increase* in corticosteroids correlating with the effectiveness of ethanol). Another reason for not concluding that hormonal differences in men and women at risk of alcoholism are not absolute is that both groups of researchers reported that the family history positive men *and* women showed significantly lower prolactin levels following alcohol challenge than did the appropriate controls (SCHUCKIT et al. 1987b; LEX et al. 1991).

Animal experiments are not much help in untangling these differences to date. Historically there have been many observations that stress will increase voluntary ethanol consumption by animals, but whether this has anything to do specifically with the HPA axis remains uncertain. In addition to the simple hypothesis that links activation or inhibition of the HPA axis to ethanol consumption, there are several other ways in which the HPA axis could be involved in susceptibility to ethanol dependence. For example there is ample evidence for effects of the neurohormone, CRH, on ethanol tolerance and dependence (CICERO 1982; KOOB 1982) and this could have a secondary effect on voluntary consumption of the drug. This also illustrates one of the problems in interfering with this system experimentally. Effects of adrenalectomy, for example, could be due to a loss of the adrenal steroid hormones or a compensatory increase in the secretion of CRH.

III. Appetitive Systems

Of all the sections in this chapter the next proved to be the most difficult to write. There is clearly an interaction between ethanol, the systems that control satiety and appetite (e.g., serotonin) and the systems that control thirst and fluid balance (e.g., renin-aldosterone and vasopressin), and this may well be important in controlling ethanol consumption. The difficulty is in making sense of the mass of conflicting evidence concerning a system which is not well understood anyway.

Serotonin may just about qualify as a neurohormone, and so be within the scope of this chapter. It certainly seems to be closely concerned with the neural substrates that control appetite and satiety and this makes it of interest here. In particular there is good experimental evidence that inhibitors of serotonin reuptake, such as fluoxetine, are effective inhibitors of voluntary ethanol consumption in animals (see AMIT and SMITH 1992) though preliminary studies in man are equivocal (GORELICK and PAREDES 1992). In any event the evidence that these drugs are effective because of an action on *serotonin* is not yet convincing but this may be because of the complexity of the system.

One of the *potential* explanations for a role of serotonin in ethanol consumption is that it causes the release of renin from the juxtaglomerular cells in the kidney (GRUPP et al. 1988). Renin is converted via a series of enzyme reactions in the plasma to angiotensin II, which then causes the release of aldosterone from the adrenal cortex. Aldosterone in turn increases Na^+/K^+ exchange and increases the reabsorption of fluid from the kidney tubule, thus having an antidiuretic effect. The reason for interest in this system is that drugs which act as angiotensin-converting enzyme (ACE) inhibitors are effective at inhibiting voluntary ethanol consumption in rodents (GRUPP et al. 1991). The inhibition of ACE by these drugs would of course reduce the effective concentration of angiotensin II in the plasma, reduce the release of aldosterone and tend to produce a diuresis. How exactly this could inhibit the "appetite" for ethanol is uncertain, but explanations are possible (GRUPP et al. 1991). This approach to modifying ethanol consumption may be of more than theoretical interest because enalapril does have some effects in man (NARANJO et al. 1991).

Another link between serotonin and fluid balance is the secretion of vasopressin from the posterior pituitary. One of the normal functions of circulating vasopressin is to allow fluid reabsorption from the collecting tubules in the kidney so that it acts as an "antidiuretic hormone". Ethanol inhibits its secretion, probably by an action in the hypothalamus (LINKOLA et al. 1974), and so provokes a diuresis. The paper cited showed a difference in the effect of ethanol in two rat lines genetically selected for differences in voluntary ethanol consumption. The inhibition of vasopressin release and the subsequent diuresis might provoke a thirst and a desire for more alcohol, but the effects of vasopressin are probably a good deal more subtle (and interesting) than this. In particular there is a great deal of evidence that vasopressin can prolong the retention of alcohol tolerance (HOFFMAN et al. 1978; HOFFMAN 1982) and this may influence alcohol consumption indirectly. A common theme in the literature on neurohormones is that their release as hormones from the pituitary is but one of their functions. Vasopressin, just as are most of the other pituitary peptides, is also found in other brain regions, where it functions as a neurotransmitter or neuromodulator. In some of these areas vasopressin can influence serotonin synthesis and release (AUERBACH and LIPTON 1982) and it may be here that any important effects on ethanol consumption occur.

C. Overall Conclusions

The literature on neurohormonal effects of ethanol is vast and confusing, but that is because it is such a vast subject and has attracted so much interest. Despite the confusion there are coherent patterns in the literature, and I have tried to emphasize these. Much of the evidence supports the idea that ethanol-induced alterations in many endocrine systems are a consequence of effects on the neurohormonal *regulation* of these systems so the title of the chapter, and the approach I have taken, is justified. However, these effects on neurohormones cannot be viewed in isolation since effects of ethanol, both above the hypothalamic level and below the endocrine gland, have a marked influence on their functional consequences. This complexity is one reason for the earlier contention that alcohol-related endocrine disturbances can mimic almost any pathological endocrine condition.

In describing the consequences of ethanol use on the neurohormonal system I have tried to stay largely within the "endocrine boundaries" because these are well-established effects of the drug. I have not attempted a detailed review of the role of neurohormones in learning and memory (or any other behaviour) simply because I do not believe there is enough evidence to make a coherent story out of the effects of ethanol on these hormones and these functions. Again, that does not mean to say that they are unimportant. The brief section on the role of neurohormones in ethanol consumption illustrates that neurohormones could have very profound effects on ethanol-induced behaviours, but it would be premature to attempt to review this now.

As far as the pathophysiological consequences of actions of ethanol on neurohormones are concerned there can be no doubt that these can be severe. These comments apply not only to direct effects of the drug on hormonal function, but also to indirect effects such as the potential role of raised glucocorticoid levels in alcohol-induced neurodegeneration. There is clearly a lot to be done in unravelling the harmful effects of ethanol on neurohormones in the adult, and there is even more potential for interactions at the level of fetal development. Since endocrine influences are so important in the normal development of the fetus any alteration in endocrines that can be produced by alcohol during pregnancy (in the mother or the fetus) must be viewed with extreme suspicion. The mechanisms for some aspects of the fetal alcohol syndrome are almost certain to be found in ethanol-induced neurohormonal disturbances.

In the sum total of human misery caused by excessive use of ethanol, effects of the drug on the neurohormonal system rank highly. Inevitably alcohol research in this area lags some way behind pure physiological research on neurohormonal regulation, but patterns are emerging of almost universally detrimental effects of the hypothalamus and pituitary. Ethanol and neurohormones seems destined to be a fruitful area of research for many years to come.

References

Abadallah RM, Starkey JR, Meadows GG (1988) Toxicity of chronic high alcohol intake on mouse NK cell activity. Res Commun Chem Pathol Pharmacol 59:254–258

Adams ML, Cicero TJ (1991) Effects of alcohol on beta-endorphin and reproductive hormones in the male rat. Alcohol Clin Exp Res 15:685–692

Adinoff B, Nemeroff CG, Bissette G et al. (1991) Inverse relationship between CSF TRH concentrations and the TSH response to TRH in abstinent alcohol dependent patients. Am J Psychiatry 148:1586–1588

Adler RA (1992) Clinically important effects of alcohol on endocrine function. J Clin Endocrinol Metab 74:957–960

Amit Z, Smith BR (1992) Neurotransmitter systems regulating alcohol intake. In: Naranjo CA, Sellers EM (eds) Novel pharmacological interventions for alcoholism. Springer, Berlin Heidelberg New York, pp 161–183

Anderson RA Jr (1981) Endocrine balance as a factor in the etiology of the fetal alcohol syndrome. Neurobehav Toxicol Teratol 3:89–104

Annunziato L, Amoroso S, Di Renzo G, Argenzio F, Aurilio C, Grella A, Quattrone A (1983) Increased responsiveness to DA receptor stimulation in alcohol addicts during the late withdrawal syndrome. Life Sci 33:2651–2655

Arnold AP, Gorski RA (1984) Gonadal steroid-induction of structural sex differences in the CNS. Annu Rev Neurosci 7:413–422

Auerbach S, Lipton P (1982) Vasopressin augments depolarization-induced release and synthesis of serotonin in hippocampal slices. J Neurosci 2:477–482

Badger TM, Ronis MJJ, Lumpkin CK, Valentine CR et al. (1993) Effects of chronic ethanol on growth hormone secretion and hepatic cytochrome p450 isozymes in the rat. J Pharmacol Exp Ther 264:435–447

Barron S, Riley EP (1985) Pup-induced maternal behavior in adult and juvenile rats exposed to ethanol prenatally. Alcohol Clin Exp Res 9:360–365

Becker U, Tomesdu H, Kaas-Claesen N, Glund C (1989) Menstrual disturbances and fertility in chronic alcoholic women. Drug Alcohol Depend 24:75–82

Blank SE, Duncan DA, Meadows GG (1991) Suppression of NK cell activity by ethanol consumption and food restriction. Alcohol Clin Exp Res 15:16–22

Borg S, Kvande H, Rydberg U (1982) Endorphin levels in human CSF during alcohol intoxication and withdrawal. Psychopharmacology (Berl) 78:101–103

Boyden TW, Parmenter RW (1983) Effects of ethanol on the male hypothalamic-pituitary-gonadal axis. Endocrinol Rev 4:389–395

Breeden J (1984) Alcohol, alcoholism and cancer. Med Clin North Am 68:163–177

Castells S, Mark E, Abaci F, Schwartz E (1981) Growth retardation in fetal alcohol syndrome. Dev Pharmacol Ther 3:232–241
Chao YB, van Thiel DH (1983) Biochemical mechanisms that contribute to alcohol-induced hypogonadism in the male. Alcohol Clin Exp Res 7:131–134
Chopra IJ, Solomon DH, Chopra K, Young RT, Cuateco GN (1974) Alterations in circulating thyroid hormones and thyrotropin in hepatic cirrhosis. Evidence for euthyroidism despite subnormal serum triiodothyronine. J Clin Endocrinol Metab 39:501–511
Chung M, Valenca M, Negro-vilar A (1988) Acute ethanol treatment lowers hypophyseal portal plasma LHRH and systemic LH levels in orchidectomised rats. Brain Res 443:325–328
Cicero TJ (1982) Alcohol effects on the endocrine system. NIAAA Alcohol Health Monogr 2:53–94
Cicero TJ, Bell RD (1980) Effects of ethanol and acetaldehyde on the biosynthesis of testosterone in the rodent testis. Biochem Biophys Res Commun 94:814–819
Cobb CF, van Thiel DH (1982) Mechanism of ethanol induced adrenal secretion. Alcohol Clin Exp Res 6:202–206
Dees WL, McArthur NH, Harris PG (1984) Effects of ethanol on LHRH in the male rat; an immunocytochemical study. Exp Brain Res 54:197–202
Dolin S, Little HJ, Hudspith M, Pagonis C, Littleton J (1987) Increased DHP-sensitive calcium channels in rat brain may underlie ethanol physical dependence. Neuropharmacology 26:275–279
Eagon PK, Porter LE, Gavaler JS, Egler KM, van Thiel DH (1981a) Effects of ethanol feeding upon a male specific estrogen binding protein: a possible mechanism of feminization. Alcohol Clin Exp Res 5:183–187
Eagon PK, Zdmek JR, van Thiel DH, Singletary BK, Egler KM, Gavaler JS, Porter LE (1981b) Alcohol induced changes in hepatic estrogen-binding proteins: a mechanism explaining feminization. Arch Biochem Biophys 211:48–54
Eden S (1979) Age and sex related differences in episodic growth hormone secretion in the rat. Endocrinology 105:555–560
Emanuele MA, Tentler J, Reda D, Kirstens L, Emanuele NV, Lawrence AM (1989) Failure of ethanol to inhibit LHRH release from hypothalamus. Alcohol 6:263–266
Emanuele MA, Tentler J, Halloran M, Emanuele NV, Keeley MR (1991) In vivo effects of acute ethanol on rat alpha and beta LH gene expression. Alcohol 8:345–348
Emanuele MA, Tentler JJ, Halloran MM, Emanuele NV, Wallock L, Kelley MR (1992) The effect of acute in vivo ethanol exposure on FSH transcription and translation. Alcohol Clin Exp Res 16:776–780
Erickson JD, Masserano JM, Zoeller RT et al. (1991) Differential responsiveness of the pituitary thyroid axis to TRH in mouse lines selected to differ in CNS sensitivity to ethanol. Endocrinology 128:3013–3020
Fazebas JG (1966) Hydrocortisone content of human blood and alcohol content of blood and urine after wine consumption. Q J Stud Alcohol 27:439–446
Froelich JC, Li T-K (1992) The enkephalinergic system and maintenance of ethanol drinking. In: Naranjo CA, Sellers EM (eds) Novel pharmacological interventions for alcoholism. Springer, Berlin Heidelberg New York, pp 135–136
Froelich JC, Harts J, Lumeng L, Li T-K (1990) Naloxone attenuates voluntary ethanol intake in rats selectively bred for high ethanol preference. Pharmacol Biochem Behav 35:385–390
Froelich JC, Zweifel M, Harts J, Lumeng L, Li T-K (1991) Importance of delta opioid receptors in maintaining alcohol drinking. Psychpharmacology (Berl) 103:467–472
Gavaler JS (1990) Effects of alcohol on endocrine function in postmenopausal women: a review. J Stud Alcohol 46:495–516
Gavaler JS, van Thiel DH (1987) Reproductive consequences of alcohol abuse. Males and females compared and contrasted. Mutat Res 186:269–277

Gavaler JS, Love K, Starzl TE, van Thiel DH (1990) Hormonal status of postmenopausal women with alcohol-induced liver disease. Hepatology 12:924
Gianoulakis C, Beliveau D, Angelogianni P (1989) Different pituitary beta-endorphin and adrenal cortisol response to ethanol in individuals with high and low risk for future development of alcoholism. Life Sci 33:1097–1109
Gianoulakis C, de Waele JP, Kiianmaa K (1992) Difference in the brain and pituitary beta-endorphin system between the alcohol-preferring AA and alcohol avoiding ANA rats. Alcohol Clin Exp Res 16:453–465
Gordon CG, Southren LA, Vittek J, Lieber CS (1979) The effect of alcohol ingestion on hepatic aromatase activity and plasma steroid hormones in the rat. Metabolism 28:20–24
Gorelick DA, Paredes A (1992) Effects of fluoxetine on alcohol consumption in male alcoholics. Alcohol Clin Exp Res 16:261–265
Gottfeld Z, Silverman PB (1990) Developmental delays associated with prenatal alcohol exposure are reversed by thyroid hormone treatment. Neurosci Lett 109:42–47
Grant KA, Valverius P, Hudspith M, Tabakoff B (1990) Ethanol withdrawal seizures and the NMDA receptor complex. Eur J Pharmacol 176:289–296
Greenspan FS (ed) (1991) Basic and clinical endocrinology, 3rd edn. Appleton and Lange, East Norwalk
Grupp LA, Perlanski E, Stewart RB (1988) Attenuation of alcohol intake by a serotonin uptake inhibitor: evidence for mediation through the renin-angiotensin system. Pharmacol Biochem Behav 30:823–827
Grupp LA, Perlanski E, Stewart RB (1991) Regulation of alcohol consumption by the renin-angiotensin system; a review of recent findings and a possible mechanism of action. Neurosci Biobehav Rev 15:265–275
Halmesmaki E, Autti I, Granstrom M-L, Heikinheino M, Raivio KO, Ylikorkala O (1987) Prediction of fetal alcohol syndrome by maternal alpha-fetoprotein, human placental lactogen and pregnancy specific beta-1 glycoprotein. Alcohol Alcohol [Suppl 1]:474–476
Hannigan JH, Bellisario RL (1990) Lower serum thyroxine levels in rats following prenatal exposure to ethanol. Alcohol Clin Exp Res 14:456–460
Hannigan J, Flood C, diCarbo J, Mizejewski G (1992a) Prenatal alcohol exposure reduces amniotic fluid levels of alpha fetoprotein in rats (Abstr). Am J Obstet Gynecol 166:351
Hannigan JH, Naber J, Martier S, Read K (1992b) Use of newborn screening data to assess possible fetal alcohol effects in infants: preliminary reports on methods and thyroxine levels (Abstr). Alcohol Clin Exp Res 16:383
Hauda RJ, McGivern RF, Noble EP, Gorski RA (1985) Exposure to alcohol in utero alters the adult pattern of Lh secretion in male and female rats. Life Sci 37:1683–1690
Hegedus L, Rasmussen N, Ravi V, Kastrup J, Krogsgaard K, Adershville J (1988) Independent effects of liver disease and chronic alcoholism on thyroid function and size; the possibility of a toxic effect of alcohol on the thyroid gland. Metabolism 37:229–233
Hein MD, Jackson IMD (1990) Review: thyroid function in psychiatric illness. Gen Hosp Psychiatry 12:232–244
Helderman JH, Vestal RE, Rowe JW, Tobin JD, Andres R, Robertson CL (1978) The response of AVP to intravenous ethanol and hypertonic saline in man; the impact of aging. J Gerontol 33:39–47
Hoffman PL (1982) Structural requirements for neurohypophyseal peptide modulation of ethanol tolerance. Pharmacol Biochem Behav 17:685–690
Hoffman PL, Ritzmann RF, Walter R, Tabakoff B (1978) Arginine-vasopressin maintains ethanol tolerance. Nature 276:614–616
Irwin MR, Hauger RL (1987) Adaptation to chronic stress: temporal pattern of immune and neuroendocrine correlates. Neuropsychopharmacology 1:239–248

Irwin MR, Caldwell C, Smith TL, Brown S, Schuckit MA, Gillin JC (1990a) Major depressive disorder, alcoholism and reduced NK cell cytotoxicity. Arch Gen Psychiatry 47:713–719
Irwin MR, Vale W, Rivier C (1990b) Central CRF mediates the suppressive effect of stress on natural killer cytotoxicity. Endocrinology 126:2837–2844
Irwin MR, Hauger RL, Brown M, Britton KT (1988) CRF activates autonomic nervous system and reduces natural killer cytotoxicity. Am J Physiol 255: R744–R747
Israel Y, Walfish PG, Orrego H (1979) Thyroid hormones in alcoholic liver disease. Effect of treatment with 6-N propylthiouracil. Gastroenterology 76:116–122
Johnson DE, Chiao Y-B, Gavaler JS et al. (1987) Inhibition of testosterone biosynthesis by ethanol and acetaldehyde. Biochem Pharmacol 30:1827–1831
Keane B, Leonard BE (1989) Rodent models of alcoholism: a review. Alcohol Alcohol 24:299–309
Kikihana R, Butte JC, Hathaway A et al. (1971) Adrenocortical response to ethanol in mice: modification by chronic ethanol consumption. Acta Endocrinol (Copenh) 67:653–664
Koob GF (1982) Interaction of vasopressin and CRF with stress: implications for alcohol tolerance research. In: Cicero TJ (ed) Ethanol tolerance and dependence: endocrinological aspects. DHHS, Rockville, pp 217–230 (NIAAA research monograph)
Kornguth DI, Rutledge JJ, Sunderland E, Seigel F, Carlsson I, Smallens J, Juhl U, Young D (1979) Impeded cerebellar development and reduced serum thyroxine levels associated with fetal alcohol intoxication. Brain Res 177:347–360
Landfield PW, Eldridge JC (1991) The glucocorticoid hypothesis of brain aging and neurodegeneration; recent modifications. Acta Endocrinol (Copenh) 125: 54–64
Lee M, Wakabayashi K (1986) Pituitary and thyroid hormones in pregnant alcohol-fed rats and their fetuses. Alcohol Clin Exp Res 10:428–431
Lee SY, Imalin T, Vale W, Rivier CL (1990) Effect of prenatal exposure to ethanol on the activity of the hypothalamic-pituitary-adrenal axis of the offspring. Mol Cell Neurosci 1:168–177
Lex BW, Ellingboe JE, Teoh SK, Mendelson JH, Rhoades E (1991) Prolactin and cortisol levels following acute alcohol challenge in women with and without a family history of alcoholism. Alcohol 8:383–387
Li X-W, Li T-K, Froelich JC (1992) The enkephalinergic system and alcohol preference (Abstr). Alcohol Clin Exp Res 16:359
Linkola J, Fyrhquist F, Forsander O (1974) Effects of ethanol on urinary arginine vasopressin excretion in two rat strains selected for their different ethanol preferences. Acta Physiol Scand 101:126–128
Loosen PT (1988) Thyroid function in affective disorders and alcoholism. Endocrinol Metab Clin North Am 17:55–82
Lukas SE, Mendelson JH (1988) Electroencephalographic activity and plasma ACTH during ethanol-induced euphoria. Biol Psychiatry 23:141–148
Matusek N, Ackenheil M, Herz M (1984) The dependence of the clonidine growth hormone test on alcohol drinking habits and the menstrual cycle. Psychoneuroendocrinology 9:173–177
McGivern RF, Yellon S (1992) Delayed onset of puberty and subtle alterations in GnRH neuronal morphology in female rats exposed prenatally to ethanol. Alcohol 9:335–340
McGivern RF, Clancy AN, Hill MA, Noble EP (1984) Prenatal alcohol exposure alters adult expression of sexually dimorphic behavior in the rat. Science 224: 896–898
McGivern RF, Raum WJ, Salido E, Redei E (1988) Lack of prenatal testosterone surge in fetal rats exposed to alcohol; alterations in testicular morphology and physiology. Alcohol Clin Exp Res 12:243–247

Meadows GG, Wallendal M, Kosugi A, Wunderlich J, Singer DS (1992) Ethanol induces marked changes in lymphocyte populations and NK cell activity in mice. Alcohol Clin Exp Res 16:474–479
Mello NK, Mendelson JH, Bree MP, Skupny AST (1986a) Alcohol effects on LHRH-stimulated LH and FSH in female rhesus monkeys. J Pharmacol Exp Ther 236:590–595
Mello NK, Mendelson JH, Bree MP, Skupny AST (1986b) Alcohol effects on LHRH-stimulated LH and FSH in ovariectomized female rhesus monkeys. J Pharmacol Exp Ther 239:693–700
Mello NK, Mendelson JH, Teoh SK (1989) Neuroendocrine consequences of alcohol abuse in women. Ann NY Acad Sci 562:211–240
Mello NK, Mendelson JH, Teoh SK (1992) Alcohol and neuroendocrine function in women of reproductive age. In: Mendelson JH, Mello NK (eds) Medical diagnosis and treatment of alcoholism. McGraw-Hill, New York
Mendelson JH, Mello NK (1988) Chronic alcohol effects on anterior pituitary and ovarian hormones in healthy women. J Pharmacol Exp Ther 245:407–412
Mendelson JH, Mello NK, Cristofaro P, Ellingboe J, Skupny A, Palmieri SL, Benedikt R, Schiff I (1987) Alcohol effects on naloxone stimulated LH, prolactin and estradiol in women. J Stud Alcohol 48:287–294
Mendelson JH, Mello NK, Teoh SK, Ellingboe J (1989) Alcohol effects on LHRH-stimulated anterior pituitary and gonadal hormones in women. J Pharmacol Exp Ther 250:902–909
Merry J, Marks V (1972) The effect of alcohol, barbiturates and diazepam on hypothalamic-pituitary-adrenal function in chronic alcoholics. Lancet 2:990–992
Miller N, Hoehe M, Klein HE et al. (1989) Endocrinological studies in alcoholics during withdrawal and after abstinence. Psychoneuroendocrinology 14:113–123
Morris DL, Harms FG, Peterson HD, McArthur NH (1989) LHRH and LH in peripubertal female rats following prenatal and/or postnatal ethanol exposure. Life Sci 44:1165–1171
Naranjo CA, Kadlec KE, Sanhueza D, Woodley-Remus D, Sellers EM (1991) Enalapril effects on alcohol intake and other consummatory behaviors in alcoholics. Clin Pharmacol Ther 50:96–106
Nomura S, Pittman CS, Chambers JB, Buck MW, Shimizu T (1975) Reduced peripheral conversion of thyroxine to triiodothyronine in patients with hepatic cirrhosis. J Clin Invest 56:643–652
Othner E, Daughaday WH, Goodwin DW, Levine WR, Malarkey WB, Freemon F, Halihas JA (1982) Sleep and growth hormone secretion in alcoholics. J Clin Psychiatry 43:411–414
Phipps WR, Lukas SE, Mendelson JH, Ellingboe J, Palmieri SL, Schiff I (1987) Acute ethanol administration enhances plasma testosterone levels following gonadotrophin stimulation in men. Psychoneuroendocrinology 12:459–465
Portoles M, Sanchis R, Guerri C (1988) Thyroid hormone levels in rats exposed to alcohol during development. Horm Metab Res 20:267–270
Redei E, Branch BJ, Taylor AN (1986) Direct effect of ethanol on ACTH release in vitro. J Pharmacol Exp Ther 237:59–64
Redei E, Branch BJ, Gholani S, Lin EYR, Taylor AN (1988) Effect of ethanol on CRF release in vitro. Endocrinology 123:2736–2743
Redmond GP (1980) Effect of ethanol on endogenous rhythms of growth hormone secretion. Alcohol Clin Exp Res 4:50–56
Rees LH, Beser GM, Jeffcoate WF, Goldie DJ, Marks V (1977) Alcohol-induced pseudo-Cushing's syndrome. Lancet 1:726–728
Riley JN, Walker DW (1978) Morphological alterations in hippocampus after long-term alcohol consumption in mice. Science 201:646–648
Rivier C, Imaki T, Vale W (1990) Prolonged exposure to alcohol: effect on CRF mRNA levels and CRF- and stress-induced ACTH secretion in the rat. Brain Res 520:1–5

Rivier C, Rivest S, Vale W (1992) Alcohol-induced inhibition of LH secretion in intact and gonadectomized male and female rats: possible mechanisms. Alcohol Clin Exp Res 16:935–941
Rudeen PK (1984) Fetal alcohol and brain differentiation. In: Miller MW (ed) Development of the CNS: effects of alcohol and opiates. Wiley-Liss, New York, pp 169–188
Sapolsky R (1990) Glucocorticoids, hippocampal damage and the glutamatergic synapse. Prog Brain Res 86:13–23
Sapolsky RM, Rivier C, Yamammoto G, Plotsky P, Vale W (1987) Interleukin I stimulates the secretion of hypothalamic DRF. Science 238:522–524
Schade RR, Bonner G, Gay VC, van Thiel DH (1983) Evidence for a direct effect of ethanol on gonadotrophin secretion at the pituitary level. Alcohol Clin Exp Res 7:150–152
Schuckit MA, Gold E, Risch C (1987a) Plasma cortisol levels following ethanol in sons of alcoholics and controls. Arch Gen Psychiatry 44:942–945
Schuckit MA, Gold E, Risch C (1987b) Serum prolactin levels in sons of alcoholics and control subjects. Am J Psychiatry 144:854–859
Scott HC, Westling E, Paull WK, Rudeen PK (1990) LHRH neuron migration in mice exposed to ethanol in utero. Soc Neurosci Abstr 16:32
Shavit Y, Lewis JW, Terman GW, Gale RP, Liebeskind JC (1984) Opioid peptides mediate the suppressive effects of stress on natural killer cell cytotoxicity. Science 223:188–191
Smals AG, Njo KT, Knoben JM, Ruland CM, Kloppenberg PW (1977) Alcohol-induced Cushingoid syndrome. J R Coll Physicians Lond 12:36–41
Teoh SK, Lex BW, Mendelson JH, Mello NK, Cochin J (1992) Hyperprolactinemia and macrocytosis in women with alcohol and polysubstance dependence. J Stud Alcohol 53:176–182
Udani M, Parker S, Gavaler J, van Thiel DH (1985) Effects of in utero exposure to alcohol on male rats. Alcohol Clin Exp Res 2:355–359
Valuniaki MR, Pelkonen M, Salasporo M, Harkonen E, Hirvonen E, Ylikahari R (1984) Sex hormones in amenorrheic women with alcoholic liver disease. J Clin Endocrinol Metab 59:133–138
Van Thiel DH (1983) Ethanol: its adverse effects on the hypothalamic-pituitary-gonadal axis. J Lab Clin Med 101:21–33
Van Thiel DH, Gavaler JS (1990) Endocrine consequences of alcohol abuse. Alcohol Alcohol 25:341–344
Van Thiel DH, Lester R, Sherins RJ (1974) Hypogonadism in alcoholic liver disease; evidence for a double defect. Gastroenterology 67:1188–1199
Van Thiel DH, Gavaler JS, Cobb CF (1982) Pseudo-Cushing syndrome and alcohol abuse. In: Cicero TJ (ed) Ethanol tolerance and dependence: endocrinological aspects. DHHS, Rockville, pp 117–126 (NIAAA research monograph)
Van Thiel DH, Gavaler JS, Cobb CF, Santucci L, Graham TO (1983) Ethanol, a Leydig cell toxin: evidence obtained in vitro and in vivo. Pharmacol Biochem Behav 18 [Suppl 1]:317–323
Vanicelli M, Nash L (1984) Effect of sex bias on women's studies on alcoholism. Alcohol Clin Exp Res 8:334–336
Wand GS (1990a) Differential regulation of anterior pituitary corticotrope function is observed in vivo but not in vitro in two lines of ethanol sensitive mice. Alcohol Clin Exp Res 14:100–106
Wand GS (1990b) Ethanol differentially regulates pro-adrenocorticotrophin-endorph n production and corticosterone secretion in LS and SS lines of mice. Endocrinology 124:518–526
Wand GS, Dobs AS (1991) Alterations in the hypothalamic-pituitary adrenal axis in actively drinking alcoholics. J Clin Endocrinol Metab 72:1290–1295
Wand GS, Levine MA (1991) Hormonal tolerance to ethanol is associated with decreased expression of the GTP-binding protein Gs-alpha and adenyl cyclase activity in ethanol-treated LS mice. Alcohol Clin Exp Res 15:705–710

Waxman DJ, Pampori NA, Ram PA, Agrawal AK, Shapiro BH (1991) Interpulse interval in circulating growth hormone patterns regulates sexually dimorphic expression of hepatic cytochrome p450. Biochemistry 88:6868–6872
Weinberg J (1988) Hyperresponsiveness to stress: differential effects of prenatal ethanol on males and females. Alcohol Clin Exp Res 12:647–652
Widdowson PS, Holman RB (1992) Ethanol-induced increase in endogenous dopamine release may involve endogenous opioids. J Neurochem 59:157–163
Widenius TV, Erickson CJP, Ylikahari R, Harkonen M (1989) Inhibition of testosterone synthesis by ethanol: role of luteinizing hormone. Alcohol 6:241–244
Wright J (1978) Endocrine effects of ethanol. Clin Endocrinol Metab 7:351–367
Zanboni A, Zecca L, Zanboni-Muciaccia W (1983) Failure of inhibition by TRH and L-DOPA stimulated GH secretion in patients with alcoholic cirrhosis of the liver. Clin Endocrinol (Oxf) 18:233–239

CHAPTER 11

Clinical Application of Findings from Animal Research on Alcohol Self-Administration and Dependence

M.K. ROMACH and D.M. TOMKINS

A. Overview

Preceding chapters have reviewed progress made in the understanding of the basic mechanisms of action of alcohol and preclinical evidence suggesting the involvement of a number of neurotransmitter and neuropeptide systems in the modulation of ethanol[1] self-administration, dependence, and tolerance. The aim of this chapter is to integrate some of these experimental findings with the clinical manifestations of alcohol abuse and dependence and to discuss their application to the development of pharmacotherapies for treatment of these disorders. It is impossible to review all the important research developments of the past several years, and we will therefore focus on those areas with direct and current relevance to the management of alcoholism. In particular, self-administration studies and their pharmacologic manipulation will be emphasized.

Since the ideal initial requirement for clinical studies of medications to alter alcohol consumption is that there should be extensive preclinical pharmacology with the compounds, we will selectively review those drugs which have been systematically evaluated preclinically and as a result showed clinical promise. The ensuing clinical work will then be described. In addition, several issues are emerging with important clinical implications, for example, comorbid psychiatric disorders in alcohol abusers. They underscore the need for new approaches to investigate clinically driven concepts in laboratory animals. These will be reviewed briefly.

B. Alcohol Consumption

I. Clinical Aspects of Alcoholism

Alcoholism is an acquired disorder determined by pharmacologic, contextual, behavioral, and genetic factors. The numerous theories put forward to explain the etiology, natural history, and consequences of alcoholism attest to our incomplete understanding of the disorder and its complexity. These

---

[1]The term ethanol will be used throughout this chapter as a synonym for both the pure psychoactive agent and beverage alcohol.

theories range from simple formulations such as the tension reduction or self-medication hypothesis (CAPPELL 1975) to integrated conceptualizations of the interactions among alcohol, the drinker, and the environment as expounded in Cloninger's neurobiological learning model (see MEYER and BABOR 1989 for a more complete listing). For clinicians and researchers, one particularly influential, comprehensive formulation about the nature of alcohol abuse and dependence is detailed in a WHO Memorandum (EDWARDS et al. 1981). This formulation effectively illustrates how various biological, psychological, and social factors may interact with alcohol ingestion in the development of alcoholism.

The ingestion of ethanol, motivated by numerous antecedents, produces a variety of effects, some rewarding and some aversive. The dose of alcohol, the genetic makeup of the individual, the environmental and behavioral circumstances, and other factors will influence the intensity and relative balance of these effects so as either to increase or reduce the probability of repeated ingestion of ethanol on other occasions (KALANT 1987). Repeated ingestion may lead to neuroadaptive changes manifested as tolerance, physical dependence, and a withdrawal syndrome and these may alter the balance of reinforcing and aversive consequences. Conditioning and learning processes are also important determinants of the disposition to use alcohol. Other factors that need to be considered include various mood states (MARLATT and GORDON 1980) and the frequent co-occurrence of other mental disorders, particularly affective and anxiety disorders (REGIER et al. 1990; ROSS et al. 1988). The interaction of alcohol use with these disorders is complex and not well understood. It has been postulated that it may involve alcohol use initiating or exacerbating the course of certain psychiatric illnesses or alternatively that certain disorders increase the likelihood of initial alcohol use, progression to dependence, or relapse after remission. Social factors such as attitudes and behaviors of an individual's peer group, parents and the larger societal millieu, the availability of alcohol, current life events, previous drug experiences, coping skills, attributions, and expectancies are also important. All of these factors form the basis of individual differences in the regulation of alcohol consummatory behavior and their diversity argues against a search for constant etiological factors.

Over the last 2 decades, our understanding of the mechanisms of action of alcohol and the factors affecting the development of tolerance and physical dependence has grown considerably. However, the role of a withdrawal state, classically defined as the demonstration of physical dependence, as a contributor to continued alcohol use is the source of some debate (GOUDIE and EMMETT-OGLESBY 1989).

It has been hypothesized that the repeated occurrence of withdrawal symptoms and their relief by further alcohol consumption results in reinforcement of drug-taking behavior and provides an opportunity for internal stimuli (moods and cognitions) to become paired with external environmental cues (SEIGEL 1989). These cues may subsequently elicit components of the withdrawal state, long after actual withdrawal, and lead to the resumption of drinking. However, physically dependent rats, monkeys, and humans experiencing withdrawal will not perform an operant response to obtain ethanol when it is available (SAMSON 1987; MELLO 1983). In general, there is not a good correlation between the development of physical dependence (withdrawal) and drug-seeking behavior in experimental animal models (SAMSON and HARRIS 1992). Furthermore, inquiries of patients as to why they drink consistently fail to provide evidence that withdrawal is an important reason for drinking (ANNIS and DAVIS 1989; ANNIS 1982). The consensus seems to be that physical dependence and withdrawal symptoms may be neither necessary nor sufficient to perpetuate the excessive consumption of alcohol, but they are likely modulators of the patterns of drinking (JAFFE 1992). The positive reinforcing effects of alcohol are viewed as proportionate to the intensity of drug-seeking behavior and the nature of these has been the focus of much research (STOLERMAN 1992). Several neurotransmitters have been implicated in certain neuronal pathways (KOOB and WEISS 1992; MILLER et al. 1987). The self-administration paradigm has been used extensively in this work. However, there is no fully satisfactory model of human alcoholism in experimental animals and this has hampered efforts to specify the neural mechanisms and behavioral components responsible for alcohol's reinforcing properties.

The diversity in explanatory theories of ethanol dependence is reflected in the eclectic nature of therapies for the management of alcohol abuse/dependence (SAUNDERS 1989). Treatment approaches are further complicated by two often conflicting philosophies of treatment, abstinence versus reduction to nonhazardous levels of drinking. Increasingly, as the outcomes of abstinence-oriented programs have shown that complete abstinence is uncommon, and as a public health perspective (i.e., decreasing individual and population risk and consequences) has been brought to treatment, the goal of reducing drinking has been more widely advocated. This goal is often more easily achieved and more generally acceptable to the very large group of individuals with mild-to-moderate alcohol dependence.

Psychosocial interventions represent the most widely used methods for the treatment of alcohol problems. A major review covering more than 900 references concluded that behavioral approaches constitute the most evaluated and best-validated psychosocial interventions (MILLER and HESTER 1986). However, a number of widely used approaches such as Alcoholics Anonymous or confrontational therapy have not been carefully appraised.

Behavioral treatments involve the application of theories of learning to change patterns of behavior (SOBELL et al. 1990). Pavlovian conditioning theory (learning by association) forms the basis for the use of relaxation training techniques, aversive conditioning, and cue exposure/response-prevention approaches. Operant conditioning theory (learning what behaviors result in rewards and punishments) promotes the identification of high-risk situations for excess alcohol consumption and encourages individuals to develop more appropriate responses to those situations, as well as to restructure the physical and social environment. Such approaches include contingency management programs, the development of social relationships with nonalcoholics, and constructive uses of leisure time. Social learning theory has become prominent recently and has a strong cognitive element. This approach enhances a person's motivation to change (i.e., thoughts are important modulators of learning and behavioral expression). It has spawned procedures such as self-monitoring records, self-selection of treatment goals, and the provision of social support for attempts to change behavior (Sellers et al. 1992).

Pharmacologic interventions, by contrast, are used far less frequently in the treatment of alcoholism. Disulfiram (Antabuse), the aldehyde dehydrogenase inhibitor, is the most commonly prescribed medication. Clinical trials of disulfiram have been fraught with methodological problems and properly controlled studies have failed to demonstrate pharmacologic efficacy (Litten and Allen 1991). For example, in a multicenter trial, Fuller et al. (1986) found no significant differences in abstinence rates between disulfiram (250 mg and 1 mg) and placebo groups, although patients taking a pharmacologic dose of disulfiram (250 mg) reported fewer drinking days than did subjects in the control group. Advocates for the use of alcohol-sensitizing drugs (Brewer 1986; Azrin et al. 1982; Wright and Moore 1989) argue for their efficacy when they are taken under supervision (e.g., by a spouse or work supervisor) as part of a broader treatment program. Clearly, therapist and family involvement and patient expectancy are the primary reasons disulfiram can "work" as part of an overall treatment program. However, disulfiram's toxicity and contraindications to use would appear to limit its utility as an expectancy enhancer.

For the development of drugs to treat alcohol abuse and dependence, the neurobiological basis of the disorders must be understood. In dissecting the neurobiology of alcoholism, a large number of neurotransmitter and neuropeptide systems have been implicated including dopamine, serotonin, GABA, and opioid peptides (Koob and Bloom 1988). The delineation of multiple, distinct receptor sites has stimulated studies investigating the role of these neurotransmitter systems in the etiology and treatment of many neuropsychiatric disorders, including alcoholism (Murphy 1990). One of the strategies that appears to be yielding promising treatment approaches is the identification of receptor subtypes that regulate alcohol consumption and the use of specific and selective medications acting at these receptors. Manipulation of these systems by receptor subtype selective agonists and antagonists has been limited primarily to the preclinical domain. Since many of these agents have not been approved for study in humans and the time required to complete an adequate clinical study is quite lengthy, progress in treatment development has been slow. Table 1 serves as a summary of the ways in which medication could be used to modify human alcohol consumption. It suggests that medications currently under investigation may have

Table 1. Targets and mechanisms by which drugs could reduce alcohol consumption (from SELLERS et al. 1992)

Target	Mechanism
Ethanol-reinforced behavior	Antagonize the reinforcing effects, e.g., ethanol-selective antagonist
	Substitute for the reinforcing effects, e.g., medication with ethanol-like properties
	Provoke an unconditioned aversive or dysphoric reaction by pairing behavior with shock or any other aversive conditioning stimulus
	Provide an alternative and dissimilar reinforcer, e.g., stimulant
	Provoke a conditioned aversive or dysphoric physiologic reaction, e.g., apormorphine or possibly disulfiram
	Induce dysphoric symptoms that produce mild malaise (in other drugs this is sometimes a side effect)
	Modify ethanol biodisposition, e.g., accelerate elimination from body
Mood, motivation, or cognitions	Treatment of primary or secondary mental disorders associated with alcohol abuse/dependence, such as major depression or chronic anxiety
	Suppress target symptoms, such as anxiety, that may prompt or sustain alcohol use or that may prevent reduction or cessation of use
	Facilitate the learning or retention of a new behavior, e.g., coping skills
	Augment self-efficacy by providing cues that active medication is part of treatment
	Threatened punishment, e.g., disulfiram
	Decrease the desire to drink (no examples known)
	Increase patient control over initiation and continuation of drinking, which may be the product of several of the other approaches listed
	Accelerate or modify the conditioned cues associated with alcohol use

multiple mechanisms by which they may demonstrate clinical utility (SELLERS et al. 1992).

II. Animal Models of Alcoholism

Various animal models have been useful for informing investigators about the core mechanisms of alcohol-consuming behavior and the relative intensity of alcohol effects as a reinforcer. The majority of research using animal models has employed a variety of self-administration paradigms, although drug discrimination, self-stimulation, and withdrawal models have also been an integral part of alcohol research. They have been valuable in the prediction and evaluation of drug treatments directed at the elemental features of alcohol self-administration. However, there are a number of constraints that limit their utility. The range of factors involved in the human situation, as outlined above, is much more complex, from cognitive determinants to complicated behaviors that cannot be mimicked in an animal paradigm. The animal models must therefore be viewed as reductionist in nature or at best simulations of isolated aspects of the human condition.

1. Self-Administration

A number of criteria have been proposed as essential for an appropriate animal model of alcoholism (CICERO 1980). These include:

1. Sufficient intake to achieve pharmacologically relevant blood alcohol levels (BAL)
2. Self-administration of ethanol by the oral route, and in preference to other solutions
3. A demonstration that ethanol is serving as a reinforcer to the animal, i.e., the animal is prepared to work for the drug
4. Manifestation of tolerance to ethanol's effects
5. Withdrawal symptoms on discontinuation indicating physical dependence
6. The development of biomedical complications (e.g., liver damage)

However, animals rarely self-administer ethanol in sufficient quantities to produce physical dependence. One exception to this appears to be selectively bred animal lines, such as the P rats (WALLER et al. 1982). Furthermore, this may be an unrealistic goal for an animal model, as many patients who are heavy drinkers are not physically dependent on alcohol. As discussed earlier, physical dependence and withdrawal symptoms may not be critical determinants of excess alcohol consumption in humans. These criteria, therefore, do not accord closely with the nature and clinical definitions of alcoholism that have evolved over the years and need to be reconsidered.

A number of self-administration paradigms have been utilized in alcohol research. These include nonoperant "passive" oral self-administration, where the animals are given access to an ethanol solution presented in a drinking tube and allowed to consume the ethanol voluntarily (LINSEMAN 1987). Rats and primates can also be trained, using operant procedures, to emit a number of responses prior to delivery of ethanol. However, unlike many drugs abused in man, such as cocaine, ethanol typically does not support high rates of responding (SAMSON et al. 1988). Using these methodologies, a number of access schedules have been employed. These generally fall into two broad categories, continuous availability or restricted access. It is not yet apparent how these different schedules may affect the profile of activity of investigational drugs on voluntary ethanol intake. Though not extensively characterized, it is nonetheless clear that both the drinking characteristics and the animals' behavior differ in the two schedules (GILL et al. 1986).

Rhesus monkeys given continuous access to alcohol develop an inconsistent profile of drinking, but generally one typified by extended periods of drinking interspersed with periods of abstinence (HENNINGFIELD and MEISCH 1979). Rats allowed continuous access to ethanol drink in short discrete bouts throughout the dark period with minimal intake during the light phase (HIGGINS et al. 1992). Although some proportion of this intake may be attributable to prandial drinking, nonprandial drinking also occurs (GILL et al. 1986). The quantities of ethanol consumed in these bouts have been shown to be sufficient to induce behavioral changes, such as anxiolysis and enhanced locomotor activity (HIGGINS et al. 1992).

Restricted access schedules result in relatively stable and consistent drinking patterns in rodents and monkeys. Restricted access animals can be trained to consume ethanol in much larger amounts in a single bout, attaining greater BAL. Another interesting feature is that the behavioral activation that occurs prior to the access-to-ethanol period correlates with the subsequent intake (GILL et al. 1986; Tomkins, unpublished). Conditioning cues are known to play an important role in the maintenance of drinking in humans, a feature that is prominent in the limited access procedure. It has recently been noted that the efficacy of the serotonergic antagonist ondansetron in attenuating ethanol intake in Wistar rats was greater when the animals were maintained on a limited, compared to a continuous, access paradigm (TOMKINS and SELLERS 1992). This suggests that these models may differ in their sensitivities to drugs, though this requires further study.

Continuous and restricted access models generally do not result in the animals becoming physically dependent. However, as noted earlier, this is not a requirement for alcohol dependence in humans. In most regards, the limited access model has the best face validity. It is learned, persists over time, results in behavioral change, and is associated with the drive to obtain the reinforcer.

2. Drug Discrimination

In this procedure, animals are trained to exhibit one response while in the drugged state and a different response following saline administration. Typically, the response takes the form of pressing a lever for a food reinforcement using an operant procedure. This procedure enables one to investigate whether a drug produces an interoceptive cue that the animal identifies as being ethanol-like or whether the drug antagonizes the ethanol-discriminative stimulus (SIGNS and SCHECTER 1988). As ethanol can produce an array of physiological, behavioral, and biochemical effects, it is difficult to ascertain which one of ethanol's many effects is producing the interoceptive cue the animal is identifying. Whereas rats can discriminate aversive effects of ethanol withdrawal (GAUVIN et al. 1992), drug discrimination does not necessarily indicate that pharmacologic agents are interacting with ethanol's reinforcing properties. It does tell us, however, the extent to which agents can modify the perceived subjective effects of ethanol, which may supplement the positive reinforcing properties of ethanol and thereby encourage drug-seeking behavior. Interestingly, Grant has recently reported that the ethanol dose used in training is important, as there appears to be some selectivity of certain drug classes to substitute for either low-, medium-, or high-ethanol dose cues (GRANT and COLOMBO 1992). These observations may further help us to understand the cue that is most salient to the animal.

3. Genetic Strains

Genetic factors are important in determining individual differences in the propensity to self-administer ethanol. Thus, genetically derived strains of animals with different susceptibilities to ethanol's effects are helpful in understanding the importance of these differences (CRABBE and BELKNAP 1992). Three rat strains exhibiting high spontaneous preference for ethanol have been extensively studied, the alcohol-preferring (P) rat and the high-alcohol-drinking (HAD) strain, both bred at Indiana University, and the alcohol-accepting (AA) rat bred at Alko Laboratories in Finland. During the selection procedure low-ethanol-consuming counterparts were also derived known as the non-preferring (NP), low-alcohol-drinking (LAD), and alcohol-non-accepting (ANA) strains, respectively. Differences in biochemical measurements and pharmacologic responsiveness reported among these genetic lines have proved useful in guiding certain aspects of research.

III. Preclinical Studies

From the vast preclinical literature published to date, numerous neurotransmitters, neuropeptides, ion channels, and hormones have been implicated in the modulation of ethanol intake. While no one system would appear unique and essential to regulation of ethanol intake, certain neurotransmitter systems appear to present promising targets for pharmacotherapies.

1. Serotonin

A substantial body of evidence supports a role for serotonin (5-HT) in the modulation of ethanol intake. Rats genetically bred for high ethanol preference, the alcohol-preferring P-line and the high-alcohol-drinking (HAD) strain, have been shown to have low forebrain 5-HT content compared to their nonpreferring (NP) and low-alcohol-drinking (LAD) counterparts (MCBRIDE et al. 1990; GONGWER et al. 1989). Pharmacologic manipulations to enhance central 5-HT activity, for example with serotonin uptake inhibitors, have consistently produced a reduction in voluntary ethanol intake in both animals and man (SELLERS et al. 1992). The 5-HT uptake inhibitors have been reported to attenuate ethanol intake in both hetereogeneous and genetically bred high-preferring rat strains following acute administration (GILL et al. 1988; MYERS and QUARFORDT 1991). However, the effects do not appear to be selective for ethanol alone as the drugs have been shown to attenuate responding for both natural and drug reinforcers (CARROLL et al. 1990). Decreases in food intake can occur, but there is evidence to suggest that these agents may exhibit preferential effects on substances having high motivational value, such as cocaine (CARROLL et al. 1990). One explanation offered for these observations is that 5-HT may modulate the salience of a specific factor or reinforcer to the animal, thereby reducing its incentive value (SOUBRIE 1986; FLETCHER and DAVIES 1990). Although this hypothesis has been explored in feeding behavior, it has received little attention from the alcohol field to date. The modest suppression in alcohol consumption following 5-HT uptake inhibitor administration may present some restriction on the clinical applicability of this class of medications. This limited effect is probably due to their mechanism of action, which appears to rely on the endogenous tone of the serotonergic system and activation of presynaptic autoreceptors to decrease 5-HT release (SELLERS et al. 1992).

Dexfenfluramine, a 5-HT releaser and uptake inhibitor, produces marked reductions in ethanol intake in both Wistar and P-rat strains (HIGGINS et al. 1992; ROWLAND and MORIAN 1992), an effect that is highly selective for alcohol consumption compared to water. Interestingly, data from our laboratory (unpublished observation) and that of ROWLAND and MORIAN (1992) suggest that tolerance to the dexfenfluramine effect occurs following chronic administration. This may be due to 5-HT depletion or pharmacologic conditioning (tolerance). These results would suggest that potent agents such as dexfenfluramine may be useful clinically as initial treatments but that over the long term the drug may need to be superceded by other therapies. However, it has been reported that less intense treatment may prevent 5-HT depletion and hence increase long-term effectiveness (HINDMARSH et al. 1993). Further chronic administration studies in animals are required to understand this phenomenon more clearly.

The involvement of different 5-HT receptor subtypes in the modulation of ethanol intake is now receiving increasing attention from both a neurochemical and a pharmacologic standpoint. Evidence from ligand binding experiments has demonstrated a higher density of postsynaptic 5-HT_{1A} receptors and a lower density of 5-HT_2 receptor sites in certain limbic structures of the alcohol P-rat compared to their nonpreferring (NP) counterparts (WONG et al. 1990). These differences may contribute to the maintenance of high alcohol drinking behavior in these strains.

The functional role of these different receptor subtypes has been investigated using pharmacologic manipulations with selective agonists and antagonists. 5-HT_{1A} partial agonists (ipsapirone, buspirone, gepirone) reduce alcohol consumption in a range of strains and species, although generally only when tested with restricted access models (KOSTOWSKI and DYR 1992; KNAPP et al. 1992). This model dependency probably is due to the short duration of action of many of these compounds. In a continuous access model, in which the temporal effects can be evaluated, the 5-HT_{1A} agonist 8-OH-DPAT ($\geq 125\,\mu g/kg$) can produce a secondary increase in ethanol intake in rats (HIGGINS et al. 1992), an effect also observed by MURPHY et al. (1987). Whether this is a drug effect or an ethanol deprivation effect is at present not known, although low doses of 8-OH DPAT can enhance ethanol intake in the absence of an initial suppression (TOMKINS et al. 1993; TOMKINS et al. 1993, submitted). These latter observations initially seem contradictory to the accepted role of 5-HT and ethanol self-administration. However, 8-OH DPAT as well as other 5-HT_{1A} agonists also activate inhibitory somatodendritic autoreceptors localized on the dorsal and median raphe nuclei, in addition to postsynaptic sites. The consequence of this is an attenuation in 5-HT neurotransmission to the areas innervated by these nuclei (HJORTH and SHARP 1991). This mechanism supports the concept of reduced 5-HT function as a factor in high ethanol intake. The nature of the effect of 5-HT_{1A} agonists on ethanol intake, enhancement versus attenuation, is still a matter of some debate, although the recent availability of more specific and selective antagonists for this receptor subtype should aid in the resolution of this issue.

Less extensively studied are agents interacting with 5-HT_2 receptors. While it has been shown that ritanserin and LY 53857, antagonists at this receptor subtype, reduce intake in high-preferring strains (MEERT et al. 1991), this has not been completely reproducible in other strains such as Wistar rats (HIGGINS et al. 1992). These differences may be due to variations in the level of baseline intake of alcohol and/or other biochemical parameters.

Finally, 5-HT_3 antagonists reduce ethanol intake in both Wistar rats and in strains bred for high preference (HIGGINS et al. 1992; TOMKINS and SELLERS 1992; FADDA et al. 1991; KNAPP and POHORECKY 1992). Unlike other 5-HT manipulations, 5-HT_3 antagonists attenuate ethanol intake in the absence of changes in food consumption (HIGGINS et al. 1992) and have little effect upon cocaine and heroin self-administration (LANE et al. 1992). Thus, these agents appear to produce selective effects on ethanol intake. Electrophysiologic techniques have been employed to demonstrate an interaction between ethanol and the 5-HT_3 receptor ion gated channel (LOVINGER 1991), which may underlie the specificity of these observations for ethanol self-administration. In contrast to both 5-HT_{1A} and 5-HT_2 agents, 5-HT_3 antagonists can block an ethanol discriminative cue (GRANT and BARRETT 1991), suggesting an alternative mechanism of action. Ethanol-induced dopamine release within the mesolimbic system, which has been implicated in mediating its reinforcing properties, is also attenuated by 5-HT_3 antagonists (CARBONI et al. 1989). These agents are effective following both acute and subchronic administration. Indeed, a number of laboratories have shown that a maximum attenuation is achieved following several drug administration procedures (FADDA et al. 1991; KNAPP and POHORECKY 1992). This suggests a learning component is involved in these effects; hence animals may require repeated or long-term pairings with the drug and ethanol to demonstrate a reduction in alcohol intake. In this regard, 5-HT_3 antagonists have been reported to improve cognitive performance in rodents and primates, including man (BARNES et al. 1990).

2. Opioids

Considerable evidence implicates endogenous opioid peptides in the modulation of ethanol's effects. Changes in opiate content both peripherally and centrally, particularly within the hypothalamus, have been reported following acute and chronic administration of ethanol in both animals and man (GIANOULAKIS et al. 1992; BORG et al. 1982). Furthermore, rats bred for alcohol preference show a relationship between high alcohol consumption and high endogenous opioid activity (FROEHLICH et al. 1988). High doses of morphine suppress drinking of alcohol in rats (SINCLAIR 1974), whereas morphine withdrawal increases voluntary consumption of alcohol (VOLPICELLI et al. 1990). Cerebrospinal fluid from alcoholic patients has been shown to enhance ethanol intake in macaque monkeys when infused intracerebroventricularly, and this effect can be partially reversed by naltrexone, an opiate antagonist (MYERS et al. 1986). The opiate antagonists naloxone and naltrexone interfere with the excitatory effect of ethanol on lateral hypothalamus stimulation (LORENS and SAINASTI 1978), attenuate ethanol-induced hyperactivity (MIDDAUGH et al. 1978), and reduce ethanol self-administration in both rodents (SAMSON and DOYLE 1985) and primates (ALTSHULER et al. 1980), suggesting that the reinforcing properties of ethanol are altered. The opioid system has been implicated in the physiologic responses to stress and can influence poststress elevations of ethanol intake in rats. Naltrexone administration attenuates poststress elevations of ethanol consumption in rats (VOLPICELLI et al. 1986). Decreases in other consummatory behaviors have been observed, leading some to question the selectivity of the opioid system for ethanol's reinforcing effects.

As in the case of serotonin, there are a number of opioid receptor subtypes to which both naloxone and naltrexone exhibit some antagonist activity, albeit in a dose-dependent manner (CHANG and CUATRECASAS 1981). Recently, there has been some attempt to dissect the relative importance of these subtypes in mediating the observed effects on alcohol intake. Of the receptor subtypes the δ-receptor, for which met- and leu-enkephalins exhibit high affinity, shows some promise as a target for potential pharmacotherapies. ICI 174864, an antagonist for this receptor subtype (COTTON et al. 1984), decreased ethanol intake without affecting that of water when HAD rats were given a free choice situation (FROEHLICH et al. 1991). It was more effective than naloxone in producing this effect. Nalmefene, a newer opiate antagonist with greater affinity for δ- and κ-receptors (MICHEL et al. 1985), reduced alcohol intake in both AA and Sprague-Dawley rat strains in a restricted access paradigm (SINCLAIR et al. 1992; HUBBELL et al. 1991). These data provide support for the hypothesis that the opioidergic system interacts with ethanol's reinforcing properties.

3. Dopamine

The mesolimbic dopamine system has been the focus of much attention in all areas of substance abuse research and ethanol is no exception. It has been shown that ethanol, at doses that increase locomotor activity in rats, can enhance dopamine (D) release in the nucleus accumbens, and this has been linked to the reinforcing properties of ethanol (DI CHIARA and IMPERATO 1985). High-preferring AA rats are more sensitive than their low-preferring ANA counterparts to this stimulatory effect on dopamine release (KIIANMAA et al. 1992). P rats have lower dopamine and dopamine metabolite content within the nucleus accumbens in addition to a lower density of D_2 receptors than NP rats (MCBRIDE et al. 1990; STEFANINI et al. 1992). This suggests that perturbation of this system may predispose to high volitional intake of ethanol. Preclinical evidence shows that enhanced dopamine availability results in an attenuation of ethanol self-administration. Thus, bromocriptine (a D_2 agonist), GBR 12909 (a D reuptake inhibitor), and amphetamine (a D releaser) significantly decreased ethanol intake (KOOB and WEISS 1992; MCBRIDE et al. 1990) in P rats. It has been proposed that these agents are producing the same net effect as ethanol and hence reducing the animals' need to respond for ethanol. Of further interest is the observation that rats trained to lever press for ethanol show elevations in accumbens dopamine during the waiting period along with behavioral activation (WEISS et al. 1992). This increase in dopamine may serve as a primer for the subsequent ethanol access period which the animal has been trained to expect. Similar observations have been reported in rats trained to expect alternative reinforcers, including saccharin and food (BLACKBURN et al. 1989). While these findings argue against a specific release of dopamine in response to expectancy for ethanol, its importance as an antecedent to drinking behavior has yet to be evaluated. Determining how pharmacologic agents interact with the anticipatory biochemical and behavioral changes described will help to clarify this process and the extent to which it may mimic craving in humans.

IV. Clinical Studies

1. Serotonin

Preclinical studies, as outlined above, have shown that central serotonin systems may play an important modulatory role in alcohol dependence through effects on behavioral control. They have provided the rationale for clinical testing of several classes of serotonergic drugs. Furthermore, there is clinical evidence indicating that serotonin and its metabolites, especially 5-hydroxyindoleacetic acid, are decreased in the cerebrospinal fluid of many alcoholics (BORG et al. 1985; LINNOILA 1990), suggesting the presence of a central 5-HT deficiency in alcohol dependence. Challenge tests with serotonergic probes looking at hormonal responses have also inferred a serotonergic abnormality, a subsensitivity of $5\text{-HT}_2/5\text{-HT}_{1C}$ receptors (LEE and MELTZER 1990). A similar 5-HT deficiency has been proposed for other disorders that, like alcoholism, are characterized by a lower degree of behavioral control such as bulimia nervosa, obsessive-compulsive disorder, aggressive behavior, and suicide (ERIKSSON and HUMBLE 1990), adding further support to this hypothesis.

The most extensively studied group of serotonergic drugs in alcohol-related disorders are the serotonin uptake inhibitors. A number of placebo-controlled clinical trials, each 2–4 weeks in duration, have been conducted (NARANJO et al. 1984, 1987, 1989, 1990) in mild to moderately alcohol-dependent individuals (mean Alcohol Dependence Scale Scores 8–12; mean daily ethanol consumption 7–8 standard drinks). Zimelidine (200 mg daily), citalopram (40 mg daily), viqualine (200 mg daily), and fluoxetine (60 mg/day) reduced alcohol intake by an average of 9%–17%. There appeared to be some minor differential effects among the drugs. Zimelidine and citalopram increased the number of abstinent days, whereas viqualine, a serotonin releaser and uptake inhibitor, and fluoxetine decreased the number of drinks on drinking days. There were also large interindividual variations in the size of the response to the active drugs. However, the observed effects were dose related, independent of the drugs' antidepressant action and rapid in onset. Side effects were mild, transient, and unrelated to response. No patient characteristics predictive of a significant decrease in alcohol consumption could be identified (NARANJO and SELLERS 1989).

Recent human experimental ethanol self-administration studies with citalopram (40 mg/day) (NARANJO et al. 1992) and fluoxetine (60 mg daily) (GORELICK and PAREDES 1992) supported the earlier findings of onset of effect within the 1st week of treatment, resulting in modest decreases in alcohol intake of 18% and 14%, respectively. Furthermore, these experimental studies reported diminution in craving and interest to use ethanol. These data support the notion that animal and human experimental studies have predictive validity for the clinical utility of such medications.

The precise mechanism of action of these serotonergic agents has not been determined, but since they have shown broad effects on consummatory behaviors (e.g., feeding and drinking) it has been proposed that they act by facilitating central satiety signals through enhancement of central serotonergic tone. Further investigation is necessary to elaborate more fully the pathophysiologic mechanisms of these responses. Moreover, longer term studies of these drugs to examine their efficacy over time and utility in relapse prevention are required. Such studies are important because they are more typical of the way in which medications would be used in a chronic, relapsing disorder. During such use, these drugs may have a cumulative effect, especially when combined with psychosocial interventions. Preliminary results from a clinical study with fluoxetine have been promising (KRANZLER et al. 1991). Alternatively, chronic use of these drugs may produce tolerance to the effects, as has been suggested by preclinical studies with other serotonergic drugs (e.g., dexfenfluramine).

Buspirone, a 5-HT_{1A} receptor partial agonist, has shown efficacy as an anxiolytic in clinical populations (GOLDBERG and FINNERTY 1982). Animal and clinical research suggested that buspirone might have a role in the treatment of alcoholism. Buspirone reduced alcohol intake in macaque monkeys (COLLINS and MYERS 1987) and in rats (PRIVETTE et al. 1988) and has shown low abuse liability in alcohol-dependent patients (GRIFFITH et al. 1986).

These favorable data and the emphasis on serotonin manipulation prompted the assessment of buspirone's efficacy in the reduction of alcohol consumption in a clinical population. BRUNO (1989) compared buspirone (15–30 mg/day) to placebo in a double-blind, 8-week trial in 50 outpatients diagnosed as mild-to-moderate alcohol abusers. The buspirone-treated group reported a significantly lower intensity of craving (desire to drink) in association with a reduction in other psychological symptoms (anxiety/depression) and a 57% decrease in alcohol consumed. There was also a lower rate of dropout from treatment in the buspirone group. The author proposed that the effectiveness of the drug might be partially related to its anxiolytic effect, which in turn produced a decrease in desire for alcohol. A subsequent open trial of buspirone in recently detoxified alcohol-dependent individuals supported this proposal (KRANZLER and MEYER 1989b). Buspirone (mean dose 40 mg) reduced both anxiety and the desire to drink in alcoholics with high levels of anxiety. Changes in alcohol consumption were not reported. The authors concluded that the association between craving and anxiety is consistent with the view that dysphoric mood states may act as conditioned cues for drinking in a subgroup of alcoholics (LUDWIG and STARK 1974). They also suggested that the apparent drug effect may have been influenced by the psychosocial component of the treatment intervention. These earlier studies, in which post hoc analyses suggested an association between anxiolysis and decreased drinking, added impetus to the subsequent exploration of treatment of comorbid psychiatric disorders (anxiety disorders, depression) as a means of reducing alcohol consumption.

A preliminary investigation with another serotonergic agent, ritanserin, a $5\text{-HT}_2/5\text{-HT}_{1C}$ antagonist with antidepressant and anxiolytic properties, has been reported (REYNTJENS et al. 1986). A 10-mg daily dose was given in a single-blind fashion for 28 days to five moderately to severely dependent alcoholics who had been abstinent for 1 month. Efficacy of the drug was measured by the magnitude of the decreases in depressive and anxiety symptoms. These patients described a concurrent reduction in their desire to drink alcohol that was evident within the 1st week of drug treatment and which persisted during the week after drug discontinuation (MONTE and ALTERWAIN 1991).

Several pieces of preclinical data from self-administration studies suggested a unique role for 5-HT_3 receptors and their antagonists in the regulation of ethanol consumption (i.e., decreasing the size of drinks, gradual onset of effect, dissociation from effects on food and water intake). These findings prompted a clinical trial to determine the efficacy of ondansetron, a 5-HT_3 antagonist, in patients with alcohol abuse or dependence (SELLERS et al. 1991). The study was a randomized, placebo-controlled trial in 71 male problem drinkers. The efficacy of ondansetron was evaluated as part of a treatment program that incorporated nonpharmacologic components including daily self-monitoring and a structured, problem-solving treatment. Ondansetron at the lower dose of 0.25 mg twice daily significantly reduced mean weekly alcohol consumption by 37% compared to baseline and 18% compared to placebo after 6 weeks of treatment. The pattern of response was particularly interesting. The onset of effect was gradual, reaching its peak 1 week following the medication period. It is possible that, given the short duration of treatment, the effects of the behavioral intervention obscured some of the effect produced by ondansetron. This slow onset was also demonstrated in animal studies and has been reported in a clinical trial of the anxiolytic effects of ondansetron (KILPATRICK 1992, unpublished). The anxiolytic effects continued to increase for 2 weeks after the medication period ended. This pattern of onset is in marked contrast to the rapid onset of reductions in alcohol consumption seen with the serotonin uptake inhibitors.

Another interesting feature of the trial of ondansetron is the fact that the greatest treatment effect was observed in the low-dose group. This is similar to the dose effect patterns seen preclinically, where an inverted U-shaped dose response curve was observed for anxiolytic and place-conditioning effects. This suggests that subsequent clinical studies should include a wider range of medication doses. Post hoc analyses of the ondansetron findings also showed a trend for patients to achieve a greater reduction in drinking based on the belief that they were receiving active medication, regardless of their actual medication condition. This interaction of pharmacologic activity of a drug with the attribution that active medication has been given emphasizes the importance of examining the relationships among patient perception of self-efficacy, the perception of effectiveness of medication, and the actual pharmacologic mechanism by which medications may modify alcohol-dependent behavior. These findings underscore the accumulating evidence that medication response may be most prominent in moderately heavy drinkers and suggests that future work on medication development should focus on this group. This would also be appropriate from a public health perspective, as the majority of individuals requiring treatment likely fall within this category.

2. Opioids

The influence of the endogenous opioid system on the intake of ethanol has been receiving increasing attention. VOLPICELLI et al. (1986) proposed several mechanisms to account for the apparent interaction of alcohol with endogenous opiates. In particular, they argued that ethanol stimulates endorphin release, although the effects depend on the duration of exposure. Decreased levels of central β-endorphin have been reported in alcohol-dependent men 10 days after detoxification (GENAZZANI et al. 1982). Therefore, if endorphin deficiency leads to voluntary alcohol consumption, then blocking opioid receptors with an antagonist should interfere with the reinforcing properties of alcohol once drinking is initiated.

Animals will reduce responding for alcohol if their opiate receptors are blocked by pretreatment with naltrexone (MYERS et al. 1986; ALTSHULER et al. 1980; VOLPICELLI et al. 1986; KORNET et al. 1991). Interestingly, in a relapse to drinking model in monkeys, naltrexone also acutely decreased alcohol consumption after an initial imposed abstinence (KORNET et al. 1991). The potential utility of naltrexone in treating alcoholism in humans was first demonstrated by VOLPICELLI et al. (1990). In a double-blind study in 30 recently detoxified alcohol-dependent males, naltrexone 50 mg was administered for 12 weeks in conjunction with a standard treatment (group therapy and individual counselling). In these patients, naltrexone decreased craving for alcohol, mean drinking days during the medication period, and rates of relapse. Naltrexone appeared to be particularly effective in reducing drinking in subjects who sampled alcohol but consequently did not "lose control" and relapse. This effect had been suggested in the animal studies. In this trial, naltrexone was well tolerated, with few side effects or significant effects on mood. The study sample was eventually enlarged to 78 patients and the findings were similar (VOLPICELLI et al. 1992). A more recent trial sought to replicate and expand on these findings (O'MALLEY et al. 1992). O'Malley and colleagues examined the effectiveness of naltrexone combined with two different manual-guided forms of psychotherapy: coping skills/relapse prevention and nondirective supportive therapy. The study was conducted in an alcohol-dependent sample that was predominantly (74%) male. Subjects had been abstinent for 7–30 days and received 50 mg naltrexone daily for 12 weeks. Those who received naltrexone drank on half as many days and drank a third of the number of standard drinks compared with subjects who received placebo. Abstinence rates were higher and relapse rates lower in the naltrexone group. Severity of alcohol problems, assessed with the Addiction Severity Index, was lower in patients on naltrexone at termination of the study, in particular with respect to employment status. Medication effect interacted with the type of psychotherapy administered, such that the cumulative rate of abstinence was highest for patients treated with naltrexone and supportive therapy. However, among patients who initiated drinking, those who received naltrexone and coping skills therapy were the least likely to relapse to heavy drinking. As with ondansetron, the naltrexone studies emphasize the importance of interactions among pharmacologic agents, psychotherapeutic strategies, and patient expectations. The nature of these interactions needs to be explored further in order to maximize therapeutic benefits.

Most recently, nalmefene, a new opiate antagonist with less liver toxicity than naltrexone, has shown promise in the treatment of alcohol dependence (MASON et al. 1992). A pilot study found a trend to decreasing number of drinks per drinking day in two dosage groups (5 and 20 mg daily), relative to placebo. In addition, the higher dose group had a significantly lower rate of relapse and a reduction in number of drinking days per week. A larger treatment trial with nalmefene, currently underway, should provide a firmer basis for evaluating the drug's effects in alcohol dependence.

3. Dopamine

The extent of preclinical investigation of the dopaminergic system has not been matched clinically. Furthermore, data with dopaminergic compounds in preclinical studies have been conflicting with agonists and antagonists producing similar behavioral effects. There is a need for more experimental work to understand this complex interaction within the dopamine system and to guide further clinical study. However, a few clinical studies have been conducted. In a double-blind, 6-month placebo-controlled study, Borg (1983) reported a reduction in desire to drink and improved psychosocial functioning in chronic alcoholics treated with bromocriptine, a dopamine agonist.

A more recent trial consisting of 8 weeks of treatment with bromocriptine (mean dose, 6.9 mg daily) and 2 months follow-up found no significant differences between the placebo and drug groups with respect to alcohol consumption and anxiety and depression scores (Dongier et al. 1991). Global psychiatric symptomatology decreased more in the bromocriptine group. Interestingly, patients in both groups reported decreasing their alcohol consumption by more than 90%. However, the dropout rate in the study exceeded 50%, making the results difficult to interpret. A larger multicenter trial looking at the long-term effectiveness of bromocriptine is in progress.

C. Alcohol Withdrawal

I. Preclinical Studies

Physical dependence is defined as the onset of a number of predictable signs and symptoms on reduction or cessation of alcohol use which may be alleviated by further alcohol intake. In the alcohol withdrawal syndrome, the major symptoms include anxiety, tremor, nausea, insomnia, perceptual disturbances, seizures, and autonomic system activation. As noted earlier, the criteria for clinical alcohol dependence do not require the occurrence of a full-blown withdrawal syndrome and there is limited evidence to support the notion that avoidance of a withdrawal state is of critical importance in maintaining drinking. However, dysphoria and anxiety states have been implicated as determinants of both the initiation and maintenance of drinking. The role of withdrawal anxiogenesis is being explored in animal models. In animals, ethanol withdrawal produces an interoceptive cue similar to that produced by the anxiogenic agent pentylenetetrazol (PTZ). Hence, animals trained to discriminate PTZ will exhibit a PTZ-like stimulus-response during ethanol withdrawal (IDEMUDIA et al. 1989; LAL et al. 1988). This withdrawal stimulus-response can be blocked by anxiolytic drugs like diazepam (LAL et al. 1988). The PTZ discrimination paradigm supports the concept of withdrawal anxiogenesis and may prove useful in determining the efficacy of certain new drugs in blocking the effects of acute and prolonged withdrawal from ethanol. Other animal models have been used to assess drug action in attenuating the behavioral consequences of ethanol withdrawal. The elevated plus maze test, the social interaction test, and the mouse black/white box system have been employed as measures of behaviors related to anxiogenic stimuli.

Several drugs have been tested for their ability to suppress withdrawal anxiogenesis. Benzodiazepines have been effective and are often used as the standard for comparison when newer anxiolytic agents are evaluated (FILE et al. 1992; COSTALL et al. 1988). Serotonergic drugs show promise. Buspirone, in a dose-dependent fashion, reversed the anxiogenic profile exhibited by rats and mice after cessation of subchronic treatment with ethanol (LAL et al. 1991). The mechanism underlying this effect is not known but may be related to activation of 5-HT_{1A} autoreceptors, thus inhibiting 5-HT release and leading to anxiolysis.

A role for 5-HT_3 receptor antagonists in alleviating the symptoms of withdrawal from various substances of abuse has also become apparent. Ondansetron ameliorates the behavioral changes seen in rodents following abrupt withdrawal from chronic benzodiazepine, cocaine, nicotine, and ethanol treatment (SELLERS et al. 1991; HIGGINS et al. 1991; REITH 1990; VAN DER HOEK and COOPER 1990; COSTALL et al. 1990). The drug has also been shown to reduce anxiety-related withdrawal behavior in alcohol-dependent marmosets (OAKLEY et al. 1988). These studies suggest potential utility in the treatment of ethanol withdrawal. Of some clinical concern is the suggestion by Grant (personal communication) that 5-HT_3 antagonists may decrease seizure threshold.

Mianserin, a clinically effective anxiolytic (BJERTNAES et al. 1982), which is believed to act as an antagonist at 5-HT_2 receptors, has also been studied in the rat elevated plus maze test. The drug prevented withdrawal-induced anxiogenesis at several time points, suggesting its effects could be longer lasting than those seen with other drugs (PRATHER et al. 1991). Many other drugs with established clinical indications other than withdrawal have been tested preclinically – calcium channel antagonists, β-blockers, etc. Their review is beyond the scope of this chapter and is covered in Chap. 14 by Anton and Becker in this volume.

Seizures are of particular concern in the management of the withdrawal syndrome. In rodents, repeated withdrawal states lead to a progressive exacerbation of the syndrome with particular effects on seizure activity (HUNT 1973; HUNTER et al. 1973; WALKER and ZORNETZER 1974). This relationship has also been substantiated clinically. A number of investigators have confirmed that the incidence of alcohol withdrawal seizures is greater in patients who have undergone repeated detoxification and a kindling model has been proposed to account for the progressive neuronal hyperexcitability (BALLENGER and POST 1978; BROWN et al. 1988). A promising line of research in understanding the mechanisms underlying this process has involved a subtype of the glutamate receptor, a major excitatory neurotransmitter in the CNS. Activation of the receptor by the ligand *N*-methyl-D-aspartate (NMDA) is associated with increased permeability of the neuron to calcium. The consequent increase in intracellular calcium levels has implicated this receptor in the development of epileptiform seizures and brain damage (ROTHAM and OLNEY 1987; DINGLEDINE et al. 1986). It has been postulated that chronic ethanol intake leads to a supersensitivity (upregulation) of NMDA receptors. On cessation of ethanol, this change may contribute to the generation of withdrawal seizures (GRANT et al. 1990; HOFFMAN et al. 1990). Following chronic ethanol treatment, an increase in the density of hippocampal NMDA receptors has been observed (GRANT et al. 1990). Whereas pretreatment with the NMDA receptor antagonist MK-801 effectively reduced, in a dose-dependent manner, the severity of ethanol withdrawal seizures in mice, other withdrawal symptoms (e.g., tremor) were unaltered. MK-801 did not cause concurrent sedation, which commonly occurs in association with the anticonvulsant effects of benzodiazepines. The underlying role of the benzodiazepine receptor complex in withdrawal seizures appears to be different from that of the NMDA receptor complex. Furthermore, NMDA antagonists have been reported to have neuroprotective qualities, thus limiting damage from kindling-induced seizures. These observations indicate that NMDA antagonists warrant further evaluation for treatment of ethanol withdrawal seizures and alcohol-induced cognitive deficits. Unfortunately, at the present time, there are no NMDA antagonists available for study in humans.

II. Clinical Studies

The alcohol withdrawal syndrome encompasses a cluster of characteristic signs and symptoms, the severity of which varies with the individual and the amount of ethanol consumed. The pathophysiology underlying the withdrawal syndrome appears to be a disturbance of several neurotransmitter systems, including increases in noradrenergic and excitatory glutaminergic (via NMDA receptors) function and alterations in the GABAergic system. Therapy is aimed at relief of symptoms, prevention, or treatment of the more serious complications, and preparation of patients for long-term rehabilitation. In patients undergoing mild-to-moderate withdrawal, nonpharmacologic detoxification is often adequate. However, those in moderate-to-severe withdrawal should be assessed for pharmacotherapy. For the past 25 years, the benzodiazepines have been the mainstay of treatment of the alcohol withdrawal syndrome (ROMACH and SELLERS 1991; LITTEN and ALLEN 1991). There is an extensive preclinical literature that predicted their utility and describes their mechanism of action through the GABA-benzodiazepine receptor complex (NUTT et al. 1989; TICKU and KULKARNIZ 1988). The reader is referred to Chaps. 5 and 13 by Ticku and by Anton and Becker in this volume for more detailed information.

Numerous compounds of this class have been used and all can be equally effective in treatment, provided appropriate dose adjustments are made for potency and kinetic differences. Diazepam is the most widely used drug, primarily because of its long half-life. Various regimens of dosing have been used, the most popular being flexible approaches, rather than the more traditional fixed dosage schedules. The loading dose technique in particular is useful because of its simplicity and efficacy, allowing for titration of dosage and gradual self-tapering of drug (SELLERS et al. 1983). The benzodiazepines are effective in diminishing the behavioral features of withdrawal, e.g., anxiety, as well as the autonomic symptoms and may be used both in the prevention and, if necessary, in the treatment of seizures if they occur. In addition, they have a wide margin of safety and are well tolerated by most patients. Despite these advantages, repeated concerns have been raised about their use in alcoholics, i.e., the risk of alcoholics becoming dependent on them (MEYER 1986). The scientific data to support this concern are very limited and yet clinically this restriction on prescribing is generally accepted. There are data to suggest that many alcoholics in remission who have used benzodiazepines for prolonged periods do so in an appropriate manner (i.e., as prescribed for approved indications) and are able to discontinue their use when the medications are no longer required (ROMACH et al. 1992).

Nevertheless, these worries have prompted clinical assessment of various other compounds either because of promising preclinical data or because of approaches targeted at specific neurotransmitter systems. Carbamazepine, an anticonvulsant (BUTLER and MESSIHA 1986; MALCOLM et al. 1989), β-blockers (SELLERS et al. 1977; KRAUS et al. 1985), clonidine (BAUMGARTNER and ROWEN 1987), bromocriptine (BURROUGHS et al. 1985; BORG and WEINHOLDT 1982), and calcium channel blockers (LITTLE et al. 1986; KOPPI et al. 1987) have been investigated. None of these agents have shown consistent superiority to the benzodiazepines and many have shown lack of effect on the behavioral components of withdrawal (e.g., anxiety and agitation) and on major symptoms such as seizures. In most studies, combination therapy has been required, increasing the risk of side effects and drug interactions.

The non-benzodiazepine anxiolytic buspirone has also been evaluated in alcohol withdrawal (DOUGHERTY and GATES 1990). This drug is of particular interest because its reported effects on diminution of craving (KRANZLER and MEYER 1989) and alcohol consumption make it a medication with therapeutic potential for different stages of treatment of alcohol abuse/dependence. In a recent open trial, 100 patients received buspirone over varying periods of time during detoxification from alcohol. Mean dose of drug required by patients was not described but, interestingly, almost a third of patients required more than 22 doses of 5 mg given every 4 h over the course of the study and several received supplementary treatment with clonidine.

The potential advantage of buspirone over benzodiazepines that the authors proposed was less sedation. However, the methodological shortcomings of the study make it difficult to draw firm conclusions about possible advantages and a persistent concern is that buspirone will not prevent withdrawal seizures. The benefit of 5-HT_{1A} agents like buspirone may lie primarily in their alleviation of anxiety symptoms both during and after withdrawal.

As mentioned earlier, other drugs that warrant clinical assessment in withdrawal are the 5-HT_3 antagonists and novel anticonvulsants like NMDA antagonists.

D. Comorbidity

I. Overview

Over the past decade, the relationship between substance abuse and comorbid psychiatric disorders has attracted considerable attention. Clinical studies have shown that significant proportions of patients seeking treatment for alcohol abuse suffer from concurrent depressive, anxiety, and personality disorders (Ross et al. 1988). Data from the NIMH Epidemiologic Catchment Area (ECA) Study (REGIER et al. 1990) showed that 37% of alcohol-abusing individuals met DSM III-R criteria for at least one other mental disorder. Although prevalent, the significance of the comorbidity is not well understood. A few studies have provided some evidence that the presence of certain comorbid psychiatric disorders in alcohol abusers adversely influences outcomes in treatment of alcoholism, although there may be gender differences (ROUNSAVILLE et al. 1987; KRANZLER et al. 1992).

Despite the frequency of comorbid disorders, there have been few well-formulated clinical treatment trials focused on this subgroup of alcoholics. The first studies were mostly intuitive in nature based on the commonly held belief that alcoholism is a form of self-medication that has become maladaptive. Certainly, the role of negative affective states such as anxiety and depression in the initiation and maintenance of maladaptive drinking patterns has been acknowledged (MARLATT and GORDON 1980; HATSUKAMI and PICKENS 1982). It has been hypothesized that pharmacologic manipulation of these dysphoric states could lead to attenuation of excessive drinking. Consequently, drugs that had primarily antidepressant effects were selected for investigation of their effects on drinking. However, few, if any, of these early studies were preceded by preclinical investigation that evaluated the drugs' effects in models of self-administration. Increasingly though, careful preclinical evaluation prior to the initiation of a clinical trial has become the preferred pattern.

II. Animal Models

Newer pharmacologic agents are being assessed in novel models of depression and anxiety as well as in alcohol self-administration paradigms. However, there are no suitable animal models to simulate the influence of these affective states on alcohol self-administration. Some work has been initiated in an effort to evaluate preclinically drugs for treatment of comorbid conditions. For example, the anxiolytic and anxiogenic actions of ethanol in a mouse model have led to speculation about the contribution of withdrawal anxiogenesis to continued drinking in humans (COSTALL et al. 1988). Similarly, selective breeding of rodent lines has produced animals that are purported to possess alcohol-preferring traits and behavioral features consistent with an animal model of depression (OVERSTREET et al. 1992). Social phobia and its relationship to alcoholism has begun to figure prominently in discussions of comorbid disorders. Most animal models use individually housed animals. Few animal models have been devised where manipulation of social settings and effects on ethanol consumption can be investigated. Those that have been proposed have focused on characterization of the factors involved rather than on drug intervention.

Generally, social isolation enhances drug-taking behavior (WOLFFGRAMM and HEYNE 1991). In self-administration models using colonies of animals to mimic social influences, one of the problems is identifying which animals in a group are drinking and the extent of their drinking. Some work has shown that, in a rat colony, the submissive or low-ranking animals consume greater quantities of ethanol than their dominant counterparts (BLANCHARD et al. 1987; ELLISON 1987). HILAKIVI-CLARKE and LISTER (1992) suggested that within rodent social colonies enhanced ethanol consumption in submissive animals was a state, as opposed to a trait, effect: namely that being the target of social and physical stress results in increased alcohol intake. Their methodology, in which they introduce a dominant male into a colony, may prove useful as a model for comorbid disorders. However, pharmacologic studies in colony groups require large numbers of animals and the instrumentation and intensity of behavioral analysis required may prevent widespread utilization of such models. Nevertheless, the elaboration of such animal models may offer an opportunity not only to discover genetic influences on neurotransmitter systems thought to underlie psychiatric disorders such as depression or anxiety and alcoholism but also to develop more effective treatment interventions. This is a vitally important area of research.

III. Clinical Studies

1. Depression

Until recently, the most widely studied psychiatric problem in alcohol abusers has been depression. Prevalence rates are generally high, but they have varied considerably owing to a number of factors (heterogeneous patient populations, diagnostic criteria used, chronology of development of the disorders, the temporal association to drinking or withdrawal) (KRANZLER and LIEBOWITZ 1988). The utility of antidepressants in treating alcohol dependence in depressed patients, both for immediate efficacy in reducing alcohol consumption and for relapse prevention, has been largely unexplored. Preclinical data have suggested that noradrenergic and serotonergic uptake inhibitors, which are effective antidepressants, can reduce the volitional intake of ethanol by rats (MCBRIDE et al. 1988; DAOUST et al. 1984; MURPHY et al. 1985). JAFFE and CIRAULO (1986) reviewed the utility of the tricyclic antidepressants for treatment of depression in alcoholics and found that, although a few showed some positive effect (imipramine, doxepin, amitriptyline), the methodology in the majority of the clinical studies was poor and limited any conclusions. However, they suggested that this category of drugs deserved another look using better methodology, including multidimensional treatment outcome measures. Recent preliminary reports once again suggest promise for imipramine (MCGRATH et al. 1991; NUNES et al. 1993) and desipramine (MASON and KOCSIS 1991). A 6-month double-blind, placebo-controlled trial of desipramine in 42 recently abstinent alcoholics (which included desipramine plasma level monitoring) showed an advantage to desipramine treatment of depression secondary to alcoholism. The results also showed a trend for the desipramine-treated subjects to maintain sobriety for a longer period of time while in the study.

The antimanic medication lithium has generated much discussion of its effectiveness in diminishing alcohol intake. Although not an antidepressant per se, it has mood-stabilizing properties, specifically in patients with bipolar disorder. In an effort to remedy an equivocal literature, DORUS et al. (1989) conducted a multisite trial in alcoholics with a history of depression and those without such a history. They found no differences between lithium and placebo conditions on a number of outcome measures including rates of abstinence, reduction in drinking, and depression. Ultimately, a role may be found for lithium in the treatment of alcoholism in subgroups of patients with psychiatric disorders known to be responsive to the drug, such as bipolar disorder, possibly attention deficit disorder, or antisocial personality disorder (KRANZLER and ORROK 1989).

Serotonergic antidepressants may hold the most promise in reversing mood disorders associated with alcohol dependence and consequently reducing drinking and relapse risks. Their effectiveness in the treatment of depressed alcoholics has not been systematically investigated. One open trial of fluoxetine (20–40 mg) in 12 detoxified alcoholics showed significant improvements in measures of depressive symptomatology and alcohol intake, suggesting that larger double-blind trials are warranted (CORNELIUS et al. 1992).

2. Anxiety

Similar to the data with depression, there is considerable evidence of a high rate of anxiety disorders in alcohol-abusing patients (ROSS et al. 1988). Of particular note is the comorbidity between phobic disorders (e.g., social phobia and agoraphobia) and alcohol problems (QUITKIN et al. 1972; MULLANEY and TRIPPETT 1979; COX et al. 1990; KUSHNER et al. 1990; WILSON 1988; SMAIL et al. 1984), suggesting that alcoholism and some forms of anxiety may be related disorders. A linkage between anxiety disorders and serotonergic function has been intensively investigated in the last few years, both in animal and human studies, and provides a theoretical basis for the association between alcoholism and anxiety. It is possible that the decrease in alcohol consumption produced by drugs acting on the serotonergic system (e.g., buspirone, ondansetron, ritanserin, and the 5-HT uptake inhibitors) may be secondary to anxiolytic effects. This hypothesis has been tested to a limited extent with buspirone. In addition to the two clinical studies mentioned earlier, in which post hoc analyses revealed anxiolytic effects, OLIVERA et al. (1990) examined the effectiveness of buspirone in 60 individuals, the majority of whom were severely dependent on a variety of substances, but specifically selected for study because of the presence of an anxiety disorder. However, confounding factors included the presence of a concurrent major affective disorder in almost half the subjects, the use of concomitant psychotropic agents (antidepressants, lithium, carbamazepine), and the diagnosis of alcohol dependence in only 43% of the sample. Nevertheless, in this open trial, the author reported a significant reduction in anxiety and overall improvement in psychosocial functioning when buspirone was combined with some form of drug rehabilitation. TOLLEFSON et al. (1992) conducted a placebo-controlled trial of buspirone in 51 patients dually diagnosed with generalized anxiety disorder and alcohol dependence. Recently abstinent subjects received from 15–60 mg buspirone daily for a minimum of 4 weeks. Once again, significant reductions in anxiety were noted in the buspirone-treated group and these were correlated with reports of diminished intensity of craving for alcohol. However, there was no significant reduction in the overall use of alcohol as measured by the Addiction Severity Index. Both studies, although suggestive of potential utility of this drug in alcohol treatment, underscore the need for refinement of methodology when evaluating the therapeutic efficacy of newer pharmacologic agents. This problem was addressed to a large extent in a trial of buspirone in inpatient alcoholics with generalized anxiety disorder carried out by MALCOLM et al. (1992). Specific outcome measures relating to both levels of anxiety and consumption of alcohol were reported. Anxiety scores declined significantly in both the buspirone and placebo groups but there were no differential group differences throughout the 6-month treatment period on a number of anxiety measures and drinking-related behaviors. The failure to find a significant drug effect may have been related to the high placebo response and early dropout rates, limiting the size of the sample for end point analysis. The authors identified a number of other variables which may have influenced the outcome and which need to be considered in future pharmacologic trials. Nevertheless, this negative result will have an inpact on clinical practice.

Recent reports describing the action of *m*-chlorophenylpiperazine (*m*-CPP), a nonselective 5-HT receptor agonist, provide intriguing data with respect to the interrelationship of alcoholism and anxiety. In animal models of anxiety, *m*-CPP has anxiogenic effects that appear to be mediated by 5-HT_{1C} receptors (Kennett and Curzon 1988; Kennett et al. 1989; Whitton and Curzon 1990; Curzon and Kennett 1990). In human experimental studies, *m*-CPP has precipitated anxiety in control subjects and in patients with anxiety disorders (panic and obsessive compulsive disorders) (Kahn and Wetzler 1991). Infusions of *m*-CPP in abstinent alcoholic patients produce subjective experiences described as a "high" and elicit craving for alcohol (George et al. 1990). It has been suggested that this craving reflects anxiety induced by *m*-CPP, which acts as an interoceptive cue for drinking.

An alternative explanation for the craving produced by *m*-CPP is that this represents a response to the perception of an ethanol-like stimulus in the absence of the expected pharmacologic effect. TFMPP (*m*-trifluoromethylphenylpiperazine), an analogue of *m*-CPP, is also a 5-HT agonist with purported selectivity for the 5-HT_{1C} receptor (Kennett and Curzon 1988; Kennett et al. 1989). TFMPP has shown similarities to the ethanol cue in a rodent drug discrimination paradigm (Signs and Schecter 1988). The clinical reports of craving are of interest because serotonergic drugs like citalopram, fluoxetine, and ondansetron may decrease the desire to drink (Naranjo et al. 1992; Naranjo and Sellers 1989; Gorelick and Paredes 1992; Sellers et al. 1991). These subjective reports may reflect an involvement of serotonin in individuals' perceptions of their ability to regulate their drinking. However, recent preclinical studies of alcohol self-administration that show *m*-CPP decreases ethanol intake (Higgins et al. 1992) confound the above clinical interpretations and suggest that current animal models may lack predictive validity for certain aspects of human alcohol abuse.

E. Conclusions

Over the past decade, considerable knowledge has accumulated about the role that pharmacotherapy can play in the treatment of alcohol abuse and dependence. New and more specific drugs show clinical promise in attenuating alcohol consumption and preventing relapse to hazardous drinking.

Preclinical investigation has driven much of the recent clinical work. This has served to emphasize the need for systematic evaluation of new pharmacologic agents to guide clinical development, from the preclinical stage to clinical trials in humans and ultimately to broader clinical application. Such an approach has allowed us to examine more critically the suitability of the animal models used to simulate alcoholism and to consider newer, alternative paradigms to approximate more closely the human condition. One example is the effort to mimic high anxiety levels and elevated alcohol consumption as seen in some comorbid conditions.

The study of temporal patterns of drinking in animals has been limited. Research into these patterns and how drugs may modify them would be extremely useful in understanding more fully the determinants of the variations in drinking behavior seen in humans. In addition to implicating specific neurotransmitters/neuropeptides and receptor subtypes, preclinical studies suggest that certain pharmacologic agents (e.g., serotonergic drugs) can affect conditioning and learning as well as behavioral control and self-administration. Further clarification of these mechanisms and their manipulation with selective agonists and antagonists could have great clinical utility.

Our review of more recent clinical studies generated by preclinical findings underscores the continued methodological shortcomings that have hampered pharmacologic treatment research. Outcome measures need to be clearly defined, with several determined concurrently, particularly in the area of comorbidity. Study samples need to be described in greater detail and future studies should aim to select more homogeneous samples. Matching specific patients with specific treatments may enhance treatment outcome. Pharmacologic treatment matching should consider several variables including severity of alcohol dependence, presence of concurrent psychopathology, number and type of previous treatments, and their success. Importantly, the desired treatment goal must be consistently defined to allow comparison of the effectiveness of different drugs. Many clinical trials have involved recently detoxified individuals in whom the treatment goal was relapse prevention. Fewer studies have looked at drug effectiveness in reducing drinking (to nonhazardous levels or abstinence) in active drinkers. Lastly, it is important to recognize that decreasing the reinforcing properties of alcohol pharmacologically does not necessarily produce a reduction in consumption. Individuals can increase their intake to offset the decreased alcohol effect or they can refuse the medication. The successful treatment of alcohol abuse/dependence will likely require the prolonged use of medication integrated with cognitive and behavioral treatments directed at controlling use over long periods of time. The potential for developing newer pharmacotherapies by extending preclinical findings to clinical application and by developing appropriate strategies for their use is extensive.

References

Altshuler HL, Phillips PE, Feinhandler DA (1980) Alteration of ethanol self-administration by naltrexone. Life Sci 26:679–688

Annis HM (1982) Inventory of drinking situations (IDS-100). Addiction Research Foundation, Toronto

Annis HM, Davis CS (1989) Relapse prevention. In: Hester RK, Miller WR (eds) Handbook of alcoholism treatment approaches. Pergamon, New York, pp 170–182

Azrin NH, Sisson RW, Meyers R, Godley M (1982) Alcoholism treatment by disulfiram and community reinforcement therapy. J Behav Ther Exp Psychiatry 13:105–112

Ballenger JC, Post RM (1978) Kindling as a model for alcohol withdrawal syndromes. Br J Psychiatry 133:1–14

Barnes JM, Costall B, Coughlan J, Domeney AM, Gerrard PA, Kelly ME, Naylor RJ, Tomkins DM, Tyers MB (1990) The effects of ondansetron, a 5-HT_3 receptor antagonist in rodents and primates. Pharmacol Biochem Behav 35: 955–962

Baumgartner GR, Rowen RC (1987) Clonidine vs chlordiazepoxide in the management of acute alcohol withdrawal syndrome. Arch Intern Med 147:1223–1226

Bjertnaes A, Block JM, Hafstad PE, Holte M, Ottemo I, Larsen T, Pinder RM, Steffensen K, Stulemeijer SM (1982) A multicentre placebo-controlled trial comparing the efficacy of mianserin and chlordiazepoxide in general practice patients with primary anxiety. Acta Psychiatr Scand 66:199–207

Blackburn JR, Phillips AG, Jakubovic A, Fibiger HC (1989) Dopamine and preparatory behaviour. II. A neurochemical analysis. Behav Neurosci 103:15–23

Blanchard RJ, Hori K, Tom P, Blanchard DC (1987) Social structure and ethanol consumption in the laboratory rat. Pharmacol Biochem Behav 28:437–442

Borg S, Kuande H, Ryberg U et al. (1982) Endorphin levels in human cerebrospinal fluid during alcohol intoxication and withdrawal. Psychopharmacology (Berl) 78:101–103

Borg S, Kuande H, Liljeberg P et al. (1985) 5 Hydroxyindoleacetic acid in cerebrospinal fluid in alcoholic patients under different clinical conditions. Alcohol 2:415–418

Borg V (1983) Bromocriptine in the prevention of alcohol abuse. Acta Psychiatr Scand 68:100–110

Borg V, Weinholdt T (1982) Bromocriptine in the treatment of the alcohol withdrawal syndrome. Acta Psychiatr Scand 65:101–111

Brewer C (1986) Supervised disulfiram in alcoholism. Br J Hosp Med 35:116–119

Brown ME, Anton RF, Malcolm R, Ballenger JC (1988) Alcohol detoxification and withdrawal seizures: clinical support for a kindling hypothesis. Biol Psychiatry 23:507–514

Bruno F (1989) Buspirone in the treatment of alcoholic patients. Psychopathology 22 [Suppl]:49–59

Burroughs AK, Morgan MY, Sherlock S (1985) Double-blind controlled trial of bromocriptine, chlordiazepoxide and chlormethiazole for alcohol withdrawal symptoms. Alcohol Alcohol 20:263–271

Butler D, Messiha FS (1986) Alcohol withdrawal and carbamazepine. Alcohol 3:113–129

Cappell H (1975) An evaluation of tension models of alcohol consumption. In: Gibbons RJ, Israel Y, Kalant H et al. (eds) Research advances in alcohol and drug problems. Wiley, New York, pp 177–210

Carboni E, Acquas E, Frau R, Di Chiara G (1989) Differential inhibitory effect of a 5-HT_3 antagonist on drug-induced stimulation of dopamine release. Eur J Pharmacol 164:515–519

Carroll ME, Lac ST, Asencio M, Klagh R (1990) Fluoxetine reduces intravenous cocaine self-administration in rats. Pharmacol Biochem Behav 35:237–244

Chang KJ, Cuatrecasas P (1981) Heterogeneity and properties of opiate receptors. Fed Proc 40:2729–2734
Cicero TJ (1980) Animal models of alcoholism. In: Sinclair JD, Kiianmaa K (eds) Animal models in alcohol research. Academic, New York, pp 99–117
Collins DM, Myers RD (1987) Buspirone attenuates volitional alcohol intake in the chronically drinking monkey. Alcohol 4:49–56
Cornelius JR, Fisher BW, Salloum IM, Cornelius MD, Ehler JG (1992) Fluoxetine trial in depressed alcoholics. Alcohol Clin Exp Res 16(2):362
Costall B, Kelly ME, Naylor RJ (1988) The anxiolytic and anxiogenic actions of ethanol in a mouse model. J Pharm Pharmacol 40:197–202
Costall B, Jones BJ, Kelly ME, Naylor RJ, Onaivi ES, Tyers MB (1990) Ondansetron inhibits behavioural consequence of withdrawing from drugs of abuse. Pharmacol Biochem Behav 36:339–344
Cotton R, Giles MG, Miller L, Shaw JS, Timms D (1984) ICI174864: a highly selective antagonist for the opioid delta receptor. Eur J Pharmacol 97:331–332
Cox BJ, Norton GR, Swinson RP, Endler NS (1990) Substance abuse and panic related anxiety: a critical review. Behav Res Ther 28:385–393
Crabbe JC, Belknap JK (1992) Genetic approaches to drug dependence. Trends Pharmacol Sci 15:212–219
Curzon G, Kennett GA (1990) *m*-CPP: a tool for studying behavioural responses associated with $5HT_{1C}$ receptors. Trends Pharmacol Sci 11:181–182
Daoust M, Saligaut C, Chadelaud M, Chretien P, Moore N, Biosmare F (1984) Attenuation by antidepressant drugs of alcohol intake in rats. Alcohol 1:379–384
Di Chiara G, Imperato A (1985) Ethanol preferentially stimulates dopamine release in the nucleus accumbens of freely moving rats. Eur J Pharmacol 115:131–132
Dingledine R, Hynes MA, King GL (1986) Involvement of N-methyl-D aspartate receptors in epileptiform bursting in the rat hippocampal slice. J Physiol (Lond) 380:175–189
Dongier M, Vachon L, Schwartz G (1991) Bromocriptine in the treatment of alcohol dependence. Alcohol Clin Exp Res 15:970–977
Dorus W, Ostrow DG, Anton R, Cushman P, Collins JF, Schaefer M, Charles HL, Desai P, Hayashida M, Malkerneker U, Willenbring M, Fiscella R, Sather MR (1989) Lithium treatment of depressed and nondepressed alcoholics. JAMA 262:1646–1652
Dougherty RJ, Gates RR (1990) The role of buspirone in the management of alcohol withdrawal. A preliminary investigation. J Subst Abuse Treatm 7:189–192
Edwards G, Arif A, Hodgson R (1981) Nomenclature and classification of drug and alcohol related problems: a WHO memorandum. Bull WHO 59:225–242
Ellison G (1987) Stress and alcohol intake: the socio-pharmacological approach. Physiol Behav 40:378–392
Eriksson E, Humble M (1990) Serotonin in psychiatric pathophysiology: a review of data from experimental and clinical research. In: Pohl R, Gershon S (eds) The biological basis of psychiatric treatment. Karger, Basel, pp 66–119 (Progress in basic clinical pharmacology, vol 3)
Fadda F, Garau B, Marchei F, Colombo G, Gessa GL (1991) MDL72222, a selective 5-HT_3 receptor antagonist, suppresses voluntary ethanol consumption in alcohol-preferring rats. Alcohol Alcohol 26:107–110
File SE, Zharkovsky A, Hitchcott PK (1992) Effects of nitrendipine chlordiazepoxide, flumazenil and baclofen on the increased anxiety resulting from alcohol withdrawal. Prog Neuropsychopharmacol Biol Psychiatry 16:87–93
Fletcher PJ, Davies M (1990) A pharmacological analysis of the eating response induced by 8-OH-DPAT injected into the dorsal raphe nucleus reveals the involvement of a dopaminergic mechanism. Psychopharmacology (Berl) 100: 188–194
Froehlich JC, Harts J, Lumeng L, Li T-K (1988) Enkephalinergic involvement in voluntary ethanol consumption. In: Kuriyama K, Takada A, Ishii H (eds)

Biomedical and social aspects of alcohol and alcoholism. Excerpta Medica, Amsterdam, pp 235–238
Froehlich JC, Zweifel M, Harts J, Lumeng L, Li T-K (1991) Importance of delta opioid receptors in maintaining high alcohol drinking. Psychopharmacology (Berl) 103:467–472
Fuller RK, Branchey L, Brightwell DR, Derman RM, Emrick CD, Iber FL, James KE, Lacoursiere RB, Lee KK, Lowenstam I, Maany I, Neiderhiser D, Nocks JJ, Shaw S (1986) Disulfiram treatment of alcoholism: a veterans administration cooperative study. JAMA 256:1449–1455
Gauvin DV, Youngblood BD, Holloway FA (1992) The discriminative stimulus properties of acute ethanol withdrawal (hangover) in rats. Alcohol Clin Exp Res 16:336–341
Genazzani AR, Nappi G, Facchinetti F, Mezzela GL, Parrini D, Sinforiani E, Petraglia F, Savoldi J (1982) Central deficiency of B-endorphin in alcohol addicts. J Clin Endocrinol Metab 55:583–586
George DT, Benkelfat C, Murphy DJ, Schmitz J, Linnoila M (1990) Ethanol-like behavioural response in abstinent alcoholics infused with the 5-HT agonist mCPP. Biol Psychiatry 27:176A
Gianoulakis C, de Waele JP, Kiianmaa K (1992) Differences in the brain and pituitary B-endorphin system between the alcohol-preferring AA and alcohol-avoiding ANA rats. Alcohol Clin Exp Res 16:453–459
Gill K, France C, Amit Z (1986) Voluntary ethanol consumption in rats: an examination of blood/brain ethanol levels and behaviour. Alcohol Clin Exp Res 4:457–462
Gill K, Amit Z, Koe BK (1988) Treatment with sertraline, a new serotonin uptake inhibitor, reduces voluntary ethanol consumption in rats. Alcohol 5:349–354
Goldberg HL, Finnerty R (1982) Comparison of buspirone in two separate studies. J Clin Psychiatry 43:87–91
Gongwer MA, Murphy JM, McBride WJ, Lumeng L, Li T-K (1989) Regional brain contents of serotonin, dopamine and their metabolites in the selectively bred high and low alcohol drinking lines of rats. Alcohol 6:317–320
Gorelick DA, Paredes A (1992) Effect of fluoxetine on alcohol consumption in male alcoholics. Alcohol Clin Exp Res 16(2):261–265
Goudie AJ, Emmett-Oglesby MW (eds) (1989) Psychoactive drugs: tolerance and sensitization. Humana, Clifton
Grant KA, Barrett JE (1991) Blockade of the discriminative stimulus effects of ethanol with 5-HT_3 receptor antagonists. Psychopharmacology (Berl) 104:451–456
Grank KA, Colombo G (1992) Ethanol-like discriminative stimulus effects of the 5-HT, agonist TFMPP: effect of training dose. Alcohol Clin Exp Res 16:368
Grant KA, Valverius P, Hudspith M, Tabakoff B (1990) Ethanol withdrawal seizures and the NMDA receptor complex. Eur J Pharmacol 176:289–296
Griffith JD, Jasinski DR, Casten GP et al. (1986) Investigation of the abuse liability of buspirone in alcohol-dependent patients. Am J Med 80:30–35
Hatsukami D, Pickens RW (1982) Post treatment depression in an alcohol and drug abuse population. Am J Psychiatry 39:1563–1566
Henningfield JE, Meisch RA (1979) Ethanol drinking by rhesus monkeys with concurrent access to water. Pharmacol Biochem Behav 10:777–782
Higgins GA, Nguyen P, Joharchi N, Sellers EM (1991) Effects of 5-HT_3 receptor antagonists on behavioural measures of naloxone-precipitated opioid withdrawal. Psychopharmacology (Berl) 105:322–328
Higgins GA, Tomkins DM, Fletcher PJ, Sellers EM (1992) Effect of drugs influencing 5-HT function on ethanol drinking behaviour in rats: studies using a drinkometer system. Neurosci Biobehav Rev 16:535–552
Hilakivi-Clarke L, Lister RG (1992) Social status and voluntary alcohol consumption in mice: interaction with stress. Psychopharmacology (Berl) 10:276–282

Hindmarsh JG, Rose S, Collins P, Jenner P (1993) Stepwise administration prevents the fenfluramine-induced depletion of brain 5-hydroxytryptamine in the rat. British Pharmacological Society, winter meeting, p 144
Hjorth S, Sharp T (1991) Effect of the 5-HT1A receptor agonist 8-OH DPAT on the release of 5-HT in dorsal and median raphe-innervated rat brain regions as measured by in vivo microdialysis. Life Sci 48:1779–1786
Hoffman PL, Rabe CS, Grant KA, Valverius P, Hudspith M, Tabakoff B (1990) Ethanol and the NMDA receptor. Alcohol 7:229–231
Hubbell CL, Marglin SH, Spitalnic SJ, Abelson ML, Wild KD, Reid LD (1991) Opioidergic, serotonergic and dopaminergic manipulations and rats' intake of a sweetened alcoholic beverage. Alcohol 8:355–367
Hunt WA (1973) Changes in the neuroexcitability of alcohol dependent rats undergoing withdrawal as measured by the pentylenetetrazole seizure threshold. Neuropharmacology 12:1097–1102
Hunter BE, Boast CA, Walker DW, Zornetzer SF (1973) Alcohol withdrawal syndrome in rats: neural and behavioural correlates. Pharmacol Biochem Behav 1:719–725
Idemudia SO, Bhadra S, Lal H (1989) The pentylenetetrazole-like interoceptive stimulus produced by ethanol withdrawal is potentiated by bicuculline and picrotoxin. Neuropsychopharmacology 2:115–122
Jaffe J, Ciraulo D (1986) Alcoholism and depression. In: Meyer RE (ed) Psychopathology and addictive disorders. Guilford, New York, pp 293–320
Jaffe JH (1992) Current concepts of addiction. In: O'Brien CP, Jaffe JH (eds) Addictive states. Raven, New York, pp 1–21
Kahn RS, Wetzler S (1991) m-Chlorophenylpiperazine as a probe of serotonin function. Biol Psychiatry 30:1139–1166
Kalant AH (1987) Current trends in biomedical research on alcohol. Alcohol Alcohol [Suppl 1]:1–12
Kennett GA, Curzon G (1988) Evidence that mCPP may have behavioural effects mediated by central $5HT_{1C}$ receptors. Br J Pharmacol 94:137
Kennett GK, Whitton P, Shah K, Curzon G (1989) Anxiogenic-like effects of mCPP and TFMPP in animal models are opposed by $5\text{-}HT_{1C}$ receptor antagonists. Eur J Pharmacol 164:445–454
Kiianmaa K, Kokkonen J, Nurmi M, Nykanen I (1992) Effect of ethanol on the release of monoamines in the nucleus accumbens of the alcohol-preferring AA and alcohol avoiding ANA rats. Soc Neurosci Abstr 18:598.6
Knapp DJ, Pohorecky LA (1992) Zacopride, a 5-HT3 receptor antagonist, reduces voluntary ethanol consumption in rats. Pharmacol Biochem Behav 41:847–850
Knapp DJ, Benjamin D, Pohorecky LA (1992) Effects of gepirone on ethanol consumption, exploratory behaviour and motor performance in rats. Drug Dev Res 26:319–341
Koob GF, Bloom FE (1988) Cellular and molecular mechanisms of drug dependence. Science 242:715–723
Koob GF, Weiss F (1992) Neuropharmacology of cocaine and ethanol dependence. Recent Dev Alcohol 10:201–233
Koppi S, Eberhardt G, Haller R, Konig P (1987) Calcium channel blocking agent in the treatment of acute alcohol withdrawal-caroverine versus meprobamate in a randomized double-blind study. Neuropsychobiology 17:49–52
Kornet M, Goosen C, Van Ree JM (1991) Effect of naltrexone on alcohol consumption during chronic alcohol drinking and after a period of imposed abstinence in free choice drinking rhesus monkeys. Psychopharmacology (Berl) 104:367–376
Kostowski W, Dyr W (1992) The effects of 5-HT1A receptor agonists on ethanol preference in the rat. Alcohol 9:283–286
Kranzler HR, Liebowitz NR (1988) Anxiety and depression in substance abuse: clinical implications. Med Clin North Am 72:867–885
Kranzler HR, Meyer RE (1989) An open trial of buspirone in alcoholics. J Clin Psychopharmacol 9(5):379–80

Kranzler HR, Orrok B (1989) The pharmacotherapy of alcoholism. In: Tasman A, Hales RE, Frances AJ (eds) Review of psychiatry, vol 8. American Psychiatric Association, Washington, pp 359–379
Kranzler H, Del Boca F, Babor T (1991) Fluoxetine as an adjunct to relapse prevention in alcoholics. Annual Meeting of the American College of Neuropsychopharmacology, San Juan/Puerto Rico
Kranzler H, Del Boca F, Rounsaville B (1992) Psychopathology as a predictor of outcome three years after alcoholism treatment. Alcohol Clin Exp Res 16(2):363
Kraus ML, Gottlieb LD, Horwitz RI, Anscher M (1985) Randomized clinical trial of atenolol in patients with alcohol withdrawal. N Engl J Med 313(15):905–909
Kulkosky PJ, Sanchez MR, Foderaro MA, Chiu N (1989) Cholecystokinin and satiation with alcohol. Alcohol 6:395–402
Kushner MG, Sher KJ, Beitman BD (1990) The relation between alcohol problems and the anxiety disorders. Am J Psychiatry 147:685–695
Lal H, Harris CM, Benjamin D, Springfield AC, Bhadra S, Emmett-Oglesby MW (1988) Characterization of a pentylenetetrazole-like interoceptive stimulus produced by ethanol withdrawal. J Pharmacol Exp Ther 247:508–518
Lal H, Prather PL, Rezazadeh SM (1991) Anxiogenic behaviour in rats during acute and protracted ethanol withdrawal: reversal by buspirone. Alcohol 8:467–471
Lane JD, Pickering CL, Hooper ML, Fagan K, Tyers MB, Emett-Oglesby MW (1992) Failure of ondansetron to block the discriminative or reinforcing stimulus effects of cocaine in the rat. Drug Alcohol Depend 30:151–162
Lee MA, Meltzer HY (1990) Neuroendocrine responses to serotonergic agents in alcoholics. Biol Psychiatry 30(10):1017–1030
Linnoila M (1990) Monoamines and impulse control. In: Swinkels JA, Blijleven W (eds) Depression, anxiety and aggression: factors that influence the course. Medidact, Houten, pp 167–169
Linseman MA (1987) Alcohol consumption in free-feeding rats: procedural, genetic and pharmacokinetic factors. Psychopharmacology (Berl) 92:254–261
Litten RZ, Allen JP (1991) Pharmacotherapies for alcoholism: promising agents and clinical issues. Alcohol Clin Exp Res 15:620–633
Little HJ, Dolin SJ, Halsey MJ (1986) Calcium channel antagonists decrease the ethanol withdrawal syndrome. Life Sci 39:2059–2065
Lorens SA, Sainasti SM (1978) Naloxone blocks the excitatory effect of ethanol and chlordiazepoxide on lateral hypothalmic self-stimulation behaviour. Life Sci 23:1359–1364
Lovinger DM (1991) Ethanol potentiation of 5-HT_3 receptor mediated ion current in NCB-20 neuroblastoma cells. Neurosci Lett 122:57–60
Ludwig AM, Stark LH (1974) Alcohol craving: subjective and situational aspects. Q J Stud Alcohol 35:899–905
Malcolm R, Ballenger JC, Sturgis ET, Anton R (1989) Double-blind controlled trial comparing carbamazepine to oxazepam treatment of alcohol withdrawal. Am J Psychiatry 146(5):617–621
Malcolm R, Anton RF, Randall CL, Johnston A, Brady K, Thevos A (1992) A placebo-controlled trial of buspirone in anxious inpatient alcoholics. Alcohol Clin Exp Res 16:1007–1013
Marlatt GA, Gordon JR (1980) Determinants of relapse: implications for the maintenance of behaviour change. In: Davidson PO, Davidson SM (eds) Behavioural medicine: changing health lifestyles. Brunner/Mazel, New York, pp 410–452
Mason BJ, Kocsis JH (1991) Desipramine treatment of alcoholism. Psychopharmacol Bull 27:155–161
Mason BJ, Ritvo EC, Welch B, Zimmer E, Salvato F, Goldberg G (1992) Repeated measures of body weight and Hamilton depression scores in alcoholics chronically treated with Nalmefene, a new opiate antagonist. Annual Meeting of the American College of Neuropsychopharmacology, San Juan/Puerto Rico
McBride WJ, Murphy JM, Lumeng L, Li T-K (1988) Effects of Ro15-4513, fluoxetine and desipramine on the intake of ethanol, water and food by the alcohol-

preferring (P) and non-preferring (NP) lines of rats. Pharmacol Biochem Behav 30:1045–1050
McBride WJ, Murphy JM, Lumeng L, Li T-K (1990) Serotonin, dopamine and GABA involvement in alcohol drinking of selectively bred rats. Alcohol 7: 199–205
McGrath PJ, Goldman D, Nunes EN, Quitkin FM, Stewart JM, Goldman R (1991) Imipramine treatment of depressed alcoholics. Annual Meeting of the American Psychiatric Association, San Francisco
Meert TF, Awouters F, Niemegeers CJE, Schellenkens KHL, Janssen PAJ (1991) Ritanserin reduces abuse of alcohol, cocaine and fentanyl in rats. Pharmacopsychiatry 24:159–163
Mello NK (1983) A behavioural analysis of the reinforcing properties of alcohol and other drugs in man. In: Kissin B, Begleiter H (eds) The biology of alcoholism: the pathogenesis of alcoholism, vol 7. Plenum, New York, pp 133–198
Meyer RE (1986) Anxiolytics and the alcoholic patient. J Stud Alcohol 47(4):269–273
Meyer RE, Babor TF (1989) Explanatory models of alcoholism. In: Tasman A, Hales RE, Frances AJ (eds) Review of psychiatry, vol 8. American Psychiatric Association, Washington, pp 273–292
Michel MK, Bolger G, Weissman B-A (1985) Binding of a new opiate antagonist, nalmefene to rat membranes. Methods Find Exp Clin Pharmacol 7:175–177
Middaugh LD, Read E, Boggan WO (1978) Effects of naloxone on ethanol induced alterations of locomotor activity in C57BL/6 mice. Pharmacol Biochem Behav 11:157–160
Miller NS, Dackis CA, Gold MS (1987) The relationship of addiction, tolerance and dependence to alcohol and drugs: a neurochemical approach. J Subst Abuse Treatm 4:197–207
Miller WR, Hester RK (1986) The effectiveness of alcoholism treatment – what research reveals. In: Miller WE, Heather N (eds) Treating addictive behaviours: processes of change. Plenum, New York, pp 121–174
Monte JM, Alterwain P (1991) Ritanserin decreases alcohol intake in chronic alcoholics. Lancet 337:60
Mullaney JA, Trippett CJ (1979) Alcohol dependence and phobias: clinical description and relevance. Br J Psychiatry 35:565–573
Murphy DL (1990) Neuropsychiatric disorders and the multiple human brain serotonin receptor subtypes and subsystems. Neuropsychopharmacology 3: 457–471
Murphy JM, Waller MB, Gatto GJ, McBride WJ, Lumeng L, Li T-K (1985) Monoamine uptake inhibitors attenuate ethanol intake in alcohol-preferring (P) rats. Alcohol 2:349–352
Murphy JM, McBride WJ, Lumeng L, Li T-K (1987) Effects of serotonergic agents on ethanol intake of the high alcohol drinking (HAD) line of rats. Alcohol Clin Exp Res 11:208
Myers MD, Quarfordt SD (1991) Alcohol drinking attenuated by sertraline in rats with 6-OH DOPA or 5,7-DHT lesions of the N-accumbens: a caloric response? Pharmacol Biochem Behav 40:923–928
Myers RD, Borg S, Mossberg R (1986) Antagonism by naltrexone of voluntary alcohol selection in the chronically drinking macaque monkey. Alcohol 3:383–388
Naranjo CA, Sellers EM (1989) Serotonin uptake inhibitors attenuate ethanol intake in problem drinkers. Recent Dev Alcohol 7:255–266
Naranjo CA, Sellers EM, Roach CA, Woodley DV, Sanchez-Craig M, Sykora K (1984) Zimelidine-induced variations in alcohol intake by non-depressed heavy drinkers. Clin Pharmacol Ther 35:374–381
Naranjo CA, Sellers EM, Sullivan JT, Woodley DV, Kadlec K, Sykora K (1987) The serotonin uptake inhibitor citalopram attenuates ethanol intake. Clin Pharmacol Ther 41:266–274

Naranjo CA, Sullivan JT, Kadlec K, Woodley-Remus DV, Kennedy G, Sellers EM (1989) Differential effects of viqualine on alcohol intake and other consummatory behaviours. Clin Pharmacol Ther 46:301–309
Naranjo CA, Kadlec K, Sanhueza P, Woodley-Remus DV, Sellers EM (1990) Fluoxetine differentially alters alcohol intake and other consummatory behaviours in problem drinkers. Clin Pharmacol Ther 47:490–498
Naranjo CA, Poulos CX, Bremner KE, Lanctot KL (1992) Citalopram decreases desirability, liking and consumption of alcohol in alcohol dependent drinkers. Clin Pharmacol Ther 51:729–739
Nunes EV, McGrath PJ, Quitkin FM, Stewart JP, Harrison W, Tricamo E, Ocepek-Welikson K (1993) Imipramine treatment of alcoholism with comorbid depression. Am J Psychiatry 150:963–965
Nutt D, Adinoff B, Linnoila M (1989) Benzodiazepines in the treatment of alcoholism. In: Galanter M (ed) Treatment research. Recent Alcohol 7:283–313
O'Malley S, Jaffe AJ, Chang G, Schottenfeld RS, Meyer RE, Rounsaville B (1992) Naltrexone and coping skills therapy for alcohol dependence: a controlled study. Arch Gen Psychiatry 49:881–887
Oakley NR, Jones BJ, Tyers MB et al. (1988) The effect of GR38032F on alcohol consumption in the marmoset. Br J Pharmacol 95:870P
Olivera AA, Sarvis S, Heard C (1990) Anxiety disorders coexisting with substance dependence: treatment with buspirone. Curr Ther Res 47:52–61
Overstreet DH, Rezvani AH, Janowsky DS (1992) Genetic animal models of depression and ethanol preference provide support for cholinergic and serotonergic involvement in depression and alcoholism. Biol Psychiatry 31:919–936
Prather PL, Rezazadeh SM, Lal H (1991) Mianserin in the treatment of ethanol withdrawal in the rat: prevention of behaviours indicative of anxiety. Psychopharmacol Bull 27:285–289
Privette TH, Hornsby RL, Myers RD (1988) Buspirone alters alcohol drinking induced in rats by tetrahydropapaveroline injected into brain monoaminergic pathways. Alcohol 5:147–152
Quitkin FM, Rifkin A, Kaplan J et al. (1972) Phobic anxiety syndrome complicated by drug dependence and addiction. Arch Gen Psychiatry 27:159–162
Regier DA, Farmer ME, Rae DS et al. (1990) Comorbidity of mental disorders with alcohol and other drug abuse. JAMA 264:2511–2518
Reith MEA (1990) 5-HT_3 receptor antagonists attenuate cocaine-induced locomotion in mice. Eur J Pharmacol 186:327–330
Reyntjens A, Gelders IG, Hoppenbrowers ML, Vanden Bussche G (1986) Thymosthenic effects of ritanserin R55667, a centrally acting serotonin S-2 receptor blocker. Drug Dev Res 8:205–211
Romach MK, Sellers EM (1991) Management of the alcohol withdrawal syndrome. Annu Rev Med 42:323–340
Romach MK, Busto U, Somer G, Kaplan HL, Sellers EM (1992) Alcoholism, mental disorders and patient's sex in benzodiazepine discontinuation. Annual Meeting of the American College of Neuropsychopharmacology San Juan/Puerto Rico
Ross HE, Glaser FB, Germanson T (1988) The prevalence of psychiatric disorders in patients with alcohol and other drug problems. Arch Gen Psychiatry 45:1023–1031
Rotham SW, Olney JW (1987) Excitotoxicity and the NMDA receptor. Trends Neurosci 10:299–302
Rounsaville BJ, Dolinsky ZS, Babor TF et al. (1987) Psychopathology as a predictor of treatment outcome in alcoholics. Arch Gen Psychiatry 44:505–513
Rowland NE, Morian KR (1992) Effect of dexfenfluramine on alcohol intake in alcohol-preferring "P" rats. Alcohol 9:559–561
Samson HH (1987) Initiation of ethanol-maintained behaviour. A comparison of animal models and their implication to human drinking. In: Thompson T, Dews

PB, Barrett JE (eds) Advances in behavioural pharmacology: neurobehavioural pharmacology, vol 6. Erlbaum, Hillsdale, pp 221–248

Samson HH, Harris RA (1992) Neurobiology of alcohol abuse. Trends Pharmacol Sci 13:206–211

Samson HH, Pfeffer AO, Tolliver GA (1988) Oral ethanol self-administration in rats: models of ethanol seeking behavior. Alcohol Clin Exp Res 12:591–598

Samson HH, Doyle TF (1985) Oral ethanol self-administration in the rat: effect of naloxone. Pharmacol Biochem Behav 22:91–99

Saunders JB (1989) The efficacy of treatment for drinking problems. Int Rev Psychiatry 1:121–138

Seigel S (1989) Pharmacological conditioning and drug effects. In: Goudie AJ, Emmett-Oglesby MW (eds) Psychoactive drugs: tolerance and sensitization. Humana Clifton, pp 115–180

Sellers EM, Zilm DH, Degani NC (1977) Comparative efficacy of proprancolol and chlordiazepoxide in alcohol withdrawal. J Stud Alcohol 38:2096–2108

Sellers EM, Naranjo CA, Harrison M, Devenyi P, Roach C, Sykora K (1983) Diazepam loading: simplified treatment of alcohol withdrawal. Clin Pharmacol Ther 34:822–826

Sellers EM, Romach MK, Frecker RC, Higgins GA (1991) Efficacy of the $5HT_3$ antagonist ondansetron in addictive disorders. Biol Psychiatry 2:894–897

Sellers EM, Higgins GA, Sobell MB (1992) 5-HT and alcohol abuse. Trends Pharmacol Sci 13:69–75

Signs SA, Schecter MD (1988) The role of dopamine and serotonin receptors in the mediation of the ethanol interoceptive cue. Pharmacol Biochem Behav 30: 55–64

Sinclair JD (1974) Morphine suppresses alcohol drinking regardless of prior alcohol access duration. Pharmacol Biochem Behav 2:409–412

Sinclair JD, Scheinin H, Lammintausta R (1992) Progressive suppression of aclohol drinking in the alcohol-preferring AA rat line with nalmefene. 22nd Annual Meeting of the Behaviour Genetics Association, Boulder

Smail P, Stockwell T, Canter S et al. (1984) Alcohol dependence and phobic anxiety states. I: A prevalence study. Br J Psychiatry 144:53–57

Sobell MB, Wilkinson DA, Sobell LC (1990) Alcohol and drug problems. In: Bellack AS, Hersen M, Kazdin AE (eds) International handbook of behaviour modification and therapy, 2nd edn. Plenum, New York, pp 415–435

Soubrie P (1986) Reconciling the role of central serotonin neurons in human and animal behaviour. Behav Brain Sci 9:319–364

Stefanini E, Frau M, Garau MG, Garau B, Fadda F, Gessa GL (1992) Alcohol-preferring rats have fewer dopamine D2 receptors in the limbic system. Alcohol 27:127–130

Stolerman I (1992) Drugs of abuse: behavioural principles, methods and terms. Trends Pharmacol Sci 13:170–176

Szabo G, Tabakoff B, Hoffman PL (1988) Receptors with V_1 characteristics mediate the maintenance of ethanol tolerance by vasopressin. J Pharmacol Exp Ther 247:536–541

Ticku MK, Kulkarniz K (1988) Molecular interactions of ethanol with GABAergic system and potential of RO-15-4513 as an ethanol antagonist. Pharmacol Biochem Behav 30:501–510

Tollefson GD, Montague-Clouse J, Tollefson SL (1992) Treatment of comorbid generalized anxiety in a recently detoxified alcoholic population with a selective serotonergic drug (buspirone). J Clin Psychopharmacol 12:19–26

Tomkins DM, Sellers EM (1992) Effect of ondansetron on ethanol intake and behaviour. Alcohol Clin Exp Res 16:368

Tomkins DM, Higgins GA, Sellers EM (1993) Enhancement of ethanol intake in Wistar rats following pretreatment with low doses of 8-OH DPAT and gepirone using a limited access paradigm. British Pharmacological Society, winter meeting, C101

Van der Hoek GA, Cooper SJ (1990) Evidence that ondansetron, a selective 5-HT_3 antagonist, reduces concaine's psychomotor stimulant effects in the rat. Psychopharmacology (Berl) 101:S59
Volpicelli JR, Davis MG, Olgin JE (1986) Naltrexone blocks the post-shock increase of ethanol consumption. Life Sci 38:841–847
Volpicelli JR, O'Brien CP, Alterman AI, Hayashida M (1990) Naltrexone and the treatment of alcohol dependence: initial observations. In: Reid LD (ed) Opioids, bulimia and alcohol abuse and alcoholism. Springer, Berlin Heibelberg New York, pp 195–214
Volpicelli JR, Alterman Al, Hayashida M, O'Brien CP (1992) Naltrexone in the treatment of alcohol dependence. Arch Gen Psychiatry 49:876–880
Walker DW, Zornetzer SF (1974) Alcohol withdrawal in mice: electroencephalographic and behavioural correlates. Electroencephalogr Clin Neurophysiol 36: 233–243
Waller MB, McBride WJ, Lumeng L, Li T-K (1982) Induction of dependence on ethanol by free-choice drinking in alcohol-preferring rats. Pharmacol Biochem Behav 16:501–507
Weiss F, Bloom FE, Koob GF (1992) Effects of ethanol self-administration on accumbens dopamine release: strain differences between alcohol-preferring (P) and genetically heterogeneous Wistar rats. Alcohol Clin Exp Res 16:367
Whitton P, Curzon G (1990) Anxiogenic-like effects of infusing 1-(3 chlorophenyl) piperazine (mCPP) into the hippocampus. Psychopharmacology (Berl) 100: 138–140
Wilson GT (1988) Alcohol and anxiety. Behav Res Ther 26:369–381
Wolffgramm J, Heyne A (1991) Social behaviour, dominance and social deprivation of rats determine drug choice. Pharmacol Biochem Behav 38:389–399
Wong DT, Threlkeid PG, Lumeng L, Li T-K (1990) Higher density of serotonin 1A receptors in the hippocampus and cerebral cortex of alcohol preferring P rats. Life Sci 46:231–235
Wright C, Moore RD (1989) Disulfiram treatment of alcoholism. Ann Intern Med 111:943–945

CHAPTER 12

Genetic Factors in Alcoholism: Evidence and Implications

J. GELERNTER

A. Introduction

In this chapter we review some of the major findings pertaining to genetic factors in alcohol abuse and dependence. First we consider a series of clinical studies which establish that genetic liability can increase risk of alcoholism; then, a series of laboratory studies that investigate the nature of the liability.

Although genes alone do not make alcoholics, it has long been appreciated that a propensity to abuse alcohol seems to run in families. Many environmental factors are also important determinants of a person's drinking behavior. There is an unavoidable incongruity in our consideration of genetic factors influencing complex behaviors such as alcohol intake: how can a gene make a person obtain alcohol and drink excessive amounts of it? It is impossible to imagine a DNA sequence that would have only this specific effect. Instead of considering the behavior as a whole, we might better consider certain aspects of the individual's experiences and perceptions that may be genetically influenced. These may influence in turn how much he desires alcohol, how he perceives its effects on his mood and thought processes, and whether alcohol easily makes him physically sick. Someone carrying a gene that made it impossible to metabolize alcohol normally, resulting in severe physical discomfort after every drink, might be at decreased risk for alcohol dependence no matter how reinforcing the CNS effects. This is a very straightforward mechanism for a genetic influence on a complex behavior: the gene makes the behavior induce physical discomfort, so the behavior is inhibited. In fact this schema is important in Asian populations, where there are relatively high frequencies of abnormal alleles of alcohol-metabolizing enzymes. We might imagine several other ways for simple genes to have effects on this complicated behavior; for example, a variant form of a brain protein (such as a dopamine receptor) might make ethanol's effects more reinforcing in those individuals carrying the variant, so those individuals might have statistically greater risk of alcohol abuse, because they enjoy it more; or another gene might make the taste of alcohol unpleasant for a certain group, making it less likely that they will drink excessively.

None of these genetic factors would necessarily be expected to turn the behavior on or off. They might, though, make the behavior more or less likely to occur in individuals carrying particular variant alleles of certain genes. Many other factors could also affect the behavior, such as degree of exposure to ethanol, stress in relationships, or adherence to religious beliefs, for example.

While this raises the prospect that single genes may have measurable effects on alcohol intake, at least in certain populations, it also illustrates the complexity of the genetic system: a complex behavior can be influenced at many points along its pathway, from the cognition preceding the behavior to varying the effect of the final behavior on the individual. Although clinical heterogeneity in itself cannot prove genetic heterogeneity, alcoholism is so common and its expression so varied that major single gene effects are thought to be relatively unlikely. Many other factors complicate study of the molecular genetics of alcohol dependence. For example, alcoholics tend to select mates with substance abuse problems or other psychiatric illnesses themselves (assortative mating), complicating linkage studies. Also, segregation analyses (ASTON and HILL 1990) have tended to suggest that the underlying genetic components are complicated. Use of a candidate gene stragegy is hampered by a large excess of candidate genes, which limits the utility of a linkage or association strategy directed by candidate gene hypotheses.

The observation of familial clustering of a set of symptoms does not prove a genetic basis. Genetic factors must be proven by other means; this can be accomplished by adoption studies, twin studies, and family studies. We will first consider studies in clinical genetics (adoption, twin, family, and epidemiological studies); then, studies in laboratory genetics (association and linkage studies). Ideas about diagnosis and methodology have evolved over time; we will not discuss some older studies that would now not be considered methodologically acceptable. Even within types, studies from different groups of investigators are rarely exactly comparable. Male subjects have often been preferred in these studies because of the higher rate of alcoholism in men.

B. Clinical Studies: Familial Patterns in Alcohol Use

I. Adoption Studies

Adoption studies provide a reliable way to prove a genetic contribution to an illness. The general paradigm is to consider rates of illness of adopted-away offspring of ill and well parents. For a disease with genetic determinants, the risk of the illness in adopted away offspring of ill parents should exceed the risk in adopted away offspring of normal parents. Excess risk of illness in children of ill individuals adopted into normal homes can only be explained by an inherited liability to develop the illness. Due to the requirement for good adoption records, many adoption studies have been carried out in Scandinavian countries where such records are maintained. Adoption studies have provided good evidence for genetic factors playing some role in alcoholism.

Goodwin et al. (1973) compared 55 sons of alcoholics who were adopted away in the first 6 weeks of life and 78 controls in a Danish sample. The adoptees received direct interviews. These investigators found an increased risk of alcohol-related problems and about a fourfold increased risk of alcoholism in the children of alcoholics, compared to the children of nonalcoholics. There were no other differences in psychopathology in the groups of adoptees, but only a limited range of psychopathology was considered (e.g., no data on drug abuse were presented). A subsequent study comparing 49 adopted away daughters of alcoholics and 47 adopted away daughters of nonalcoholics from the same large sample (Goodwin et al. 1977) failed to show any increased risk in biological relatives of alcoholics (but there were only 2 alcoholics in each group).

Schuckit et al. (1972) studied children of absent alcoholics or nonalcoholics who were raised for at least 6 years by nonalcoholic parents, and children of alcoholics or nonalcoholics raised with alcoholics, in a study design using half siblings. It was thus possible to investigate the contribution of alcoholic or nonalcoholic parents who were not part of the household towards the outcome of their biological children. The idea is similar (but not identical) to that of an adoption study. They found that genetic influences were more important that environmental influences for the development of alcoholism.

Cadoret and Gath (1978), who studied a sample of 84 adoptees in Iowa (mostly biological children of nonalcoholics), also found that alcoholism was more common in adopted-away biological offspring of alcoholics that in offspring of normal controls. Cadoret et al. (1980) extended their findings with a larger sample of male adoptees who had alcoholics first- or second-degree relatives. Their data were most consistent with genetic determinants only (i.e., no environmental determinants) of alcoholism in adoptees (although a subsequent study by the same group did show environmental effects; see below).

Cloninger et al. (1981) studied a Swedish sample of 862 male adoptees. [Bohman et al. (1981) completed a parallel study with 913 women discussed below.] The evaluations of both adoptees and relatives in this study were more extensive than in the earlier studies. Eighty-seven of the subjects had fathers with moderate or severe alcohol abuse; 151 of the adoptees themselves had alcohol abuse histories. Overall, adoptees biologically related to alcoholics had increased risk for alcoholism. Based on discriminant analysis of the data, the authors hypothesized the existence of two main types of alcoholism; type 1 ("milieu-limited," more common, either mild or severe, and associated with mild alcohol abuse in both parents); and type 2 ("male-limited," less common, of moderate severity or recurrent, and associated with severe paternal alcohol abuse and criminality but no maternal pathology; expressed only in men). Type 2 alcoholism was postulated to be much more heritable (i.e., 90% heritability) than type 1 alcoholism, the latter type being much more likely to be influenced by environment. [The hypothesis was developed further by CLONINGER (1987).] Differing observations of environmental and genetic effects could be explained in terms of this hypothesized genetic heterogeneity in forms of alcohol abuse. BOHMAN et al. (1981), in their study of 913 female adoptees, 31 of whom were alcohol abusers, found evidence of transmission of alcoholism from mothers to daughters, but only from fathers to daughters when the fathers had mild alcoholism; this was felt to confirm the predictions of the hypothesized type 1/type 2 distinction. Alcoholism was found to be less heritable in women than in men (BOHMAN et al. 1987).

Although aspects of the CLONINGER et al. (1981) type 1/type 2 hypothesis are very attractive, that work has been criticized (e.g., on the basis of statistical methods and sample composition by LITTRELL 1988; refuted by CLONINGER et al. 1988; see also SCHUCKIT and IRWIN 1989) and still awaits clear confirmation. Segregation analysis of pedigree data by the same group did not demonstrate statistically significant heterogeneity between the families of male and female alcoholics (GILLIGAN et al. 1987), but in the same study etiologic heterogeneity was demonstrated between "male-like" and "female-like" families (with a major gene effect seen only in the "male-like" families). CLONINGER et al. (1988) summarized further studies by his group.

CADORET et al. (1986) demonstrated through an adoption study conducted in Iowa that parental alcoholism can result in increased incidence of both alcohol abuse and substance abuse in offspring. They studied a total of 443 adoptees; 39 had first-degree relatives with alcohol abuse. Alcoholism in a first-degree relative resulted in an odds ratio of 5.9 for alcoholism and 4.3 for substance abuse in the adoptee. Antisocial personality in first-degree relatives also increased risk of alcoholism in adoptees; and alcohol problems in the adoptive family were found to increase risk of alcohol abuse in the adoptees (odds ratio 2.6). A subsequent study (CADORET et al. 1987) showed that alcoholism and antisocial personality in first-degree relatives specifically increased risk of those same disorders in adoptees, and that alcohol problems in adoptive families (i.e., environmental effects) also increased risk of alcoholism in adoptees.

In summary, adoption studies have consistently found increased risk of alcoholism in the adopted-away children of alcoholics compared to the adopted-away children of controls. This establishes that vulnerability to alcoholism can be increased by genetic factors. Most studies have also found environmental factors to have some influence on the development of alcoholism in adoptees.

II. Twin Studies

Twin studies can be used to demonstrate the relative importance of genetic and environmental contributors to illness by comparing concordance rates for illness in monozygotic (MZ) and dizygotic (DZ) twins. Since MZ twins share 100% of their DNA and DZ twins share only 50% of theirs, MZ concordance is expected to be greater than DZ concordance for a disorder with a genetic contribution. However, MZ twins look alike and also tend to share similar behavior (BOUCHARD et al. 1990). Consequently, even when the subjects studied are pairs of MZ and DZ twins for whom each pair was raised together, the effects of shared environment may be somewhat greater for MZ twins because of their greater similarity in other spheres. Several groups have used the twin study paradigm to investigate heritability of alcoholism.

HRUBEC and OMENN (1981) investigated the inheritance of alcoholism and cirrhosis using a twin registry of almost 16000 US veterans. Information about the subjects was obtained from the US Department of Veterans Affairs medical records and questionnaire self-reports. MZ concordance for alcoholism was 26% whereas DZ concordance was 12%. MZ twins also had greater concordance for cirrhosis than DZ twins, a result that could not be explained solely on the basis of alcoholism.

PICKENS et al. (1991) studied 81 MZ and 88 DZ (same sex) male and female twin pairs, and examined concordance for alcohol dependence and other diagnoses. Differences in concordance supporting genetic factors for alcohol dependence were found in both male and female twin pairs; probandwise concordance rates for alcohol dependence in male twins were 0.59 for MZ twins and 0.36 for DZ twins; for female twins, the corresponding rates were 0.25 and 0.05. The difference for alcohol abuse was significant only for the male twins. Estimated heritability for alcohol abuse and dependence for the male subjects was 0.36, and for the female subjects 0.26. A substantial amount of liability variance (50%) was due to shared environmental factors (which contrasts with the study of KENDLER et al. 1992, discussed below). A lower heritability of alcoholism in females was also demonstrated in an overlapping twin study that also considered male-female DZ twin pairs (McGUE et al. 1992). In that study, elevated cross-sex rates of alcoholism were observed, a result that was attributed to environmental factors. HEATH et al. (1989) studied a sample of female twins by questionnaire. They noted that MZ twins (data from 1233 twin pairs) were highly correlated for alcohol use.

KENDLER et al. (1992) have provided the largest direct interview twin study so far, evaluating a sample of 1030 female twin prais. Rather than a treatment sample, they used a population sample from the Virginia Twin Registry, which included twin pairs where both cotwins were affected, where one twin was affected, and where neither was affected. Correlation of liability rather than probandwise concordance was used as the major method of analysis (although MZ twins had greater probandwise concordance for alcoholism than DZ twins in all cases). According to these authors, depending on the disease definition of alcoholism used, the genetic contribution to liability to develop alcoholism in women was between 50% (for narrowly defined alcoholism corresponding to DSM-III-R alcoholism plus tolerance or dependence) and 60% (using a broader disease definition). Surprisingly, the rest of the variance was accounted for by individual environmental differences, as opposed to family differences (shared environmental factors). These estimates of genetic and environmental contributions to alcoholism in females differ from those derived from adoption studies (e.g., GOODWIN et al. 1977; BOHMAN et al. 1981) and prior twin studies (e.g., PICKENS et al. 1991). One possible explanation for the different findings is sample size; KENDLER et al. (1992) examined the largest sample.

MULLAN et al. (1986) studied a series of 56 MZ and DZ twin pairs, and evaluated comorbid diagnoses and neuroticism to investigate whether some of these characteristics represented cause or effect with respect to alcoholism. Alcoholic twins (both MZ and DZ) had higher neuroticism scores than their nonalcoholic cotwins. The comorbid psychopathology was therefore interpreted as being a consequence of the alcoholism. MZ twins can also be used to assess the effect of two different environments on what, from a genetic standpoint, is essentially the same person. GURLING et al. (1991) studied 25 MZ twin pairs discordant for alcoholism; the non-alcohol-abusing twin represented, in effect, the developmental pathway the alcohol-abusing twin might have taken had he not abused alcohol. [This method is supported by the similarity normally demonstrated by cotwins on behavioral measures; BOUCHARD et al. (1990).] Studying discordant MZ cotwins uses genetic knowledge to obtain, in effect, a perfectly matched control group for a group of alcohol abusers. In this study, the alcohol-abusing twins had inferior performance on cognitive testing. This decrement in performance was most likely an effect of alcohol abuse, rather than a cause; all of the twins started with the same genetic potential for a certain level of cognitive function. However, unidentified environmental events (e.g., obsterical complications) also could have increased the risk of both alcoholism and a decrement in cognitive performance in the affected twins.

Thus there is greater concordance for alcoholism in MZ than in DZ twins, replicated over several samples, providing further support for genetic factors in the etiology of alcoholism. The conclusions of KENDLER et al. (1992), supporting very significant genetic effects in females, contrast with those from most previous studies assessing genetic contributions to alcoholism in females. These results may have been observed only in the study by KENDLER et al. (1992) because these investigators had a much larger sample than the others.

III. Other Clinical Data Bearing on Genetics

Family studies, not discussed here, demonstrate the range of other psychiatric disorders that may be genetically related to alcoholism. FRANCES et al. (1980, 1984) compared symptomatology and treatment outcome in male alcoholic inpatients with or without a first-degree alcohol-abusing relative, and found that those with ill relatives had more severe symptoms and worse outcome, the worst outcome correlated with the greatest number of ill first-degree relatives.

IV. Trait Markers and Differences Between Familial and Nonfamilial Alcoholism

If specific differences between that type of alcoholism that clearly runs in families, and "sporadic," or seemingly nonfamilial, alcoholism could be identified, such a heuristic could allow identification of a more homogeneous subgroup of alcohol abusers. Factors that may differentiate between alcoholics with a family history of alcoholism and those without a family history include, e.g., EEG findings (PROPPING et al. 1981), course of illness (discussed above), response to alcohol challenge (reviewed by NEWLIN and THOMSON 1990), and alexithymia (FINN et al. 1987). If clear trait markers could distinguish between alcoholics and nonalcoholics, this could lead to identification of a genetic mechanism. Several trait markers have been proposed for alcoholism (reviewed by TABAKOFF and HOFFMAN 1988; CRABB 1990). These include, e.g., EEG differences (POLLOCK et al. 1988), monoamine oxidase inhibition by ethanol and adenylate cyclase activity after stimulation (TABAKOFF et al. 1988), and variation in the alcohol-metabolizing enzymes (discussed below).

C. Laboratory Studies: Search for a Molecular Basis for a Complex Behavior

Since roughly 1980, the introduction of molecular DNA markers such as restriction fragment length polymorphisms (RFLPs) has revolutionized the search for genes that predispose to illness, especially using genetic linkage studies. In a genetic linkage study, a gene can be assigned to a certain chromosomal location (i.e., mapped) by demonstrating nonrandom segregation between the gene studied and other markers of known map location. Linkage is accepted given (a) a lod score statistic (*l*og of the *od*ds ratio) >3 and (b) replication. Increasing sophistication in methods of manipulating DNA in the laboratory has made it possible, in theory, to demonstrate genetic linkage with a disease susceptibility locus anywhere in the genome (NIH/CEPH Collaborative Mapping Group 1992; WEISSENBACH et al. 1992).

Genetic linkage has been extremely useful in medicine in general, with many notable successes, among them, identifying the gene for Huntington's disease (Huntington's Disease Collaborative Research Group 1993), a disorder that is relatively straightforward in genetic terms, and locating a gene that seems responsible for a large proportion of early onset familial Alzheimer's disease (SCHELLENBERG et al. 1992), a disorder that is genetically heterogeneous. However, in psychiatric illness, the situation has been quite different, with many highly publicized reports of linkage failing to ever find confirmation [e.g., a proposed chromosome 11 gene for bipolar affective disorder (EGELAND et al. 1987) and a proposed chromosome 5 gene for schizophrenia (SHERRINGTON et al. 1988)]. These difficulties underscore the problems of working with phenotypes that are actually complex sets of behavior. Molecular genetic studies may eventually be expected to help clarify some of these complicated situations but may also increase confusion if they are attempted without sufficient prior knowledge about the disorder. Molecular genetic studies of alcohol-metabolizing enzymes have proven to be clarifying and useful, but in most other cases these kinds of studies applied to the problem of alcohol dependence have led to controversy.

Inasmuch as linkage and association studies may eventually lead to the identification of specific genes where mutations may give rise to either increased or decreased risk for alcoholism, these avenues of research may be most useful in the long run for devising new pharmacotherapies to treat alcohol dependence. For example, suppose a certain receptor is found to be twice as sensitive to its naturally occurring ligand in alcohol-dependent individuals than in normals. Then, if this defect really increases vulnerability to alcoholism, a medication that partially blocks that receptor might be an effective treatment. No such defect has been demonstrated to date (although claims, based on indirect evidence, have been made for a defect at D_2 dopamine receptor, discussed below).

I. Linkage Studies

Several possible linkages with alcoholism have been reported (TANNA et al. 1988; HILL et al. 1988), but none reached statistical significance. Linkage studies can be a way to test for etiological involvement of candidate genes in a disorder. The candidate gene is tested for linkage; if the lod score allows rejection of linkage (with a value <-2), the candidate gene cannot be the disease gene because it cannot be at the same chromosomal location as the disease gene. Linkage studies have allowed exclusion of linkage with several interesting candidate genes including the D_2 dopamine receptor gene (BOLOS et al. 1990).

II. Association Studies

In a genetic association study, allele frequencies at candidate genes are compared in ill and well groups of unrelated individuals. The best-known genetic association studies in alcoholism are those with the alcohol-metabolizing enzymes, ADH and ALDH, and those with DRD2 D_2 dopamine receptor, both discussed below. The basic physical reality underlying an association study for a genetically simple (homogeneous) illness is quite straightforward (GEJMAN and GELERNTER 1993). All genetic effects can ultimately be traced to variation in DNA sequence. A group of unrelated people vary in most genetic characteristics (because they are unrelated) – including eye color, Rh type, and blood groups, but also including other measurable genetic polymorphism, such as other protein variants and DNA polymorphisms. This is true unless the study group is selected for homogeneity for one of these types of DNA vairation. But a group of unrelated people ill with a genetic disease *is* selected for homogeneity for a DNA variant: the DNA variant that causes the disease. Therefore, if a specific DNA variant is found to be more common in a group of ill people than in a group of well people who have no other reason to share alleles at the locus being studied, the reason must be that that DNA variant has something to do with the illness. The DNA variant responsible for causing the illness should be the only such polymorphism shared at a greater rate among ill individuals than among unrelated individuals without the illness. This is accurate in the absence of ethnic effects (e.g., all ill people deriving from an ethnic group sharing other DNA variants for reasons unrelated to illness). The assumptions do not apply if some of the individuals are related; that gives them another reason to share DNA variants at a rate greater than population expectations. There is also one other major consideration, linkage disequilibrium: it could be that the polymorphism associated with illness is not the disease gene itself, but is very close to it.

Numerous association studies have been published, where various markers have been found to be overrepresented among alcoholics (e.g., HLA antigens: CÓRSICO et al. 1988). PELCHAT and DANOWSKI (1992) compared PROP tasting ability in alcoholics, nonalcoholics, and their children, and found more PROP nontasters in the children of alcoholics than in the children of nonalcoholics. This finding, if confirmed, has interesting implications. It would represent another example (besides alcohol-metabolizing enzymes, discussed below) of a gene probably not expressed in brain affecting a complex behavior, perhaps in this case by altering the taste of alcohol-containing beverages. Association studies are generally problematic, however (GELERNTER et al. 1993), and results need to be replicated if they are to be accepted.

1. Alcohol-Metabolizing Enzymes

Although a gene affecting liability to develop a certain illness may exert its effect by increasing risk, it may also work by decreasing risk for illness. One example of this mechanism is the effect of alcohol-metabolizing enzymes on risk of alcohol dependence (reviewed in detail by EDENBERG 1991). Alcohol dehydrogenase (ADH) oxidizes alcohol to acetaldehyde (LI et al. 1977), and aldehyde dehydrogenase (ALDH) oxidizes acetaldehyde to acetate. Both of these enzymes are polymorphic in the human population, and the polymorphic forms have differing levels of biological activity (HARADA et al. 1982; YOSHIDA et al. 1984). An abnormal form of ADH is common in Asian populations but not among Caucasians; STAMATOYANNOPOULOS et al. (1975) found that 85% of Japanese carried an abnormal ADH allele. The allele frequencies of both ADH and ALDH variants depend on the population studied (GOEDDE et al. 1992).

Individuals with less active or inactive forms of ADH and ALDH are protected from developing alcoholism to some extent, presumably because when they drink alcohol they develop high blood levels of acetaldehyde (GOEDDE et al. 1979; HARADA et al. 1981; THOMASSON et al. 1991), which causes flushing and discomfort. In the case of ALDH, a mutation that inactivates a metabolic enzyme, or decreases its function, reduces the risk for alcoholism in the individual who carries it (HARADA et al. 1982; ENOMOTO et al. 1991).

Abnormalities of ALDH have been shown to affect development of alcoholism in Asian populations (GOEDDE et al. 1979; HARADA et al. 1982; SHIBUYA and YOSHIDA 1988), but not, so far, in other populations (HARADA et al. 1982; GOEDDE et al. 1979). THOMASSON et al. (1991) demonstrated associations between alcoholism, ALDH and also ADH subunits (ADH2, ADH3, and ALDH2) in a series of alcoholic and nonalcoholic Chinese men. Defective ALDH accounted for most of the abnormal physiological effects of alcohol seen in these populations. These differences are not observed in Caucasians from the United Kingdom and Ireland (GILDER et al. 1993).

Although the basic idea is straightforward – mutations in the enzymes that metabolize alcohol cause the buildup of toxic metabolites when the individual carrying those mutations drinks alcohol – the genetic situation is complex. Genetic associations exist at three different genetic loci, but only in certain populations. This situation also illustrates an exception to the usual presumption that candidate genes that affect behavior are likely to be those that code for brain proteins. These are the only findings in the molecular genetics of alcoholism that have unambiguously been replicated, have a well-understood physiological basis, and are generally accepted.

2. D_2 Dopamine Receptor Gene

Reports of a genetic association between alleles at the D_2 dopamine receptor gene (*DRD2*) and alcoholism (BLUM et al. 1990) have generated considerable interest and controversy. When genomic human DNA is digested with the restriction endonuclease enzyme *Taq*I, electrophoresed in agarose gels, and transferred to a fixed support (Southern blotted), a probe to D_2 dopamine receptor (genetic locus *DRD2*) hybridized with the Southern blots and autoradiographed reveals an RFLP. The marker system studied most frequently, known as the "A" system (GRANDY et al. 1989), corresponds to a single base mutation about 10 kilobases 3′ to the coding region of the gene (HAUGE et al. 1991). This mutation is not in the D_2 dopamine receptor gene itself. Early association studies with this marker for *DRD2* and alcoholism were promising (BLUM et al. 1990, 1991; PARSIAN et al. 1991), but there were also conflicting results from very early on (with studies showing no association from BOLOS et al. 1990; GELERNTER et al. 1991; SCHWAB et al. 1991; TURNER et al. 1992; COOK et al. 1992; GOLDMAN et al. 1992, 1993; ARINAMI et al. 1993) (see Table 1 for data summary). A meta-analysis of all *DRD2* association data available demonstrated that the positive data from the original group accounted for all of the difference between alcoholics and controls in the cumulative literature (GELERNTER et al. 1993).

DRD2 alleles have also been reported to be associated with substance abuse (SMITH et al. 1992), Tourette's syndrome (COMINGS et al. 1992), and obesity (COMINGS et al. 1992). All of these studies (including the alcoholism studies also) have shared certain design features which could have rendered them vulnerable to false-positive results. These features include lack of matching for ethnicity between alcoholic and control groups, multiple statistical tests, and also other factors (discussed in GELERNTER et al. 1991, 1993). A wide range of allele frequencies at *DRD2* among individuals of different ethnic backgrounds has been demonstrated (GOLDMAN et al. 1992, 1993; BARR and KIDD 1992), such that small differences in the ethnic makeup of ill and well groups could account for a significant difference in allele frequencies between the groups. A debate also arose over the suitability of using a random control group vs. a screened control group (to eliminate

Table 1. Allele frequencies for the *Taq*I "A" marker for the *D2* dopamine receptor gene (*DRD2*) in alcoholics and controls. Note that no study drawing alcoholics and controls from a single ethnic group showed a significant difference between alcoholics and controls

Investigators	Alcoholics		Controls		Population
	Total	f (A1)	Total	f (A1)	
BLUM et al. (1990)	35	0.37	35	0.13	Mixed
BLUM et al. (1991)	96	0.29	43	0.10	Caucasian
BOLOS et al. (1990)	40	0.22	127	0.18	Caucasian
COMINGS et al.* (1991)	104	0.23	39	0.08	Caucasian
COOK et al. (1992)	20	0.15	20	0.15	Caucasian
GELERNTER et al. (1991)	44	0.23	68	0.20	Caucasian
GRANDY et al. (1989)	–	–	43	0.24	Caucasian
PARSIAN et al. (1991)	32	0.20	25	0.06	Caucasian
TURNER et al. (1992)	47	0.10	–	–	Caucasian
GOLDMAN et al. (1993)	23	0.74	24	0.73	Cheyenne
GOLDMAN et al. (1992)	46	0.15	36	0.21	Finn
SCHWAB et al. (1991)	45	0.12	69	0.22	German
ARINAMI et al. (1993)	78	0.40	100	0.42	Japanese

* Excluding CEPH controls included with Bolos et al.

individuals from the control group who might have the illness) (SMITH et al. 1991; NOBLE and BLUM 1991). We have demonstrated that using a random control group entails little loss of power for association studies and have modeled this particular (*DRD2*) system (GELERNTER et al. 1991). The screening process may itself introduce other less quantifiable artifacts into the control group.

The reports of an association are problematic for other reasons too. For example, if there really is a genetic association between *DRD2* alleles and alcohol dependence, there should be a mutation somewhere in or near the *DRD2* gene. However, despite many attempts to demonstrate such a mutation, none has been found. SARKAR et al. (1991) sequenced a series of the expressed part of *DRD2* genes without finding any mutations that would be reflected in altered protein structure. Results from a recent comprehensive mutation search study, using denaturing gradient gel electrophoresis to examine all coding *DRD2* exons (GEJMAN et al. 1994), are in agreement on the major point, i.e., there are no common mutations in the coding region of *DRD2* that could account for the association findings. NOBLE et al. (1991) have reported an alteration in D_2 dopamine receptor properties in post-mortem brain samples from alcoholics, but the changes observed could have been a consequence of alcohol use rather than a result of a genetic mutation.

CLONINGER (1991) has concluded that a relationship between *DRD2* alleles and alcoholism has been proven; UHL et al. (1993) also concluded that there is an effect of *DRD2* alleles on alcoholism and substance abuse. While CLONINGER (1991; PARSIAN et al. 1991) stated that association was demonstrated in the absence of linkage, HODGE (1993) has shown that what PARSIAN et al. (1991) described as a negative test of linkage was actually itself a test of association, so "merely *undermines* their previous finding of an association" (HODGE 1993, p. 376). We (GELERNTER et al. 1993) and others have concluded that no association between *DRD2* alleles and alcohol dependence has been proven and that other explanations more conservative than an allelic association of *DRD2* with alcoholism could account for the positive findings of some groups. Other analyses have led to similar conclusions, i.e., that a physiological association should not be accepted at this point (KARP 1992; STONE and GOTTESMAN 1993). A true association of physiological significance would be not be the most likely explanation of the data.

D. Conclusions

The data supporting a genetic contribution to the development of alcoholism are very strong. In one specific case and only in specific populations, there is also strong evidence pointing to one part of the genetic basis of predisposition, i.e., mutations in the alcohol-metabolizing enzymes that provide a protective effect. However, this is not a factor in most populations, so most aspects of genetic predisposition to alcoholism are unexplained. The characteristic difficulties inherent in the study of genetically complex disorders such as alcoholism all but guarantee that the search for genes accounting for most of the variance will be a difficult and time-consuming task. Given the increasingly powerful methods of genetic analysis under development (e.g., advances in sequencing techniques and progress in the Human Genome project) (GEJMAN and GELERNTER 1994), it seems likely that more genes contributing to the development of alcoholism will eventually be identified.

How might improved knowledge of genetics lead to new developments in pharmacotherapy for alcohol dependence? The case of the alcohol-metabolizing enzymes provides a good starting point. We know that mutations in these enzymes inhibit alcohol abuse in individuals who harbor those mutations. Suppose a certain medication could interfere with the metabolism of alcohol in a similar way; it might also inhibit alcohol abuse. In fact, disulfiram works by this mechanism; it is a widely used (though problematic) pharmacotherapy for alcohol dependence. Although the treatment did not derive from the genetic knowledge in this case, it illustrates a potential interaction between genetics and pharmacotherapy. New insights into other genetic mechanisms might lead to similar insights into potential pharmacotherapies.

Acknowledgements. This work was supported in part by funds from the U.S. Department of Veterans Affairs [the National Center for PTSD Research, the VA-Yale Alcoholism Research Center, and the VA Medical Research Program (Merit Review grant to J.G.)], and NIMH SDA-C grant MH00931 to JG.

References

American Psychiatric Association (1987) Diagnostic and statistical manual of mental disorders, 3rd edn (DSM-IIIR). American Psychiatric Association, Washington

Arinami T, Itokawa M, Komiyama T, Mitsushio H, Mori H, Mifune H, Hamaguchi H, Toru M (1993) Association between severity of alcoholism and the A1 allele of the dopamine D2 receptor gene TaqI A RFLP in Japanese. Biol Psychiatry 33:108–114

Aston CE, Hill SY (1990) Segregation analysis of alcoholism in families ascertained through a pair of male alcoholics. Am J Hum Genet 46:879–887

Barr CL, Kidd KK (1992) Population frequencies of the TaqI A-system alleles at the dopamine D_2 receptor locus. Am J Hum Genet 51:A145

Blum K, Noble EP, Sheridan PJ, Montgomery A, Ritchie T, Jagadeeswaran P, Nogami H, Briggs AH, Cohn JB (1990) Allelic association of human dopamine D_2 receptor gene in alcoholism. JAMA 263:2055–2060

Blum K, Noble EP, Sheridan PJ, Finley O, Montgomery A, Ritchie T, Ozkaragoz T, Fitch RJ, Sadlack F, Sheffield D, Dahlmann T, Halbardier S, Nogami H (1991) Association of the A1 allele of the D_2 dopamine receptor gene with severe alcoholism. Alcohol 8:409–416

Bohman M, Sigvardsson S, Cloninger CR (1981) Maternal inheritance of alcohol abuse. Cross-fostering analysis of adopted women. Arch Gen Psychiatry 38: 965–969

Bohman M, Cloninger R, Sigvardsson S, von Knorring AL (1987) The genetics of alcoholism and related disorders. J Psychiatry Res 21:447–452

Bolos AM, Dean M, Lucas-Derse S, Ramsburg M, Brown GL, Goldman D (1990) Population and pedigree studies reveal a lack of association between the dopamine D_2 receptor gene and alcoholism. JAMA 264:3156–3160
Bouchard TJ Jr, Lykken DT, McGue M, Segal NL, Tellegen A (1990) Sources of human psychological differences: the Minnesota Study of Twins Reared Apart. Science 250:223–228
Cadoret RJ, Gath A (1978) Inheritance of alcoholism in adoptees. Br J Psychiatry 132:252–258
Cadoret RJ, Cain CA, Grove WM (1980) Development of alcoholism in adoptees raised apart from alcoholic biologic relatives. Arch Gen Psychiatry 37:561–563
Cadoret RJ, Troughton E, O'Gorman TW, Heywood E (1986) An adoption study of genetic and environmental factors in drug abuse. Arch Gen Psychiatry 43: 1131–1136
Cadoret RJ, Troughton E, O'Gorman TW (1987) Genetic and environmental factors in alcohol abuse and antisocial personality. J Stud Alcohol 48:1–8
Cloninger CR (1987) Neurogenetic adaptive mechanisms in alcoholism. Science 236:410–416
Cloninger CR (1991) D2 dopamine receptor gene is associated but not linked with alcoholism (Editorial). JAMA 266:1833–1834
Cloninger CR, Bohman M, Sigvardsson S (1981) Inheritance of alcohol abuse. Arch Gen Psychiatry 38:861–868
Cloninger CR, Sigvardsson S, von Knorring AL, Bohman M (1988) The Swedish studies of the adopted children of alcoholics: a reply to Littrell. J Stud Alcohol 49:500–509
Comings DE, Comings BG, Muhleman D, Dietz G, Shahbahrami B, Tast D, Knell E, Kocsis P, Baumgarten R, Kovacs BW, Levy DL, Smith M, Borison RL, Evans D, Klein DN, MacMurray J, Tosk JM, Sverd J, Gysin R, Flanagan SD (1991) The dopamine D2 receptor locus as a modifying gene in neuropsychiatric disorders. JAMA 266:1793–1800
Comings DE, MacMurray J, Dietz G, Muhleman D, Knell E, Flanagan S, Gysin R, Ask M, Johnson J (1992) Dopamine D_2 receptor as a major gene in obesity. Am J Hum Genet 51:A211
Cook BL, Wang ZW, Crowe RR, Hauser R, Freimer M (1992) Alcoholism and the D_2 receptor gene. Alcohol Clin Exp Res 4:806–809
Córsico R, Pessino OL, Morales V, Jmelninsky A (1988) Association of HLA antigens with alcoholic disease. J Stud Alcohol 49:546–550
Crabb DW (1990) Biological markers for increased risk of alcoholism and for quantitation of alcohol consumption. J Clin Invest 85:311–315
Edenberg HJ (1991) Molecular biological approaches to the studies of alcohol-metabolizing enzymes. In: Carbbe JC, Harris RA (eds) The genetic basis of alcohol and drug actions. Plenum, New York, pp 165–223
Egeland JA, Gerhard DS, Pauls DL, Susses JN, Kidd KK, Allen CR, Hostetter AM, Housman DE (1987) Bipolar affective disorder linked to DNA markers on chromosome 11. Nature 325:783–787
Enomoto N, Takase S, Yasuhara M, Takada A (1991) Acetaldehyde metabolism in different aldehyde dehydrogenase-2 genotypes. Alcohol Clin Exp Res 15: 141–144
Falk CT, Rubinstein P (1987) Haplotype relative risks: an easy reliable way to construct a proper control sample for risk calculations. Ann Hum Genet 51: 227–233
Finn PR, Martin J, Pihl (1987) Alexithymia in males at high genetic risk for alcoholism. Psychother Psychosom 47:18–21
Frances RJ, Timm S, Bucky S (1980) Studies of familial and nonfamilial alcoholism, I: demographic studies. Arch Gen Psychiatry 37:564–566
Frances RJ, Bucky S, Alexopoulos GS (1984) Outcome study of familial and non-familial alcoholism. Am J Psychiatry 141:1469–1471

Gejman PV, Gelernter J (1993) Mutational analysis of candidate genes in psychiatric disorders. Am J Med Genet 48:184–191
Gejman PV, Ram A, Gelernter J, Friedman E, Cao Q, Pickar D, Blum K, Noble EP, Kranzler H, O'Malley S, Hamer DH, Whitsitt F, Rao P, DeLisi LE, Virkkunen M, Linnoila M, Goldman D, Gershon ES (1994) No structural mutation in the dopamine D_2 receptor gene in alcoholism or schizophrenia. JAMA 271:204–208
Gelernter J, O'Malley S, Risch N, Kranzler H, Krystal J, Merikangas K, Kennedy J, Kidd KK (1991) No association between an allele at the D2 dopamine receptor gene (DRD2) and alcoholism. JAMA 266:1801–1807
Gelernter J, Goldman D, Risch N (1993) The A1 allele at the D_2 dopamine receptor gene and alcoholism: a reappraisal. JAMA 269:1673–1677
Gilder FJ, Hodgkinson S, Murray RM (1993) ADH and ALDH genotype profiles in Caucasians with alcohol-related problems and controls. Addiction 88:383–388
Gilligan SB, Reich T, Cloninger CR (1987) Etiologic heterogeneity in alcoholism. Genet Epidemiol 4:395–414
Goedde HW, Harada S, Agarwal DP (1979) Racial differences in alcohol sensitivity: a new hypothesis. Hum Genet 51:331–334
Goedde HW, Agarwal DP, Fritze G, Meier-Rackmann D, Singh S, Beckmann G, Bhatia K, Chen LZ, Fang B, Lisker R, Paik YK, Rothhammer F, Saha N, Segal B, Srivastava LM, Czeizel A (1992) Distribution of ADH_2 and ALDH2 genotypes in different populations. Hum Genet 88:344–346
Goldman D (1988) Molecular markers for linkage of genetic loci contributing to alcoholism. Recent Dev Alcohol 6:333–349
Goldman D, Dean M, Brown GL, Bolos AM, Tokola R, Virkkunen M, Linnoila M (1992) D_2 dopamine receptor genotype and cerebrospinal fluid homovanillic acid, 5-hydroxyindoleacetic acid and 3-methoxy-4-hydroxyphenylglycol in Finnish and American alcoholics. Acta Psychiatr Scand 86:351–357
Goldman D, Brown GL, Albaugh B, Robin R, Goodson S, Trunzo M, Akhtar L, Lucas-Derse S, Long J, Linnoila M, Dean M (1993) DRD2 dopamine receptor genotype, linkage disequilibrium and alcoholism in American Indians and other populations. Alcohol Clin Exp Res 17:199–204
Goodwin DW, Schulsinger F, Hermansen L, Guze SB, Winokur G (1973) Alcohol problems in adoptees raised apart from alcoholic biological parents. Arch Gen Psychiatry 28:238–243
Goodwin DW, Schulsinger F, Knop J, Mednick S, Guze SB (1977) Alcoholism and depression in adopted-out daughters of alcoholics. Arch Gen Psychiatry 34: 751–755
Grandy DK, Litt M, Allen L, Bunzow JR, Marchionni M, Makam H, Reed L, Magenis RE, Civelli O (1989) The human dopamine D2 receptor gene is located in chromosome 11 at q22–q23 and identifies a Taq I polymorphism. Am J Hum Genet 45:778–785
Gurling HMD, Curtis D, Murray RM (1991) Psychological deficit from excessive alcohol consumption: evidence from a co-twin control study. Br J Alcohol 86:151–155
Harada S, Agarwal DP, Goedde HW (1981) Aldehyde dehydrogenase deficiency as cause of facial flushing reaction to alcohol in Japanese. Lancet 2:982
Harada S, Agarwal DP, Goedde HW, Tagaki S, Ishikawa B (1982) Possible protective role against alcoholism for aldehyde dehydrogenase deficiency in Japan. Lancet 2:827
Hauge XY, Grandy DK, Eubanks JH, Evans GA, Civelli O, Litt M (1991) Detection and characterization of additional DNA polymorphisms in the dopamine D2 receptor gene. Genomics 10:527–530
Heath AC, Jardine R, Martin NG (1989) Interactive effects of genotype and social environment on alcohol consumption in female twins. J Stud Alcohol 50:38–48

Hill SY, Aston C, Rabin B (1988) Suggestive evidence of genetic linkage between alcoholism and the MNS blood group. Alcohol Clin Exp Res 12:811–814
Hodge SE (1993) Linkage analysis versus association analysis: distinguishing between two models that explain disease-marker associations. Am J Hum Genet 53: 367–384
Huntington's Disease Collaborative Research Group (1993) A novel gene containing a trinucleotide repeat that is expanded and unstable on Huntington's disease chromosomes. Cell 72:971–983
Hrubec Z, Omenn GS (1981) Evidence of genetic predisposition to alcoholic cirrhosis and psychosis: twin concordances for alcoholism and its biologic end-points by zygosity among male veterans. Alcohol Clin Exp Res 5:207–215
Karp RW (1992) D_2 or not D_2? Alcohol Clin Exp Res 16:786–787
Kendler KS, Heath AC, Neale MC, Kessler RC, Eaves LJ (1992) A population-based twin study of alcoholism in women. JAMA 268:1877–1882
Li T-K, Bosron WF, Dafeldecker WP, Lange LG, Vallee BL (1977) Isolation of π-alcohol dehydrogenase of human liver: is it a determinant of alcoholism? Proc Natl Acad Sci USA 74:4378–4381
Littrell J (1988) The Swedish studies of the adopted children of alcoholics. J Stud Alcohol 49:491–499
McGue M, Pickens RW, Svikis DS (1992) Sex and age effects on the inheritance of alcohol problems: a twin study. J Abnorm Psychol 101:3–17
Mullan MI, Gurling HMD, Oppenheim BE, Murray RM (1986) The relationship between alcoholism and neurosis: evidence from a twin study. Br J Psychiatry 148:435–441
Newlin DB, Thomson JB (1990) Alcohol challenge with sons of alcoholics: a critical review and analysis. Psychol Bull 108:383–402
NIH/CEPH Collaborative Mapping Group (1992) A comprehensive genetic linkage map of the human genome. Science 258:67–86
Noble EP, Blum K (1991) (Letter). JAMA 268:2667
Noble EP, Blum K, Ritchie T, Montgomery A, Sheridan PJ (1991) Allelic association of the D_2 dopamine receptor gene with receptor-binding characteristics in alcoholism. Arch Gen Psychiatry 48:648–654
Parsian A, Todd RD, Devor E, O'Malley KL, Suarez BK, Reich T, Cloninger CR (1991) Alcoholism and alleles of the human dopamine D2 receptor locus: studies of association and linkage. Arch Gen Psychiatry 48:655–663
Pelchat ML, Danowski S (1992) A possible genetic association between PROP-tasting and alcoholism. Physiol Behav 51:1261–1266
Pickens RW, Svikis DS, McGue M, Lykken DT, Heston LL, Clayton PJ (1991) Heterogeneity in the inheritance of alcoholism: a study of male and female twins. Arch Gen Psychiatry 48:19–28
Pollock VE, Gabrielli WF, Mednick SA, Goodwin DW (1988) EEG identification of subgroups of men at risk for alcoholism? Psychiatry Res 26:101–114
Propping P, Kruger J, Mark N (1981) Genetic predisposition to alcoholism. An EEG stu in alcoholics and their relatives. Hum Genet 59:51–59
Risch N (1987) Assessing the role of HLA-linked and unlinked determinants of disease. Am J Hum Genet 40:1–14
Sarkar G, Kapelner S, Grandy DK, Marchionni M, Civelli O, Sobell J, Heston L, Sommer SS (1991) Direct sequencing of the dopamine D2 receptor (DRD2) in schizophrenics reveals three polymorphisms but no structural change in the receptor. Genomics 11:8–14
Schellenberg GD, Bird TD, Wijsman EM, Orr HT, Anderson L, Nemens E, White JA, Bonnycastle L, Weber JL, Alonso ME, Potter H, Heston LL, Martin GM (1992) Genetic linkage evidence for a familial Alzheimer's disease locus on chromosome 14. Science 258:668–671
Schuckit MA, Irwin M (1989) An analysis of the clinical relevance of Type 1 and Type 2 alcoholics. Br J Addict 84:869–876

Schuckit MA, Goodwin DW, Winokur G (1972) A study of alcoholism in half siblings. Am J Psychiatry 128:1132–1136
Schwab S, Soyka M, Niederecker M, Ackenheil M, Scherer J, Wildenaurer DB (1991) Allelic association of human D2-receptor DNA polymorphism ruled out in 45 alcoholics. (A1094) Am J Hum Genet 49 [Suppl 4]:203
Sherrington R, Brynjolfsson J, Petursson H, Potter M, Dudleston K, Barraclough B, Wasmuth J, Dobbs M, Gurling H (1988) Localization of a susceptibility locus for schizophrenia on chromosome 5. Nature 336:164–167
Shibuya A, Yoshida A (1988) Genotypes of alcohol-metabolizing enzymes in Japanese with alcohol liver disease a strong association of the usual caucasian-type aldehyde dehydrogenase gene ($ALDH_2^1$) with the disease. Am J Hum Genet 43:744–748
Smith SS, Gorelick DA, O'Hara BF, Uhl GR (1991) (Letter). JAMA 268:2667–2668
Smith SS, O'Hara BF, Persico AM, Gorelick DA, Newlin DB, Vlahov D, Solomon L, Pickens R, Uhl GR (1992) Genetic vulnerability to drug abuse; the D_2 dopamine receptor TaqI B1 restriction fragment length polymorphism appears more frequently in polysubstance abusers. Arch Gen Psychiatry 49:723–727
Stamatoyannopoulos G, Chen SH, Fukui F (1975) Liver alcohol dehydrogenase in Japanese: high population frequency of the atypical form and its possible role in alcohol sensitivity. Am J Hum Genet 27:789–796
Stone WS, Gottesman II (1993) A perspective on the search for the causes of alcoholism: slow down the rush to genetical judgements. Neurol Psychiatry Brain Res I:123–132
Tabakoff B, Hoffman PL (1988) Genetics and biological markers of risk for alcoholism. Public Health Rep 103:690–698
Tabakoff B, Hoffman PL, Lee JM, Saito T, Willard B, De Leon-Jones F (1988) Differences in platelet enzyme activity between alcoholics and nonalcoholics. N Engl J Med 318:134–139
Tanna VL, Wilson AF, Winokur G, Elston RC (1988) Possible linkage between alcoholism and esterase-D. J Stud Alcohol 49:472–476
Thomasson HR, Edenberg JH, Crabb DW, Mai XL, Jerome RE, Li TK, Wang SP, Lin YT, Lu RB, Yin SJ (1991) Alcohol and aldehyde dehydrogenase genotypes and alcoholism in Chinese men. Am J Hum Genet 48:677–681
Turner E, Ewing J, Shilling P, Smith TL, Irwin M, Schuckit M, Kelsoe JR (1992) Lack of association between an RFLP near the D_2 dopamine receptor gene and severe alcoholism. Biol Psychiatry 31:285–290
Uhl G, Blum K, Noble E, Smith S (1993) Substance abuse vulnerability and D_2 receptor genes. TINS 16:83–88
Weissenbach J, Gyapay G, Dib C, Vignal A, Morisette J, Millasseau P, Vaysseix G, and Lathrop M (1992) A second-generation linkage map of the human genome. Nature 359:794–801
Yoshida A, Huang I-V, Ikawa M (1984) Molecular abnormality of an inactive aldehyde dehydrogenase variant commonly found in Orientals. Proc Natl Acad Sci USA 81:258–261

CHAPTER 13

Pharmacotherapy and Pathophysiology of Alcohol Withdrawal

R.F. ANTON and H.C. BECKER

A. Introduction

The alcohol withdrawal syndrome has received increased attention over the past 40 years and it is now well established that the abrupt reduction, or total cessation, of chronic alcohol use will lead to a set of predictable signs and symptoms. Although it was suspected in the latter half of the nineteenth century that alcohol withdrawal could produce seizures (HUSS 1852), it was only in the middle of the twentieth century that the alcohol withdrawal syndrome (AWS) was fully described (VICTOR and ADAMS 1953) and demonstrated experimentally by Isbell and colleagues (ISBELL et al. 1955) and later by MENDELSON and LADOU (1963). Since that time numerous animal models have been developed to examine the pathophysiology and impact of varying conditions of alcohol intake and pharmacologic manipulation on the AWS. Considerable progress has been made in understanding the pathophysiologic mechanisms that underlie the AWS. Knowledge gained from animal models, along with clinical experience and methodologic developments, have led to advances in the treatment of the AWS.

Despite these relatively rapid advances, much still remains to be understood about the treatment of the acute AWS and especially about the subacute and chronic effects of alcohol withdrawal (AW). Some of the issues that are beginning to receive greater attention, and where much more information is needed to guide clinical practice, are the "protracted" AWS, the effects of AW on subsequent AW symptomatology ("the kindling effect"), the effects of comorbid psychiatric conditions (e.g., anxiety, depression) on the severity and course of AW symptoms, craving for alcohol during the course of AW, and the prognostic significance of an AW-related seizure for future neurological sequelae. The impact of a variety of pharmacologic treatments on these processes and on the propensity of the alcoholic to relapse following detoxification are also areas where further knowledge and clinical guidance is necessary.

In this chapter we will review what is known about the AWS with attention directed primarily towards many of the above issues, pharmacologic treatments, and mechanisms of action. Many excellent reviews of the AWS and its treatment have been written in recent years (GORELICK and WILKINS 1986; LISKOW and GOODWIN 1987; CASTANEDA and CUSHMAN 1989;

McMICKEN 1990; VICTOR 1990; SCHULTZ 1991). The reader will be referred to these, where appropriate, and a greater emphasis will be placed on newer information that may begin to guide clinical practice in the years ahead.

B. Clinical Syndrome

I. Phenomenology and Description

There is considerable interindividual variability in the presentation of AW symptoms. It appears that some alcoholics will have more severe symptomatology than others despite having ingested comparable amounts of alcohol prior to cessation of drinking. Despite this variability, AW affects the same body systems, to a greater or lesser extent, in all alcoholics in withdrawal. The symptoms which occur after a single heavy drinking episode, commonly referred to as a "hangover," are qualitatively similar to but quantitatively less than those present in AW. In addition, more serious and intense symptoms occur during withdrawal as a consequence of the chronic high level of alcohol use that characterizes most alcoholics. The primary organ systems involved are the peripheral and central nervous systems, which then affect other end organ systems such as the cardiovascular, gastrointestinal, and dermatologic systems. Some, or all, of the following symptoms are present during the AWS and, must in fact, be present for the diagnosis to be made (SELLERS et al. 1991): nausea and vomiting, tremor, sweating, anxiety and irritability, motor arousal (agitation), skin sensations, heightened sensitivity to light and sound, headache, and problems with concentration and orientation. Increases in blood pressure and pulse are often observed. In addition, most have a decreased appetite and an abnormal sleep architecture which, at the extreme, is manifest as complete insomnia.

It has been well documented that there is a temporal relationship between the cessation of drinking and the emergence of withdrawal symptoms (VICTOR and ADAMS 1953; ISBELL et al. 1955; SELLERS and KALANT 1976; VICTOR 1990). AW symptomatology first begins to appear between 6 and 8 h after the cessation of drinking (MENDELSON and LADOU 1963), often prior to the blood alcohol level reaching zero. The symptoms intensify and then diminish over the next 24–48 h. During more serious AW there are more significant visual and auditory disturbances that take the form of transient visual hallucinations (e.g., faces, bugs) as well as auditory hallucinations (e.g., voices). The auditory hallucinations seem to be less elaborate than those found in psychotic disorders, in that patients describe these as less recognizable, shorter, perhaps more focused, and less bizarre. The most intense and serious syndrome, which is characterized by increased agitation and tremulousness, autonomic instability, hyperpyrexia, persistent visual and auditory hallucinations, and disorientation, has been labeled delirium tremens (DT). Should this occur, as it does in approximately 5% of alcoholics in withdrawal, it usually becomes manifest between 48 and 96 h after the cessation of drinking (SELLERS and KALLANT 1976; VICTOR 1990). DT is more often manifest in medical settings since other organ pathology such as pancreatitis, pneumonia, and hepatitis may predispose an individual to develop this more severe form of AWS (THOMPSON 1975). When death is directly attributable to AW it generally occurs in patients with DT. Deaths due to DT have been greatly decreased in recent years secondary to early recognition and aggressive medical and pharmacologic management. Therefore, patients that have signs and symptoms suggestive of the development of DT need emergency intensive medical care. This includes hydration, correction of electrolyte and metabolic disturbances, and aggressive pharmacologic management.

Recognition of DT in the medically or neurologically compromised patient poses a challenge to the clinician. Patients in trauma and burn units, where there are high rates of alcohol abuse and dependence, run the greatest risk of a serious AWS. Unfortunately, these patients are the very same ones that may be unable or unwilling to provide adequate and valid information regarding alcohol ingestion. Use of admission blood alcohol measurement, laboratory markers of chronic abusive drinking, collateral reports, and structured questionnaires may aid in the identification of chronic, heavy alcohol use in these patients.

Another major complication of AW is seizures. Although it is estimated that seizures occur in up to 25% of all alcoholics in withdrawal (CHAN 1985), a more accurate estimate is probably between 5% and 15% (VICTOR and BRAUSCH 1967; HILLBROM and HJELM-JAGER 1983). Several studies have suggested that nearly half of all patients presenting to an emergency room for treatment of a seizure have a history of alcohol abuse (EARNEST and YARNELL 1976; HILLBOM 1980). Data from numerous animal studies and clinical information strongly suggest that increasing levels of alcohol intake increase the risk of seizures (NG et al. 1988). The relative risk of seizures rose from 3-fold at intakes of 51–100 g ethanol/day, to 8-fold at intakes of 101–200 g ethanol/day, and to almost 20-fold when the intake of ethanol per day was greater than 200 g. These data suggest that the abrupt cessation of drinking following chronic ingestion of large quantities of alcohol is sufficient to cause seizures independent of any other predisposing factors. It is also clear that individuals who have other risk factors for the development of seizures, such as idiopathic epilepsy, head trauma, or metabolic disturbances, or who are ingesting drugs that lower the seizure threshold are at even greater risk for AWS. It has been suggested that alcoholics with frontal cortical atrophy are at higher risk for the development of seizures and that the elderly in particular may develop seizures during a relative decrease in alcohol intake because of cortical atrophic changes related to the aging process (AVDALOFF 1979).

Though seizures in the context of AW usually occur within the first 24 h after the cessation of alcohol intake, they can occur up to 5 days later

(VICTOR and BRAUSCH 1967; VICTOR 1990). Alcohol withdrawal seizures are most often grand mal in type except where there has been a previous focal neurological insult, such as trauma, in which case focal seizures may be observed. Generally, there will be one or a few seizures over several hours. The electroencephalogram will be normal except in close temporal proximity to the seizure and most of the time the seizures will occur prior to the onset of DT. Although in the majority of patients the syndrome will not progress to DT, in a significant number [up to one-third in one study (VICTOR 1990)] the syndrome will progress in this manner. Experience suggests that, when seizures occur in an alcoholic more than 1 week after cessation of drinking, one should be suspicious of other etiologic factors such as metabolic disturbances, head trauma, sedative-hypnotic dependence, or the iatrogenic use of medications that might lower the seizure threshold (such as phenothiazines, tricyclic antidepressants, xanthine derivatives). A discussion of the prevention and treatment of alcohol withdrawal seizures will be presented later in this chapter.

II. Drinking Variables Related to the Alcohol Withdrawal Syndrome

Efforts to use clinical or historical data to predict the severity of the AWS have met with some success, though the predictive validity of these measures is not adequate for clinical decision making. For example, a blood alcohol concentration (BAC) measured on admission was shown to be positively but weakly correlated ($r = 0.26$) with the severity of withdrawal symptoms over the first 48 h (VINSON and MENEZES 1991). Additionally, it has been reported that withdrawal symptoms correlated with the amount of alcohol ingested in the 7 days prior to admission ($r = 0.58$) and that both total amount of alcohol consumed ($r = 0.56$) and the number of days drinking ($r = 0.42$) in the 4 weeks prior to admission correlated with AW symptoms (HERSHON 1977). Finally, there is a suggestion that longer duration of problem drinking, greater alcohol consumption during the month prior to admission, and a higher BAC at the time of admission discriminated a group of patients who needed pharmacologic treatment of their AW symptoms from patients who did not require such treatment (BENZER 1990). These findings, though of limited clinical utility, highlight the potential value in this approach. Perhaps with better predictors variables (e.g., laboratory tests), more valid reports of recent and lifetime alcohol use and through the application of multivariate statistics this approach may ultimately be clinically useful.

The issue of AW in the elderly needs more study, especially as the number of older people in our population increases. There is some evidence that older alcoholics (greater than age 58 years) have more severe withdrawal symptoms. Though this may be partially attributed to cumulative years of problem drinking, it also appears to be directly related to age (LISKOW et al. 1989). In that study it was observed that the relationship between the amount of alcohol consumed in the 30 days prior to admission and the severity of withdrawal symptoms was present only in the older age group (as above) and that older patients needed more chlordiazepoxide to treat these symptoms. More information is needed to draw firm conclusions concerning the phenomenology and treatment of AW in the elderly.

III. Protracted Alcohol Withdrawal

Although there have been a number of anecdotal reports and case studies that have alluded to an extended period of physiologic and psychologic dysregulation that occurs in alcoholics after the initiation of abstinence, there have been few systematic studies of this issue. The nature of this phenomenon, therefore, remains ill defined. There is clear clinical evidence, however, of a very high rate of relapse in patients during the first few months following detoxification, which is followed by a gradual decline over the next 6–12 months. This pattern of relapse is consistent with the notion that a protracted abstinence syndrome exists and is most prominent during the early months following detoxification.

In some elegant basic studies, BEGLEITER and PORJESZ (1979) found that rodents and monkeys given alcohol over prolonged periods of time showed an increase in brain neuronal "excitability" that persisted after the cessation of alcohol exposure. When rechallenged with alcohol, these animals showed a greater neuronal response than animals who had not been previously exposed to chronic alcohol. GILLIN et al. (1990) showed that alcoholics who had been abstinent for 17 days showed abnormalities in electroencephalographic (EEG) sleep recordings (GILLIN et al. 1990). These patients exhibited difficulty falling asleep, less overall sleep, and poor sleep efficiency and had reduced rapid eye movement (REM), stage 2, stage 4, and delta sleep when compared to age-matched, nonalcoholic controls. For at least some chronic alcoholics, disturbances in sleep may persist for months to years after the initiation of abstinence (ADAMSON and BURDICK 1973).

In a very recent study, positron emission tomography (PET) scans were done on recently abstinent (6–32 days) chronic alcoholics and healthy controls (VOLKOW et al. 1992). These investigators found that alcoholics had lower overall brain metabolic rates which were correlated with the time since alcohol discontinuation. Some patients who were studied 2–3 weeks after alcohol cessation appeared to have lower brain metabolism than the average control patient. Interestingly, this temporal period overlaps with the timing of those abnormalities in EEG sleep recordings described by GILLIN et al. (1990). Taken together, these studies strongly suggest that at least 3–4 weeks may be needed to normalize brain function after abrupt cessation of chronic alcohol consumption and that this high vulnerability period for relapse may have a significant biologic substrate.

Behavioral correlates of brain dysregulation also deserve more extensive study. However, the data that exist suggest that symptoms related to "brain hyperexcitability," such as fatigability, inner tension, insomnia, and somatic pain, may persist up to 5 weeks after the cessation of drinking (ALLING et al. 1982) and that some aspects of mood also continue to improve during the first 7 weeks of sobriety (BOKSTROM et al. 1989).

One of the greatest challenges in the treatment of alcoholics is to distinguish between protracted AW symptoms and those psychiatric symptoms that warrant specific therapeutic intervention. The signs and symptoms of schizophrenia and mania are well circumscribed and historical information lends support to the diagnoses. It is in the area of depressive and anxiety disturbances that greater difficulty exists. As many of these symptoms overlap with those of AW (including protracted withdrawal), it is particularly difficult to distinguish their etiology.

It is heuristically appealing to assume that in fact there may be a large gray area of symptomatology that would better fit a stress diathesis or vulnerability model. Patients with a genetic or biologic vulnerability to experience mood or anxiety disorders may actually experience various symptoms of these disorders during alcohol withdrawal in a more intense fashion, and for a longer period, after the cessation of drinking. Since dysphoria, and tension (anxiety), may predispose to alcoholic relapse, the identification and treatment of these symptoms may reduce the risks in particularly vulnerable individuals. However, recent attempts to treat depressive and anxiety syndromes in the immediate detoxification period with targeted pharmacologic agents have met with mixed results (DORUS et al. 1989; MALCOLM et al. 1992; TOLLEFSON et al. 1990; MASON and KOSCIS 1991; KRANZLER and MEYER 1989; KRANZLER and ORROK 1989).

IV. Psychiatric Comorbidity and Alcohol Withdrawal

Much has been written about the relationship of depressive symptomatology and AW. Good empirical evidence (SCHUCKIT 1983; DORUS et al. 1989; WILLENBRING 1986) indicates that symptoms of depression are largely a manifestation of AW that is time-limited. It is thought that most depressive symptoms will abate by 10–14 days after the last drink. Treatment for a major depressive syndrome should be withheld until enough time has elapsed to rule out the possibility of spontaneous recovery during alcohol detoxification. Theoretically, though to our knowledge unstudied, patients with bipolar affective illness who present with depression during detoxification would be less likely to recover spontaneously and may need more aggressive pharmacologic management of their depressive symptoms. On the other hand, bipolar alcoholic patients who present with manic symptoms pose a particular diagnostic and therapeutic challenge. Presenting as they do with marked agitation, insomnia, possible disorientation, and psychotic symptoms, these patients generally need simultaneous treatment for both AW and mania. As we will discuss later in this chapter, carbamazepine may be the treatment of choice for these individuals.

There is also a high prevalence of anxiety disorders in alcoholics (Reiger et al. 1990; Kushner et al. 1990). It has been suggested that the pathophysiology and symptomatology of AW and panic disorder are very similar and that repeated episodes of AW may sensitize (i.e., "kindle") an individual to develop panic-like symptoms during both recovery and relapse (George et al. 1990). This hypothesis, though as yet unsupported empirically, would suggest that AW symptomatology might best be treated aggressively in those individuals with a previous history of panic attacks or, possibly, a family history of panic disorder.

While considerable thought has been given to the impact of AW on psychiatric pathology, less study has been devoted to the impact of psychopathology on AW symptomatology. To our knowledge, data on the impact of depressive illness on AW have not been reported. However, those individuals that have had hallucinations as part of an AW syndrome, quite independent of DT, blackouts, or seizures, have more intense and persistent depressive symptoms and dysphoria after the detoxification period (Hayne and Louks 1991). Johnston et al. (1991) examined AW symptomatology in alcoholics with a comorbid diagnosis of generalized anxiety disorder. They

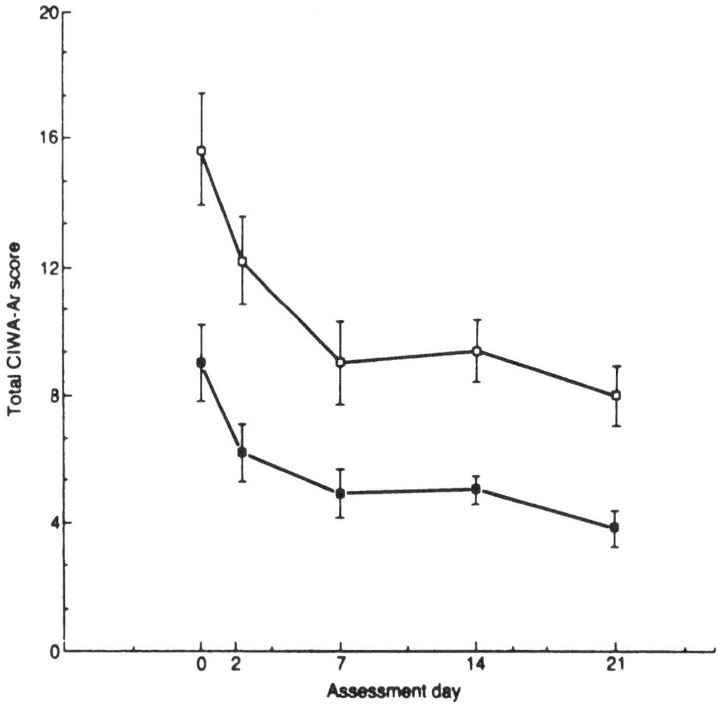

Fig. 1. Mean (± SEM) total CIWA-Ar scores for the dual-diagnosed (anxiety and alcohol) group (n = 17), shown as *open squares* (□), and alcohol-only group (n = 16), shown as *solid squares* (■), on assessment days 0, 2, 7, 14, and 21. (From Johnston et al. 1991)

found that the comorbid group exhibited higher levels of AW symptoms, which persisted over the course of the detoxification period (Fig. 1). These patients were also found to have higher levels of anxiety, which diminished to some degree during the course of AW, but remained elevated relative to the patients without comorbid anxiety disorders (THEVOS et al. 1991). The findings lend some support to the hypothesis that there may be a common pathophysiology between the anxiety disorders and the AWS and that these conditions may have an additive effect on symptomatology. Other more recent data suggest that alcoholics with social phobia may delay their presentation for treatment (ANTON 1992), which may have implications for the severity of their withdrawal symptoms once they do seek detoxification. Further investigation of the relationship between psychopathology and symptomatology is warranted, given the clinical implications of these findings.

C. Pathophysiology of Alcohol Dependence and Withdrawal

This section will provide a general overview of neurochemical changes related to the symptoms of the AWS and which, thus, represent potential targets for pharmacotherapy.

I. GABA-A Receptor System

Ethanol shares a similar pharmacologic profile with benzodiazepines (BDZs) and barbiturates. Following chronic exposure, both cross-tolerance and cross-dependence have been demonstrated to exist among these compounds (e.g., CHAN et al. 1986, 1988; KALANT et al. 1971; LE et al. 1986). In addition, all three classes of drugs have been shown to facilitate the inhibitory effects of gamma-amino butyric acid (GABA) neurotransmission (e.g., YU and HO 1990). Thus, it is not surprising that drugs which elevate GABA activity are most effective in alleviating symptoms of the AWS. Indeed, BDZs are currently the drugs of choice in treating AW (CASTANEDA and CUSHMAN 1989; SCHUCKIT 1989).

While enhancement of GABA-mediated neurotransmission has been demonstrated following acute treatment with ethanol, chronic exposure to ethanol has been shown to result in a functional downregulation or desensitization of GABA activity (for reviews, see HUNT 1983; KULONEN 1983; TICKU 1989; TICKU and KULKARNI 1988; TICKU et al. 1992). Reduced activity of GABA, the major inhibitory neurotransmitter in the brain, is thought to underlie, at least in part, the state of CNS hyperexcitability associated with AW (ALLAN and HARRIS 1987a; FRYE 1990). Support for this idea comes from animal and clinical studies demonstrating that manipulation of the GABA system reliably influences the ethanol withdrawal response. That is, drugs with an intrinsic positive modulatory effect at the GABA/benzodiazepine/

chloride (GABA/BDZ/Cl) channel complex (e.g., GABA agonists, BDZ agonists, barbiturates, as well as ethanol) facilitate GABA-mediated Cl-conductance, possess anxiolytic and anticonvulsant activity, and are effective in ameliorating AW symptomology (AARONSON et al. 1984; COOPER et al. 1979; FADDA et al. 1985; FRYE et al. 1983, 1986; GOLDSTEIN 1973; KRAMP and RAFAELSON 1978; NOBLE et al. 1976; SELLERS and KALANT 1976). Conversely, drugs with intrinsic negative modulatory action at the GABA/BDZ/Cl-channel receptor complex (e.g., GABA antagonists, BDZ inverse agonists, Cl-channel blockers) inhibit GABA-operated Cl-conductance, exhibit an anxiogenic and (pro)convulsant profile, and exacerbate the severity of the ethanol withdrawal response (BECKER and ANTON 1989; GOLDSTEIN 1973; LISTER and KARANIAN 1987).

A third class of agents that bind to the BDZ site are relatively devoid of intrinsic activity, and, hence, are behaviorally inert. These BDZ antagonists (e.g., flumazenil) are capable of antagonizing the effects of BDZ agonists and inverse agonists. Flumazenil has been found to be ineffective against ethanol withdrawal seizures (ADINOFF et al. 1986; LITTLE et al. 1985), but is efficacious in attenuating the anxiety component of the withdrawal syndrome (FILE et al. 1989).

While these above studies may be taken as indirect support for GABA involvement in AW, the mechanism underlying a purported decrease in functional GABA activity is unclear at present (FRYE 1990). Chronic treatment with ethanol has not been shown to consistently influence the binding properties of the various domains of the GABA/BDZ/Cl channel complex (e.g., see BUCK and HARRIS 1990; FRYE 1990; TICKU 1989). Similarly, prolonged ethanol treatment does not apparently alter presynaptic GABA activity (FRYE 1990; HUNT 1983). GABA levels in plasma, CSF, and postmortem bran tissue from alcoholics have varied relative to controls (COFFMAN and PETTY 1985; FREUND and BALLINGER 1988; GOLDMAN et al. 1981; PETTY 1992; ROY et al. 1990; TRAN et al. 1981). Chronic ethanol exposure has been shown to alter mRNA for various subunits of the GABA/BDZ/Cl channel complex (e.g., MONTPIED et al. 1991; MORROW et al. 1990; TICKU et al. 1992), but the functional significance of these changes is unknown. One of the most consistent findings is that the ethanol-induced enhancement of GABA-stimulated Cl-flux is diminished in brain tissue from animals chronically exposed to ethanol (ALLAN and HARRIS 1987b; MORROW et al. 1988). However, contrary to what might be expected, most studies have demonstrated no alteration in GABA-stimulated Cl-flux following chronic ethanol exposure (ALLAN and HARRIS 1987b; BUCK and HARRIS 1990; FRYE et al. 1991). This has led to the suggestion that chronic ethanol treatment may modify coupling of the chloride channel to the various binding domains of the receptor complex (ALLAN and HARRIS 1987b; BUCK and HARRIS 1990).

Support for this notion comes from a study demonstrating that chronic exposure to ethanol reduced the ability of a BDZ agonist to augment muscimol-stimulated Cl-uptake, but enhanced the action of BDZ inverse agonists to inhibit GABA-operated Cl-channels in mouse cortical microsacs (BUCK and HARRIS 1990). Further, decreased sensitivity to BDZ agonists (e.g., CHAN et al. 1988; CRISWELL and BREESE 1989; WOO and GREENBLATT 1979) and enhanced sensitivity to BDZ inverse agonists and Cl-channel blockers (BECKER and ANTON 1989; LISTER and KARANIAN 1987; MEHTA and TICKU 1989; TICKU et al. 1992) have been observed following prolonged ethanol exposure. These studies suggest that chronic ethanol exposure results in reduced coupling between BDZ agonist sites and the Cl-channel along with enhanced coupling between BDZ inverse agonist sites and the Cl-channel. This combined effect may be involved in the development of ethanol dependence, as well as contributing to the anxiety and seizure susceptibility associated with the AWS (BUCK and HARRIS 1990).

The increased sensitivity to BDZ inverse agonists following chronic ethanol is interesting in that many of the symptoms of the AWS (e.g., increased anxiety, reduced seizure threshold) are similar to those produced by compounds such as the BDZ inverse agonists and Cl-channel blockers (e.g., DORROW et al. 1983; MICZEK and WEERTS 1987). In fact, in a drug discrimination paradigm, animals have been shown to generalize the subjective effects (interoceptive cues) of AW to systemic administration of the Cl channel blocker pentylenetetrazol (IDEMUDIA et al. 1989; LAL et al. 1988). The possibility that enhanced sensitivity to BDZ inverse agonists plays a role in ethanol dependence and symptoms associated with the withdrawal syndrome supports the suggested existence of endogenous BDZ ligands with inverse agonist properties. Indeed, a number of putative endogenous BDZ receptor ligands with inverse agonist activity have been isolated, including diazepam-binding inhibitor (GUIDOTTI et al. 1983), octadecaneuropeptide (FERRERO et al. 1986), and ethyl-*b*-carboline-3-carboxylate (PENA et al. 1986). In addition, plasma levels of one such compound (norharman) have been reported to be elevated in chronic alcoholics relative to controls (ROMMELSPACHER et al. 1991). Additional research is necessary to determine whether these substances are produced in physiologically relevant concentrations, as well as to determine what, if any, role they play in ethanol dependence and withdrawal.

Furthermore, while the reduction in sensitivity to a BDZ agonist was transitory (normalizing within 24 h after the cessation of alcohol intake), augmented sensitivity to BDZ inverse agonists was more durable, lasting for up to 8 days following withdrawal from chronic ethanol in mice (BUCK and HARRIS 1990). Thus, this biochemical event may be involved in protracted withdrawal, the propensity to relapse, and withdrawal kindling. In this light, it is interesting that Ticku and his colleagues (MHATRE et al. 1988; MHATRE and TICKU 1989; TICKU et al. 1992) reported that chronic ethanol exposure resulted in an upregulation in BDZ inverse agonist-binding sites during withdrawal. In contrast, the binding of BDZ agonists, BDZ antagonists, and ligands that bind to a channel (picrotoxin) site have typically been found to be unaltered following similar treatment. Hence, it may be that during repeated episodes of AW, repeated upregulation of BDZ inverse agonist-binding sites (and/or Cl-channel coupling), along with enhanced sensitivity to the anxiogenic and (pro)convulsant properties of BDZ inverse agonists, results in an increased propensity to subsequent withdrawal occurrences.

These studies may have important and practical clinical relevance. For example, tolerance to the anticonvulsant properties of BDZ (e.g., SCHNEIDER and STEPHENS 1988), along with decreased sensitivity that may result from ethanol withdrawal "kindling," may compromise the usefulness of these drugs in treatment of AW. In this regard, it is interesting that in one study a large proportion of patients presenting with seizures during AW had received BDZ treatment during previous withdrawal episodes and many were receiving BDZ for AW at the time of the index seizure (BROWN et al. 1988).

II. NMDA Receptor System

Glutamate is the major excitatory amino acid neurotransmitter in the brain and recent studies have demonstrated an interaction between ethanol and the NMDA (*N*-methyl-D-aspartate) receptor system (DANIELL 1992; see Chap. 4, by Hoffman, this volume). While acute treatment with ethanol has been shown to have an inhibitory effect on NMDA-mediated neurotransmission, there is some evidence to suggest that chronic ethanol exposure may result in a compensatory upregulation of the NMDA receptor complex. For instance, radioligand binding studies have demonstrated an increase in the number of NMDA receptor-linked channels, as measured by [^{3}H]MK-801 binding, in several brain regions following chronic ethanol treatment (GRANT et al. 1990). Moreover, while binding returned to control levels within 24 h following withdrawal, the increase in [^{3}H]MK-801 binding in hippocampus was found to mirror the time course of ethanol withdrawal-related seizures (GULYA et al. 1991). In addition, both competitive (LILJEQUIST 1991) and noncompetitive (GRANT et al. 1990, 1992; MORRISETT et al. 1990) NMDA receptor antagonists have been shown to reduce the severity of AW seizures. Conversely, administration of NMDA during AW increased the severity of withdrawal-related seizures (GRANT et al. 1990). Further, mice selectively bred to be prone to AW seizures evidenced a greater number of hippocampal NMDA receptors both prior to and following chronic ethanol treatment in comparison to mice that were selectively bred for resistance to AW seizures (VALVERIUS et al. 1990). Collectively, these studies suggest that the compensatory upregulation in NMDA receptor density induced by chronic ethanol exposure contributes, at least in part, to the observed increase in seizure susceptibility during AW.

It is also of interest that electrolyte imbalance is common in the alcoholic, insofar as a number of cations influence NMDA receptor function. For example, zinc acts as a noncompetitive NMDA antagonist (PETERS et al. 1987). Mice prone to withdrawal seizures were found to have lower hippocampal levels of zinc compared to the withdrawal seizure resistant line of mice (FELLER et al. 1991). Similarly, magnesium ions are known to block the NMDA-gated cation channel in a concentration-dependent fashion (NOWAK et al. 1984). Hypomagnesemia is relatively common in alcoholics, is exacerbated during withdrawal, and may be associated with risk of seizure development (WOLFE and VICTOR 1969). In addition, an inverse relationship exists between the severity of withdrawal symptoms and the level of circulating magnesium in alcoholics (FLINK 1986). Thus, vitamin and nutritional supplementation during detoxification may be important, particularly given the influence of cations such as magnesium on NMDA receptor function.

In summary, a compensatory upregulation of the NMDA receptor complex following chronic ethanol exposure may represent an important neuroadaptive response that contributes to increased seizure susceptibility during withdrawal. NMDA receptors are also involved in long-term potentiation (COLLINGRIDGE and BLISS 1987), a process not only critical for learning and memory, but essential for the development of kindling (McNAMARA et al. 1988; VEZZANI et al. 1988; YEH et al. 1989). Moreover, given the excitotoxic effects of NMDA overactivity (ROTHMAN and OLNEY 1990), repeated increases in NMDA receptor function that accompany episodes of withdrawal may result in cell death in corticolimbic structures, and thus contribute to alcohol-related dementia. While we know of no reports of NMDA antagonists used clinically for treatment of ethanol withdrawal, preclinical studies have demonstrated MK-801 (dizocilpine) to be an effective antiepileptic agent (McNAMARA et al. 1988; SATO et al. 1988; WILLIAMSON and LOTHMAN 1989). However, this and other related compounds have a number of motor and psychotogenic side effects that may preclude their clinical application. Recent preclinical studies have indicated that some NMDA receptor antagonists may be effective in ameliorating AW symptoms (particularly tremor and seizures) with a relatively high therapeutic index and minimal side effects compared to dizocilpine (GRANT et al. 1992; LILJEQUIST 1991). Whether such drugs can be used for preventing the development of kindling and/or cognitive dysfunction in alcoholics remains to be determined.

III. Voltage-Operated Calcium Channels

In addition to neuroadaptive changes in receptor-gated ion channel function, chronic ethanol exposure has also been demonstrated to evoke alterations in a number of voltage-dependent ion channels (ANANTHARAM and TREISTMAN 1992; Chap. 3, this volume). The effects of chronic ethanol on voltage-gated calcium channels have been studied in the greatest detail (GREENBERG and CHAN 1992; HAWTHORN 1992; LESLIE et al. 1990). Indeed, voltage-operated calcium channels play an important role in neuronal excitability (GREENBERG 1987), and it has been suggested that CNS hyperexcitability during ethanol withdrawal may represent, at least in part, the manifestations of a compensatory upregulation in neuronal calcium channel activity (GREENBERG et al. 1990; LITTLETON 1990). More specifically, enhanced neuronal excitability, presumably related to an increase in the number of L-type calcium channels, is thought to represent a neuroadaptive response to prolonged ethanol exposure that is only revealed when the drug is abruptly removed (LITTLETON 1989).

Support for this notion comes from a number of studies using cell culture systems, as well as the intact animal. For example, while acute exposure to ethanol produced a concentration-dependent decrease in depolarization-induced calcium uptake in cultured PC12 (rat neural crest-derived pheochromocytoma) cells, chronic ethanol exposure resulted in a reciprocal increase in calcium uptake (GREENBERG et al. 1987; MESSING et al. 1986). The most potent calcium channel antagonists are the 1,4-dihydropyridines (DHPs). These drugs selectively bind with high affinity to the L-type calcium channels and radioligand binding studies indicated an increase in the number of DHP-sensitive binding sites following prolonged ethanol exposure (MESSING et al. 1986). Similar results have been obtained in studies with cultured bovine adrenal chromaffin cells (BRENNAN et al. 1989; HARPER and LITTLETON 1987). Recently, it has been suggested that second messenger systems play a role in mediating the upregulation of L-type calcium channels induced by chronic ethanol exposure (BRENNAN and LITTLETON 1990; MESSING et al. 1990). The time course of this increase in DHP-sensitive binding sites was similar in each of the cultured cell types, with the maximal effect reached after about 6 days of growth in ethanol-containing medium. Further, the enhanced uptake of calcium "normalized" within 16 h after removal of ethanol from the culture medium (MESSING et al. 1986), which is the time period during which withdrawal symptoms are typically most intense.

Similarly, an increase in the number of DHP-sensitive calcium channels has been demonstrated in brain tissue from ethanol-dependent rats (DOLIN et al. 1987; GUPPY and LITTLETON 1987). Moreover, the increase in brain DHP-binding sites was reported to be more than threefold greater in selectively bred withdrawal seizure-prone mice in comparison to those selectively bred for withdrawal seizure resistance (BRENNAN et al. 1990).

In some cases, however, chronic ethanol exposure has not been found to alter the number and/or function of DHP-sensitive channels. For example, in a synaptosomal preparation derived from the striatum of rats chronically fed an ethanol-containing diet, tolerance to the inhibitory effects of acute ethanol on depolarization-induced calcium uptake was demonstrated in the absence of an increase in the density of L-type calcium channels (WOODWARD et al. 1990). In contrast, while chronic ethanol exposure does not result in tolerance to the inhibitory effects of acute ethanol on calcium uptake in PC12 cells, an increase in the number of L-type channels was associated with an enhancement of potassium-stimulated calcium uptake (GREENBERG et al. 1987; MESSING et al. 1986). Studies with cultured chromaffin cells have demonstrated an enhancement of catecholamine release corresponding to the upregulation in DHP-sensitive calcium channels (HARPER and LITTLETON 1987; LYNCH and LITTLETON 1983). These secondary effects downstream from altered calcium channel activity may, in turn, underlie a variety of symptoms (autonomic and epileptogenic) of the withdrawal syndrome. Thus, while the discrepant results mentioned above are most likely due to differences in the assay conditions for calcium uptake, level of ethanol exposure, and cell preparation, the functional significance of these calcium channel changes remains to be determined.

Perhaps the strongest evidence in support of chronic ethanol-induced neuroadaptation of voltage-dependent calcium channels comes from studies examining the effects of DHP-sensitive calcium channel antagonists on ethanol withdrawal. In fact, a number of DHP calcium channel antagonists (e.g., nitrendipine, nimodipine, nifedipine) have been shown to attenuate ethanol withdrawal seizures (Little et al. 1986; LITTLETON et al. 1990; PUCILOWSKI et al. 1989). This effect was stereoselective and was blocked by the calcium channel agonist BAY K 8644 (LITTLETON et al. 1990). In addition, nitrendipine treatment was shown to prevent the development of ethanol dependence when administered chronically along with ethanol (WHITTINGTON and LITTLE 1988). Although nitrendipine was not found to be efficacious in the treatment of ethanol withdrawal-related anxiety in rats (FILE et al. 1989), these studies suggest that calcium channel antagonists may be clinically useful in the prevention and/or treatment of withdrawal seizures, and possibly autonomic hyperactivity.

While clinical studies of the effectiveness of calcium channel antagonists in the treatment of ethanol withdrawal are few in number, positive results have been obtained from two European studies using nimodipine (NICKEL and SCHMICKALY 1988) and a non-DHP antagonist, caroverine (KOPPI et al. 1987). Thus, although these drugs have a number of side effects (particularly cardiovascular), their utility in the treatment of AW remains to be more thoroughly evaluated.

IV. Monoamine Systems

Monoamines are involved in a number of ethanol actions, including the withdrawal syndrome (NUTT and GLUE 1986; Chap. 7, this volume). Indeed, a number of withdrawal symptoms reflect increased sympathetic activity, including hypertension, tachycardia, tremor, agitation, anxiety, and general heightened autonomic reactivity. This overactivity of brain and peripheral noradrenergic systems during withdrawal has been demonstrated by increased plasma and CSF levels of norepinephrine and its major metabolite 3-methoxy-4-hydroxyphenylethyleneglycol (MHPG) following withdrawal from acute intoxication (BORG et al. 1981) and chronic ethanol exposure (HAWLEY et al. 1985; SMITH et al. 1990). The severity of withdrawal symptoms was found to be positively correlated with levels of plasma and CSF norepinephrine (HAWLEY et al. 1985; SMITH et al. 1990). Furthermore, the level of sympathetic activity (plasma MHPG levels) was positively correlated with the number of years that patients experienced withdrawal symptoms, as well as the number of previous treatments for withdrawal (NUTT 1987). These results suggest that withdrawal symptoms related to heightened noradrenergic activity may be susceptible to a kindling process.

There is also evidence that increased noradrenergic activity during withdrawal is due to reduced α_2-adrenoceptor autoreceptor inhibition. More specifically, subsensitivity of α_2-adrenoceptors has been demonstrated by blunted physiologic (blood pressure and body temperature) and neuroendocrine (growth hormone secretion) responses following clonidine challenge (GLUE et al. 1988; NUTT et al. 1988). While downregulation of platelet α_2-adrenoceptors has been found in one study (SMITH et al. 1990), another study found no change in density of α_2-adrenoceptors in platelets from alcoholic patients during withdrawal (NUTT et al. 1987). Some impairments of α_2-adrenoceptor function have been shown to persist during continued abstinence (GLUE et al. 1989).

Given the number of withdrawal symptoms that are presumably reflective of autonomic hyperactivity, it is not surprising that several sympatholytic agents (β-adrenergic blockers and α_2-adrenoceptor agonists) have been used in the treatment of withdrawal. Indeed, the β-adrenergic blockers propranolol and atenolol (CARLSSON and JOHANSSON 1971; KRAUS et al. 1985; ZILM 1975) and the α_2-adrenoceptor agonists clonidine and lofexidine (BAUMGARTNER and ROWEN 1987; BJORKQUIST 1975; CUSHMAN et al. 1985; MANHEM et al. 1985; WALINDER et al. 1981; WILKINS et al. 1983) have been shown to be effective in ameliorating some withdrawal symptoms. However, these drugs may produce a number of potentially dangerous side effects and they lack anticonvulsant activity (BOGIN et al. 1987; ROBINSON et al. 1989; SELLERS et al. 1977). Further, while animal studies have yielded mixed results (HEMMINGSEN et al. 1984; PARALE and KULKARNI 1986), in one study clonidine was found to enhance AW symptoms in mice (BLUM et al. 1983). Thus, these drugs are of limited use in severe withdrawal where seizures and delirium tremens are likely to occur.

There is also evidence for altered dopamine activity following chronic ethanol exposure. Preclinical studies have demonstrated both enhanced (ENGEL and LILJEQUIST 1976; GONZALEZ et al. 1988; LILJEQUIST 1978) and attenuated (HOFFMAN and TABAKOFF 1977; HUNT 1981; TABAKOFF et al. 1978) dopamine receptor sensitivity during ethanol withdrawal. Two clinical studies have shown increased dopamine receptor sensitivity during the abstinence syndrome, as measured by apomorphine-induced growth hormone release (ANNUZIATO et al. 1983; BALLDIN et al. 1985). It has been suggested that hallucinations associated with severe withdrawal (particularly delirium tremens) may be reflective of dopaminergic dysfunction (GLUE and NUTT 1990; NUTT and GLUE 1990). In fact, neuroleptics have been shown to be efficacious in treating psychotic-like symptoms of severe withdrawal (PALESTINE 1973; RITTER and DAVIDSON 1971). However, these drugs are not recommended as the sole means of pharmacotherapy because they lack anticonvulsant activity and, in some cases, may lower seizure threshold (McCUTCHEN 1990).

V. Second Messenger Systems and Adenosine

Second messenger systems play an important role in signal transduction processes following activation of certain neurotransmitter and hormone receptors. These receptor systems comprise at least three linked membrane components: the receptor (recognition site for neurotransmitter of hormone); adenylate cyclase (the enzyme involved in the production of cAMP); and Gs (a stimulatory guanine nucleotide regulatory protein that couples receptor activation with adenylate cyclase stimulation) (STRYER and BOURNE 1986). Both acute and chronic ethanol exposure have been shown to influence these G protein-coupled receptor systems. For example, acute ethanol enhances receptor-stimulated cAMP production in cultured neural (GORDON et al. 1986) and lymphoma (BODE and MOLINOFF 1988) cells, as well as in cortical brain tissue (SAITO et al. 1987; VALVERIUS et al. 1987). Conversely, chronic ethanol exposure has been shown to decrease receptor-dependent cAMP levels in these systems. Moreover, lymphocytes from alcoholic patients were found to exhibit a fourfold decrease in basal and adenosine (A_2) receptor-stimulated cAMP levels in comparison to normal controls and patients with nonalcoholic liver disease (DIAMOND et al. 1987; NAGY et al. 1988). Similarly, platelets from alcoholics were more resistant to prostaglandin E_1 (PGE_1) receptor-stimulated cAMP production in comparison to controls (TABAKOFF et al. 1988). These latter findings suggest that this measure (receptor-stimulated cAMP production) may be useful as a marker for alcoholism.

Since chronic ethanol decreases cAMP production following activation of adenosine (A_2) receptors (GORDON et al. 1986), PGE_1 receptors (RICHELSON 1986), and β-adrenergic receptors (SAITO et al. 1987; VALVERIUS et al. 1987), it has been suggested that this effect is due to heterologous desensitization of receptors coupled to Gs protein (DIAMOND et al. 1990; MOCHLY-ROSEN et al. 1988). Further, studies have demonstrated that the reduced cAMP production following chronic ethanol exposure results from altered Gs protein synthesis and function. More specifically, decreased receptor-stimulated cAMP production in chronic ethanol-exposed cultured cells (neuroblastoma × glioma hybrid cells) was found to be accompanied by a 30% decrease in the amount of mRNA for the α-subunit of Gs, a 39% reduction in the amount of the Gs α-subunit protein, and a 29% decrease in the functional activity of the α-subunit of Gs (DIAMOND et al. 1990; MOCHLY-ROSEN et al. 1988).

It is interesting that recovery of normal adenosine receptor-stimulated cAMP production occurs 48 h after ethanol is withdrawn from the culture medium. This time course corresponds to the period of most severe AW symptoms. Adenosine is an inhibitory neuromodulator with sedative and anticonvulsant properties. Hence, it has been suggested that desensitization of adenosine receptors (particularly the A_2 subtype which is linked to Gs) may underlie the hyperexcitability and lowered seizure threshold associated with the AWS (DIAMOND et al. 1990).

VI. Summary

It is clear that many neurochemical systems in the brain are perturbed by acute ethanol exposure. In addition, a number of neuroadaptive changes are set in motion in response to continued exposure to the drug. These compensatory changes typically serve to mitigate the effects of ethanol (tolerance), and, while not apparent in the presence of the drug, these changes and their behavioral manifestations are revealed upon abrupt withdrawal of ethanol. Hence, the neurochemical events observed during the withdrawal state are taken as indices of neuroadaption in response to chronic exposure to ethanol. The extent to which these changes reflect the processes underlying the development of dependence (as opposed to being expressed secondary to the withdrawal reaction) remains a difficult issue to resolve. Nevertheless, as outlined above, elucidation of neurochemical changes during withdrawal from chronic ethanol exposure has greatly enhanced our knowledge of the biologic basis of the various facets of the withdrawal syndrome and, importantly, continues to guide the development of new and more efficacious medications.

A summary of the prominent changes in neurochemistry observed during ethanol withdrawal, along with their possible correspondence to specific withdrawal symptoms, is presented in Table 1. In addition, as noted by others (e.g., GLUE and NUTT 1990; NUTT and GLUE 1990), there are numerous interactions among these neurochemical systems that almost certainly contribute to the AWS. For example, increased intracellular

Table 1. Summary of neurochemical disturbances following chronic ethanol exposure, their possible relationship to withdrawal symptoms, and corresponding appropriate pharmacotherapy

Neurotransmitter disturbances	Withdrawal symptoms	Appropriate pharmacotherapy
↓ GABA	Anxiety, seizures	GABA agonists, benzodiazepines, barbiturates
↑ NMDA	Seizures	NMDA receptor antagonists
↑ Calcium channels	Seizures	Calcium channel antagonists
↑ Norepinephrine	Anxiety, autonomic hyperactivity	Sympatholytics: β-adrenergic blockers, α_2-adrenoceptor agonists
↑ Dopamine	Hallucinations, delirium tremens	Neuroleptics

calcium concentrations that result from NMDA receptor activation have been shown to inhibit GABA activity (STELTZER et al. 1987). Conversely, reduced GABA neurotransmission may enhance NMDA receptor activity (e.g., HERRON et al. 1985). Thus, enhanced NMDA function and depressed GABA activity may complement and facilitate one another during AW, resulting in an overall hyperexcitable state. Increased neuronal excitability (depolarization) due to NMDA receptor-stimulated calcium influx may also activate voltage-dependent calcium channels (HOFFMAN and TABAKOFF 1991). Further, there is evidence to suggest that both reduced GABA activity and enhanced NMDA receptor function may result in noradrenergic and dopamine overactivity (ARANEDA and BUSTOS 1989; LALIES et al. 1988). Augmented noradrenergic activity may, in turn, stimulate the hypothalamic-pituitary-adrenal axis through activation of corticotropin-releasing factor (CRF) (ALONSO et al. 1986; KITAZAWA et al. 1987), a peptide found in the hypothalamus and other brain regions. In addition to stimulating pituitary adrenocorticotrophic hormone (ACTH) release, CRF has been shown to possess anxiogenic and proconvulsant properties (KOOB and BLOOM 1985; EHLERS et al. 1983; WEISS et al. 1986). Thus, the AWS appears to be the manifestation of numerous interactions among central neurochemical and neuroendocrine systems.

D. Repeated Episodes of Alcohol Withdrawal: The "Kindling" Hypothesis

The question of whether an episode of AW influences subsequent AW symptoms has received an increasing amount of attention in recent years. The issue is of clinical significance in that periodic interruptions in alcohol consumption occur commonly during the course of alcoholism (HILLBOM 1990). While the amount of alcohol consumed and the duration of intoxication prior to cessation represent important determinants of the severity of the withdrawal reaction, it has been suggested that the prior episodes of AW can sensitize an individual to subsequent episodes of withdrawal (GROSS et al. 1972). Indeed, early episodes of AW are typically characterized as being relatively mild. However, continued alcohol abuse results in a progressive increase in the severity of symptoms associated with additional subsequent episodes.

BALLENGER and POST (1978) hypothesized that the progressive intensification of the withdrawal syndrome following repeated episodes of alcohol intoxication and withdrawal may represent the manifestations of a "kindling" mechanism. The term "kindling" was first introduced by GODDARD et al. (1969). It refers to the phenomenon wherein low-level electrical stimulation of discrete brain regions, which are initially insufficient to produce electrographic or overt behavioral effects, come to evoke major motor seizures upon repeated intermittent application. Extending this phenomenon to AW, it was postulated that each episode of CNS hyperexcitability that normally accompanies AW (BEGLEITER and PLATZ 1972; MENDELSON 1971) may support the "kindling" process. This "kindling" process, then, may underlie the commonly observed progression of withdrawal symptoms, from relatively minor responses characteristic of early withdrawal episodes (e.g., irritability, tremors) to more severe symptoms associated with later withdrawal episodes (e.g., seizures, delerium tremens) (BALLENGER and POST 1978).

Several clinical reports have provided support for the "kindling" hypothesis of AW. For example, in a small study, patients with a history of prior dependence were found to exhibit a more severe withdrawal reaction following consumption of alcohol in comparison to a group without prior dependence (MENDELSON et al. 1969). More recently, BROWN et al. (1988) found that individuals who had experienced a seizure during AW were more likely to have undergone previous alcohol detoxification than individuals who did not have an AW seizure during the index episode (Fig. 2). LECHTENBERG and WORNER (1991) reported an increased risk of AW seizures associated with greater numbers of both "nondetoxification" and, more importantly, "detoxification" hospital admissions. This relationship appeared to be stronger for women than men. To our knowledge, this is one of the few observations that has suggested that women may have a different sensitivity

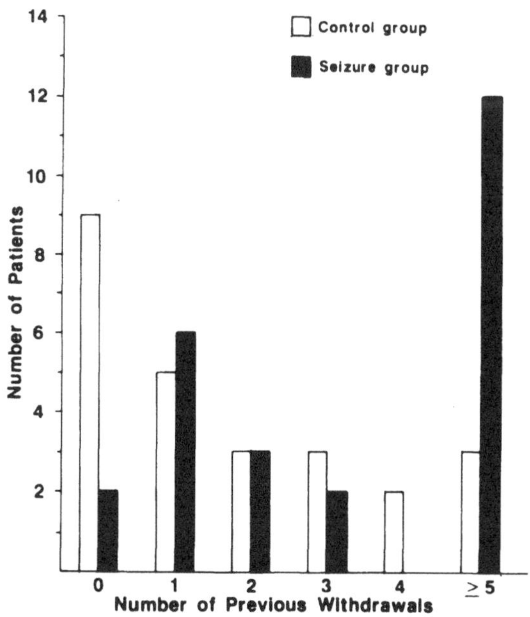

Fig. 2. Number of previous withdrawals (detoxifications) in the seizure and control groups. More alcoholics with withdrawal seizures had undergone five or more detoxifications ($\chi^2 = 7.13$, $df = 2$, $p > 0.05$). (From BROWN et al. 1988)

to AW. This suggests that the "telescoping" phenomenon described in women (PIAZZA et al. 1989), which refers to a quicker onset of alcohol-related medical problems such as liver pathology (FREZZA et al. 1990), may also be operable in the brain, leading to an accelerated "kindling" of AW. If validated, these findings would have significant theoretical and therapeutic implications.

Although the body of clinical data supporting the "kindling" hypothesis is growing, many of the concurrent and intervening variables associated with the phenomenon cannot be adequately controlled in humans. The fact that there are animal models for the "kindling" phenomenon enhances the validity of the concept. Indeed, a number of studies have demonstrated an intensification of the withdrawal reaction in animals with prior AW experience. The potentiated withdrawal response has been measured as changes in motor activity, as well as increased incidence of spontaneous audiogenic- and chemoconvulsant-induced spontaneous seizures (BAKER and CANNON 1979; BRANCHEY et al. 1971; CLEMMESEN and HEMMINGSEN 1984; MAIER and POHORECKY 1989; OLSEN et al. 1992; WALKER and ZORNETZER

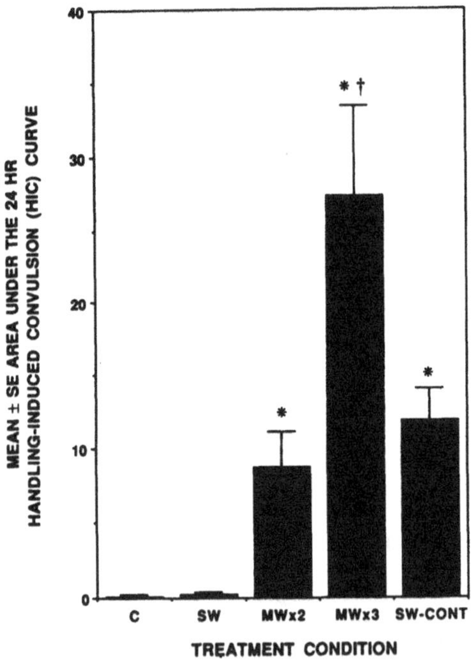

Fig. 3. Severity of withdrawal seizures as a function of treatment condition (*C*, control group; *SW*, single withdrawal group; *MWx2*, multiple withdrawal group (two cycles); *MWx3*, multiple withdrawal group (three cycles); *SW-CONT*, single withdrawal group after continuous ethanol exposure; $N = 8$/group – see text for details). Blood ethanol levels just prior to withdrawal testing were similar for all ethanol-exposed groups (135–155 mg/dl). *significantly differs from C and SW group ($p < 0.05$); †significantly differs from C, SE, SW-CONT groups ($p < 0.05$)

1974). Recent reports have also indicated that some symptoms of the AWS (e.g., startle and seizure responsiveness) may be more susceptible to "kindling" than others (POHORECKY and ROBERTS 1991; ULRICHSEN et al. 1992).

BECKER and HALE (1992, 1993) recently developed a mouse model of AW that is sensitive to the effects of prior withdrawal experience. In this model, mice are continuously exposed to ethanol vapor for varying periods in inhalation chambers. As shown in Fig. 3, a positive relationship exists between the number of previous AW episodes and the severity of subsequent withdrawal seizures. Moreover, this potentiated withdrawal response was observed even when the total amount of ethanol exposure was equated across groups. Importantly, the more severe withdrawal response exhibited by animals undergoing repeated episodes of withdrawal compared with mice that experienced withdrawal only once cannot be explained by differences in blood alcohol levels at the time of withdrawal or differences in the rate of ethanol elimination (BECKER 1994).

The exacerbated behavioral symptoms observed in animals undergoing repeated episodes of withdrawal have been shown to be accompanied by progressively greater changes in EEG (POLDRUGO and SNEAD 1984; WALKER and ZORNETZER 1974), as well as alterations in local metabolic activity (i.e., glucose consumption) in cortical and limbic brain regions (CLEMMESEN et al. 1988). Additionally, more severe cognitive/memory deficits have been exhibited by animals (BOND 1979; FREUND 1971) and humans (GLENN et al. 1988) that have experienced multiple episodes of ethanol withdrawal, compared with those withdrawn from alcohol only once. Studies with mice (BECKER and HALE 1992; WALKER and ZORNETZER 1974), rats (BAKER and CANNON 1979; BRANCHEY et al. 1971), and primates (ELLIS and PICK 1970) have also shown that a prior history of withdrawal episodes decreases the duration or extent of intoxication necessary to provoke a subsequent withdrawal response upon cessation of alcohol use. Finally, animal studies have shown that kindling (whether produced by electrical stimulation or repeated administration of chemoconvulsants) potentiates the symptoms of subsequent AW (PINEL 1980; PINEL and VAN OOT 1975, 1978; PINEL et al. 1975). Conversely, AW accelerates the development of electrical kindling in various limbic brain structures (CARRINGTON et al. 1984; MCCOWAN and BREESE 1990). Thus, there is a growing body of clinical and experimental evidence that supports the "kindling" hypothesis of AW.

The mechanism(s) involved in this "kindling" phenomenon associated with AW are presently unknown. Neurochemical systems that are not only highly sensitive to ethanol's action, but are also important in brain epileptogenic activity, may be particularly significant with regard to the "kindling" phenomenon. Presumably, changes in any number of these systems that progressively intensify with each withdrawal episode may culminate in a persistent state of CNS hyperexcitability that is manifested as a "kindled" AW response. For example, a decrease in the CNS inhibitory effects of GABA following chronic alcohol exposure and withdrawal might represent one such mechanism. There is some evidence that chronic ethanol exposure results in decreased GABA activity by altering the binding capacity and actions of ligands that allosterically modulate the ability of GABA to increase permeability and transport of chloride ions across the neuronal membrane. Both increased sensitivity to BDZ inverse agonists (BUCK and HARRIS 1990) and upregulation of BDZ inverse agonist-binding sites (MHATRE and TICKU 1989; MHATRE et al. 1988) have been reported in animals chronically treated with alcohol. It may be that "kindling" is a function of increasing sensitivity to putative endogenous BDZ inverse agonists and/or repeated upregulation of binding sites for these ligands concomitant with repeated episodes of AW, which sensitizes an individual to subsequent seizure-inducing events, including AW.

The CNS hyperexcitability associated with AW may also result from a variety of other neurochemical effects, including enhancement of the excitatory effects of glutamate at NMDA receptors. Thus, repeated occurrences of an increase in the number of MK-801-binding sites that have been reported in animals chronically exposed to ethanol (GRANT et al. 1990; GULYA et al. 1991) may contribute to the progessive intensification of the withdrawal syndrome. This hypothesis is supported by the fact that NMDA-type glutamate receptors are known to play an important role in neuronal plasticity, including both long-term potentiation (COLLINGRIDGE and BLISS 1987; COLLINGRIDGE and DAVIES 1989) and seizure production by electrical kindling (WILSON et al. 1989; YEH et al. 1989). Finally, other consequences of chronic ethanol treatment that increase CNS excitability, such as an increase in the density of dihydropyridine-sensitive calcium channels (BRENNAN et al. 1990; DOLIN et al. 1987), desensitization of A_2 adenosine receptors (DIAMOND et al. 1990), functional subsensitivity and receptor downregulation of α_2-adrenoceptor systems (GLUE et al. 1989; LINNOILA et al. 1987), and excessive secretion of CRF (ADINOFF et al. 1990) may, upon repeated occurrence, progressively intensify and contribute to "kindling" of the AWS. The brain mechanisms underlying AW-induced "kindling" deserve additional experimental attention.

In view of the data supporting the existence of AW-induced kindling, there is ongoing debate as to whether all alcohol-dependent patients should be aggressively treated during detoxification. Mild withdrawal symptoms have been shown to be adequately managed through the use of nonpharmacologic supportive care (NARANJO et al. 1983). While BDZ remain the drugs of choice in the pharmacologic treatment of AW, there is some concern over their use (given their potential for abuse) in patients with a substance use disorder.

On the other hand, the growing body of clinical and experimental evidence supporting the AW-induced "kindling" hypothesis underscores the negative consequences that may occur if withdrawal episodes (even those that are relatively mild) are not treated aggressively. This has led some to argue that all patients undergoing alcohol detoxification should be medically treated to prevent kindling (LINNOILA et al. 1987). In support of this idea, a recent study demonstrated that prevention of withdrawal seizures in rats by treatment with phenobarbital thwarted the development of "kindling" such that subsequent untreated withdrawal episodes were much less severe than those in animals that received no medication during prior withdrawal episodes (ULRICHSEN et al. 1992). Furthermore, untreated episodes of withdrawal may result in repeated states of hypercortisolemia that may, in turn, facilitate the development of neuropathologic changes (e.g., hippocampal lesions) that underlie alcohol-related cognitive dysfunction and dementia (ADINOFF et al. 1990; SAPOLSKY and PULSINELLI 1985). New research in this area may provide guidance in the development of more effective strategies for the treatment of AW and improvements in the long-term management of alcoholism.

E. Treatment of Alcohol Withdrawal

While it is generally accepted that severe AW requires pharmacologic intervention (and possibly hospitalization), clinical practice differs when it comes to the treatment of mild-to-moderate AW. Most physicians would agree that delirium tremens merits aggressive medical and pharmacologic treatment and that the prevention of AW seizures is a worthwhile goal of treatment. Other goals of pharmacologic treatment of AW are to increase the patient's level of comfort, reduce dysphoria and anxiety which may lead to alcohol-seeking behavior, and relieve physical symptoms so that the process of rehabilitation may begin as soon as possible. Countervailing views make note of the fact that seizures occur in only a very small number of individuals in AW and that there are abundant data attesting to the efficacy of psychosocial support alone in the treatment of uncomplicated AW. Furthermore, many clinicians believe that the use of a sedative medication during withdrawal reinforces the need for self-medication and supports dependence on substances rather than on internal strength and social support. Data will subsequently be presented to address this controversy. However, prior to that discussion, some description of efforts to quantify the AWS is in order.

F. Measurement of the Alcohol Withdrawal Syndrome

The ability to measure and quantify AW symptoms is useful for a number of reasons. First, it enhances clinical care by standardizing measurement and thereby providing a firmer basis for communication among clinicians. Second, the reliable and valid measurement of the severity of AW is essential for meaningful research on the efficacy of different treatments for AW. Third, standardization of the measurement of the AWS provides a basis for the scientific study of the pathophysiology of the AWS.

GROSS and colleagues (1973) are credited with having developed the first scales for measurement of AW symptoms. The Total and Selective Severity Assessment Scales laid the foundation on which future scale development was built. The main problem with these scales was that they were too lengthy for clinical application. They also were intended for use in rating symptoms on a daily basis rather than at multiple time points in a given day. A modified version of the original scale of GROSS and colleagues (1973) called the modified, or Milwaukee, Selective Severity Assessment (MSSA) was developed by BENZER (1990). They found it to be useful in the selection of patients who are in need of pharmacologic management (BENZER 1990).

SHAW et al. (1981) adapted the work of Gross and colleagues to yield a 15-item scale called the Clinical Institute Withdrawal Assessment for Alcohol (CIWA-A). Insofar as this scale could be administered every half-hour, it was useful for the examination of rapid changes in symptom level. Consequently, the CIWA-A could be used to guide treatment and in comparing the efficacy of different treatments. The CIWA-A was utilized to validate the DSM-III-R classification for uncomplicated alcohol withdrawal (SELLERS et al. 1991). The CIWA-A was revised and shortened by SULLIVAN and associates (1989) to 10 items (CIWA-Ar), with no loss of clinical utility, reliability, or validity, despite greater efficiency in its application. Both the CIWA-A and the CIWA-Ar have been used clinically to monitor AW symptoms and to guide pharmacologic dosing, with excellent results (WARTENBERG et al. 1990; SULLIVAN et al. 1991). Other data suggest that patients who eventually progress to complicated AW (seizures or confusion) score higher on the CIWA prior to the development of these complications (FOY et al. 1988).

It is essential that research on the treatment of the AWS employ a standardized rating scale, in order that the results be widely interpretable. It would also appear to be desirable for all clinical detoxification units to adopt a standard AW scale. As the studies above indicate, this procedure has enhanced the standard of care in a number of clinical settings and should be strongly advocated.

G. Nonpharmacologic Treatment of Alcohol Withdrawal

A study of 1114 patients suggested that the overwhelming majority (1024) could be safely and effectively treated with "social setting" detoxification alone (WHITFIELD et al. 1978). Several patients with clear evidence of delirium tremens and 38 patients with "hallucinosis" responded to this approach. However, the authors also reported that 12 other patients developed one or more seizures during the treatment period. While the results of this study suggest that a nonpharmacologic approach may work for the majority of admissions to detoxification centers, the results are global in nature and random assignment to treatment alternatives was not employed. While the authors pointed out that "much staff time is needed" to deliver treatment using this approach, detailed information on the cost of the treatment was not provided.

Subsequently, SHAW and colleagues (1981) reported that 75% of hospitalized chronic alcoholics could be successfully treated with supportive care, which included evaluation of their symptoms, nursing support, and fluids every 30 min. Patients in this study were in moderate-to-severe withdrawal and those who responded did so within the first 4–8 h after the initiation of the supportive care. One patient developed a seizure and another manifested hallucinations subsequent to showing an initial response. As with the earlier study by WHITFIELD et al. (1978), this study did not employ an adequate control treatment or random assignment to conditions.

NARANJO and colleagues (1983) conducted a small, but well-designed, study of alcoholics who presented to an emergency room with mild-to-moderate AW; patients in the study were randomly assigned to receive supportive care with or without lorazepam. Though a few of the patients who received medications showed greater overall improvement, and the majority receiving the drug showed a more rapid decline in symptoms, there was no overall group difference in symptom reduction over the 7-h assessment period. As in previous studies, this study showed that, while a few patients benefit from pharmacologic treatment, the majority of patients in mild-to-moderate AW do well with supportive care alone. However, to date, studies in this area have been limited in the number of patients studied, the duration of treatment, the use of a controlled experimental design, and the period over which outcome is assessed. Until studies meeting these qualifications have been conducted, the appropriate role of pharmacologic treatment in the majority of patients with AW symptoms will continue to be debated.

H. Pharmacologic Treatment

I. Complicated Alcohol Withdrawal

As discussed previously, there seems to be near universal agreement that complicated AW (that is when seizures, hallucinations, or delirium tremens occur during the AW) should be treated pharmacologically. There is also agreement that patients should receive adequate hydration, thiamine, and close observation of their vital signs, behavior, and cognition during all phases of AW. We will cover the prevention and treatment of AW seizures in a subsequent section. However, the aggressive treatment of impending, or full-blown, delirium tremens will be mentioned at this time. Delirium tremens (DT) occurs in less than 5% of alcoholics who abruptly cease chronic ingestion. The rate varies with the treatment setting. Alcoholics who are debilitated, or medically ill, have higher rates of DT than the usual patient who presents to primary detox and rehabilitation facilities. As noted earlier, the syndrome starts sometime later than the more common symptoms of alcohol withdrawal and may in fact appear as these are waning. It is marked, in its full-blown form, by hyperpyrexia, tachycardia, hallucinosis, disorientation, tremulousness, incontinence of urine, agitation and aggressivity, and inability to take fluids and nutrition.

In a well-designed study, THOMPSON (1975) randomly assigned patients who had severe DT, with or without additional medical illnesses, to treatment with either paraldehyde or intravenous diazepam. They found that diazepam reduced symptoms and returned the patients to normal more rapidly than did paraldehyde. The average length of treatment necessary to achieve complete symptom resolution was 56 hours, irrespective of the presence of medical complications. However, medically ill patients needed significantly more medication to alleviate symptoms. Diazepam was initially given at a dosage of 10 mg, with additional doses of 5 mg given every 5 min until a calming effect was achieved. Doses of 5–10 mg were then given every 1–4 h to maintain the calming effect. Patients without medical illness needed 10–160 mg diazepam to produce a calming effect, while those with medical illness needed 35–215 mg and twice as long to achieve a calm state. Importantly, all of the serious complications, including death in two patients, the need for resuscitation in two others, and serious injuries to patients and staff, occurred in the paraldehyde group. These data as well as accumulated clinical experience have led clinicians to use BDZs, in general, and diazepam, in particular, as the treatment of choice for serious alcohol withdrawal symptoms. In general, as the severity of DT increases, more medication and greater intensity of supportive care are necessary. The need for repletion of electrolytes, treatment of comorbid medical conditions, and rehydration cannot be overemphasized. With this type of aggressive medical care becoming the norm, the mortality rate of DT has declined significantly.

To our knowledge there are no controlled studies that satisfactorily address the treatment of patients who experience hallucinations in the absence of specific symptoms of DT. It is clear that a number of patients who experience illusions, or low-grade visual or acoustic hallucinations, will benefit and recover with sedative-hypnotic (e.g., benzodiazapine) treatment. There are a few patients, however, who experience more discrete, intense, and distressing hallucinations. These patients appear to benefit more from the use of antipsychotic medications. Haloperidol seems to be the most widely used in this regard, possibly because it has minimal active metabolites which might be increased by alcoholic liver disease, and also because it may reduce seizure threshold less than phenothiazine antipsychotics. SOYKA and colleagues (1992) have shown haloperidol to be safe and efficacious for the treatment of AW hallucinosis. Many clinicians prescribe haloperidol on an "as needed" basis during acute alcohol withdrawal at the first sign of hallucinatory symptoms. This strategy remains to be validated empirically.

II. Uncomplicated Alcohol Withdrawal

A variety of strategies have been employed in the treatment of uncomplicated alcohol withdrawal, which occurs upon abrupt cessation of chronic heavy drinking. These strategies range from psychosocial support to standardized medical detoxification. However, a number of questions remain unanswered concerning the treatment of AW, including whether medication is useful in the treatment of mild-to-moderate AW and, if so, what type of medication should be used? The remainder of this section will review the information available that bears on these issues. The focus will be on studies conducted during the last 10 years. In a review of studies published prior to that time MOSKOWITZ and colleagues (1983) concluded that most studies on the pharmacotherapy of AW were deficient in some manner. According to these authors, the only justified conclusion is that benzodizapines are superior to placebo in the treatment of AW. During the past decade a number of methodologically sound studies of AW have been conducted, which have yielded useful data on the clinical management of AW.

III. Benzodiazapines

In most treatment settings benzodiazepines (BDZs) are the drugs of choice for the treatment of uncomplicated AW. SELLERS and colleagues (1983) employed a standardized evaluation technique (i.e., the CIWA-A scale), and found that diazepam was more efficacious than supportive care in rapidly reducing symptoms of moderate-to-severe AW. All of the serious complications in that study occurred in the nonmedicated supportive care group. A number of studies comparing newer BDZ preparations with placebo or standard drugs (e.g., diazepam, chlordiazepoxide) are summarized in Table 2. In general, the newer benzodiazapines have been found to have efficacy similar to that of the standard compounds. It is also clear that the rate of placebo response is substantial, especially in the less severe forms of AW. Therefore, studies that utilize a standard drug as a comparison are primarily useful in examining different side effect profiles and less useful in evaluating efficacy. As the severity of the AWS in the study sample increases this becomes less of a problem, since more severe withdrawal symptoms are less likely to respond to placebo treatment.

The accumulated evidence to date suggests that BDZs, when utilized in equivalent pharmacologic doses, are all equally useful in the treatment of uncomplicated AWS. Some suggest that those with longer half-lives (e.g., diazepam and chlordiazepoxide) are more useful, since fewer doses are needed and their pharmacokinetic profile allows for "autotitration" or "self-tapering." Others suggest that shorter half-life drugs (e.g., oxazepam, lorazepam) are more useful, given that their more rapid metabolism and excretion limit drug accumulation and associated adverse effects. Also, these drugs have fewer active metabolites to accumulate in patients with

Table 2. Studies published on the pharmacologic treatment of alcohol withdrawal during the last 10 years

Drug class	Author	Design	Comparison drugs	*N*	Duration (days)	Efficacy	Side effects
Benzodiazepines							
Diazepam	Sellers et al. (1983)	RCT	Diazepam 20 mg q2 h vs. placebo	50	2	Diazepam produced more rapid and complete improvement	Four patients on placebo had AW seizure during Rx
Clobazam (CLB)	Mukherjee (1983)	RCT	Clobazam 20 mg tid vs. chlordiazepoxide (CDZ)	40	14	Both drugs produced significant improvement; trend for CLB > CDZ	None severe No group differences
Halazepam (HAL)	Mendels et al. (1985)	RCT	Halazepam 160–480 mg day 1 vs. CDZ 100–300 mg day 1	80	5	Both drugs produced sigificant improvement. HAL = CDZ	None severe No group differences
Alprazolam (ALPZ)	Wilson and Vulcand (1985)	RCT	Alprazolam $\bar{x}$ = 3.8 mg day 1 vs. CDZ $\bar{x}$ = 197 mg day 1	100	5	Both drugs produced significant improvement. ALPZ = CDZ	Nine ALPZ and 4 CDZ patients had AW seizure during Rx; 3 ALPZ and 3 CDZ patients developed DT
Alprazolam	Turbridy (1988)	RCT	Alprazolam 4 mg day 1 vs. chlormethiazole 3 mg day 1	90	5	Both drugs produced significant improvement. ALPZ = chlormethiazole	None severe No group differences
α-Adrenergic blockers							
Clonidine (CLON)	Wilkins et al. (1983)	Single Subject A–B	Clonidine 5 μg/kg vs. placebo	11	4 h	Clonidine improved VS and withdrawal symptoms	None

Clonidine	Manhem et al. (1985)	RCT	Clonidine 0.6–1.2 mg/day vs. chlormethiazole 2–4 g/day	20	4	Both drugs produced significant improvement. CLON > chlormethiazole for BP and HR decrease	None noted
Clonidine	Baumgartner and Rowen (1987)	RCT	Clonidine 0.2–0.6 mg/day vs. CDZ 50–150 mg/day	47	4	Clonidine better than CDZ in withdrawal score. Improvement primary because of decreased HR and systolic BP	Confounded with withdrawal symptoms; less nausea/vomiting in clonidine group
Clonidine	Robinson et al. (1989)	RCT	Clonidine 0.6 mg tid vs. chlormethiazole 2 mg tid	32	4	Chlormethiazole significantly better than clonidine; 8 patients on clonidine withdrawn from study because of worse withdrawal	All in clonidine group: 2 pts. – hallucinations 2 pts. – seizure 3 pts. – decreased BP 1 pt. – drowsiness
Clonidine	Baumgartner (1991)	RCT	Clonidine transdermal patch No. 2; patches on each arm vs. CDZ 150 mg day 1	50	4	Clonidine better than CDZ in withdrawal symptoms and vital signs	Clonidine less fatigue; CDZ less nausea, no DT or seizures
Lofexidine	Cushman et al. (1985)	RCT	Lofexidine 1.6 mg day 1 vs. placebo	63	4	Lofexidine superior to placebo on BP and tremor	Lofexidine patients had significantly more hypotension and dry mouth. Six placebo patients worsened
Lofexidine	Brunning et al. (1986)	Open trial	Lofexidine $\bar{x}$ = 1.5 mg day 1	28	6	Symptoms rapidly decreased over 36 h	One patient had seizure
Lofexidine	Cushman and Sowers (1989)	RCT	Lofexidine 1.6 or 2.4 mg day 1 vs. placebo	23	4	Lofexidine superior to placebo during first 48 h esp. on HR, BP, and tremor	Two lofexidine patients had hypotension

Table 2. *Continued*

Drug class	Author	Design	Comparison drugs	*N*	Duration (days)	Efficacy	Side effects
Beta-adrenergic blockers							
Timolol	Potter et al. (1984)	RCT	Timolol 10 mg bid vs. placebo	18	4	Timolol patients required less ancillary chlormethiazole	None reported
Atenolol	Kraus et al. (1985)	RCT	Atenolol 50–100 mg/day vs. placebo	120	9	Atenolol superior to placebo in normalization of VS, tremor and less use of ancillary oxazepam	None severe or drug specific
Atenolol	Horwitz et al. (1989)	RCT (outpatients)	Atenolol 50–100 mg/day vs. placebo	180	14	Atenolol superior to placebo in normalization of VS, dropout rate, and craving. Similar use of ancillary oxazepam	One atenolol patient had AW seizure. No other difference between groups
Anticonvulsants							
Carbamazepine (CARB)	Flygenring et al. (1984)	RCT	Carbamazepine 400–4200 mg total dose vs. barbital 250–5250 mg total dose	72	7	Both groups responded well. CARB = barbital	Eight patients in CARB group reported "dizziness." Six patients in barbital group had CNS side effects

Carbamazepine	MALCOLM et al. (1989)	RCT	Carbamazepine 200 mg qid vs. oxazepam 30 mg qid	63	7	Both groups responded well. CARB = oxazepam	CARB patients had less overall dysphoria or anger during withdrawal
Carbamazepine and Valproic acid (VAL)	HILLBOM et al. (1989)	RCT	Carbamazepine 600 bid vs. valproic acid 600 bid vs. placebo	138	4	Side effects in both groups limited efficacy analysis. One VAL, two CARB, and three placebo patients had AW seizure	Of CARB group, 56% terminated because of vertigo, nausea, vomiting, diplopia, and rash; 32% of VAL group terminated because of gastric distress, nausea, and vomiting
Carbamazepine	STUPPAECK et al. (1992)	RCT	Carbamazepine 200 mg qid X3D, bid X4D vs. oxazepam 120 mg X3D, 90 mg X4D	58	7	CARB = OZP at 5 days; CARB > OZP day 6–7; 2 oxazepam had DT; 1 oxazepam had seizure	Three CARB patients withdrew because of nausea and vomiting
Ca^{2+} Channel Blocker							
Caroverine (CAROV)	KOPPI et al. (1987)	RCT	Caroverine 120 mg/day vs. meprobamate 2400 mg/day	20	5	Both groups responded well. CAROV = meprobamate	Meprobamate caused more sedation than caroverin
Nimodipine (NIMOD)	BANGER et al. (1992)	RCT	Nimodine 60 mg X4 doses vs. placebo	32	3	Chlormethiazole prn doses equal in both groups. NIMOD = placebo	None severe reported; two seizures; one DT

Table 2. *Continued*

Drug class	Author	Design	Comparison drugs	N	Duration (days)	Efficacy	Side effects
Miscellaneous							
Magnesium sulfate	WILSON and VULCANO (1984)	RCT	Magnesium sulfate 2 g i.m. q6 h vs. placebo	100	5	Both groups needed a similar amount of chlordiazepoxide ($\bar{x} = 92 \pm 107$ mg on day 1) to control withdrawal symptoms. No group differences	Six patients developed DT and five patients developed AWS. No group differences
γ-Hydroxy-butyric acid (GHB)	GALLIMBERTI et al. (1989)	RCT	GHB = 50 mg/kg (one dose) vs. placebo	23	7 h RTC then 7 days open	GHB markedly better than placebo in relieving withdrawal symptoms	Self-limited dizziness reported by GHB-treated patients
Buspirone (BUS)	DOUGHERTY and GATES (1990)	Open trial	Buspirone 5 mg q4 h prn	118	6	All patients had an "effective" control of withdrawal symptoms. Eight patients needed ancillary clonidine for increased BP	Global staff and patient impression was that BUS was less sedating than historical experience with benzodiazepines

RCT, randomized control trial; VS, vital signs; BP, blood pressure; HR, heart rate.

alcoholic liver disease. It is argued that short-acting BDZs afford the physician greater control in the treatment of AW. If anything, the longer-acting drugs pose a danger of toxicity, while the shorter-active ones pose a danger of inadequate treatment.

Given their efficacy and safety, one might ask why drugs other than the BDZs are under investigation for the treatment of AW. There are, in fact, several reasons. First, critics of BDZ use for treatment of AW worry about the substitution of "solid alcohol" for liquid alcohol. They fear the continued reinforcement of substance use, especially one that allows the alcoholic to feel and behave as if he or she were using alcohol. The perpetuation of dependence on a brain depressant is clearly an issue that should be taken seriously since these drugs have been shown to have reinforcing effects and abuse potential in the recently detoxified alcoholic (JAFFE et al. 1983). Second, it is not completely clear if BDZs completely prevent the development of AW seizures, especially in patients who have experienced seizures in the past. Third, if the "kindling hypothesis" of AW is correct, it is not at all clear whether BDZs will inhibit or enhance the development of this process. From the clinical data that are available it does not appear that BDZs inhibit the development of kindling, while in animal studies they clearly do not block kindling as well as some of the anticonvulsants (e.g., carbamazepine). Fourth, there are multiple BDZ receptors in the brain that have different anatomical distributions and functions. BDZ compounds that are clinically available do not selectively bind to these different types of receptors. Hence, the therapeutic effects of these medications are accompanied by a number of unwanted effects. Theoretically, greater selectivity would enhance clinical efficiency. Therefore, the search for new pharmacotherapeutic agents for use in AW appears justified.

IV. Alpha- and Beta-Adrenergic Blockers

Considerable research has examined the utility of drugs that bind to brain and peripheral autonomic adrenergic receptors in the treatment of AW. This line of investigation has intrinsic appeal, since a number of studies (see the section on the biology of AW) have suggested that both norepinephrine and epinephrine are elevated in AW. Furthermore, the symptoms of AW, such as increased heart rate and blood pressure, are those typically associated with the stimulation of adrenergic systems. Hypothetically, drugs that either inhibit the release of norepinephrine by presynaptic agonist action (e.g., α_2-agonists such as clonidine and lofexidine) or those that block the postsynaptic binding of epinephrine (e.g., β-blockers such as propranolol and atenolol) should be useful in alleviating AW symptoms. Table 2 lists studies that have been reported in this area and their outcomes.

In general, research has supported the utility of α_2-agonist drugs in the treatment of withdrawal symptoms. Though these drugs have some sedative effects, these are not the primary basis of their activity. Rather, the majority of their utility is accounted for by the lowering of heart rate and blood pressure. Common side effects associated with their use include orthostatic hypotension, fatigue, and dry mouth. However, ROBINSON and associates (1989) reported that a number of patients being treated with clonidine experienced side effects serious enough to require discontinuation, including hallucinations, seizures, hypotension, and drowsiness. Particularly in patients with moderate-to-severe withdrawal, risk of AW seizures is of concern when α_2-adrenoceptor agonists are used as monotherapy.

The β-adrenergic blocker atenolol has been found to be useful in the treatment of inpatient and outpatient alcohol withdrawal symptoms. It too appears to be most effective in the reduction of cardiovascular symptoms. It is more difficult to evaluate the results of the few studies conducted with atenolol since ancillary oxazepam was also allowed (KRAUS et al. 1985; HORWITZ et al. 1989). It is noteworthy, however, that although in an outpatient study (HORWITZ et al. 1989) there was no significant difference in oxazepam use between the treatment groups, those patients treated with atenolol remained in treatment longer and experienced less craving than did placebo-treated patients. Nevertheless, long-term outpatient treatment with atenolol was no better than placebo in preventing relapse.

Taken together, it would appear that drugs that decrease adrenergic tone are useful in the treatment of mild to moderate withdrawal. These drugs may be most useful in an outpatient setting, where the abuse liability of BDZs is of more concern and where the severity of withdrawal is generally less. On the other hand, it would not appear wise to utilize these drugs alone when more severe withdrawal symptomatology is present or anticipated. This is especially true in patients that have previously experienced AW seizures or delirium tremens. Obviously, attention to the underlying cardiovascular health of the patients is important since these drugs could have grave consequences for some medically compromised individuals.

V. Anticonvulsants

Much of the work on the use of anticonvulsants, particularly carbamazepine, for AW was done in Scandinavia during the late 1970s. The topic was comprehensively reviewed by BUTLER and MESSIHA (1986). In the last 10 years there have been several well-controlled trials of carbamazepine for the treatment of AW (see Table 2).

There are a number of potential advantages to the use of anticonvulsants, for the treatment of AW. First, seizures are one of the more serious complications of AW. Consequently, antiseizure medications would be expected to decrease the probability of experiencing a seizure and carbamazepine has ben shown in animal studies to inhibit the development of AW seizures (CHU 1979). Second, these drugs have been shown to block kindling in brain cells. Third, anticonvulsants have no apparent abuse liability. Fourth, they have been found to be useful in the treatment of mood disorders, which share with AW such symptoms as depression, irritability, and anxiety.

Carbamazepine was shown by BUTLER and MESSIHA (1986) to be efficacious in the treatment of AW. Presently, it is widely used in Europe for that indication. More recently, a randomized, double-blind trial showed carbamazepine to be comparable to oxazepam both in safety and efficacy (MALCOLM et al. 1989). Carbamazepine (200 mg four times a day) was well tolerated. In addition, patients on carbamazepine appeared to be less angry and dysphoric than those treated with oxazepam, particularly during the latter part of the 7-day treatment period. These results have been replicated in a very similar study in Austria (STUPPAECK et al. 1992). In contrast, HILLBOM et al. (1989) found that side effects limited the usefulness of both carbamazepine and valproic acid, and 56% and 32% of the treatment groups respectively terminated the study early. The high rate of intolerable side effects may have been due to the high dosages used in this study. Carbamazepine and valproic acid were given in 600 mg doses twice daily. When the dose of carbamazepine was reduced to 400 mg twice a day there were significantly fewer side effects and the dropout rate was reduced to 17%. Clearly, if anticonvulsants are to be used for the treatment of AW, dosage must be adjusted to minimize adverse effects.

There have been no studies reported that specifically examine the effect of treatment with anticonvulsant on the incidence of AW seizures. In the studies described above (MALCOLM et al. 1989; STUPPAECK et al. 1992; HILLBOM et al. 1989), a total of three AW seizures were reported: one occurred in a patient treated with oxazepam (STUPPAECK et al. 1992) and two others occurred in patients within 4–5 h of receiving a first dose of carbamazepine (HILLBOM et al. 1989). Additional research is needed before firm conclusions can be drawn regarding the most effective treatment to prevent AW seizures.

Since both carbamazepine and valproic acid are metabolized by the liver and have been implicated in the causation of liver pathology, it would be prudent to use these drugs cautiously in alcoholics with significant liver damage.

VI. Miscellaneous Treatments

The calcium channel blocking drug caroverine was found to be less sedative but equal in efficacy to the sedative-hypnotic meprobamate (KOPPI et al. 1987). However, in a study whose findings may be limited because of poor design, the calcium antagonist nimodipine was found to be no better than placebo in the treatment of AW (BANGER et al. 1992). Based on preclinical efficacy and one positive clinical trial, calcium channel blockers would seem worthy of further clinical evaluation.

Though magnesium sulfate is used routinely with the expectation that it will prevent AW symptoms, particularly seizures, a randomized control trial suggests that there is no utility in this practice (WILSON and VULCANO 1984). Magnesium sulfate given intramuscularly on admission was no better than placebo injections in terms of subsequent chlordiazepoxide use or in the number of patients who experienced seizures or DT.

In an interesting early report, GALLIMBERTI et al. (1989) administered γ-hydroxybutyric acid (GHB), an analog of the brain neurotransmitter GABA, or placebo to patients and evaluated them over a 7-h period. They found a markedly better response to GHB with minimal side effects. The efficacy and side effect profile of this drug will need to be compared with other medications, including both anticonvulsant and BDZs.

A large open trial of the serotonin 5HT1a agonist buspirone suggested that it may be useful in the management of withdrawal symptoms, with perhaps less sedation that BDZs (DOUGHERTY and GATES 1990). If confirmed, buspirone's lack of both abuse potential and additive sedative effects with alcohol would make it particularly suitable for use in outpatient detoxification.

Finally, nitrous oxide gas inhalation has been found to be efficacious in the treatment of mild-to-moderate AW (GILLMAN and LICHTIGFELD 1990). The inhalation of nitrous oxide gas in concentrations up to 70% (i.e., an analgesic, but not anesthetic, dose) for up to 40 min was found to be both well tolerated and immediately efficacious in 7000 patients in South Africa. Though 50%–60% of patients respond to the administration of air or oxygen (placebo treatment), the administration of nitrous oxide was effective in up to 90% of patients who failed to respond to these treatments. If further study supports its utility, the use of nitrous oxide may have both practical and theoretical implications.

I. Treatment of Alcohol Withdrawal Seizures

The prevention and treatment of AW seizures remains controversial. In fact, the prevention of a first AW seizure in any given episode of AW may be qualitatively different than the prevention of subsequent seizures once the first has occurred. The main controversy surrounds the use of diphenylhydantoin (DPH), or dilantin, for both of these indications. The most useful evidence reported on this issue is summarized in Table 3.

The evidence suggests that if the population under treatment does not have a history of previous seizures, then BDZ treatment is all that is needed to prevent the majority of AW seizures (ROTHSTEIN 1973). In patients with a history of a seizure occurring in adult life, irrespective of whether it was an AW seizure, the combined use of DPH and a BDZ may help to prevent an initial AW seizure (SAMPLINER and IBER 1974). However, in a patient who has already experienced an AW seizure during the index episode, it appears that DPH will not prevent the occurrence of a subsequent seizure (ALLDREDGE et al. 1989; CHANCE 1991).

Table 3. Summary of studies done on the prevention and treatment of alcohol withdrawal (AW) seizures with diphenylhydantoin (DPH)

Authors	Design	Comparison drugs	N	Duration	Efficacy	Comments
Rothstein (1973)	Open random assignment	Chlordiazepoxide $\bar{x}$ = 360 mg day 1 vs. chlordiazepoxide plus DPH 200 mg bid	200	5 days	No AW seizure in either group	Mild DT in 4 CDZ alone patients and in 5 CDZ + DPH patients
Sampliner and Iber (1974)	RCT	Chlordiazepoxide x = 400 mg day 1 vs. chlordiazepoxide plus DPH 100 mg tid	136	5 days	No AW seizure in DPH + CDZ group; 11 AWS in CDZ alone group	All patients had a history of previous adult seizures
Alldredge et al. (1989)	RCT	DPH 1000 mg i.v./20 min vs. placebo infusion	90	12 h	13% in each group had an AW seizure	All patients had a history of previous AW seizure in index episode
Chance (1991)	RCT	DPH 15 mg/kg i.v./25 min vs. placebo infusion	55	6 h	21% in DPH group had an AW seizure: 19% in placebo group had an AW seizure	All patients had a history of previous AW seizure in index episode

RCT, randomized control trial; CDZ, chlordiazepoxide.

There are limited data on the utility of other anticonvulsants and BDZs in the prevention or treatment of AW seizures. Studies in the literature are generally not well controlled and consequently provide little guidance in clinical management. There are conflicting reports on the utility of carbamazepine (see BUTLER and MESSIHA 1986), while the use of phenobarbital has been advocated (YOUNG 1987). However, more studies are needed in this area, particularly focusing on the efficacy of BDZs (particularly those with the most anticonvulsant potential, such as clonazepam) as well as carbamazepine and valproic acid.

In addition, not much is known about the longitudinal outcome of patients who experience one or more AW seizures. As mentioned in an earlier section, there is some indication that patients with AW seizures are at higher risk for future withdrawal seizures and possibly other, more serious, neurologic sequelae such as earlier onset of dementia and organic psychotic symptoms.

J. Biologic Aspects of Alcohol Withdrawal in Man

Much of the research on the biologic and physiologic basis of alcohol withdrawal has been done in animals, particularly the rodent. These data have been reviewed in a previous section. Data from human studies, though not as extensive as animal data, have generally been congruent with the animal data. In fact, AW may be one area of alcohol research where animal models are particularly useful for understanding the effects of alcohol in the human.

The adrenergic neurotransmitter system has received the most extensive study during AW in humans, in large part because it is both heuristically appealing and the measurement technology is well developed. A number of studies have shown elevated levels of catecholamines in cerebrospinal fluid (CSF) during AW (FUJIMOTO et al. 1983; BORG et al. 1983; HAWLEY et al. 1981). HAWLEY and colleagues (1985) found that, during the early stages of withdrawal, alcoholics have higher CSF levels of MHPG, the major central metabolite of norepinephrine, than controls. More importantly, the level of MHPG was highly correlated with clinical signs and symptoms of AW (e.g., blood pressure, heart rate, tremor, and sweating). In addition, the higher the initial level of MHPG, the longer the patient experienced AW symptomatology. These investigators also found a normalization of CSF in MHPG levels following recovery from AW. Taken together, these data offer strong support to the notion that a perturbation in the noradrenergic system is responsible for some of the symptomatology observed in AW. Convergent validity for this concept comes from the studies that have used α_2-adrenergic agonists (such as clonidine) in the treatment of this condition. In these studies (see Table 2), the clinical signs and symptoms most responsive to these compounds were those (normalization of blood pressure, heart rate, tremor, sweating) that were correlated with noradrenergic activity in the study by HAWLEY and colleagues (1981). Whereas noradrenergic neurotransmission has clearly been implicated in the pathophysiology of AW, many other systems serve to regulate noradrenergic output, including the GABAergic, opioid, serotonergic, NMDA, and CRF systems. Animal studies have shown that most of these systems are affected by alcohol consumption and its abrupt cessation after chronic use (see section on pathophysiology).

Another area of interest in relationship to AW in humans is the hypothalamic-pituitary-adrenal (HPA) axis. Cortisol secretion is increased in the period immediately following the cessation of chronic alcohol consumption (ADINOFF et al. 1991; ANTON et al. 1985). It also appears that the perturbation in the HPA axis is present for at least 3 weeks after the cessation of drinking as indicated by decreased release of ACTH by the pituitary following administration of corticotropin-releasing factor (CRF). Abnormalities in the HPA axis may be evident in some people for as long as 6 months following AW (ADINOFF et al. 1990). It has been hypothesized that the blunted ACTH response to CRF stimulation may be caused by higher than normal release of CRF during the withdrawal period. CRF is a neuropeptide that may influence noradrenergic and perhaps other neurotransmitter activity. It has also been implicated in the development of kindling (see GLUE and NUTT 1990).

The HPA axis may also be involved in the pathophysiology of alcohol-related cognitive impairment. Cortisol and other glucocorticoids have been shown to damage hippocampal cells when present in high levels over a prolonged period of time. Glucocorticoids have been shown to increase susceptibility of hippocampal cells to excitatory amino acid exposure (PACKAN and SAPOLSKY 1990). The increase in excitatory amino acids, especially NMDA, combined with hypercortisolemia, may have a significant negative impact on hippocampal cells during AW. If the kindling mechanism is also operative (as potentially mediated by CRF and the excitatory amino acids), the initial subclinical impact could eventually lead to major cellular destruction during repeated withdrawal episodes. The clinical impact could either take the form of memory loss, limbic system dysregulation (anxiety, panic, depression, irritability), or the production of seizure-like discharges. In relation to the last point, it has been reported that mice who were genetically inbred for susceptibility to AW seizures are sensitized by glucocorticoids to experience greater withdrawal seizures while, in contrast, if glucocorticoid synthesis is blocked, neuronal excitability is diminished (ROBERTS et al. 1991).

In conclusion, AW is a complex disorder involving perturbations in number of neurotransmitter and neuroendocrine systems. Ultimately, it may be best to think of the AW syndrome as a pathophysiologic cascade, similar to those associated with inflammation, and other acute pathologic states. The removal of alcohol from a dependent animal, as the initial event, sets in motion a variety of subsequent processes that affect numerous tissues, organs, and systems, sometimes with an amplification that is logarithmic. The goal of treatment in these conditions, in most instances, is the interruption of the pathophysiologic progression at one, or perhaps several, entry points. This model explains why many different drugs, working on many different receptors and neurotransmitter systems, serve to ameliorate the AWS. The earlier the pathophysiologic cascade can be interrupted the better the outcome. The prompt initiation of treatment during an episode of AW may be expected to have beneficial longer-term effects as well. By interrupting the cascade of events associated with acute AW, it is possible that the chronic effects of alcohol dependence, including cognitive decline, can also be avoided.

References

Aaronson LM, Hinman DJ, Okamoto M (1984) Effects of diazepam on ethanol withdrawal. J Pharmacol Exp Ther 221:319–325
Adamson J, Burdick JA (1973) Sleep of dry alcoholics. Arch Gen Psychiatry 28:146–149
Adinoff B, Majchorwicz E, Martin PR, Linnoila M (1986) The benzodiazepine antagonist RO15-1788 does not antagonize the ethanol withdrawal syndrome. Biol Psychiatry 21:643–649
Adinoff B, Martin PR, Bone GHA, Eckardt MJ, Roehrich L, George DT, Moss HB, Eskay R, Linnoila M, Gold PW (1990) Hypothalamic-pituitary-adrenal axis functioning and cerebrospinal fluid corticotropin releasing hormone and corticotropin levels in alcoholics after recent and long-term abstinence. Arch Gen Psychiatry 47:325–330
Adinoff B, Risher-Flowers D, De Jong J, Ravitz B, Bone GHA, Nutt DJ, Roehrich L, Martin PR, Linnoila M (1991) Disturbances of hypothalamic-pituitary-adrenal axis functioning during ethanol withdrawal in six men. Am J Psychiatry 148: 1023–1025
Avdaloff W (1979) Alcoholism, seizures and cerebral atrophy. Adv Biol Psychiatry 3:20–32
Allan AM, Harris RA (1987a) Involvement of neuronal chloride channels in ethanol intoxication, tolerance and dependence. In: Galanter M (ed) Recent developments in alcoholism. Plenum, New York, p 313
Allan AM, Harris RA (1987b) Acute and chronic ethanol treatments alter GABA receptor-operated chloride channels. Pharmacol Biochem Behav 27:665–670
Alldredge BK, Lowenstein DH, Simon RP (1989) Placebo-controlled trial of intravenous diphenylhydantoin for short-term treatment of alcohol withdrawal seizures. Am J Med 87:645–648
Alling C, Balldin J, Bokstrom K, Gottfries CG, Karlsson I, Langstrom G (1982) Studies on duration of a late recovery period after chronic abuse of ethanol. Acta Psychiatr Sand 66:384–397
Alonso G, Szafarczyk A, Balmefrezol M, Assenmacher I (1986) Immunocytochemical evidence for stimulatory control by the ventral noradrenergic bundle of parvocellular neurons of the paraventricular nucleus secreting corticotropin releasing hormone and vasopressin in rats. Brain Res 397:297–307
Anantharam V, Treistman SN (1992) Effects of ethanol on neuronal voltage-gated ion channels. In: Watson RR (ed) Alcohol and neurobiology: receptors, membranes, and channels. CRC Press, Boca Raton, p 13

Annuziato L, Amoroso S, Di Renzo G, Argenzio F, Aurilio C, Grella A, Quattrone A (1983) Increased GH responsiveness to dopamine receptor stimulation in alcohol addicts during the late withdrawal syndrome. Life Sci 33:2651–2655

Anton RF (1992) Buspirone in the treatment of anxious alcoholics with social phobia. Presented at the Research Society of Alcoholism meeting, San Diego, CA

Anton RF, Malcolm R, Thevos A (1985) Urinary free cortisol and clinical correlates during alcohol withdrawal. Alcohol Clin Exp Res 9(2):205

Araneda R, Bustos G (1989) Modulation of dendritic release of dopamine by N-methyl-D-aspartate receptors in rat substantia nigra. J Neurochem 52:962–970

Baker TB, Cannon DS (1979) Potentiation of ethanol withdrawal by prior dependence. Psychopharmacology 60:105–110

Balldin J, Alling C, Gottfris CG, Lindstedt G, Langstrom G (1985) Changes in dopamine receptor sensitivity in humans after heavy alcohol intake. Psychopharmacology 86:142–146

Ballenger JC, Post RM (1978) Kindling as a model for alcohol withdrawal syndromes. Br J Psychiatry 133:1–14

Banger M, Benkert O, Roschke J, Herth T, Hebenstreit M, Philipp M, Aldenhoff JB (1992) Nimodipine in acute alcohol withdrawal state. J Psychiatr Res 26:117–123

Baumgartner GR, Rowen RC (1987) Clonidine vs chlordiazepoxide in the management of acute alcohol withdrawal syndrome. Arch Intern Med 147:1223–1226

Baumgartner GR, Rowen RC (1991) Transdermal clonidine versus chlordiazepoxide in alcohol withdrawal: A randomized, controlled clinical trial. South Med J:84:312–321

Becker HC (1994) Positive relationship between the number of prior ethanol withdrawel episodes and the severity of subsequent withdrawal seizures. Psychopharmacology (in press)

Becker HC, Anton RF (1989) The benzodiazepine inverse agonist RO15-4513 exacerbates, but does not precipitate, ethanol withdrawal in mice. Pharmacol Biochem Behav 32:162–167

Becker HC, Hale RL (1992) Further characterization of an animal model of ethanol withdrawal "kindling". Soc Neurosci Abstr 18:1073

Becker HC, Hale RL (1993) Repeated episodes of ethanol withdrawal potentiate the severity of subsequent withdrawal seizures: an animal model of alcohol withdrawal "kindling". Alcohol Clin Exp Res 17:94–98

Begleiter H, Platz A (1972) The effects of alcohol on the central nervous system in humans. In: Kissin B, Begleiter H (eds) The biology of alcoholism, vol 2. Plenum, New York, p 293

Begleiter H, Porjesz B (1979) Persistence of a "subacute withdrawal syndrome" following chronic ethanol intake. Drug Alcohol Depend 4:353–357

Benzer DG (1990) Quantification of the alcohol withdrawal syndrome in 487 alcoholic patients. J Subst Abuse Treat 7:117–123

Bjorkquist SE (1975) Clonidine in alcohol withdrawal. Acta Psychiatr Scand 52:256–263

Blum K, Briggs AH, De Lallo L (1983) Clonidine enhancement of ethanol withdrawal in mice. Subst Alcohol Actions Misuse 4:59–63

Bode DC, Molinoff PB (1988) Effects of ethanol in vitro on the beta adrenergic receptor-coupled adenylate cyclase system. J Pharmacol Exp Ther 246:1040–1047

Bogin TM, Nostrant TT, Young MI (1987) Propranolol for the treatment of the alcoholic hangover. Am J Drug Alcohol Abuse 13:175–180

Bokstrom K, Balldin J, Langstrom G (1989) Alcohol withdrawal and mood. Acta Psychiatr Scand 80:505–513

Bond NW (1979) Impairment of shuttlebox avoidance learning following repeated alcohol withdrawal episodes in rats. Pharmacol Biochem Behav 11:589–591

Borg S, Kvande H, Sedvall G (1981) Central norepinephrine metabolism during alcohol intoxication in addicts and healthy volunteers. Science 213:1135–1137
Borg S, Czarnecka A, Kvande H, Mossberg D, Sedvall G (1983) Clinical conditions and concentrations of MOPEG in the cerebrospinal fluid and urine of male alcoholic patients during withdrawal. Alcoholism 7:411–415
Branchey M, Rauscher G, Kissin B (1971) Modifications in the response to alcohol following the establishment of physical dependence. Psychopharmacologia 22:314–322
Brennan CH, Littleton JM (1990) Second messengers involved in genetic regulation of the number of calcium channels in bovine adrenal chromaffin cells in culture. Neuropharmacology 29:689–693
Brennan CH, Lewis A, Littleton LM (1989) Membrane receptors involved in up-regulation of calcium channels in bovine adrenal chromaffin cells chronically exposed to ethanol. Neuropharmacology 28:1303–1307
Brennan CH, Crabbe JC, Littleton JM (1990) Genetic regulation of dihydropyridine-sensitive calcium channels in brain may determine susceptibility to physical dependence on alcohol. Neuropharmacology 29:429–432
Brown ME, Anton RF, Malcolm R, Ballenger JC (1988) Alcohol detoxification and withdrawal seizures: clinical support for a kindling hypothesis. Biol Psychiatry 23:507–514
Brunning J, Mumford JP, Keaney FP (1986) Lofexidine in alcohol withdrawal states. Alcohol Alcohol 21:167–170
Buck KJ, Harris RA (1990) Benzodiazepine agonist and inverse agonist actions on GABA-A receptor-operated chloride channels. II. Chronic effects of ethanol. J Pharmacol Exp Ther 253:713–719
Butler D, Messiha FS (1986) Alcohol withdrawal and carbamazepine. Alcohol 3:113–129
Carlsson C, Johansson T (1971) The psychological effects of propanolol in the abstinence phase of chronic alcoholism. Br J Psychiatry 119:605–606
Carrington CD, Ellinwood EH, Krishnan RR (1984) Effects of single and repeated alcohol withdrawal on kindling. Biol Psychiatry 19:525–537
Castaneda R, Cushman P (1989) Alcohol withdrawal: a review of clinical management. J Clin Psychiatry 50:278–284
Chan AWK (1985) Alcoholism and epilepsy. Epilepsia 26:323–333
Chan AWK, Langan MC, Leong FW, Penetrante ML, Schanely DL, Aldrich-Castanik L (1986) Substitution of chlordiazepoxide for ethanol in alcohol-dependent mice. Alcohol 3:309–316
Chan AWK, Langan MC, Leong FW, Schanley DL, Penetrante ML (1988) Does chronic ethanol intake confer full cross-tolerance to chlordiazepoxide? Pharmacol Biochem Behav 30:385–389
Chance JF (1991) Emergency department treatment of alcohol withdrawal seizures with phenytoin. Ann Emerg Med 20:520–522
Chu N-S (1979) Carbamazepine: prevention of alcohol withdrawal seizures. Neurology 29:1397–1401
Clemmesen L, Hemmingsen R (1984) Physical dependence on ethanol during multiple intoxication and withdrawal episodes in the rat: evidence of a potentiation. Acta Pharmacol Toxicol 55:345–350
Clemmesen L, Ingvar M, Hemmingsen R, Bolwig TG (1988) Local cerebral glucose consumption during ethanol withdrawal in the rat: effects of single and multiple episodes and previous convulsive seizures. Brain Res 453:204–214
Coffman JA, Petty F (1985) Plasma GABA levels in chronic alcoholics. Am J Psychiatry 142:1204–1205
Collingridge GL, Bliss TVP (1987) NMDA receptors – their role in long-term potentiation. Trends Neurosci 10:288–293
Collingridge GL, Davies SN (1989) NMDA receptors and long-term potentiation in the hippocampus. In: Watkins JC, Collingridge GL (eds) The NMDA receptor. Oxford University Press, Oxford, p 123

Cooper BR, Viik K, Ferris RM, White HL (1979) Antagonism of the enhanced susceptibility to audiogenic seizures during alcohol withdrawal in the rat by γ-aminobutyric acid (GABA) and "GABAmimetic" agents. J Pharmacol Exp Ther 209:396–403
Criswell HE, Breese GR (1989) A conflict procedure not requiring deprivation: evidence that chronic ethanol treatment induces tolerance to the anticonflict action of ethanol and chlordiazepoxide. Alcohol Clin Exp Res 13:680–685
Cushman P, Sowers JR (1989) Alcohol withdrawal syndrome: clinical and hormonal responses to α_2-adrenergic agonist treatment. Alcohol Clin Exp Res 13:361–364
Cushman P, Forbes R, Lerner W, Stewart M (1985) Alcohol withdrawal syndromes: clinical management with lofexidine. Alcohol Clin Exp Res 9:103–108
Daniell LC (1992) Ethanol effects on central N-methyl-D-aspartate receptors. In: Watson, RR (ed) Alcohol and neurobiology: receptors, membranes, and channels. CRC Press, Boca Raton, p 13
Diamond I, Wrubel B, Estrin W, Gordon AS (1987) Basal and adenosine receptor-stimulated levels of cAMP are reduced in lymphocytes from alcoholic patients. Proc Natl Acad Sci USA 84:1413–1416
Diamond I, Mochly-Rosen D, Gordon AS (1990) Reduced adenosine receptor activation in alcoholism: implications for alcohol withdrawal seizures. In: Porter RJ, Mattson RH, Cramer JA, Diamond I (eds) Alcohol and seizures: basic mechanisms and clinical concepts. Davis, Philadelphia, p 79
Dolin SI, Little HJ, Hudspith M, Pagonis C, Littleton JM (1987) Increased dihydropyridine-sensitive calcium channels in rat brain may underlie ethanol physical dependence. Neuropharmacology 26:275–279
Dorrow R, Horowski R, Paschelke G, Amin M, Braestrup C (1983) Severe anxiety induced by FG 7142, a β-carboline ligand for the benzodiazepine receptor. Lancet ii:98–99
Dorus W, Ostrow DG, Anton RF, Cushman P, Collins JF, Schaefer M, Charles HL, Desai P, Hayashida M, Malkerneker U, Willenbring M, Fiscella R, Sather MR (1989) Evaluation of lithium carbonate treatment of depressed and non-depressed alcoholics in a double-blind placebo controlled study. JAMA 262(12):1646–1652
Doughty RJ, Gates RR (1990) The role of buspirone in the management of alcohol withdrawal: a preliminary investigation. J Subst Abuse Treat 7:189–192
Earnest MP, Yarnell PR (1976) Seizure admissions to a city hospital: the role of alcohol. Epilepsia 17:387–393
Ehlers CL, Henriksen S, Wang M, Rivier J, Vale W, Bloom FE (1983) Corticotropin-releasing factor increases brain excitability and convulsive seizures in the rat. Brain Res 278:332–336
Ellis FW, Pick JR (1970) Experimentally induced ethanol dependence in rhesus monkeys. J Pharmacol Exp Ther 175:88–93
Engel J, Liljequist S (1976) The effect of long-term ethanol treatment on the sensitivity of the dopamine receptors in the nucleus accumbens. Psychopharmacol 49:253–257
Fadda F, Mosca E, Meloni R, Gessa GL (1985) Suppression by progabide of ethanol withdrawal syndrome in rats. Eur J Pharmacol 109:321–325
Feller DJ, Tso-Olivas DY, Savage DD (1991) Hippocampal mossy fiber zinc deficit in mice genetically selected for ethanol withdrawal seizure susceptibility. Brain Res 545:73–79
Ferrero PA, Vaccarino F, Alho H, Mellstrom B, Costa E, Guidotti A (1986) Study of an octadecaneuropeptide derived from diazepam binding inhibitor (DBI): biological activity and presence in rat brain. Proc Natl Acad Sci USA 83:827–831
File SE, Baldwin HA, Hitchcott PK (1989) Flumazenil but not nitrendipine reverse the increased anxiety during ethanol withdrawal in the rat. Psychopharmacology 98:262–264
Flink EB (1986) Magnesium deficiency in alcoholism. Alcohol Clin Exp Res 10:590–594

Flygenring J, Hansen J, Holst B, Petersen E, Sorensen A (1984) Treatment of alcohol withdrawal symptoms in hospitalized patients. Acta Psychiatr Scand 69:398–408
Foy A, March S, Drinkwater V (1988) Use of an objective clinical scale in the assessment and management of alcohol withdrawal in a large general hospital. Alcohol Clin Exp Res 12:360–364
Freund G (1971) Alcohol, barbiturate, and bromide withdrawal in mice. In: Mello N, Mendelson J (eds) Recent advances in studies of alcoholism. US Government Printing Office, Washington DC, p 453
Freund G, Ballinger WE (1988) Decrease of benzodiazepine receptors in frontal cortex of alcoholics. Alcohol 5:275–282
Frezza M, DiPadova C, Pozzato G, Terpin M, Baraona E, Lieber CS (1990) High blood alcohol levels in women: the role of decreased gastric alcohol dehydrogenase activity and first-pass metabolism. N Engl J Med 322:95–99
Frye GD (1990) Gamma-aminobutyric acid (GABA) changes in alcohol withdrawal. In: Porter RJ, Mattson RH, Cramer JA, Diamond I (eds) Alcohol and seizures: basic mechanisms and clinical cocepts. Davis, Philadelphia, p 87
Frye GD, McCown T, Breese G (1983) Differential sensitivity of ethanol withdrawal signs in the rat to gamma-aminobutyric acid (GABA) mimetics: blockade of audiogenic seizures but not forelimb tremors. J Pharmacol Exp Ther 226:720–725
Frye GD, McCown T, Breese GR, Petersen SL (1986) GABAergic modulation of inferior colliculus excitability: role in the ethanol withdrawal audiogenic seizures. J Pharmacol Exp Ther 237:478–485
Frye GD, Matthew J, Trzeciakowski JP (1991) Effect of ethanol dependence on GABA-A antagonist-induced seizures and agonist-stimulated chloride uptake. Alcohol 8:453–459
Fujimoto A, Nagao T, Ebara T, Sato M, Otsuki S (1983) Cerebrospinal fluid monoamine metabolites during alcohol withdrawal syndrome and recovered state. Biol Psychiatry 18:1141–1152
Gallimberti L, Gentile N, Cibin M, Fadda F, Canton G, Ferri M, Ferrara SD, Gessa GL (1989) Gamma-hydroxybutyric acid for treatment of alcohol withdrawal syndrome. Lancet 30:787–789
George DT, Nutt DJ, Dwyer BA, Linnnoila M (1990) Alcoholism and panic disorder: is it comorbidity more than coincidence? Acta Psychiatr Scand 81: 97–107
Gillin JC, Smith TL, Irwin M, Kripke DF, Schuckit M (1990) EEG sleep studies in "pure" primary alcoholism during subacute withdrawal: relationships to normal controls, age and other clinical variables. Biol Psychiatry 27:477–488
Gillman MA, Lichtigfeld FJ (1990) Analgesic nitrous oxide for alcohol withdrawal: a critical appraisal after 10 years' use. Postgrad Med J 66:543–546
Glenn SW, Parsons OA, Sinha R, Stevens L (1988) The effects of repeated withdrawals from alcohol on the memory of male and female alcoholics. Alcohol Alcohol 23:337–342
Glue P, Nutt DJ (1990) Overexcitement and disinhibition: dynamic neurotransmitter interactions in alcohol withdrawal. Br J Psychiatry 157:491–499
Glue P, Sellman JD, Joyce PR, Nicholls MG, Nutt DJ (1988) The hypothermic response to clonidine is absent in alcohol withdrawal but returns in abstinence. Biol Psychiatry 24:102–104
Glue P, Sellman JD, Nicholl MG, Abbott R, Joyce PR, Nutt DJ (1989) Studies of alpha-2-adrenoceptor function in abstinent alcoholics. Br J Addict 84:97–102
Goddard GV, McIntyre DC, Leech CK (1969) A permanent change in brain function resulting from daily electrical stimulation. Exp Neurol 25:295–330
Goldman GD, Volicier L, Gold BI, Roth RH (1981) Cerebrospinal fluid GABA and cyclic nucleotides in alcoholics with and without seizures. Alcohol Clin Exp Res 5:431–434
Goldstein DB (1973) Alcohol withdrawal reactions in mice: effects of drugs that modify neurotransmission. J Pharmacol Exp Ther 186:1–9

Gonzalez LP, Colbourn DL, Czachura JF (1988) Apomorphine-induced hypothermia increases during ethanol withdrawal. Alcohol 5:107–109

Gordon AS, Collier K, Diamond I (1986) Ethanol regulation of adenosine receptor-stimulated cAMP levels in a clonal neural cell line: an in vitro model of cellular tolerance to ethanol. Proc Natl Acad Sci USA 83:2105–2108

Gorelick DA, Wilkins JN (1986) Special aspects of human alcohol withdrawal. In: Galanter M (ed) Recent development in alcoholism, vol 4. Plenum, New York, pp 283–305

Grant KA, Valverius P, Hudspith M, Tabakoff B (1990) Ethanol withdrawal seizures and the NMDA receptor complex. Eur J Pharmacol 176:289–296

Grant KA, Snell LD, Rogawski MA, Thurkauf A, Tabakoff B (1992) Comparison of the effects of the uncompetitive N-methyl-D-aspartate antagonist (±)-5-aminocarbonyl-10,11-dihydro-5H-dibenzo[a,d]cyclohepten-5,10-imine (ADCL) with its structural analogs dizocilpine (MK-801) and carbamazepine on ethanol withdrawal seizures. J Pharmacol Exp Ther 260:1017–1022

Greenberg DA (1987) Calcium channels and calcium channel antagonists. Ann Neurol 21:317

Greenberg DA, Chan J (1992) Ethanol, calcium channels, and intracellular calcium in cultured neural cells. In: Watson RR (ed) Alcohol and neurobiology: receptors, membranes, and channels. CRC Press, Boca Raton, p 141

Greenberg DA, Carpenter CL, Messing RO (1987) Ethanol-induced component of $^{45}Ca^{2+}$ uptake in PC12 cells is sensitive to Ca^{2+} channel modulating drugs. Brain Res 410:143–146

Greenberg DA, Messing RO, Marks SS, Carpenter CL (1990) Calcium channel changes during alcohol withdrawal. In: Poter RJ, Mattson RH, Cramer JA, Diamond I (eds) Alcohol and seizures: basic mechanisms and clinical concepts. Davis, Philadelphia, p 60

Gross MM, Rosenblatt SM, Lewis E, Chartoff S, Malenowski B (1972) Acute alcoholic psychoses and related syndromes: psychosocial and clinical characteristics and their implications. Br J Addict 67:15–31

Gross M, Leis E, Nagareijan M (1973) An improved quantitative system for assessing the acute alcohol psychoses and related states (TSA and SSA). In: Gross MM (ed) Alcohol intoxication and withdrawal experimental studies, V-35. Plenum, New York, pp 365–376

Guidotti A, Forchetti CM, Corda MG, Kondel D, Bennett CD, Costa E (1983) Isolation, characterization and purification to homogeneity of an endogenous polypeptide with agonistic action on benzodiazepine receptors. Proc Natl Acad Sci USA 80:3531–3533

Gulya K, Grant KA, Valvarius P, Hoffman PL, Tabakoff B (1991) Brain regional specificity and time course of changes in the NMDA receptor-ionophore complex during ethanol withdrawal. Brain Res 547:129–134

Guppy LJ, Littleton JM (1987) Increased [^{3}H]dihydropyridine binding sites in brain, heart, and smooth muscle of ethanol dependent rats. Br J Pharmacol 92: 662P

Harper JC, Littleton JM (1987) Putative alcohol dependence in adrenal cell cultures: relation to calcium channel activity. Br J Pharmacol 92:661P

Hawley RJ, Major LF, Schulman EA, Lake CR (1981) CSF levels of norepinephrine during alcohol withdrawal. Arch Neurol 38:289–292

Hawley RJ, Major LF, Schulman E, Linnoila M (1985) Cerebrospinal fluid 3-methoxy-4-hydroxyphenylglycol and norepinephrine levels in alcohol withdrawal. Arch Gen Psychiatry 42:1056–1062

Hawthorn MH (1992) Alcohol and the voltage-dependent calcium channels. In: Watson RR (ed) Alcohol and neurobiology: receptors, membranes, and channels. CRC Press, Boca Raton, p 109

Hayne CH, Louks JL (1991) Dysphoria in male alcoholics with a history of hallucinations. J Nerv Ment Dis 179:415–419

Hemmingsen R, Clemmesen L, Barry DI (1984) Blind study of the effect of the alpha adrenergic agonists clonidine and lofexidine on alcohol withdrawal in the rat. J Stud Alcohol 45:310–315
Herron CE, Williamson R, Collingridge GL (1985) A selective N-methyl-D-aspartate antagonist depresses epileptiform activity in rat hippocampal slices. Neurosci Lett 61:255–260
Hershon HI (1977) Alcohol withdrawal symptoms and drinking behavior. J Stud Alcohol 38:953–971
Hillbom ME, Hjelm-Jäger M (1984) Should alcohol withdrawal seizures be treated with anti-epileptic drugs? Acta Neurol Scand 69:39–42
Hillbom M, Tokola R, Kuusela V, Karkkainen P, Kalli-Lemma L, Pilke A, Kaste M (1989) Prevention of alcohol withdrawal seizures with carbamazepine and valproic acid. Alcohol 6:223–226
Hillbom ME (1980) Occurrence of cerebral seizures provoked by alcohol abuse. Epilepsia 21:459–466
Hillbom ME (1990) Alcohol withdrawal seizures and binge versus chronic drinking. In: Porter RJ, Mattson RH, Cramer JA, Diamond I (eds) Alcohol and seizures: basic mechanisms and clinical concepts. Davis Company, Philadelphia, p 206
Hoffman PL, Tabakoff B (1977) Alterations in dopamine receptor sensitivity by chronic ethanol treatment. Nature 268:551–553
Hoffman PL, Tabakoff B (1991) The contribution of voltage-gated and NMDA receptor-gated calcium channels to ethanol withdrawal seizures. Alcohol Alcohol [Suppl 1]:171–175
Horwitz RI, Gottlieb LD, Kraus ML (1989) The efficacy of atenolol in the outpatient management of the alcohol withdrawal syndrome. Arch Intern Med 149:1089–1093
Hunt WA (1981) Neurotransmitter function in the basal ganglia after acute and chronic ethanol treatment. Fed Proc 40:2077–2081
Hunt WA (1983) The effect of ethanol on GABAergic transmission. Neurosci Biobehav Rev 7:87–95
Huss M (1852) Chronische alcohol – Krankheit. Fritze, Stockholm
Idemudia SO, Bhadra S, Lal H (1989) The pentylenetetrazol-like interoceptive stimulus produced by ethanol withdrawal is potentiated by bicuculline and picrotoxin. Neuropsychopharmacology 2:115–122
Isbell H, Fraser HF, Wikler A, Belleville RE, Eisenman AJ (1955) An experimental study of the etiology of "rum fits" and delirium tremens. Quant J Stud Alcohol 16:1–33
Jaffe JH, Ciraulo DA, Nies A, Dixon RB, Monroe LL (1983) Abuse potential of halazepam and of diazepam in patients recently treated for acute alcohol withdrawal. Clin Pharmacol Ther 34:623–630
Johnston AL, Thevos AK, Randall CL, Anton RF (1991) Increased severity of alcohol withdrawal in in-patient alcoholics with co-existing anxiety diagnosis. Br J Addict 86:719–725
Kalant H, LeBlanc AE, Gibbins RJ (1971) Tolerance to, and dependence on, some nonopitae psychotropic drugs. Pharmacol Rev 23:135–191
Kitazawa S, Shioda S, Nakai Y (1987) Catecholaminergic innervation of neurons containing corticotropin-releasing factor in the paraventricular nucleus of the rat hypothalamus. Acta Anat 129:337–343
Koob GF, Bloom FE (1985) Corticotropin-releasing factor and behavior. Fed Proc 44:259–263
Koppi S, Eberhardt G, Haller R, Konig P (1987) Calcium-channel-blocking agent in the treatment of acute alcohol withdrawal-caroverine versus meprobamate in a randomized double-blind study. Neuropsychobiology 17:49–52
Kramp P, Rafaelson OJ (1978) Delirium tremens: a double-blind comparison of diazepam and barbital treatment. Acta Psychiatr Scand 58:174–190

Kranzler HR, Meyer RE (1989) An open trial of buspirone with alcoholics. J Clin Psychopharmacol 9:379–380
Kranzler HR, Orrok B (1989) The pharmacotherapy of alcoholism. In: Tasman A, Hales RE, Frances AJ (eds) Review of psychiatry, vol 8. American Psychiatric Association Washington DC, pp 359–379
Kraus ML, Gottlieb LD, Horwitz RI, Anscher M (1985) Randomized clinical trial of atenolol in patients with alcohol withdrawal. N Engl J Med 313:905–909
Kril JJ, Gundlach AL, Dodd PR, Johnston GAR, Harper CG (1989) Cortical dihydropyridine binding sites are unaltered in human alcoholic brain. Ann Neurol 26:395–397
Kulonen E (1983) Ethanol and GABA. Med Biol 61:147–157
Kushner MG, Sher KJ, Beitman BD (1990) The relationship between alcohol problems and the anxiety disorders. Am J Psychiatry 6:685–695
Lal H, Harris CM, Benjamin D, Springfield AC, Bhadra S, Emmet-Oglesby MW (1988) Characterization of a pentylenetetrazol-like interoceptive stimulus produced by ethanol withdrawal. J Pharmacol Exp Ther 247:508–518
Lalies M, Middlemiss DN, Ransom R (1988) Stereoselective antagonism of NMDA-stimulated noradrenaline release from rat hippocampal slices by MK-801. Neurosci Lett 91:339–342
Le AD, Khanna JM, Kalant H, Grossi F (1986) Tolerance to and cross-tolerance among ethanol, pentobarbital and chlordiazepoxide. Pharmacol Biochem Behav 24:93–98
Lechtenberg R, Worner TM (1991) Relative kindling effect of detoxification and non-detoxification admissions in alcoholics. Alcohol Alcohol 26:221–225
Leslie SW, Brown LM, Dildy JE, Sims JS (1990) Ethanol and neuronal calcium channels. Alcohol 7:233–236
Liljequist S (1978) Changes in the sensitivity of dopamine receptors in the nucleus accumbens and in the striatum induced by chronic ethanol administration. Acta Pharmacol 43:19–28
Liljequist S (1991) The competitive NMDA receptor antagonist, CGP 39551, inhibits ethanol withdrawal seizures. Eur J Pharmacol 192:197–198
Linnoila M, Mefford I, Nutt DJ (1987) Alcohol withdrawal and noradrenergic function. Ann Intern Med 108:875–889
Liskow BI, Goodwin DW (1987) Pharmacological treatment of alcohol intervention, withdrawal and dependence: a critical review. J Stud Alcohol 48:356–370
Liskow BI, Rinck C, Campbell J, DeSouza C (1989) Alcohol withdrawal in the elderly. J Stud Alcohol 50:414–421
Lister RG, Karanian JW (1987) RO15-4513 induces seizures in DBA/2 mice undergoing alcohol withdrawal. Alcohol 4:409–411
Little HJ, Taylor SC, Nutt DJ, Cowen PJ (1985) The benzodiazepine antagonist RO15-1788 does not decrease ethanol withdrawal convulsions in rats. Eur J Pharmacol 107:375–377
Little HJ, Dolin SJ, Halsey MJ (1986) Calcium channel antagonists decrease the ethanol withdrawal syndrome. Life Sci 39:2059–2065
Littleton JM (1989) Alcohol intoxication and physical dependence: a molecular mystery tour. Br J Addict 84:267–276
Littleton JM (1990) Calcium channel activity in alcohol dependency and withdrawal seizures. In: Porter RJ, Mattson RH, Cramer JA, Diamond I (eds) Alcohol and seizures: basic mechanisms and clinical concepts. Davis, Philadelphia, p 51
Littleton JM, Little HJ, Whittington MA (1990) Effects of dihydropyridine calcium channel antagonsits in ethanol withdrawal: doses required, stereospecificity and actions of BAY K8644. Psychopharmacology 100:387–392
Lynch MA, Littleton JM (1983) Possible association of alcohol tolerance with increased synaptic Ca^{2+} sensitivity. Nature 303:175–176

Maier DM, Pohorecky LA (1989) The effect of repeated withdrawal episodes on subsequent withdrawal severity in ethanol-treated rats. Drug Alcohol Depend 23:103–110
Maki T, Heikkonen E, Harkonen T, Kontula K, Harkonen M, Ylikahri R (1990) Reduction of lymphocyte β-adrenoceptor level in chronic alcoholics and rapid reversal after ethanol whthdrawal. Eur J Clin Invest 20:313–316
Malcolm R, Anton RF, Randall CL, Johnston A, Brady K, Thevos A (1992) A placebo-controlled trial of buspirone in anxious inpatients alcoholics. Alcohol Clin Exp Res 16:1007–1013
Malcolm RJ, Sturgis E, Ballenger JC, Anton RF (1989) A double-blind trial of comparing carbamazepine to oxazepam treatment of alcohol withdrawal. Am J Psychiatry 146:617–621
Manhem P, Nilsson LH, Moberg AL, Wadstein J, Hokfelt B (1985) Alcohol withdrawal: effects of clonidine treatment on sympathetic activity, the renin-aldosterone system, and clinical symptoms. Alcohol Clin Exp Res 9:238–243
Marks SS, Watson DL, Carpenter CL, Messing RO, Greenberg DA (1989) Calcium channel antagonist receptors in cerebral cortex from alcoholic patients. Brain Res 478:196–198
Mason BJ, Koscis JH (1991) Desepramine treatment of alcholism. Psychopharm Bull 27(2):155–161
McCowan TJ, Breese GR (1990) Multiple withdrawals from chronic ethanol "kindles" inferior collocular seizure activity: evidence for kindling of seizures associated with alcoholism. Alcohol Clin Exp Res 14:394–399
McCutchen CB (1990) Treatment of alcohol withdrawal seizures with other drugs. In: Porter RJ, Mattson RH, Cramer JA, Diamond I (eds) Alcohol and seizures: basic mechanisms and clinical concepts. Davis, Philadelpia, p 321
McMicken DB (1990) Alcohol withdrawal syndromes. Emerg Med Clin North Am 8:805–819
McNamara JO, Russell RD, Rigsbee L, Bonhaus DW (1988) Anticonvulsant and antiepileptogenic actions of MK-801 in the kindling and electroshock models. Neuropharmacology 27:563–568
Mehta AK, Ticku MK (1989) Chronic ethanol treatment alters the behavioral effects of RO15-4513, a partially negative ligand for benzodiazepine bindings sites. Brain Res 489:93–100
Mendels J, Wasserman TW, Michals TJ, Fine EW (1985) Halazepam in the managment of acute alcohol withdrawal syndrome. J Clin Psychiatry 46:172–174
Mendelson JH (1971) Biochemical mechanisms of alcohol addiction. In: Kissin B, Begleiter H (eds) The biology of alcoholism, vol 1. Plenum, New York, p 513
Mendelson JH, LaDou J (1963) Experimentally induced chronic intoxication and withdrawal in alcoholics, part 2. Psychophysiological findings. Quart J Stud Alcohol [Suppl] 2:14
Mendelson JH, Stein S, McGuire MT (1966) Comparative psychophysiological studies of alcoholic and nonalcoholic subjects undergoing experimentally induced ethanol intoxication. Psychosom Med 28:1–12
Messing RO, Carpenter CL, Diamond I, Greenberg DA (1986) Ethanol regulates calcium channels in clonal neural cells. Proc Natl Acad Sci USA 83: 6213–6215
Messing RO, Sneade AB, Savidge B (1990) Protein kinase C participates in up-regulation of dihydropyridine-sensitive calcium channels by ethanol. J Neurochem 55:1383–1389
Mhatre M, Ticku MK (1989) Chronic ethanol treatment selectively increases the binding of inverse agonists for benzodiazepine binding sites in cultured spinal cord neurons. J Pharmacol Exp Ther 251:164–168
Mhatre M, Mehta AK, Ticku MK (1988) Chronic ethanol administration increases the binding of the benzodiazepine inverse agonist and alcohol antagonist [3H]RO15-4513 in rat brain. Eur J Pharmacol 153:141–145
Miczek KA, Weerts EM (1987) Seizures in drug-treated animals. Science 235:1127

Mochly-Rosen D, Chang FH, Cheever L, Kim M, Diamond I, Gordon AS (1988) Chronic ethanol causes heterologous desensitization of receptors by reducing alpha-s messenger RNA. Nature 333:848–850
Montpied P, Morrow AL, Karanian JW, Ginns EI, Martin BM, Paul SM (1991) Prolonged ethanol inhalation decreases GABA-A receptor α-subunit mRNAs in the rat cerebral cortex. Mol Pharmacol 39:157–163
Morrisett RA, Rezvani AH, Overstreet D, Janowsky DS, Wilson WA, Swartzwelder S (1990) MK-801 powerfully inhibits alcohol withdrawal seizures in rats. Eur J Pharmacol 176:103–105
Morrow AL, Suzdak PD, Karanian JW, Paul SM (1988) Chronic ethanol administration alters γ-aminobutyric acid, pentobarbital and ethanol-mediated $^{36}Cl^{-}$ uptake in cerebral cortical synaptoneurosomes. J Pharmacol Exp Ther 246: 158–164
Morrow AL, Montpied P, Lingford-Hughes A, Paul SM (1990) Chronic ehtanol and pentobarbital administration in the rat: effects on GABA-A receptor function and expression in brain. Alcohol 7:237–244
Moskowitz G, Chalmers TC, Sacks HS, Fagerstrom RM, Smith H (1983) Deficiencies of clinical trials of alcohol withdrawal. Alcohol Clin Exp Res 7:42–46
Mukherjee PK (1983) A comparison of the efficacy and tolerability of clobazam and chlordiazepoxide in the treatment of acute withdrawal from alcohol in patients with primary alcoholism. J Intern Med Res 11:205–211
Nagy LE, Diamond I, Gordon AS (1988) Cultured lymphocytes from alcoholic subjects have altered cAMP signal transduction. Proc Natl Acad Sci USA 85: 6973–6976
Naranjo CA, Sellers EM, Chater K, Iversen P, Roach C, Sykora K (1983) Nonpharmacologic intervention in acute alcohol withdrawal. Clin Pharmacol Ther 34:214–219
Ng SKC, Hauser WA, Brust JCM, Susser M (1988) Alcohol consumption and withdrawal in new-onset seizures. N Engl J Med 319:666–716
Nickel B, Schmickaly R (1988) Calcium channel blockers in the treatment of alcohol withdrawal syndromes. Alcohol Alcohol 23:A60
Noble EP, Gillies R, Vigran R, Mandel P (1976) The modification of the ethanol withdrawal syndrome in rats by di-n-propylacetate. Psychopharmacology 46: 127–131
Nowak L, Bregestovski P, Ascher P, Herbert A, Proachiantz A (1984) Magnesium gates glutamate-activated channels in mouse central neurones. Nature 307:462–465
Nutt DJ (1987) Alpha-2-adrenoceptor function during ethanol withdrawal, pp 880–884. In: Linnoila M (ed) Alcohol withdrawal and noradrenergic function. Ann Intern Med 107:875–889
Nutt DJ, Glue P (1986) Monoamines and alcohol. Br J Addict 81:327–338
Nutt DJ, Glue P (1990) Neuropharmacological and clinical aspects of alcohol withdrawal. Ann Med 22:275–281
Nutt DJ, Glue P, Stewart A (1987) Platelet monoamine binding in alcoholics during withdrawal. Br J Addict 82:1253–1255
Nutt DJ, Glue P, Molyneux S, Clark E (1988) Alpha-2-adrenoceptor activity in alcohol withdrawal: a pilot study of the effects of i.v. clonidine in alcoholics and normals. Alcohol Clin Exp Res 12:14–18
Olsen RW, Kokka N, Sapp DW (1992) The GABA-A receptor complex in alcohol dependence. In: Watson RR (ed) Alcohol and neurobiology: receptors, membranes, and channels. CRC Press, Boca Raton, p 37
Packan DR, Sapolsky RM (1990) Glucocorticoid endangerment of the hippocampus: tissue, steroid and receptor specificity. Neuroendocinology 51:613–618
Palestine ML (1973) Drug treatment of the alcohol withdrawal syndrome and delirium tremens. Quart J Stud Alcohol 34:185–193
Parale MP, Kulkarni SK (1986) Studies with alpha-2-adrenoceptor agonists and alcohol abstinence syndrome in rats. Psychopharmacology 88:237–239

Pena C, Medina JH, Novas ML, Paladini AC, DeRobertis E (1986) Isolation and identification in bovine cerebral cortex of n-butyl-b-carboline-3-carboxylate, a potent benzodiazepine binding ihibitor. Proc Natl Acad Sci USA 83:4952–4956
Peters S, Koh J, Choi DW (1987) Zinc selectively blocks the action of N-methyl-D-aspartate on cortical neurons. Science 236:589–593
Petty F (1992) Plasma GABA as a biological marker for alcoholism. In: Watson RR (ed) Alcohol and neurobiology: receptors, membranes, and channels. CRC Press, Boca Raton, p 1
Piazza NJ, Vrbka JL, Yeager RD (1989) Telescoping of alcoholism in women alcoholics. Int J Addict 24:19–28
Pinel JPJ (1980) Alcohol withdrawal seizures: implications of kindling. Pharmacol Biochem Behav 13:225–231
Pinel JPJ, Van Oot PH (1975) Generality of the kindling phenomenon: some clinical implications. Can J Neurol Sci 2:467–475
Pinel JPJ, Van Oot PH (1978) Increased susceptibility to the epileptic effects of alcohol withdrawal following periodic electroconvulsive shocks. Biol Psychiatry 13:353–368
Pinel JPJ, Van Oot PH, Mucha RF (1975) Intensification of the alcohol withdrawal syndrome by repeated brain stimulation. Nature 254:510–511
Pohorecky LA, Roberts P (1991) Development of tolerance to and physical dependence on ethanol: daily versus repeated cycles treatment with ethanol. Alcohol Clin Exp Res 15:824–833
Poldrugo F, Snead OC (1984) Electroencephalographic and behavioral correlates in rats during repeated ethanol withdrawal syndromes. Psychopharmacology 83: 140–146
Potter JF, Bannan LT, Beevers DG (1984) The effect of a non-selective lipohilic beta-blocker on the blood pressure and noradrenaline, vasopressin, cortisol and renin release during alcohol withdrawal. Clin Exp Hyper Theor Pract A6(6): 1147–1160
Pucilowski O, Krzascik P, Trzaskowska E, Kostowski W (1989) Different effect of diltiazem and nifedipine on some central actions of ethanol in the rat. Alcohol 6:165–168
Reiger DA, Farmer ME, Rae DS, Locke BZ, Keith SJ, Judd LL, Goodwin FK (1990) Comorbidity of mental disorders with alcohol and other drug abuse. Results from the epidemiologic catchment area (ECA) study. JAMA 264:2511–2518
Richelson E (1986) Effects of chronic exposure to ethanol on the prostaglandin E1 receptor-mediated response and binding in a murine neuroblastoma clone (NIE-115). J Pharmacol Exp Ther 239:687–692
Ritter RM, Davidson DE (1971) Haloperidol for acute psychiatric emergencies: a double-blind comparison with periphenazine in acute alcoholic psychosis. South Med J 64:249–250
Roberts AJ, Chu H-P, Crabbe JC, Keith LD (1991) Differential modulation by the stress axis of ethanol withdrawal seizure expression in WSP and WSR mice. Alcohol Clin Exp Res 15:412–417
Robinson BJ, Robinson GM, Maling JB, Johnson RH (1989) Is clonidine useful in the treatment of alcohol withdrawal? Alcohol Clin Exp Res 13:95–98
Rommelspacher H, Schmidt LG, May T (1991) Plasma norharman (β-carboline) levels are elevated in chronic alcoholics. Alcohol Clin Exp Res 15:553–559
Rothman SM, Olney JW (1990) Excitotoxicity and the NMDA receptor. Trends Neurosci 10:216–222
Rothstein E (1973) Prevention of alcohol withdrawal seizures: the roles of diphenylhydantoin and chlordiazepoxide. Am J Psychiatry 130:1381–1432
Roy A, DeJong J, Ferraro T, Adinoff B, Ravitz B, Linnoila M (1990) CSF γ-aminobutyric acid in alcoholic and control subjects. Am J Psychiatry 147:1294–1296

Saito T, Lee JM, Hoffman PL, Tabakoff B (1987) Effects of chronic ethanol treatment on the beta adrenergic receptor-coupled adenylate cyclase system of mouse cerebral cortex. J Neurochem 48:1817–1822

Sampliner R, Iber FL (1974) Diphenylhydantoin control of alcohol withdrawal seizures. JAMA 230:1430–1432

Sapolsky RM, Pulsinelli WA (1985) Glucocorticoids potentiate ischemic injury to neurons: therapeutic implications. Science 229:1397–1400

Sato K, Morimoto K, Okamoto M (1988) Anticonvulsant action of a non-competitive antagonist of NMDA receptors (MK-801) in the kindling model of epilepsy. Brain Res 463:12–20

Schneider HH, Stephens DN (1988) Co-existence of kindling induced by the β-carboline, FG 7142, and tolerance to diazepam following chronic treatment in mice. Eur J Pharmacol 154:35–45

Schuckit M (1983) Alcoholic patients with secondary depression. Am J Psychiatry 140:711–713

Schuckit MA (1989) Drug and alcohol abuse: a clinical guide to diagnosis and treatment. Plenum, New York, p 77

Schultz TK (1991) Alcohol withdrawal syndrome: clinical features, pathophysiology, and treatment. In: Miller NS (ed) Comprehensive handbook of drug and alcohol addiction. Dekker, New York, pp 1091–1112

Sellers EM, Kalant H (1976) Alcohol intoxication and withdrawal. N Engl J Med 294:757–762

Sellers EM, Zilm DH, Degani DC (1977) Comparative efficacy of propranolol and chlordiazepoxide in alcohol withdrawal. J Stud Alcohol 38:2096–2108

Sellers EM, Naranjo CA, Harrison M, Devenyi P, Roach C, Sykora K (1983) Diazepam loading: simplified treatment of alcohol withdrawal. Clin Pharmacol Ther 34:822–926

Sellers EM, Sullivan JT, Somer G, Sykora K (1991) Characterization of DSM-III-R criteria for uncomplicated alcohol withdrawal provides an empirical basis for DSM-IV. Arch Gen Psychiatry 48:442–447

Shaw JM, Kolesar GS, Sellers EM, Kaplan HL, Sandors P (1981) Development of optimal treatment tactics for alcohol withdrawal. I. Assessment and effectiveness of supportive care. J Clin Psychopharmacol 1:382–387

Smith AJ, Brent PJ, Henry DA, Foy A (1990) Plasma noradrenaline, platelet alpha-2-adrenoceptors, and functional scores during ethanol withdrawal. Alcohol Clin Exp Res 14:497–502

Soyka M, Botschev C, Volcker A (1992) Neuroleptic treatment in alcohol hallucinosis – no evidence for increased seizure risk. J Clin Psychopharmacol 12:66–67

Stelzer A, Slater NT, Bruggencate G (1987) Activation of NMDA receptors blocks GABAergic ihibition in an in vitro model of epilepsy. Nature 326:698–701

Stryer L, Bourne HR (1986) G proteins: a family of signal transducers. Annu Rev Cell Biol 2:391–419

Stuppaeck CH, Pycha R, Miller C, Whitworth AB, Oberbauer H, Fleischhacker WW (1992) Carbamazepine versus oxazepam in the treatment of alcohol withdrawal: a double-blind study. Alcohol Alcohol 27:153–158

Sullivan JT, Swift RM, Lewis DC (1989) Assessment of alcohol withdrawal: the revised clinical institute withdrawal assessment for alcohol scale (CIWA-Ar). Br J Addict 84:1353–1357

Sullivan JT, Swift RM, Lewis DC (1991) Benzodiazepine requirements during alcohol withdrawal syndrome: clinical implications of using a standardized withdrawal scale. J Clin Psychopharmacol 11:291–295

Tabakoff B, Hoffman PL, Ritzman RF (1978) Dopamine receptor function after chronic ingestion of ethanol. Life Sci 23:643–648

Tabakoff B, Hoffman PL, Lee JM, Saito T, Willard B, De Leon-Jones F (1988) Differences in platelet enzyme activity between alcoholics and nonalcoholics. N Engl J Med 318:134–139

Thevos AK, Johnston AL, Latham PK, Randall CL, Adinoff B, Malcolm R (1991) Symptoms of anxiety in inpatient alcoholics with and without DAMIII-R anxiety diagnoses. Alcohol Clin Exp Res 15:102–105
Thompson WL (1975) Management of alcohol withdrawal syndromes. Arch Intern Med 138:278–283
Ticku MK (1989) Ethanol and the benzodiazepine-GABA receptor-ionophore complex. Experientia 45:413–418
Ticku MK, Kulkarni SK (1988) Molecular interactions of ethanol with GABAergic system and potential of RO15-4513 as an ethanol antagonist. Pharmacol Biochem Behav 30:501–510
Ticku MK, Mhatre M, Mehta AK (1992) Modulation of GABAergic transmission by ethanol. In: Biggio G, Concas A, Costa E (eds) GABAergic synaptic transmission: molecular, pharmacological, and clinical aspects. Raven, New York, p 255
Tollefson GD, Montagu-Clouse J, Lancaster SP (1990) Buspirone in comorbid alcohol dependency and generalized anxiety disorders. Drug Ther Bull [Suppl]: 35–50
Tran VT, Snyder SH, Major LF, Hawley RJ (1981) GABA receptors are increased in brains of alcoholics. Ann Neurol 9:289–292
Turbridy P (1988) Alpraxolam versus chlormethiazole in acute alcohol withdrawal. Br J Addict 83:581–585
Ulrichsen J, Clemmesen L, Hemmingsen R (1992) Convulsive behavior during alcohol dependence: discrimination between the role of intoxication and withdrawal. Psychopharmacology 107:97–102
Valverius P, Hoffman PL, Tabakoff B (1987) Effect of ethanol on mouse cerebral cortical beta-adrenergic receptors. Mol Pharmacol 32:217–222
Valverius P, Crabbe JC, Hoffman PL, Tabakoff B (1990) NMDA receptors in mice bred to be prone or resistant to ethanol withdrawal seizures. Eur J Pharmacol 184:185–189
Vezzani A, Wu H-Q, Montena E, Samanin R (1988) Role of the N-methyl-D-aspartate-type receptors in the development and maintenance of hippocampal kindling in rats. Neurosci Lett 87:63–68
Victor M (1990) Alcohol withdrawal seizures: an overview. In: Porter R (ed) Alcohol and seizures. Davis, New York
Victor M, Adams RD (1953) The effect of alcohol on the nervous system. Res Publ Assoc Res Nerv Ment Dis 32:526
Victor M, Brausch C (1967) The role of abstinence in the genesis of alcoholic epilepsy. Epilepsia 8:1–20
Vinson DC, Menezes M (1991) Admission alcohol level: a predictor of the course of alcohol withdrawal. J Fam Prac 33:161–167
Volkow ND, Hitzemann R, Wang G-J, Fowler JS, Burr G, Pascani K, Dewey SL, Wolf AP (1992) Decreased brain metabolism in neurologically intact healthy alcoholics. Am J Psychiatry 149:1016–1022
Walinder J, Balldin J, Bokstrom K, Karlsson I, Lundstrom B (1981) Clonidine suppression of the alcohol withdrawal syndrome. Drug Alcohol Depend 8:345–348
Walker DW, Zornetzer SF (1974) Alcohol withdrawal in mice: electroencephalographic and behavioral correlates. Electroencephalogr Clin Neurophysiol 36: 233–243
Wartenberg AA, Nirenberg TD, Liepman MR, Silvia LY, Begin AM, Monti PM (1990) Detoxification of alcoholics: improving care by symptom-triggered sedation. Alcohol Clin Exp Res 14:71–75
Weiss SRB, Post RM, Gold PW, Chrousos G, Sullivan TL, Walker D, Pert A (1986) CRF-induced seizures and behavior: interaction with amygdala kindling. Brain Res 372:345–351
Whitfield CL, Thompson G, Lamb A, Spencer V, Pfeifer M, Browning-Ferrando M (1978) Detoxification of 1024 alcoholic patients without psychoactive drugs. JAMA 239:1409–1410

Whittington MA, Little HJ (1988) Nitrendipine prevents the ethanol withdrawal syndrome when administered chronically with ethanol prior to withdrawal. Br J Pharmacol 94:385P

Wilkins AJ, Jenkins WJ, Steiner JA (1983) Efficacy of clonidine in treatment of alcohol withdrawal state. Psychopharmacology 81:78–80

Wilkins JN, Gorelick DA (1986) Clinical neuroendocrinology and neuropharmacology of alcohol withdrawal. In: Galanter M (ed) Recent development in Alcoholism, vol 4. Plenum, New York, pp. 241–262

Willenbring ML (1986) Measurement of depression in alcoholics. J Stud Alcohol 47:367–371

Williamson JM, Lothman EW (1989) The effect of MK-801 on kindled seizures: implications for use and limitations as an antiepileptic drug. Ann Neurol 26: 85–90

Wilson A, Vulcano B (1984) A double-blind, placebo-controlled trial of magnesium sulfate in the ethanol withdrawal syndrome. Alcohol Clin Exp Res 8:542–549

Wilson A, Vulcano BA (1985) Double-blind trial of alprazolam and chlordiazepoxide in the management of the acute ethanol withdrawal syndrome. Alcohol Clin Exp Res 9:23–27

Wilson WA, Stasheff S, Swartzwelder S, Clark S, Anderson WW, Lewis D (1989) The NMDA receptors in epilepsy. In: Watkins JC, Collingridge GL (eds) The NMDA receptor. Oxford University Press, Oxford, p 167

Wolfe SM, Victor M (1969) The relationship of hypomagnesaemia and alkalosis to alcohol withdrawal symptoms. Ann NY Acad Sci 162:973–984

Woo E, Greenblatt DJ (1979) Massive benzodiazepine requirements during acute alcohol withdrawal. Am J Psychiatry 136:821–823

Woodward JJ, Machu T, Leslie SW (1990) Chronic ethanol treatment alters ω-conotoxin and BAY K 8644 sensitive calcium channels in rat striatal synaptosomes. Alcohol 7:279–284

Yeh G-C, Bonhaus DW, Nadler JV, McNamara JO (1989) N-Methyl-D-aspartate receptor plasticity in kindling: quantitative and qualitative alterations in the N-methyl-D-aspartate receptor-channel complex. Proc Natl Acad Sci USA 86: 8157–8160

Young GP (1987) Intravenous phenobarbital for alcohol withdrawal and convulsion. Ann Emerg Med 16:847

Yu S, Ho IK (1990) Effects of acute barbiturate administration, tolerance and dependence on brain GABA system: comparison to alcohol and benzodiazepines. Alcohol 7:261–272

Zilm DH (1975) Propranolol effect on tremor in alcoholic withdrawal. Ann Intern Med 83:234–236

CHAPTER 14

Drugs to Decrease Alcohol Consumption in Humans: Aversive Agents

R.K. FULLER and R.Z. LITTEN

A. Introduction

Tetraethylthiuram (disulfiram) is used for the treatment of alcoholism. Its use is intended to prevent relapse from abstinence by deterring impulsive drinking. This rationale is based on the fact that an intense adverse reaction, the disulfiram-ethanol reaction (DER), occurs when alcohol is ingested subsequent to the administration of disulfiram. Within 5–10 min after ingesting an alcoholic beverage, peripheral cutaneous vasodilation manifested by flushing and tachycardia occurs, and an intense throbbing is often felt in the head and neck. Vertigo, nausea, vomiting, chest pain, and hypotension may also occur. Fatal DERs have been reported. Most of these occurred with the higher dosages that were used when disulfiram was initially introduced into clinical practice. Because of its action, disulfiram belongs to a category of drugs referred to as alcohol-deterrent, alcohol-sensitizing, or antialcohol drugs.

The DER occurs primarily because disulfiram inhibits the oxidation of acetaldehyde, the first metabolic product of ethanol. The impaired oxidation is a result of disulfiram inhibiting the aldehyde dehydrogenase (ALDH) enzymes which promote the catabolism of acetaldehyde to acetic acid. This inhibition results in an accumulation of circulating acetaldehyde.

Development of medications can be conceptualized as a series of steps from drug discovery to animal testing to preliminary human studies to controlled clinical trials. The history of disulfiram is interesting in this regard. Disulfiram was used as an antioxidant in the rubber industry. It was observed that workers exposed to disulfiram became sick if they drank, and the suggestion was made that disulfiram should be tried as a treatment for alcoholism (WILLIAMS 1937). However, this idea was ignored until two Danish investigators who had taken disulfiram in their search for an effective antihelminthic agent became ill when they drank. In 1948 they published a study which showed the potential usefulness of disulfiram as a treatment for alcohol dependence (HALD and JACOBSEN 1948).

While disulfiram has been used clinically for over 40 years, much is still not known about the drug. For example, is the inhibition of ALDH activity the only basis for the disulfiram-ethanol reaction? Disulfiram inhibits other enzymes, and that inhibition also may play a role in the DER. Is disulfiram or one of its metabolites primarily responsible for the DER? Is it effective, and, if so, how should it be used in alcoholism treatment programs? Our ignorance stems, in large part, because disulfiram was introduced into clinical practice without having been evaluated by some of the steps in medications development that are considered standard today.

B. Absorption, Metabolism, and Excretion

Ninety percent of an oral dose of disulfiram is absorbed from the gastrointestinal tract (IBER et al. 1977). After absorption, the catabolism of disulfiram results in several metabolites. First, disulfiram is reduced to diethyldithiocarbamate (DDTC). Disulfiram is the disulfide dimer of DDTC. Diethyldithiocarbamic acid both forms mixed disulfides with serum protein sulfhydryl groups and is catabolized to a variety of metabolites. For example, it forms a bis-diethyldithiocarbamato copper complex (JOHANSSON and STANKIEWICZ 1985), because DDTC is a chelator and the copper complex results from the thiol group of DDTC reacting with cupric ions bound to plasma proteins. It is degraded nonenzymatically to diethylamine, which is excreted in the urine (NIEDERHISER 1983), and carbon disulfide, which is expired in the breath (IBER et al. 1977). In addition, DDTC is conjugated with glucuronic acid and is methylated by microsomal *S*-methyltransferase to form a methyl ester (GESSNER and JAKUBOWSKI 1972). The DDTC-methyl ester is metabolized to diethylthiocarbamic (DTC) methyl ester by the cytochrome P-450 oxidase system and to thiocarboxylic acid by thioesterase. Recently, HART et al. (1990) have described another metabolite, a sulfoxide derivative of the DTC-methyl ester.

IBER et al. (1977) studied the absorption and excretion of a 250-mg dose of ^{35}S-labeled disulfiram in 20 male alcoholics, 10 with and 10 without liver disease. The intersubject variability in the urinary excretion of radioactivity ranged from 12% to 90%. Two-thirds of the total urinary excretion occurred within 24 h of ingesting disulfiram. It then declined rapidly during the next 2 days, with 97% of the total urinary radioactivity being excreted by 72 h. The excretion of radioactivity in the breath was also greatest during the 1st day, but those subjects with relatively low urinary excretion of radioactivity had higher expiration in the breath. There was no difference in urine, breath, and fecal excretion between subjects with and those without liver disease.

C. Pharmacokinetics

There is limited information on the pharmacokinetics of disulfiram. FAIMAN et al. (1984) used high-performance liquid chromatography to measure disulfiram and its metabolites after a single oral 250-mg dose and again after 12 days of dosing in 15 male alcoholics. Average time to maximal plasma concentration after either single or repeated dosing was 8–10h for disulfiram, DDTC, DDTC-methyl ester, and diethylamine and CS_2 in the breath. The apparent half-lives for disulfiram, DDTC, DDTC-methyl ester, diethylamine, and CS_2 were 7.3, 15.5, 22.1, 13.9, and 8.9h, respectively. However, marked intersubject variability was noted in the plasma levels of disulfiram and its metabolites.

D. Mechanism of Disulfiram: Ethanol Reaction

The mechanism by which disulfiram inhibits ALDH is not fully elucidated. Disulfiram inhibits ALDH in vitro. However, this may not explain the DER in vivo, which may result from the inhibition of ALDH by one of its metabolites. The initial metabolite of disulfiram (DDTC) does not produce inhibition in vitro unless nonphysiological doses are used (DIETRICH and HELLERMAN 1963). However, DDTC does inhibit ALDH in vivo. When DDTC is administered to the rat in doses equivalent to disulfiram, it is as effective as disulfiram in inhibiting ALDH activity (DIETRICH and ERWIN 1971). The inhibition of ALDH by either disulfiram or DDTC in vivo is irreversible and requires new protein synthesis (DIETRICH and ERWIN 1971).

Since DDTC does not inhibit ALDH in vitro but does in vivo, this suggests that a metabolite of DDTC is responsible for the inhibition of ALDH. In the rat, the DDTC-methyl ester is more potent and has a more rapid onset of action than either disulfiram or DDTC (YOURICK and FAIMAN 1989). Methyl-DDTC is oxidized to methyl-diethylthiocarbamate (methyl-DTC) and it has been suggested that methyl-DTC is the immediate inhibitor of ALDH (HART et al. 1990; JOHANSSON and STANKEWICZ 1989). However, KITSON (1991) found that $10\,\mu M$ methyl-DTC produces only 2% inhibition of ALDH. HART et al. (1990) proposed that the sulfoxide derivative of methyl DTC, which has been identified, is the key inhibitor of ALDH.

Further complicating our understanding of the DER is that disulfiram inhibits other enzymes, most notably dopamine-β-hydroxylase. While most of the symptoms and signs of the DER likely result from inhibition of ALDH and the resulting accumulation of acetaldehyde, it is possible that the inhibition of other enzymes plays a role in the complete clinical presentation of the DER.

E. Clinical Use

The studies of HALD and JACOBSEN (1948) led to disulfiram being marketed. In the United States, it is sold under the trade name "Antabuse." Initially, high doses (1000–3000mg daily) of the drug were used. This is in contrast to the 250-to 500-mg daily doses used today. Also, in the past, physicians conducted one or more test sessions in which the patient was given alcohol to drink after taking disulfiram for a few days so that he experienced the reaction. This is not done today. Current practice involves providing the patient with an explanation of the DER before the drug is prescribed.

The current dosage of 250–500 mg daily was arrived by empircally balancing deterrence against toxicity. This dose level has been challenged as being insufficient to cause a DER (BREWER 1984). However, a recent, carefully conducted dose-effect study resulted in the conclusion that, while a daily dose of 200 mg disulfiram resulted in a DER in nonalcoholic volunteers after an alcohol challenge, 300 mg might be required to produce severe enough symptomatology to deter subsequent drinking (CHRISTENSEN et al. 1991).

F. Efficacy

Initial clinical studies of the efficacy of disulfiram reported excellent outcomes with the use of the drug. However, disulfiram was incorporated into clinical practice at about the time that the concepts inherent to randomized clinical trials were beginning to be developed. Consequently, many years elapsed before the efficacy of disulfiram was appropriately tested. This may account, in part, for the wide variation in the use of disulfiram among treatment programs. A recent study (FRIEDMAN and FULOP 1988) examined alcoholism treatment centers in the New York City metropolitan area and found that the percentages of patients receiving disulfiram ranged from 0% to 97%. In this study, there was no significant difference between hospital-based programs and freestanding programs. The percentage of hospital-based programs that use disulfiram was 27% compared to 34% in free-standing programs.

A quarter of a century elapsed following the introduction of disulfiram into clinical practice before the appearance of two reviews that criticized the reports supporting the efficacy of disulfiram because those studies had serious flaws in their design (LUDWALL and BAEKELAND 1971; MOTTIN 1973). Most (90%) of the studies were uncontrolled, and the criteria for a successful outcome were often vague or ill-defined. FULLER and ROTH (1979) identified other methodological deficiencies: nonrandom assignment if control groups were used, lack of "blinding," high attrition, measurement of treatment outcome based solely on the patient's self-report, failure to measure adherence to the disulfiram regimen, and absence of statistical analysis.

GALLANT and his colleagues (1968) did a randomized controlled study of disulfiram in skid-row alcoholics. These men were recruited in a municipal court in New Orleans and were randomly assigned to disulfiram, group therapy, disulfiram and group therapy, or routine sentencing. While there was a slight trend in favor of disulfiram, recidivism was very high in all groups.

The largest well-designed clinical trial of disulfiram therapy was supported by the Veterans Administration Cooperative Studies Program

(FULLER et al. 1986). This was a randomized, controlled, blinded multicenter study. Response to treatment was assessed bimonthly for 1 year. Cohabiting relatives or friends and blood and urine alcohol tests were used to corroborate patients' reports.

The active treatment group was prescribed 250 mg disulfiram to take daily. There were two control groups. One group received 1 mg disulfiram. This dose is not sufficient to cause a DER. As GORDIS and PETERSON (1977) wrote, "... it is probable that it is the patient's belief that he is taking disulfiram (whether or not he actually is) that is therapeutic and not the action of the drug itself...." The 1-mg dose was the control for the implied threat (expectancy) of the DER. Recipients of the 250-mg and 1-mg disulfiram doses were told they were receiving disulfiram. The second control group received a vitamin (50 mg riboflavin) rather than disulfiram. These patients were told they were not receiving disulfiram. This group was a control for counseling that all patients received without the adjunctive use of disulfiram. Riboflavin (50 mg) was also incorporated into the two disulfiram dosage forms. Urine specimens were collected at both clinic visits and assessment interviews and analyzed for riboflavin. These analyses were used to measure adherence to the medication regimen.

At the end of 1 year of follow-up, there were sufficient data to categorize 91.6% of the 605 subjects as continuously abstinent or not for 1 year. The continuous abstinence rate for the 250-mg, 1-mg, and no-disulfiram groups were 18.8%, 22.5%, and 16%, respectively. These differences are not significant. The sample sizes of the three groups were sufficient to have a power of 91% to detect a 12% difference among the groups. Similarly, there were no significant differences among the groups on time to first drink by life table analysis, employment, and social stability. However, among the patients who drank and had a complete set of assessment interviews, those assigned to 250 mg disulfiram reported significantly fewer drinking days during the year than those in the control groups, and these findings were corroborated by the cohabiting relative or friend. The men who drank and provided all seven assessment interviews (about half of those who drank) were slightly older and had longer residential stability than those who drank and provided fewer interviews. The two groups did not differ significantly on marital status, employment status, or level of education.

There was also a significant relationship between compliance with the medication regimen and continuous abstinence, which was present for all three groups. Forty-three percent of those judged to exhibit "good" compliance were abstinent compared with 9% of those who did not adhere to a medication regimen. Does good compliance result in more abstinence, or does a third factor such as motivation result in both abstinence and compliance? The answer to this question cannot be determined by this study.

What can one conclude from this important, well-designed, and well-executed study? It would appear that disulfiram is not the panacea that the initial clinical studies claimed. However, there may be a subgroup of patients who will benefit from disulfiram to the extent that they drink less frequently. A reasonable clinical approach may be to not recommend disulfiram to patients who are entering treatment for the first time. However, for those patients who relapse a discussion concerning the potential utility of taking disulfiram is appropriate. If a patient decides to take disulfiram, the next decision to be made is whether the medication should be ingested under supervision. This is discussed in the next session.

G. Enhancing Disulfiram Compliance

Poor compliance with the medication regimen limits the efficacy of disulfiram. Several strategies to improve compliance have been devised. These include incentives, contracts, supervised administration, and implants. Preliminary studies have shown that incentives, such as cash incentives and availability of methadone treatment, may increase disulfiram compliance and improve treatment outcome (ALLEN and LITTEN 1992). These preliminary studies, while interesting, need replication and validation.

Disulfiram contracts and supervised administration may have promise for increasing the effectiveness of disulfiram. KEANE et al. (1984) employed a contract in which it was agreed that disulfiram would be taken in the presence of the significant other. Both the patient and the significant other signed the contract. While compliance with the disulfiram regimen was higher in the contract group (88%) than in the no-contract group (56%), there was no difference in abstinence between the two groups.

AZRIN (1976) combined a supervised disulfiram procedure, i.e., disulfiram taken in the presence of a significant other, with a broad-spectrum treatment approach that uses behavioral techniques to achieve an alcohol-free lifestyle. Clients and their counselors signed a contract for all the procedures. Alcoholics receiving the combined supervised disulfiram-behavioral treatment drank fewer days, were employed more, spent more days at home, and spent fewer days institutionalized than the control group which had received only standard treatment. Unfortunately, the effects of supervised disulfiram could not be separated from the behavioral program.

In a subsequent study, AZRIN et al. (1982) contrasted the following interventions: (1) traditional treatment without supervised disulfiram; (2) traditional treatment with disulfiram taken in the presence of a significant other; and (3) a behavioral program including supervised use of disulfiram. After 6 months, the traditionally treated group without supervised disulfiram was drinking on most days and no longer taking the disulfiram. The traditional treatment combined with supervised use of disulfiram achieved abstinence for almost all of the married clients, but was considerably less successful for those who were single. Finally, the combined behavioral program with supervised use of disulfiram produced fewer drinking days and less alcohol consumption in both the single and married subjects.

Two mandatory supervised disulfiram studies have also yielded favorable results. BREWER and SMITH (1983) conducted a pilot study of 16 "habitual drunken" offenders who agreed to take disulfiram under supervision as a condition of probation. Twelve of the 16 either abstained or had brief and "comparatively harmless" lapses. However, they did not have a concurrent control group.

In another study, SERENY et al. (1986) initiated a mandatory supervised disulfiram program with patients who had relapsed at least three times. Patients were required to take disulfiram three times per week under supervision for a minimum of 6 weeks. If they missed more than two sessions, they would be discharged from the clinic. A contract containing these provisions was signed by the patient, physician, nurse, and counselor. The investigators found that 60% of the patients who agreed to these provisions obtained a level of sobriety not previously achieved by them. However, the absence of both random assignment and a control group limits the validity of these results.

KOFOED (1987) conducted a very interesting randomized, three-group study in which subjects were given disulfiram to take at home. Two groups had their compliance with disulfiram monitored by testing the breath of subjects for carbon disulfide at weekly clinic visits. The case managers for the members of one of these two groups were told the results of the breath tests and were encouraged to discuss the results with the patients, whereas the case managers for the members of the other group were not told the results of the breath tests. A third group did not have their adherence to the disulfiram regimen monitored by the breath tests. The compliance rate was significantly greater for the group in which the case managers were informed of the breath test results (71%) compared to the group in which the case managers did not know the results (44%). However, the three groups did not differ on retention in treatment or on a global measure of severity of alcohol-related problems. Specific drinking variables such as days of abstinence, frequency of drinking, or quantity of alcohol consumed were not reported. Nevertheless, it is significant that improved compliance in this study was not reflected in better outcome.

CHICK et al. (1992) recently conducted a randomized, single-blind, 6-month study of supervised disulfiram as an adjunct to outpatient treatment or alcoholics. Patients received either 200 mg disulfiram or 100 mg vitamin C daily in the presence of an informant (usually the spouse, though occasionally another relative, colleague, or clinic staff member). Patients were informed of the type of medication and, consequently, were not blinded to treatment they received. However, the independent assessor was blinded to the medication received. There were no written contracts or penalties for not taking the medication. Given that there were no penalties, this trial was a supervised study, in contrast to the mandatory supervised studies conducted by BREWER and SMITH (1983) and SERENY et al. (1986). In addition to the medications, patients also received a variety of treatments including

Alcoholics Anonymous (AA), marital therapy, relaxation therapy, and supportive group therapy. The patients treated with disulfiram had more days of abstinence and less alcohol consumption than did patients assigned to the vitamin C group. In addition, follow-up serum levels of γ-glutamyl transferase (GGT) were lower in the disulfiram group but higher in the vitamin C group, thus supporting a decrease in total alcohol intake in the supervised disulfiram patients. These results suggest that supervised disulfiram in conjunction with the usual verbal therapies improves outcome.

Finally, disulfiram implants have been used over the past 30 years to increase patient compliance. Unfortunately, most of the early studies evaluating the efficacy of disulfiram implants had methodological flaws. In addition, a frequent complication was the formation of sterile abscesses at the site of implantation. Recently, better designed studies have been conducted, thus allowing more accurate appraisal of the effectiveness of disulfiram implants.

JOHNSEN et al. (1987) conducted a double-blind, placebo-controlled trial in which ten 100-mg doses of disulfiram or ten 100-mg doses of calcium phosphate were implanted subcutaneously into 11 and 10 alcoholics, respectively. Twenty weeks after implantation the investigators found no differences between the two groups in alcohol consumption, rate of weekly abstinence, and days to first drink. Also, there were no differences in serum levels of GGT, alanine aminotransferase, and aspartate aminotransferase between the two groups. Finally, neither group reported a DER nor had differences in blood acetaldehyde levels following acute alcohol challenge (at a dosage of 0.4 g/kg i.v.).

The inability of disulfiram implants to elicit a DER has been observed in other studies as well. BERGSTROM et al. (1982) implanted 1 g disulfiram (10 × 100 mg) into 11 chronic alcoholics and orally administered 15 g alcohol 4–6 weeks later. No DER was observed.

JOHNSEN et al. (1990) conducted a double-blind, placebo-controlled study of subcutaneous disulfiram implants in nonalcoholic volunteers. Each participant received one pre-implant and six post-implant (1–18 weeks following implementation) sessions of intravenous alcohol challenges (0.4 g/kg) as well as one oral alcohol challenge (0.8 g/kg). These investigators found no clinical signs of a DER. In addition, the intravenous infusion of alcohol produced no changes in blood acetalde-hyde concentration in either group. The oral challenge did, however, produce higher blood acetaldehyde levels in the disulfiram-implanted group compared to the placebo-implanted group, though acetaldehyde levels were insufficient to cause a DER.

Finally, JOHNSEN and MORLAND (1991) recently conducted a randomized, controlled, double-blind study of disulfiram implants in alcohol-dependent patients. Forty subjects received subcutaneous abdominal implantation of ten 100-mg disulfiram tablets and 36 subjects received nine 100-mg calcium phosphate tablets plus a tablet containing 1 mg disulfiram. After a study period of 300 days there were no differences between the two groups in days to first drink, level of weekly abstinence, alcohol consumption, and overall psychosocial and physical functioning. However, both groups reduced alcohol consumption, suggesting a psychological deterrent effect rather than a pharmacological effect. It is unknown from this study if sufficient disulfiram was released into the bloodstream to produce a DER.

These studies show that disulfiram tablets implanted subcutaneously are not effective. This is most likely because a DER does not occur when alcohol is ingested because of poor absorption from the tablets. Improvements in the technology of disulfiram implants may correct this problem. Once developed, improved implants would have to be assessed in randomized, controlled studies. One potential problem with effective implants would be how to treat a DER if a person with an implant drinks, given the long duration of action of implant disulfiram.

H. Toxicity

An issue that has to be considered in recommending disulfiram to a patient is the potential toxicity of the compound. One side effect to disulfiram is drowsiness, which can ususally be avoided if the medication is taken at bedtime. It also has to be used with caution by individuals who work around machinery, ladders, or in other settings that require alertness. Other side effects, e.g., impotence, have been reported, but the VA Cooperative Study found drowsiness to be the only side effect to be reported more frequently by those assigned to the 250-mg dose compared to 1 mg or no disulfiram. Another controlled study also found no significant differences in side effects between patients receiving disulfiram and those receiving a placebo (Christensen et al. 1984).

A worrisome adverse reaction from disulfiram is hepatoxicity, which is rare but can be fatal. Severe disulfiram-induced fulminant hepatitis usually occurs within the first 2 months of initiating treatment (Wright et al. 1988). Since it is idiosyncratic, it is impossible to predict who will develop this serious toxic reaction. The VA Cooperative Study obtained periodic liver tests to monitor for hepatoxicity. In this study an increase in liver test values from baseline almost always indicated a resumption of drinking and not hepatoxicity (Iber et al. 1987). Nevertheless, it is important to obtain liver tests frequently during the first 2 months of treatment. Wright et al. (1988) recommend that liver function tests should be obtained before treatment, at 2-week intervals for 2 months, and at 3- to 6-month intervals thereafter. If there is an increase in the liver function tests, a careful interview with the patient is necessary, and the medication should be stopped unless it is evident that the patient has been drinking.

Another uncommon but serious adverse reaction is peripheral neuropathy (Frisoni and DiMonda 1989). Its most serious form is optic neuritis, which can cause blindness. When higher doses of disulfiram were used, psychotic reactions and major depression were frequently observed. At the doses used today, these serious adverse reactions are rare. The VA Cooperative Study did not find a greater incidence among those who received disulfiram than placebo or no disulfiram (BRANCHEY et al. 1987), but these investigators excluded patients with major affective disorders and schizophrenia.

I. Contraindications to Disulfiram Treatment

Because hypotension can occur during the DER, patients with cardiovascular or cerebrovascular disease should not be prescribed disulfiram. Patients with organic brain disease should not be given disulfiram because they are unable to fully comprehend the potentially dangerous consequences of the DER. Patients with schizophrenia or a major affective disorder should not receive disulfiram because these illnesses may be exacerbated with the use of disulfiram. Pregnant women should not receive the drug because disulfiram has been reported to cause serious fetal abnormalities (NORA et al. 1977). Since disulfiram may lower the seizure threshold, patients with an idiopathic seizure disorder should not take the drug.

Disulfiram should be used carefully in patients taking certain other medications because it interferes with the biotransformation of such drugs as phenytoin, warfarin, isoniazid, rifampin, diazepam, chlordiazepoxide, imipramine, and desipramine.

J. Calcium Carbamide

Although currently not available in the United States, calcium carbamide is another alcohol-sensitizing agent. Its onset of action is much more rapid than disulfiram (1 h versus 12 h), and the duration of the DER-like reaction after termination of the drug is considerably shorter (24 h versus several days) (ANNIS and PEACHEY 1992).

PEACHEY et al. (1989a) conducted a double-blind, crossover study of calcium carbamide in alcohol-dependent patients. Study completers in both groups were judged to be highly compliant as determined by daily self-reports, pill counts, and urine levels of the riboflavin medication marker. Frequency and total volume of alcohol consumed were diminished significantly but equivalently in both groups. Side effects and medical problems of the calcium carbamide-treated subjects were essentially the same in both groups. In addition, thyroid function was not affected by calcium carbamide and no signs of liver toxicity were observed (PEACHEY et al. 1989b).

ANNIS and PEACHEY (1992) recently conducted a clinical trial of calcium carbamide in combination with a psychosocial therapy. Alcoholics were randomly assigned to either relapse prevention therapy or counseling by a physician. All were encouraged to reduce pill intake over time. Both groups demonstrated substantial improvement in amount of alcohol consumed at 6-, 12-, and 18-month follow-up. The relapse prevention group exhibited lower drinking at the 18-month follow-up. This study suggests that integration of alcohol-sensitizing agents with psychosocial therapies may be promising in the treatment of alcoholism.

K. Future Research Directions

Disulfiram treatment is likely to be improved if the results of CHICK et al. (1992) are confirmed, because these results indicate that supervised administration should be used rather than giving the patient disulfiram to take at his/her discretion. Disulfiram treatment would also be improved if a compound or dosage form with fewer side effects were developed. Perhaps one of the disulfiram metabolites would be as effective as the parent compound in causing a DER but have fewer side effects. Lastly, combining specific psychosocial therapies with disulfiram may also improve treatment outcome.

References

Allen JP, Litten RZ (1992) Techniques to enhance compliance with disulfiram. Alcohol Clin Exp Res 16:1035–1041

Annis HM, Reachey JE (1992) The use of calcium carbamide in relapse prevention counselling: results of a randomized controlled trial. Br J Addict 87:63–72

Azrin NH (1976) Improvements in the community-reinforcement approach to alcoholism. Behav Res Ther 14:339–348

Azrin NH, Sisson RW, Meyers R, Godley M (1982) Alcoholism treatment by disulfiram and community reinforcement therapy. J Behav Ther Exp Psychiatry 13:105–112

Bergstrom B, Ohlin H, Lindblom PE, Wadstein J (1982) Is disulfiram implantation effective? Lancet 1:49–50

Branchey L, Davis W, Lee KK, Fuller RK (1987) Psychiatric complications following disulfiram treatment. Am J Psychiatry 144:1310–1312

Brewer C (1984) How effective is the standard dose of disulfiram? A review of the alcohol-disulfiram reaction in practice. Br J Psychiatry 144:200–202

Brewer C, Smith J (1983) Probation linked supervised disulfiram in the treatment of habitual drunken offenders: results of a pilot study. Br Med J 287:1282–1283

Chick J, Gough K, Falkowski W et al. (1992) Disulfiram treatment of alcoholism. Br J Psychiatry 161:84–89

Christensen JK, Ronsted P, Vaag UH (1984) Side effects after disulfiram: comparison of disulfiram and placebo in a double-blind multicentre study. Acta Psychiatr Scand 69:265–273

Christensen JK, Moller IW, Ronstad P et al. (1991) Dose-effect relationship of disulfiram in human volunteers. I. Clinical studies. Pharmacol Toxicol 68:163–165

Dietrich RA, Erwin VG (1971) Mechanism of the inhibition of aldehyde dehydrogenase in vivo by disulfiram and diethyldithiocarbamate. Mol Pharmacol 7:301–307

Dietrich RA, Hellerman L (1963) Diphosphopyridine nucleotide-linked aldehyde dehydrogenase. J Biol Chem 238:1683–1689

Faiman MD, Jensen CJ, Lacoursiere RB (1984) Elimination kinetics of disulfiram in alcoholics after single and repeated doses. Clin Pharmacol Ther 36:520–526

Friedman TC, Fulop G (1988) Disulfiram use at hospital-based and free-standing alcoholism treatment centers. J Subst Abuse Treatment 5:139–143
Frisoni GB, DiMonda V (1989) Disulfiram neuropathy: a review (1971–1988) and report of a case. Alcohol Alcohol 24:429–437
Fuller RK, Roth HP (1979) Disulfiram for the treatment of alcoholism. An evaluation in 128 men. Ann Intern Med 90:901–904
Fuller RK, Branchey L, Brightwell DR et al. (1986) Disulfiram treatment of alcoholism: a Veterans Administration Cooperative Study. J Am Med Assoc 256:1449–1455
Gallant DM, Bishop MP, Faulkner MA et al (1968) A comparative evaluation of compulsory (group therapy and/or antabuse) and voluntary treatment of the chronic alcoholic municipal court offender. Psychosomatics 9:306–310
Gessner T, Jakubowski M (1972) Diethyldithiocarbamic acid methyl ester: a metabolite of disulfiram. Biochem Pharmacol 21:219–230
Gordis E, Peterson K (1977) Disulfiram therapy in alcoholism: patient compliance studied with a urine detection procedure. Alcohol Clin Exp Res 1:213–216
Hald J, Jacobsen E (1948) A drug sensitizing the organism to ethyl alcohol. Lancet 2:1001–1003
Hart BW, Yourick JJ, Faiman MD (1990) S-Methyl-N,N-diethylthiocarbamate: a disulfiram metabolite and potent rat liver mitochondrial low K_m aldehyde dehydrogenase inhibitor. Alcohol 7:165–169
Iber FL, Dutta S, Shamszad M, Krause S (1977) Excretion of radioactivity following the administration of ^{35}Sulfur-labeled disulfiram in man. Alcohol Clin Exp Res 1:359–364
Iber FL, Lee KK, Lacoursier RB, Fuller RK (1987) Liver toxicity encountered in the Veterans Administration trial of disulfiram in alcoholics. Alcohol Clin Exp Res 11:301–304
Johansson B, Stankiewicz Z (1985) Bis-(diethyldithiocarbamato) copper complex: a new metabolite of disulfiram? Biochem Pharmacol 34:2989–2991
Johansson B, Stankiewicz Z (1989) Inhibition of erythrocyte aldehyde dehydrogenase activity and elimination kinetics of diethyldithiocarbamic acid methyl ester and its monothio analogue after administration of single and repeated doses of disulfiram to man. Eur J Clin Pharmacol 37:133–130
Johnsen J, Morland J (1991) Disulfiram implant: a double-blind placebo controlled follow-up on treatment outcome. Alcohol Clin Exp Res 15:532–536
Johnsen J, Stowell A, Bache-Wiig JE et al. (1987) A double-blind placebo controlled study of male alcoholics given a subcutaneous disulfiram implantation. Br J Addict 82:607–613
Johnsen J, Stowell A, Stensrud T et al. (1990) A double-blind placebo controlled study of healthy volunteers given a subcutaneous disulfiram implantation. Pharmacol Toxicol 66:227–230
Keane TM, Foy DW, Nunn B, Rychtarik RG (1984) Spouse contracting to increase Antabuse compliance in alcoholic veterans. J Clin Psychol 40:340–344
Kitson TM (1991) Effect of some thiocarbamate compounds on aldehyde dehydrogenase and implications for the disulfiram ethanol reaction. Biochem J 278:189–192
Kofoed LL (1987) Chemical monitoring of disulfiram compliance: a study of alcoholic outpatients. Alcohol Clin Exp Res 11:481–485
Lundwall L, Baekeland F (1971) Disulfiram treatment of alcoholism: a review. J Nerv Ment Dis 153:381–394
Mottin JL (1973) Drug-induced attentuation of alcohol consumption: a review and evaluation of claimed, potential or current therapies. Q J Stud Alcohol 34:444–472
Neiderhiser DH, Wych G, Fuller RK (1983) The metabolic fate of double labelled disulfiram. Alcohol Clin Exp Res 7:199–202
Nora AH, Nora JJ, Blu J (1977) Limb-reduction anomalies in infants born to disulfiram-treated alcoholic mothers. Lancet 2:664

Peachey JE, Annis HM, Bornstein ER et al. (1989a) Calcium carbamide in alcoholism treatment, part 1: a placebo-controlled, double-blind clinical trial of short-term efficacy. Br J Addict 84:877–887
Peachey JE, Annis HM, Bornstein ER et al. (1989b) Calcium carbamide in alcoholism treatment, part 2: medical findings of a short-term, placebo-controlled, double-blind clinical trial. Br J Addict 84:1359–1366
Sereny G, Sharma V, Holt J, Gordis E (1986) Mandatory supervised Antabuse therapy in an outpatient alcoholism program: a pilot study. Alcohol Clin Exp Res 10:290–292
Williams EE (1937) Effects of alcohol on workers with carbon disulfide. J Am Med Assoc 109:142–143
Wright C, Vafier JA, Lake R (1988) Disulfiram-induced fulminating hepatitis: guideliness for liver-panel monitoring. J Clin Psychiatry 49:430–434
Yourick JJ, Faiman MD (1989) Comparative aspects of disulfiram and its metabolites in the disulfiram-ethanol reaction in the rat. Biochem Pharmacol 38:413–421

CHAPTER 15

Drugs Attenuating Alcohol Consumption in Humans Through Effects on Various Neurotransmitter Systems*

C.A. NARANJO and K.E. BREMNER

A. Introduction

Several pharmacologic treatments for decreasing alcohol consumption and related problems have recently been identified and tested in clinical studies. These developments have been a consequence of increased understanding of the biological mechanisms regulating ethanol intake. Research studies suggest a role for several neurotransmitter systems and hormones in regulating the initiation, maintenance, and cessation of alcohol drinking in animals (NARANJO and SELLERS 1992). Recently serotonin, dopamine, and endogenous opioids have been shown to be particularly important in alcohol consummatory behavior (NARANJO and SELLERS 1992; NARANJO et al. 1992c; AMIT and SMITH 1992). Most studies, both preclinical and clinical, have concentrated on a single neurotransmitter system or mechanism, but these independent lines of research will, most likely, be integrated in the future (AMIT and SMITH 1992). The interactions of systems and the potential synergisms of various drugs must be assessed since few medications have a sufficiently potent effect. This paper will review recent clinical research into neuropharmacologic agents and their effects on alcohol consumption. Some of the major findings are summarized in Table 1.

B. Serotonin

The relationship between central serotonergic neurotransmission and alcohol intake was first determined in preclinical studies. Many agents have been tested and the results in general suggest that medications that increase serotonin tone decrease alcohol intake (NARANJO and BREMNER 1992a; MCBRIDE et al. 1992; ENGEL et al. 1992; HIGGINS et al. 1992). For example, serotonin uptake inhibitors were found to decrease alcohol intake in rats (AMIT and SMITH 1992; LAWRIN et al. 1986). These medications were developed primarily as antidepressants and were available for testing in

* Parts of this manuscript have been published in abstracts and papers by the authors.

Table 1. Drugs to reduce alcohol consumption

Medication	Comments
Serotonin uptake inhibitors (e.g., citalopram, fluoxetine)	Consistent decreases of 20% in alcohol intake in mildly/moderately dependent alcoholics, perhaps by reducing desire to drink. Clinical testing continues
Serotonin antagonists (e.g., ondansetron, ritanserin)	Inconsistent results in clinical trials. Further evidence required to confirm effectiveness
Serotonin agonists (e.g., buspirone)	Reduced anxiety and alcohol craving, but not alcohol intake
Dopamine agonists (e.g., bromocriptine)	Alleviated withdrawal symptoms, alcohol intake, and/or psychopathology. Testing continues
Opiate antagonists (e.g., naltrexone)	Preliminary results indicate naltrexone reduces relapse rate in detoxified alcoholics
GABA agents (e.g., acamprosate)	May reduce relapse in detoxified alcoholics. More studies are needed

humans. Therefore, we developed a research program to evaluate systematically the clinical effects of serotonin uptake inhibitors on ethanol intake and other consummatory behaviors in humans (NARANJO and BREMNER 1992b). Details of the methodology and the main results of our initial studies with zimeldine, citalopram, viqualine, and fluoxetine are described elsewhere (NARANJO et al. 1984b, 1987, 1989, 1990a). All were randomized, placebo-controlled, double-blind studies. Subjects had mild-to-moderate alcohol dependence (American Psychiatric Association 1980, 1987), and were nondepressed, socially stable, and did not abuse or only occasionally abused other drugs (other than cigarette smoking). They reported drinking 28 or more standard drinks (each containing 13.6 g absolute ethanol) per week before enrollment in the studies.

After the 2-week baseline periods, subjects were randomized to receive a serotonin uptake inhibitor (zimeldine 200 mg/day, citalopram 20 or 40 mg/day, viqualine 100 or 200 mg/day, or fluoxetine 40 or 60 mg/day) or placebo for 2 or 4 weeks according to the study protocols. They attended weekly/biweekly assessments, but no other treatment or advice was offered. Several objective measures confirmed compliance with medication and accurate reporting of alcohol consumption. None of the lower doses of serotonin uptake inhibitor significantly changed alcohol intake compared with placebo or baseline. Zimeldine 200 mg/day and citalopram 40 mg/day, viqualine 200 mg/day, and fluoxetine 60 mg/day all decreased short-term (2–4 weeks) alcohol intake by averages of 14%–20% from baseline, with some subjects decreasing their alcohol intake by up to 60%. Placebo treatment did not change alcohol intake significantly from baseline (less than 2% average).

Consistent results in several human studies are required to establish the potential clinical usefulness of a new medication. Our results with zimeldine were confirmed in another study (AMIT et al. 1985), in which 12 social drinkers reduced their alcohol intake during inpatient experimental drinking sessions. Although zimeldine was withdrawn from the market because of hepatic and neurologic toxicity (FAGIUS et al. 1985; NARANJO et al. 1990b), other serotonin uptake inhibitors, such as fluoxetine, citalopram, sertraline and paroxetine, have been approved for clinical use as antidepressants in many countries. More recent studies have been conducted with these drugs. For example, the effects of fluoxetine on alcohol consumption were tested in a double-blind, placebo-controlled inpatient study (GORELICK and PAREDES 1992). Twenty alcohol-dependent males were housed in a locked hospital ward and had measured alcoholic drinks available to them in a fixed interval drinking decision paradigm 13 times each day. Subjects received fluoxetine 20–60 mg/day ($n = 10$) or placebo daily ($n = 10$) for 28 days after a 3-day baseline period. The fluoxetine group decreased their alcohol intake by 14% from baseline during the 1st week of treatment only. During the last 3 weeks, alcohol intake was not significantly different from baseline. The placebo group showed no significant changes at any time period (less than 4% from baseline).

The mechanism of the effect of serotonin uptake inhibitors on alcohol intake is not fully understood. Subjects in our studies were not clinically depressed or anxious and changes in depression or anxiety were not observed. No consistent temporal relationship between the few side effects and decreases in drinking could be discerned. In addition, serotonin uptake inhibitors do not produce an alcohol-sensitizing reaction and have not been found to have any adverse pharmacokinetic interactions with alcohol (NARANJO and BREMNER 1992b; NARANJO et al. 1984a; LADER et al. 1986; SULLIVAN et al. 1989; SHAW et al. 1989; LEMBERGER et al. 1985).

The effect of serotonin uptake inhibitors appears to be fairly specific to alcohol intake. For example, there is no evidence from the studies described that these drugs decrease all fluid intake or other consummatory behaviors, because nonalcoholic drinks and cigarette smoking did not change (NARANJO and BREMNER 1992b; SELLERS et al. 1987). Weight loss was observed in most of the male heavy drinkers who received citalopram, viqualine, and fluoxetine. The average weight loss per 2 weeks was 0.3–1.2 kg and could not be accounted for by decreased calories from alcohol. Moreover, the lower doses of serotonin uptake inhibitors decreased body weight, but not alcohol intake. Changes in body weight and alcohol intake did not correlate, suggesting that these effects are independent (NARANJO and BREMNER 1992b). As loss of appetite was frequently reported during citalopram 40 mg/day and fluoxetine 60 mg/day (p's < 0.05 compared with placebo), food intake probably decreased. Therefore, serotonin uptake inhibitors may decrease appetite, or desire to eat, and also decrease desire to drink alcohol. In outpatient studies, increases in abstinent days were frequently observed

(NARANJO et al. 1984b, 1987), indicating that these drugs may exert their main effect before the initiation of drinking. Decreases in desire to drink were reported with zimeldine (AMIT et al. 1985), fluoxetine (GORELICK and PAREDES 1992), and viqualine (NARANJO et al. 1989).

In order to determine the importance of this mechanism, the effects of citalopram (40 mg/day) and fluoxetine (60 mg/day) on desire to drink were directly assessed in two recent studies, in which drinking sessions were conducted in an experimental bar immediately following one or two out-patient weeks of serotonin uptake inhibitor and placebo treatments. Subjects ($n = 16$ per study) were offered alcoholic minidrinks at 5-min intervals during the experimental drinking sessions. Subjects consumed the maximum number of minidrinks (i.e., up to 18, which is equivalent to 6 standard drinks) that they could and rated their desire for alcohol, intoxication, and mood. There was some indication that citalopram decreased the desirability of alcohol early in the experimental drinking session (NARANJO et al. 1992d). Fluoxetine almost completely suppressed desire for alcohol ratings in the experimental drinking sessions compared with placebo (NARANJO et al. 1992b). Also, decreases in liking for alcohol were detected. These results indicate that decreases in interest and desire (urge to drink), and liking (reinforcing effects) of alcohol may mediate the effects of serotonin uptake inhibitors.

This mechanism of action should be considered in the development of treatments to reduce alcohol intake and prevent relapse. Serotonin uptake inhibitors could be combined with a psychological treatment for alcohol abuse in order to maximize the drug effect and assess long-term effects of the combination of treatments. In a recent study 62 mildly/moderately dependent alcoholics received, double-blind, citalopram 40 mg/day ($n = 31$) or placebo ($n = 31$) in conjunction with a brief cognitive-behavioral treatment program (SANCHEZ-CRAIG and WILKINSON 1987). The short-term reduction in alcohol intake by citalopram ($p < 0.05$ compared with placebo) was potentiated by the nonpharmacologic intervention (NARANJO et al. 1992a). Citalopram had no overall significant effect, compared with placebo, on alcohol intake during the entire 12-week treatment period. However, among the 31 subjects who received citalopram, the males responded significantly better than the females. The average decreases from baseline alcohol intake were 44% for the males ($n = 16$) and 26% for the females ($n = 15$). The reductions in alcohol intake were maintained into the 5th–8th week post-treatment for subjects in both treatment groups; average decreases from baseline were 46% (post-citalopram) and 53% (post-placebo). Concomitant decreases in alcohol dependence (Alcohol Dependence Scale scores) and alcohol-related problems (MAST scores) were observed (NARANJO and BREMNER 1993).

Recently, several serotonin receptor subtypes have been identified and serotonergic agents have been classified according to the one(s) upon which they act (WATSON and GIRDLESTONE 1993). Many drugs, both agonists and antagonists, have been tested in animals and humans in order to determine which receptor subtypes are most important in the regulation of alcohol intake (SELLERS et al. 1992). In preclinical studies, the decrease in alcohol intake observed during administration of the serotonin releaser and uptake inhibitor dexfenfluramine was reversed by pretreatment with the $serotonin_{1/2}$ antagonist metergoline and the $serotonin_2$ antagonist ritanserin, but not by the $serotonin_3$ antagonist ondansetron, suggesting that central $serotonin_{1/2}$ receptors may mediate the effect (SELLERS et al. 1992). Further research is required to determine the relevance of these observations to humans. Buspirone, a $serotonin_{1A}$ agonist, has been studied in humans. In one study, buspirone (average dose 20.5 mg/day) reduced alcohol craving and anxiety, but not alcohol intake, compared with placebo in 50 mild-to-moderate alcohol-dependent subjects, 40% of whom also had evidence of mild-to-moderate anxiety (BRUNO 1989). Similarly, buspirone (average dose 42.25 mg/day) reduced anxiety and desire for alcohol in 42 patients with a dual diagnosis of generalized anxiety/alcohol abuse or dependence who had abstained from alcohol for 30–90 days (TOLLEFSON et al. 1992). Thus, its effects on alcohol craving may be secondary to the alleviation of anxiety, and buspirone may be efficacious only in anxious alcoholics (KRANZLER and MEYER 1989). However, it may be useful therapeutically in these patients. A $serotonin_{1c}$ and $serotonin_3$ agonist, mCPP, decreased oral ethanol self-administration in rats, but it increased craving for alcohol and the urge to drink, and produced ethanol-like effects in abstinent alcoholics in two studies (BENKELFAT et al. 1991; KRYSTAL et al., in press).

Some selective serotonin antagonists have been tested for their direct effects on alcohol intake. The $serotonin_3$ antagonist ondansetron decreased alcohol intake in rats (SELLERS et al. 1992) and was studied in 71 male alcohol abusers. Patients were randomly assigned to receive ondansetron 0.25 mg or 2 mg or placebo p.o. b.i.d. for 6 weeks after a 2-week baseline period. All patients also received a brief psychosocial intervention for alcoholism. Ondansetron 0.25 mg b.i.d. reduced alcohol intake from baseline, compared with ondansetron 2 mg b.i.d. or placebo, with the effect increasing in the second half of the treatment period (SELLERS et al. 1991, 1992). Studies with the $serotonin_2$ antagonist ritanserin have yielded no conclusive results. In rats, it has been reported to reduce alcohol intake and preference in a dose-dependent manner (MEERT and JANSSEN 1991), and to reduce alcohol intake and measures of alcohol withdrawal (MEERT et al. 1993), but it had no effect in another study (SELLERS et al. 1992). In a small, uncontrolled, open clinical study with five male abstinent alcoholics, ritanserin 10 mg/day reportedly reduced desire to drink and helped to maintain abstinence during the 28-day treatment period and a 15-day follow-up placebo period (medication administered single-blind) (MONTI and ALTERWAIN 1991). However, in a double-blind placebo-controlled study with 39 mildly/moderately dependent alcoholics ritanserin 5 mg/day decreased desire and craving for alcohol compared with baseline, but not compared with ritanserin 10 mg/day or placebo treatments. Neither dose of ritanserin significantly reduced alcohol intake (NARANJO et al. 1993).

In summary, many serotonergic drugs show considerable promise as novel pharmacotherapies for alcohol dependence. Further research is required to enhance the size and duration of their effects (e.g., by combining them with psychosocial therapies and with other medications) and to determine the precise neuropharmacologic mechanisms underlying their clinical effects.

C. Dopamine

Alcohol induces the release of dopamine in some of the regions of rat brain in which reinforcement is mediated; this effect is more marked in high alcohol-preferring than in low alcohol-preferring rats (ENGEL et al. 1992). A low endogenous level of brain dopamine has been found in alcohol-preferring rats and increasing dopamine tone with L-dopa and carbidopa reduced alcohol consumption in these rats (GEORGE et al. 1993). However, other studies do not provide clear evidence for the role of dopamine in ethanol consumption. For example, there are contradictory reports of reductions in ethanol intake in rats following treatment with both bromocriptine, a dopamine agonist (WEISS et al. 1990), and haloperidol, a dopamine antagonist (SAMSON et al. 1990). A possible interpretation of these results is that dopamine blockade reduces alcohol's reinforcing effects (and therefore intake) while dopamine receptor activation substitutes for alcohol intake (NARANJO et al. 1992c).

The importance of the dopaminergic system in human alcohol dependence and consumption has yet to be confirmed by clinical studies. The dopamine agonist bromocriptine alleviated alcohol withdrawal symptoms (BORG and WEINHOLDT 1982) and possibly decreased alcohol consumption in humans (BORG 1983). In a more recent study, severely dependent alcoholics were randomly assigned to receive bromocriptine 2.5 mg t.i.d. ($n = 20$) or placebo ($n = 18$) for 7 weeks following a 1-week baseline period and a 9-day titration of bromocriptine from 1.25 mg/day (placebo was dosed similarly). Significant decreases in alcohol consumption were observed with both bromocriptine and placebo treatments, compared with baseline, but there were significant differences in favor of bromocriptine in various psychopathological measures such as the SCL-90-R global severity index, interpersonal sensitivity, and anger/hostility (DONGIER et al. 1991). A large multicenter trial to determine the effects of bromocriptine on alcohol dependence is underway.

D. Opioids

A genetically determined deficiency of opioid function may predispose to alcoholism. Alcohol-preferring mice have low opioid tone, perhaps due to enhanced enkephalin degradation (GEORGE et al. 1991). Both prevention of enkephalin degradation with the enkephalinase inhibitor kelatorphan and increasing opioid activity with a μ-receptor opioid agonist decreased alcohol consumption in these mice (GEORGE et al. 1991). An alternative hypothesis, that a surfeit of opioid activity potentiates alcohol drinking, is based on the observation that alcohol consumption in animals increases with the prior administration of a small dose of an opiate agonist such as morphine and decreases when opiate receptors are blocked by opiate antagonists such as naltrexone (HUBBELL and REID 1990; BERG et al. 1992; REID and HUBBELL 1992).

Naltrexone, an opiate antagonist, was tested in two clinical studies using a variety of subjective and objective measures to assess alcohol intake and craving for alcohol. Seventy detoxified male alcoholics received 12 weeks of naltrexone 50 mg/day ($n = 14$) or placebo ($n = 16$), following a 1-week placebo baseline. Naltrexone significantly reduced alcohol intake, drinking days, and craving for alcohol compared with placebo (VOLPICELLI et al. 1992). Preliminary results from this initial study led to a larger study with 97 alcohol-dependent male and female patients. Patients received either naltrexone 50 mg/day or placebo and either weekly individual coping skills relapse prevention sessions or supportive psychotherapy sessions, for 12 weeks. They were required to abstain from alcohol for 7–30 days before beginning treatment. Naltrexone was better than placebo on a number of measures related to alcohol use and medication interacted with the type of psychotherapy on some measures. For example, 61% of the patients who received naltrexone and supportive therapy and 28% of those who received naltrexone and coping skills therapy remained abstinent for 12 weeks compared with 21% of the placebo/coping skills patients and 19% of the placebo/supportive therapy patients. The naltrexone-treated patients drank on an average of 4.3% of the study days, and drank a total of 13.7 drinks during the trial on the average, while the placebo-treated patients drank on 9.9% of the study days and a total of 38.0 drinks. Ratings of craving for alcohol were lowest for the naltrexone/coping skills patients. Naltrexone was well tolerated and was also associated with better patient retention than placebo (O'MALLEY et al. 1992).

E. Medications Acting on Other Neurotransmitter Systems

Recent research has focused on the role of the inhibitory neurotransmitter γ-aminobutyric acid (GABA) in the regulation of alcohol intake (AMIT and SMITH 1992). The administration of a GABA agonist, calcium acetylhomotaurinate, reduced voluntary alcohol consumption in rats (BOISMARE et al. 1984). This drug, named acamprosate, has been reported to reduce relapse in recently detoxified alcoholics. In a double-blind, placebo-controlled, randomized, multicenter study, 181 of the 279 subjects who were randomized to acamprosate (1.3 g/day) completed the 3-month treatment period. These subjects had significantly lower plasma γ-glutamyl transpeptidase (the major efficacy criterion used as an indicator of recent alcohol ingestion) than the 175 placebo subjects who completed the study (of 290 randomized to placebo) (LHUINTRE et al. 1990). Amounts of alcohol consumed were not reported. In a 6-month multicenter study in 102 recently detoxified alcoholics, calcium acetyl-homotaurinate (1999 mg/day) improved compliance to follow-up, length of abstinence, and clinical global assessment compared with placebo (PELC et al. 1992). This medication shows promise but more studies are needed to determine the role of GABA in the moderation of alcohol consumption and alcohol-induced effects.

F. Conclusions

Much progress has been achieved in the identification and testing of neuropsychopharmacologic agents to decrease alcohol consumption. New experimental paradigms have been developed (e.g., experimental drinking sessions) to test specific hypotheses related to drug effects. The quality of clinical studies has been improved by better descriptions of patients and identification of subgroups, the use of larger sample sizes and placebo-controlled designs, and isolation and characterization of the pharmacologic effect (e.g., size, possible mechanisms, dose-response relationships). Agents acting on serotonin (e.g., serotonin uptake inhibitors), endogenous opioids (e.g., naltrexone), and dopamine (e.g., bromocriptine) have been tested and research continues to explore systematically their therapeutic potentials. In particular, we need to determine better ways to produce a synergistic effect of pharmacologic and nonpharmacologic treatments, since human drinking behavior is regulated by many types of factors, some of which may be amenable to psychosocial or cognitive-behavioral therapies. Because the effect of the nonpharmacologic interventions may interact with or mask that of a medication, large-scale studies (e.g., multicenter trials) using standardized nonpharmacologic treatments and adequate control groups may be required to provide a firm basis for pharmacologic efficacy. Similarly, since it is unlikely that a single neurotransmitter system regulates desire for alcohol, and its intake and effects, and since experimental evidence suggests the involvement of several neurotransmitter systems, studies to test combinations of medications may be warranted. Issues of drug interactions and safety in such studies, however, must be carefully scrutinized.

In conclusion, considerable progress has been made in the neuropharmacologic treatment of alcohol dependence, and the days in which disulfiram and similar medications were the only options seem far away. Researchers continue to investigate new pharmacotherapies for alcohol abuse in both basic and clinical studies. Better use of collaborative networks will accelerate progress.

References

American Psychiatric Association (1980) Diagnostic and statistical manual of mental disorders, 3rd edn. Washington DC

American Psychiatric Association (1987) Diagnostic and statistical manual of mental disorders, 3rd edn, revised. Washington DC

Amit Z, Smith BR (1992) Neurotransmitter systems regulating alcohol intake. In: Naranjo CA, Sellers EM (eds) Novel pharmacological interventions for alcoholism. Springer, Berlin Heidelberg New York, pp 161–183

Amit Z, Brown Z, Sutherland A, Rockman G, Gill K (1985) Reduction in alcohol intake in humans as a function of treatment with zimelidine: implications for treatment. In: Naranjo CA, Sellers EM (eds) Research advances in new psychopharmacological treatment for alcoholism. Elsevier Science, Amsterdam, pp 189–198

Benkelfat C, Murphy DL, Hill JL et al. (1991) Ethanollike properties of the serotonergic partial agonist m-chlorophenylpiperazine in chronic alcoholic patients. Arch Gen Psychiatry 48:383

Berg BJ, Volpicelli JR, Alterman AI, O'Brien CP (1992) The relationship of alcohol drinking and endogenous opioids: the opioid compensation hypothesis. In: Naranjo CA, Sellers EM (eds) Novel pharmacological interventions for alcoholism. Springer, Berlin Heidelberg New York, pp 137–141

Boismare F, Daoust M, Moore N et al. (1984) A homotaurine derivative reduces the voluntary ethanol intake by rats: are cerebral GABA receptors involved? Pharmacol Biochem Behav 21:787–789

Borg V (1983) Bromocriptine in the prevention of alcohol abuse. Acta Psychiatr Scand 68:100–110

Borg V, Weinholt T (1982) Bromocriptine in the treatment of the alcohol withdrawal syndrome. Acta Psychiatr Scand 65:101–111

Bruno F (1989) Buspirone in the treatment of alcoholic patients. Psychopathology 22 [Suppl 1]:49–59

Dongier M, Vachon L, Schwartz G (1991) Bromocriptine in the treatment of alcohol dependence. Alcohol Clin Exp Res 15:970–977

Engel JA, Enerback C, Fahlke C et al. (1992) Serotonergic and dopaminergic involvement in ethanol intake. In: Naranjo CA, Sellers EM (eds) Novel pharmacological treatments for alcoholism. Springer, Berlin Heidelberg New York, pp 68–82

Fagius J, Osterman PO, Siden A, Wiholm BE (1985) Guillain-Barre syndrome following zimeldine treatment. Neurol Neurosurg Psychiatry 48:65–69

George SR, Roldan L, Lui A, Naranjo CA (1991) Endogenous opioids are involved in the genetically determined high preference for ethanol consumption. Alcohol Clin Exp Res 15:668–672

George SR, Fan T, Jun Ng G, Naranjo CA (1993) Low endogenous dopamine function in brain predicts high alcohol preference and consumption: reversal by increasing dopamine neurotransmission (abstract). Alcohol Clin Exp Res 17:482

Gorelick DA, Paredes A (1992) Effects of fluoxetine on alcohol consumption in male alcoholics. Alcohol Clin Exp Res 16:261–265

Higgins GA, Lawrin MO, Sellers EM (1992) Serotonin and alcohol consumption. In: Naranjo CA, Sellers EM (eds) Novel pharmacological interventions for alcoholism. Springer, Berlin Heidelberg New York, pp 83–91

Hubbell CL, Reid LD (1990) Opioids modulate rats' intakes of alcohol beverages. In: Reid LD (ed) Opioids, bulimia and alcohol abuse and alcoholism. Springer, Berlin Heidelberg New York, pp 145–174

Kranzler HR, Meyer RE (1989) An open trial of buspirone in alcoholics. J Clin Psychopharmacol 9:379–380

Krystal JH, Webb E, Cooney N et al. (in press) Specificity of ethanol-like effects elicited by serotonergic and noradrenergic mechanisms: m-CPP and yohimbine effects in recently detoxified alcoholics. Arch Gen Psychiatry

Lader M, Melhuish A, Frcka G (1986) The effect of citalopram in single and repeated doses and with alcohol on physiological and psychological measures in healthy subjects. Eur J Clin Pharmacol 31:183–190
Lawrin MO, Naranjo CA, Sellers EM (1986) Identification and testing of new drugs for modulating alcohol consumption. Psychopharmacol Bull 22:1020–1025
Lemberger L, Rowe H, Bergstrom RF et al. (1985) Effect of fluoxetine on psychomotor performance, physiologic response and kinetics of ethanol. Clin Pharmacol Ther 37:658–664
Lhuintre JP, Moore N, Tran G et al. (1990) Acamprosate appears to decrease alcohol intake in weaned alcoholics. Alcohol Alcohol 25:613–622
McBride WJ, Murphy WJ, Lumeng L, Li T-K (1992) Serotonin and alcohol consumption. In: Naranjo CA, Sellers EM (eds) Novel pharmacological interventions for alcoholism. Springer, Berlin Heidelberg New York, pp 59–67
Meert TF, Janssen PAJ (1991) Ritanserin, a new therapeutic approach for drug abuse, part I: effects on alcohol. Drug Dev Res 24:235–249
Meert TF, Huysmans H, Melis W, Clincke GHC (1993) Ritanserin and other 5-HT agents on alcohol preference and alcohol withdrawal in rats (abstract). College on Problems of Drug Dependence, 55th annual scientific meeting, Toronto, Canada
Monti JM, Alterwain P (1991) Ritanserin decreases alcohol intake in chronic alcoholics. Lancet 337:60
Naranjo CA, Bremner KE (1992a) Clinical pharmacology of serotonin-altering medications for decreasing alcohol consumption. 6th congress of the ISBRA: symposium on role of serotonin in alcohol dependence, Bristol, UK, 22 June 1992. Alcohol Alcohol 27 [Suppl 1]:9
Naranjo CA, Bremner KE (1992b) Evaluation of the effects of serotonin uptake inhibitors in alcoholics: a review. In: Naranjo CA, Sellers EM (eds) Novel pharmacological interventions for alcoholism. Springer, Berlin Heidelberg New York, pp 105–117
Naranjo CA, Bremner KE (1993) Treatment related attenuations of alcohol intake, alcohol dependence (AD) and problems (abstract). Clin Pharmacol Ther 53:176
Naranjo CA, Sellers EM (eds) (1992) Novel pharmacological interventions for alcoholism. Springer, Berlin Heidelberg New York
Naranjo CA, Sellers EM, Kaplan HL et al. (1984a) Acute kinetic and dynamic interactions of zimelidine with ethanol. Clin Pharmacol Ther 36:654–660
Naranjo CA, Sellers EM, Roach CA et al. (1984b) Zimelidine-induced variations in alcohol intake by non-depressed heavy drinkers. Clin Pharmacol Ther 35:374–381
Naranjo CA, Sellers EM, Sullivan JT et al. (1987) The serotonin uptake inhibitor citalopram attenuates ethanol intake. Clin Pharmacol Ther 41:266–274
Naranjo CA, Sullivan JT, Kadlec KE et al. (1989) Differential effects of viqualine on alcohol intake and other consummatory behaviors. Clin Pharmacol Ther 46:301–309
Naranjo CA, Kadlec KE, Sanhueza P et al. (1990a) Fluoxetine differentially alters alcohol intake and other consummatory behaviours in problem drinkers. Clin Pharmacol Ther 47:490–498
Naranjo CA, Lane D, Ho-Asjoe M, Lanctôt KL (1990b) A Bayesian assessment of idiosyncratic adverse reactions to new drugs: Guillain-Barre syndrome and zimeldine. J Clin Pharmacol 30:174–180
Naranjo CA, Bremner KE, Lanctôt KL (1992a) Short- and long-term effects of citalopram (C) combined with a brief psychosocial intervention (BPI) for alcoholism (abstract). Clin Pharmacol Ther 51(2):168
Naranjo CA, Bremner KE, Poulos CX, Lanctôt KL (1992b) Fluoxetine decreases desire for alcohol (abstract). Clin Pharmacol Ther 51(2):168
Naranjo CA, George SR, Bremner KE (1992c) Novel neuropharmacological treatments of alcohol dependence. Clin Neuropharmacol 15 [Suppl 1, A]:74A–75A

Naranjo CA, Poulos CX, Bremner KE, Lanctôt KL (1992d) Citalopram decreases desirability, liking and consumption of alcohol in alcohol-dependent drinkers. Clin Pharmacol Ther 51:729–739

Naranjo CA, Poulos CX, Umana M, Lanctôt KL, Bremner KE (1993) Effects of ritanserin (R) on desire to drink and consummatory behaviors (CB) in heavy alcohol drinkers (abstract). Clin Pharmacol Ther 53:176

O'Malley SS, Jaffe A, Chang G et al. (1992) Naltrexone and coping skills therapy for alcohol dependence: a controlled study. Arch Gen Psychiatry 49:881–887

Pelc I, Le Bon O, Verbanck P et al. (1992) Calcium-acetylhomotaurinate for maintaining abstinence in screened alcoholic patients: a placebo-controlled double-blind multicentre study. In: Naranjo CA, Sellers EM (eds) Novel pharmacological interventions for alcoholism. Springer, Berlin Heidelberg New York, pp 348–352

Reid LD, Hubbell CL (1992) Opioids modulate rats' propensities to take alcoholic beverages. In: Naranjo CA, Sellers EM (eds) Novel pharmacological interventions for alcoholism. Springer, Berlin Heidelberg New York, pp 121–134

Samson HH, Tolliver GA, Schwarz-Stevens K (1990) Oral ethanol self-administration: a behavioral pharmacologic approach to CNS control mechanisms. Alcohol 7:187–191

Sanchez-Craig M, Wilkinson DA (1987) Theory and methods for secondary prevention of alcohol problems: a cognitively based approach. In: Cox WM (ed) Treatment and prevention of alcohol problems: a resource manual. Academic, New York, pp 287–331

Sellers EM, Naranjo CA, Kadlec K (1987) Do serotonin uptake inhibitors decrease smoking? Observations in a group of heavy drinkers. J Clin Psychopharmacol 7:417–420

Sellers EM, Romach MK, Frecker RC, Higgins GA (1991) Efficacy of the 5-HT_3 antagonist ondansetron in addictive disorders, In: Racagni G, Brunello N, Fukudo T (eds) Proceedings of the 5th world congress of biological psychiatry, vol 2. Elsevier Science, Amsterdam, pp 894–897

Sellers EM, Higgins GA, Tomkins DM, Romach MK (1992) Serotonin and alcohol drinking. US Government Printing Office, Washington DC, pp 141–145 (NIDA research monograph series 119)

Shaw CA, Sullivan JT, Kadlec KE et al. (1989) Ethanol interactions with serotonin selective and non-selective antidepressants: fluoxetine and amitriptyline. Hum Psychopharmacology 4:113–120

Sullivan JT, Naranjo CA, Shaw CA (1989) Kinetic and dynamic interactions of oral viqualine and ethanol in man. Eur J Clin Pharmacol 36:93–96

Tollefson GD, Montague-Clouse J, Tollefson SL (1992) Treatment of comorbid generalized anxiety in a recently detoxified alcoholic population with a selective serotonergic drug (Buspirone). J Clin Psychopharmacol 12:19–26

Volpicelli JR, Alterman AI, Hayashida M, O'Brien CP (1992) Naltrexone in the treatment of alcohol dependence. Arch Gen Psychiatry 49:876–880

Watson S, Girdlestone D (1993) Receptor nomenclature supplement. Trends Pharmacol Sci 14:21–22

Weiss F, Mitchiner M, Bloom FE, Koob GF (1990) Free-choice responding for ethanol versus water in alcohol preferring (P) and unselected Wistar rats is differentially modified by naloxone, bromocriptine, and methysergide. Psychopharmacology 101:178–186

CHAPTER 16

Pharmacology of Gastrointestinal Comorbidity in Alcoholics

T.M. WORNER and P. MANU

A. Introduction

In this chapter, we will examine published data regarding patient cohorts assembled for specific pharmacologic interventions addressed to specified medical complications of heavy alcohol consumption. The review will evaluate only peer-assessed publications that: (1) included full disclosure of their methodology; (2) identified the proportion of alcoholics among heterogeneous patient groups with a given clinical diagnosis (e.g., pancreatitis); (3) used standard diagnostic criteria; and (4) employed a control group whose clinical characteristics matched those of the study group. Whenever possible, our presentation will describe two categories of facts: (1) confirmations (i.e., similar findings obtained by at least two independent research groups) and (2) contradictions (i.e., unequivocally opposite findings in independent studies with similar methodology).

The literature regarding pharmacologic approaches to the medical complications of alcoholism was reviewed by accessing the Medline database of the National Library of Congress available in the form of compact disk storage of the abstracts of all pertinent articles published from January 1988 through June 1992. The analysis of these abstracts did not reveal significant pharmacologic advances obtained by well-controlled studies of cardiovascular (e.g., alcohol-related hypertension, alcoholic cardiomyopathy, and arrhythmias), neurologic (e.g., Wernicke's encephalopathy, cerebellar degeneration, and peripheral neuropathy), renal (i.e., alcohol-induced rhabdomyolysis with renal failure), endocrinologic (e.g., testicular atrophy, amenorrhea, hypercortisolism), hematologic (i.e., pancytopenia, macrocytosis), and metabolic complications of alcoholism (i.e., magnesium and phosphate deficiency, alcoholic ketosis). When progress was made, the pharmacologic advance was due to work in the general area of, for example, the management of arterial hypertension, rather than in the specific field of alcohol-induced changes in blood pressure. However, substantial results have been published regarding the pharmacologic treatment of alcohol-induced gastrointestinal, pancreatic, and hepatologic complications; these advances will be described in detail. We will not address the medical managment of the alcohol withdrawal syndrome and its complications (alcohol withdrawal seizures, alcohol withdrawal delirium, and alcoholic hallucinosis) because pharmacologic advances in these areas are presented in detail by Anton and Becker (Chap. 14, this volume).

B. Drug Therapy for Alcohol-Induced Liver Disease

I. Steatosis

Hepatic steatosis is generally considered a benign stage of alcoholic liver disease which is reversible if the offending toxin (alcohol) is removed. A prospective Danish study suggests, however, that moderate-to-severe steatosis increases the risk of development of cirrhosis, even in the absence of alcoholic hepatitis (SØRENSEN et al. 1984).

In view of the demonstrated progression of steatosis to cirrhosis, one might consider the use of hepatoprotective drugs for the most severe stages of steatosis. Hepatoprotective agents such as (+)-cyanindanol-3, a free radical scavenger, however, have not proved efficacious in placebo-controlled studies (WORLD et al. 1987; ANONYMOUS 1982).

The bioflavonoid silymarin, an active extract of *Silbyium marinanum*, has been evaluated in a double-blind, randomized controlled trial in 97 young soldiers with persistent transaminase elevation. Steatosis was present in almost half of these recruits. Transaminases were significantly decreased in the actively treated patients when compared to controls. Seventy-three percent of treated patients versus 29% of controls showed histologic improvement on rebiopsy (SALMI and SARNA 1982). The clinical use of this agent awaits the results of other randomized controlled trials.

Protein, methionine, and choline deficiencies have been implicated in the pathogenesis of alcoholic liver disease. *S*-Adenosyl-L-methionine (SAMe), the activated form of methionine, is depleted in the livers of baboons fed alcohol chronically. SAMe supplementation results in improvement in the severity of histologic damage in the baboon (LIEBER 1990). By contrast, a 1-month inpatient, randomized, placebo-controlled trial of intravenous SAMe in human alcoholics did not demonstrate a beneficial effect of SAMe (T.M. Worner and C.S. Lieber, unpublished observations).

II. Alcoholic Hepatitis

Rapid deterioration is characteristic of the patient with severe alcoholic hepatitis. Mortality may approach 50% during the first 30 days of hospitalization (MADDREY 1988; SHERLOCK 1990). Malnutrition is common in patients with alcoholic hepatitis and has been correlated with mortality (MENDENHALL et al. 1986). Studies aimed at nutritional repletion of subjects with alcoholic hepatitis have, therefore, been undertaken. Randomized controlled trials of amino acid infusions, though initially encouraging, have produced contradictory results. Most, but not all, of the recent studies demonstrate improvement in biochemical and nutritional parameters, but no improvement in mortality (NASRALLAH and GALAMBOS 1980; CALVEY et al. 1985; DIEHL et al. 1985; ACHORD 1987; SIMON and GALAMBOS 1988; MEZEY et al. 1991; BIRD and WILLIAMS 1990). In view of the cost, as well as the associated morbidity, this form of nutritional supplementation cannot currently be recommended.

1. Insulin and Glucagon

In an attempt to stimulate hepatic regeneration, other investigators undertook randomized controlled trials of insulin and glucagon infusions. The first of these trials, published in 1981 (BAKER et al. 1981), demonstrated an improvement in biochemical parameters in the treated group, but no difference in mortality between treated subjects and controls. In the most recent studies, subjects were randomized to daily 12-h infusions of insulin and glucagon for 3 weeks or until hospital discharge, if it occurred earlier (TRINCHET et al. 1992; BIRD et al. 1991). Neither study demonstrated an improved outcome in the treated group.

2. Corticosteroids

Numerous other pharmacologic therapies have been evaluated. Corticosteroids have been the most extensively studied, with the first controlled trials having been reported in 1971. Though corticosteroids are theoretically advantageous due to their antiinflammatory and antifibrotic effects, contradictory results have been reported in the 12 published randomized controlled trials (BIRD and WILLIAMS 1990; REYNOLDS 1990). Eleven randomized studies of corticosteroids in the management of alcoholic hepatitis (10 of which were placebo-controlled) were published between 1971 and 1989. The results of these studies were evaluated by meta-analysis (IMPERIALE and MCCULLOUGH 1990). The authors concluded that corticosteroids decreased short-term mortality in those subjects with alcoholic hepatitis complicated by encephalopathy. These results were limited to studies in which acute gastrointestinal bleeding was an exclusion criterion for study entry. One randomized controlled trial has subsequently been published (RAMOND et al. 1992). Sixty-one alcoholics with biopsy-proven alcoholic hepatitis and either hepatic encephalopathy or a discriminant function value greater than 32 (an empirically derived value which identified patients who had a 50% risk of dying within 2 months) were randomly assigned to either placebo or prednisolone. The majority of subjects also had cirrhosis. Nearly 90% of the treated subjects were alive at month 2 as compared with only 45% of control subjects. Thus, for a subgroup of patients with alcoholic hepatitis, corticosteroids appear to produce short-term benefits.

3. Anabolic Steroids

Anabolic steroids have been thought to promote hepatic regeneration and to stimulate anabolism. Contradictory results, however, have been reported in randomized controlled trials of these drugs published prior to 1986 (BIRD and WILLIAMS 1990). A subsequent, small, randomized controlled trial of parenteral nutrition and oxandrolone has recently been reported (BONKOVSKY et al. 1991a). Of the 43 subjects recruited for the study, two exsanguinated prior to randomization, one signed out against medical advice, and one developed acute renal failure and sepsis. Of the remaining 39 subjects, 11 were Child's class B and 28 were class C. Subjects underwent 17 days of observation and specialized testing prior to enrollment into the 21-day treatment protocol. All subjects improved with time, but the group treated with nutritional supplementation and/or oxandrolone improved the most. Though the results of this study provide evidence of the short-term benefits of parenteral nutrition and/or oxandrolone on nitrogen metabolism and metabolic balance, they are not readily generalized to other patient samples (BONKOVSKY et al. 1991b). The extremely low mortality in this sample is at variance with other randomized controlled trials of alcoholic hepatitis. Despite the histologic documentation of hepatitis in most subjects, they were all well enough to participate in a 17-day protocol prior to enrollment into the treatment arm of the study. Thus, the use of anabolic steroids in the management of alcoholic hepatitis requires evaluation in a sample that includes more severely ill patients.

4. Colchicine

Due to the contradictory results obtained in controlled clinical trials with steroids, alternative treatments of alcoholic hepatitis have been examined. Colchicine, due to its known antiinflammatory effects and favorable effects on granulocytes, was prospectively employed in a randomized, 1-month placebo-controlled trial (AKRIVIADIS et al. 1990). Approximately 20% of each group died during the study period. There was no significant difference between groups in laboratory parameters or in the degree of neutrophilic infiltration found on repeat liver biopsy. Results of a 6-month, randomized placebo-controlled trial of colchicine also demonstrated no beneficial effect of the active drug in 67 subjects, one-third of whom also had cirrhosis (TRINCHET et al. 1989). However, more than one-fourth of the subjects were lost to follow-up at 3 months and over one-half of the subjects could not be located for the 6-month follow-up. Thus, at the present time, colchicine should be used only in controlled clinical trials.

5. Propylthiouracil

Another alternative to steroids has been propylthiouracil. The rationale for use of this drug, first tested in rats fed alcohol chronically, is that it would ameliorate the hypermetabolic state in the liver, which is associated with an increased rate of oxygen consumption (ISRAEL et al. 1975). One hundred and thirty-three inpatient alcoholics were randomly assigned to placebo or propylthiouracil. The treatment protocol was continued until the subjects were stable enough to be discharged from the hospital, with a predetermined maximum treatment period of 46 days. Treatment response was measured with a composite clinical and laboratory index. Short-term improvement was observed only in those subjects with documented alcoholic hepatitis. There was no difference in mortality between the two groups (ORREGO et al. 1979). Another 6-week, double-blind placebo-controlled trial of propylthiouracil, which employed a similar study design, demonstrated no advantage to propylthiouracil either on biochemical parameters or on mortality (HALLÉ et al. 1982). A 2-year study, in which subjects were randomly assigned to propylthiouracil or placebo, demonstrated a striking reduction in mortality in the active treatment group (ORREGO et al. 1987). Pending further efficacy studies, propylthiouracil will remain an experimental drug for the treatment of alcoholic hepatitis (MADDREY 1979; KAPLOWITZ 1982; ANONYMOUS 1988b).

III. Cirrhosis

Management of cirrhosis has focused primarily on the modification of etiologic factors such as heavy alcohol consumption or on the treatment of major complications (vide infra). Attempts to reverse the fibrotic process have been limited. Though numerous drugs are capable of modifying collagen metabolism, only colchicine has been evaluated in humans with alcoholic cirrhosis. Among its many actions, colchicine interferes with microtubular assembly, inhibits collagen synthesis and secretion, and enhances collagenase activity (DIEGELMAN and PETERKOFSKY 1972; ERLICH and BORNSTEIN 1972; ERLICH et al. 1974; HARRIS and KRANE 1971). One hundred patients in a randomized placebo-controlled trial of colchicine have been followed for up to 14 years (KERSHENOBICH et al. 1979, 1988). Colchicine-treated patients had longer overall survival (by 7.5 years) and a rate of mortality from hepatic failure that was 9% lower than the placebo-treated patients. In addition, 9 of 30 subjects in the colchicine-treated group underwent repeat liver biopsy: all showed histologic improvement. By contrast, no such improvement was observed in 14 placebo-treated subjects. The results of this trial must be interpreted cautiously, however, for several reasons. First, random assignment resulted in a more favorable patient profile in the treated group. Second, survival outcome could not be determined in almost 20% of subjects. Third, compliance with the treatment regimen was not documented (BOYER and RANSOHOFF 1988). At this time, colchicine can only be recommended for treatment as part of a controlled clinical trial.

1. Ascites

A number of theories, including the "underfill" and "overflow" and more recently the "peripheral arterial vasodilation" theories (SCHRIER et al. 1988) of ascites formation, continue to be debated. Despite a limited understanding of the pathophysiology of ascites formation, there is general agreement that the renal regulation of salt and water is impaired. Ultimately, sodium is conserved. Thus, initial treatment includes sodium restriction, bed rest, and, if indicated, potassium repletion. Water restriction, however, remains controversial (FRAKES 1980). Weight loss of 0.5–1 kg/day is considered optimal, though more rapid weight loss is safe in those subjects with peripheral edema. Approximately 5%–15% of cirrhotics will respond to this regimen (ROCCO and WARE 1986). If less than 0.5 kg/day weight loss occurs over 3–4 days, spironolactone in lower dosage (e.g., 25 mg q.i.d.) may be added to the treatment regimen. As reviewed by BOYER and WARNOCK (1983), up to 90% of cirrhotics will experience a satisfactory diuresis when the dosage of spironolactone is increased to 300–600 mg/day. For the nonresponders, a loop diuretic can be added.

Until the introduction of diuretics in the 1950s, therapeutic paracentesis was a standard treatment modality. Concerns about adverse effects, including deterioration in renal function, hemodynamic compromise, encephalopathy, and bacterial peritonitis caused paracentesis to fall into disfavor by the 1960s (REYNOLDS 1987; KELLERMAN and LINAS 1990). Recent controlled studies have shown that large volume paracentesis (4–6 l/day), together with intravenous albumin infusion, was as effective as diuretics in reducing both edema and the duration of the inpatient hospital stay (QUINTERO et al. 1985; GINÈS et al. 1987). More recently, it has been suggested that tense ascites can be treated by "total paracentesis" followed by diuretic administration (spironolactone and furosemide) to prevent recurrence of ascites (TITÓ et al. 1990). It has now been suggested that large volume paracentesis be considered the initial treatment for ascites (ANONYMOUS 1988a; KELLERMAN and LINAS 1990; EPSTEIN 1989). Alternatively, for selected patients with massive, refractory ascites, peritovenous shunting remains an option (GINÈS et al. 1991; SHOCKET 1992; STANLEY et al. 1989).

2. Encephalopathy

For discussions of pathogenesis, diagnostic evaluation, and general management of hepatic encephalopathy, the reader is referred to recent overviews (GAMMAL and JONES 1989; SZAUTER and MULLEN 1989). Several controlled studies of the efficacy of branched-chain amino acids or enteral nutrition in the management of encephalopathy have been published (EGBERTS et al. 1985; HORST et al. 1984; CABRE et al. 1990; KEARNS et al. 1992). The results of these trials have been contradictory (ERIKSSON and CONN 1989; NAYLOR et al. 1989; BERLIN and CHALMERS 1989). The use of branched-chain amino acids cannot be recommended as standard therapy for hepatic encephalopathy at this time.

Lactulose has been used for at least 20 years in the management of encephalopathy and is considered by many to be the drug of choice. Though the exact mechanism of action of the drug is unknown, the improvement in encephalopathy has been attributed to its cathartic properties, which reduce the time available both for the production and absorption of toxins. Lactulose also lowers colonic pH, thus suppressing both the generation and absorption of ammonia. Dosage of the drug is titrated to achieve two to four loose bowel movements each day. Toxic effects include hypernatremia, hypomagnesemia, and perianal irritation. More recently, lactitol, a disaccharide analogue of lactulose, has been proposed as an alternative treatment. The onset of this drug's action is more rapid than lactulose and its side effect profile is more favorable (ANONYMOUS 1987).

With the widespread acceptance of lactulose, the use of broad-spectrum antibiotics has fallen into some disfavor due to the serious side effects associated with their use, including nephrotoxicity, ototoxicity, and antibiotic-associated colitis. However, for selected patients, neomycin, 4–6 g q.d. in divided doses is effective. It is thought to improve encephalopathy by reducing the concentration of urease-containing bacteria in the gut, thereby decreasing ammonia production. Alternatively, metronidazole, which is active against *Bacteroides* and other anaerobes, is also effective. Since metronidazole is metabolized by the liver, a dosage lower than the 500–750 mg/day that is generally prescribed is required.

The most recent addition to the pharmacologic armamentarium in the treatment of hepatic encephalopathy is flumazenil, a benzodiazepine receptor antagonist. Traditionally, ammonia, multiple neurotoxins, false neurotransmitters, or an imbalance of amino acids have been implicated in the pathogenesis of hepatic encephalopathy. More recently, two additional hypotheses have been propounded. The first of these is the highly controversial true neurotransmitter or GABA hypothesis (SCHAFER and JONES 1982). Several lines of evidence support this hypothesis. First, when GABA, the principal inhibitory neurotransmitter of mammalian brain, was applied to the hippocampal region of conscious rabbits, spontaneous motor activity declined and was associated with the development of spreading δ-wave activity on the electroencephalogram. Similar electroencephalographic observations have been reported in man and rabbits with hepatic encephalopathy. Second, in the rabbit model of galactosamine-induced hepatic failure, as the liver fails, plasma levels of GABA rise for several hours prior to the onset of encephalopathy. Third, enteric flora, particularly *Escherichia coli*, are a source of GABA. Fourth, based on studies with ^{14}C-α-aminoisobutyric acid, it has been hypothesized that GABA enters the brain during liver failure. Fifth, an increased number of GABA receptors have been observed during liver failure. Based on these observations, Schafer and Jones have hypothesized that "in liver failure gut derived GABA passes through a permeable blood-brain barrier and induces its own receptors on postsynaptic neural membranes, possibly increasing the sensitivity of the brain to GABA-ergic neural inhibition." This hypothesis has subsequently been expanded to include other neurotransmitter systems. The second hypothesis is the endogenous benzodiazepine hypothesis, an extension of the GABA hypothesis, whih proposes that an endogenous substance with benzodiazepine agonist-like properties augments GABA-ergic neurotransmission (MULLEN et al. 1988). This hypothesis stresses the events occurring at the benzodiazepine-binding site, which ultimately potentiate the action of GABA. Both elevations of concentrations of 1,4-benzodiazepines and "endogenous" benzodiazepines have been reported in humans (BASILE et al. 1991; MULLEN et al. 1990). In two small clinical trials encephalopathy has been reported to improve with varying doses of intravenous flumazenil (GRIMM et al. 1988; BANSKY et al. 1989). There is one case report of the successful long-term use of flumazenil (FERENCI et al. 1989). To date, no controlled, double-blind studies of flumazenil for treatment of hepatic encephalopathy have been published.

3. Esophageal Varices

Management of esophageal varices remains controversial (TERBLANCHE et al. 1989a,b). There have been numerous randomized clinical trials to assess the efficacy of β-blockers, sclerotherapy, and/or surgical shunts to prevent eigher a first or recurrent variceal hemorrhage. The reader is referred to recent controlled trials of surgical shunts or sclerotherapy (CELLO et al. 1987; SAUERBRUCH et al. 1988; O'CONNOR et al. 1989; BURROUGHS et al. 1989; HENDERSON et al. 1990; VA COOPERATIVE VARICEAL SCLEROTHERAPY GROUP 1991; PLANAS et al. 1991; STIEGMANN et al. 1992).

A meta-analysis of 9 randomized controlled trials of β-blockers and 19 comparable trials of sclerotherapy concluded that β-blockers are effective for the prevention of a first bleeding episode in cirrhotics with "high-risk" varices. The use of sclerotherapy as prophylaxis remains to be demonstrated (PAGLIARO et al. 1992). Other meta-analyses on the use of β-blockers have reached similar conclusions (POYNARD et al. 1991; HAYES et al. 1990; PAGLIARO et al. 1990). Though controlled studies of β-blockers have demonstrated a reduction in first or recurrent bleeding episodes, the reduction in mortality has been variable (PASCAL et al. 1987; COLOMBO et al. 1989; GARDEN et al. 1990; ANDREANI et al. 1990; ROSSI et al. 1991). Since only one-third of cirrhotics with esophageal varices hemorrhage, prophylaxis with propranolol is recommended only for the high-risk subgroup (POLIO and GROSZMANN 1989; DE FRANCHIS et al. 1988). Propranolol therapy can be started at 40 mg p.o. b.i.d. and titrated upward until the resting heart rate has been decreased by 25%.

IV. Liver Transplantation

Orthotopic liver transplantation in the alcoholic has become a reality in several major centers in the United States. Long-term outcome data indicate that survival is comparable to a nonalcoholic control group if the alcoholics are carefully screened prior to transplantation (Van Thiel et al. 1991; Lucy et al. 1992). Transplant candidates must be able to undergo the rigorous treatment schedule of chronic immunosuppressive therapy, which currently includes cyclosporin, prednisone, and azathioprine (Starzl et al. 1988). Intravenous bolus methylprednisone, followed by a tapering steroid schedule, is used to treat acute rejections. For persistent rejections, antilymphocyte globulin, antithymocyte globulin, or the monoclonal OKT_3 immunoglobulin is added (Dindzans et al. 1989).

C. Drug Therapy of Acute Alcoholic Pancreatitis

1. Cimetidine

Cimetidine, an H_2-receptor antagonist widely used for the treatment of peptic ulcer disease and other hypersecretory states of the gastrointestinal tract, has been tried in patients with acute alcoholic pancreatitis on the basis of encouraging preclinical investigations (Hadas et al. 1978; Joffe and Lee 1978). The rationale for this therapeutic approach is to diminish the release of pancreatic stimulating hormones by decreasing the entry of acid gastric secretions into the duodenum.

The efficacy of cimetidine was well tested in two studies (Meshkinpour et al. 1979; Goff et al. 1982). The first study used a placebo-controlled, double-blind design to study patients presenting for evaluation with a chief complaint of upper abdominal pain. Patients also had a confirmed history of excessive alcohol use for many years, alcohol ingestion during the week preceding onset of the abdominal pain, and either abnormally high serum lipase levels or high serum amylase levels with an abnormal amylase-creatinine clearance ratio (Meshkinpour et al. 1979). Patients with hypotension, oliguria, and other indications of severe hemodynamic abnormalities were excluded. The studied cohort consisted of 27 patients, 26 of whom were male. Fourteen randomly selected patients (13 men and 1 woman, mean age 43.1 years) received 300 mg cimetidine hydrochloride diluted in 20 ml 0.15 *M* NaCl solution intravenously. The same volume of 0.15 *M* NaCl solution served as placebo and was administered to 13 patients (all men, mean age 45.4 years). The patients started the protocol within 12 h of admission and received cimetidine or placebo intravenously at 6-h intervals until one of the following three end points was reached: pain-free status lasting 24 h, or the patient's condition required a surgical intervention, or 120 h had elapsed since the initial dose. Clinical status was assessed by a physician unaware of whether the patient was given cimetidine or placebo.

The results indicated no statistically significant difference between the cimetidine and placebo groups. Further stratification of the patient groups according to the severity of the clinical condition also failed to demonstrate a beneficial effect of treatment with cimetidine. A trend toward more persistent hyperamylasemia was noted among those patients receiving cimetidine.

These results were confirmed by a second study, a large randomized trial comparing nasogastric suction to cimetidine in acute pancreatitis (GOFF et al. 1982). Although nasogastric suction has traditionally been considered part of the primary therapeutic regimen for patients with acute alcoholic pancreatitis, recent studies have raised doubts about its efficacy (LEVANT et al. 1974; NAEIJE et al. 1978; FIELD et al. 1979). Moreover, nasogastric suction may create substantial physical and psychological discomfort; the potential for hypochloremia, metabolic alkalosis, and other acid-base and electrolyte abnormalities; and the risk of mechanical injury to the esophagogastric mucosa. In the study by GOFF and colleagues (1982), 95 patients with 103 episodes of abdominal pain and elevated serum or urine amylase levels were randomized to receive either continuous nasogastric suction or 300 mg cimetidine intravenously every 6 h. Cimetidine was used in 57 cases and nasogastric suction in 46 cases; 86.4% of episodes of pancreatitis were alcohol-related. The trial showed a significantly shorter duration of hospitalization for the episodes of pancreatitis in the patients treated with cimetidine (mean 6.8 days), as compared with the group treated with nasogastric suction (mean 8.5 days); all other clinical variables were similar in the two groups. Therefore, both studies reviewed here indicate that treatment with cimetidine does not affect the outcome of acute alcoholic pancreatitis.

2. Somatostatin

Somatostatin, a hormone that inhibits pancreatic enzyme secretion (DOLLINGER et al. 1976), has been the object of controlled clinical trials. These trials followed work with animal models that showed the drug's ability to prevent experimentally induced pancreatitis and to decrease the mortality of established pancreatitis (SCHWEDES et al. 1979; BAXTER et al. 1985).

In a prospective study carried out in Hong Kong, 71 patients with acute pancreatitis (etiologically linked to cholelithiasis in 37 cases, to alcohol abuse in 11 cases, to ascaris in 1 case, and idiopathic in 22 patients) were included in a clinical trial to evaluate the efficacy of somatostatin (CHOI et al. 1989). All patients were treated with nasogastric suction, intravenous hydration, and cephoperazone for 5 days. Immediately after the diagnosis of acute pancreatitis was made, 35 randomly selected patients also received an intravenous bolus of 250 μg somatostatin, followed by a continuous infusion of 100 μg/h for 48 h. The effect of somatostatin was compared with the control group on the following variables: mortality, severity of pancreatitis at admission and 48 h after admission, and the incidence and type of complications.

One patient in the somatostatin group and two patients in the control group died. Five patients from the control group and one from the somatostatin group had an increase in the severity of their pancreatitis within 2 days after admission. Thirteen patients in the control group, but only five patients in the somatostatin group developed complications during their hospital stay. The most frequent complication in this series was "inflammatory pancreatic swelling" detected by abdominal ultrasound and/or computed tomography. After the exclusion of this item from the list of complications, the difference between the two groups appears much smaller (three versus seven cases) and is not statistically significant.

A multicenter study conducted in Italy evaluated the efficacy of somatostatin and total parenteral nutrition in 164 patients with acute pancreatitis. Approximately one-fourth of the study cohort had alcoholic pancreatitis (D'AMICO et al. 1990). The main study group comprised 60 patients given total parenteral nutrition and somatostatin (3.5 μg/kg per hour) by continuous intravenous infusion for between 72 h and 120 h. These patients were compared to a control group of 60 patients given total parenteral nutrition plus symptomatic treatment. Two smaller groups, each comprising 22 patients, required emergency surgical treatment in addition to total parenteral nutrition and either somatostatin or symptomatic therapy.

Two of the 82 patients given somatostatin died, as compared to seven control patients. Nine patients from the somatostatin group and 10 from the control group developed major complications (i.e., pancreatic abscess, pancreatic pseudocyst, upper gastrointestinal bleeding, or multiorgan failure). Neither of these differences reached statistical significance. Recovery, defined as clinical improvement and normalization of the biochemical parameters at the end of treatment, was more frequently recorded in the somatostatin group (60 of the 82 patients) than in the control group (44 patients). Statistically significant differences favoring somatostatin were observed in the frequency of recovery among nonsurgical cases with acute edematous pancreatitis and in the risk of death among surgical cases with acute necrotic-hemorrhagic pancreatitis. From these two studies, we conclude that, despite modest reductions in morbidity and mortality associated with local complications, somatostatin does not provide convincing advantages over standard symptomatic medical therapy of acute pancreatitis.

3. Pirenzepine

Cholinergic receptors influence the exocrine function of the acinar cells of the pancreas (VALENZUELA et al. 1986) and conventional anticholinergic agents, such as atropine, have been used in acute pancreatitis. However, these drugs have never gained popularity given their many unpleasant or life-threatening side effects (CAMERON et al. 1979). Pirenzepine, a selective

M1-receptor antagonist, has a much safer antimuscarinic profile (Hammer et al. 1980). It was recently tested in a double-blind clinical trial that included 115 patients with acute pancreatitis (Moreno-Otero et al. 1989). Pancreatitis was alcohol-related in 29 (25%) of patients. The study cohort was randomly divided into three groups: two treatment groups, receiving either 10 or 20 mg pirenzepine intravenously every 12 h, and a control group. All patients were treated with nasogastric suction and intravenous hydration. Anticholinergic side effects attributable to pirenzepine were quite rare: blurred vision in three cases, urinary retention in three men with benign prostatic hypertrophy, and mental confusion, diplopia, and dryness of the mouth in one patient each.

Eight of the 115 patients died; four were from the control group and three were from the study group receiving the higher dose of pirenzepine. Pancreatic abscesses were diagnosed in three patients receiving pirenzepine and in none of the control subjects. Trends toward fewer major systemic complications (e.g., renal failure, adult respiratory distress syndrome, disseminated intravascular coagulation) and a shorter duration of abdominal pain were noted in the study group, but were not statistically significant. The therapeutic effect, if any, did not seem to be dose dependent.

D. Drug Therapy of Chronic Alcoholic Gastritis

Alcoholism is often associated with clinical symptoms of gastritis and histologic evidence of hyperemia and erosions affecting predominantly the antrum of the stomach (Dinoso et al. 1972; Pitchumoni and Glass 1976; Gottfried et al. 1978). A new development in the drug therapy of this entity is based on the realization that antral gastritis is often caused by a localized infection with *Helicobacter pylori* (Marshall 1983). In a recent study (Uppal et al. 1991), 18 alcoholics with a history of upper gastrointestinal complaints underwent endoscopic evaluation of the stomach with antral biopsies. The endoscopic appearance was graded according to standard criteria (Lanza et al. 1984) to record the severity of edema, erythema, hemorrhages, and erosions. The presence and severity of histologic gastritis was assessed by counting the number of inflammatory cells in representative fields of multiple biopsy samples. *Helicobacter pylori* infection was diagnosed if the bacterial agent was identified in at least two separate biopsy specimens.

A large majority (14 of 18 cases) of alcoholics enrolled in the study showed evidence of infection with *H. pylori*. The endoscopic appearance of gastric mucosa and the symptoms of patients positive or negative for *H. pylori* were similar. However, histologic signs of moderate or severe inflammation were found only in the patients positive for *H. pylori*. A subgroup of ten patients positive for *H. pylori* were treated with bismuth subsalicylate, amoxacillin, and metronidazole; the remaining four patients positive for *H. pylori* received only antacids. After 3 weeks, only the patients receiving triple therapy showed significant clinical improvement. Although the sample was small, this study suggests that alcoholics may constitute a population with a high risk of gastric colonization with *H. pylori*. Specific antibacterial treatment appears efficacious in alleviating the dyspeptic symptoms of alcoholics. Further evaluation of this promising preliminary finding appears warranted.

E. Summary

We have reviewed the peer-assessed publications which addressed specific pharmacologic interventions for the medical complications of alcoholism. Accessing the Medline database of the National Library of Congress revealed no significant recent pharmacologic advances in the management of cardiovascular, neurologic, renal, endocrine, hematologic, or metabolic complications of alcoholism.

Based on the recent literature, no new pharmacologic interventions are currently recommended for the treatment of hepatic steatosis. Reduced alcohol consumption remains the only effective intervention.

For subjects with alcoholic hepatitis, insulin and glucagon cannot currently be recommended. When patients have alcoholic hepatitis complicated by encephalopathy but without associated gastrointestinal bleeding, corticosteroids appear to produce short-term benefits. The use of anabolic steroids, propylthiouracil, and colchicine require further study before they can be recommended for treatment of patients with alcoholic hepatitis.

For treatment of cirrhotic patients, colchicine can only be recommended as part of controlled clinical trials. Likewise, the use of flumazenil for the management of hepatic encephalopathy awaits the results of controlled, double-blind clinical trials. β-Blockers are effective for the prevention of a first bleeding episode in cirrhotics with "high-risk" varices.

For those subjects with pancreatitis, neither somatostatin nor pirenzepine provides convincing advantages over standard symptomatic therapy.

Based on a small study, alcoholics with gastritis complicated by *Helicobacter pylori* infection have resolution of their dyspeptic symptoms when treated with a triple antibiotic regimen.

In summary, while considerable recent investigations have focused on a variety of treatments for gastrointestinal complications of heavy alcohol consumption, the therapeutic armamentarium remains quite limited.

References

Achord JL (1987) A prospective randomized clinical trial of peripheral amino acid-glucose supplementation in acute alcoholic hepatitis. Am J Gastroenterol 82: 871–875

Akriviadis EA, Steindel H, Pinto PC, Fong TL, Kanel G, Reynolds TB and Gupta S (1990) Failure of colchicine to improve short-term survival in patients with alcoholic hepatitis. Gastroenterology 99:811–818

Andreani T, Poupon RE, Balkau BJ et al. (1990) Preventive therapy of first gastrointestinal bleeding in patients with cirrhosis: results of a controlled trial comparing propranolol, endoscopic sclerotherapy and placebo. Hepatology 12: 1413–1419
Anonymous (1982) (+)-Cyanindanol-3. Lancet 1:549
Anonymous (1987) Lactitol. Lancet 2:81–82
Anonymous (1988a) Diuretics or paracentesis for ascites? Lancet 2:775–776
Anonymous (1988b) Propylthiouracil and alcoholic liver disease. Lancet 1:450
Baxter AL, Jaspan JB, Haines NW, Hatfield GE, Krager PS, Schneider JF, University of Chicago Medical Housestaff (1981) A randomized clinical trial of insulin and glucagon infusion for treatment of alcoholic hepatitis: progress report in 50 patients. Gastroenterology 80:1410–1414
Bexter JN, Jenkins SA, Day DW (1985) Effects of somatostatin and a long-acting somatostatin analogue on the prevention and treatment of experimentally induced acute pancreatitis in the rat. Br J Surg 72:382–385
Bansky G, Meier PJ, Riederer E, Walser H, Ziegler WH, Schmid M (1989) Effects of the benzodiazepine receptor antagonist flumazenil in hepatic encephalopathy in humans. Gastroenterology 97:744–750
Baraldi M, Zeneroli ML, Ventura E et al. (1984) Supersensitivity of benzodiazepine receptors in hepatic encephalopathy due to fulminant hepatic failure in the rat: reversal by a benzodiazepine antagonist. Clin Sci 67:167–175
Basile AS, Hughes RD, Harrison PM, Murata Y, Pannell L, Jones EA, Williams R, Skolnick P (1991) Elevated brain concentrations of 1,4-benzodiazepines in fulminant hepatic failure. N Engl J Med 325:473–478
Berlin JA, Chalmers TC (1989) Meta-analysis of branched-chain amino acids in hepatic encephalopathy. Gastroenterology 97:1043–1045
Bird GLA, Williams R (1990) Treatment of advanced alcoholic liver disease. Alcohol Alcohol 25:197–206
Bird G, Lau JYN, Koskinas J, Wicks C, Williams R (1991) Insulin and glucagon infusion in acute alcoholic hepatitis: a prospective randomized controlled trial. Hepatology 14:1097–1101
Bonkovsky HL, Fiellen DA, Smith GS, Slaker DP, Simon D, Galambos JT (1991a) A randomized, controlled trial of treatment of alcoholic hepatitis with parenteral nutrition and oxandrolone. I. Short-term effects on liver function. Am J Gastroenterol 86:1200–1208
Bonkovsky HL, Singh RH, Jafri IH, Fiellin DA, Smith GS, Simon D, Cotsonis GA, Slaker DP (1991b) A randomized controlled trial of treatment of alcoholic hepatitis with parenteral nutrition and oxandrolone. II. Short-term effects on nitrogen metabolism, metabolic balance, and nutrition. Am J Gastroenterol 86:1209–1218
Boyer JL, Ransohoff DF (1988) Is colchicine effective therapy for cirrhosis? N Engl J Med 318:1751–1752
Boyer TD, Warnock DG (1983) Use of diuretics in the treatment of cirrhotic ascites. Gastroenterology 84:1051–1055
Burroughs AK, Hamilton G, Phillips A, Mezzanotte G, McIntyre N, Hobbs KEF (1989) A comparison of sclerotherapy with staple transection of the esophagus for the emergency control of bleeding from esophageal varices. N Engl J Med 321:857–862
Cabre E, Gonzalez-Huix F, Abad-LaCruz A, Esteve M, Acero D, Fernandez-Bañares F, Xiol X, Gassull MA (1990) Effect of total enteral nutrition on the short-term outcome of severely malnourished cirrhotics. Gastroenterology 98: 715–720
Calvey H, Davis M, Williams R (1985) Controlled trial of nutritional supplementation, with and without branched chain amino acid enrichment, in treatment of acute alcoholic hepatitis. J Hepatol 1:141–151
Cameron JL, Mehigan D, Zuidema GD (1979) Evaluation of atropine in acute pancreatitis. Surg Gynecol Obstet 148:206–208

Cello JP, Grendell JH, Crass RA, Weber TE, Trunkey DD (1987) Endoscopic sclerotherapy versus portacaval shunt in patients with severe cirrhosis and acute variceal hemorrhage: long-term follow-up. N Engl J Med 316:11–15
Choi TK, Mok F, Zhan WH, Fan ST, Lai ECS, Wong J (1989) Somatostatin in the treatment of acute pancreatitis: a prospective randomized controlled trial. Gut 30:223–227
Colombo M, De Franchis R, Tommasini M, Sangiovanni A, Dioguardi N (1989) β-Blockade prevents recurrent gastrointestinal bleeding in well-compensated patients with alcoholic cirrhosis: a multicenter randomized controlled trial. Hepatology 9:433–438
D'Amico D, Favia G, Biasiato R, Casaccia M, Falcone F, Fersini M, Marrano D, Napolitano F, Oliviero S, Rodolico A (1990) The use of somatostatin in acute pancreatitis: results of a multicenter trial. Hepatogastroenterology 37:92–98
De Franchis R, North Italian Endoscopic Club for the Study and Treatment of Esophageal Varices (1988) Prediction of the first variceal hemorrhage in patients with cirrhosis of the liver and esophageal varices. N Engl J Med 319:983–989
Diegelman RF, Peterkofsky B (1972) Inhibition of collagen secretion from bone and cultured fibroblasts by microtubular disruptive drugs. Proc Natl Acad Sci USA 69:892–896
Diehl AM, Boitnott JK, Herlong HF et al. (1985) Effect of parenteral amino acid supplementation in alcoholic hepatitis. Hepatology 5:57–63
Dindzans VJ, Gavaler JS, Tarter RE, Van Thiel DH (1989) Liver transplantation. Contemp Gastroenterol 2:9–16
Dinoso VP, Chey WY, Braverman SP, Rosen AR, Ottenberg D, Lorber SH (1972) Gastric secretion and gastric mucosal morphology in chronic alcoholics. Arch Intern Med 130:715–719
Dollinger HC, Raptis S, Pfeiffer EF (1976) Effects of somatostatin on exocrine and endocrine pancreatic function stimulated by intestinal hormones in man. Horm Metab Res 8:74–78
Egberts EH, Schomerus H, Hamster W, Jürgens P (1985) Branched chain amino acids in the treatment of latent portosystemic encephalopathy. Gastroenterology 88:887–895
Epstein M (1989) Treatment of refractory ascites. N Engl J Med 321:1675–1677
Eriksson LS, Conn HO (1989) Branched-chain amino acids in the management of hepatic encephalopathy: an analysis of variants. Hepatology 10:228–246
Erlich HP, Bornstein P (1972) Microtubules in transcellular movement of procollagen. Nature 238:257–260
Erlich HP, Ross R, Bornstein P (1974) Effects of antimicrotubular agents on the secretion of collagen. J Cell Biol 62:390–405
Ferenci P, Grimm G, Meryn S, Gangl A (1989) Successful long-term treatment of portal-systemic encephalopathy by the benzodiazepine antagonist flumazenil. Gastroenterology 96:240–243
Field BW, Heppner GW, Shabot MM, Schwartz AA, State D, Worthen N, Wilson R (1979) Nasogastric suction in alcohol pancreatitis. Dig Dis Sci 24:339–344
Frakes JT (1980) Physiologic considerations in the medical management of ascites. Arch Intern Med 140:620–623
Gammal SH, Jones EA (1989) Hepatic encephalopathy. Med Clin North Am 73: 793–813
Garden OJ, Mills PR, Birnie GG, Murray GD, Carter DC (1990) Propranolol in the prevention of recurrent variceal hemorrhage in cirrhotic patients. Gastroenterology 98:185–190
Ginès P, Arroyo V, Quintero E, Planas R, Bory F, Cabrera J, Rimola A, Viver J, Camps J, Jiménez W, Mastai R, Gaya J, Rodés J (1987) Comparison of paracentesis and diuretics in the treatment of cirrhotics with tense ascites. Gastroenterology 93:234–241
Ginès P, Arroyo V, Vargas V, Planas R, Casafont F, Panés J, Hoyos M, Viladomiu L, Rimola A, Morillas R, Salmerón JM, Ginès A, Esteban R, Rodés J (1991)

Paracentesis with intravenous infusion of albumin as compared with peritoneovenous shunting in cirrhosis with refractory ascites. N Engl J Med 325:829–835
Goff JS, Feinberg LE, Brugge WR (1982) A randomized trial comparing cimetidine to nasogastric suction in acute pancreatitis. Dig Dis Sci 27:1085–1088
Gottfried EB, Korsten MA, Lieber CS (1978) Alcohol-induced gastric and duodenal lesions in man. Am J Gastroenterol 70:587–592
Grimm G, Ferenci P, Katzenschlager R, Madl C, Schneeweiss B, Laggner AN, Lenz K, Gangl A (1988) Improvement of hepatic encephalopathy treated with flumazenil. Lancet 2:1392–1394
Hadas N, Wapnick S, Groskero SJ (1978) Cimetidine in pancreatitis (letter). N Engl J Med 299:487
Hallé P, Paré P, Kaptein E, Kanel G, Redeker AG, Reynolds TB (1982) Double-blind, controlled trial of propylthiouracil in patients with severe acute alcoholic hepatitis. Gastroenterology 82:925–931
Hammer R, Berrie CP, Birdsall NJM, Burgen ASV, Hulme EC (1980) Pirenzepine distinguishes between different subclasses of muscarinic receptors. Nature 383: 90–92
Harris ED, Krane SM (1971) Effects of colchicine on collagenase in cultures of rheumatoid synovium. Arthritis Rheum 14:669–684
Hayes PC, Davis JM, Lewis JA, Bouchier IAD (1990) Meta-analysis of value of propranolol in prevention of variceal haemorrhage. Lancet 336:153–156
Henderson JM, Kutner MH, Millikan WJ, Galambos JT, Riepe SP, Brooks WS, Bryan FC, Warren WD (1990) Endoscopic variceal sclerosis compared with distal splenorenal shunt to prevent recurrent variceal bleeding in cirrhosis. Ann Intern Med 112:262–269
Horst D, Grace ND, Conn HO, Schiff E, Schenker S, Viteri A, Law D, Atterbury CE (1984) Comparison of dietary protein with an oral, branched chain-enriched amino acid supplement in chronic portal-systemic encephalopathy: a randomized controlled trial. Hepatology 4:279–287
Imperiale TF, McCullough AJ (1990) Do corticosteroids reduce mortality from alcoholic hepatitis? Ann Intern Med 113:299–307
Israel Y, Videla L, Bernstein J (1975) Liver hypermetabolic state after chronic ethanol consumption: hormonal interrelationships and pathogenic implications. Fed Proc 34:2052–2059
Joffe SN, Lee FD (1978) Acute pancreatitis after cimetidine administration in experimental duodenal ulcer (letter). Lancet 1:383
Kaplowitz N (1982) Propylthiouracil treatment for alcoholic hepatitis: should it and does it work? Gastroenterology 82:1468–1472
Kearns PJ, Young H, Garcia G, Blaschke T, O'Hanlon G, Rinki M, Sucher K, Gregory P (1992) Accelerated improvement of alcoholic liver disease with enteral nutrition. Gastroenterology 102:200–205
Kellerman PS, Linas SL (1990) Large-volume paracentesis in treatment of ascites. Ann Intern Med 112:889–891
Kershenobich D, Uribe M, Suárez GI, Mata JM, Pérez-Tamayo R, Rojkind M (1979) Treatment of cirrhosis with colchicine. Gastroenterology 77:532–536
Kershenobich D, Vargas F, Garcia-Tsao G, Perez-Tamayo R, Gent M, Rojkind M (1988) Colchicine in the treatment of cirrhosis of the liver. N Engl J Med 318:1709–1713
Lanza FL, Nelson RS, Rack MF (1984) A controlled endoscopic study comparing the toxic effects of sulindac, naproxen, aspirin, and placebo on the gastric mucosa of healthy volunteers. J Clin Pharmacol 24:89–95
Levant JA, Secrist DM, Resin H, Sturdevant RAL, Guth PH (1974) Nasogastric suction in the treatment of alcoholic pancreatitis. A controlled study. JAMA 229:51–52
Lieber CS (1990) Interaction of alcohol with other drugs and nutrients. Drugs 40 [Suppl 3]:23–44

Lucey MR, Merion RM, Henley KS, Campbell DA, Turcotte JG, Nostrant TT, Blow FC, Beresford TP (1992) Selection for and outcome of liver transplantation in alcoholic liver disease. Gastroenterology 102:1736–1741

Maddrey WC (1979) Propylthiouracil treatment of alcoholic hepatitis. Gastroenterology 76:218–219

Maddrey WC (1988) Alcoholic hepatitis: clinicopathologic features and therapy. Semin Liver Dis 8:91–102

Marshall BJ (1983) Unidentified curved bacilli in the stomach of patients with active chronic gastritis. Lancet 2:1273–1275

Mendenhall CL, Tosch T, Weesner RE, Garcia-Pont P, Goldberg SJ, Kiernan T, Seeff LB, Sorrell M, Tamburro C, Zetterman R, Chedid A, Chen T, Rabin L (1986) VA cooperative study on alcoholic hepatitis II: prognostic significance of protein-calorie malnutrition. Am J Clin Nutr 43:213–218

Meshkinpour H, Molinari MD, Gardner L, Berk JE, Hoehler FK (1979) Cimetidine in the treatment of acute alcoholic pancreatitis. A randomized, double-blind study. Gastroenterology 77:687–690

Mezey E, Caballería J, Mitchell MC, Parés A, Herlong HF, Rodés J (1991) Effect of parenteral amino acid supplementation on short-term and long-term outcomes in severe alcoholic hepatitis: a randomized controlled trial. Hepatology 14: 1090–1096

Moreno-Otero R, Rodriguez S, Carbo J, Garcia-Buey L, Pajares JM (1989) Double-blind trial of pirenzepine in acute pancreatitis. Digestion 42:51–56

Mullen KD, Martin JV, Mendelson WB, Bassett ML, Jones EA (1988) Could an endogenous benzodiazepine ligand contribute to hepatic encephalopathy? Lancet 1:457–459

Mullen KD, Szauter KM, Kaminsky-Russ K (1990) "Endogenous" benzodiazepine activity in body fluids of patients with hepatic encephalopathy. Lancet 336: 81–83

Naeijie R, Salingret E, Clumeck N, DeTroyer A, Devis G (1978) Is nasogastric suction necessary in acute pancreatitis? Br Med J 2:659–660

Nasrallah SM, Galambos JT (1980) Aminoacid therapy of alcoholic hepatitis. Lancet 2:1276–1277

Naylor CD, O'Rourke K, Detsky AS, Baker JP (1989) Parenteral nutrition with branched-chain amino acids in hepatic encephalopathy. Gastroenterology 97: 1033–1042

O'Connor KW, Lehman G, Yune H, Brunelle R, Christiansen P, Hast J, Compton M, McHenry R, Klatte E, Cockerill E, Holden R, Becker G, Kopecky K, Hawes R, Pound D, Rex D, Lui A, Snodgrass P,Weddle R, Crabb D, Lumeng L (1989) Comparison of three nonsurgical treatments for bleeding esophageal varices. Gastroenterology 96:899–906

Orrego H, Kalant H, Israel Y, Blake J, Medline A, Rankin JG, Armstrong A, Kapur B (1979) Effect of short-term therapy with propylthiouracil in patients with alcoholic liver disease. Gastroenterology 76:105–115

Orrego H, Blake JE, Blendis LM, Compton KV, Israel Y (1987) Long-term treatment of alcoholic liver disease with propylthiouracil. N Engl J Med 317: 1421–1427

Pagliaro L, Burroughs AK, Sorensen TIA, Lebrec D, Morabito A, Amico GD, Tine F (1990) Beta-blockers for preventing variceal bleeding. Lancet 336:1001–1002

Pagliaro L, D'Amico G, Sörensen TIA, Lebrec D, Burroughs AK, Morabito A, Tiné F, Politi F, Traina M (1992) Prevention of first bleeding in cirrhosis. Ann Intern Med 117:59–70

Pascal JP, Cales P, Multicenter Study Group (1987) Propranolol in the prevention of first upper gastrointestinal tract hemorrhage in patients with cirrhosis of the liver and esophageal varices. N Engl J Med 317:856–861

Pitchumoni CS, Glass GBJ (1976) Patterns of gastritis in alcoholics. Biol Gastroenterol 9:11–16

Planas R, Boix J, Broggi M, Cabré E, Gomes-Vieira MC, Morillas R, Armengol M, de León R, Humbert P, Salvá JA, Gassull MA (1991) Portacaval shunt versus endoscopic sclerotherapy in the elective treatment of variceal hemorrhage. Gastroenterology 100:1078–1086
Polio J, Groszmann RJ (1989) Bleeding esophageal varices. Pract Gastroenterol 13:41–50, 72
Poynard T, Calès P, Pasta L, Ideo G, Pascal JP, Pagliaro L, Lebrec D, Franco-Italian Multicenter Study Group (1991) Beta-adrenergic-antagonist drugs in the prevention of gastrointestinal bleeding in patients with cirrhosis and esophageal varices. N Engl J Med 324:1532–1538
Quintero E, Ginès P, Arroyo V, Rimola A, Bory F, Planas R, Viver J, Cabrera J, Rodés J (1985) Paracentesis versus diuretics in the treatment of cirrhotics with tense ascites. Lancet 1:611–612
Ramond MJ, Polynard T, Rueff B, Mathurin P, Theódore C, Chaput JC, Benhamou JP (1992) A randomized trial of prednisolone in patients with severe alcoholic hepatitis. N Engl J Med 326:507–512
Reynolds TB (1987) Therapeutic paracentesis. Gastroenterology 93:386–388
Reynolds TB (1990) Corticosteroid therapy of alcoholic hepatitis: How many studies will it take? Hepatology 12:619–621
Rocco VK, Ware AJ (1986) Cirrhotic ascites. Ann Intern Med 105:573–585
Rossi V, Calès P, Burtin P, Charneau J, Person B, Pujol P, Valentin S, D'Aubigny N, Joubaud F, Boyer J (1991) Prevention of recurrent variceal bleeding in alcoholic cirrhotic patients: prospective controlled trial of propranolol and sclerotherapy. J Hepatol 12:283–289
Salmi HA, Sarna S (1982) Effect of silymarin on chemical, functional and morphological alterations of the liver. A double-blind controlled study. Scand J Gastroenterol 17:517–521
Sauerbruch T, Wotzka R, Köpcke W, Härlin M, Heldwein W, Bayerdörffer E, Sander R, Ansari H, Starz I, Paumgartner G (1988) Prophylactic sclerotherapy before the first episode of variceal hemorrhage in patients with cirrhosis. N Engl J Med 319:8–15
Schafer DF, Jones EA (1982) Hepatic encephalopathy and the γ-aminobutyric-acid neurotransmitter system. Lancet 1:18–20
Schrier RW, Arroyo V, Bernardi M, Epstein M, Henriksen JH, Rodés J (1988) Peripheral arterial vasodilation hypothesis: a proposal for the initiation of renal sodium and water retention in cirrhosis. Hepatology 8:1151–1157
Schwedes M, Althoff PH, Klempa L (1979) Effects of somatostatin on bile induced acute haemorrhagic pancreatitis in the dog. Horm Metab Res 11:647
Sherlock S (1990) Alcoholic hepatitis. Alcohol Alcohol 25:189–196
Shocket ID (1992) Paracentesis or peritoneovenous shunting for ascites? ACP J Club 116 [Suppl 1]:10
Simon D, Galambos JT (1988) A randomized controlled study of peripheral parenteral nutrition in moderate and severe alcoholic hepatitis. J Hepatol 7:200–207
Sørensen TIA, Orholm M, Bentsen K, Høybye G, Eghøje K, Christoffersen P (1984) Prospective evaluation of alcohol abuse and alcoholic liver injury in men as predictors of development of cirrhosis. Lancet 2:241–244
Stanley MM, Ochi S, Lee KK, Allen MJ, Baum RA, Gadacz TR, Camara DS, Caruana JA, Schiff ER, Livingstone AS, Samanta AK, Najem AZ, Glick ME, Juler GL, Adham N, Baker JD, Cain GD, Jordan PH, Wolf DC, Fulenwider JT, James KE, VA Cooperative Study on Treatment of Alcoholic Cirrhosis with Ascites (1989) Peritoneovenous shunting as compared with medical treatment in patients with alcoholic cirrhosis and massive ascites. N Engl J Med 321: 1632–1638
Starzl TE, Van Thiel D, Tzakis AG, Iwatsuki S, Todo S, Marsh JW, Koneru B, Staschak S, Stieber A, Gordon RD (1988) Orthotopic liver transplantation for alcoholic cirrhosis. JAMA 260:2542–2544

Stiegmann GV, Goff JS, Michaletz-Onody PA, Korula J, Lieberman D, Saeed ZA, Reveille RM, Sun JH, Lowenstein SR (1992) Endoscopic sclerotherapy as compared with endoscopic ligation for bleeding esophageal varices. N Engl J Med 326:1527–1532

Szauter KM, Mullen KD (1989) Management of hepatic encephalopathy. Pract Gastroenterol 13:40–48, 68

Terblanche J, Burroughs AK, Hobbs KEF (1989a) Controversies in the management of bleeding esophageal varices. N Engl J Med 320:1393–1398

Terblanche J, Burroughs AK, Hobbs KEF (1989b) Controversies in the management of bleeding esophageal varices. N Engl J Med 320:1469–1475

Titó L, Ginès P, Arroyo V, Planas R, Panés J, Rimmola A, Llach J, Humbert P, Badalamenti S, Jiménez W, Rodés J (1990) Total paracentesis associated with intravenous albumin management of patients with cirrhosis and ascites. Gastroenterology 98:146–151

Trinchet JC, Beaugrand M, Callard P, Hartmann DJ, Gotheil C, Nusgens BV, Lapiere CM, Ferrier JP (1989) Treatment of alcoholic hepatitis with colchicine. Gastroenterol Clin Biol 13:551–555

Trinchet JC, Balkau B, Poupon RE, Heintzmann F, Callard P, Gotheil C, Grange JD, Vetter D, Pauwels A, Labadie H, Chazouilleres O, Mavier P, Desmorat H, Zarski JP, Barbare JC, Chambre JF, Pariente EA, Roulot D, Beaugrand M (1992) Treatment of severe alcoholic hepatitis by infusion of insulin and glucagon: a multicenter sequential trial. Hepatology 15:76–81

Uppal R, Lateef SK, Korsten MA, Paronetto F, Lieber CS (1991) Chronic alcoholic gastritis. Role of alcohol and Helicobacter pylori. Arch Intern Med 151:760–764

VA Cooperative Variceal Sclerotherapy Group (1991) Prophylactic sclerotherapy for esophageal varices in men with alcoholic liver disease. N Engl J Med 324: 1779–1784

Valenzuela JE, Weiner K, Saad C (1986) Cholinergic stimulation of human pancreatic secretion. Dig Dis Sci 31:615–619

Van Thiel DH, Carr B, Iwatsuki S, Tzakis A, Fung JJ, Starzl TE (1991) Liver transplantation for alcoholic liver disease, viral hepatitis, and hepatic neoplasms. Transplant Proc 23:1917–1921

World MJ, Ryle PR, Aps EJ, Shaw GK, Thomson AD (1987) Palmitoyl-catechin for alcoholic liver disease: results of a three-month clinical trial. Alcohol Alcohol 22:331–340

CHAPTER 17

Drugs for the Treatment of Psychiatric Comorbidity in Alcoholics: Recent Developments

M.J. BOHN and D. HERSH

A. Introduction

Alcoholism has long been associated with other psychiatric *symptoms*, including anxiety, depression, and hallucinations. Hippocrates advised that "wine drunk with an equal quantity of water puts away anxiety and terrors." As distinct *syndromes*, coexisting psychiatric disorders are also common among alcoholics. Anxiety disorders, affective disorders, antisocial personality disorder (ASP), and drug use disorders are particularly prevalent comorbid disorders. These coexisting psychiatric conditions generally confer a poor prognosis following psychosocial treatment of the alcoholic (MCLELLAN et al. 1983a; ROUNSAVILLE et al. 1987; HELZER and PRYZBECK 1988).

Traditionally, alcoholism rehabilitation has employed a variety of psychosocial treatment approaches. In a study of predictors of response to substance abuse treatment, MCLELLAN and collaborators (1983a) found that patients with low severity of psychiatric problems improved in each of six programs, while those with high severity of psychiatric problems showed virtually no improvement in any program. A patient-treatment matching study conducted in the early 1980s attempted to match alcoholics with psychiatric problems to psychiatric treatment programs (MCLELLAN et al. 1983b). The investigators hypothesized that alcoholics with severe psychiatric problems would benefit most from treatment in psychiatric (and not substance abuse) treatment centers. However, they were unable to arrange treatment of alcoholics in such psychiatric treatment programs, so that this patient-treatment matching hypothesis could not be evaluated. Recently, the efficacy of two distinct psychosocial group treatments in the care of alcoholics following inpatient rehabilitation was compared in a randomized clinical trial. Results suggest that structured relapse prevention treatment, which employs cognitive-behavioral strategies for management of negative affects and related psychologic distress, are more effective for alcoholics with moderate-to-severe psychologic distress, including those with ASP (COONEY et al. 1991; LITT et al. 1992). In contrast, patients with less severe psychiatric symptoms had better drinking outcomes following interactive group treatment.

A large number of controlled clinical trials have demonstrated the effectiveness of a variety of antidepressant, mood-stabilizing, anxiolytic, and antipsychotic agents in the treatment of *nonalcoholic* patients with specific psychiatric syndromes. Pharmacologic treatment of coexisting psychiatric syndromes or symptoms may also benefit the *alcoholic* by facilitating abstinence or reduced drinking, or by reducing the severity of distress associated with psychiatric symptoms. Medications may also enhance the rate of retention and/or efficacy of psychosocial treatments for alcoholism. They may reduce the desire to drink or the occurrence of alcohol-associated psychiatric symptoms and associated social or occupational dysfunction. Recent trials of pharmacotherapy for alcoholism have addressed many important issues in the measurement of these outcomes and have employed compliance assessment and drug level monitoring to assure the adequacy of the medication treatment. Investigators have selected alcoholic subgroups whose symptoms are likely to respond to pharmacotherapy and for whom there is relatively low risk of adverse effects. These efforts will, we hope, overcome traditional barriers to the appropriate psychopharmacologic treatment of alcoholism.

In this chapter, we will evaluate the current status of pharmacologic treatment of selected psychiatric disorders that occur commonly among alcoholics, thereby highlighting areas in which future research efforts might be focused. We will concentrate on the treatment of major depression, anxiety disorders, bipolar affective disorder, and schizophrenia. While alcoholics with antisocial personality traits appear to respond preferentially to cognitive-behavioral treatment (Litt et al. 1992), specific pharmacotherapy for ASP does not now exist. Alcoholics with opiate dependence can be treated successfully with opiate agonists such as methadone (Liebson et al. 1973), and we refer the reader to other sources for a review of this topic (Lowinsohn et al. 1992).

For each disorder, we will first review its relationship to alcohol abuse and dependence. We will then discuss diagnostic issues important to the conduct of pharmacotherapy trials among alcoholics with these disorders. Finally, we will systematically evaluate published studies assessing the efficacy and safety of pharmacologic treatment of alcoholic patients with these disorders. Given the relative dearth of published experimental literature addressing these issues, we will call attention to areas in which future research efforts might be appropriately directed.

B. Depression

Because of the high prevalence of depressive symptoms among alcoholics and the demonstrated responsiveness of these symptoms to pharmacotherapy, we begin with a discussion of the pharmacologic treatment of depression in the alcoholic patient. We will review clinical and epidemiologic evidence for the relationship between alcoholism and depression, and will discuss the pretreatment evaluation of the depressed alcoholic. We will then discuss the results of clinical trials of a variety of classes of antidepressant medications in alcoholics, emphasizing the potential efficacy of these medications in reducing depressive symptoms and drinking behavior.

Nearly all alcoholics report dysphoric affects, such as low mood, anhedonia, and anxiety (KEELER et al. 1979). Simultaneously, there may be depressive cognitions such as guilt and suicide wishes, with neurovegetative symptoms such as low energy, sleep or appetite disturbance, and impaired concentration or memory. Longitudinal epidemiologic studies support the association between depression and alcohol use (ANESHENSEL and HUBA 1983). While there is a bidirectional relationship between depression and alcoholism, current evidence suggests that depression is a more frequent *consequence* of drinking than it is a *cause* of alcoholism. Experimental administration of alcohol to healthy nonalcoholics and to alcoholics produces an initial period of euphoria, often followed by depressed mood and guilt after prolonged drinking (TAMERIN and MENDELSON 1969; NATHAN et al. 1970). Transient grief and remorse commonly occur in the early days following spontaneous cessation of drinking, as well as during periods of pharmacologic detoxification from alcohol. The severity of depressive symptoms can be reliably rated using self-report rating scales, such as the Beck Depression Inventory (BDI, BECK et al. 1961), as well as observer-scored rating scales, such as the Hamilton Depression Rating Scale (HDRS, HAMILTON 1960). Over 80% of depressed inpatient alcoholics show a reduction in scores on these scales following 2–4 weeks of abstinence (SCHUCKIT 1979; NAKAMURA et al. 1983; DORUS et al. 1987; BROWN and SCHUCKIT 1988), without specific treatment of depressive symptoms.

In contrast to depressive *symptoms*, depressive *syndromes* are relatively stable conditions that can be reliably ascertained, have a familial or heritable predisposition, and have a well-defined natural history and response to specific treatment. In the DSM-III-R diagnostic classification system, four depressive syndromes are identified (AMERICAN PSYCHIATRIC ASSOCIATION 1987). Major depressive disorder (MDD) involves temporal clustering of several persistent depressive symptoms for a minimum of 2 weeks. Untreated, an episode of MDD typically lasts 6 or more months (AMERICAN PSYCHIATRIC ASSOCIATION 1993a). If depressive symptoms persist in milder form for a period of 2 years, dysthymia may be present. Organic mood disorders have symptoms that can be identical to MDD or dysthymia, but are caused or maintained by a physical condition such as intoxication or withdrawal. Adjustment disorders with depressed mood occur following stressful events, are less severe and prolonged than major depressive episodes, and remit in the absence of pharmacotherapy.

The prevalence of current and lifetime categoric diagnoses of major depression has been estimated among alcoholics in both community samples and treatment samples. In the largest published community survey, the Epidemiologic Catchment Area Study (ECA), over 20000 respondents in five United States sites were interviewed using the Diagnostic Interview

Schedule (DIS, Robins et al. 1989) to determine the prevalence of DSM-III psychiatric diagnoses. While the DIS interview inquired about whether alcohol or drug use was associated with the particular depressive and other psychiatric symptoms, the interviewers were not able to examine the respondent for organic conditions unknown to the respondent. Thus, the DIS may have underestimated the relative prevalence of organic disorders and overestimated the relative prevalence of "functional" psychiatric disorders. The weighted, 1-month prevalence of DSM-III MDD in the ECA was 2.2%. The lifetime prevalence of DSM-III alcoholism was 13.7%, and the lifetime prevalence of major depression and dysthymic disorder were 5.1% and 1.5%, respectively (Robins and Regier 1990). Among all respondents with a *lifetime history of alcoholism*, there was a 1.7-fold increased risk of *lifetime major depression* and a 1.8-fold increased risk of dysthymic disorder; among female alcoholics, the risk of major depression was increased 2.7-fold (Helzer and Pryzbeck 1988).

A practical approach to the reliable classification of depressive disorders among alcoholics is the primary–secondary distinction, which is based on the order of onset of the two disorders. While useful, it is subject to recall bias, which can be substantial in the case of affective disorders (Zimmerman and Coryell 1986). Use of a collateral historian and ascertainment of the age at which diagnostic syndrome criteria (in contrast to symptom presence or absence) were met may enhance the reliability of this approach (Schuckit 1979). There was a large sex differential in the order of onset of major depression and alcoholism in the ECA. In men, alcoholism preceded the onset of depression in 78% of cases, while in women this pattern occurred in 66% of cases; in both men and women, antecedent depression was associated with fewer symptoms of alcoholism. For those with *current alcoholism*, the rates of *lifetime* major depression and dysthymia were approximately 2% for men and 13% for women (Robins and Regier 1990).

Rates of depressive disorders have been reported to vary widely among clinical samples of alcoholics. In part, this may be due to differences in the clinical samples. Additionally, the method of ascertainment of MDD may exert a powerful impact on the apparent prevalence of this disorder among recently abstinent alcoholics. Given the high rates of remission of depressive *symptoms* among recently abstinent alcoholics, it is likely that most cases of depressive *disorders* among actively drinking and recently abstinent alcoholics could be classified as organic mood or adjustment disorders, rather than MDD. Dackis and colleagues (1986) reevaluated 49 alcoholics who met symptomatic criteria for MDD that had originally been ascertained within the first 5 days of abstinence. Without treatment, over 80% failed to meet MDD criteria at re-interview following 2 weeks of abstinence. Those who did remain depressed had a higher prevalence of familial affective disorder. Jaffe and Ciraulo (1986) have cited several causes of depressive symptoms in alcoholics, including malnutrition, alcohol withdrawal, direct effects of alcohol on the brain, hepatic dysfunction, head trauma, social losses, use of other drugs, personality disorder, and an independent mood disorder.

Pooled data from two groups who interviewed treatment samples of alcoholics following 2 weeks of abstinence (HESSELBROCK et al. 1985; ROSS et al. 1988) to systematically evaluate DSM-III disorders indicate the high prevalence of these disorders in clincal treatment programs. Affective disorders were more common in these samples than in the ECA. Lifetime MDD was found in 29%, current MDD in 17%, and dysthymic disorder in 15% of alcoholics. Consistent with results of the ECA, a substantial proportion of these depressed alcoholics, i.e., 41% of men and 65% of women, developed secondary alcoholism after primary MDD. Alcoholism was primary in 47% of men and 22% of women, while about 13% of both men and women had onset of both alcoholism and MDD within the same year (HESSELBROCK et al. 1985).

Clinically, a concurrent major depressive episode may have a significant effect on outcomes following psychosocial treatment for alcoholism. Among active and abstinent alcoholics, depression and other negative affects are frequently cited as potent stimuli for initiation of drinking bouts (HATSUKAMI and PICKENS 1982). For males, MDD was associated with poorer 1-year outcomes following inpatient treatment for alcoholism, while for females major depression was associated with better drinking-related outcome measures (SCHUCKIT and WINOKUR 1972; ROUNSAVILLE et al. 1987). This difference in outcomes may be due to the fact that among males alcoholism is often primary while among females alcoholism is more often secondary. Most alcoholics who commit suicide are males, and major depression is the principal factor in a majority of alcoholic suicides (MURPHY 1992).

I. Assessment

Since most depressed alcoholics undergo a marked reduction in depressive symptom severity (or frank remission of the depressive disorder) soon after initiating abstinence, the reliability of ascertainment of MDD will be increased by assessments that are deferred or repeated following 2–4 weeks of abstinence. Such a waiting period can occasionally be difficult for the patient to tolerate; ressurance and psychosocial treatments often suffice during this interval. If the MDD is primary, if the alcoholic had prior episodes of MDD during prolonged (e.g., 3- to 6-month) periods of abstinence, or if there is a strong family history of MDD, earlier initiation of pharmacotherapy may be justified. To our knowledge, published studies have not addressed the issue of whether earlier initiation of pharmacologic treatment may improve outcomes for alcoholics with *symptoms* of depression.

Given the above, all alcoholics should be assessed for current and past affective disorders in both the treatment and research clinical setting. The

BDI, Zung Depression Scale (ZUNG 1965), and HDRS are not particularly sensitive or specific in identifying patients with MDD or other depressive disorder (GRANT et al. 1989). Reliable ascertainment of psychiatric diagnoses will be enhanced by adherence to specified diagnostic criteria such as the DSM-III-R (AMERICAN PSYCHIATRIC ASSOCIATION 1987), DSM-IV (AMERICAN PSYCHIATRIC ASSOCIATION 1993b), or ICD-10 (WORLD HEALTH ORGANIZATION 1990a) criteria, and by the use of structured or semistructured diagnostic interviews such as the Diagnostic Interview Schedules for DSM-III-R (DIS-R, ROBINS et al. 1989), the Structured Clinical Interview for DSM-III-R (SCID, SPITZER et al. 1992), or the Comprehensive International Diagnostic Interview (CIDI, WORLD HEALTH ORGANIZATION 1990b). Use of these instruments reduces diagnostic errors among dually diagnosed substance abuse patients (BRYANT et al. 1992; WEISS et al. 1992).

II. Treatment

Among patients without concurrent alcoholism, a variety of antidepressant medications, including cyclic antidepressants (CAs), monoamine oxidase inhibitors (MAOIs), serotonin reuptake inhibitors (SUIs), lithium, and other medications are effective in treatment of MDD and in the prevention of depressive relapses (KUPFER et al. 1992; AMERICAN PSYCHIATRIC ASSOCIATION 1993a). In clinical trials of nonalcoholics with MDD, two-thirds of those treated with antidepressants experienced significant reduction of depressive symptoms; only one-third of patients treated with placebo experienced similar improvement (AMERICAN PSYCHIATRIC ASSOCIATION 1993a). Diurnal symptom worsening, marked psychomotor disturbance, sleep disturbance, poor energy, and appetite disturbance are symptoms that are particularly predictive of antidepressant medication responsiveness in MDD (AMERICAN PSYCHIATRIC ASSOCIATION 1993a). In addition, adequate serum levels of antidepressant medications are strongly predictive of antidepressant efficacy (AMERICAN PSYCHIATRIC ASSOCIATION 1993a). There is typically a 2- to 3-week interval between the achievement of therapeutic serum antidepressant levels and the reduction in depressive symptoms. Among mildly-to-moderately depressed patients, cognitive and interpersonal psychotherapy enhances the effectiveness of antidepressant medications in the treatment of a current major depression and in the prevention of recurrences of major depression (ELKIN et al. 1989).

III. Pharmacotherapy

Antidepressant medications carry risks in any patient, particularly those who have underlying illnesses or who are elderly. These may include sedation, confusion, anxiety, insomnia, and other symptoms that could lessen the effectiveness of psychosocial treatments for alcoholism and/or depression. In themselves, these adverse effects of antidepressant medications may mimic intoxication or withdrawal states or, if alcohol is consumed, lead to enhanced toxicity. However, given the negative impact of comorbid depressive disorders on outcome following alcoholism rehabilitation, including the risk of suicide, we conclude that carefully monitored psychopharmacologic treatment of depressed alcoholics is currently justified in the clinical setting. Such treatment should be used in conjunction with effective psychotherapy.

We now review the major classes of antidepressants whose safety and efficacy have been studied in alcoholics, and discuss the potential utility and limitations of these medications in treatment of the depressed alcoholic.

1. Cyclic Antidepressants

The cyclic antidepressants (CAs) have been the focus of several clinical treatment trials for depressed alcoholics conducted since the 1960s. In most studies, they have been compared with benzodiazepines or placebo for treatment of depressive symptoms occurring in the immediate post-withdrawal period.

Several methodologic flaws make it difficult to assess the results of the earlier studies (JAFFE 1987). First, in these studies there was no effort to measure the blood levels of the CA in order to assure adequate dosage and compliance. The oral CA doses employed (75–150 mg imipramine, 25–80 mg amitriptyline, or 25–150 mg doxepin daily) were lower than those now known to be required for effective treatment of depressed nonalcoholics (150–200 mg imipramine or equivalent; AMERICAN PSYCHIATRIC ASSOCIATION 1993a). Blood levels of the CAs imipramine (IMI), amitriptyline (AMI), and, to a lesser extent, desipramine (DMI) are typically lower among alcoholics than among nonalcoholics, and the clearance of these drugs is substantially increased in recently detoxified alcoholics, compared with nonalcoholic controls (SANDOZ et al. 1983; CIRAULO et al. 1982, 1988a). This is probably due to recent alcohol and cigarette tar ingestion, which results in hepatic microsomal enzyme induction, enhancing oxidation and elimination of CAs and other drugs. The failure to measure blood CA levels makes it difficult to determine if the lack of treatment efficacy in these studies was due to pharmacokinetic abnormalities or to a lack of pharmacodynamic potency. Second, CA treatment was often initiated within 2 weeks after cessation of drinking, which is a period associated with high rates of spontaneous symptom remission. Coupled with the typical lag time for CAs to exert an antidepressant effect in nonalcoholics, the lack of superiority of active medication may have been an artifact of a high placebo response rate. Third, it is unclear how many of the enrolled subjects had either MDD or dysthymia, and whether they had symptom patterns that may have been expected to respond favorably to CA treatment. Fourth, the type and extent of psychosocial treatment, including psychotherapies specific for depression, were rarely specified. Particularly powerful or highly variable nonpharmacologic treatment may have obscured effects of active medication treatment. Fifth, depressive symptoms and drinking patterns were not simultaneously reported in these studies. Since recent drinking is correlated with depressive symptoms, concurrent measurement of these two outcomes is critical.

Two recently reported trials illustrate how many of these concerns can be appropriately addressed. MASON and KOCSIS (1991) conducted a 6-month, placebo-controlled, double-blind trial of desipramine (DMI) in a sample of 16 depressed and 26 nondepressed alcoholics. MDD diagnoses were made using structured interviews following 3 weeks of abstinence. All subjects received weekly therapy (the nature of which was unspecified) and were encouraged to participate in Alcoholics Anonymous groups. Thirty-one patients completed the trial; most of the 11 patients who dropped out did so during the initial portion of the study. Among the depressed subjects, DMI was associated with significant improvements (compared with placebo) in treatment retention and with a significant reduction of depressive symptoms (as assessed by the HDRS). DMI levels (mean oral dose, 275 mg/day; serum level, 141 ng/ml) were in the usual therapeutic range. Compared with placebo, DMI was not associated with an increased risk of adverse effects among the depressed alcoholics. Several nondepressed subjects receiving DMI withdrew from the study due to bothersome side effects. There was a marginally significant trend for depressed subjects taking desipramine to remain abstinent for a longer period of time.

NUNES and colleagues (1993) recently reported results of a 12-week, open-label trial of imipramine (IMI) treatment of 85 adult alcoholics with dysthymia or primary MDD. Subjects met weekly with a psychiatrist and alcoholism counselor for an unspecified type of alcoholism counseling. Sixty subjects completed at least 4 weeks of treatment at a final IMI dosage of at least 150 mg/day, with mean blood IMI levels of 368 ± 264 ng/ml. Twenty-seven (45%) had substantial improvement, i.e., both reduced HDRS scores *and* abstinence or substantially reduced drinking. Three additional subjects responded following an increase in their IMI dosage. Finally, five subjects responded to the addition of disulfiram. Twenty-three of the 35 responders later completed a randomized discontinuation trial. Of those continuing imipramine, 31% relapsed, while 70% of placebo-treated subjects relapsed, a marginally significant difference ($p = 0.09$). Of those who relapsed upon placebo substitution, resumption of IMI improved mood and lowered drinking significantly in 70%, while the remaining 30% had a favorable mood response but no reduction in drinking. Though uncontrolled, these results suggest that IMI may yield a significant reduction in both depression and alcoholism symptoms in selected depressed alcoholics.

Cyclic antidepressants can produce a variety of adverse effects, and can be lethal in overdose. This is particularly relevant in depressed alcoholics who, compared with depressed nonalcoholics, are at increased risk of suicide (MURPHY 1992). They should not be dispensed a lethal quantity (typically more than 1 g) of a CA, and their suicide potential should be frequently assessed. Psychiatric hospitalization may be necessary if the patient is psychotic, actively drinking, has a suicide plan, or has other risk factors for imminent suicide. Because of potential toxicity, CAs are contraindicated in patients with underlying seizure disorders, cardiac failure, heart block, hepatic or renal disease, glaucoma, impaired balance, bladder outlet obstruction, or dementia. Sedation and confusion may occur following CA treatment, particularly if potent anticholinergic CAs such as amitriptyline, imipramine, or doxepin are used. Elderly patients are at particular risk. Such cognitive impairment may interfere with successful psychosocial treatment of both alcoholism and depression. By inhibition of hepatic microsomal oxidase activity, disulfiram can increase the half-life of CAs (CIRAULO et al. 1985), which may exaggerate adverse CA effects. Thus, initiation of CA treatment should occur at lower initial doses in alcoholics taking disulfiram. Oral doses of CAs may need to be adjusted after determination of serum CA levels.

2. Monoamine Oxidase Inhibitors

In adequate dosages (e.g., typically ≥60 mg/day phenelzine), monoamine oxidase inhibitors (MAOIs) are at least as effective as CAs in treatment of patients with major depression, especially those with bipolar affective disorder (AMERICAN PSYCHIATRIC ASSOCIATION 1993a). There have been reports that a subgroup of alcoholics with early onset and high degrees of impulsivity have low platelet levels of MAO activity (VON KNORRING et al. 1985). In a placebo-controlled trial, the MAOI isocarboxazide was not superior to placebo in facilitating abstinence among depressed alcoholics (KISSIN 1975). While the effect of this treatment on depressive symptoms was not reported, the doses were low and an unknown proportion of subjects had MDD. SCHOTTENFELD et al. (1989), despite a well-designed trial to test the efficacy of the MAOI phenelzine in depressed abstinent alcoholics, found that too few alcoholics were free of medical contraindications, willing to comply with the dietary restrictions, and able to tolerate the side effects of this agent to permit an adequate evaluation of the medication's potential efficacy. The efficacy of MAOIs in treating depressed alcoholics has yet to be meaningfully tested.

To reduce the risk of hypertension and intracerebral hemorrhage associated with the ingestion of vasopressors, alcoholics taking MAOIs should adhere to a low-tyramine diet and avoid use of sympathomimetics, including cocaine and amphetamines. Because alcoholics display impaired judgement during periods of intoxication and may be unable to avoid vasopressors, use of MAOIs in treatment of depressed alcoholics should be approached with caution. MAOIs are also associated with postural hypotension, tremor, weight gain, and sexual dysfunction. A variety of medication interactions can occur with MAOIs. Selective MAO-A inhibitors such as clorgyline appear to have strong antidepressant efficacy. They do not inhibit gut MAO as completely as do nonselective MAO inhibitors, and may be less likely to cause severe hypertension following intake of tyramine. Their efficacy in treatment of depressed alcoholics has not been reported.

3. Serotonin Uptake Inhibitors

Recent work on alcoholic subtypes has suggested that a subtype exists which is characterized by early onset, impulsivity, sociopathic behavior, and prominent depressive and anxiety symptoms (BABOR et al. 1992). Deficits in serotonergic activity have been found among a similar subgroup of alcoholics (BUYDENS-BRANCHEY et al. 1989), which may be related to the genesis or maintenance of excessive drinking. As reviewed by Naranjo in Chap. 16 of this volume, serotonin uptake inhibitors (SUIs) facilitate reduction of drinking or delay relapses in abstinent alcoholics. Fluoxetine, sertraline, and other SUIs are effective antidepressants.

Serotonin reuptake inhibitors have relatively few adverse effects at usual therapeutic doses, but can cause headache, insomnia, anxiety, tremor, akathisia, nausea, sexual dysfunction, and diarrhea. With the exception of fluvoxamine (KRANZLER et al. 1990), they are well tolerated by depressed and alcoholic patients. SUI overdose is rarely lethal, and SUIs do not accentuate alcohol-induced cognitive or psychomotor impairment.

In an 8-week, open-label study of the effectiveness of 20–40 mg/day fluoxetine in 12 alcoholics with major depression (assessed following detoxification and a 1-week washout period), significant improvements were noted in both drinking and depressive symptoms (CORNELIUS et al. 1992). Placebo-controlled studies are currently in progress to evaluate the efficacy of a variety of SUIs in depressed and nondepressed alcoholics.

4. Lithium

At adequate serum levels, lithium is effective in treatment of acute mania, prevention of relapses of both mania and depression, and treatment of depression (JEFFERSON et al. 1987; GOODWIN and JAMISON 1990). In a double-blind crossover study, euthymic alcoholics maintained on lithium for several weeks reported feeling less intoxicated, had less desire for a second drink, and evidenced less cognitive impairment following an alcohol challenge (JUDD and HUEY 1984). JAFFE and CIRAULO (1986) reviewed early studies of lithium treatment of alcoholics, noting that treatment was begun in recently detoxified alcoholics, dropout rates were high (47%–83%), and depressive diagnoses were based not on MDD diagnostic criteria, but on severity ratings, which are known to be poorly related to diagnoses among alcoholics (DORUS et al. 1987). They concluded that these flawed studies could not properly evaluate the efficacy of lithium in alcoholism treatment.

More recently, in a 1-year, placebo-controlled, double-blind trial of lithium carbonate treatment of 104 alcoholics, 88% met criteria for current major depression similar to those in the DSM-III-R, and 17% met criteria for primary major depression on diagnostic interview (FAWCETT et al. 1987). Compliance with either lithium or placebo was associated with superior outcomes for both drinking and depressive symptoms among both non-depressed and depressed alcoholics. Adequate serum lithium levels were not associated with a significant reduction in depressive symptom severity or with improvement in the social adjustment of these alcoholics. A placebo-controlled trial of lithium treatment of 457 male alcoholics, which specifically excluded those with bipolar affective disorder, was conducted by the VA Cooperative Studies Program, and yielded similar results (DORUS et al. 1989). Only 280 patients completed the year-long study, but neither depression nor lithium treatment affected study completion rates. Medication compliance, but not the presence of MDD, lithium treatment, or adequate serum lithium levels resulted in a significant reduction in drinking and depressive symptoms among these alcoholics. A third study reported similar results (DE LA FUENTE et al. 1989). While many bipolar patients drink less following lithium treatment (GOODWIN and JAMISON 1990), controlled studies evaluating the efficacy of lithium in adult alcoholics with bipolar disorder have not been reported. Preliminary results from an ongoing study of adolescents with bipolar disorder and substance abuse indicate that lithium treatment can both reduce alcohol use and normalize mood (GELLER et al. 1992).

Based on these findings, lithium is not warranted for treatment of alcoholism, except when bipolar affective disorder is also present. Furthermore, lithium has a number of potential adverse effects that must be considered in treatment of the alcoholic patient, including nausea, diarrhea, tremor, hypothyroidism, and cerebellar and other neurotoxic effects. Alcohol-induced diuresis can lower serum lithium levels, decreasing its potential efficacy (TYRER et al. 1990). In overdose, lithium can produce lethal toxicity (JEFFERSON et al. 1987).

In summary, pharmacotherapy is indicated for alcoholics who have primary MDD and those who meet criteria for an MDD following 3 or 4 weeks of abstinence. Given in doses adequate to achieve therapeutic serum drug levels, cyclic antidepressants are indicated for treatment of depressive symptoms. The adverse effects of these medications may limit their acceptability. SUIs have a more favorable side effect profile and may have more general efficacy in reducing drinking relapses. At present, there are insufficient data to warrant the use of MAOIs or lithium in treatment of depressive disorders in alcoholics.

C. Anxiety Disorders

Symptoms of anxiety are common among alcoholics; yet most of these symptoms (especially generalized anxiety and panic symptoms) abate, without specific pharmacologic or psychosocial treatment, during the first

2–4 weeks of abstinence (Brown et al. 1991). Regular intake of even moderate amounts of alcohol exerts anxiogenic effects (Tamerin and Mendelson 1969). Many anxious alcoholics report that they frequently drink to alleviate their anxiety symptoms. Even alcoholics without anxiety disorders frequently cite anxiety and associated symptoms of sympathetic hyperactivity as potent stimuli for continued drinking and relapse to drinking (Schuckit et al. 1990).

In the ECA survey, anxiety disorders were the only group of disorders whose prevalence was higher than alcoholism (Robins and Regier 1990). Simple phobias had 1-month and lifetime prevalences of approximately 6.2% and 12.5%, respectively. Obsessive-compulsive disorder (OCD) and panic disorder (PD) had a lifetime prevalence of 2.5% and 2.0%, respectively. Generalized anxiety disorder (GAD), in which excessive worrying is accompanied by chronic symptoms of motor tension, autonomic hyperactivity, and cognitive vigilance, occurred among 5% of ECA respondents. Social phobia was found to occur among 1%–2% of ECA respondents. Alcoholism increased the lifetime risk of PD and OCD more than twofold each, while phobias were increased 1.4-fold by alcoholism. Posttraumatic stress disorder (PTSD) was found among 1.4% of ECA respondents; its prevalence was not increased among heavy drinkers (Cottler et al. 1992). Among treated alcoholics assessed by structured interviews, the prevalence of DSM-III anxiety disorders appears higher than that observed in such community samples. In two studies, the mean current (lifetime) rates of these disorders were 20(26)% for phobias, 3(9)% for PD, 5(10)% for OCD, 9(12)% for social phobia, and 26(52)% for GAD, respectively (Hesselbrock et al. 1985; Ross et al. 1988). In clinical populations, the prevalence of alcoholism among patients with PTSD was 22%–86%, with higher rates reported among Vietnam War veterans with PTSD (Cottler et al. 1992). Thus alcoholism and anxiety disorders frequently co-occur in both the general population and in clinical samples.

In contrast to the case with MDD, a substantial minority of alcoholics with anxiety disorders develop alcoholism *following* the onset of a primary anxiety disorder. More than half of the alcoholics with PD or phobias, and over one-third of those with OCD or GAD, had secondary alcoholism (Hesselbrock et al. 1985; Ross et al. 1988). Thus, anxiety disorders may lead to the development of alcoholism. Conversely, repeated episodes of alcohol withdrawal or panic symptoms may result in "kindling" within limbic and other subcortical systems, yielding chronic elevation in sympathetic tone (George et al. 1988), which maintains drinking behavior and anxiety symptoms (Ballenger and Post 1978; Brown et al. 1990). Eventually, repeated cycles of drinking and withdrawal may lower the seizure threshold, producing withdrawal (or spontaneous) seizures.

Considering the high prevalence of anxiety during periods of active drinking and withdrawal, it is possible that many cases of apparent anxiety disorders, including PD and GAD, are actually organic anxiety disorders which, like organic affective disorders, usually resolve with abstinence.

I. Assessment

Assessment of an anxiety disorder in an alcoholic patient is enhanced using the same methods recommended for assessment of depression in this patient group, i.e., use of reliable criterion sets, structured psychiatric interviews, and assessment following 2–4 weeks of abstinence (MANUZZA et al. 1989; BRYANT et al. 1992). Given their common co-occurrence, anxiety and depressive disorders should be assessed simultaneously. This is particularly relevant for alcoholics with PD, because PD confers an increased risk of suicide in depressed alcoholics (HENRIKSSON et al. 1993).

II. Treatment

As with depressive symptoms, abstinence appears to reduce substantially the severity of anxiety symptoms. Cognitive and behavioral therapies are effective for phobias, PD, OCD, PTSD, and GAD among nonalcoholics (MARKS 1987). Therefore, behavioral or cognitive treatment would appear to be warranted following detoxification. We know of no research studies that have assessed whether pharmacotherapy of alcoholics is more successful than cognitive–behavioral treatments alone, or whether the combination of psychosocial and pharmacologic treatment is more effective than either single treatment modality.

III. Pharmacotherapy

Benzodiazepines and related sedatives are effective in the treatment of GAD, PD, and other anxiety disorders in patients with substance use disorders, yet pose serious potential risks for a minority of alcoholics. These anxiolytics carry at least a moderate risk of abuse or dependence (WOODS et al. 1987). Oxazepam may have lower abuse liability than the lipophilic benzodiazepines such as diazepam and alprazolam (BLIDING 1978; CIRAULO et al. 1988b). Even in the absence of alcohol dependence, diazepam causes significant impairment of motor performance in conjunction with alcohol. Diazepam has amnestic properties that may impede the learning of new information, which may be particularly problematic in cognitive–behavioral treatments for alcoholism (WOODS et al. 1987). Paradoxical agitation or violent disinhibition may also occur with benzodiazepine treatment and may interfere with the successful psychosocial rehabilitation of alcoholics. The risk of these adverse effects has not been reported in studies of less-lipophilic benzodiazepines such as clonazepam, which is frequently used to treat patients with panic disorder.

Many clinicians have suggested that prescribed benzodiazepines (particularly lipophilic agents such as diazepam or alprazolam) may increase the risk of relapse among alcoholics, even among alcoholics whose medication use and treatment responsiveness is carefully monitored by physicians. We know of no well-designed studies that have tested this hypothesis. For the present, however, it appears prudent to employ non-benzodiazepines (e.g., CAs or buspirone) as the primary agents for treatment of anxiety disorders in abstinent alcoholics.

Beta blockers such as atenolol and the MAOI phenelzine are effective in the treatment of social phobia, particularly for patients with readily identified, specific, fear-inducing situations (LIEBOWITZ et al. 1988). To our knowledge, there are no published studies evaluating the efficacy of these or other drugs in the treatment of alcoholics with simple or social phobia.

In nonalcoholics, CAs and MAOIs are effective in the treatment of PD, particularly when combined with behavioral treatment (MARKS 1987). No studies have been published to address the effectiveness of these agents in alcoholics with PD, though at least on study is currently underway to address this this issue.

Serotonergic medications, such as the CA clomipramine and the SUIs fluvoxamine, fluoxetine, and sertraline, are effective in treatment of obsessions and compulsions in nonalcoholics with OCD (BLACK 1992). Because of the risk of seizures at higher doses, daily doses of clomipramine above 250 mg should be avoided. KRANZLER and collaborators (1990) found that fluvoxamine was poorly tolerated by abstinent alcoholics, none of whom had OCD. The safety and efficacy of these medications for treatment of alcoholics with OCD has not been documented in published reports.

Symptoms of GAD respond to treatment with buspirone, a serotonergic anxiolytic that, in contrast to benzodiazepines, does not enhance alcohol-induced impairment of psychomotor speed or reactivity. It also appears to have little abuse potential. In an open trial, buspirone treatment was associated with decreased anxiety and desire to drink among anxious alcoholics (KRANZLER and MEYER 1989). An early double-blind, placebo-controlled trial in alcoholics found buspirone to be associated with significantly improved treatment retention, lower levels of alcohol craving, decreased anxiety, and decreased depression (BRUNO 1989). In a placebo-controlled trial among alcoholics with GAD, buspirone treatment was associated with significantly lower levels of premature treatment withdrawal and greater reduction in anxiety than was placebo (TOLLEFSON et al. 1992). Effects on drinking were not reported for this study, in which no specific psychosocial treatment was provided beyond referral to Alcoholics Anonymous. A study of alcoholic male veterans with GAD who had recently completed inpatient detoxification from alcohol found that buspirone treatment (45–60 mg/day) was not superior to placebo on measures of anxiety or drinking (MALCOLM et al. 1993). Of note, subjects received only supportive physician contact in this study, and no specific psychotherapy was provided to address alcoholism or anxiety symptoms.

Kranzler and colleagues recently reported the results of a 12-week clinical trial comparing buspirone and placebo as an adjunct to structured relapse prevention therapy in a group of anxious alcoholics (Kranzler et al., unpublished manuscript). Buspirone treatment was associated with significantly greater treatment retention and a significantly longer interval to relapse to heavy drinking (i.e., ≥5 drinks/occasion). Buspirone treatment also produced lower rates of drinking during both the treatment and the 6-month follow-up period. While 36%–40% of subjects had GAD, all subjects had persistently elevated anxiety ratings at the time of entering the study. Among subjects with the highest pretreatment anxiety levels, buspirone treatment was associated with a significant reduction in anxiety. No significant adverse effects of buspirone were noted at the dosage (40–60 mg/day) employed in this study. The finding that buspirone was superior to placebo in this study, but not in previous studies, may be attributable to several factors. The subjects in this study had lower levels of alcohol dependence and greater social stability than did the subjects studied by Malcolm and colleagues (1993). In addition, the concomitant psychosocial treatment provided in this study may have enabled subjects to remain in treatment and cope more effectively with anxiety symptoms and drinking situations.

At present, buspirone appears indicated for abstinent alcoholics with significant levels of persistent generalized anxiety. Clinical experience indicates that abstinent alcoholics with panic disorder may benefit from avoidance of caffeine and treatment with CAs. Similarly, alcoholics with OCD may benefit from SRI treatment. However, no published data are available to support these recommendations. Pharmacologic treatment of other anxiety disorders is not clearly indicated. Benzodiazepines should be avoided whenever possible in treatment of alcoholics.

D. Bipolar Affective Disorder

Manic episodes are distinct periods, usually lasting at least 1 week, of uncharacteristically and markedly elevated, expansive, euphoric, or irritable mood, together with decreased subjective need for sleep, increased energy and activity, distractibility, rapid or pressured speech, and/or a subjective sense of racing thoughts. Many manic patients display grandiosity and take unusual risks, are physically reckless, spend wildly, or are physically aggressive, though these behaviors are rare once the mania resolves. Psychosis may be present. In mixed states, manic and depressive symptoms occur together and suicide risk may be high (Goodwin and Jamison 1990). Bipolar II disorder is a milder form in which recurrent depressive episodes occur, together with periods of hypomania (American Psychiatric Association 1993b). Cyclothymia is another mild form characterized by recurrent periods of depressive symptoms which are of insufficient severity to qualify as a major depressive episode, together with periods of hypomania (AMERICAN PSYCHIATRIC ASSOCIATION 1987, 1993b).

Alcoholism has been associated with manic and depressive states since ancient times. Plato described alcoholism as a cause of mania. In the ECA, the lifetime prevalence of mania was 0.4% in the general population, but the rate of mania among alcoholics was sixfold greater (HELZER and PRYZBECK 1988). Among those with bipolar disorder, 15% were alcohol abusers and 31% had a lifetime history of alcohol dependence; this represents a 3.5-fold increased lifetime risk of alcoholism for those with bipolar I disorder compared to those without it (GOODWIN and JAMISON 1990). Alcoholism is common among patients with bipolar disorder. It is estimated that least 15%–65% of patients with bipolar I, bipolar II, and cyclothymic disorders abuse alcohol (GOODWIN and JAMISON 1990).

It is, however, relatively uncommon for bipolar patients to be treated in alcoholism rehabilitation programs. Among alcoholics studied in a variety of treatment settings, 2%–8% met criteria for bipolar disorder (HESSELBROCK et al. 1985; ROSS et al. 1988). Heavy drinking frequently contributes to symptomatic exacerbations of mania, mixed mania, or depression, resulting in hospitalization of bipolar disordered patients (GOODWIN and JAMISON 1990). Drinking patterns appear to be different between manic and depressive phases, with reduced or periodic excessive drinking predominating during the depressive phase and chronic excessive drinking occurring during manic periods (GOODWIN and JAMISON 1990). These patterns may partly reflect the social withdrawal typical of bipolar depression and the increased psychomotor activity, impulsiveness, and efforts to heighten euphoria typical of manic episodes. Drinking by manic and depressed bipolar patients increases their risk of attempted and successful suicide (MURPHY 1992).

I. Assessment

Mania and bipolar disorder can be difficult to identify in their atypical forms, particularly when the clinician is inexperienced or when the patient is intoxicated or in alcohol withdrawal. Patients often are unable to recall diagnostically important features of past episodes of mania and depression (ZIMMERMAN and CORYELL 1985), even when there is no history of a substance use disorder. Depressed or manic patients may not provide reliable information regarding past symptoms. However, when respondents are not manic and when a standardized psychiatric interview is employed, bipolar disorder is recognized with good reliability among substance abuse patients, though it is occasionally confused with schizophrenia (BRYANT et al. 1992). Longitudinal history, e.g., a history of episodic mood disturbance with little dysfunction between such episodes, would support a diagnosis of bipolar disorder and enhance diagnostic reliability. Lower levels of reliability are achieved with bipolar II and cyclothymic disorders than with classic bipolar disorder;

their identification may require greater interviewer skill, often supplemented by information provided by collateral informants or treatment records.

Differentiation of symptoms secondary to antisocial personality disorder, underlying alcohol or drug use, or bipolar disorder can be difficult. It is important, therefore, to obtain a detailed chronology of the onset and duration of characteristic symptoms of each disorder, with particular emphasis on their occurrence during periods of abstinence and their association with periods of alcohol or drug use or withdrawal. In all cases where a diagnosis of bipolar affective disorder is considered, other organic causes of mania, such as hyperthyroidism, must be excluded. Like alcoholism, bipolar affective disorder is transmitted within families. An alcoholic with a clear history of bipolar disorder in a first-degree family member is at increased risk of manic and depressive episodes. Family history may aid in the differential diagnosis of psychiatric symptoms suggestive of bipolar disorder in alcoholics.

II. Treatment

Successful pharmacotherapeutic treatment of cyclic mood disorders in the alcoholic is enhanced by appropriate psychosocial treatment of the mood disorder itself (Goodwin and Jamison 1990). Psychiatric hospitalization may be necessary for accurate diagnosis, detoxification, safekeeping, or behavioral stabilization, particularly if the risk of suicide or violence is high, or if complicated withdrawal, psychosis, or marked impairment in judgment are present. Counseling helps the patient and family reduce stresses and behaviors (e.g., drinking, sleep deprivation) that may precipitate a relapse, recognize signs of impending relapse, cope with manic or depressive episodes, and deal with interpersonal, occupational, and other consequences of past affective episodes (Goodwin and Jamison 1990).

III. Pharmacotherapy

As discussed above, lithium carbonate is effective for the acute treatment of mania and depression and for prevention of relapse in patients with bipolar affective disorder (Jefferson et al. 1987). The efficacy of lithium in treatment of bipolar disorder is contingent on maintenance of therapeutic serum lithium levels. For treatment of acute mania, lithium levels of 0.9–1.2 mEq/l are typically effective, whereas levels of 0.7–1.0 mEq/l are effective for prevention of relapses (Jefferson et al. 1987).

Preliminary results from an ongoing clinical trial of lithium treatment of adolescents with comorbid bipolar and substance use disorders suggest that lithium at serum levels of 0.9–1.3 mEq/l is effective in treatment of both the cyclic mood disorder and alcohol and/or drug use disorder (Geller et al. 1992). To our knowledge, studies of the efficacy of lithium in adults with both bipolar disorder and alcoholism have not been published.

Adverse effects of lithium include increased weight, nausea, diarrhea, increased urination, acne, and a fine tremor of the hands that is worsened by caffeine (Jefferson et al. 1987). Some patients report drinking more alcohol in response to lithium-induced increases in thirst, but this is uncommon. At high serum levels, lithium causes ataxia, slurred speech, vomiting, confusion, coma, and death. Beyond intentional overdose, high lithium levels can arise due to renal disease, concurrent hyponatremia, or concurrent use of medications such as nonsteroidal antiinflammatory agents or thiazide diuretics. Bipolar alcoholics are at substantial risk of intentional overdose, particularly during depressive and drinking relapses, and following serious interpersonal losses such as separation or divorce (Murphy 1992). Chronic lithium treatment is associated with an increased risk of hypothyroidism, nephrogenic diabetes insipidus, and impaired renal filtration. Thus, pretreatment evaluation of the lithium candidate should include assessment of serum creatinine, electrolytes, and thyroid indices, as well as evaluation of urine specific gravity. Periodic monitoring of these indices and of serum lithium levels is also indicated during a course of lithium treatment.

Noncompliance with prescribed lithium is often a problem, and can contribute to manic or depressive relapses and increased drinking. Noncompliance is most commonly associated with the patient's desire to lessen the adverse effects of lithium, psychosocial stressors, alcohol or drug use, boredom, and a desire to avoid the stigma of having a chronic mental illness. Efforts to reduce the severity of adverse effects of lithium by dose modification, use of sustained-release lithium preparations, reduction in caffeine use, or addition of propranolol for treatment of lithium-induced tremor (Jefferson et al. 1987) may be helpful. Psychosocial interventions can improve family and other interpersonal relationships and employment, and can help reduce the stigma of mental illness and feelings of boredom that may occur following resolution of a hypomanic episode (Goodwin and Jamison 1990).

Carbamazepine (CBZ) is also effective in treatment of acute manic and depressive episodes and in prevention of relapses in patients with bipolar disorder. It may be particularly effective for patients with mixed or rapid-cycling bipolar disorder (Post et al. 1989). As reviewed by Anton and Becker (Chap. 13) in this volume, CBZ is effective in treatment of alcohol withdrawal. To our knowledge, there are no published studies of the effectiveness of CBZ in the prevention of relapses among alcoholics with bipolar affective disorder, nor are there published studies that compare CBZ and lithium in the rehabilitation of bipolar alcoholics.

Carbamazepine is oxidized by hepatic enzymes that are, themselves, induced by CBZ, alcohol, and tobacco smoke. Thus, the half-life of CBZ is reduced following chronic use, and by concurrent heavy drinking or smoking. While a clear dose-response relationship for CBZ in prophylaxis of mania and depression has not been established, experienced psychopharmacologists suggest that clinicians titrate oral doses to achieve a serum level of CBZ in the antiepileptic range, i.e., 8–12 μg/ml (Post et al. 1989).

Carbamazepine has a number of potential adverse effects (Pose et al. 1989). It can cause leukopenia and, rarely, agranulocytosis, thrombocytopenia, eosinophilia, or aplastic anemia. It causes hyponatremia in a substantial minority of cases, particularly in elderly patients treated at high dosages. Rarely, it causes cholestatic hepatitis or hepatocellular damage. At high serum levels, it can lead to nausea or vomiting. CBZ levels, liver function tests, and blood counts should be monitored periodically in patients treated with CBZ. CBZ is neurotoxic, producing sedation and cerebellar symptoms, which may be exacerbated by concurrent use of alcohol or other drugs such as phenytoin. In overdose, CBZ can cause significant cardiovascular and neurologic toxicity. Alcohol ingestion may also increase the risk of overdose and suicide.

As with lithium treatment, CBZ treatment of bipolar patients is often accompanied by noncompliance, which can be associated with symptomatic relapses.

Other antiepileptics, including valproic acid and the benzodiazepine clonazepam, have efficacy in bipolar disorder (Post et al. 1989). Their use in treatment of the alcoholic with bipolar disorder has not been reported. There is a risk of abuse of benzodiazepines by patients with alcoholism, but the relative lack of euphoria produced and long half-life of clonazepam should theoretically reduce its abuse potential. Systematic evaluation of the abuse potential of clonazepam has not been reported.

Pharmacologic treatment of alcoholics with bipolar disorder using adequate dosages of lithium is indicated, both to reduce the risk of manic and depressive relapses and to facilitate social stability, which, in turn, will likely reduce the risk of alcoholic relapses. CBZ may be appropriate for bipolar patients who do not respond to lithium maintenance treatment. Because of the need for long-term pharmacotherapy of patients with bipolar disorder, treatment must address issues of medication compliance.

E. Schizophrenia

Most schizophrenics living in the community drink alcohol, and many spend substantial portions of their limited incomes on alcohol and low-cost recreational drugs, which they often use with their peers (Test et al. 1989; Soni and Brownlee 1991; Westermeyer 1992). Alcohol use often contributes to medication noncompliance, homelessness, and rehospitalization of schizophrenics (Drake et al. 1989, 1990, 1991).

Schizophrenic disorders are characterized by functional decline, psychotic symptoms, and, in most cases, chronicity (American Psychiatric Association 1987, 1993b). In addition to psychotic symptoms, patients with schizophrenic disorders often display odd behaviors, blunted affect, impoverished speech, poor hygiene, social isolation, and limited social skills. These "negative" symptoms can be gravely disabling and often render schizophrenic patients ill-suited to treatments offered in traditional substance abuse treatment programs. Organic mental conditions must be considered in evaluation of the potentially schizophrenic patient.

Schizophrenia is a relatively rare disorder. In the ECA, the lifetime prevalence of schizophrenia was approximately 1%, but alcoholism was associated with a fourfold increased risk of schizophrenia (Helzer and Pryzbeck 1988). Conversely, alcoholism was common among schizophrenics. Schizophrenia was associated with a 30% prevalence of current alcoholism, which was eight- to tenfold higher than the prevalence in the general population (Robins and Regier 1990). In clinical samples, the lifetime prevalence of alcoholism was 31%–59% among schizophrenics in the United States (Drake et al. 1990; Mueser et al. 1992). As in the ECA, schizophrenic alcoholics in treatment are most often male, single, and unemployed.

I. Assessment

Because schizophrenics often demonstrate cognitive impairment, it is important that collateral information from reliable informants be employed in making a substance use disorder diagnosis (Test et al. 1989; Drake et al. 1990). In evaluating the alcoholic patient with schizophrenia for possible pharmacologic treatment, several other concerns should also be addressed. First, it is important to evaluate the presence of extrapyramidal and other adverse effects of prescribed antipsychotics. Akathisia, for example, may be mistaken for anxiety or psychotic decompensation, while bradykinesia may mimic depression. Reduction in antipsychotic dosage, use of a less potent agent, or (for treatment-resistant patients or those who develop tardive dyskinesia) replacement with atypical antipsychotics such as clozapine, or use of anticholinergic agents or beta blockers may lessen these effects. Such treatment may lessen the patient's motivation to drink or stop taking his or her antipsychotic medication. Second, it is important to review the context of drinking, motivations for drinking, and effects of drinking reported by the patient. Schizophrenics lead isolated lives, have poor social skills, and spend time with others who have similar problems, including alcohol-related problems (Dixon et al. 1990). They report drinking to relieve boredom, to socialize, to feel relaxed, to sleep better, to make side effects of medications more tolerable, and to decrease hallucinations (Test et al. 1989; Dixon et al. 1990; Noordsy et al. 1991). These symptoms may be lessened by a variety of psychosocial interventions (described below), in addition to adjustments in psychotropic medications. Occasionally, use of benzodiazepines for treatment of akathisia or marked anxiety may be indicated. Third, concurrent depression should be evaluated and treated, as described earlier in this chapter. Fourth, the temporal relationship between alcohol use, antipsychotic medication noncompliance, and psychosis or other adverse symptoms of schizophrenia should be reviewed with the patient. Avoidance of adverse consequences may motivate the schizophrenic patient to comply with an antipsychotic medication regimen and to reduce alcohol and drug use.

II. Treatment

Psychosocial interventions, including social skills training, family psychoeducation, assistance with the needs of daily living (such as employment, housing, money management), and nonconfrontational alcoholism counseling (Test et al. 1989; Drake et al. 1991, 1993), complement pharmacotherapy and appear useful in treating the schizophrenic with alcoholism. Controlled studies of psychosocial treatment of schizophrenics with concurrent alcoholism are in progress.

III. Pharmacologic Treatment

Antipsychotics remain the mainstay in the pharmacologic treatment of schizophrenics, including those who abuse alcohol. A variety of antipsychotics are effective in the treatment of acute psychotic episodes and in the prevention of relapses among schizophrenics, including typical antipsychotics and newer, atypical antipsychotic agents such as clozapine and risperidone. Side effect profiles generally guide the choice of an antipsychotic. Low-potency antipsychotics such as chlorpromazine are more sedating and have more anticholinergic effects, but are less often associated with extrapyramidal effects, while high-potency agents such as haloperidol produce the opposite pattern. Monitoring of drug levels is most practical with haloperidol and fluphenazine, and may be useful, particularly as alcohol may lower fluphenazine serum levels (Soni and Brownlee 1991). Noncompliant schizophrenics may benefit from fluphenazine or haloperidol decanoate injections every 2–4 weeks. Atypical antipsychotics may lessen negative or positive symptoms, and may occasionally contribute to abstinence. Clozapine and resperidone currently are the only atypical antipsychotic approved in the United States. Risperidone has been shown in Europe and elsewhere to be effective and well tolerated. There is a 0.5%–1% risk of agranulocytosis in patients treated with clozapine, and weekly blood counts are required of all United States patients receiving this medication. To our knowledge, no published studies have compared the effectiveness of these oral antipsychotics, depot antipsychotics, or atypical antipsychotics in the reduction or prevention of drinking or psychotic relapses among schizophrenics with alcoholism.

In summary, antipsychotic medications are clearly indicated in the treatment of alcoholics with schizophrenia. The choice of a particular agent and dosage form (oral or depot) will depend on a variety of factors, including past antipsychotic response, adverse effects, compliance, and convenience. Ongoing treatment of the schizophrenic alcoholic with antipsychotics requires regular assessment of the patient for signs of tardive dyskinesia and other drug side effects, as well as counseling to enhance medication compliance and to reduce alcohol and drug use.

F. Conclusions

Antidepressant medications, beta blockers, and buspirone are effective for appropriately selected alcoholics whose depressive or anxiety symptoms are persistent and/or meet criteria for a depressive or anxiety disorder. In addition, a minority of alcoholics have persistent mental illnesses such as bipolar affective disorder or schizophrenia. Relapses among patients with these disorders are often associated with substance abuse. While studies have not addressed the specific issues of pharmacotherapy of these two groups of disorders in alcoholics, lithium, carbamazepine, and neuroleptics are effective in treating a majority of patients with these conditions. Patients receiving such treatment must be closely monitored.

It should be clear from this discussion that clinical researchers have only begun to evaluate the potential for effective pharmacologic treatment of coexisting psychiatric disorders in alcoholics. Improved clinical care of alcoholics will be enhanced by carefully conducted studies designed to evaluate the efficacy of psychotropic medications, specific psychosocial therapies, and the combinations of these treatments in alcoholics with specific concurrent psychiatric diagnoses. The impact of these treatments on psychiatric symptom profiles, drinking behavior, and functional capacity of alcoholics should be further evaluated.

Several specific questions should be addressed in future research:

Are there subtypes of affective, anxiety, and other comorbid psychiatric disorders that are more responsive to pharmacotherapeutic treatment than other subtypes? Among nonalcoholics with MDD, evidence suggests that more severe MDD characterized by marked psychomotor retardation and persistent abnormality in the hypothalamic–pituitary–adrenal axis is particularly responsive to antidepressant medications (American Psychiatric Association 1993a). Given the large number of subtypes of depression among alcoholics, it is possible that the various depressive subtypes will respond differently to various antidepressant medications. Similar subtyping may apply to anxiety and psychotic disorders.

Are there particular characteristics of the alcoholic that predict both a reduction in psychiatric symptoms and a lower rate of drinking during or following pharmacotherapy? There are empiric subtypes of alcoholics who display differential responsiveness to distinct psychosocial treatments (Litt et al. 1992). At least one of these subtypes may display a neurochemical abnormality, i.e., low serotonin activity (Buydens-Branchey et al. 1989). Do depressed or anxious alcoholics of this type experience a reduction in anxiety or affective symptoms, as well as a lower rate of relapse, following treatment with a serotonergic agent such as buspirone?

Is there a maximally effective duration of pharmacotherapy for coexisting psychopathology in alcoholics? Considerable evidence suggests that maintenance treatment for bipolar disorder reduces rates of manic and depressive relapses (Goodwin and Jamison 1990), and that prolonged antidepressant treatment reduces relapses to MDD (American Psychiatric Association 1993a), but these conclusions are based on studies of nonalcoholics. While maintenance antipsychotic treatment may be well justified for alcoholics with schizophrenia, we do not know if the same is true for anxious or depressed alcoholics treated with buspirone or cyclic antidepressants.

We hope that by investigating the answers to these and other questions clinical researchers can find methods of treating alcoholics more effectively with medications.

References

American Psychiatric Association (1987) Diagnostic and statistical manual for psychiatric disorders, 3rd edn, revised. American Psychiatric Association, Washington DC

American Psychiatric Association (1993a) Practice guideline for major depressive disorder in adults. American Psychiatric Association, Washington DC

American Psychiatric Association (1993b) DSM-IV draft criteria. American Psychiatric Association, Washington DC

Aneshensel CS, Huba GJ (1983) Depression, alcohol use, and smoking over one year: a four-wave longitudinal causal model. J Abnorm Psychol 92:134–150

Babor TF, Hoffman M, Del Boca FK, Tennen H, Hesselbrock VM, Meyer RE (1992) Types of alcoholics I: evidence for an empirically derived typology based on indicators of vulnerability and severity. Arch Gen Psychiatry 49:599–608

Ballenger JC, Post RM (1978) Kindling as a model for alcohol withdrawal syndromes. Br J Psychiatry 133:1–14

Beck AT, Ward CH, Mendelson M (1961) An inventory for measuring depression. Arch Gen Psychiatry 4:461–471

Black JL (1992) Obsessive-compulsive disorder: a clinical update. Mayo Clin Proc 67:266–275

Bliding A (1978) The abuse potential of benzodiazepines with special reference to oxazepam. Acta Psychiatr Scand 24:111–116

Brown ME, Anton RF, Malcolm R, Ballenger JC (1990) Alcohol detoxification and withdrawal seizures: clinical support for a kindling hypothesis. Biol Psychiatry 23:507–514

Brown SA, Schuckit MA (1988) Changes in depression among abstinent alcoholics. J Stud Alcohol 49:412–417

Brown SA, Irwin M, Schuckit MA (1991) Changes in anxiety among abstinent male alcoholics. J Stud Alcohol 52:55–61

Bruno F (1989) Buspirone in the treatment of alcoholic patients. Psychopathology 22 [Suppl]:49–59

Bryant KJ, Rounsaville B, Spitzer RL, Williams JBW (1992) Reliability of dual diagnosis: substance dependence and psychiatric disorders. J Nerv Ment Dis 180:251–257

Buydens-Branchey L, Branchey MJ, Noumair D (1989) Age of alcoholism onset II. Relationship to susceptibility to serotonin precursor availability. Arch Gen Psychiatry 46:231–236

Ciraulo DA, Alderson LM, Chapron DJ (1982) Imipramine disposition in alcoholics. J Clin Psychopharmacol 2:2–7

Ciraulo DA, Barnhill J, Boxenbaum HG (1985) Pharmacokinetic interaction of disulfiram and antidepressants. Am J Psychiatry 142:1373–1374
Ciraulo DA, Barnhill JG, Jaffe JH (1988a) Clinical pharmacokinetics of imipramine and desipramine in alcoholics and normal volunteers. Clin Pharmacol Ther 43:509–518
Ciraulo DA, Barnhill J, Greenblatt DJ, Shader RE, Ciraulo AM, Tarmey MF, Molloy MA, Foti ME (1988b) Abuse liability and clinical pharmacokinetics of alprazolam in alcoholic men. J Clin Psychiatry 49:333–337
Cooney NL, Kadden RM, Litt MD, Getter H (1991) Matching alcoholics to coping skills or interactional therapies: two-year follow-up results. J Consult Clin Psychol 59:598–601
Cornelius JR, Fisher BW, Salloum IM, Cornelius MD, Ehler JG (1992) Fluoxetine trial in depressed alcoholics. Alcohol Clin Exp Res 16:362
Cottler LB, Compton WM, Mager D, Spitznagel EL, Janca A (1992) Posttraumatic stress disorder among substance users from the general population. Am J Psychiatry 149:664–670
Dackis CA, Gold MS, Pottash ALC, Sweeney DR (1986) Evaluating depression in alcoholics. Psychiatry Res 17:105–109
de la Fuente J-R, Morse RM, Niven RG, Ilstrup DM (1989) A controlled study of lithium carbonate in the treatment of alcoholism. Mayo Clin Proc 64:177–180
Dixon L, Haas G, Weiden P, Sweeney J, Frances A (1990) Acute effects of drug abuse in schizophrenic patients: clinical observations and patients' self-reports. Schizophr Bull 16:69–79
Dorus W, Kennedy J, Gibbons RD, Raavi SD (1987) Symptoms and diagnosis of depression in alcoholics. Alcohol Clin Exp Res 11:150–154
Dorus W, Ostrow DG, Anton R, Cushman P, Collins JF, Schaefer M, Charles HL, Desai P, Hayashida M, Malkernecker U, Willenbring M, Fiscella R, Sather MR (1989) Lithium treatment of depressed and nondepressed alcoholics. JAMA 262:1646–1652
Drake RE, Osher FC, Wallach MA (1989) Alcohol use and abuse in schizophrenia: a prospective community study. J Nerv Ment Dis 177:408–414
Drake RE, Osher FC, Noordsy DL (1990) Diagnosis of alcohol use disorders in schizophrenia. Schizophr Bull 16:57–67
Drake RE, Wallach MA, Teague GB (1991) Housing instability and homelessness among rural schizophrenic patients. Am J Psychiatry 148:330–336
Drake RE, McHugo GJ, Noordsy DL (1993) Treatment of alcoholism among schizophrenic outpatients: 4-year outcomes. Am J Psychiatry 150:328–329
Elkin I, Shea T, Watkins JT, Imber SD, Sotsky SM, Collins JF, Glass DR, Pilkonis PA, Leber WR, Docherty JP, Fiester SJ, Parloff MB (1989) National Institute of Mental Health treatment of depression collaborative research program: general effectiveness of treatments. Arch Gen Psychiatry 46:971–982
Fawcett J, Clark DC, Aagesen CA (1987) A double-blind, placebo-controlled trial of lithium carbonate therapy for alcoholism. Arch Gen Psychiatry 44:248–256
Geller B, Cooper TB, Watts HE, Cosby CM, Fox LW (1992) Early findings from a pharmacokinetically designed, double-blind, placebo-controlled study of lithium for adolescents with comorbid bipolar and substance dependence disorders. Prog Neuropsychopharmacol Biol Psychiatry 16:281–299
George DT, Nutt DJ, Dwyer BA, Linnoila M (1988) Alcoholism and panic disorders: is the comorbidity more than coincidence? Acta Psychiatr Scand 81:97–107
Goodwin FK, Jamison KR (1990) Manic depressive illness. Oxford University Press, New York
Grant BF, Hasin DS, Harford TC (1989) Screening for major depression among alcoholics: an application of receiver operating characteristic analysis. Drug Alcohol Depend 23:123–131
Hamilton M (1960) A rating scale for depression. J Neurol Neurosurg Psychiatry 23:56–62

Hatsukami D, Pickens RW (1982) Post-treatment depression in an alcohol and drug abuse population. Am J Psychiatry 139:1563–1566
Helzer JE, Pryzbeck TR (1988) The co-occurrence of alcoholism with other psychiatric disorders in the general population and its impact on treatment. J Stud Alcohol 49:219–224
Henriksson MM, Aro HM, Marttunen MJ, Heikkinen ME, Isometsa ET, Kuoppasalmi KI, Lonnqvist JK (1993) Mental disorders and comorbidity in suicide. Am J Psychiatry 150:935–940
Hesselbrock MN, Meyer RE, Keener JJ (1985) Psychopathology in hospitalized alcoholics. Arch Gen Psychiatry 42:1050–1055
Jaffe JH (1987) Alcoholism and affective disturbance: current drugs and current shortcomings. In: Edwards G, Littleton J (eds) Pharmacologic treatments for alcoholism. Methuen, New York, pp 463–490
Jaffe JH, Ciraulo DA (1986) Alcoholism and depression. In: Meyer RE (ed) Psychopathology and addictive disease. Guilford, New York, pp 293–320
Jefferson JW, Greist JH, Ackerman DL, Carroll JA (1987) Lithium encyclopedia for clinical practice, 2nd edn. American Psychiatric Press, Washington DC
Judd, LL, Huey LY (1984) Lithium antagonizes ethanol intoxication in alcoholics. Am J Psychiatry 141:1517–1521
Keeler MH, Taylor I, Miller WC (1979) Are all recently detoxified alcoholics depressed? Am J Psychiatry 136:586–588
Kissin B (1975) The use of psychoactive drugs in the long-term treatment of chronic alcoholics. Ann NY Acad Sci 252:385–395
Kranzler HR, Meyer RE (1989) An open trial of buspirone in alcoholics. J Clin Psychopharmacol 9:379–380
Kranzler HR, Del Boca FK, Korner P, Brown J (1990) Fluvoxamine is poorly tolerated by alcoholics. In: Naranjo CA, Sellers EM (eds) Novel pharmacological interventions for alcoholism. Springer, Berlin Heidelberg New York, pp 304–308
Kranzler HR, Burleson JA, Del Boca FK, Babor TF, Korner P, Brown J, Bohn MJ (unpublished manuscript) Placebo-controlled trial of buspirone as an adjunct to relapse prevention in anxious alcoholics
Kupfer DJ, Frank E, Perel JM, Cornes C, Mallinger AG, Thase ME, McEachran AB, Grochocinski VJ (1992) Five-year outcome for maintenance therapies in recurrent depression. Arch Gen Psychiatry 49:769–773
Liebowitz MR, Gorman JM, Fyer AJ, Campeas R, Levin AP, Sandberg D, Hollander E, Papp L, Goetz D (1988) Pharmacotherapy of social phobia: an interim report of a placebo-controlled comparison of phenelzine and atenolol. J Clin Psychiatry 49:252–257
Liebson I, Bigelow G, Flamer R (1973) Alcoholism among methadone patients: a specific treatment method. Am J Psychiatry 130:483–485
Litt MD, Babor TF, Del Boca FK, Kadden RM, Cooney NL, Getter H (1992) Types of alcoholics: application of an empirically derived typology to treatment matching. Arch Gen Psychiatry 49:609–614
Lowinsohn JH, Ruiz P, Millman RB, Langrod JG (1992) Substance abuse: a comprehensive textbook, 2nd edn. Williams and Wilkins, Baltimore
Malcolm R, Anton RF, Randall CL, Johnston A, Brady K, Thevos A (1993) A placebo-controlled trial of buspirone in anxious inpatient alcoholics. Alcohol Clin Exp Res 16:1007–1014
Manuzza S, Fyer AJ, Martin LY, Gallops MS, Endicott J, Gorman J, Liebowitz MR, Klein DF (1989) Reliability of anxiety assessment. I. Diagnostic agreement. Arch Gen Psychiatry 46:1093–1101
Marks IM (1987) Fears, phobias, and rituals: panic, anxiety, and their disorders. Oxford University Press, New York
Mason BJ, Kocsis JH (1991) Desipramine treatment of alcoholism. Psychopharmacol Bull 27:155–161

McLellan AT, Luborsky L, Woody GE, O'Brien CP, Druley KA (1983a) Predicting response to alcohol and drug abuse treatments: role of psychiatric severity. Arch Gen Psychiatry 40:620–625
McLellan, AT, Woody GE, Luborsky L, O'Brien CP, Druley KA (1983b) Increased effectiveness of substance abuse treatment: a prospective study of patient–treatment "matching". J Nerv Ment Dis 171:597–605
Mueser KT, Bellack AS, Blanchard JJ (1992) Comorbidity of schizophrenia and substance abuse: implications for treatment. J Consult Clin Psychol 60:845–856
Murphy GE (1992) Suicide in alcoholism. Oxford University Press, New York
Nakamura MM, Overall JE, Hollister LE, Radcliffe E (1983) Factors affecting outcome of depressive symptoms in alcoholics. Alcohol Clin Exp Res 7:188–193
Nathan PE, Titler NA, Lowenstein LM, Solomon P, Rossi AM (1970) Behavioral analysis of chronic alcoholism: interaction of alcohol and human contact. Arch Gen Psychiatry 22:419–430
Noordsy DL, Drake RE, Teague GB (1991) Subjective experiences related to alcohol use among schizophrenics. J Nerv Ment Dis 179:410–414
Nunes EV, McGrath PJ, Quitkin FM, Stewart JP, Harrison W, Tricamo E, Ocepek-Welikson K (1993) Imipramine treatment of alcoholism with comorbid depresion. Am J Psychiatry 150:963–965
Post RM, Trimble MR, Pippinger C (1989) Clinical use of anticonvulsants in psychiatric disorders. Demos, New York
Robins LN, Regier DA (1990) Psychiatric disorders in America. Free Press, New York
Robins LN, Cottler L, Keating S (1989) NIMH Diagnostic Interview Schedule, Version III, Revised. Washington University Department of Psychiatry, St Louis
Ross HE, Glaser FB, Germanson T (1988) The prevalence of psychiatric disorders in patients with alcohol and other drug problems. Arch Gen Psychiatry 45:1023–1031
Rounsaville BJ, Dolinsky ZS, Babor TF, Meyer RE (1987) Psychopathology as a predictor of treatment outcome in alcoholics. Arch Gen Psychiatry 44:505–513
Sandoz M, Vandel S, Vandel B, Bonin B, Allers G, Volmot R (1983) Biotransformation of amitriptyline in alcoholic depressive patients. J Clin Pharmacol 24:615–621
Schottenfeld RS, O'Malley SS, Smith L, Rounsaville BJ, Jaffe JH (1989) Limitation and potential hazards of MAOIs for the treatment of depressive symptoms in abstinent alcoholics. Am J Drug Alcohol Abuse 15:339–344
Schuckit M (1979) Alcoholism and affective disorder: diagnostic confusion. In Goodwin DW, Erickson CK (eds) Alcoholism and affective disorders. Spectrum, New York, pp 9–19
Schuckit MA, Winokur G (1972) A short-term follow-up of women alcoholics. Dis Nerv Syst 33:672–678
Schuckit MA, Irwin M, Brown SA (1990) History of anxiety symptoms among 171 primary alcoholics. J Stud Alcohol 51:34–41
Soni SD, Brownlee M (1991) Alcohol abuse in chronic schizoprenics: implications for management in the community. Acta Psychiatr Scand 84:272–276
Spitzer RL, Williams JBW, Gibbon M, First MB (1992) The Structured Clinical Interview for DSM-III-R (SCID). I. History, rationale, and description. Arch Gen Psychiatry 49:624–629
Tamerin JS, Mendelson JH (1969) The psychodynamics of chronic inebriation: observations of alcoholics during the process of drinking in an experimental group setting. Am J Psychiatry 125:58–71
Test MA, Wallisch LS, Allness DJ, Ripp K (1989) Substance use in young adults with schizophrenic disorders. Schizophr Bull 15:465–476
Tollefson GD, Montague-Clouse J, Tollefson SL (1992) Treatment of comorbid generalized anxiety in a recently detoxified alcoholic population with a selective serotonergic drug (buspirone). J Clin Psychopharmacol 12:19–26

Tyrer SP, Peat MA, Minty PSB (1990) The effect of alcohol on lithium kinetics. Lithium 1:163–168

Von Knorring A-L, Bohman M, Von Knorring L (1985) Platelet MAO-activity as a biological marker in subgroups of alcoholism. Acta Psychiatr Scand 72:51–58

Weiss RD, Mirin SM, Griffin ML (1992) Methodological considerations in the diagnosis of coexisting psychiatric disorders in substance abusers. Br J Addict 87:179–187

Westermeyer J (1992) Schizophrenia and substance abuse. In: Tasman A, Riba M (eds) Annual review of psychiatry, vol 11. American Psychiatric Press, Washington DC, pp 379–401

Woods JH, Katz JL, Winger G (1987) Abuse liability of benzodiazepines. Pharmacol Rev 39:254–413

World Health Organization (1990a) International classification of diseases - 10, chapter V: mental and behavioural disorders: diagnostic criteria for research. World Health Organization, Geneva

World Health Organization (1990b) Composite international diagnostic interview, core Version 1.0 World Health Organization, Geneva

Zimmerman M, Coryell W (1986) Reliability of follow-up assessments of depressed inpatients. Arch Gen Psychiatry 43:468–470

Zung WWK (1965) A self-rating depression scale. Arch Gen Psychiatry 12:63–70

CHAPTER 18

Clinical Markers of Alcohol Abuse

H. Rommelspacher and C. Müller

A. Introduction

I. Definition of the Term "Marker"

The term "marker", which was originally introduced in molecular genetics and is now also established in research on psychoses and alcoholism, has the same meaning as "indicator", at least as long as no true genetic markers of alcoholism have been detected (Hill et al. 1987; Rommelspacher 1992; Schmidt and Rommelspacher 1990). Thus, researchers in the field of alcoholism use the term "marker" synonymously with "indicator". It should be noted that the commonly used markers refer to superficial phenomena and do not permit conclusions to be drawn about the underlying disease structure. Maas and Katz (1992) have pointed out that "data suggesting relationships between brain serotonin (5-HT) systems and the impulsive expression of aggression are much more persuasive than any data linking 5-HT to a particular psychiatric diagnosis." Thus, as long as relatively little is known of the neurobiological causes of alcoholism, the findings from marker studies should be interpreted with caution with respect to the information they provide about the underlying disease.

Several attempts have been undertaken to categorize the markers for alcoholism. Gordis proposed a schema involving three groups: genetic markers, diagnostic markers, and clinical markers. *Genetic markers* are those that are positive in nondrinking subjects who are biologically at risk for the development of alcoholism if they begin to drink. A *diagnostic marker* is a test for alcoholism independent of the patient's "recovery," and indicates whether the patient has ever been engaged in protracted heavy drinking. The test is positive even if the patient has been abstinent for many months at the time of sampling. *Clinical markers* correlate strongly with the patient's mean blood alcohol concentration (BAC) during the preceding week(s). The test is positive even if BAC is zero at time of sampling. Furthermore, Gordis described the uses of a "clinical" marker: collection of accurate epidemiological data on drinking, evaluation of alcoholism treatment outcome, advising patients of the health risks of excessive drinking, and resolution of clinical diagnostic problems. Gordis also elaborated the desirable characteristics of a clinical marker. It should not depend on the presence of organ damage; false positives should not occur in other medical conditions; its sensitivity to both intake drinking pattern and total intake should be known; it should be easily performed in a clinical setting; and it should be inexpensive (presented at International Society for Biomedical Research on Alcoholism (ISBRA) meeting by Fuller 1992). Begleiter and Porjesz (1988) elaborated the following criteria as being necessary for the identification of a potential *biological marker*: the trait can be reliably measured and is stable over time, it is genetically transmitted, the "abnormal" trait has a low base rate, and it identifies individuals at risk. Studies should show that the trait is prevalent in the patient population, is present during symptom remission, occurs among the first-degree relatives of the index case at a rate higher than that of the normal population, and segregates with the illness in affected relatives. The authors presented some neurophysiological measures that meet the criteria for a marker. These tests, however, are beyond the scope of this chapter.

Our approach to categorizing markers has been influenced in several respects by other authors. *Trait markers* are time-invariant indicators that are present during one's entire lifetime. They can vary with respect to the degree of expression. Trait markers are not specific for a certain disease, especially where some personality traits are concerned. *State markers* are present during the acute disease but not before or after. They vary with conditions and time. Although this classification is useful in psychosis research, it is not sufficient for alcoholism because in the latter disease the compound ethanol plays an important role and the course of the disease is characterized by specific phases.

Psychic alcohol dependence does not require a predisposing factor for its development. However, predisposing factors, which may be either genetic or acquired (such as in utero ethanol exposure), increase the likelihood of its development. Psychic dependence becomes manifest after a period of alcohol misuse lasting many years, followed by a transition phase (point of no return; Coper et al. 1991).

The cellular factors that play a role in the pathophysiology of psychic alcohol dependence during the transition phase, as well as those neuronal mechanisms crucial for the development of dependence symptomatology, are largely unknown. It appears that during the transition phase, the central control mechanisms increasingly lose their effectiveness. Furthermore, during this time the modulation of drinking by social factors is reduced or suspended and a compulsive desire for alcohol (craving), among other factors, induces the individual to abuse the substance. Following withdrawal, a high risk of relapse is characteristic of the disease.

The features of substance abuse make it necessary to introduce the term "*intoxication marker*" for those measures indicative of tissue damage by the substance. The term "*residual marker*" refers to those measures that are demonstrable only after the disease is manifest (and are, therefore, comparable to the "diagnostic markers" of Gordis).

The transition from state markers to intoxication markers can proceed smoothly for some measures. Intoxication markers can persist beyond the period of intoxication through the phase of tissue regeneration. Residual markers appear during the transition phase and reflect the changes in neuronal mechanisms that underlie addiction. The variable extent of their expression is comparable to that of trait markers. As elaborated above, the individual determinants of relapse are residual markers. These determinants contribute to the loss of control and to craving, to mention just two states. Distortions of thinking, most notably denial, are other residual markers. If, as claimed by Alcoholics Anonymous, the disease of alcoholism is not reversible, residual markers should show lifelong persistence.

It is important for the understanding of residual markers that they should be seen not simply as more strongly expressed disposition markers, but as representative of an additional dimension in the disease process. They can be distinguished from trait markers by their occurrence only when the disease is established. Examples of trait markers include the dopamine D_2 receptor allele *A1*. Provided the *A1* allele represents a marker of disposition (trait marker), it in some way exacerbates the course of the disorder (see Chapter 12 of this handbook for a discussion of this and other genetic issues). It is noteworthy that one must not necessarily search for trait markers using genetic methods. Examples of a different strategy are the synapsin II variants of synaptic vesicles, which may be a trait marker for alcoholism (GREBB and GREENGARD 1990), or the evaluation of subjective feelings, motor performance, and endocrinologic measures following the intake of a test dose of ethanol in a certain population (high-risk sons; SCHUCKIT 1992).

In this chapter no endocrinologic findings will be presented. This topic will be dealt with in Chapter 10 of this handbook. Psychophysiologic findings are also not included because such investigations are generally not of pharmacologic interest (for a review see BEGLEITER and PORJESZ 1988).

B. Ethanol, Acetaldehyde, and Acetate

I. Ethanol

Blood ethanol levels are the most direct test for the initial diagnosis of alcohol abuse. However, blood ethanol has low sensitivity as a clinical marker because of its rapid elimination and enhanced elimination in subjects with chronic intake of alcoholic beverages. Although the presence of alcohol does not distinguish acute from chronic alcohol intake, under certain circumstances it can be highly suggestive of abusive alcohol consumption. Thus, a blood alcohol level higher than 300 mg/dl recorded at any time or a level higher than 100 mg/dl (1‰ = 80 mg/dl = 17.4 mmol/l) recorded during a routine medical examination can be regarded as a strong indicator of alcoholism (NATIONAL COUNCIL ON ALCOHOISM 1972). The measurement of ethanol in urine might be a more sensitive marker for acute and chronic intake (TANG 1987). Tang found that ethanol concentrations in the urine of 11 alcoholics after 14 days of abstinence were at least seven times higher than those of social drinkers (excluding heavy drinkers) and ten times higher than those of nonalcoholics who had not consumed ethanol during the 7 days before the determination. In the morning following the day of admission, alcoholics had an average urine ethanol level nearly 160 times that found in light drinkers. The ability to detect ethanol in the urine even after 2 weeks of abstinence greatly improves the sensitivity of the test as an indicator of heavy ethanol consumption. The author speculated that the source of ethanol in the urine might have been ethanol conjugates.

II. Acetaldehyde

Because of the artifactual formation of acetaldehyde during the extraction procedure, the true levels of blood acetaldehyde, free and bound, with or without ethanol intake, remain to be established. The true levels in alcoholics and nonalcoholics are below $0.5\,\mu M$. Eriksson underlined in his 1992 summary that low concentrations do appear (~4 μM) in blood from the right atrium; and up to $68\,\mu M$ has been recorded in blood from the hepatic vein using reliable methods. This means that even during normal ethanol oxidation substantial acetaldehyde gradients exist in the circulation. The hepatic oxidation of acetaldehyde has been determined by Eriksson to be roughly 99% under normal conditions. Thus, aldehyde dehydrogenase deficiency causes a dramatic increase in acetaldehyde in the hepatic vein ($\sim 100\,\mu M$). The same has been reported for dehydrogenase inhibitors (calcium carbimide $\sim 210\,\mu M$, chlorpropamide $\sim 10\,\mu M$, disulfiram $\sim 130\,\mu M$, nitrefazole $\sim 150\,\mu M$). Increasing the ethanol oxidation rate by the ingestion of fructose also causes higher acetaldehyde levels ($\sim 38\,\mu M$) (NUUTINEN et al. 1984), and acetaldehyde has been reported to exist in different bound forms, e.g., bound to blood proteins, to hemoglobin, and to erythrocytes (COLLINS 1985). Recently, free and/or loosely bound blood acetaldehyde levels in women were obtained with methods emphasizing an appropriate correction for artifactual formation.

Acetaldehyde was found only at lower ethanol concentrations ($<15\,\mu M$), whereas no acetaldehyde was observed at higher ethanol concentrations (FUKUNAGA et al. 1993).

III. Acetaldehyde-Hemoglobin Adduct

The binding of acetaldehyde to proteins depends upon the concentration of acetaldehyde, the duration of the reaction, and the presence of reducing agents (TUMA and SORRELL 1985). Acetaldehyde-protein adducts have been reported not only for hemoglobin (STEVENS et al. 1981) but also for actin, spectrin (GAINES et al. 1977), RNAase, lactate dehydrogenase, and glucose-6-phosphate dehydrogenase (TUMA and SORRELL 1987). It should be kept in mind that acetaldehyde binds reversibly until fixed by covalent bonding due to chemical reduction. This reversible binding of acetaldehyde to erythrocytes can be regarded as a form of transport. The reversibly bound concentration of acetaldehyde could be more than ten times as high as in plasma (BARAONA et al. 1987; ROMMELSPACHER et al. 1985). Binding kinetics were consistent with at least two sites, the one with the highest affinity and lower capacity for acetaldehyde corresponding to hemoglobin. It is likely that erythrocyte proteins are particularly vulnerable to the toxic effects of acetaldehyde. Affinity and B_{max} did not differ between alcoholics and controls. The low-affinity site(s) was enhanced in alcoholics and comprised 56% of the overall binding capacity compared with it was 20% in controls. The authors suggest that L-cysteine binds the acetaldehyde-forming thiazolidine-4-carboxylic acids. Levels of red blood cell cysteine and acetaldehyde remained high for 2 weeks after withdrawal; these changes may provide new markers of alcoholism (HERNANDEZ-MUNOS et al. 1989). The formation of 1-methyl-thiazolidine-4-carboxylic acid from L-cysteine and acetaldehyde has been demonstrated in liver and brain homogenate (WLODEK et al. 1993).

The sites to which acetaldehyde binds in hemoglobin may be the same as those in other carbonyl-group-containing compounds (e.g., glucose). The level of HbA_{1ACH} (hemoglobin fraction in cation exchange chromatography, measured as a percentage of total hemoglobin) is elevated in heavy drinkers, alcoholics, and diabetics compared with teetotalers and social drinkers. The HbA_{1ACH} and the HbA_{1ACH}/HbA_{1C} ratio seem of value in the detection of excessive alcohol consumption in its early phase (33% of heavy drinkers and 25% of alcoholics were detected) (SILLANAUKEE et al. 1991). HbA_{1ACH} evidently gave information on the effective acetaldehyde concentration in cases when the amount of glycated hemoglobin was normal. The sensitivity and specificity of the method could be improved if better suited HPLC columns were available.

Immunological measures utilizing antibodies against acetaldehyde-modified protein epitopes seem the most sensitive and specific markers of alcohol misuse (ISRAEL et al. 1986; NIEMELÄ et al. 1987). Antibodies raised against hemoglobin-acetaldehyde adducts detected alcoholic patients with sensitivities from 43%–78%, depending on the antibody. These findings merited further investigation (LIN et al. 1992). As pointed out by these authors, antibodies raised with protein-acetaldehyde adduct immunogens produced under different conditions recognize different epitopes.

Others have found that hemoglobin-acetaldehyde adducts were increased in 50% of alcohol abusers, whereas 24% of social drinkers exceeded the reference interval obtained from abstaining controls. Adducts were also increased in 23% of the hospitalized controls, seven of whom could retrospectively be verified as heavy drinkers. Upon abstinence from ethanol, the adducts decreased during a period of 1–3 weeks. High titers of circulating antibodies against acetaldehyde adducts were frequently observed in sera of chronic alcoholics with low adduct values (NIEMELÄ and ISRAEL 1992). Combined use of the adduct and the corresponding antibody measurements may prove to be useful as markers of excessive alcohol consumption (clinical markers according to Gordis-state markers according to our nomenclature).

IV. Acetate

Measurement of serum acetate has little value as a screening method for early identification of alcoholism because it depends on the presence of ethanol. It could be used for the screening of problem drinkers among drunken drivers (ROINE et al. 1988). Tests in which abnormal values revert toward normal values within 1 or 2 weeks of abstinence may not be sufficiently sensitive and specific to identify binge drinkers or alcoholics in recent remission. It may be a good marker for monitoring abstinence in alcoholics during treatment (KORRI et al. 1985; NUUTINEN et al. 1985; SALASPURO et al. 1987).

C. Monoamine Oxidase

I. Regulation of Enzyme Activity

The enzyme monoamine oxidase (MAO; EC 1.4.3.4) occurs in the outer mitochondrial membrane and several extramitochondrial subcellular sites (ARNOLD et al. 1990). It oxidizes a number of important biogenic and xenobiotic amines. Monoclonal antibodies specific for human MAO-A and -B and immunocytochemical techniques were used to demonstrate that subtype B but not subtype A occurs in platelets and lymphocytes (THORPE et al. 1987). MAO-B in lymphocytes has received little attention in clinical studies. However, MAO-B in human platelets is often used as a peripheral marker of vulnerability in psychiatric disorders such as schizophrenia, depression, and alcoholism (FARAJ et al. 1987; MAJOR et al. 1981).

Several studies suggest that the levels of platelet MAO-B are determined genetically. Interclass correlation coefficients are 0.76–0.88 for monozygotic twins and 0.39–0.45 for dizygotic twins (NIES et al. 1973; WYATT et al. 1975); heritability has been estimated at ~0.8 (NIES et al. 1973). A single major locus appears to determine heritable activity but there is no evidence that this is the X-linked MAO-B locus itself (RICE et al. 1984). The differences in activity correlate directly with levels of active enzyme molecules. The other genes involved in the regulation of levels of active enzymes are thought to include those genes affecting transcription, posttranslational modification, and control of the microenvironment

(BREAKEFIELD and EDELSTEIN 1980). No differences were observed in K_m values or isoelectric points (GILLER et al. 1982). In fact, some variation in the catalytic activity of enzyme molecules among individuals has been demonstrated (ROSE et al. 1986). Furthermore, thermolabile forms have been described (BRIDGE et al. 1981). The potency of a number of alcohols to inhibit MAO-B was linearly related to the logarithm of the alcohol-water-partition coefficient and to the chain length of the alcohols. This suggests that the lipid-perturbing properties of alcohols play a significant role in the inhibition of MAO activity (TABAKOFF et al. 1985). Furthermore, dietary studies with rats have demonstrated that a fat-free diet leads to a 50% reduction in liver MAO-B and subsequent supplementation of fats restored activities to normal levels (KANDASWAMI and D'IORIO 1979). Other studies with phosphatidyl serine (50% reduction of MAO-B activity) and other phospholipids underline the effect of the lipid environment on MAO-B activity (BUCKMAN et al. 1983).

No correlation has been found for levels of MAO-B activity in platelets and brain from the same individual (ORELAND 1979; YOUNG et al. 1986). However, a positive correlation has been found between levels of 5-hydroxyindoleacetic acid (5-HIAA) in cerebrospinal fluid and platelet MAO in volunteers and in chronic pain patients (VON KNORRING et al. 1986). Furthermore, there is a strong connection between personality traits such as aggressive behavior, sensation-seeking violence, and impulsiveness, which are supposed to be linked to low serotonergic activity and low MAO activity, with the nosological entity playing a secondary role (LIDBERG et al. 1985; BELFRAGE et al. 1992). These observations support the theory that several genetic and environmental factors regulate levels and activity of MAO-B in human tissues.

II. In Vitro Effects of Ethanol

Ethanol inhibits the activity of the B-form of MAO in membranes obtained from human platelets. Using a concentration of β-phenylethylamine (PEA) of 50 nM, which is reported to be present in the blood of humans, TABAKOFF et al. (1985) found an inhibition of 22% of platelet MAO-B at an ethanol concentration of 75 mM. The K_i value was 167 mM with PEA as substrate and 338 mM with p-dimethylaminobenzylamine (DAB; competitive inhibition). We found a K_i value of 272 mM and a Hill coefficient of 1.17 with kynuramine as substrate (with 25, 50, 100, and 200 mM ethanol competitive inhibition, MAY and ROMMELSPACHER 1994).

III. Enzyme Activity in Alcoholics and Nonalcoholics

Chronic ethanol intake has been associated with decreased activity of MAO-B in platelets from recently abstinent alcoholics. Although this finding has also been linked to other psychiatric disturbances, perhaps the most consistent findings have been in patients with alcoholism (BROWN 1977; FOWLER et al. 1982; MURPHY et al. 1982; AGARWAL et al. 1983).

Recovering alcoholics with 2–10 years of abstinence showed a low MAOB V_{max} ($p < 0.01$). Greater than 95% of the alcoholics had V_{max} values lower than the smallest value of control subjects (FARAJ et al. 1987). A more detailed analysis of the time course revealed a normalization of MAO-B activity at day 5 postwithdrawal followed again by lowered levels at days 12–14 and 21–27 (MAJOR et al. 1981). We were able to confirm the time course of the V_{max} in the postwithdrawal period. At day 1 the activity was reduced (25%; $p = 0.055$), at day 8 and after 3 and 6 months there were no changes (MAY and ROMMELSPACHER 1994). This confirms other studies, which have found no difference between alcoholics who had abstained from alcohol for from 5 days to several months and controls (TABAKOFF et al. 1988b). One explanation for the postwithdrawal normalization at days 5–8 is the liberation of newer platelets which are larger and contain more MAO activity (MURPHY et al. 1978). An increased percentage of newer platelets would result in a higher MAO activity.

Monoamine oxidase B activity is higher in female than in male subjects. The reason for this is not clear since testosterone and progesterone have no impact, whereas the plasma concentration of estrogen is negatively correlated with enzyme activity (POIRIER et al. 1985). YOUDIM (1991a) observed no hormonal regulatory mechanism for MAO-B activity in any cell type examined so far, and presented interesting evidence that MAO-B requires a genetic promoter to be expressed in the cell. Female alcoholic platelet MAO levels were significantly lower than those of controls and no different from activity levels in male alcoholics (YATES et al. 1990). Others have found a higher mean platelet MAO activity in female alcoholics than in male alcoholics (PANDEY et al. 1988).

Using the typology of Cloninger, male alcoholics are divided into type 1 and type 2 subgroups: type 2 is male-limited and characterized by high heritability, early onset, frequent social complication and mixed misuse, and high impulsiveness and sensation-seeking behavior scores; type 1 alcoholics (milieu-limited) have a later onset, lower degrees of heritability, and rarely misuse illegal drugs. MAO levels in type 1 alcoholics were lower than in controls and levels in type 2 alcoholics did not differ from type 1 levels (YATES et al. 1990). These findings are in contrast to those of other studies, which reported lower MAO activity in type 2 alcoholics when compared with both healthy controls and type 1 alcoholics. Also, the type 1 alcoholics had lower platelet MAO activity than the controls (ORELAND et al. 1985; VON KNORRING et al. 1991; PANDEY et al. 1988; SULLIVAN et al. 1990). The mean platelet volume has been found to be essentially the same for controls and type 1 and type 2 alcoholics (SULLIVAN et al. 1990).

In line with these findings is the low platelet MAO activity demonstrated in teenage boys with mixed alcohol and illegal drug use (VON KNORRING et al. 1987). It should be pointed out, however, that the temperament linked to low platelet MAO activity also makes low-MAO subjects overrepresented in such groups as mountaineers and violent offenders (VON KNORRING et al. 1991). MAO activity was low in both alcoholics and their first-degree relatives (ALEXOPOULOS et al. 1983). Others have found no difference or only a trend toward reduced platelet MAO activity in men with a positive family history of alcoholism, compared with men with a negative family history of alcoholism (SCHUCKIT 1984). Thus, the enzyme activity as assessed by the V_{max} value seems to be a trait marker, at least in a subgroup of alcoholics (type 2 alcoholics). In addition, our data and those of others (low levels during intoxication), suggest intoxication marker characteristics (ROMMELSPACHER et al. 1994).

IV. In Vitro Studies of Platelets from Alcoholics

Ethanol (200 mM) causes a highly significant increase in K_m (decrease in substrate affinity) but does not influence the V_{max} (maximum enzyme activity) of platelet MAO from both controls and alcoholics. Ethanol is obviously a competitive inhibitor of the enzyme. The extent of the reduction of affinity by ethanol in vitro is not influenced by acute intake of ethanol (1 g/kg, blood withdrawn 4 h after intake) in control subjects but is significantly reduced (15%; $p < 0.01$) in intoxicated alcoholics and in alcoholics 8 days after the last intake of ethanol. The findings suggest a reduced sensitivity of the enzyme to the effects of ethanol in vitro in chronic alcoholics and a recovery from the adaptive changes after 3–6 months (tolerance; marker of intoxication; ROMMELSPACHER et al. 1994). These observations are in contrast to results of TABAKOFF et al. (1988), who found an increased inhibition by ethanol in vitro from 6.1% in control subjects to 12% in sober alcoholics; it is therefore a more sensitive enzyme in alcoholics. The male alcoholics had abstained from alcohol for at least 5 days. The difference was found with a high saturating, but not with a low, substrate concentration. Further examination of the time course is warranted.

D. Guanine Nucleotide Binding Proteins and Adenylylcyclase

Guanine nucleotide binding proteins and the enzyme adenylylcyclase (EC 4.6.1.1) are part of the signal-transduction system in cell membranes including those from lymphocytes and thrombocytes. Cyclic AMP, the product of the enzymatic reaction, serves as intracellular second messenger. It has been known for many years that ethanol not only affects proteins such as neurotransmitter receptors but also phospholipids. Thus, the signal-transduction system was chosen as a model for studying the changes in the interaction of proteins with phospholipids of the membranes and those between several proteins: namely, the receptor, the G proteins, and the adenylylcyclase induced by acute and chronic ethanol exposure.

I. Acute Effects of Ethanol

Ethanol had little effect on basal adenylylcyclase activity in lymphocytes, platelets, and granulocytes, but enhanced the response of cyclic AMP to hormones, drugs, and guanine nucleotides, respectively, in reasonably low concentrations (20 m*M*) (TABAKOFF et al. 1988a). At least in cerebral cortical tissue, a primary site of action of ethanol within the receptorcoupled adenylylcyclase system is G_S (increased efficacy of Gpp(NH)p and Mg^{2+} to stimulate adenylylcyclase activity (HOFFMAN and TABAKOFF 1990); the major site of action is located distal to the guanine nucleotide-G-protein interaction (LUTHIN and TABAKOFF 1984). In human platelets only very high concentrations of ethanol (200–400 m*M*) were found to increase guanine nucleotide-stimulated and basal phospholipase C activity, respectively (RUBIN and HOEK 1988).

II. Chronic Effects of Ethanol

Little is known about the in vitro effects of ethanol on adenylylcyclase despite the interesting observations concerning monoamine oxidase. We have found that ethanol (200 m*M*) facilitates the stimulating effect of Gpp(NH)p (10 μM) on adenylylcyclase activity in platelet membranes. There is no difference between the platelets from nonalcoholics, 4 h after a load with ethanol (1 g/kg body weight), and alcoholics (intoxicated, 8 days after withdrawal, 3 and 6 months after withdrawal). Thus, there is no tolerance with respect to the in vitro facilitating effect of ethanol in platelet membranes.

In a study of ten alcoholics, ten matched normal individuals, and ten patients with nonalcoholic liver disease, basal and adenosine receptor-stimulated cAMP levels were reduced by 75% in lymphocytes from alcoholic subjects. Also, there was a 76% reduction in ethanol stimulation of cAMP accumulation in lymphocytes from alcoholics. Similar results were demonstrated in isolated T cells. The authors speculated that these measurements were due either to chronic alcohol abuse or to genetic predisposition unique to alcoholic subjects (DIAMOND et al. 1987). In a study with 94 alcoholics no change in basal adenylylcyclase activity of platelet membranes was measured (TABAKOFF et al. 1988b). We found a 30% reduction in basal activity in platelets from intoxicated alcoholics. One week after with-drawal the activity was back to normal levels. Since the patients in the study of Tabakoff and coworkers abstained for at least 5 days, both studies agreed that a change in enzyme activity is an indicator of alcohol intoxication (marker of intoxication).

Stimulation of adenylylcyclase activity by Gpp(NH)p, fluoride ion, and PGE_1 was lower in the platelets of alcoholics than in those of controls.

These differences were evident among inpatient alcoholic subjects with long-term abstinence, compared with controls. Smoking had no significant effect on enzyme activity. Among the alcoholics, there was a statistically significant decrease in fluoride-stimulated activity with duration of abstinence. However, this negative correlation did not remain statistically significant when the alcoholics with long-term abstinence were excluded from analysis (TABAKOFF et al. 1988b). We found a reduced activation by Gpp(NH)p during intoxication only, with normal stimulation during abstinence. The effects of sodium fluoride (10 m*M*), forskolin (5 μ*M* to determine affinity changes and 100 μ*M* to assess maximum activity), and PGE_1 (1 μ*M*) in platelets were not different from those in alcoholics and controls. Thus, the data support the theory that activation of the adenylylcyclase in platelet membranes is a marker of intoxication.

A combination of Gpp(NH)p and forskolin results in ~50% inhibition of forskolin-induced activation, probably due to an action of Gpp(NH)p on G*i* protein, which is present in the membranes of platelets. The inhibiting effect did not differ between alcoholics and nonalcoholics. Despite these negative findings, there is increasing evidence for a marker function of Gi (LICHTENBERG-KRAAG et al. 1994). The G_s-subunit in a neuroblastoma cell line (NG 108–15) is reduced by 38% and mRNA by 30% in cells treated with 100 m*M* ethanol for 48 h (MOCHLY-ROSEN et al. 1988). The ethanol concentration in the in vitro experiments was much higher than under in vivo conditions.

In a study to assess the effect of genetic factors, particularly in male-limited alcoholism, lymphocytes were cultured from alcoholics for 7–8 days without ethanol and then a search made for differences in receptor-mediated cAMP accumulation. After four to six generations in culture, lymphocytes from alcoholic subjects had 2.8-fold higher adenosine-receptor-mediated cAMP levels than cells from nonalcoholic subjects. Thus, a difference in cAMP signal transduction is demonstrable in cells from alcoholic subjects grown without ethanol. To determine whether alcoholic subjects have altered sensitivity to ethanol, lymphocytes were challenged with 100 m*M* ethanol for 24 h. Under these conditions, adenosine-receptor-stimulated cAMP levels did not change in cells from nonalcoholic subjects. However, lymphocytes from alcoholic subjects showed a significant decrease in adenosine-agonist-stimulated cAMP levels (DIAMOND et al. 1991). These data suggest that the regulation of cAMP signal transduction is altered in alcoholic subjects. It remains to be established whether these findings represent a trait marker.

E. Serum Enzymes

Elevated activity of enzymes in human serum are better suited to tracing specific cell systems or organs that have been irritated or damaged than to identifying the responsible mechanisms. Consequently, no single "marker enzyme" has been found that reliably (1) detects excessive chronic ethanol consumption, (2) discerns an ethanol-related etiology of organ damage, or (3) quantitates the organ damage. With the host of possible non-ethanol-related influences on serum enzyme activity, reports on sensitivities and specificities of marker enzymes vary considerably, as evidenced in recent in-depth reviews on the subject (SALASPURO 1987; MIHAS and TAVASSOLI 1992). Within the limited space of this review only new data on established or new promising markers will be discussed.

I. γ-Glutamyltransferase

The enzyme γ-glutamyl transferase (GGT) is associated with the membrane fraction of many tissues, including liver, biliary tract, kidney, pancreas, and heart. Isoenzyme patterns have been detected but they have not found a diagnostic application. Primarily, GGT is a prominent tool in the evaluation of hepatobiliary and pancreatic diseases, but it has also found wide application as a screening and follow-up marker in alcoholism. Mechanisms that may lead to increased serum GGT levels upon chronic ethanol ingestion include induction of hepatic GGT synthesis, impairment of biliary GGT secretion, and interference with hepatic GGT clearance. Diagnostic sensitivities between 30% and 85% have been reported, but not enough attention has been paid to specificity (SALASPURO 1987; MIHAS and TAVASSOLI 1992).

In more recent studies using carefully evaluated ethanol consumers and control groups of patients with non-ethanol-related liver diseases, sensitivities of 69%, 59%, and 59%, with specificities between 50% and 59%, respectively, have been reported for serum GGT (KWOH-GAIN et al. 1990; STIBLER et al. 1988; BEHRENS et al. 1988). In a study of pregnant ethanol abusers, 33% had elevated GGT values; within the subgroup of women that gave birth to infants with fetal alcohol effects, 52% had above-normal GGT serum activities (HALMESMÄKI et al. 1992). WEILL et al. (1988) made the interesting proposal that the sensitivity of GGT in detecting alcohol abuse can be raised to 90% when a decrease in GGT activity, irrespective of the initial value, is detected after abstinence for 7 days. Finally, GGT appears to be a very sensitive relapse marker: a study group of moderate ethanol abusers with elevated serum GGT activities showed a marked increase in GGT values within 24–48 h when challenged with a single oral load of ethanol of 1 g/kg after an abstinence period of 4 weeks (NEMESANSZKY et al. 1988).

II. Aspartate Aminotransferase

The enzyme aspartate aminotransferase (AST) (or glutamate oxalacetate transaminase, GOT) is abundant in liver, heart, and skeletal muscle and serves as a sensitive indicator of hepatocyte necrosis. A cytoplasmic and a mitochondrial isoenzyme of AST are known that are immunochemically distinct and differ in kinetic properties. With total AST as a marker of chronic ethanol ingestion, problems of diagnostic accuracy are even more pronounced than with GGT, since liver damage is required for abnormal values to be detected (lower overall sensitivity) and specificity is reduced by the ubiquity of the enzyme. Sensitivities between 18% and 100% have been reported (SALASPURO 1987; MIHAS and TAVASSOLI 1992).

The mitochondrial AST (mAST) or the mAST/AST ratio has been evaluated in recent years, since it was noted that mAST levels were higher in alcoholic hepatitis than expected from total AST (MIHAS and TAVASSOLI 1992). Using physicochemical or immunochemical methods to separate and quantitate the isoenzymes, the mAST/AST ratio has been claimed to discriminate more efficiently chronic alcoholics from patients with non-ethanol-related liver diseases (OKUNO et al. 1988). KWOH-GAIN et al. (1990) determined a sensitivity of 92% and a specificity of 70% in a study with carefully selected comparison groups. Other recent studies have failed to support the utility of mAST as a screening parameter for excessive ethanol consumption (SCHIELE et al. 1989; CHAN et al. 1989; NILSSEN et al. 1992).

III. Other Enzymes, Indices, and Test Combinations

Various combinations of markers, ratios, or other indices have been tried. Besides mAST/AST, the ratio of AST to alanine aminotransferase has been widely used to diagnose an ethanol-related etiology of liver diseases (SALASPURO 1987). Also, the ratio of GGT to alkaline phosphatase has been proposed as a sensitive indicator of excessive ethanol ingestion (MIHAS and TAVASSOLI 1992). Of the various additive test panels proposed, the combination of carbohydrate-deficient transferrin (CDT) and GGT seems the most promising, since these tests have been shown to behave independently in the evaluation of ethanol consumers (STIBLER et al. 1988; BEHRENS et al. 1988; NILSSEN et al. 1992; NYSTRÖM et al. 1992).

Among the host of other enzymes suggested, serum β-hexosaminidase (β-HEX; also: *N*-acetyl-β-D-glucosaminidase) may deserve some attention. In several studies it was shown that β-Hex is increased in more than 80% of heavy drinkers and that it rapidly and significantly decreases with abstinence (KÄRKKÄINEN et al. 1990a,b). The sensitivity of this marker in detecting chronic excessive ethanol ingestion could be increased by measuring the isoenzyme B of β-Hex with an enzyme immunoassay (HULTBERG et al. 1991).

F. Tetrahydroisoquinolines and β-Carbolines

The search for naturally addictive compounds in mammals has involved two strategies: one that could be called the "peptide strategy," and the other the "alkaloid strategy." The peptide strategy has identified the existence of the three families of endorphins and other peptides with potent opioid activity called exorphins in the normal diet [fragments of the milk protein casein (casomorphins) of wheat gluten and of chocolate] (MAX 1992). The alkaloid strategy has led to the discovery of morphinanes (morphine, 6-acetylmorphine, and codeine) in bovine hypothalami and rat brain (DONNERER et al. 1986, 1987; WEITZ et al. 1986, 1987). Some evidence has been presented that morphine-like immunoreactivity occurs in human cerebroventricular fluid (SHORR et al. 1978; CARDINALE et al. 1987). Furthermore, morphine, codeine, and tetrahydropapaveroline have been detected at picomole per milliliter levels in the urine of humans (MATSUBARA et al. 1992). The methods available are not sensitive enough for the quantification of morphine alkaloids. Whether these compounds are disposition markers remains to be established.

I. Tetrahydroisoquinolines

The "alkaloid strategy" has revealed the natural presence of tetrahydroisoquinolines (TIQs) and β-carbolines (BCs) in mammals, including man (BROSSI 1991; ROMMELSPACHER et al. 1991b). Among the TIQs the compound salsolinol has been most widely investigated. The (R)-enantiomer of this compound predominates in human urine (STROLIN BENEDETTI et al. 1989), while the (S)-enantiomer seems to be formed by subjects who drink a substantial amount of alcoholic beverages. This may be because the acetaldehyde produced during ethanol metabolism favors the production of racemic salsolinol (DOSTERT et al. 1988). Some authors reported an increased salsolinol level in the urine of alcoholics (COLLINS et al. 1979; DOSTERT et al. 1991), whereas others have not detected such a difference (CLOW et al. 1985; VERNAY et al. 1989; FEEST et al. 1991). These differences could be explained in part by different circumstances (ethanol present/not present, time elapsed since last ethanol intake, etc.) and by the disturbed formation of salsolinol from its precursor 1,2-dehydrosalsolinol in some alcoholics (DOSTERT et al. 1991).

FARAJ et al. (1989) showed that alcoholics had significantly elevated blood plasma concentrations of salsolinol sulfate compared with controls. The two enantiomers were not separated in this study. Others detected no salsolinol in blood plasma from controls, parkinsonian patients, or alcoholics (VERNAY et al. 1989). However, salsolinol was present in cerebrospinal fluid and a slight increase was found in alcoholics (SJÖQUIST et al. 1981a,b; VERNAY et al. 1989). Further studies are warranted to assess the marker function of salsolinol in alcoholics. With respect to the pathogenesis of alcoholism it should be noted that salsolinol and its precursor 1-carboxysalsolinol cannot be converted to morphine. Salsolinol is methylated in position 7 (salsoline) of the catechol ring, whereas the biosynthesis of morphine requires methylation in position 6 (BROSSI 1991).

The availability of pyruvate and thiamine, the redox equilibrium in the cell, and many other factors influence the biosynthesis of salsolinol, which suggests an intoxication marker function for the compound.

II. β-Carbolines

β-Carbolines with a methyl substituent in position 1 have been the subject of interest and controversy in alcoholism research for several decades owing to the possibility of a precursor role for acetaldehyde, the metabolite of ethanol (COLLINS 1988). In human urine, 1-methyltetrahydro-β-carboline is present normally (ALLAN et al. 1980; MATSUBARA et al. 1986). Some authors, however, have found the BC to be present only following ethanol ingestion by the subject (PEURA et al. 1980; ROMMELSPACHER et al. 1980). Others have not observed an effect of ethanol ingestion on the amount excreted in the urine (MATSUBARA et al. 1986). 7-Hydroxy-1-methyltetrahydro-β-carboline was identified in human urine with different portions of the (S)(−)- and (R)(+)-enantiomers (BECK et al. 1986). The effect of ethanol ingestion on the presence of this compound was not investigated.

The fully aromatic β-carboline with the methyl substituent in position 1 is designated harman. It is formed from tryptamine and pyruvic acid. Acetaldehyde presumably only plays a role as precursor in chronically intoxicated subjects ingesting large amounts of ethanol (for discussion of the topic see ROMMELSPACHER et al. 1991b).

Harman normally occurs in blood, urine, and in 1 out of 24 cerebrospinal fluid samples (ROMMELSPACHER et al. 1984, 1985, 1990; BIDDER et al. 1979; SCHOUTEN and BRUINVELS 1985; BOSIN et al. 1989). In nonalcoholics a slight increase was measured in erythrocytes following an ethanol load (1 g/kg; ROMMELSPACHER et al. 1990). An increased harman concentration was observed in some alcoholics (ROMMELSPACHER et al. 1991b). Therefore, harman can function as a state marker and its levels were normalized 3 months after withdrawal (unpublished data).

The unsubstituted fully aromatic β-carboline, namely norharman, seems to be a better marker for alcoholism than other β-carbolines. It can be formed from tetrahydro-β-carboline and from tryptamine and formaldehyde (SCHOUTEN and BRUINVELS 1985; ROMMELSPACHER et al. 1991a). The levels of norharman in blood plasma showed a much greater variation in nonalcoholics following a load with ethanol (1 g/kg) than without ethanol loading. Even 6 months after withdrawal the levels remained significantly elevated in the blood plasma of alcoholics. The findings suggest that norharman could serve as a disposition marker for alcoholism or as a residual marker (ROMMELSPACHER et al. 1991a and unpublished results). Alcoholics with a high anxiety score had higher levels of norharman than those with other predominant psychopathological traits (ROMMELSPACHER et al. 1992). The sensitivity and specificity of norharman level elevations in alcoholics remain to be determined.

G. Immune System

Considerable evidence has accumulated documenting a wide array of immunosuppressive effects of both acute and chronic ethanol administration (BALLARD 1989; YOUNG and MACGREGOR 1989). Studies linking alcoholism with color blindness, ABO blood groups, HLA antigens, complement C3 β-glycoprotein, and other measures have been reported (HILL et al. 1975).

I. Blood Groups

HILL and coworkers (1975) studied association and linkage of ABO, MNSs, Rh, Kell, Duffy, and Xg in alcoholics and their nonalcoholic first-degree relatives. Except for a higher frequency of homozygous recessive *ss* (MNSs system) in the nonalcoholic family members than in the alcoholics, no significant relationship between blood groups and alcoholism was noted. Other investigators could not demonstrate any link between the ABO system and alcoholism (BUCKWALTER et al. 1964; CAMPS et al. 1969; SWINSON and MADDEN 1973).

II. Immunoglobulins

Elevated serum IgA and IgE levels are often seen in chronic alcoholics (HALLGREN and LUDIN 1983; ITURRIAGA et al. 1977) possibly due to an increased rate of synthesis (KALSI et al. 1983). Many authors agree that elevated levels of IgA can be observed consistently only after the development of liver fibrosis or cirrhosis. Thus, IgA could serve as an intoxication marker. Although IgA may also be increased in nonalcoholic liver diseases, one study showed that the IgA/IgG ratio can be used to differentiate liver damage of alcoholic and nonalcoholic origins (ITUARRIAGA et al. 1977). However, the specificity of this ratio has been disputed by many other studies (e.g. SOPENA et al. 1993).

III. Human Leukocyte Antigens

Significant associations between various human leukocyte antigens (HLAs), such as HLA-B8, HLA-B13, HLA-B15, HLA-B40, HLA-DR3, alcoholic hepatitis, and alcoholic cirrhosis have been observed in several studies (DICK et al. 1982; SAUNDERS et al. 1982; ROBERTSON et al. 1984; WILSON et al. 1984; DOFFOEL et al. 1986). However, in other studies no such association was demonstrated (SCOTT et al. 1977; MELENDES et al. 1979; RADA et al. 1981). Thus, the findings are primarily confined to demonstrating a higher frequency of certain HLA antigens in patients with alcoholic cirrhosis compared to healthy controls. Only the report by HRUBEC and OMENN (1981) provided empirical support for a genetic basis of alcoholic cirrhosis. These investigators found a twin concordance rate for alcoholic cirrhosis of 14.6%

for monozygotic twins and 5.4% for dizygotic twins from a study of more than 15000 male twin pairs.

IV. Cell-Mediated Immune Function

Natural killer cell activity was significantly lower in alcoholics (IRWIN et al. 1990a). Some disturbances were also demonstrated in T-helper and T-suppressor cells (NOURI-ARIA et al. 1986).

H. Alcohol Intake and Serum Trace Elements

I. Trace Elements and Electrolytes

Serum selenium (AASETH et al. 1982; KORPELA et al. 1985; RISSANEN et al. 1987; VÄLIMÄKI et al. 1983), zinc (VALLEE et al. 1957; KIILERICH et al. 1980; MCCLAIN et al. 1986), and magnesium (FLINK 1986; MCCOLLISTER et al. 1963; WOLFE and VICTOR 1969) have been found to be lower in alcoholics than in nonalcoholics, whereas serum iron was higher (CHAPMAN et al. 1982; HEIDEMANN et al. 1981; KRISTENSON et al. 1981). In a study correlating alcohol intake and serum trace element levels in a population showing no obvious nutritional deficiencies, serum copper correlated positively with alcohol intake per drinking day whereas selenium and magnesium showed the reverse relationship. No significant correlations were found between ethanol intake and serum zinc or iron levels (KÄRKKÄINEN et al. 1988).

Decreased selenium levels in serum and erythrocytes were found even in alcoholics without severe malnutrition (DWORKIN et al. 1988). A rapid trend toward the normalization of blood selenium levels was observed within 14 days (GIRRE et al. 1990) and levels were found to be normal after 1 month of abstinence (JOHANSSON et al. 1986). These results are of importance because selenium is a cofactor of glutathione peroxidase, the key enzyme in defense against oxidative injury. The findings support the presence of an intoxication marker characteristic for the trace elements and electrolytes.

II. Thiamine (Vitamin B_1)

Thiamine (T) and its phosphorylated derivatives TP, TPP, and TPPP have attracted a lot of interest due to the observation that T-deficiency is associated with Wernicke's encephalophathy and beriberi. The concentrations of T and TPP in blood have been utilized as markers for alcoholism. Chronic alcohol consumption can create thiamine deficiency by decreased absorption due to mucosal atrophy, and decreased synthesis of the biologically active thiamine diphosphate. TPP is a cofactor of the pyruvate dehydrogenase complex. Decreased activity of pyruvate dehydrogenase leads to an increase in pyruvate levels that equilibrate with lactate (ROMMELSPACHER et al. 1991b). TPP represents about 80% of the total thiamine concentration in human tissues and in blood (HOYUMPA 1983). The intracellular blood T concentration is >100 times higher than the extracellular concentration (BAINES et al. 1988). Serum TPP and blood TPP concentration did not differ between alcoholics and nonalcoholics regardless of gender.

The extent of changes following thiamine treatment also did not differ. TP was the only significantly decreased compound in serum and blood both before and after treatment. TP is formed solely by the dephosphorylation of TPP to T (DAVIS and ICKE 1983). An explanation for the low TP levels might be a decreased thiamine pyrophosphatase and a partly increased thiamine monophosphatase due to chronic alcohol ingestion (LAFORENZA et al. 1990). Because of the low TP concentrations (close to the detection limit of available analytical methods), TP cannot be utilized as a marker of thiamine deficiency in serum or whole blood (TALLAKSEN et al. 1992). A recent review of the changes in vitamin metabolism related to ethanol (HALSTED and HEISE 1988) lead to the conclusion that none of them can be used as a marker.

I. Carbohydrate-Deficient Transferrin

I. Biochemistry

Human serum transferrin, when analyzed with high-resolution techniques, shows a microheterogeneity related to differences in amino acid composition, iron saturation, and carbohydrate side chain structures (DE JONG et al. 1990). Since genetic variants are rare and readily detectable, isotransferrin patterns seen upon isoelectric focusing (IEF) of iron-saturated samples will depend solely on carbohydrate variations, especially on the content of negatively charged terminal sialic acid molecules. Two N-linked oligosaccharide side chains with variable degrees of branching may result in up to nine isotransferrins containing from zero to eight terminal sialic acid molecules. The abundant transferrin isoform in human serum contains two two-branched carbohydrate side chains with a total of four sialic acid residues and has an isoelectric point (pI) of 5.4–5.5. Removal (addition) of a sialic acid residue raises (lowers) the pI of this transferrin by an increment of about 0.1 (PETREN and VESTERBERG 1989; DE JONG et al. 1990; STIBLER 1991).

STIBLER et al. (1978) were the first to identifying an abnormal isotransferrin in the serum of alcoholics. Present at pI 5.7, the protein disappeared with abstinence. When the decrease in the number of substituted sialic acids of these isotransferrins was confirmed, desialylated transferrin became a descriptive term. To illustrate the fact that with the sialic acid residues some of the neutral carbohydrates constituting the terminal trisaccharides are also diminished in alcohol-abusing subjects, STIBLER et al. (1986) proposed the term "carbohydrate-deficient transferrin" (CDT), which is now in general use.

The pathobiochemical mechanisms that lead to the imbalance in human serum isotransferrins with chronic excessive ethanol consumption are still obscure. Several reports described evidence of reduced glycosyl transferase activity in ethanol-treated rats (MALAGOLINI et al. 1989; GUASCH et al. 1992) and in alcohol-abusing humans (STIBLER and BORG 1991); other authors demonstrated ethanol-enhanced cleavage of sialoconjugates in mice (MATHEW and KLEMM 1989), ethanol-promoted desialylation of transferrin by rat liver endothelium (MIHAS and TAVASSOLI 1991), or altered elimination and degradation of asialoglycoproteins in the perfused liver of ethanol-fed rats (CASEY et al. 1990). Reports on the suitability of rat models for the study of CDT are controversial (STIBLER et al. 1989; BATEY and Patterson 1991). Besides any direct effects of ethanol, acetaldehyde as the reactive intermediate in ethanol metabolism may play a crucial role in the generation of CDT (STIBLER 1991).

II. Methodological Approaches

Despite many favorable reports on the discriminatory power of CDT for the assessment of chronic ethanol abuse, widespread use of this marker was hampered due to the unavailability of a method suited for routine measurement. Electrophoretic approaches favor isoelectric focussing (IEF): (1) in polyacrylamide gels followed by immunofixation (IF) and densitometry (POUPON et al. 1989); (2) in polyacrylamide/immobiline gels followed by crossed immunoelectrophoresis (VAN EIJK et al. 1987); (3) in agarose gels followed by direct IF/Coomassie staining and visual inspection (KAPUR et al. 1989); (4) in agarose gels followed by zone immunoelectrophoresis (PETREN and VESTERBERG 1988; CHAN et al. 1989); and (5) in agarose gels followed by western blotting/immunostaining and densitometric scanning (XIN et al. 1991). A single report used conventional native polyacrylamide electrophoresis followed by western blotting/immunostaining and visual inspection (REISINGER and SOYKA 1990).

Another more promising approach has been anion exchange chromatography followed by sensitive immunoassays for the detection of eluted isotransferrins: (1) a patented method eluating transferrin species with pI above pH 5.65 that are measured with transferrin RIA (STIBLER et al. 1986); initial kits distributed by Pharmacia were favorably evaluated (BEHRENS et al. 1988; SCHELLENBERG et al. 1989); (2) a slightly modified microcolumn approach using DEAE-Sephacel at pH 5.65 and a transferrin RIA (KWOH-GAIN et al. 1990); and (3) a revised version by STIBLER et al. (1991) using ionic strength elution from an anion exchange microcolumn and a transferrin RIA for the quantitation of asialotransferrin. The latter should behave more stably under routine work conditions.

Either CDT concentrations, various ratios of isotransferrins, or the ratio of CDT to total transferrin have been used as discriminators. The ratio of isoforms has been preferred with electrophoretic/immunostaining techniques in order to circumvent calibration problems, but XIN et al. (1991) used absolute calibration by the inclusion of a CDT standard set with each IEF run. In this study, absolute CDT proved to be the more accurate marker in terms of specificity and sensitivity. With anion exchange chromatography, KWOH-GAIN et al. (1990) found a slightly better performance with a CDT to total transferrin ratio, whereas STIBLER et al. (1986) and BEHRENS et al. (1988) with larger study populations found CDT to be unrelated to total transferrin and to be a better discriminator on its own.

III. Performance: Sensitivity and Specificity

In view of the host of methodological variations adopted for the determination of CDT or ratio values it is obvious that results of clinical investigations regarding the discriminatory power of this marker vary. Variables to be considered include (1) the practicality, precision, and reliability of the method in question; (2) the choice of the discriminator and its threshold value; (3) the definition and recruitment of alcohol abusers in terms of duration and amount of ethanol ingestion as well as duration of eventual abstinence; and (4) the inclusion of patients with non-alcohol-related diseases among the controls.

Studies performed with the original anion exchange microcolumn technique (STIBLER et al. 1986) make up the largest group of individuals examined with a consistent quantitative method: sensitivities for ethanol abuse ranged from 91% for alcoholic patients without clinical signs of liver cirrhosis (STIBLER et al. 1986) to 87% in alcoholic patients with biopsy-verified liver diseases (STIBLER and HULTCRANTZ 1987), to 83% in alcoholic women (STIBLER et al. 1988), to more than 80% in black, Puerto Rican, and white alcoholics (BEHRENS et al. 1988), to 76% in 160 alcoholics without clinical signs of cirrhosis (SCHELLENBERG et al. 1989). Sensitivities of 81% have been repeatedly reported for a slightly modified microcolumn technique (KWOH-GAIN et al. 1990; FLETCHER et al. 1991). For the electrophoretic techniques, recent studies using IEF/western blotting resulted in a sensitivity of 75% (XIN et al. 1992).

In general, these data relate to individuals that have consumed more than 60 g ethanol/day regularly until at least 2 days before sampling. As a consequence of the short half-life of CDT (see below), calculated sensitivities in some studies may be obscured by variable periods of abstinence before sampling (CHAN et al. 1989). Conflicting reports have been made regarding the performance of the revised anion exchange microcolumn method based on ionic strength elution. This method should work even slightly better than its predecessor (STIBLER et al. 1991); however, sensitivities determined with initial kits distributed by Pharmacia were around 60% (XIN et al. 1992) or even lower than 30% (NILSSEN et al. 1992; NYSTRÖM et al. 1992).

Published results regarding specificity of CDT are unanimous: it is near to 100%. Compiling data from some 20 investigations, STIBLER (1991) counted 40 individual false-positive CDT results among more than 1000 healthy controls or patients with non-alcohol-related diseases. The majority of these false-positives were from patients with cirrhosis caused by chronic active hepatitis or from patients with primary biliary cirrhosis. Other patients with non-alcohol-related liver diseases have CDT values within the reference range (BEHRENS et al. 1988; SCHELLENBERG et al. 1989; FLETCHER et al. 1991; XIN et al. 1992). Other causes of false-positive results relate to either rare D-variants (abundant is the C-variant) or a recently discovered inherited disorder, the carbohydrate-deficient glycoprotein syndrome (STIBLER 1991).

It has been claimed that CDT is as sensitive and specific in female ethanol abusers as it is in males (STIBLER et al. 1988), but again this has been questioned in two recent studies where the sensitivity of CDT to detect female heavy drinkers was greatly reduced (NILSSEN et al. 1992; NYSTRÖM et al. 1992). It is generally accepted that method-specific cutoff values for CDT are slightly higher in female than in male reference populations (STIBLER 1991). A slight but significant correlation between reported regular ethanol ingestion and CDT values has been reported by several groups (STIBLER et al. 1986; POUPON et al. 1989; XIN et al. 1991) and there is general accord that initial CDT levels decline, upon abstinence, with a half-life of 10–14 days.

In summary, CDT appears to be the best single biochemical state marker for accurate detection of chronic excessive ethanol abuse. It is suited for screening and monitoring of healthy individuals at risk and for the detection of involvement of ethanol abuse in various disease states. However, a practical method adapted to the requirements of a standard clinical laboratory without losses in overall accuracy is still awaited.

J. Conclusion

There are a number of definitions for the clinical markers of alcoholism: trait markers (synonym, genetic markers), state markers, intoxication markers, and residual markers (synonym, diagnostic markers). The term "clinical marker" comprises both state markers and intoxication markers. Only in recent years have many researchers become aware of the necessity of categorizing the clinical characteristics of alcoholism into such constructs as markers. An obstacle to categorizing these characteristics is the dual character of many parameters. For example, monoamine oxidase activity is reduced in intoxicated alcoholics, suggesting it is an intoxication marker. The activity tends to remain low even after several weeks of abstinence, suggesting it is a trait and/or a residual marker. In individuals with certain personality traits, the activity is especially low, suggesting a trait character and calling into question the specificity of this marker for alcoholism. Another example of the difficulties in categorizing markers is the adenylylcyclase in blood platelets and lymphocytes. The enzyme is an intoxication marker. While detailed investigations with lymphocytes cultivated for several generations have revealed a possible trait marker quality, such a measure is not suited for routine laboratory assessment. Markers of special interest for an understanding of the disease are the residual markers. The determinants of this type of marker are responsible for distortions of thinking and affect, as well as the lapse/relapse liability of abstaining alcoholics.

Another problem is the lack of field studies demonstrating the degree of specificity and sensitivity of a certain marker in a larger population. The exception is the SVALBARD study of 1992. Nevertheless, the present review demonstrates that in many areas tremendous progress has been made in recent years with respect to the search for suitable markers of alcoholism. The efforts of the many laboratories involved hold future promise that more systematic studies will be performed as a prerequisite to the discovery of more elaborate markers of alcoholism.

References

Aaseth J, Alexander J, Thomassen Y, Blomihoff JP, Skrede P (1982) Serum selenium levels in liver diseases. Clin Biochem 15:281–282

Agarwal DP, Philippe G, Milech U, Goedde HW, Schrappe O (1983) Platelet monoamine oxidase activity in alcoholics. Mod Probl Pharmacopsychiatry 19:260–264

Alexopoulos G, Lieberman K, Frances R (1983) Platelet MAO activity in alcoholic patients and their first-degree relatives. Am J Psychiatry 140:1500–1504

Allen JRF, Beck O, Borg S, Skroeder R (1980) The analysis of 1-methyl-1,2,3,4-tetrahydro-β-carboline in human urine and cerebrospinal fluid by gas chromatography-mass spectrometry. Eur J Mass Spectr Biochem Med Environ Res 1:171–177

Arnold JR, Hysmith RM, Boor PJ (1990) The ultracytochemical demonstration of extramitochondrial monoamine oxidase activity. Biogen Amines 7:127–143

Baines M, Bligh JG, Madden JS (1988) Tissue thiamin levels of hospitalised alcoholics before and after oral or parenteral vitamins. Alcohol Alcohol 23:49–52

Ballard HS (1989) Hematological complications of alcoholism. Alcohol Clin Exp Res 13:706–720

Baraona E, Di Pova C, Tabasco J, Lieber CS (1987) Transport of acetaldehyde in red blood cells. Alcohol Alcohol [Suppl 1]:203–206

Batey R, Patterson F (1991) Carbohydrate-deficient transferrin in the ethanol-consuming rat model. Alcohol 8:487–490

Beck O, Repke DB, Faull KF (1986) 6-Hydroxy-methyltryptoline is naturally occurring in mammalian urine: identification by combined chiral capillary gas chromatography and high resolution mass spectrometry. Biomed Environ Mass Spectr 13:469–472

Begleiter H, Porjesz (1988) Potential biological markers in individuals at high risk for developing alcoholism. Alcohol Clin Exp Res 12:488–493

Behrens UJ, Worner TM, Braly LF, Schaffner F, Lieber CS (1988) Carbohydrate-deficient transferrin, a marker for chronic alcohol consumption in different ethnic populations. Alcohol Clin Exp Res 12:427–432

Belfrage H, Lidberg L, Oreland L (1992) Platelet monoamine oxidase activity in mentally disordered violent offenders. Acta Psychiatr Scand 85:218–221

Bidder TG, Schoemaker DW, Böttger HG, Evans M, Cummins JT (1979) Harman in human platelets. Life Sci 25:157–164

Bosin TR, Borg S, Faull KF (1989) Harman in rat brain, lung and human CSF: effect of alcohol consumption. Alcohol 5:501–511
Breakefield XO, Edelstein SB (1980) Inherited levels of A and B types of monoamine oxidase activity. Schizophr Bull 6:282–288
Bridge TP, Wise DC, Potkin SG, Chelps BH, Wyatt RJ (1981) Platelet monoamine oxidase: studies of activity and thermolability in a general population. In: Gershon ES, Matthysse S, Breakefield XO (eds) Genetic strategies in psychopharmacology. Boxwood, Pacific Centre, CA, pp 95–105
Brossi A (1991) Mammalian alkaloids: conversions of tetrahydoisoquinoline-1-carboxylic acids derived from dopamine. Planta Med 57:S93–S100
Brown JB (1977) Platelet MAO and alcoholism. Am J Psychiatr 134:206–207
Buckman TD, Eiduson S, Sutohin MS, Chang R (1983) Selective effects on catalysis by the multiple forms of monoamine oxidase produced by interactions of acidic phospholipids with mitochondrial membranes. J Biol Chem 258:8670–8676
Buckwalter JA, Pollock CB, Hasleton G, Krohn JA, Nance MJ, Ferguson JL, Bondi RL, Jacobsen JJ, Lubin AH (1964) The Iowa blood type disease research project II. J Iowa State Med Soc 54:58–66
Camps FE, Dodd BE, Lincoln PJ (1969) Frequencies of secretors and nonsecretors of ABH group substances among 1000 alcoholic patients. Br Med J 4:457–459
Cardinale GJ, Donnerer J, Finck AD, Kantrowitz JD, Oka K, Spector S (1987) Morphine and codeine are endogenous components of human cerebrospinal fluid. Life Sci 40:301–306
Casey C, Volentine G, Jankovich C, Kragskow S, Tuma D (1990) Effect of chronic ethanol administration on the uptake and degradation of asialoglycoproteins by the perfused rat liver. Biochem Pharmacol 40:1117–1123
Chan AWK, Leong FL, Schanley DL, Welte JW, Wieczorek W, Rej R, Whitney RB (1989) Transferrin and mitochondrial aspartate aminotransferase in young adult alcoholics. Drug Alcohol Depend 23:13–18
Chapman RW, Morgan MY, Laulicht M, Hofbrand AV, Sherlock S (1982) Hepatic iron stores and markers of iron overload in alcoholics and patients with ideopathic hemochromatosis. Dig Dis Sci 27:909–916
Clow A, Topham A, Saunders JB, Murray R, Sandler M (1985) The role of salsolinol in alcohol intake and withdrawal. In: Collins MA (ed) Aldehyde adducts in alcoholism. Liss, New York, pp 101–113
Collins MA (1985) Aldehyde adducts in alcoholism. Liss, New York
Collins MA (1988) Acetaldehyde and its condensation products as markers in alcoholism. In: Galanter M (ed) Recent developments in alcoholism, vol 6. Plenum, New York, pp 387–403
Collins MA, Nijm WP, Borge GF, Teas G, Goldfarb C (1979) Dopamine-related tetrahydroisoquinolines: significant urinary excretion by alcoholics after alcohol consumption. Science 206:1184–1186
Coper H, Rommelspacher H, Wolffgramm J (1990) The "point of no return" as a target of experimental research on drug dependence. Drug Alcohol Depend 25:129–134
Davis RE, Icke GC (1983) Clinical chemistry of thiamin. Adv Clin Chem 23:93–138
De Jong G, van Dijk JP, van Eijk HG (1990) The biology of transferrin. Clin Chim Acta 190:1–46
Diamond J, Wrubel B, Estrin W, Gordon A (1987) Basal and adenosine receptor-stimulated levels of cAMP are reduced in lymphocytes from alcoholic patients. Proc Natl Acad Sci USA 84:1413–1416
Diamond J, Nagy L, Mochly-Rosen D, Gordon A (1991) The role of adenosine and adenosine transport in ethanol-induced cellular tolerance and dependence. Ann NY Acad Sci 625:473–487
Dick HM, MacSween RNM, Hislop S, Mills P (1982) HLA antigens and alcohol liver disease. Lancet ii:325–326

Doffoel M, Tongio MM, Gut JP, Ventre G, Charrault A, Vetter D, Ledig M, North ML, Mayer S, Bockel R (1986) Relationship between 34 HLA-A, HLA-B and HLA-DR antigens and three serological markers of viral infections in alcoholic cirrhosis. Hepatology 6:457–463

Donnerer J, Oka K, Brossi A, Rice KC, Spector S (1986) Presence and formation of codeine and morphine in the rat. Proc Natl Acad Sci USA 83:4566–4567

Donnerer J, Cardinale G, Coffey J, Lisek CA, Jardine I, Spector S (1987) Chemical characterization and regulation of endogenous morphine and codeine in the rat. J Pharmacol Exp Ther 242:583–587

Dostert P, Strolin Benedetti M, Dordain G (1988) Dopamine-derived alkaloids in alcoholism and in degenerative diseases. J Neural Transm 74:61–74

Dostert P, Strolin Benedetti M, Dordain G, Vernay D (1991) Urinary elimination of salsolinol enantiomers in alcoholics. J Neural Transm 85:51–59

Dworkin MB, Rosenthal WS, Stahl RE, Panesar NK (1988) Decreased hepatic selenium content in alcoholic cirrhosis. Dig Dis Sci 33:1213–1217

Eriksson CJP (1992) Human blood acetaldehyde (update 1992). Alcohol Alcohol 27 [Suppl 1]:2

Faraj BA, Lenton JD, Kutner M, Camp VM, Stammers TW, Lee SR, Lolies PA, Chandora D (1987) Prevalence of low monoamine oxidase function in alcoholism. Alcohol Clin Exp Res II:464–467

Faraj BA, Camp VM, Davis DC, Lenton JD, Kutner M (1989) Elevation of plasma salsolinol sulfate in chronic alcoholics as compared to nonalcoholics. Alcohol Clin Exp Res 13:155–163

Feest U, Kemper A, Nickel B, Rabe H, Koalik F (1991) Comparison of salsolinol excretion in alcoholics and nonalcoholic controls. Alcohol 9:49–52

Fletcher LM, Kwoh-Gain I, Powell EE, Powell LW, Halliday JW (1991) Markers of chronic alcohol ingestion in patients with nonalcoholic steatohepatitis: an aid to diagnosis. Hepatology 13:455–459

Flink EB (1986) Magnesium deficiency in alcoholism. Alcohol Clin Exp Res 10: 590–594

Fowler CJ, Tipton KF, Mackay AVP, Youdim MBH (1982) Human platelet monoamine oxidase – a useful enzyme in the study of psychiatric disorders? Neuroscience 7:1577–1584

Fukunaga T, Sillanaukee P, Eriksson CJP (1993) Occurrence of blood acetaldehyde in women during ethanol intoxication: preliminary findings. Alcohol Clin Exp Res 17:1198–1200

Fuller RK (1992) Overview: problems, pitfalls and opportunities in monitoring alcohol consumption. Alcohol Alcohol 27:S10–1

Gaines KG, Salhany JM, Tuma DJ, Sorrell MF (1977) Reaction of acetaldehyde with human erythrocyte membrane proteins. FEBS Lett 75:115–119

Giller E Jr, Castiglione C, Wojciechowski J, Breakefield XO (1982) Molecular properties of platelet MAO in psychiatric patients and controls. In: Usdin E, Hanin I (eds) Biological markers in psychiatry and neurology. Pergamon, New York, pp 111–121

Girre C, Hispard E, Therond S, Guedj S, Bourdon R, Dally S (1990) Effect of abstinence from alcohol on the depression of glutathione peroxidase activity and selenium and vitamin E levels in chronic alcoholic patients. Alcohol Clin Exp Res 14:909–912

Grebb JA, Greengard P (1990) An analysis of synapsin II, a neuronal phosphoprotein, in postmortem brain tissue from alcoholic and neuropsychiatric ill adults and medically ill children and young adults. Arch Gen Psychiatry 47: 1149–1156

Guasch R, Renau-Piqueras J, Guerri C (1992) Chronic ethanol consumption induces accumulation of proteins in the liver Golgi apparatus and decreases galactosyltransferase activity. Alcohol Clin Exp Res 16:942–948

Hallgen R, Ludin L (1983) Increased total serum IgE in alcoholics. Acta Med Scand 213:99–103

Halmesmäki E, Roine R, Salaspuro M (1992) Gammaglutamyltransferase, aspartate and alanineaminotransferase and their ratio, mean cell volume and urinary dolichol in pregnant alcohol abusers. Br J Obstet Gynaecol 99:287–291
Halsted CH, Heise CL (1988) Ethanol and vitamin metabolism. Pharmacol Ther 34:453–464
Heidemann E, Nerke O, Waller HD (1981) Alkoholtoxische Veränderungen der Hämatopoese. Klin Wochenschr 59:1303–1312
Hernandez-Munos R, Baraona E, Blacksberg I, Lieber CS (1989) Characterization of the increased binding of acetaldehyde to red blood cells in alcoholics. Alcohol Clin Exp Res 13:654–659
Hill SY, Goddwin DM, Cadoret R, Osterland K, Doner SM (1975) Association and linkage between alcoholism and eleven serological markers. J Stud Alcohol 36:918–982
Hill SY, Steinhauer SR, Zubin J (1987) Biological markers for alcoholism: a vulnerability model conceptualization. In: Rivers PC (ed) Alcohol and addictive behavior: Nebraska symposium on motivation 1986. University of Nebraska Press, Lincoln, pp 207–256
Hoffman PL, Tabakoff B (1990) Ethanol and guanine nucleotide binding proteins: a selective interaction. FASEB J 4:2612–2622
Hoyumpa AM (1983) Alcohol and thiamine metabolism. Alcohol Clin Exp Res 7:11–14
Hrubec Z, Omenn GS (1981) Evidence for genetic predisposition to alcoholic cirrhosis and psychosis: twin concordances for alcoholism and its biological end points by zygosity among male veterans. Alcoholism 5:207–215
Hultberg B, Isaksson A, Berglund M, Moberg AL (1991) Serum β-hexosaminidase isoenzyme: a sensitive marker for alcohol abuse. Alcohol Clin Exp Res 15:549–552
Irwin M, Caldwell C, Smith TL, Brown S, Schuckit MA, Gillin C (1990a) Major depressive disorder, alcoholism, and reduced natural killer cell cytotoxicity. Arch Gen Psychiatry 47:713–719
Irwin M, Patterson T, Smith TL, Caldwell C, Brown SA, Gillin JC, Grant J (1990b) Reduction of immune function in life stress and depression. Biol Psychiatry 27:22–30
Israel Y, Hurwitz E, Niemelä O, Arnon R (1986) Monoclonal and polyclonal antibodies against acetaldehyde-containing epitopes in acetaldehyde-protein adducts. Proc Natl Acad Sci USA 83:7923–7927
Iturriaga H, Pereda T, Etevez A, Ugrate G (1977) Serum immunoglobulin A changes in alcoholic patients. Ann Clin Res 9:39–43
Johansson, U, Johnsson F, Joelsson B, Berglund M, Akesson B (1986) Selenium status in patients with liver cirrhosis and alcoholism. Br J Nutr 55:227–233
Kalsi J, Delacroix DL, Hedgson HJF (1983) IgA in alcoholic cirrhosis. Clin Exp Immunol 52:499–504
Kandaswami C, D'Iorio A (1979) On hepatic mitochondrial monoamine oxidase activity in lipid deficiency. Can J Biochem 57:588–594
Kapur A, Wild G, Milford-Ward A, Triger DR (1989) Carbohydrate deficient transferrin: a marker for alcohol abuse. Lancet 299:427–431
Kärkkäinen P, Mussalo-Ranhamaa H, Poikolainen K, Lehto J (1988) Alcohol intake correlated with serum trace elements. Alcohol Alcohol 23:279–282
Kärkkäinen P, Jokelainen K, Roine R, Suokas A, Salaspuro M (1990a) The effects of moderate drinking and abstinence on serum and urinary β-hexosaminidase levels. Drug Alcohol Depend 25:35–38
Kärkkäinen P, Poikalainen K, Salaspuro (1990b) Serum β-hexosaminidase as a marker of heavy drinking. Alcohol Clin Exp Res 14:187–190
Kiilerich S, Dietrichson O, Loud FB, Naestoft J, Christofferson P, Juhl E, Kjems G, Christiansen C (1980) Zinc depletion in alcoholic liver disease. Scand J Gastroenterol 15:363–367

Korpela H, Kumpulainen J, Luoma P, Arranto A, Sotaniemi E (1985) Decreased serum selenium in alcoholics as related to liver structure and function. Am J Clin Nutr 42:147–151

Korri U-M, Nuutinen H, Salaspuro M (1985) Increased blood acetate: a new laboratory marker of alcoholism and heavy drinking. Alcohol Clin Exp Res 9:468–471

Kristenson H, Fex G, Trell E (1981) Serum ferritin, gamma-glutamyltransferase and alcohol consumption in healthy middle-aged men. Drug Alcohol Depend 8:43–50

Kwoh-Gain I, Fletcher LM, Price J, Powell LW, Halliday JW (1990) Desialylated transferrin and mitochondrial aspartate aminotransferase compared as laboratory markers of excessive alcohol consumption. Clin Chem 36:841–845

Laforenza U, Patrini C, Gastaldi G et al. (1990) Effects of acute and chronic ethanol administration on thiamine metabolizing enzymes in some brain areas and other organs of the rat. Alcohol Alcohol 25:591–603

Lichtenberg-Kraag B, May T, Schmidt LG, Rommelspacher H (1994) changes of G protein levels in platelet membranes from alcoholics during short-term and long-term abstinence. Alcohol Alcohol (in press)

Lidberg L, Modin I, Oreland L, Tuck JR, Gillner A (1985) Platelet monoamine oxidase activity and psychopathy. Psychiatry Res 16:339–343

Lin RC, Shahidi S, Kelly TJ, Lumeng C, Lumeng L (1992) Measurement of hemoglobin-acetaldehyde adducts (Hb-AA) in alcoholic patients. Alcohol Alcohol 27 [Suppl 1]:81

Luthin GR, Tabakoff B (1984) Activation of adenylate cyclase by alcohol requires the nucleotide-binding protein. J Pharmacol Exp Ther 228:579–587

Maas JW, Katz MM (1992) Neurobiology and psychopathological states: are we looking in the right place? Biol Psychiatry 31:757–758

Major LF, Goyer PF, Murphy DL (1981) Changes in platelet monoamine oxidase in brain tissues. Biochem Pharmacol 17:1285–1297

Malagolini N, Dall'Olio F, Serafini-Cessi F, Cessi C (1989) Effect of acute and chronic ethanol administration on rat liver α_2,6-sialyltransferase activity responsible for sialylation of serum transferrin. Alcohol Clin Exp Res 13:649–653

Mathew J, Klemm WR (1989) Ethanol promotes hydrolysis of ^{3}H-labeled sialoconjugates from brain of mice in vitro. Alcohol 5:499–503

Matsubara K, Fukushima S, Akane A, Hama K, Fukui Y (1986) Tetrahdro-β-carbolines in human urine and rat brain – no evidence of formation by alcohol drinking. Alcohol Alcohol 21:339–345

Matsubara K, Fukushima S, Akane E, Kobayashi M, Shiono H (1992) Increased urinary morphine, codeine and tetrahydropapaveroline in parkinsonian patients undergoing L-3,4-dihydroxyphenylalanine therapy: a possible biosynthetic pathway of morphine form L-3,4-dihydroxyphenylalanine in humans. J Pharmacol Exp Ther 260:974–978

Max B (1992) This and that. An artefactual alkaloid and its peptide analogs. TIPS 13:341–345

May T, Rommelspacher H (1994) Monoamine oxidase (MAO; EC 1.4.3.4) characteristics of platelets influenced by in vitro and in vivo ethanol on alcoholics and on control subjects. J Neural Transm 41:69–73

McClain CJ, Antonow DR, Cohen DA, Shedlofsky SI (1986) Zinc metabolism in alcoholic liver disease. Alcohol Clin Exp Res 10:582–589

McCollister RJ, Flink EB, Lewis MD (1963) Urinary excretion of magnesium in man following the ingestion of ethanol. Am J Clin Nutr 12:415–420

Melendez M, Vargas-Tank K, Fuentes C (1979) Distribution of HLA histocompatibility antigens. ABO blood groups and Rh antigens in alcoholic liver disease. Blut 20:288–290

Mihas AA, Tavassoli M (1991) The effect of ethanol on the uptake, binding, and desialylation of transferrin by rat liver endothelium: implications in the pathogenesis of alcohol-associated hepatic siderosis. Am J Med Sci 301:299–304

Mihas AA, Tavassoli M (1992) Laboratory markers of ethanol intake and abuse: a critical appraisal. Am J Med Sci 303:415–428

Mochly-Rosen D, Chang F-H, Cheever L, Kim M, Diamond J, Gordon AS (1988) Chronic ethanol causes heterologous desensitization of receptors by reducing α_s messenger RNA. Nature 333:848–850

Murphy DL, Costa JL, Schafer B, Corash L (1978) Monoamine oxidase activity in different density gradient fractions of human platelets. Psychopharmacology 59:193–197

Murphy DL, Coursey RD, Taenel T, Aloi J, Buchsbaum MS (1982) Platelet monoamine oxidase as a biological marker in the affective disorders and alcoholism. In: Usdin E, Hanin I (eds) Biological markers in psychiatry and neurology. Pergamon, New York, pp 123–136

National Council on Alcoholism, Criteria Committee (1972) Criteria for the diagnosis of alcoholism. Am J Psychiatry 129:127–135

Nemesanszky E, Lott JA, Arato M (1988) Changes in serum enzymes in moderate drinkers after an alcohol challenge. Clin Chem 34:525–527

Niemelä O, Israel Y (1992) Acetaldehyde adducts in human alcohol abusers. Alcohol Alcohol 27 [Suppl 1]:82

Niemelä O, Klajner F, Orrego H, Vidins E, Blendis L, Israel Y (1987) Antibodies against acetaldehyde-modified protein epitopes in human alcoholics. Hepatology 7:1210–1214

Nies A, Robinson DS, Lamborn KR, Lampert RP (1973) Genetic control of platelet and plasma monoamine oxidase activity. Arch Gen Psychiatry 28:834–838

Nilssen O, Huseby NE, Hoyer G, Brenn T, Schirmer H, Forde OH (1992) New alcohol markers – how useful are they in population studies: the Svalbard study 1988–1989. Alcohol Clin Exp Res 16:82–86

Nouri-Aria KT, Alexander GJM, Portmann BC, Hegarty JE, Eddleston ALWF, Williams K (1986) T and B cell function in alcoholic liver disease. J Hepatol 2:195–207

Nuutinen HU, Salaspuro MP, Valle M, Lindros KO (1984) Blood acetaldehyde concentration gradient between hepatic and antecubital venous blood in ethanol-intoxicated alcoholics and controls. Eur J Clin Invest 14:306–311

Nuutinen HU, Lindros K, Hekali P, Salaspuro M (1985) Elevated blood acetate as an indicator of fast ethanol elimination in chronic alcoholics: Alcohol 2:623–626

Nyström M, Peräsalo J, Salaspuro M (1992) Carbohydrate-deficient transferrin (CDT) in serum as a possible indicator of heavy drinking in young university students. Alcohol Clin Exp Res 16:93–97

Okuno F, Ishii H, Kashiwazaki K, Takagi S, Shigeta Y, Arai M, Takagi T, Ebihara Y, Tsuchiya M (1988) Increase in mitochondrial GOT (m-GOT) activity after chronic alcohol consumption: clinical and experimental observations. Alcohol 5:49–53

Oreland L (1979) The activity of human brain and thrombocyte mononamine oxidase (MAO) in relation to various psychiatric disorders. In: Singer TP, von Korff RW, Murphy DL (eds) Monamine oxidase: structure, function and altered function. Academic, New York, pp 379–388

Oreland L, von Knorring L, von Knorring A-L, Bohman M (1985) Studies on the connection between alcoholism and low platelet monoamine oxidase activity. In: Parvez S, Burns E, Burrow Y, Parvez H (eds) Progress in alcohol research. VNU Science, Utrecht, pp 83–117

Pandey GN, Fawcett J, Gibbons R, Clark DC, Davis JM (1988) Platelet monoamine oxidase in alcoholism. Biol Psychiatry 24:15–24

Petren S, Vesterberg O (1988) Concentration differences in isoforms of transferrin in blood from alcoholics during abuse and abstinence. Clin Chim Acta 175: 183–188

Petren S, Vesterberg O (1989) The N-acetylneuraminic acid content of five forms of human transferrin. Biochim Biophys Acta 994:161–165

Peura P, Kari I, Airaksinen MM (1980) Identification by selective ion monitoring of 1-methyl-1,2,3,4-tetrahydro-beta-carboline in human platelets and plasma after ethanol intake. Biomed Mass Spectr 7:553–555
Poirier MF, Loô H, Lefur G, Mitrani N (1985) Monoamine oxidase plaquettaire et hormones sexuelles dans une population de déprimés. Encephale 11:125–129
Poupon RE, Schellenberg F, Nalpas B, Weill J (1989) Assessment of the transferrin index in screening heavy drinkers from a general practice. Alcohol Clin Exp Res 13:549–553
Rada RT, Knodell RG, Troup GM, Kellner R, Hermanson SM, Richards M (1981) HLA antigen frequencies in cirrhotic and noncirrhotic male alcoholics: a controlled study. Alcohol Clin Exp Res 5:188–191
Reisinger PWM, Soyka M (1990) The diagnosis of alcoholism on the basis of detection of a transferrin variant by polyacrylamide gel electrophoresis and immunoblotting. Blutalkohol 27:427–433
Rice J, McGuffin O, Goldin LR, Shaskan EG, Gershon ES (1984) Platelet monoamine oxidase (MAO) activity: evidence for a single major locus. Am J Hum Genet 36:36–43
Rissanen A, Sarlio-Lähteenkorva S, Alfthan G, Gref C-G, Keso L, Salaspuro M (1987) Employed problem drinkers: a nutritional risk group? Am J Clin Nutr 45: 456–461
Robertson DM, Morse RM, Moore SB, O'Fallon WM, Hurt RD (1984) A study of HLA antigens in alcoholism. Mayo Clin Proc 59:243–246
Roine RP, Korri U-M, Ylikahri R, Pentilla A, Pilekarainen J, Salaspuro M (1988) Increased serum acetate as a marker of problem drinking among drunken drivers. Alcohol Alcohol 23:123–126
Rommelspacher H (1992) Marker des Alkoholismus. Psycho 18:165–177
Rommelspacher H, Strauss S, Lindemann J (1980) Excretion of tetrahydroharmane and harmane into the urine of man and rat after a load with ethanol. FEBS Lett 109:209–212
Rommelspacher H, Damm H, Strauss S, Schmidt G (1984) Ethanol induces an increase of harman in the brain and urine of rats. Arch Pharmacol 327:107–113
Rommelspacher H, Damm H, Schmidt L, Schmidt G (1985) Increased excretion of harman by alcoholics depends on events of their life history and the status of the liver. Psychopharmacology 87:64–68
Rommelspacher H, Damm H, Lutter S, Schmidt LG, Otto M, Sachs-Ericsson N, Schmidt G (1990) Harman (1-methyl-β-carboline) in blood plasma and erythrocytes of non-alcoholics following ethanol loading. Alcohol 7:27–31
Rommelspacher H, Schmidt LG, May T (1991a) Plasma norharma (β-carboline) levels are elevated in chronic alcoholics. Alcohol Clin Exp Res 15:553–559
Rommelspacher H, May T, Susilo R (1991b) β-Carbolines and tetrahydroisoquinolines: detection and function in mammals. Planta Med S85–S92
Rommelspacher H, Schmidt LG, Leitner A, Lesch O (1992) Are β-carbolines biochemical marker of subgroups of chronic alcoholics? Alcohol Alcohol 27 [Suppl 1]:93
Rommelspacher H, May T, Dufeu P, Schmidt LG (1994) Longitudinal observations of monoamine oxidase in alcoholics: Differentiation of marker characteristics. Alcohol Clin Exp Res (in press)
Rose RM, Castellani S, Boeringa JA, Malek-Ahmadi P, Lankford Bessmann JD, Fritz RR, Denney CB, Denney RM, Abell CW (1986) Platelet MAO concentration and molecular activity: II Comparison of normal and schizophrenic populations. Psychiatry Res 17:141–151
Rubin R, Hoek J (1988) Alcohol-induced stimulation of phospholipase C in human platelets requires G protein activation. Biochem J 254:147–153
Salaspuro M (1987) Use of enzymes for the diagnosis of alcohol-related organ damage. Enzyme 37:87–107

Salaspuro M, Korri U-M, Nuutinen H, Roine R (1987) Blood acetate and urinary dolichols – new markers of heavy drinking and alcoholism. Prog Clin Biol Res 241:231–240

Saunders JB, Wodak AD, Haines A, Powell-Jackson PR, Portmans B, Davis M, Williams R (1982) Accelerated development of alcoholic cirrhosis in patients with HLA-B8. Lancet i:1381–1384

Schellenberg F, Benard JY, Le Goff AM, Bourdin C, Weill J (1989) Evaluation of carbohydrate-deficient transferrin compared with Tf index and other markers of alcohol abuse. Alcohol Clin Exp Res 13:605–610

Schiele F, Artur Y, Varasteh A, Wellman M, Siest G (1989) Serum mitochondrial aspartate aminotransferase activity: not useful as a marker of excessive alcohol consumption in an unselected population. Clin Chem 35:926–930

Schmidt LG, Rommelspacher H (1990) Biologische Marker des Alkoholismus. Nervenarzt 61:140–147

Schouten MR, Bruinvels J (1985) High-performance liquid chromatography of tetrahydro-β-carbolines extracted from plasma and platelets. Anal Biochem 147:401–409

Schouten MR, Bruinvels J (1986) Endogenously formed norharman (β-carboline) in platelet rich plasma obtained from porphyric rats. Pharmacol Biochem Behav 24:1219–1223

Schuckit MA (1984) Genetic and biochemical factors in the etiology of alcoholism. In: Grinspoon L (ed) Psychiatry update, vol III. American Psychiatry Press, Washington DC, pp 320–327

Schuckit MA (1992) Reaction to alcohol as a predictor of alcoholism. Alcohol Alcohol 27:S17–4

Scott BB, Rajah SM, Losowsky MS (1977) Histo-compatibility antigens in chronic liver disease. Gastroenterology 72:122–125

Shorr Y, Foley K, Spector S (1978) Presence of a non-peptide morphine-like compound in human cerebrospinal fluid. Life Sci 23:2057–2062

Sillanaukee P, Seppä K, Koivula T (1991) Effect of acetaldehyde on hemoglobin: HbA_{IACH} as a potential marker of heavy drinking. Alcohol 8:377–381

Sjöquist B, Borg S, Kvande H (1981a) Catecholamine derived compounds in urine and cerebrospinal fluid from alcoholics during and after longstanding intoxication. Subst Alcohol Act Misuse 2:63–72

Sjöquist B, Borg S, Kvande H (1981b) Salsolinol and methylated salsolinol in urines and cerebrospinal fluid from healthy volunteers. Subst Alcohol Act Misuse 2:73–77

Sopena B, Martinez-Vasquez C, de la Fuente J, Fernandez C, Rivera A (1993) Serum levels of immunoglobulins and complement in alcoholic liver disease. Rev Clin Esp 193:419–423

Stevens VJ, Fault WJ, Newman CB, Sims RV, Cerami A, Peterson CM (1981) Acetaldehyde adducts with hemoglobin. J Clin Invest 67:361–369

Stiber H (1991) Carbohydrate-deficient transferrin in serum: a new marker of potentially harmful alcohol consumption reviewed. Clin Chem 37:2029–2037

Stibler H, Borg S (1991) Glycoprotein glycosyltransferase activities in serum in alcohol-abusing patients and healthy controls. Scand J Lab Invest 51:43–51

Stibler H, Hultcrantz R (1987) Carbohydrate-deficient transferrin in serum in patients with liver diseases. Alcohol Clin Exp Res 11:468–473

Stibler H, Allgulander C, Borg S, Kjellin KG (1978) Abnormal microheterogeneity of transferrin in relation to alcohol consumption. Acta Med Scand 204:49–56

Stibler H, Borg S, Joustra M (1986) Micro anion exchange chromatography of carbohydrate-deficient transferrin in serum in relation to alcohol consumption (Swedish patent 8400587-5). Alcohol Clin Exp Res 10:535–544

Stibler H, Dahlgren L, Borg S (1988) Carbohydrate-deficient transferrin (CDT) in serum of women with early alcohol addiction. Alcohol 5:393–398

Stibler H, Beauge F, Bjorneboe A, Aufrere G (1989) Transferrin microheterogeneity in rats treated chronically with ethanol. Pharmacol Toxicol 64:383–385
Stibler H, Borg S, Joustra M (1991) A modified method for the assay of carbohydrate-deficient transferrin (CDT) in serum. Alcohol Alcohol [Suppl 1]:451–454
Strolin Benedetti M, Bellotti U, Pianezzola E, Moro E, Carminati P, Dostert P (1989) Ratio of the R and S enantiomers of salsolinol in food and human urine. J Neural Transm 77:47–53
Sullivan JL, Baenziger JC, Wagner DL, Rauscher FP, Nurnberger JJ Jr, Holmes JS (1990) Platelet MAO in subtypes of alcoholism. Biol Psychiatry 27:911–922
Swinson RP, Madden JS (1973) ABO blood groups and ABH substance secretion in alcoholics. Q J Stud Alcohol 34:64–70
Tabakoff B, Lee JM, De Leon-Jones F, Hoffman PL (1985) Ethanol inhibits the activity of the B form of monoamine oxidase in human platelet and brain tissue. Psychopharmacology 87:152–156
Tabakoff B, Hoffman PL, McLaughlin A (1988a) Is ethanol a discriminating substance? Semin Liver Dis 8:26–35
Tabakoff B, Hoffman PL, Lee JM, Saito T, Willard B, De Leon-Jones F (1988b) Differences in platelet enzyme activity between alcoholics and nonalcoholics. N Engl J Med 318:134–139
Tallaksen CME, Bohmer T, Bell H (1992) Blood and serum thiamin and thiamin phosphate ester concentrations in patients with alcohol dependence syndrome before and after thiamin treatment. Alcohol Clin Exp Res 16:320–325
Tang BK (1987) Detection of ethanol in urine of abstaining alcoholics. Can J Physiol Pharmacol 65:1225–1227
Thorpe LW, Westlund KN, Kochersperger LM, Abell CW, Denney RM (1987) Immunocytochemical localization of monoamine oxidase A and B in human peripheral tissues and brain. J Histochem Cytochem 35:23–32
Tuma DJ, Sorrell MF (1985) Hypothesis: alcoholic liver injury and the covalent binding of acetaldehyde. Alcohol Clin Exp Res 9:306–309
Tuma DJ, Sorrell MF (1987) Functional consequences of acetaldehyde binding to proteins. Alcohol Alcohol [Suppl 1]:61–66
Välimäki MJ, Harju KJ, Clikahri RH (1983) Decreased serum selenium in alcoholics – a consequence of liver dysfunction. Clin Chim Acta 130:291–296
Vallee BL, Wacker WEC, Bartholomay AF, Hoch FL (1957) Zinc metabolism in hepatic dysfunction. N Engl J Med 257:1055–1065
van Eijk HG, van Noort WL, de Jong G, Koster JF (1987) Human serum sialotransferrins in diseases. Clin Chim Acta 165:141–145
Vernay D, Eschalier A, Durif F, Aumaitre O, Rigal B, Ben Sadoun A, Fialip J, Marty H, Philip E, Bourgerolle AM, Dostert P, Strolin Benedetti M, Dodrain G (1989) Le salsolinol, molécule endogène. Encephale XV:511–516
von Knorring A-L, Oreland L, Häggendal J, Magnusson T, Almay B, Johansson F (1986) Relationship between platelet MAO activity and concentrations of 5-HIAA and HVA in cerebrospinal fluid in chronic pain patients. J Neural Transm 66:37–46
von Knorring L, Oreland L, von Knorring A-L (1987) Personality traits and platelet MAO activity in alcohol and drug abusing teenage boys. Acta Psychiat Scand 75:307–314
von Knorring A-L, Hallman J, von Knorring L, Oreland L (1991) Platelet monoamine oxidase activity in type 1 and type 2 alcoholism. Alcohol Alcohol 26:409–416
Weill J, Schellenberg F, Le Goff AM, Benard JY (1988) The decrease of low serum gamma glutamyl transferase during short-term abstinence. Alcohol 5:1–3
Weitz CJ, Lowney LJ, Faull KJ, Feistner G, Goldstein A (1986) Morphine and codeine from mammalian brain. Proc Natl Acad Sci USA 83:9784–9788
Weitz CJ, Faull KF, Goldstein A (1987) Synthesis of the skeleton of morphine molecule by mammalian liver. Nature 330:674–677

Wilson JS, Gossat D, Tait A, Rouse S, Juan XJ, Pirola RC (1984) Evidence for an inherited predisposition to alcoholic pancreatitis. A controlled HLA-typing study. Dig Dis Sci 29:727–730

Wlodek L, Rommelspacher H, Susilo R, Radomski J, Höfle G (1993) Thiazolidine derivatives as source of free l-cysteine in rat tissue. Biochem Pharmacol 46: 1917–1928

Wolfe SM, Victor M (1969) The relationship of hypomagnesemia and alkalosis to alcohol withdrawal symptoms. Ann NY Acad Sci 162:973–984

Wyatt RJ, Belmaker R, Murphy DL (1975) Low platelet monoamine oxidase and vulnerability to schizophrenia. In: Mendlewicz J (ed) Modern problems of pharmacopsychiatry, vol 10. Karger, Basel, pp 38–56

Xin Y, Lasker JM, Rosman AS, Lieber CS (1991) Isoelectric focusing/western blotting: a novel and practical method for quantitation of carbohydrate-deficient transferrin in alcoholics. Alcohol Clin Exp Res 15:814–821

Xin Y, Rosman AS, Lasker JM, Lieber CS (1992) Measurement of carbohydrate-deficient transferrin by isoelectric focusing/western blotting and by micro anion-exchange chromatography/radioimmunoassay: comparison of diagnostic accuracy. Alcohol Alcohol 27:425–433

Yates WR, Wilcox J, Knudson R, Myers C, Kelly MW (1990) The effect of gender and subtype on platelet MAO in alcoholism. J Stud Alcohol 51:463–467

Youdim MB (1991a) Modulation of monoamine oxidase A activity in PC12 cells by steroids. Eur J Pharmacol 192:201–202

Youdim MB (1991b) PC12 cells as a window for the differentiation of neural crest into adrenergic nerve ending and adrenal medulla. J Neural Transm [Suppl 334]:61–67

Young CL, MacGregor RR (1989) Alcohol and host defenses: infections consequences. Infect Med Sept: 163–175

Young WF Jr, Laws ER Jr, Sharbrough FW, Weinshilboum RM (1986) Human monoamine oxidase – lack of brain and platelet correlation. Arch Gen Psychiatry 43:604–609

CHAPTER 19

Interaction of Alcohol With Therapeutic Drugs and Drugs of Abuse

B.F. SANDS, C.M. KNAPP, and D.A. CIRAULO

A. Introduction

In societies where alcoholic beverages are widely consumed, a large number of individuals are exposed to the effects of drug-alcohol combinations. Such exposures can lead to a broad range of medical problems. Of great concern are ethanol-sedative drug combinations, which may produce psychomotor impairment, which, in turn, can be the cause of motor vehicle and heavy machinery accidents (LANDAUER et al. 1969; LINNOILA and HAKKNEW 1974; MILNER 1969; MILNER and LANDAUER 1971; SEPPALA 1977; SEPPALA et al. 1975, 1979). Severe, sometimes lethal, central nervous system depression may occur when ethanol and sedative agents such as barbiturates are used concurrently (MILNER 1970). The use of alcohol can often complicate the pharmacological management of chronic diseases such as diabetes, epilepsy, and psychiatric disorders.

The effects that result from drug-alcohol interactions may be produced by pharmacokinetic and/or receptor-mediated mechanisms or, in some cases, by the physiologic actions of ethanol itself. Pharmacokinetic mechanisms may involve alcohol-induced changes in drug bioavailability or distribution or in some instances alterations in alcohol bioavailability as a consquence of drug action. Often more problematic are the large and sometimes unexpected alterations in drug or ethanol levels that can result from the acute inhibitory effects of these agents on the activity of hepatic drug-metabolizing enzyme systems. Undesirable fluctuations in drug concentrations may also result from the enzyme-inducing effects of chronic ethanol administration on the activity of drug-metabolizing systems. Many of the pharmacodynamic mechanisms that play a role in the effects of drug-ethanol interactions are not well understood. It is becoming increasingly evident, however, that ethanol can interact with other agents by specifically acting on a number of neurotransmitter and second messenger systems. There is now considerable evidence that, in addition to its nonspecific effects on cell membranes, alcohol can alter the activity of many neurotransmitter systems, including GABAergic, dopaminergic, and serotoninergic receptor systems and also some calcium channels. These effects are discussed in detail in a number of other chapters in this volume. In addition, alcohol has a variety of physiologic actions including effects on glucose metabolism, alcohol dehydrogenase (ADH) secretion, and, particularly at higher doses, CNS regulation of respiration and cardiovascular system functioning (RALL 1990). The effectiveness and sometimes the safety of many agents can be altered by ethanol-induced physiologic changes. Ethanol-induced alterations in glucose metabolism, for example, may lead to hypoglycemia in patients being treated with antihyperglycemic agents (BARUH et al. 1973).

The prolonged ingestion of ethanol frequently leads to the development of pathologic changes, foremost among which are alcoholic liver disease and cardiac disease. These changes cause alterations in organ function that can then affect the actions of many drugs. As will be discussed below, alcoholic liver disease can produced a reduction in the binding of drugs to plasma proteins. Severe liver damage may result in decreases in the rates of metabolism of many compounds. The clearance of both alprazolam (JUHL et al. 1984) and lidocaine (COLLI et al. 1988), for instance, has been found to be reduced in individuals diagnosed as suffering from alcoholic cirrhosis.

This chapter will begin with an overview of ethanol drug interaction pharmacology and conclude with a discussion of individual drugs.

B. Pharmacokinetic Mechanisms

I. Bioavailability and Absorption

Ethanol may have modest effects on the absorption of some drugs. Gastric emptying time may be delayed when alcoholic solutions are ingested (COOKE 1972). The ingestion of diazepam concurrent with moderate doses of diazepam seems to slow the rate of diazepam absorption, leading to a delay in the appearance of peak diazepam levels and a reduction in the magnitude of the peak (DIVOLL and GREENBLATT 1981; GREENBLATT et al. 1978). Consumption of larger doses of alcohol prior to the ingestion of diazepam may result in elevation of diazepam blood concentrations (MACLEOD et al. 1977; LAISI et al. 1979). Whether or not this ethanol-induced increase in diazepam levels can be attributed to alterations in the absorption process or to some change in the rate of clearance of diazepam remains unclear. There is evidence that ethanol can inhibit the metabolism of diazepam (SELLERS et al. 1980).

Treatment with ethanol may enhance the bioavailability of at least some compounds that are subject to significant first pass effects. Both propoxyphene (GRAM et al. 1979) and amitriptyline (DORIAN et al. 1983) undergo extensive first pass metabolism. The administration of ethanol significantly enhances the extent of absorption of both of these agents (DORIAN et al. 1983; GIRRE et al. 1991). This effect may be attributable to inhibition of hepatic drug-metabolizing enzymes by ethanol. The observation that levels of the propoxyphene metabolite norpropoxyphene were lower when propoxyphene was administered with ethanol than when given alone supports the hypothesis that alcohol can enhance drug bioavailability by inhibiting drug-metabolizing enzymes (GIRRE et al. 1991).

The absorption of ethanol may be delayed by the concurrent administration of drugs such as the anticholinergics, which reduce gastric emptying rates. Evidence of this is provided by the finding that treatment with desipramine, which has anticholinergic properties, resulted in the delayed appearance of peak ethanol concentrations (HALL et al. 1976). Cimetidine and other H_2 receptor antagonists such as ranitidine and nizatidine that contain pyrazole-like structures may selectively inhibit the activity of gastric ADH (LIEBER 1990). It has been suggested that ADH in the gastric mucosa can metabolize substantial amounts of ethanol and alterations in the activity of the enzyme affect the bioavailability of ethanol (LIEBER 1990). Pretreatment with cimetidine was found too elevate alcohol blood concentrations significantly above control levels in subjects treated with comparatively small oral doses of ethanol (CABALLERIA et al. 1989b). However, no changes in blood alcohol levels were observed in individuals treated with combinations of high oral doses of ethanol (0.8 g/kg) and either cimetidine or ranitidine (JONSSON et al. 1992), suggesting that large doses of ethanol may overwhelm the capacity of the gastric ADH system.

II. Distribution

Drug distribution can be altered by changes in such factors as plasma and tissue protein binding and cardiac output. Most of the alcohol-related alterations in drug distribution that have been described thus far can be attributed to the pathologic changes and physiologic effects associated with ethanol ingestion. Plasma protein levels, especially albumin levels, are frequently abnormally low in chronic alcoholics, in many cases due to malnutrition and the effects of alcoholic liver disease. The extent of plasma protein binding of tolbutamide and diazepam was significantly lower than control levels in alcoholics who had albumin concentrations less than 3.5 g/dl (THIESSEN et al. 1976). Patients with alcoholic liver disease, almost all of whom also had hypoalbuminemia, were reported to have significant elevations in the percentage of unbound fractions of several basic drugs including quinidine, dapsone, and triamterence (AFFRIME and REIDENBERG 1975). Binding of the acidic drug phenytoin was within normal limits in these subjects. The protein binding of the opiate drug alfentanil also was significantly below control levels in alcoholic patients (BOWER et al. 1992). This reduction in alfentanil binding correlated with a decrease in the concentrations of the plasma protein α-1-acid glycoprotein.

In alcoholics, the plasma concentrations of α-1-glycoprotein may increase during the early stages of ethanol withdrawal and then return to lower concentrations after several days (SANDOR et al. 1983). In patients who had recently been withdrawn from alcohol, free fractions of propanolol and warfarin decreased over time, while the unbound fraction of diazepam increased (SANDOR et al. 1983). Changes in the percentage free fraction of propanolol were inversely related to changes in α-1-acid glycoprotein plasma concentrations (SANDOR et al. 1983). Similarly, abnormally high levels of α-1-glycoprotein in alcoholics who did not show signs of hypoalbuminemia were associated with free fractions of desipramine and imipramine that were lower than those found in non-alcoholic subjects (CIRAULO and BARNHILL 1986).

III. Metabolism

Ethanol is metabolized primarily by hepatic alcohol dehydrogenase into acetaldehyde (HAWKINS and KALANT 1972). This product is then converted into acetic acid by aldehyde dehydrogenase. Substances which include pyrazole and related analogues (KISSIN 1974), chlorpromazine (KISSIN 1974), and the chloral hydrate metabolite trichlorethanol (SELLERS et al. 1971a) can inhibit the activity of liver ADH and, as a result, potentiate ethanol's intoxicating effects (KOFF and FITTS 1972). Human liver ADH catalyzes the oxidation of the pharmacologically active metabolites of digoxin and digitoxin. Ingestion of ethanol can reduce the rate of the oxidation of these compounds by acting as competitive substrate for liver ADH (FREY and VALLEE 1980).

The actions of aldehyde dehydrogenase can be blocked by disulfiram (SANNY and WEINER 1987) and other compounds such as metronidazole. Acetaldehyde levels may greatly increase in individuals who ingest ethanol while taking an aldehyde dehydrogenase blocking agent. This increase in acetaldehyde levels may lead to the appearance of a distince syndrome which is characterized by unpleasant flushing, vomiting, hypotension, sweating, and headache.

Within the liver an oxidizing system, referred to as the microsomal ethanol oxidizing system (MEOS), has been identified that may serve as a second pathway through which alcohol is metabolized (LIEBER and DE CARLI 1972). The MEOS uses cytochrome P450, NADPH, and oxygen to oxidize ethanol. The acute administration of drugs such as verapamil (BAUER et al. 1992) and possibly high-dose cimetidine leads to a decrease in the rate of clearance of alcohol, an effect that may result from the competition of these drugs with ethanol for binding sites on cytochrome P450. The acute administration of ethanol may also inhibit the metabolism of several agents including pentobarbital (RUBIN et al. 1970), methadone, and many other oxidatively metabolized drugs (BOROWSKY and LIEBER 1978). This action is presumed to involve interference by ethanol with the interactions that occur between cytochrome P450 systems and drug substrates. The precise identity of the forms of cytochrome P450 that are involved in acute ethanol-drug interactions have, for the most part, yet to be determined. In rabbits the cytochrome P450II plays a role in the metabolism of both ethanol and acetaminophen (MORGAN et al. 1983).

The chronic administration of alcohol can result in enhanced rates of clearance of many other agents. This enhancement of drug-metabolizing capacity probably results from the induction of cytochrome P450-metabolizing enzymes by alcohol. Animal studies indicate that prolonged ethanol treatment leads to an increase in hepatic smooth endoplasmic recticulum and drug-metabolizing enzymes (RUBIN et al. 1970). The activity of the MEOS, in vitro, is increased by prolonged treatment with either ethanol or barbiturates such as hexobarbital (LIEBER and DECARLI 1972). Chronic ethanol administration in rats can produce a ninefold increase in cytochrome P450j (JOHANSSON et al. 1988). Also in the rat, ethanol treatment will cause the induction of proteins that belong to two distinct cytochrome P450II gene subfamilies, IIB and IIE (JOHANSSON et al. 1988). Exposure of cultured rat hepatocytes to ethanol has produced increases in cytochrome P450IIE, IIB1/2, and IIIA (SINCLAIR et al. 1991). A human cytochrome P450J isoenzyme has been identified which is analogous to rat cytochrome P450J (WRIGHTON et al. 1986); NEBERT et al. (1987) has suggested that this enzyme be designated cytochrome P450IIE1.

Induction of the cytochrome P450 mixed function oxidizing enzyme system may cause an increase in the production of reactive intermediate metabolites. The enhanced generation of these reactive metabolites may lead to the development of liver toxicity. In the mouse, the hepatotoxic effects of cocaine are dependent upon the induction of hepatic cytochrome P450 mixed function oxidase systems (SMITH et al. 1981) and may be related to enhanced transformation of cocaine into reactive metabolites by these systems. The hepatotoxic effects of acetaminophen are greatly increased in both humans (SEEFF et al. 1986) and rodents (SATO et al. 1981a) by the chronic consumption of ethanol. While the primary route of acetaminophen metabolism involves conjugation with glucuronide and sulfate, a portion of this drug is converted by mixed function oxidase into reactive metabolites (MORGAN et al. 1983; SATO et al. 1981a). These metabolites may lose their toxicity by combining with glutathione (SATO et al. 1981a). In the rat, the prolonged ingestion of ethanol appears to lead to decreased levels of hepatic glutathione in animals that have been treated with acetaminophen (SATO et al. 1981a). The induction of mixed function oxidase enzymes by ethanol may result in an increased rate of production of reactive acetaminophen metabolites. If the rate of reactive metabolite production is high then, glutathione stores may become depleted, leaving excess reactive metabolites to bind to hepatic proteins. The covalent binding of reactive acetaminophen metabolities to proteins is greatly elevated following the addition of acetaminophen to in vitro preparations of microsomes obtained from ethanol-treated animals (SATO et al. 1981a).

Some of the ethanol effects on drug metabolism cannot be directly attributed to its interaction with microsomal oxidizing systems. Ethanol at low concentrations reduced the rates of glucuronidation of phenolphthalein, harmol, and other compounds in in vitro hepatocyte preparations (MOLDEUS et al. 1978). Results consistent with a decrease of the glucuronidation of morphine in vitro have also been reported (BODD et al. 1986). Alcohol may inhibit glucuronidation by reducing the rate of UDP-glucuronic acid synthesis (MOLDEUS et al. 1978).

Ethanol may enhance rates of drug acetylation. Treatment with ethanol appears to increase the acetylation rate of procainamide in humans (OLSEN and MORLAND 1982). The enzyme *N*-acetyltransferase uses acetyl-coenzyme A (CoA) to acetylate this compound. The metabolism of ethanol by ADH leads to the production of acetic acid, which is converted into acetyl-CoA. The increase in the availability of acetyl-CoA which may result from the metabolism of alcohol may allow rates of drug acetylation to increase.

C. Pharmacodynamic Mechanisms

Alcohol has been shown to interact with a variety of neurotransmitter and second messenger systems. Here the interactions of ethanol with GABAergic, noradrenergic, serotonergic, dopaminergic, opioidergic systems, and calcium channels will be briefly reviewed.

The interactions between ethanol and two classes of sedative drugs, the barbiturates and the benzodiazepines, have been well documented. In classic studies alcohol has been shown to potentiate the anesthetic and other effects of pentobarbital (KISSIN 1974). It has also been shown in numerous studies that alcohol can act synergisticallly with benzodiazepines to produce sedation and psychomotor impairments (ARANKO et al. 1985; LINNOILA et al. 1977d; MORLAND et al. 1974). While there is clear evidence that pharmacokinetic factors are often involved in the interactions that occur between ethanol and the benzodiazepines or the barbiturates, pharmacodynamic factors also play a major role. The observation that barbiturates and benzodiazepines can be used to prevent the signs of alcohol withdrawal from appearing suggest that these agents share some common mechanism of action. The finding that cross-tolerance can develop among ethanol, barbiturates, and benzodiazepines to the sedative and motor coordination impairing effects produced by these agents may also indicate that these agents act through some common mechanism (LAU et al. 1981; LÊ et al. 1986).

Animal studies indicate that ethanol-induced impairment of motor coordination and narcosis can be potentiated by treatment with GABA receptor agonists and attenuated or blocked by the administration of GABA antagonists (FRYE and BREESE 1982; LILJEQUIST and ENGEL 1982; MARTZ et al. 1982). Under certain experimental conditions ethanol can potentiate the inhibitory actions of GABA on cortical neuronal activity (NESTOROS 1979) and stimulate the GABA receptor mediated uptake of chloride into brain synaptosomes (MORROW et al. 1988; SUZDDAK et al. 1986) and cultured hippocampal and cortical neurons (AGUAYO 1990). These results suggest that many actions of ethanol may be caused by the enhancement of GABA receptor activity. The precise mechanism through which ethanol may act to augment the actions of GABA is not at present known. There is evidence, however, that a particular γ-subunit of the GABA receptor is required for such augmentation to occur (WAFFORD et al. 1991).

Like ethanol, the barbiturates and the benzodiazepines can enhance GABAergic system activity (STUDY and BARKER 1981). These agents may act on two distinct receptor sites that exist in close association with the GABA A receptor/chloride channel complex. Activation of the benzodiazepine receptor site may lead too an increase in the affinity of the GABA receptor while barbiturates may act on a site that is directly involved in the regulation of the flow of ions through the chloride channel (MATHERS 1985; TWYMAN et al. 1989). Genetic studies suggest that differences in the sensitivity of certain strains of mice to alcohol's effects on sleep may be related to the extent to which there is a difference in the coupling that occurs between benzodiazepine receptors and chloride ion channels among these strains (HARRIS and ALLAN 1989).

The long-term administration of alcohol seems to alter the functioning of GABAergic systems. Chronic ethanol treatment, for example, decreases GABA levels in the brains of animals and increases the activity of the GABA-synthesizing enzyme glutamate decarboxylase in the cerebellum and cortex of rats (SYTINSKY et al. 1975). Chronic treatment with alcohol or with benzodiazepines may lead to a decrease in the ability of these agents to potentiate GABA-mediated chloride uptake (MORROW et al. 1988; BUCK and HARRIS 1992; ALLAN et al. 1992). These changes may play a role in the development of tolerance to effects of both ethanol and the benzodiazepines. In the mouse, prolonged ethanol administration can lead to a loss in flunitrazepam-induced enhancement of the effects of the GABA agonist muscimol on chloride uptake (BUCK and HARRIS 1992), while chronic treatment with the benzodiazepine lorazepam can result in a decrease in both ethanol and flunitrazepam augmentation of GABA-mediated chloride up take (ALLAN et al. 1992). This evidence suggests that ethanol-benzodiazepine cross-tolerance can develop at the level of the GABA receptor/chloride channel complex.

The development of tolerance to the effects of ethanol does not appear to depend solely on changes in the functioning of GABAergic systems. The selective destruction of brain noradrenergic systems in mice blocks the development of functional tolerance to the hypnotic and hypothermic effects of ethanol (MELCHIOR and TABAKOFF 1981). Tolerance to the effects of barbiturates is also prevented by the selective depletion of norepinephrine from the brains of mice (TABAKOFF and RITZMANN 1977; TABAKOFF et al. 1978). Similarly the selective depletion of brain serotonin has been shown to delay the development of tolerance to the motor function impairing and hypnotic effects of both ethanol and the barbiturates (LEDUNG et al. 1980; FRANKEL et al. 1975; KHANNA et al. 1980). These results suggest that common noradrenergic and serotonergic systems are involved in the development of tolerance to actions of both the barbiturates and ethanol.

In animals administration of the calcium channel antagonist nifedipine causes a consistent reduction in the motor activity of animals treated with a range of doses of alcohol (WHITE and SMITH 1992). The duration of calcium action potentials in dorsal root ganglion cell was decreased by treatment with ethanol (OAKES and POZOS 1982). Exposure to ethanol can suppress calcium currents in cultured neuroblastoma cells (TWOMBLY et al. 1990) and synaptosomes (HARRIS and HOOD 1980). These results suggest that ethanol can affect calcium currents and can interact with calcium channel antagonists.

The symptoms of ethanol withdrawal can be attenuated by the administration of calcium channel blockers. Following withdrawal from alcohol, the number of convulsions in mice were found to be reduced by treatment with the calcium channel antagonists nimodipine and nitrendipine (LITTLETON et al. 1990; WHITTINGTON et al. 1991). Dihydropyridine-binding sites were increased in the cerebral cortical tissue of rats exposed chronically to ethanol (DOLIN and LITTLE 1989). Treatment with the calcium antagonist nitredipine prevents ethanol-induced increases in the density of dihydropyridine-binding sites (DOLIN and LITTLE 1989; WHITTINGTON et al. 1991).

Dihydropyridine-binding sites were increased in cultured bovine adrenal chromaffin cells bathed in medium containing ethanol, alprazolam, or buspirone, agents which were all found to inhibit basal and carbachol-stimulated catecholamine release from these cells (BRENNAN and LITTLETON 1991). Some common adaptative mechanism may exist in these cells which is activated in response to substances which cause a reduction in their excitabilty. The extent to which a similar adaptive mechanism may exist within the central nervous system is not yet clear. Treatment with nitrendipine decreased the incidence of seizures induced by the partial inverse benzodiazepine agonist FG7142 in mice withdrawn from flurazepam but it did not affect the development of tolerance to the ataxic effects of flurazepam. Chronic flurazepam treatment did not alter either the K_d or B_{max} of (^{3}H)-nimodipine binding in the mouse whole brain and cerebral cortex, which may indicate that treatment with benzodiazepines does not readily lead to an upregulation of dihydopyridine receptor sites in the brain (DOLIN et al. 1990).

Brain serotonin, opioid, and dopamine systems may all be involved in the processes that give rise to the reinforcing effects of ethanol. Patterns of ethanol consumption can be altered by drugs that interact with these systems. The results of several studies indicate that serotonin uptake blocking agents such as sertraline and tianeptine may decrease the intake of ethanol by animals (DAOUST et al. 1992; SELLERS et al. 1992). The serotonin uptake blocker fluoxetine has been shown to produced modest decreases in the consumption of ethanol by alcoholics (NARANJO et al. 1990). Ethanol consumption was reduced in animals treated with the 5-HT$_3$ antagonists ondansetron (SELLERS et al. 1992) and zacopride (KNAPP and POHORECKY

1992). The administration of ondansetron produced a slight decrease in the alcohol intake of alcohol-dependent patients (SELLERS et al. 1991).

Mesolimbic dopaminergic systems play a critical role in the neural substrates associated with the production of the reinforcing effects of the opiates and the psychomotor stimulants (NESTLER 1992). Alcohol also appears to interact with these systems. Ethanol administration increases dopamine turnover in several areas of the brain, including the caudate nucleus and the olfactory tubercle (FADDA et al. 1980, 1988), enhances the firing rate of dopaminergic neurons located in the ventral tegmental area (BRODIE et al. 1990; GEESA et al. 1985), and enhances the release of dopamine in the nucleus accumbens (DICHIARA and IMPERATO 1985; YOSHIMOTO et al. 1992). Evidence that ethanol can alter the actions of psychomotor stimulants within central reward systems includes the findings that the administration of ethanol can potentiate the brain reward stimulation threshold-lowering effects of the psychomotor stimulants cocaine and *d*-amphetamine and that chronic administration of ethanol can potentiate the threshold-lowering actions of cocaine when this drug is administered alone (MOOLTEN and KORNETSKY 1992). Ethanol may also potentiate the euphoric effects of cocaine in human subjects (MCCANCE-KATZ et al., in press). Manipulations of dopaminergic systems such as the selective depletion of dopamine from the nucleus accumbens (LYNESS and SMITH 1992) and treatments with some dopamine agonists or antagonists (LINESEMAN 1990) have failed to lead to significant changes in either alcohol self-administration or consumption. Consequently, doubts about the involvement of the dopaminergic system in the production of ethanol's reinforcing effects still exist.

Evidence that ethanol may interact with opioid systems within the brain include the findings that cross-tolerance can occur between the hypothermic effects of morphine and ethanol (KHANNA et al. 1979). Futhermore, alcoholic patients exhibit tolerance to the anethestic effects of alfantenil (LEMMENS et al. 1989). Studies of the effects of prolonged exposure of different neuroblastoma cell lines to ethanol indicate that alcohol treatment can lead to the upregulation of δ-opioid receptors and to an increase in the sensitivity of cells to opioid inhibition of adenyl cyclase activity (CHARNESS 1989).

Opioid systems exist in close association with dopaminergic systems within the mesolimbic system. The reinforcing actions of opiates appear to be dependent upon dopaminergic systems for their expression (NESTLER 1992). Animals studies indicate that the administration of the opioid antagonist naloxone leads to a reduction in the consumption of ethanol (MARFAING-JELLET et al. 1983; SAMSON and DOYLE 1985). Alcohol-dependent subjects treated with the opioid antagonist naltrexone were found to have a lower rate of relapse than were those who were treated with placebo (VOLPICELLI et al. 1992; O'MALLEY et al. 1992).

D. Individual Drugs

I. Acetaminophen

Chronic use of ethanol enhances acetaminophen hepatotoxicity in humans (SEEFF et al. 1986) and in rodents (SATO et al. 1981a). In rodents, the mechanism is increased accumulation of toxic metabolites of acetaminophen (SATO et al. 1981a; WALKER et al. 1983). Acetaminophen is metabolized in the liver via two pathways. Usually, over 90% is conjugated with glucuronic acid or sulfate and less than 5% is oxidized by hepatic cytochrome P450 to a reactive electrophilic intermediate (MORGAN et al. 1983; SATO et al. 1981a). This toxic intermediate is then conjugated with glutathione, excreted into bile, further metabolized by gut or kidneys to mercapturate derivatives, and finally excreted into urine. Chronic ethanol exposure decreases glutathione stores, resulting in decreased elimination of the intermediate (SATO et al. 1981a) and may enhance the cytochrome P450 route, resulting in increased production of the intermediate.

Acute administration of ethanol reduces hepatotoxicity in fasted Sprague-Dawley rats (SATO et al. 1981b). In this animal model, the mechanism appears to be through re uced production of reactive intermediates (SATO and LIEBER 1981), presumably by competition for microsomal cytochrome P450.

II. Angiotensin-Converting Enzyme Inhibitors

1. Enalapril

Angiotensin-converting enzyme inhibitors may decrease alcohol consumption in rodents (LINGHAM et al. 1990; SPINOZA et al. 1988) but 10 or 20 mg enalapril/day did not have an overall effect on alcohol consumption in normotensive alcoholics with normal renin levels (NARANJO et al. 1991).

III. Anticonvulsants

1. Phenytoin

Chronic ethanol ingestion increases clearance of phenytoin (SANDOR et al. 1981). Phenytoin clearance was higher during alcohol withdrawal than during a period of concurrent ethanol ingestion, reflecting the enzyme inhibition that occurs during acute ethanol administration and the enzyme induction that predominates in recently abstinent alcoholics without liver disease (SANDOR et al. 1981).

IV. Antidepressants

1. Amitriptyline

Moderate drinking (more than the equivalent of 12 fl. oz. beer/day) did not enhance amitriptyline metabolism in one study (LINNOILA et al. 1981). However, in another study, the rate of demethylation of amitriptyline in depressed subjects was greater in 10 alcoholics than in 11 controls (SANDOZ et al. 1983). Therefore, in alcoholics but not in moderate drinkers, enhanced clearance would be expected.

Acute ethanol ingestion may result in decrements in tests of motor functioning which are at least additive to those of amitriptyline. In healthy subjects given one, two, or no doses of amitriptyline 0.8 mg/kg followed by enough ethanol to bring their BAC up to 0.8 mg/l, performance on three motor skill tests thought relevent to driving were significantly impaired (LANDAUER et al. 1969). In another placebo-controlled study, volunteers were given amitriptyline 40 mg/day for 1 week followed by 60 mg/day for a 2nd week. When these and placebo-treated subjects were then given ethanol 0.5 g/kg, those treated with amitriptyline had increased cumulative choice reaction times and impaired coordination as compared with controls (SEPPALA et al. 1975). While ethanol (enough to maintain a BAC of 17–22 m*M*) had no effect on the pharmacokinetics of amitripyline 50 mg/day taken for 14 days, ethanol increased free amitriptyline concentrations brought about by a decrease in first pass extraction resulting from decreased hepatic clearance (DORIAN et al. 1983). Although it appears that changes in distribution may therefore play a role, the interaction is also likely to be on a pharmacodynamic basis. There is one case report of amitriptyline used by an alcoholic to prolong and enhance the euphoria from ethanol and also decrease the amount of ethanol needed to "get high" (HYATT and BIRD 1987).

2. Imipramine

Chronic ethanol administration increases clearnance of imipramine and its primary hydroxylated metabolite (CIRAULO et al. 1988). The pharmacokinetics of imipramine (and desipramine) were studied in 15 men recently detoxified from alcohol compared with 14 healthy volunteers. The alcoholic group had significantly greater total body clearance of imipramine (0.93 vs. $0.48\,l\,h^{-1}\,kg^{-1}$) and greater intrinsic clearance of unbound imiramine (19.8 vs. $6.56\,l\,h^{-1}\,kg^{-1}$). Mean elimination half-life was decreased in the alcoholic group (8.7 h after i.v. infusion and 10.9 h after oral administration) as compared with 19.9 and 19.6 h, respectively, in the control group (CIRAULO et al. 1988). In another study, apparent oral imipramine clearance was 2½ times higher in alcoholics than in controls (CIRAULO et al. 1982). The primary hydroxylated metabolite of imipramine is 2-hydroxyimipramine. Terminal elimination half-life of this substance was also increased (7.07

vs. 10.12 h) in seven recently abstinent alcoholics as compared with eight weight- and age-matched controls (CIRAULO et al. 1990).

3. Clomipramine

Acute ethanol administration results in at least additive clomipramine-induced decrements in tests of motor functioning. In double-blind, crossover trials, volunteers were given 30 mg/day for 1 week followed by either 75 mg/day or placebo for 1 week. An ethanol test dose of 0.5 g/kg resulted in only "slight" impairment of psychomotor skills in the clomipramine group vs. the placebo group (SEPPALA et al. 1975).

4. Desipramine

Chronic ethanol consumption increases elimination of desipramine, but less so than with imipramine. The pharmacokinetics of desipramine (and imipramine) were studied in 15 recently detoxified men compared with 14 healthy volunteers (CIRAULO et al. 1988). The alcoholic group showed greater total body clearance (1 vs. 0.62 $l\,h^{-1}\,kg^{-1}$) and greater intrinsic clearance of unbound desipramine (14.52 vs. 9.05 $l\,h^{-1}\,kg^{-1}$) as compared with the control group. Mean elimination half-life was decreased in alcoholics as compared to controls (16.5 vs. 22.4 h for the i.v. route and 15.12 vs. 21.5 h for oral administration). However, there was much greater interindividual variation with the oral route.

5. Doxepin

In tests of motor functioning, acute ethanol results in decrements that are at least additive to those produced by doxepin (MILNER and LANDAUER 1973; STROMBERG et al. 1988). In volunteers given doxepin 40 mg/day for 1 week followed by doxepin 60 mg/day for 1 week, compared with placebo, an ethanol test dose of 0.5 g/kg increased cumulative choice reaction times, inaccuracy of reactions, and impaired coordination. In another study, 12 healthy volunteers were given doxepin 25 mg alone and with alcohol (1 g/kg) in a double-blind, crossover trial. Doxepin increased alcohol-induced decrements in several objective tests of psychomotor performance including tracking, choice reaction, and body sway (STROMBERG et al. 1988).

6. Nortriptyline

Chronic ethanol does not appear to alter nortriptyline metabolism. Moderate drinking (more than the equivalent of 12 fl. oz. beer/day) did not enhance metabolism (LINNOILA et al. 1981).

Acute ethanol resulted in decrements in tests of motor function which are at least additive to those produced by nortriptyline. In volunteers given nortriptyline 40 mg/day for 1 week followed by nortriptyline 60 mg/day for 1 week, an ethanol test dose of 0.5 g/kg resulted in only "slight" impairment of psychomotor skills (SEPPALA et al. 1975).

V. Serotonin Selective Reuptake Inhibitors

Acute ethanol does not appear to interact pharmacokinetically with fluoxetine (SHAW et al. 1989) or fluvoxamine (VAN HARTEN et al. 1992). Fluoxetine has been shown to inhibit cytochrome P450 2D6 activity (OTTON et al. 1993) and ethanol has been shown specifically to induce cytochrome P450 IIE1 (NEBERT et al. 1987). The lack of interaction of ethanol with fluoxetine may be explained by these differences in P450 isozymes. As mentioned above, some serotonin selective reuptake inhibitors may decrease relapse in abstinent alcoholics, but this is an early finding.

Acute ingestion of alcohol interacts with tricyclic antidepressants to produce decrements in psychomotor function which are at least additive, but this does not seem to be the case with the serotonin selective reuptake inhibitors fluoxetine and fluvoxamine. Chronic alcohol intake enhances biotransformation of tricyclic antidepressants with effects on tertiary amines being much more marked than with secondary amines. With the secondary amines, and possibly with the tertiary amines, the increased biotransformation can result in clinically significant decreases in serum drug levels in alcoholics. These kinetic effects do not appear to obtain with the serotonin selective reuptake inhibitors fluoxetine or fluvoxamine.

VI. Antipyrine

Chronic ethanol administration may enhance antipyrine elimination. In six human subjects (prisoners in solitary confinement) who were given 1 ml/kg of 95% ethanol for 21 days, reductions from 4.2% to 37.7% were seen in antipyrine half-life as compared with half-life prior to ethanol administration (VESELL et al. 1971).

VII. Ascorbic Acid

Ascorbic acid causes enhanced ethanol clearance. In one human study, 1 g ascorbic acid daily for 2 weeks increased ethanol clearance. The same authors found that leukocyte ascorbic acid was correlated with activities of hepatic ADH in 12 human subjects with clinical or biochemical evidence of non-alcoholic liver disease, suggesting that ascorbic acid may increase ethanol clearnace by enhancing the activity of hepatic ADH (KRASNER et al. 1974).

VIII. Barbiturates

1. Phenobarbital

The most significant acute interaction between ethanol and phenobarbital is at least additive CNS depression and toxicity (GUPTA and KOFOLD 1966; BROUGHTON et al. 1956). This is a clinically important and long-recognized interaction. Phenobarbital also increases ethanol clearance. In four male chronic alcoholic patients given phenobarbital 240 mg/day for 6 days, ethanol clearance went from 15.7 ± 3.6 mg/100 ml serum per hour prior to the 6 days of phenobarbital administration to 20.1 ± 2.4 mg/100 ml serum per hour following it (MEZEY and ROBLES 1974).

2. Pentobarbital

Chronic ethanol adminstration decreases pentobarbital half-life. For 1 month, four volunteers with no history of alcoholism received ethanol as an isocaloric substitute for carbohydrates (up to 46% of total calories). Pentobarbital half-life decreased from 35.1 ± 6.1 to 26.3 ± 6.3 h at least in part due to induction of hepatic microsomal drug metabolism (MISRA et al. 1971).

For barbiturates in general, there is a highly significant, synergistic, acute interaction seen with increased CNS depression and enhanced toxicity when administered with ethanol. Barbiturates promote cytochrome P450 systems, resulting in increased ethanol clearance. Chronic ethanol administration can increase the clearance of barbiturates. However, these effects are less important clinically than are the CNS effects.

IX. Benzodiazepines

1. Chlordiazepoxide

Acute ethanol admistration impairs chlordiazepoxide elimination and also lowers plasma binding. In five healthy male volunteers, ethanol 0.8 g/kg P.O. was given 1 h before 0.6 mg/kg i.v. chlordiazepoxide. Plasma concentrations of ethanol were maintained from 50–150 mg/100 ml for 32 h by administration of ethanol 0.5 g/kg every 5 h. As compared with measurements following administration of chloridiazepoxide alone in these subjects, plasma clearance of chlordiazepoxide fell from 26.6 ± 2.6 ml/min to 16.6 ± 3.1 ml/min, with no change in the volume of distribution. Plasma binding of chlordiazepoxide was lowered by ethanol from 94.7 ± 6% to 93.4 ± 1.3% and the plasma clearance of unbound chlordiazepoxide feel from 468 ± 51 ml/min to 264 ± 98 ml/min (DESMOND et al. 1980).

2. Diazepam

The pharmacokinetics of diazepam in chronic alcoholics are not well understood. In recently detoxified alcoholics given oral and intravenous diazepam, plasma diazepam levels and AUC were decreased as compared with controls (SELLMAN et al. 1975a,b). Interpretation of the data is compromised, however, because the alcoholics were all also receiving other medications, and there was wide scatter in the results. In another study, the pharmacokinetics of intravenous diazepam (10 mg) were examined in seven chronic alcoholics on the 1st day of detoxification (accomplished with hydroxyzine) and on the 6th–8th days of detoxificaton (POND et al. 1979). In comparing values at 6–8 days vs. the 1st day of detoxification, terminal elimination half-life, clearance, and volumes of distribution changed in individual patients, but mean values for the group did not change. The only difference noted was that the initial exponential decline of plasma diazepam was more rapid during, as compared with after, withdrawal. The authors concluded that the only kinetic effect was on initial distribution (POND et al. 1979).

Conflicting evidence exists as to the presence of a pharmacokinetic interaction with concurrent or other acute use, but pharmacodynamic effects on motor performance are at least additive. A number of studies have included measures of one or both of these parameters. Diazepam (0.7 mg/kg) was given orally with water or with 30 ml 50% ethanol to seven normal volunteers, with the ethanol group attaining a serum diazepam level of 373 ng/ml vs. 197 ng/ml for the water group. The authors suggested that enhanced absorption was the mechanism for this effect (HAYES et al. 1977). In another study, diazepam 10 mg i.v. was given over 20 min, preceded by 60 min with ethanol 0.7 mg/kg orally and followed for 8 h by ethanol $0.15\,g/kg^{-1}/h^{-1}$ to maintain blood alcohol levels between 0.8 and 1 g/l. As compared with 10 mg diazepam i.v. alone, AUC increased from 1771 ± 277.6 to 2320 ± 484 ng h/ml, AUC_{free} increased from 23.87 ± 3.09 to 30 ± 4.97, but AUC of *N*-desmethyldiazepam fell by 11.7%. The authors concluded that the increase in diazepam AUC was due to impaired hepatic metabolism. In this same study, performance on six measures of psychomotor performance were not significantly changed with diazepam alone, but were significantly affected with the combination of diazepam and ethanol (SELLERS et al. 1980). In another study, 20 healthy students ingested diazepam 10 mg 30 min after administration of different forms of ethanol 0.8 g/kg. In this study, beer and white wine elevated diazepam level for 2 h, as compared with 90 min for whiskey, and there was no effect on diazepam level from red wine. Tests of psychomotor performance were impaired more with the ethanol-diazepam combination than with diazepam alone, and, for nystagmus, red wine had the greatest effect with diazepam. The authors concluded that while simultaneous ingestion of alcohol and diazepam accelerates absorption, combined psychomotor effects were mainly due to pharmacodynamic interactions (LAISI et al. 1979). However, in two other studies, ethanol slowed the rate of diazepam absorption. Six healthy volunteers ingested diazepam 5 mg with 1.5 fl. oz. 80% proof vodka combined with 4 fl. oz. orange juice. The time of peak serum diazepam concentration was lengthened from 0.79 h without ethanol to 1.79 h with ethanol, but the completeness of absorption was not significantly affected (GREENBLATT et al. 1978). In another study by the same group, 5 mg diazepam was given orally to six healthy volunteers in a randomized three-way crossover study, where diazepam was either preceded by water, preceded by 30 min with 60 ml vodka, or given concurrently with 60 ml vodka. AUC, diazepam elimination half-life, and rate and extent of desmethyldiazepam formation were not significantly different between the three test groups but the time to peak concentration increased from 0.71 to 1.5 h (DIVOLL and GREENBLATT 1981).

3. Flurazepam

Acute ethanol interacted synergistically with flurazepam to produce decrements of psychomotor function in a rodent model. ICR albino mice showed synergism in loss of righting reflex, light sedation, and rotorod performance, but not for LD_{50} or anesthesia. Dose ranges were 0–5 g/kg ethanol and 0–325 mg/kg flurazepam (HU et al. 1986).

4. Lorazepam

Acute ethanol adminstration decreases clearance of lorazepam in human and animal models (HOYUMPA et al. 1981), causes at least additive decrements in psychomotor performance, and increases anxiety in humans (LISTER and FILE 1983). Ethanol administered i.v. to five healthy dogs (3 g/kg) resulted in a 52% decrease in first pass metabolism of lorazepam 20 mg administered orally 1 h later (HOYUMPA et al. 1981). In a human study by the same authors, seven healthy subjects were given 0.8 g/kg of P.O. ethanol followed by lorazepam 2 mg i.v. 1 h later. Ethanol was repeated 0.5 mg/kg every 5 h for hour more doses. There was an 18% decrease in systemic clearnce of the i.v. lorazepam. The mean peak plasmal level of conjugated lorazepam was lower and time to attain peak concentration was increased following ethanol ingestion, but neither measure reached statistical significance (HOYMPA et al. 1981). In another study, 17 human subjects received 1 mg lorazepam and 23.7 or 71 ml vodka in a double-blind, placebo-controlled design. Self-ratings of sedation were significantly increased by both lorazepam and ethanol alone, and this effect increased at least additively when the two drugs were combined. Interestingly, while both drugs alone were anxiolytic, when combined they increased self-reported anxiety (LISTER and FILE 1983).

5. Triazolam

There is contradictory evidence on the influence of acute alcohol adminstration on the kinetics of triazolam. There is, however, agreement that combined administration of these drugs has at least additive pharmacodynamic effects. In one placebo-controlled study (DORIAN et al. 1985), six normal subjects were given ethanol sufficient to maintain breath concentrations of 800–950 mg/l, followed 1 h later by an oral dose of 0.25 mg triazolam. In that study, total AUC of triazolam increased by 21% ± 18%, as compared with subjects who were not given ethanol. Subjects given the combination also showed greater impairment on measures of free recall, postural stability, and hand-eye coordination, as compared with the effects of each drug alone. However, in another study where five healthy volunteers received 0.5 mg triazolam alone and with 60 ml "commercial vodka," there were no significant differences in peak triazolam level, time of peak level, elimination half-life, or apparent oral clearance (Ochs et al. 1984).

Benzodiazepines and ethanol interact at least additively in producing decrements in psychomotor function, increasing CNS depression and increasing the risk of toxicity. This is quite significant clinically, since when administered alone benzodiazepines have a very high therapeutic index. The interaction is largely a pharmacodynamic one involving the GABA benzodiazepine-chloride channel complex. Especially with triazolam-ethanol combinations, anterograde memory loss may be profound, resulting in what has been termed "traveler's amnesia." Pharmacokinetic interactions are less significant, with the effects varying between drugs and different studies showing conflicting findings.

X. Bromocriptine

In a small number of case reports, alcohol worsened abdominal distress caused by bromocriptine (Wass et al. 1977; Ayres and Maisey 1980).

XI. Ca^{2+} Blockers

1. Nifedipine

Acute ethanol administration increased the elimination half-life and AUC of i.v. nifedipine in a rodent model (Boje et al. 1984). In humans, pretreatment with 10–20 mg nifedipine administered 30 min before ethanol 0.85 g/kg failed to antagonize the inebriating effects of ethanol, including decrements in short-term memory and psychomotor performance (Perez-Reyes et al. 1992).

2. Nimodipine

In a rodent model, nimodipine increased motor incoordination and hypothermia caused by acute ethanol, possibly by synergistic vasodilator effects (Issacson et al. 1985).

3. Verapamil

In humans, pretreatment with single doses of verapamil 80–160 mg, administered 90 min before ethanol 0.85 g/kg, failed to antagonize the inebriating effects of ethanol or ethanol-induced impairment of short-term memory and psychomotor performance (PEREZ-REYES and JEFFCOAT 1992). In another study, ten healthy men were given verapamil 240 mg/day or placebo for 6 days in a double-blind, randomized crossover design. On the 6th day, ethanol 0.8 g/kg was administered orally. Compared with placebo, peak ethanol concentrations and AUC were significantly increased, as was the subjective sense of intoxication as measured by a visual analog scale (BAUER et al. 1992).

Calcium channels paly a role in the development of pharmacodynamic ethanol tolerance (DOLIN et al. 1987; BRENNAN and LITTLETON 1991) and further study may reveal clinical uses, possibly blocking the development of tolerance. Verapamil in humans and nifedipine in rodents can decrease ethanol clearance but whether this results in clinically significant elevations of serum ethanol levels has yet to be determined.

XII. Cannabis

Ethanol administered acutely in combination with cannabis results in decrements in tests of motor function that are at least additive. In a test of the effects of 0.64 g/kg ethanol and orally administered 143 μg/kg δ^9 tetrahydrocannabinol (THC), decrements of psychomotor skills related to driving were considered to be "at least additive" (FRANKS et al. 1975). In another study where volunteers received either 0, 2.5, or 5 mg of Δ^9 THC alone or in combination with 0.4 ml/lb ethanol, decrements in performance were seen as additive, but not dose dependent (MANNO et al. 1970). In a study where driving a motor vehicle through a prepared course was observed, ethanol and cannabis (smoked) decreased performance in an additive way on most measures, but seemed not to interact at all on others (HANSTEEN et al. 1976).

XIII. Cocaine

In one study of acute ethanol administration, cocaine insufflation did not alter blood ethanol levels or ratings of intoxication, but did cause a significant increase in plasma cocaine concentration, ratings of cocaine "high," and heart rate. Eleven male volunteers were given ethanol 0.85 g/kg or placebo over 30 min. Cocaine HCl (1.25 and 1.9 mg/kg) was given 15 min later by nasal insufflation. Peak cocaine concentrations in subjects given ethanol, as compared with placebo, were 156.0 ± 17.4 ng/ml vs. 114.8 ± 14.0 for the low-dose group and 312.0 ± 31.8 vs. 238.1 ± 29.2 for the high-dose group. AUC was also increased, with 13281 ± 1510 ng/ml min vs. 9480 ± 950 in the low-dose group and 26265 ± 2830 vs. 20530 ± 2457 in the high-dose group (HIGGINS et al. 1991).

In nine healthy male volunteers given ethanol and cocaine alone, heart rate was increased by up to 6 beats/min, with no effect on blood pressure.

Ethanol 38.7 or 51.8 g followed by cocaine 96 mg increased heart rate by 20 beats/min, which would indicate a synergistic effect. In nine healthy male volunteers given ethanol and cocaine alone, heart rate was increased by up to 6 beats/min, with no effect on blood pressure. Ethanol 38.7 or 51.8 g, followed by cocaine 96 mg, increased heart rate by 20 beats/min, which would indicate a synergistic effect (FOLTIN and FISCHMAN 1988). In another study in nine healthy adults, cocaine alone and in combination with alcohol significantly increased blood pressure and heart rate, with a greater increase in heart rate from the combination, compared with either drug alone (FARRE et al. 1991).

Alcohol's enhancement of the euphoric effect of cocaine has been attributed to cocaethylene, a metabolite that is formed by transesterification of cocaine in the presence of ethanol, and which has been found in the urine and plasma of patients who have used ethanol and cocaine concurrently (SEEF et al. 1986; SMITH 1984). This transesterification in the presence of alcohol is thought to be the result (in mice) of hepatic and possibly renal carboxyesterases present in the endoplasmic reticulum (BOYER and PETERSON 1992). Cocaethylene is equipotent with cocaine in inhibiting binding to the dopamine transporter site in neural membranes from humans (HEARN et al. 1991) and rats (JATLOW et al. 1991b). It is also equipotent with cocaine in causing increased locomotor activity and rearing in mice (JATLOW et al. 1991b) and in maintaining self-administration in primates (JATLOW et al. 1991a). However, in a study involving 11 male volunteers who were given ethanol 0.85 g/kg or placebo over 30 min followed by cocaine HCl (1.25 and 1.9 mg/kg) given 15 min later by nasal insufflation, cocaethylene levels rose after peak subjective (euphoric) effects (PEREZ-REYES and JEFFCOAT 1992). The authors of this study concluded that, although cocaethylene was a potent behavioral reinforcer, the potentiation of cocaine effect by ethanol was due to ethanol increasing the bioavailability of cocaine. Other data suggest that ethanol may influence brain dopamine systems and interact with cocaine on that level (see above).

Chronic ethanol does not alter metabolism of cocaine in a rodent model, but increases cocaine brain-to-plasma ratio. In rats given chronic ethanol treatment (2.5 g/kg twice orally for 14 days), there was no change in cocaine metabolism, but a significantly higher cocaine brain-to-plasma ratio was seen, when compared with controls (VADLAMANI et al. 1984)

Ethanol enhances cocaine-induced hepatotoxicity in animal models and probably in man, but human data are less certain. In mice, cocaine alone is hepatotoxic, with fatty degeneration, parenchymal necrosis, and elevated transaminase levels (SHUSTER et al. 1977; THOMPSON et al. 1979; KLOSS et al. 1984). Ethanol enhances cocaine-induced hepatotoxicity in animal models (BOYER and PETERSON 1990; PETERSON et al. 1983), presumably through induction of the cytochrome P450 monooxygenases, which also produce hepatotoxic metabolites of cocaine (THOMPSON et al. 1979). However, clinical evidence in humans is scant: several case reports suggest cocaine-induced hepatotoxicity is exacerbated by ethanol and barbiturates (WANLESS et al. 1990; GUBBINS et al. 1990). One study using cultured human hepatocytes showed that cocaine potentiated ethanol-induced impairment of hepatocyte functioning (JOVER et al. 1991). After 48 h of pretreatment with 50 m*M* ethanol, a concentration of cocaine (0.6 m*M*) which previously had no effects on hepatocyte metabolism, then caused a 20% inhibition of the urea synthesis rate, a 40% degradation of glycogen stores, and a 30% reduction in glutathione content. The authors concluded that ethanol enhanced the toxic effects of cocaine on human hepatocytes by a factor of 10. However, in a clinical population of nonparenteral cocaine users, there was no significant difference in liver function tests between those who also used alcohol and those who did not (TABASCO-MINGUILLAN et al. 1990).

XIV. Sodium Warfarin

Chronic administration of ethanol may decrease the anticoagulant effect of sodium warfarin (Coumadin) by induction of hepatic microsomal enzymes (O'REILLY 1986, 1987). Acute ethanol administration increased the anticoagulant effect of sodium warfarin, which presumably resulted from decreased sodium warfarin clearance secondary to competition for cytochrome P450 (O'REILLY 1986). However, other studies with wine (O'REILLY 1979) and fortified wine (O'REILLY 1981) showed no effect. All of these studies looked at effects on coagulation, rather than sodium warfarin levels.

XV. Cromoglycate

Acute ethanol administration may interact with cromoglycate, but not to a clinically significant degree. In 17 human subjects, disodium cromogylcate 40 mg was given alone and in combination with ethanol 0.75 g/kg in a double-blind, crossover study. On several measures of psychomotor performance, there was slight synergism, which was thought by the authors to be clinically insignificant (CRAWFORD et al. 1976).

XVI. Disulfiram

The combination of disulfiram and ethanol produces the well-known "disulfiram ethanol reaction." When subjects ingest ethanol while taking disulfiram, blood acetaldehyde concentration is increased to five to ten times higher than in subjects not treated with disulfiram, and a number of symptoms follow (RITCHIE 1980). These symptoms include the sensation of heat in the face, flushing of the face, intense throbbing, and, in some cases, a throbbing headache. The initial symptoms are followed by diaphoresis, blurred vision, nausea, vomiting, respiratory difficulties, and hypotension. The mechanism of these effects is irreversible inhibition of hepatic aldehyde dehydrogenase by disulfiram, resulting in acetaldehyde accumulation (Kitson 1977).

XVII. H_2 Blockers

1. Cimetidine

Cimetidine increases ethanol absorption (Feely and Wood 1982; DiPadova et al. 1992; Seitz et al. 1983) by inhibiting gastric alcohol dehydrogenase (Caballaria et al. 1989a; Hernandez-Munoz et al. 1990). The metabolism of ethanol by gastric ADH has been called gastric first pass metabolism by Lieber (1991). Cimetidine 800 mg/day administered for 7 days significantly increased AUC after orally administered ethanol at 0.15 g/kg but not after the same amount of ethanol administered intravenously. However, subjects given higher dose ethanol (0.8-g/kg) had no changes in blood alcohol levels (Jonsson et al. 1992), suggesting that large doses of ethanol may overwhelm the capacity of the gastric ADH system. This effect is apparently dose related, as a higher cimetidine dose (1200 mg/day) increased peak plasma ethanol levels resulting from the same higher dose of ethanol (Feely and Wood 1982). Cimetidine 1200 mg/day or placebo was administered for 7 days to six healthy male subjects followed by a single 0.8-g/kg oral dose of ethanol. As compared to placebo, cimetidine increased peak plasma ethanol level (146 ± 5.2 to 163 ± 7.6 mg/dl), AUC (717 ± 17 to 771 ± 44 mg/dl h), and subjective feeling of intoxication on a visual analog scale (Feely and Wood 1982).

2. Famotidine

Famotidine has no significant effect on ethanol bioavailability (Hernandez-Munoz et al. 1990). Famotidine does not inhibit gastric alcohol dehydrogenase (Hernandez-Munoz et al. 1990).

3. Ranitidine

Ranitidine increases ethanol bioavailability (DiPadova et al. 1992; Guram and Holt 1991), which has been attributed to decreased gastric first pass metabolism of ethanol secondary to ranitidine-induced inhibition of gastric alcohol dehydrogenase (DiPadova et al. 1992). As with cimetidine, when subjects were given ranitidine administered with higher dose ethanol no changes in blood alcohol levels were observed (Jonsson et al. 1992).

Some H_2 antagonists with a five-member ring structure such as cimetidine, ranitidine, and nizatidine inhibit gastric ADH and can result in greater peripheral blood levels of alcohol (Guram and Holt 1991). This effect occurs at levels of ethanol that would be seen in low-to-moderate levels of drinkng, but may not occur with higher levels of ethanol consumption. Other H_2 antagonists such as famotidine and omeprazole do not have any effect on gastric ADH and do not affect absorption (ROINE et al. 1990a; HERNANDEZ-MUNOZ et al. 1990).

XVIII. Lithium Carbonate

Lithium carbonate may antagonize decrements in cognitive and psychomotor function. The combined effects of lithium and alcohol on a series of psychomotor tests in normal subjects showed antagonistic and opposite effects on response orientation and information retrieval as compared to both drugs given alone (LINNOILA et al. 1974a). In another study, in which 23 normal male subjects maintained on lithium for 2 weeks (mean level of 0.91 mEq/l) were given 1.32 ml/kg ethanol, lithium attenuated alcohol-induced cognitive impairment (JUDD et al. 1977). In a third study with 35 detoxified alcoholics, lithium also antagonized ethanol-induced decrements in cognitive and perceptual-motor performance (JUDD and HUEY 1984).

In addition to effects on cognitive and psychomotor performance, one study of 35 detoxified alcoholics found that lithium treatment (14 days with a mean level of 0.89 mEq/l) resulted in decreased feelings of intoxication and a decrease in desire to drink after ethanol administration (JUDD and HUEY 1984). However, another study with 23 nonalcoholic subjects showed no effect on ethanol-induced "high" (JUDD et al. 1977). In a number of studies attempting to show whether lithium can decrease drinking in alcoholics, patients drank while maintained on lithium with no adverse consequences noted (POND et al. 1981; FAWCETT et al. 1987; KLINE et al. 1974). Lithium has been examined as an aid to relapse prevention, but outcomes are mixed.

XIX. Methylxanthines

1. Caffeine

Acute ethanol administation impairs hepatic metabolism of caffeine and decreases clearance (MITCHELL et al. 1983). Administration of 0.8 g/kg ethanol resulted in a decrease in plasma clearance of orally administered caffeine from 96.6 ± 13.4 ml/min to 60.7 ± 10.5 ml/min in healthy volunteers (MITCHELL et al. 1983). In dogs given 200 mg caffeine i.v. after 3.0 g/kg oral ethanol, clearance was also decreased significantly (MITCHELL et al. 1983).

2. Theophylline

Acute ethanol administration does not alter theophylline levels (KOYSOOKO et al. 1975). Theophylline administered orally along with ethanol resulted in serum theopylline levels that were not significantly different than theophylline administered in aqueous solution (KOYSOOKO et al. 1975). There was no difference in serum theophylline levels when the drug was administered orally together with ethanol compared with an aqueous solution (KOYSOOKO et al. 1975).

XX. Metoclopramide

Metaclopramide increases the speed of ethanol absorption from the oral route and is at least additive in producing sedation (BATEMAN et al. 1978). In seven normal male volunteers given metoclopramide 10 mg i.v. followed by ethanol, there was a significant decrease in time to peak ethanol concenptration, but no significant change in peak plasma concentrations. A significant level of sedation occurred with the combination as compared with metoclopramide alone at low ethanol concentrations [12.69 ± 3.37 mg% in one trial and 11.3 ± 3.56 mg% in another] (BATEMAN et al. 1978).

XXI. Neuroleptics

1. Chlorpromazine

Acute ethanol adminstration decreases clearance of chlorpromazine in animals (KOFF and FITTS 1972) and humans (FORREST et al. 1972). Twelve human subjects who were considered to be on "long-term" therapy with chlorpromazine (600–1200 mg/day) were administered 50 or 75 ml ethanol orally. Chlorpromazine urinary excretion was measured. Up to a 33% decrease in urinary excretion was noted in 54% of subjects (FORREST et al. 1972). In this same study, alcohol concentrations in blood and urine after adminstration of 50–75 ml ethanol orally were no different in the chlorpromazine-treated patients than in controls. In an animal study in which rats were given 30 mg/kg chlorpromazine 1 h before gavage with 4.8 g/kg ethanol, blood ethanol levels were significantly elevated in pretreated animals as compared with controls (KOFF and FITTS 1972).

2. Remoxipride

Acute ethanol adminstration results in at least additive decrements in tests of motor function (MATTILA et al. 1988). In 12 healthy volunteers given single oral doses of remoxipride 100 mg alone and in combination with 0.8 g/kg ethanol, psychomotor testing showed deficits that, in the view of the authors, were "at least additive" (MATTILLA et al. 1988).

3. Thioridazine

Alcohol 0.8 g/kg administered to five healthy male subjects in a placebo-controlled, double-blind, crossover study had no effect on the kinetics of a 25-mg dose of thioridazine (LINNOILA et al. 1974a).

The most clinically significant interaction of ethanol with neuroleptics is at least additive impairment in psychomotor function. Effects of ethanol on the pharmacokinetics of neuroleptics are slight, implying that the interaction is likely a pharmacodynamic one (LINNOILA et al. 1974a).

XXII. Nitrates

1. Nitroglycerin

Acute ethanol administration enhanced nitroglycerin-induced hypotension (KAPARI et al. 1984). Nitroglycerin-induced reductions in left ventricular preload and afterload were not affected by 1 g/kg ethanol immediately preceding sublingual nitroglycerin. However, afterload was significantly decreased, as compared to the sober state, when nitroglycerin was administered 1 h after ethanol (KAPARI et al. 1984).

XXIII. Opioids

Acute ethanol administration causes inhibition of opiate binding in animal models, but the evidence in humans is inconclusive. One animal study (using mouse caudate membranes) showed increased ^{3}H-dihydromorphine binding at the physiologically attainable ethanol concentration of 50 m*M*, but decreased ^{3}H-dihydromorphine binding at higher ethanol concentrations of 250–1000 m*M* (TABAKOFF and HOFFMAN 1983). In another study using the mouse neuroblastoma-rat glioma hybrid cell line, NG108-15, ethanol at 200 m*M* inhibited opiate receptor binding acutely (CHARNESS et al. 1983). The inhibitory effects on opiate binding in these two studies occurred only at ethanol levels that were far above what could be accomplished in a living animal, so they are probably not clinically significant.

CUSHMAN and colleagues (1987) reviewed 12 studies of endogenous opiate peptide involvement in ethanol intoxication or withdrawal in humans using naloxone as a probe. They concluded that, although the data are inconclusive, they do not appear to support a major modification of ethanol intoxication by opiate antagonists (CUSHMAN et al. 1978). Any purported effect of naloxone on reversal of ethanol overdose is probably due to improved circulatory status and decreased brain ischemia, rather than to direct antagonism of ethanol.

More recent data suggest that naltrexone can significantly decrease relapse in alcoholics (VOLPICELLI et al. 1992; O'MALLEY et al. 1992). In one double-blind, placebo-controlled study in which 70 male alcoholics were given naltrexone 50 mg/day as an adjunct to psychosocial treatment, 23% of subjects taking naltrexone relapsed as compared with 54.3% of subjects taking placebo (VOLPICELLI et al. 1992). In placebo-treated patients, 95% of patients who sampled alcohol relapsed, while only 50% of the naltrexone-treated subjects relapsed after alcohol exposure. The authors state that the effect of naltrexone was in the patients who sampled alcohol, suggesting that naltexone altered some aspect of the experience of drinking.

1. Methadone

Chronic ethanol adminstration increases hepatic metabolism of methadone in rodent models (BOROWSKY and LIEBER 1978). Chronic administration of ethanol resulted in decreased levels of unmetabolized methadone in brain and liver of live rats, and in vitro studies of hepatic microsomes showed increased N-demethylation of methadone (BOROWSKY and LIEBER 1978).

Acute ethanol administration decreases hepatic metabolism of methadone in rodent models (BOROWSKY and LIEBER 1978) and more variably in humans (CUSHMAN et al. 1987; KREEK 1984). Oral administration of 90 ml of a 50% solution of ethanol (approximately three standard drinks) to subjects on a stable dose of methadone produced no significant changes in plasma levels of methadone and no significant changes in ethanol metabolism (CUSHMAN et al. 1987), but preliminary data from the same group suggest that higher doses of ethanol may inhibit methadone biotransformation (KREEK 1984). In rats, acute administration of ethanol resulted in increased brain and liver concentrations of methadone, and decreased biliary output of pharmacologically active metabolites (BOROWSKY and LIEBER 1978). In vitro, ethanol inhibited N-demethylation of methadone by microsomes from livers of naive rats (BOROWSKY and LIEBER 1978). There may be clinically relevant inhibition of methadone metabolism that results from alcohol intoxication, which is not uncommon among methadone maintenance patients.

2. Propoxyphene

Acute ethanol administration may increase toxicity of propoxyphene. There are no human kinetic studies, but a review of 1000 propoxyphene-related deaths showed that alcohol was involved in 42% of cases (FINKLE et al. 1976). One animals study in rats suggests that ethanol may increase systemic availability of propoxyphene by inhibition of presytemic hepatic biotransformation (OGUMA and LEVY 1981).

XXIV. Oral Contraceptives

Conflicting data exist concerning the effects of acute ethanol on the kinetics of oral contraceptives, but limited evidence suggests that oral contraceptives may antagonize ethanol-induced decrements in psychomotor function (JONES and JONES 1984; HOBBES et al. 1975). Decreased ethanol elimination rate (105 vs. 121 $mg\,kg^{-1}\,h^{-1}$) was seen in women decribed as "light to moderate social drinkers" taking oral contraceptives, as compared with healthy controls (JONES and JONES 1984). However, in another study of women described as light to moderate drinkers, there was no difference in peak plasma ethanol concentration, mean time to peak, mean AUC, or mean rate of ethanol disappearance between groups taking and not taking estrogen-containing oral contraceptives, but the oral contraceptive group had less impairment on two psychomotor tests (HOBBES et al. 1975).

XXV. Oral Hypoglycemics

1. Chlorpropramide

Acute ethanol administration may cause moderate flushing reactions (FITZGERALD et al. 1968) thought to be secondary to inhibition of aldehyde dehydrogenase by chlorpropramide (PODGAINY and BRESSLER 1968).

2. Tolbutamide

Tolbutamide elimination half-life is decreased in alcoholics (232 ± 40 vs. 384 ± 76 min) as compared with controls (CARULLI et al. 1971). This decrease has been demonstrated in other studies (KATER et al. 1969; SHAH et al. 1972) and normalizes over time with abstinence (SHAH et al. 1972).

On the other hand, tolbutamide elimination half-life is increased by acute ethanol administration. In nondiabetics, tolbutamide elimination half-life was increased by 39.8% during i.v. ethanol infusion (CARULLI et al. 1971). Diabetics treated with tolbutamide showed an increased rate of ethanol clearance (9 ± 5.7 vs. 11.5 ± 2.1 mg/100 ml blood/h) as compared with nondiabetic controls (CARULLI et al. 1971).

XXVI. Sedative-Hypnotics

1. Chloral Hydrate

Acute ethanol administration alters chloral hydrate biotransformation, resulting in higher levels of the active metabolite trichloroethanol, and the combination produces decrements of psychomotor function that are at least additive. In several human studies, ethanol caused earlier and higher peak concentrations of trichloroethanol, the active metabolite of chloral hydrate produced by alcohol dehydrogenase (KAPLAN et al. 1967; SELLERS et al. 1972a) and decreased levels of trichloroacetic acid (SELLERS et al. 1972a). The mechanism appears to involve, through ethanol stimulation of NADH, production which increases the rate of chloral hydrate reduction to trichloroethanol by liver alcohol dehydrogenase (SELLERS et al. 1972a). First-order elimination of trichloroethanol was unaffected by ethanol (SELLERS et al. 1972a). Chloral hydrate ingestion also increases peak blood ethanol concentration (SELLERS et al. 1972a) although at least one earlier study did not find this (KAPLAN et al. 1967). Increased peak blood alcohol levels may result from trichloroethanol impairment of the metabolism of ethanol to acetaldehyde. Ethanol and chloral hydrate are at least additive, if not synergistic, in their impairment of complex motor tasks and auditory vigilance (SELLERS et al. 1972b).

2. Glutethimide

Glutethimide increases levels of acutely administered ethanol and causes at least additive decrements in tests of motor function. Glutethimide 250 mg given with ethanol (100 ml vodka = 40% ethanol w/v) to six healthy subjects increased serum ethanol levels from 11% to 30%, while ethanol in the same paradigm lowered glutethimide levels (MOULD et al. 1972). Reaction time tests were affected when the drugs were given in combination, but not when they were given alone. Tracking and finger tapping tests were impaired more with glutethimide alone than with ethanol alone or in combination (MOULD et al. 1972).

3. Meprobamate

Chronic ethanol administration increases meprobamate clearance through induction of hepatic microsomal enzymes. For 1 month, eight volunteers (four alcoholics and four nonalcoholics) received ethanol as an isocaloric substitute for carbohydrates (up to 46% of total calories). Meprobamate clearance increased from pre-ethanol feeding of 16.7 ± 2.5 mg/100 ml/h for the alcoholics and 13.7 ± 1.0 for the nonalcoholics to 18.5 ± 2 and 23.8 ± 2, respectively (MISRA et al. 1971).

Acute ethanol administration with meprobamate results in at least additive decrements in tests of motor function. In 22 normal subjects given meprobamate 1200 mg/day for 1 week, ethanol (sufficient to produce a serum concentration of 0.05%) was given alone, and in combination with a 300-mg meprobamate tablet. Ethanol and meprobamate were at least additive in their impairment of performance on eight psychological tests (ZIRKLE et al. 1960). In another series of studies, prolonged administration of meprobamate had an ameliorative effect on alcohol-induced changes in a single motor coordination task (CARPENTER et al. 1975). This is consistent with cross-tolerance between meprobamate and ethanol.

XXVII. Salicylates

Acutely administered ethanol potentiates aspirin-induced prolongation of bleeding time. In nine subjects given 50 g ethanol (the equivalent of two drinks of 86% proof whiskey) and 325 mg aspirin, bleeding time was prolonged (DEYKIN et al. 1982). Aspirin increases bioavailability of ethanol in the fed state (ROINE et al. 1990b) but not in the fasted state (LINNOILA et al. 1974c; TRUITT et al. 1987) and aspirin decreases ADH activity in human and rat gastric mucosa (ROINE et al. 1990b).

XXVIII. Tobacco

There are no direct studies on interactions between ethanol and tobacco, but many studies have looked at how drinking ethanol influences smoking.

In one study based on questionnaires sent to 103 alcoholic outpatients in Ontario, 93% of male alcoholics and 91% of female alcoholics smoked cigarettes as compared to 63% and 33%, respectively, in the general population (DREHER and FRASER 1967). In another study comparing 57 hospitalized alcoholics and 50 nonhospitalized control subjects, the alcoholics smoked an average of 48.7 cigarettes/day as compared with 12.5/day in the control group. The influence of hospitalization was not accounted for (MALETZKY and KLOTTER 1974). A study conducted in a residential laboratory also found that ethanol significantly increased cigarette consumption and that this was independent of socialization (GRIFFITHS et al. 1976). Other researchers have also noted covariation between drinking and cigarette smoking (MELLO and MENDELSON 1988a,b).

In summary, ethanol interacts with a large number of other drugs, and, given the significant prevalence of alcohol consumption in our society, these interactions are of great clinical importance. Long-term ethanol use often has effects opposite to short-term use, and both must be considered in the selection of appropriate pharmaceutical agents and in the education of patients.

References

Affrime M, Reidenberg MM (1975) The protein binding of some drugs in plasma from patients with alcoholic liver disease. Eur J Clin Pharmacol 8:267–269

Aguayo LG (1990) Ethanol potentiates the $GABA_A$-activated Cl^- current in mouse hippocampal and cortical neurons. Eur J Pharmacol 187:127–130

Allan AM, Baier LD, Zhang X (1992) Effects of lorazepam tolerance and withdrawal on $GABA_A$ receptor-operated chloride channels. J Pharmacol Exp Ther 261:395–402

Aranko K, Seppala T, Pellinen J, Mattila MJ (1985) Interaction of diazepam or lorazepam with alcohol. Eur J Clin Pharmacol 28:559–565

Ayres J, Maisey MN (1980) Alcohol increases bromocriptine's side effects (letter to the editor). NEJM 302:806

Baruh S, Sherman L, Kolodny HD, Singh AJ (1973) Fasting hypoglycemia. Med Clin North Am 57:1441–1462

Bateman DN, Kahn C, Mashiter K, Davies DS (1978) Pharmacokinetic and concentration-effect studies with intravenous metoclopramide. Br J Clin Pharmacol 6:401–407

Bauer LA, Schumock G, Horn J, Opheim K (1992) Verapamil inhibits ethanol elimination and prolongs the perception of intoxication. Clin Pharmacol Ther 52:6–10

Bodd E, Drevon CA, Kveseth N, Olse H, Morland J (1986) Ethanol inhibition of codeine and morphine metabolism in isolated rat hepatocytes. J Pharmacol Exp Ther 237:260–264

Boje KM, Dolce JA, Fung HL (1984) Oral ethanol ingestion altered ethanol pharmacokinetics in the rat: a preliminary study. Res Commun Chem Pathol Pharmacol 46:219

Borowsky SA, Lieber CS (1978) Interaction of methadone and ethanol metabolism. J Pharmacol Exp Ther 207:123–129

Bower S, Sear JW, Ro RC, Carter RF (1992) Effects of different hepatic pathologies on disposition of alfentanil in anaesthetized patients. Br J Anaesth 68:462–465

Boyer CS, Petersen DR (1990) Potentiation of cocaine-mediated hepatotoxicity by acute and chronic ethanol. Alcohol Clin Exp Res 14:28–31
Boyer CS, Peterson DR (1992) Enzymatic basis for the transesterification of cocaine in the presence of ethanol: evidence for the participation of microsomal carboxylase. J Pharmacol Exp Ther 260:939–946
Brennan CH, Littleton JM (1991) Chronic exposure to anxiolytic drugs, working by different mechanisms causes up-regulation of dihydropyridine binding sites on cultured bovine adrenal chromaffin cells. Neuropharmacology 30:199–205
Brodie MS, Shefner SA, Dunwiddie TV (1990) Ethanol increases the firing rate of dopamine neurons of the rat ventral tegmental area in vitro. Brain Res 508: 65–69
Broughton PM, Higgins G, O'Bien JRP (1956) Acute barbiturate poisoning. Lancet 270:180–184
Buck J, Harris RA (1990) Benzodiazepine agonist and inverse agonist actions on $GABA_A$ receptor-operated chloride channels. II Chronic effects of ethanol. J Pharmacol Exp Ther 253:713–719
Caballeria J, Baraona E, Rodamilans M, Lieber CS (1989a) Cimetidine and alcohol absorption. Gastroenterology 97:1067–1068
Caballeria J, Baraona E, Rodamilans M, Lieber CS (1989b) Effects of cimetidine on gastric alcohol dehydrogenase activity and blood ethanol levels. Gastroenterology 96:388–392
Carpenter J, Gibbins R, Marshman (1975) Drug interactions (the effects of alcohol and meprobamate applied singly and jointly in human subjects). J Stud Alcohol 36:54–139
Carulli N, Manenti F, Gallo M, Salvioli (1971) Alcohol-drugs interaction in man: alcohol and tolbutamide. Eur J Clin Invest 1:421–424
Charness ME (1989) Ethanol and opioid receptor signalling. Experientia 45:418–435
Charness ME, Gordon AS, Diamond I (1983) Ethanol modulation of opiate receptors in cultured neural cells. Science 222:1246–1248
Ciraulo DA, Barnhill J (1986) Pharmacokinetic mechanisms of ethanol-psychotropic drug interactions. NIDA Res Monogr 68:73–88
Ciraulo D, Alderson L, Chapron D, Jaffe J, Subbarao B, Kramer P (1982) Imipramine disposition in alcoholics. J Clin Psychopharmacol 2:2–7
Ciraulo D, Barnhill J, Jaffe J (1988) Clinical pharmacokinetics of imipramine and desipramine in alcoholics and normal volunteers Clin Pharmacol Ther 43: 509–518
Ciraulo DA, Barnhill JG, Jaffe JH, Ciraulo AM, Tarmey MF (1990) Intravenous pharmacokinetics of 2-hydroxyimipramine in alcoholics and controls. J Stud Alcohol 51:366–372
Colli A, Buccino G, Cocciolo M, Parravicini R, Scaltrini G (1988) Disposition of a flow-limited drug (lidocaine) and a metabolic capacity limited drug (theophylline) in liver cirrhosis. Clin Pharmacol Ther 44:6429
Cooke AR (1972) Ethanol and gastric function. NEJM 62(3):501–502
Crawford WA, Franks HM, Hensley VR, Hensley WJ, Starmer GA, Teo RKC (1976) The effect of disodium cromoglycate on human performance, alone and in combination with ethanol. Med J Aust 1:997–999
Cushman P, Kreek MJ, Gordis E (1987) Ethanol and methadone in man: a possible drug interaction. Drug Alcohol Depend 3:35–42
Daoust M, Compagno P, Legran E, Mocaer E (1992) Tianeptine, a specific serotonin uptake enhancer, decreases ethanol intake in rats. Alcohol Alcoholism 27:15–17
Desmond P, Patwardhan R, Schenker S, Hoyumpa A (1980) Short-term ethanol administration impairs the elimination of chlordiazepoxide (librium) in man. Eur J Clin Pharmacol 18:275–278
Deykin D, Janson P, McMahon L (1982) Ethanol potentiation of aspirin-induced prolongation of the bleeding time. NEJM 306:852–854
DiChiara G, Imperato A (1985) Ethanol preferentially stimulates dopamine release in the nucleus accumbens of freely moving rats. Eur J Pharmacol 115:131–132

DiPadova C, Roine R, Frezza M, Gentry T, Baraona E, Lieber C (1992) Effects of ranitidine on blood alcohol levels after ethanol ingestion (comparison with other h2-receptor antagonists). JAMA 267:84–86
Divoll M, Greenblatt DJ (1981) Alcohol does not enhance diazepam absorption. Pharmacology 22:263–268
Dolin S, Little H, Hudspith M, Pagonis C, Littleton J (1987) Increased dihydrophyridine-sensitive calcium channels in rat brain may underlie ethanol physical dependence. Neuropharmacology 26:275–279
Dolin S, Little HJ (1989) Changes in neuronal calcium channels involved in ethanol tolerance. J Pharmacol Exp Ther 25:985–991
Dolin SJ, Patch TL, Rabbani M, Siarey RJ, Bowhay AR, Little HJ (1990) Nitrendipine decreases benzodiazepine withdrawal seizures but not the development of benzodiazepine tolerance or withdrawal signs. Br J Pharmacol 101:691–697
Dorian P, Sellers EM, Reed KL, Warsh JJ, Hamilton C, Kaplan HL, Fan T (1983) Amitriptyline and ethanol: pharmacokinetic and pharmacodynamic interaction. Eur J Clin Pharmacol 25:325–331
Dorian P, Sellers EM, Kaplan H, Hamilton C, Greenblatt D, Abernethy D (1985) Triazolam and ethanol interaction: kinetic and dynamic consequences. Clin Pharmacol Ther 37:558–562
Dreher K, Fraser J (1967) Smoking habits of alcoholic out-patients. Int J Addict 2:259–270
Fadda F, Argiolas A, Melis MR, Serra G, Gessa GL (1980) Differential effect of acute and chronic ethanol on dopamine metabolism in frontal cortex, caudate nucleus and substantia nigra. Life Sci 27:979–986
Fadda F, Mosca E, Colombo G, Gessa GL (1988) Effect of spontaneous ingestion of ethanol on brain dopamine metabolism. Life Sci 44:281–287
Farre M, Llorente, M, Ugena B, Lamas X, Cami J (1991) Interaction of cocaine with ethanol. NIDA Res Monogr 105:570–571
Fawcett J, Clark D, Aagesen C, Pisani V, Tilkin J, Sellers D, McGuire M, Gibbons R (1987) A double-blind, placebo-controlled trail of lithium carbonate therapy for alcoholism. Arch Gen Psychiatry 44:248–256
Feely J, Wood A (1982) Effects of cimetidine on the elimination and actions of ethanol. JAMA 247:2819–2821
Finkle BS, McCloskey KL, Kiplinger GF, Bennett IF (1976) A national assessment of propoxyphene in post-mortem medicolegal investigation, 1972–1975. J Forensic Sci 21:706–742
Fitzgerald MG, Gaddie R, Malins JM, O'Sullivan DJ (1968) Alcohol sensitivity in diabetics receiving chlorpropramide. Diabetes 11:40–43
Foltin RW, Fischman MW (1988) Ethanol and cocaine interactions in humans: cardiovascular consequences. Pharmacol Biochem Behav 31:887–833
Forrest F, Forrest I, Finkle B (1972) Alcohol-chlorpromazine interaction in psychiatric patients. Agressologie 13:67–74
Frankel D, Khanna JM, Leblanc AE, Kalant H (1975) Effect of *p*-chlorophenylalanine on the acquisition of tolerance to ethanol and pentobarbital. Psychopharmacologia 44:247–252
Franks H, Starmer G, Chesher G, Jackson D, Hensley V, Hensley W (1975) The interaction of alcohol and tetrahydrocannabinol in man: effects on psychomotor skills related to driving. In: Alcohol drugs, and traffic safety. Addiction, Toronto, pp 461–466
Frey WA, Vallee BL (1980) Digitalis metabolism and human liver alcohol dehydrogenase. Proc Natl Acad Sci USA 77(2):924–927
Frye GD, Breese GR (1982) GABAergic modulation of ethanol-induced motor impairment. J Pharmacol Exp Ther 223:750–756
Gessa GL, Muntoni F, Collu M, Vargiu L, Mereu G (1985) Low doses of ethanol activate dopaminergic neurons in the ventral tegmental area. Brain Res 348: 201–203

Gilman AG, Rall TW, Nies AS, Taylor B (1990) Hypnotics and sedatives: ethanol. In: Rall TW (ed) The pharmacological basis of therapeutics, 8th edn. Pergamon, New York, pp 345–382

Girre C, Hirschhorn M, Bertaux L, Palombo S, Dellatolas F, Ngo R, Moreno M, Fournier PE (1991) Enhancement of propoxyphene bioavailability by ethanol. Eur J Clin Pharmacol 41:147–152

Gram LF, Schou J, Way WL, Heltberg J, Bodin NO (1979) d-Propoxyphene kinetics after single oral and intravenous doses in man. Clin Pharmacol Ther 26:473–482

Greenblatt DJ, Shader RI, Weinberger DR, Allen MD, MacLaughlin DS (1978) Effect of a cocktail on diazepam absorption. Psychopharmacol 57:199–203

Griffiths R, Bigelow G, Liebson I (1976) Facilitation of human tobacco self-administration by ethanol: a behavioral analysis. J Exp Anal Behav 25:279–292

Gubbins, Guillermo P, Schiffman RM, Ravindra S, Alapati, Batra SK (1990) Cocaine-induced hepatonephrotoxicity: a case report. Henry Ford Hospital Med J 38:55–56

Guram M, Holt S (1991) Are ethanol-H_2-receptor antagonist interactions 'relevant'? Gastroenterology 100:A749

Gupta RC, Kofold J (1966) Toxicological statistics for barbiturates, sedatives, and tranquilizers in Ontario: a ten year survey. Can Med Assoc J 94:863–865

Hall C, Brown D, Carter R, Kendall MJ (1976) The effect of desmethylimipramine on the absorption of alcohol and paracetamol. Postgrad Med J 52:139–142

Hansteen R, Miller R, Lonero L, Reid L, Jones B (1976) Effects of cannabis and alcohol on automobile driving and psychomotor tracking. Ann NY Acad Sci 282:240–256

Harris RA, Allan AM (1989) Genetic differences in coupling of benzodiazepine receptors to chloride channels. Brain Res 490:26–32

Harris RA, Hood WF (1980) Inhibition of synaptosomal calcium uptake by ethanol. J Pharmacol Exp Ther 213:562–568

Hawkins RD, Kalant H (1972) The metabolism of ethanol and its metabolic effects. Pharmacol Rev 24:68–157

Hayes S, Pablo G, Radomski T, Palmer R (1977) N Eng J Med 296:186–189

Hearn WL, Flynn DD, Hime GW, Rose S, Cofino JC, Mantaro-Atienza E, Wetli CV, Mash DC (1991) Cocaethylene: a unique cocaine metabolite displays high affinity for the dopamine transporter. J Neurochem 56:698–701

Hernandez-Munoz R, Caballeria J, Baraona E, Uppal R, Greenstein R, Lieber CS (1990) Human gastric alcohol dehydrogenase: its inhibition by H_2 receptor antagonists and its effects on the bioavailability of alcohol. Alcohol Clin Exp Res 14:946–950

Higgins ST, Bickel WK, Hughes JR, Lynn M, Capeless MA (1991) Behavioral and cardiovascular effects of cocaine alcohol combinations in humans. NIDA Res Monogr 105:501

Hobbes J, Boutagy J, Shenfield G (1975) Interactions between ethanol and oral contraceptive steroids. Clin Pharmacol Ther 38:371–380

Hoyumpa AM, Patwardhan R, Maples M, Desmond PV, Johnson RF, Sinclair AP, Schenker S (1981) Effect of short-term ethanol administration on lorazepam clearance. Hepatology 1:47–53

Hu WY, Reiffenstein RJ, Wong L (1986) Interaction between flurazepam and ethanol. Alcohol Drug Res 7:107–117

Hyatt M, Bird M (1987) Amitriptyline augments and prolongs ethanol-induced euphoria. J Clin Psychopharmacol 7:277–278

Issacson RL, Molina JC, Draski LJ, Johnston JE (1985) Nimodipine's interaction with other drugs. I Ethanol. Life Sci 36:2195

Jatlow P, Elsworth JD, Bradberry CW, Winger G, Taylor JR Russell R, Roth RH (1991a) Cocaethylene: a neuropharmacologically active metabolite associated with concurrent cocaine-ethanol ingestion. Life Sci 48:1787–1794

Jatlow P, Hearn WL, Elsworth JD, Roth RH, Bradberry CW, Taylor JR (1991b) Cocaethylene inhibits uptake of dopamine and can reach high plasma concentrations following combined cocaine and ethanol use. NIDA Res Monogr 105: 572–573
Johansson I, Ekstrom G, Scholte B, Puzycki D, Jornvall H, Ingelman-Sundberg M (1988) Ethanol-, fasting-, and acetone-inducible cytochromes P-450 in rat liver: regulation and characteristics of enzymes belonging to the IIB and IIE gene subfamilies. Biochemistry 27:1925–1934
Jones M, Jones B (1984) Ethanol metabolism in women taking oral contraceptives. Alcohol Clin Exp Res 8:24–28
Jonsson KA, Jones AW, Bostrom H, Andersson T (1992) Lack of effect of omeprazole, cimetidine, ranitidine on the pharmacokinetics of ethanol in fasting male volunteers. Eur J Clin Pharmacol 42:209–212
Jover R, Ponsoda X, Gomez-Lechon MJ, Herrero C, del Pino J, Castell JV (1991) Potentiation of cocaine hepatotoxicity by ethanol in human hepatocytes. Toxicol Appl Pharmacol 107:526–534
Judd L, Huey L (1984) Lithium antagonizes ethanol intoxication in alcoholics. Am J Psychiatry 141:1517–1521
Judd L, Hubbard R, Huey L, Attewell P, Janowsky D, Takahashi K (1977) Lithium carbonate and ethanol induced "highs" in normal subjects. Arch Gen Psychiatry 34:463–467
Juhl RP, Van Thiel DH, Dittert LW, Smith RB (1984) Alprazolam pharmacokinetics in alcoholic liver disease. J Clin Pharmacol 24:113–119
Kapari M, Heikkila J, Ylikahri R (1984) Does alcohol intensify the hemodynamic effects of nitroglycerin? Clin Cardiol 7:382–386
Kaplan H, Forney R, Hughes F, Jain N, Crim D (1967) Chloral hydrate and alcohol metabolism in human subjects. Forensic Sci 12:295–304
Kater RMH, Tobon F, Iber FL (1969) Increased rate of tolbutamide metabolism in alcoholic patients. JAMA 207:363–365
Khanna JM, Le AD, Kalant H, Leblanc (1979) Cross-tolerance between ethanol and morphine with respect to their hypothermic effects. Eur J Pharmacol 59:145–149
Khanna JM, Lê AD, Meyer J, LeBlank A (1980) The effect of p-chlorophenylalanine on the acquisition of tolerance to the hypnotic effects of pentobabital, barbital, and ethanol. Can J Physiol Pharmacol 58:1031–1041
Kissin B (1974) Interactions of ethyl alcohol and other drugs. In: Kissin B, Kissin H (eds) The biology of alcoholism. Plenum, New York, pp 109–116
Kitson TM (1977) The disulfiram-ethanol reaction: a review. J Stud Alcohol 38(1): 96–113
Kline N, Wren J, Cooper T, Varga E, Canal O (1974) Evaluation of lithium therapy in chronic and periodic alcoholism (abstract). Am J Med Sci 268:15–22
Kloss M, Rosen GM, Rauckman EJ (1984) Cocaine-mediated hepatotoxicity. A critical review. Biochem Pharmacol 33:169–173
Knapp DJ, Pohorecky LA (1992) Zacopride, a 5-HT_3 receptor antagonist, reduces voluntary ethanol consumption in rats. Pharmacol Biochem Behav 41:847–850
Koff R, Fitts J (1972) Chlorpromazine inhibition of ethanol metabolism without prevention of fatty liver. Biochem Med 6:77–81
Koysooko R, Ellis E, Levy G (1975) Effect of ethanol on theophylline absorption in humans. J Pharm Sci 64:299–301
Krasner N, Dow J, Moore MR, Goldberg A (1974) Ascorbic acid saturation and ethanol metabolism. Lancet 2:693–694
Kreek M (1984) Opioid interactions with alcohol. Adv Alcohol Subst Abuse 3:35–46
Laisi U, Linnoila M, Seppala T, Himberg J, Mattila M (1979) Pharmacokinetic and pharmacodynamic interactions of diazepam with different alcoholic beverages. Eur J Clin Pharmacol 16:263–270
Landauer AA, Milner G, Patman J (1969) Alcohol and amitriptyline effects on skills related to driving behavior. Science 163:1467–1468

Lau CE, Tang M, Falk JL (1981) Cross-tolerance to phenobarbital following chronic ethanol polydipsia. Pharmacol Biochem Behav 15:471–475
Lê AD, Khanna JM, Kalant H, Grossi F (1986) Tolerance to and cross-tolerance among ethanol, pentobarbital and chlordiazepoxide. Pharmacol Biochem Behav 24:93–98
Le Dung A, Khanna JM, Kalant H, LeBlanc AE (1980) Effect of 5,7-dihydroxytryptamine on the development of tolerance to ethanol. Psychopharmacology 67: 143–146
Lemmens HJM, Bovill JG, Hennis PJ, Gladines MPRR, Burm AGL (1989) Alcohol consumption alters the pharmacodynamics of alfentanil. Anesthesiology 71:669–674
Lieber CS (1990) Interaction of alcohol with other drugs and nutrients. Drugs 40 [Suppl 3]:23–44
Lieber CS, DeCarli LM (1972) The role of the hepatic microsomal ethanol oxidizing system (MEOS) for ethanol metabolism in vivo. J Pharmacol Exp Ther 181: 279–287
Liljequist S, Engel J (1982) Effects of GABAergic agonists and antagonists on various ethanol-induced behavioral changes. Psychopharmacology 78:71–75
Lingham T, Perlansky E, Grupp LA (1990) Angiotensin converting enzyme inhibitors reduce alcohol consumption: some possible mechanisms and important conditions for its therapeutic use. Alcohol Clin Exp Res 14:92–99
Linnoila M, Hakkinen S (1974) Effects of diazepam and codeine, alone and in combination with alcohol, on simulated driving. Clin Pharmacol Ther 15:368–373
Linnoila M, Otterstrom S, Anttila M (1974a) Serum chlordiazepoxide, diazepam and thioridazine concentrations after the simultaneous ingestion of alcohol or placebo drink. Ann Clin Res 6:4–6
Linnoila M, Saario I, Maki M (1974b) Effect of treatment with diazepam or lithium and alcohol on psychomotor skills related to driving. Eur J Clin Pharmacol 7:337–342
Linnoila M, Seppala T, Mattila MJ (1974c) Acute effect of antipyretic analgesics, alone or in combination with alcohol, on human psychomotor skills related to driving. Br J Clin Pharmacol 1:477–484
Linnoila M, Saario I, Maki M (1974d) Effect of treatment with diazepam or lithium and alcohol on psychomotor skills related to driving. Eur J Clin Pharmacol 7:337–342
Linnoila M, George L, Guthrie S, Leventhal B (1981) Effect of alcohol consumption and cigarette smoking on antidepressant levels of depressed patients. Am J Psychiatry 138:841–842
Linsempn MA (1990) Effects of dopaminergic agents on alcohol consumption by rats in a limited access paradigm. Psychopharmacol 10:195–200
Lister R, File S (1983) Performance impairment and increased anxiety resulting from the combination of alcohol and lorazepam. J Clin Psychopharmacol 3: 66–70
Littleton JM, Little HJ, Whittington MA (1990) Effects of dihydropyridine calcium channel antagonists in ethanol withdrawal; doses required, stereospecificity and actions of Bay K 8644. Psychopharmacology 100:387–392
Lyness WH, Smith FL (1992) Influence of dopaminergic and serotonergic neurons on intravenous ethanol self-administration in the rat. Pharmacol Biochem Behav 42:187–192
MacLeod M, Giles HG, Patzalek G, Thiessen JJ, Sellers EM (1977) Diazepam actions and plasma concentrations following ethanol ingestion. Eur J Clin Pharmacol 11:345–349
Maletzky B, Klotter J (1974) Smoking and alcoholism. Am J Psychiatry 131:445–447
Manno J, Kiplinger G, Scholz N, Forney R, Haine S (1970) The influence of alcohol and marihuana on motor and mental performance. Clin Pharmacol Ther 12: 202–211

Marfaing-Jallat P, Miceli D, Le Magen J (1983) Decrease in ethanol consumption by naloxone in naive and dependent rats. Pharmacol Biochem Behav 18 [Suppl]: 537–539

Martz A, Deitrich RA, Harris RA (1983) Behavioral evidence for the involvement of gamma-aminobutyric acid in the actions of ethanol. Eur J Pharmacol 89:53–62

Mathers DA (1985) Pentobarbital promotes bursts of gamma-aminobutyric acid-activated single channel currents in cultured mouse central neurons. Neurosci Lett 60:121–126

Mattila M, Mattila ME, Konno K, Saarialho-Kere U (1988) Objective and subjective effects of remoxipride, alone and in combination with ethanol or diazepam, on performance in health subjects. J Psychopharmacol 2:138–149

McCance-Katz EF, Price LH, McDougle EJ. Concurrent cocaine-ethanol ingestion in humans: pharmacology, physiology, behavior, and the role of cocaethylene. NIDA Res Monogr (in press)

Melchior CL, Tabakoff B (1981) Modification of environmentally cued tolerance to ethanol in mice. J Pharmacol Exp Ther 219(1):175–180

Mello N, Mendelson J (1988a) Concurrent alcohol and tobacco use by women. NIDA Res Monogr 81:26–32

Mello N, Mendelson J (1988b) Cigarette smoking: interactions with alcohol, opiates, and marijuana. NIDA Res Monogr 81:154–180

Mezey E, Robles E (1974) Effects of phenobarbital administration on rates of ethanol clearance and on ethanol-oxidizing enzymes in man. Gastroenterology 66:248–253

Milner G (1969) Drinking and driving in 753 general practice and psychiatric patients on psychotropic drugs. Br J Psychiatry 115:99–100

Milner G (1970) Interaction between barbiturates, alcohol and some psychotropic drugs. Med J Aust 13:1204–1207

Milner G, Landauer AA (1971) Alcohol, thioridazine and chlorpromazine effects on skills related to driving behavior. Br J Psychiatry 118:3512

Milner G, Landauer AA (1973) The effects of doxepin, alone and together with alcohol, in relation to driving safety. Med J Aust 28:837–841

Misra PS, Lefevre A, Ishii H, Rubin E, Lieber CS (1971) Increase of ethanol, meprobamate and pentobarbital metabolism after chronic ethanol administration in man and rats. Am J Med 51:346–351

Mitchell M, Hoyumpa A, Schenker S, Johnson R, Nichols S, Patwardhan R (1983) Inhibition of caffeine elimination by short-term ethanol administration. J Lab Clin Med 101:826–834

Moldeus P, Andersson B, Norling A (1978) Interaction of ethanol oxidation with glucuronidation in isolated hepatocytes. Biochem Pharmacol 27:2583–2588

Moolten M, Kornetsky C (1990) Oral self-administration of ethanol and not experimentally administered ethanol facilitates rewarding electrical brain stimulation. Alcohol 7(3):221–225

Morgan ET, Koop DR, Coon MJ (1983) Comparison of six rabbit liver cytochrome P-450 isozymes in formation of a reactive metabolite of acetaminophen. Biochem Biophys Res Commun 112:8–13

Morland JFW, Setekleiv J, Haffner JFW, Stromsaether CE, Danielsen A, Weth GH (1974) Combined effects of diazepam and ethanol on mental and psychomotor functions. Acta Pharmacol Toxicol 34:5–15

Morrow AL, Suzdak PD, Karanian JW, Paul SM (1988) Chronic ethanol administration alters γ-aminobutyric acid, pentobarbital and ethanol-mediated ^{36}Cl-uptake in cerebral cortical synaptoneurosomes. J Pharmacol Exp Ther 24:158–164

Mould G, Curry S, Binns T (1972) Interaction of glutethimide and phenobarbitone with ethanol in man. J Pharm Pharmacol 24:894–899

Naranjo CA, Kadlec KE, Sanhueza P, Woodley-Remus D, Sellers EM (1990) Fluoxetine differentially alters alcohol intake and other consummatory behaviors in problem drinkers. Clin Pharmacol Ther 47:490–498

Naranjo CA, Kadlec KE, Sanhueza P, Woodley-Remus D, Sellers EM (1991) Enalapril effects on alcohol intake and other consummatory behavior in alcoholics. Clin Pharmacol Ther 50:96–106
Nebert DW, Adesinik M, Coon MJ, Estabrook RW, Gonzalez FJ et al. (1987) The P450 gene superfamily: recommended nomenclature. DNA 6:1–11
Nestler EJ (1992) Molecular mechanisms of drug addiction. J Neurosci 12:2439–2450
Nestoros JN (1979) Ethanol specifically potentiates GABA-mediated neurotransmission in feline cerebral cortex. Science 209:708–710
Oakes SG, Pozos RS (1982) Electrophysiologic effects of acute ethanol exposure II Alterations in the calcium component of action potentials from sensory neurons in dissociated culture. Brain Res 281:251–255
Ochs H, Greenblatt D, Arendt R, Hubbel W, Shader R (1984) Pharmacokinetic noninteraction of triazolam and ethanol. J Clin Psychopharmacol 4:106–107
Oguma T, Levy G (1981) Acute effect of ethanol on hepatic first-pass elimination of propoxyphene in rats. J Pharmacol Exp Ther 219(1):7–13
O'Malley SS, Jaffe AJ, Chang G, Schottenfeld RS, Meyer RE, Rounsaville B (1992) Naltrexone and coping skills therapy for alcohol dependence: a controlled study. Arch Gen Psychiatry 49:881–887
O'Reilly RA (1979) Lack of effect of mealtime wine on the hypoprothrombinemia of oral anticoagulants. Am J Med Sci 277:189
O'Reilly RA (1981) Lack of effect of fortified wine ingested during fasting and anticoagulant therapy. Arch Intern Med 141:458
O'Reilly RA (1986) Drug-induced vitamin K deficiency, resistance and drug interactions. In: Seegers WH, Walz DA (eds) Prothrombin and other vitamin K proteins, vol II. CRC Press, Boca Raton
O'Reilly R (1987) Warfarin metabolism and drug-drug interactions. Adv Exp Med Biol 214:205–212
Olsen H, Morland J (1982) Ethanol-induced increase in procainamide acetylation in man. Br J Clin Pharmacol 13:203–208
Otton SV, Wu D, Joffe RT, Cheung SW, Sellers EM (1993) Inhibition by fluoxetine of cytochrome P450 2D6 activity. Clin Pharmacol Ther 53:401–409
Perez-Reyes M, Jeffcoate AR (1992) Ethanol/cocaine interaction: cocaine and cocaethylene plasma concentrations and their relationship to subjective and cardiovascular effects. Life Sci 51:553–563
Perez-Reyes M, White WR, Hicks RE (1992) Interaction between ethanol and calcium channel blockers in humans. Alcohol Clin Exp Res 16:769–775
Peterson FJ, Knodell RG, Lindemann NJ, Steele NM (1983) Prevention of acetaminophen and cocaine hepatotoxicity in mice by cimetidine treatment. Gastroenterology 85:122–129
Podgainy H, Bressler R (1968) Biochemical basis of the sulfonylurea-induced antabuse syndrome. Diabetes 17:679–682
Pond S, Phillips M, Benowitz N, Galinsky R, Tong T, Becker C (1979) Diazepam kinetics in acute alcohol withdrawal. Clin Pharmacol Ther 25:832–836
Pond S, Becker C, Vandervoort R, Phillips M, Bowler R, Peck C (1981) An evaluation of the effects of lithium in the treatment of chronic alcoholism. Alcohol Clin Exp Res 5:247–251
Ritchie JM (1980) The aliphatic alcohols. In: Gilman AG, Goodman LS, Gilman A (eds) The pharmacologic basis of therapeutics. MacMillan, New York, p 387
Roine R, Dipadova C, Frezza M, Hernandez-Munoz R, Baraona E, Lieber CS (1990a) Effects of omeprazole, cimetidine, and ranitidine on blood ethanol concentrations. Gastroenterology 98:114
Roine R, Gentry RT, Hernandez-Munoz R, Baraona E, Lieber CS (1990b) Aspirin increases blood alcohol concentrations in humans after ingestion of ethanol. JAMA 264:2406–2408
Rubin E, Gang H, Misra PS, Lieber CS (1970) Inhibition of drug metabolism by acute ethanol intoxication. Am J Med 49:801–806

Samson HH, Doyle TF (1985) Oral ethanol self-administration in the rat: effect of naloxone. Pharmacol Biochem Behav 22:91–99
Sandor P, Sellers E, Dumbrell M, Khouw V (1981) Effect of short- and long-term alcohol use on phenytoin kinetics in chronic alcoholics. Clin Pharmacol Ther 30:390–397
Sandor P, Naranjo CA, Khouw V, Sellers EM (1983) Variations in drug free fraction during alcohol withdrawal. Br J Clin Pharmacol 15:481–486
Sandoz M, Vandel S, Vandel B, Bonin B, Allers G, Volmat R (1983) Biotransformation of amitriptyline in alcoholic depressive patients. Eur J Clin Pharmacol 24:615–621
Sato C, Lieber CS (1981) Mechanism of the preventive effect of ethanol on acetaminophen-induced hepatotoxicity. J Pharmacol Exp Ther 218:811–815
Sato C, Matsuda Y, Lieber CS (1981a) Increased hepatotoxicity of acetaminophen after chronic ethanol consumption in the rat. Gastroenterology 80:140–148
Sato C, Nakano M, Lieber CS (1981b) Prevention of acetominophen-induced hepatotoxicity by acute ethanol administration in the rat: comparison with carbon-tetrachloride-induced hepatotoxicity. J Pharmacol Exp Ther 218:805–810
Seef LB, Cuccherini BA, Zimmerman HJ, Adler E, Benjamin SB (1986) Acetaminophen hepatotoxicity in alcoholics. Ann Intern Med 104:399–404
Seitz HK, Bosche J, Czygan P, Veith S, Simon B, Kommerell B (1983) Increased ethanol levels following cimetidine but not ranitidine. Lancet 1:760–761
Sellers EM, Lang M, Koch-Weser J, LeBlanc E, Kalant H (1972a) Interaction of chloral hydrate and ethanol in man: I. metabolism. Clin Pharmacol Ther 13:37–49
Sellers EM, Carr G, Bernstein JG, Sellers S, Koch-Weser J (1972b) Interaction of chloral hydrate and ethanol in man: II Hemodynamics and performance. Clin Pharmacol Ther 13:50–58
Sellers EM, Naranjo CA, Giles HG, Frecker RC, Beechin M (1980) Intravenous diazepam and oral ethanol interaction. Clin Pharmacol Ther 28:638–645
Sellers EM, Romach MK, Toneatto T, Sobell LC, Somer GR, Sobell MB (1991) Efficacy of ondansetron, a 5-HT_3 antagonist, in alcoholism treatment. Biol Psychiatry 29:495s
Sellers EM, Higgins GA, Sobell MB (1992) 5-HT and alcohol abuse. TIPS 13:69–75
Sellman R, Kanto J, Raijola E, Pekkarinen A (1975a) Human and animal study on elimination from plasma and metabolism of diazepam after chronic alcohol intake. Acta Pharmacol Toxicol (Copenh) 36:33–38
Sellman R, Pekkarinen A, Kangas L, Raijola E (1975b) Reduced concentrations of plasma diazepam in chronic alcoholic patients following an oral administration of diazepam. Acta Pharmacol Toxicol (Copenh) 36:25–32
Seppala T (1977) Psychomotor skills during acute and two-week treatment with mianserin (ORG GB 94) and amitriptyline, their combined effects with alcohol. Ann Clin Res 9:66–72
Seppala T, Linnoila M, Elonen E, Mattila MJ, Maki M (1975) Effect of tricyclic antidepressants and alcohol on psychomotor skills related to driving. Clin Pharmacol Ther 17:515–522
Seppala T, Linnoila M, Mattila MJ (1979) Drugs, alcohol and driving. Drugs 17: 389–408
Shah M, Clancy B, Iber F (1972) Comparison of blood clearance of ethanol and tolbutamide and the activity of hepatic ethanol-oxidizing and drug-metabolizing enzymes in chronic alcoholic subjects. Am J Clin Nutr 25:135–139
Shaw C, Sullivan J, Kadlec K, Kaplan H, Naranjo C, Sellers E (1989) Ethanol interactions with serotonin uptake selective and non-selective antidepressants: fluoxetine and amitriptyline. Human Psychopharmacol 4:113–120
Shuster L, Quimby F, Bates A, Thompson ML (1977) Liver damage from cocaine in mice. Life Sci 20:1035–1042
Sinclair F, McCaffrey J, Sinclair PR, Bement WJ, Lambrecht LK, Wood SG, Smith EL, Schenkman JB, Guzelian PS, Park SS, Gelboin HV (1991) Ethanol

increases cytochromes P450IIE, IIB1/2, IIIA in cultured rat hepatocytes. Arch Biochem Biophys 284:360–365

Smith AC, Freeman RW, Harbison RD (1981) Ethanol enhancement of cocaine induced hepatotoxicity. Biochem Pharmacol 30:453–458

Smith RM (1984) Ethyl esters of arylhydroxy and arylhydromethycocaines in the urine of simultaneous cocaine and ethanol users. J Anal Toxicol 8:38–42

Spinoza G, Perlanski E, Leenan FHH, Stewart RB, Grupp LA (1988) Angiotensin converting enzyme inhibitors: animal experiments suggest a new pharmacological treatment for alcohol abuse in humans. Alcohol Clin Exp Res 12:65–70

Stromberg C, Seppala T, Mattila MJ (1988) Acute effects of maprotiline, doxepin, and zimelidine with alcohol in healthy volunteers. Arch Int Pharmacodyn Ther 291:217–228

Study RE, Barker JL (1981) Diazepam and (−)-pentobarbital: fluctuation analysis reveals different mechanisms for potentiation of γ-aminobutyric acid responses in cultured central neurons. Proc Natl Acad Sci USA 78:7180–7184

Suzdak PD, Schwartz RD, Skolnick P, Paul SM (1986) Ethanol stimulates gamma-aminobutyric acid receptor-mediated chloride transport in rat brain synaptoneurosomes. Proc Natl Acad Sci USA 83:4071–4075

Sytinsky IA, Guzikov BM, Gomanko MV, Eremin VP, Konovalova NN (1975) The gamma-aminobutyric acid (GABA) system in brain during acute and chronic ethanol intoxication. J Neurochem 25:43–48

Tabakoff B, Hoffman P (1983) Alcohol interactions with brain opiate receptors. Life Sci 32:197–204

Tabakoff B, Ritzmann RF (1977) The effects of 6-hydroxydopamine on tolerance to and dependence on ethanol. J Pharmacol Exp Ther 203:319–331

Tabakoff B, Yanai J, Ritzmann RF (1978) Brain noradrenergic systems as a prerequisite for developing tolerance to barbiturates. Science 200:449–451

Tabasco-Minguillan J, Novick DM, Kreek MJ (1990) Liver function tests in nonparenteral cocaine users. Drug Alcohol Depend 26:169–174

Thiessen JJ, Sellers EM, Denbeigh P, Dolman L (1976) Plasma protein binding of diazepam and tolbutamide in chronic alcoholics. J Clin Pharmacol 16:345–351

Thompson ML, Shuster L, Shaw K (1979) Cocaine-induced necrosis in mice: the role of cocaine metabolism. Biochem Pharmacol 28:2389–2395

Truitt EB, Gaynor CR, Mehl DL (1987) Aspirin attenuation of alcohol induced flushing and intoxication in oriental and occidental subjects. Alcohol 22 [Suppl 1]:595–599

Twombly DA, Herman MD, Kye CH, Narahashi T (1990) Ethanol effects on two types of voltage-activated calcium channels. J Pharmacol Exp Ther 254:1029–1037

Twyman RE, Rogers CJ, MacDonald RL (1989) Pentobarbital and picrotoxin have reciprocal actions on single $GABA_A$ receptor channels. Neurosci Lett 96: 89–95

Vadlamani NL, Pontani RB, Misra AL (1984) Effect of diamorphine, 9-tetrahydrocannabinol and ethanol on intravenous cocaine disposition. J Pharm Pharmacol 36:552–554

van Harten J, Stevens L, Raghoebar M, Holland RL, Wesnes K, Cournot A (1992) Fluvoxamine does not interact with alcohol or potentiate alcohol-related impairment of cognitive function. Clin Pharmacol Ther 52:427–435

Vesell ES, Page JG, Passananti GT (1971) Genetic and environmental factors affecting ethanol metabolism in man. Clin Pharmacol Ther 12:192–201

Volpicelli JR, Alterman AI, Hayashida M, O'Brein CP (1992) Naltrexone in the treatment of alcohol dependence. Arch Gen Psychiatry 49:876–880

Wafford KA, Burnett DM, Leidenheimer NJ, Burt DR, Wang JB, Kofuji P, Dunwiddie TV, Harris RA, Sikela JM (1991) Ethanol sensitivity of the $GABA_A$ receptor expressed in Xenopus oocytes requires 8 amino acids contained in the γ2L subunit. Neuron 7:27–33

Walker RM, McElligott TF, Power EM, Massey TE, Racz WJ (1983) Increased acetominophen-induced hepatotoxicity after chronic ethanol consumption in mice. Toxicology 28:193–206
Wanless IR, Dore S, Gopinath N, Tan J, Cameron R, Heathcote EJ, Blendis LM, Levy G (1990) Histopathology of cocaine hepatotoxicity. Report of four patients. Gastroenterology 98:497–501
Wass JAH, Thorner MO, Morris DV, Rees LH, Mason AS, Jones AE, Besser GM (1977) Long-term treatment of acromegaly with bromocriptine. Br Med J 1:875–878
White JM, Smith AM (1992) Modification of the behavioural effects of ethanol by nifedipine. Alcohol Alcohol 27(2):137–141
Whittington MA, Dolin SJ, Patch TL, Siarey RJ, Butterworth AR, Little HJ (1991) Chronic dihydropyridine treatment can reverse the behavioural consequences of, and prevent adaptations to, chronic ethanol treatment. Br J Pharmacol 103:1669–1676
Wrighton SA, Campanile C, Thomas PE, Maines SL, Watkins PB, Parker G, Mendez-Picon G, Haniu M, Shively JE, Levin W (1986) Identification of a human liver cytochrome P-450 homologous to the major isosafrole-inducible cytochrome P-450 in the rat. Mol Pharmacol 29:405–410
Yoshimoto K, McBride WJ, Lumeng L, Li TK (1992) Ethanol enhances the release of dopamine and serotonin in the nucleus accumbens of HAD and LAD lines of rats. Alcohol Clin Exp Res 16:781–785
Zirkle G, McAtee O, King P, Van Dyke R (1960) Meprobamate and small amounts of alcohol (effects on human ability, coordination and judgment). JAMA 173: 121–123

CHAPTER 20
Pharmacotherapies for Alcoholism: Theoretical and Methodological Perspectives

H.R. KRANZLER, A.T. McLELLAN, and M.J. BOHN

A. Pharmacologic Approaches to Relapse Prevention in Alcoholics

"Historically, psychotropic drug discoveries have been made, for the most part, by observant and experienced clinicians making serendipitous findings in the course of trying a candidate drug molecule on patients suffering from a variety of clinical complaints and using traditional interview and observational techniques rather than studying homogeneous groups of patients with formal assessment instruments" (LASAGNA 1991, p. 263). In addition to serendipity, developments in the pharmacotherapy of alcoholism have also depended on the availability of medications for other clinical uses, rather than on a particular theoretical rationale or compelling preclinical data. However, in recent years preclinical investigation of the effects of a variety of drugs on alcohol consumption has increasingly set the stage for clinical trials in alcoholism. This evolution is apparent with the opioidergic, serotonergic, and GABAergic compounds that are the focus of recent clinical investigation in alcoholism. Future clinical developments are likely to be informed also by molecular investigation. A promising approach might involve the identification of agents that act selectively in those precise regions of the brain that control alcohol consumption and dependence formation. Molecular neurobiologic techniques are likely to be instrumental both in the identification of these regions and in the testing of agents that modify subcellular responses to alcohol.

In the context of a more rational approach to medications development, MEYER (1989, 1992) has elaborated six major areas in which initiatives in the pharmacotherapy of alcoholism are most promising. These include the amelioration of protracted abstinence; the reduction of desire to drink as part of a general effect on consummatory behavior; the remediation of alcohol-induced cognitive impairment; the blocking of alcohol's reinforcing effects; the production of aversive reactions as a consequence of alcohol consumption (without the problems associated with disulfiram therapy); and the treatment of psychopathology that is comorbid with alcoholism.

These approaches to relapse prevention have largely been covered in the preceding chapters of this volume. The remainder of this chapter will, therefore, be devoted to three other considerations. These include matching pharmacotherapies to specific patient characteristics in an effort to enhance treatment outcome; methodologic issues in the assessment of pharmacotherapies for relapse prevention in alcoholics; and speculation concerning what the future may hold for developments in the pharmacotherapy of alcoholism.

B. Treatment Matching

I. Matching Based Upon Comorbid Psychopathology

Given evidence that heavy drinking or drug use represents an effort on the part of some individuals to self-medicate underlying psychiatric symptoms, one strategy for preventing relapse in these patients is the treatment of comorbid psychiatric disorders (MEYER 1986a, 1989). Inherent in this effort is the notion of treatment matching (MEYER 1992). Thus one practical and promising approach to matching particular kinds of alcoholism treatment to particular "types" of alcoholic patients is based on the presence of specific psychopathology.

There is both a clinical and research basis to recommend this approach. It is well known in the clinical setting and in the clinical literature that the presence of symptoms of anxiety, depression, paranoia, etc. are intimately and interactively associated with the use of alcohol and drugs. The ongoing debate regarding the etiological relationship between psychiatric illness and substance abuse (MEYER 1986b; SCHUCKIT 1994) may never be completely resolved. Nonetheless, it is clear that symptoms of psychiatric illness can serve as powerful triggers to relapse in abstinent individuals and that substance abuse can produce significant psychiatric symptoms (KRANZLER and LIEBOWITZ 1988). Thus, on clinical grounds, it is reasonable to consider the treatment of comorbid psychiatric symptoms as a means of indirectly reducing the risk of relapse to substance abuse.

There are also data indicating that the severity of psychiatric symptoms among substance abuse patients at the start of treatment is predictive of outcome following that treatment. This has been shown repeatedly with opiate-dependent patients, alcohol-dependent patients (MCLELLAN et al. 1983, 1985; ROUNSAVILLE et al. 1986, 1987), and more recently with cocaine-dependent individuals seeking treatment (Havassy et al., in press). Among the most common coexisting psychiatric disorders in alcoholics are mood disorders, particularly major depression, and anxiety disorders (Bohn and Hersh, this volume). Both groups of disorders respond well to medications in nonalcoholic patients. Thus in the text that follows, we examine the efficacy of pharmacotherapy for comorbid anxiety and depressive symptoms/disorders in alcoholics, as a means of reducing drinking.

1. Anxiety and Alcoholism

KISSIN (1975) listed three optimal qualities for an anxiolytic in the treatment of chronic alcoholism: (1) it should be effective in maintaining individuals in treatment; (2) it should have a low abuse potential; and (3) it should not potentiate the effects of alcohol. He reviewed a number of studies that showed chlordiazepoxide to be effective in retaining alcoholics in long-term outpatient treatment. Though benzodiazepines are the most widely studied and prescribed anxiolytics, their potential both for abuse by alcoholics and for potentiating the sedative and disinhibiting effects and motor impairment produced by alcohol would appear to contraindicate their prolonged use in this patient population (CIRAULO et al. 1988a). In light of these considerations, a non-benzodiazepine anxiolytic such as buspirone hydrochloride (TAYLOR et al. 1985) may be of particular utility in the treatment of comorbid anxiety and alcoholism. In contrast to benzodiazepines, buspirone does not enhance alcohol-induced impairment of psychomotor skills (MATTILA et al. 1982; SEPPALA et al. 1982), nor is it known to have abuse liability (BALSTER 1990; COLE et al. 1982; GRIFFITH et al. 1986).

A number of studies of the efficacy of the non-benzodiazepine anxiolytic buspirone in alcoholics are reviewed in detail in the chapters in this volume by Romach and Tomkins and by Bohn and Hersh. In general, these studies have shown buspirone to be safe and effective in reducing anxiety when used in a well-selected subgroup of alcoholics with substantial comorbid anxiety. Perhaps the most consistent and potent effect of the medication has been enhanced retention of alcoholics in treatment. However, there are limited data concerning the impact of treatment with buspirone on risk of relapse to heavy drinking. Future investigations should systematically evaluate the efficacy of buspirone, in combination with different psychotherapeutic interventions, in reducing both anxiety symptoms and drinking in groups of abstinent, anxious alcoholics. Skills training for management of both anxiety symptoms and drinking might, for example, be compared with supportive therapy using a factorial (i.e., medication × psychotherapy) study design.

2. Depression and Alcoholism

Tricyclic antidepressants (TCAs) have been widely used in the treatment of depression in alcoholic patients. Controlled trials of the TCAs have been conducted in heterogeneous groups of alcoholics and reviewers are generally negative in their assessments (PATTISON 1979; SCHUCKIT 1979). However, most studies have employed inadequate oral dosage of the TCA, ignoring the findings that both cigarette smoking and heavy drinking can stimulate drug metabolism, potentially yielding plasma levels that are ineffective for treatment of depression (CIRAULO et al. 1982, 1988b). Studies of doxepin and amitriptyline in alcoholics used dosages that would now be considered barely adequate for nonalcoholic depressed patients. Despite these limitations, one study of doxepin and one of amitriptyline in unselected alcoholics showed some positive effects on mood (CIRAULO and JAFFE 1981).

A recent placebo-controlled trial of desipramine in both depressed and nondepressed alcoholics addressed many of the methodologic shortcomings of prior studies (MASON and KOSCIS 1991). In this study depressed alcoholics treated with desipramine had significantly greater reductions in depressive symptoms than did depressed alcoholics treated with placebo. Furthermore, there was a trend for desipramine-treated alcoholics to have a lower rate of relapse and longer periods of sobriety than placebo-treated alcoholics, independent of depression status (MASON and KOSCIS 1991). NUNES and colleagues (1993) used a discontinuation design to compare imipramine with placebo in the treatment of depressed alcoholics. These investigators found that treatment with the TCA resulted in somewhat fewer relapses to both depression and heavy drinking. Even if one accepts the view that most instances of postwithdrawal depression will spontaneously remit within a few days to several weeks following the initiation of abstinence (BROWN and SCHUCKIT 1988; SCHUCKIT 1983), there are still some patients (in some samples as many as 20%) with severe and persistent depression requiring treatment. In these cases, treatment with a TCA may be effective.

More recent studies have provided some support for the use of fluoxetine for treatment of depression in alcoholics (CORNELIUS et al. 1992; Kranzler et al., to be published). However, these studies are preliminary and more definitive data are needed before this serotonin uptake inhibitor (SUI) can be recommended for widespread clinical use in alcoholics.

II. Matching Based Upon Other Patient Characteristics

Alcoholism is both complex and multidimensional (BABOR et al. 1988, 1992a), so much so that JACOBSON (1976) used the term "alcoholisms" to capture the variety and complexity so often observed. Clinicians and investigators have sought to reduce and explain this variability by focusing on single dimensions or factors (e.g., etiological variables, presenting symptoms, drinking patterns) to identify alcoholic subtypes with different etiologies, courses, and treatment responses (BABOR and LAUERMAN 1986). Unfortunately, single-dimension typologies have not been shown to be useful for predicting outcome status following treatment (BABOR et al. 1988, 1992a). This recognition, combined with the development of a greater data base on alcoholism and more sophisticated multivariate statistical techniques, has led to efforts to examine *multiple* factors concurrently to derive alcoholic subtypes. The aim of these efforts has been to identify alcoholic subtypes with different etiologies, co-occurring personality or temperamental features that might explain the course of alcoholism, and, most importantly, differential responses to both psychotherapy and pharmacotherapy. In the text that follows we examine some recent work that has led to two multidimensional typologies of alcoholism.

Table 1. Comparative features of two typologies of alcoholism

Feature	"Late onset"		"Early onset"	
	Type 1[a]	Type A[b]	Type 2[a]	Type B[b]
Age at onset (years)	After 25	Mean >30	Before 25	Mean <22
Gender specificity	Male or female	Male : female = 0.8	Male limited	Male : female = 1.7
Sociopathy	Low	Low	High	High
Binge drinking	Frequent	Infrequent	Infrequent	Frequent
Inability to abstain	Uncommon	Uncommon	Common	Common
Comorbid depression	High	Low	Low	High
Heritability	Low	Probably low	High	Probably high

[a] Cloninger (1987).
[b] Babor et al. (1992b).

The best known of the multidimensional personality-based typologies of alcoholism is the "Type 1/Type 2" distinction, developed by Cloninger et al. (1981). More recently, Babor and colleagues (1992b) derived a dichotomous typology ("Type A/Type B") that, though similar in many respects to the one proposed by Cloninger (1987), differs in some important ways (Table 1). Together, these two approaches illustrate current thinking about the construction of empirically derived typologies that are useful for predicting treatment outcome. These typologies also offer some potentially useful ideas for matching alcoholics to specific pharmacotherapies. However, though these typologic approaches are of clear heuristic value, their utility for understanding the etiology, natural history, and response to treatment in alcoholism requires substantially greater empirical evaluation (Kranzler and Anton, in press).

1. Cloninger's Typology

The "Type 1/Type 2" distinction was developed by Cloninger et al. (1981) from studies of adopted sons of Swedish alcoholics. Differences in the two subtypes are thought to result from differences in three basic personality (i.e., temperament) traits, each of which has a unique neurochemical and genetic substrate (Cloninger 1987). Type 1 alcoholics are characterized by high reward dependence, high harm avoidance, and low novelty seeking. In contrast, Type 2 alcoholics are characterized by high novelty seeking, low harm avoidance, and low reward dependence.

The hypothesis by Cloninger (1987), that specific neurotransmitter systems underlie a tridimensional personality structure, is of heuristic value. Specifically, dopamine is hypothesized to modulate novelty seeking, and is characterized by frequent exploratory behavior and intensely pleasurable responses to novel stimuli. Dopaminergic neurons predominate in those brain regions (such as the caudate and nucleus accumbens) that subserve both behavioral activation and the reinforcing effects of a number of appetitive behaviors, including the self-administration of a variety of drugs of abuse (Di Chiara and Imperato 1988). Serotonin (5-HT) is hypothesized to modulate harm avoidance, which is characterized by a tendency to respond intensely to aversive stimuli and their conditioned signals. Serotonin also appears to play a role in the control of mood, impulsivity, aggression, and alcohol preference and consumption (Coccaro and Murphy 1990; Naranjo et al. 1986; Roy et al. 1990). Finally, norepinephrine is hypothesized to modulate reward dependence, or the resistance to extinction of previously rewarded behavior.

Though recent work by Schuckit and colleagues (1990) and Glenn and Nixon (1991) has failed to provide empiric support for this tridimensional personality scheme, Cloninger's typology has generated substantial research. This explanatory model of personality has also been extended to include elements of character which, together with the elements of temperament described above, yield a seven-factor model of personality (Cloninger et al. 1993). To the degree that these or other personality dimensions are replicable and can be linked to neurobiologic substrates underlying alcoholic subtypes, they provide a useful approach to the development of medications for relapse prevention.

2. The "Type A/Type B" Distinction of Babor and Colleagues

Important features of the typology of Babor et al. (1992b) are that it was derived from a clinical sample of inpatient alcoholics using an empirical clustering technique and it has been successful in predicting the response to alcoholism treatment (Babor et al. 1992b; Litt et al. 1992). Litt et al. (1992) found an effect of matching subtypes of alcoholic patients to psychotherapeutic treatments, suggesting that empirically derived, multivariate typological classifications may provide a useful basis for selecting treatment.

Gerra and colleagues (1992) prospectively matched alcoholics (based only on parental history of alcoholism) to pharmacologic treatments. In that study, alcoholics with a parental history of alcoholism responded preferentially to fluoxetine. However, Kranzler et al. (to be published) were unable to replicate these findings. Insofar as typologies based on a single dimension (including family history of alcoholism) have been shown to be poor independent predictors of outcome status (Babor et al. 1988; Babor et al. 1992a), it is unclear how useful they may be for matching alcoholic patients with specific pharmacologic treatments.

Biological and clinical data from work by Buydens-Branchey and colleagues (1989a, 1989b) may help to clarify some differences that exist between the multivariate typologies discussed above. These investigators distinguished between early-onset and late-onset alcoholism in a sample of patients consecutively admitted to an inpatient rehabilitation program. The typology described by these investigators resembles the typology of BABOR et al. (1992b), with early-onset alcoholics having a greater family history of alcoholism, more antisocial behavior, and more comorbid mood disorder. Among the patients studied by BUYDENS-BRANCHEY et al. (1989a), those with early-onset alcoholism were found to have been incarcerated more frequently for violent crimes, to have made more suicide attempts, and to have been depressed more often than patients with later onset of their alcoholism. Furthermore, among the early-onset group, there was an inverse relationship between a measure of central serotonergic activity and measures of depression and aggressivity (BUYDENS-BRANCHEY et al. 1989b). This, together with increased depressive symptoms in early-onset alcoholics, suggests that drugs that enhance 5-HT function, such as the SUIs, may be particularly useful in the treatment of this alcoholic subtype. To date, however, there are no published studies in which the efficacy of an SUI has been compared in early-onset and late-onset alcoholics.

In an effort to assess the role of serotonergic neurotransmission in alcoholism, BENKELFAT and colleagues (1991) conducted a pharmacologic challenge study in alcoholics using *m*-chlorophenylpiperazine (m-CPP), a serotonergic partial agonist. They found that the drug elicited euphoria and craving for alcoholics in early-onset, but not late-onset alcoholics. KRYSTAL et al. (1992) compared subjective and neuroendocrine responses to m-CPP, yohimbine (an α_2-adrenoceptor antagonist) and placebo in 11 inpatient male alcoholics and 10 healthy control subjects. All subjects completed three test days in a randomized, double-blind design. m-CPP produced discriminative stimulus effects similar to alcohol (i.e., alcohol-like effects) among the alcoholics (whose alcoholism was primarily early-onset), but not among the controls. There was also a significant blunting of both plasma cortisol and prolactin responses to m-CPP among the alcoholics, compared with controls.

LEE and MELTZER (1991) examined neuroendocrine responses to the serotonergic agents L-5-hydroxytryptophan (L-5-HTP, a precursor of 5-HT) and MK-212 (a direct $5\text{-HT}_2/5\text{-HT}_{1C}$ receptor agonist) in alcoholics and normal controls. They found a blunted cortisol response to L-5-HTP and a blunted prolactin response to MK-212 in the alcoholics, who had mainly early-onset disorder. In addition, the alcoholics reported more "alcohol-like" effects of MK-212 and more restlessness, irritability, and anxiety than did the controls.

One interpretation of these findings is that alcoholics (especially those with early onset) have an abnormally low capacity to synthesize 5-HT. The presence of this chronic deficiency might then result in an adaptive up-regulation of 5-HT receptors, making the system more sensitive to serotonin agonist drugs. Together, these findings are consistent with the earlier work of BALLENGER et al. (1979), who demonstrated decreased 5-HT turnover in alcoholics, and BANKI (1981), who found that the decrease in 5-HT turnover in alcoholics correlated with the duration of their abstinence from alcohol.

These data are also consistent with the argument by CLONINGER (1987) that low harm avoidance, which is hypothesized to result from decreased serotonergic tone in the central nervous system, is a central element in early-onset alcoholism.

In summary, there is growing empirical evidence for a dichotomous typology of alcoholism, of which one subtype may be characterized by earlier onset of heavy drinking, greater depression, and antisocial behavior. Furthermore, the neuropharmacologic substrates for these pathologic mood and behavioral dimensions may be linked through abnormalities in serotonergic neurotransmission. Should serotonergic medications prove to be efficacious in the treatment of early-onset alcoholism, as might be predicted based on pharmacologic challenge studies (BENKELFAT et al. 1991; KRYSTAL et al. 1992; LEE and MELTZER 1991), the extent to which this effect is mediated by primary effects on mood, rather than on alcoholism per se, needs to be considered. In contrast, opioid antagonists have been shown to influence the intensity of drinking, with minimal effects on mood (O'MALLEY et al. 1992; VOLPICELLI et al. 1992). The likelihood of differential effects resulting from the use of opioid antagonists in different subtypes of alcoholics is unclear, but also warrants empirical evaluation.

C. Methodologic Issues in Clinical Trials with Alcoholics

Clinical assessment and the randomized clinical trial remain the standards for evaluating the safety and efficacy of prospective psychotropic medications (LEVINE and BAN 1987). In this section, we will discuss in some depth those methodologic issues that, while perhaps not unique to clinical trials with alcoholics, are of particular relevance to this patient group. Clinical trials methodology has developed substantially in recent years and is described in detail in works such as those by MEINERT (1986), SPILKER (1991), and SPRIET and SIMON (1985). Specific adaptations of clinical trials methodology for use in the evaluation of psychiatric medications have also been described (GOLDBERG 1987; LEVINE and BAN 1987; OVERALL 1987; OVERALL and RHOADES 1987; PRIEN and ROBINSON 1994). Recently, attention has been focused on issues specific to clinical trials in alcoholics (MEYER 1992; SELLERS and SOBELL 1992).

Prior to initiation of a clinical trial of a medication for use in alcoholics, there should be sufficient preclinical evidence that the medication modifies behaviors relevant to the hypothesized clinical effect. Thus, for a medication to be tested for relapse prevention, evidence should exist that it reduces alcohol self-administration, discriminative stimulus properties, or place preference in a relevant animal species (SELLERS and SOBELL 1992). Medications that are hypothesized to reduce alcohol consumption indirectly through direct effects on mood (e.g., antidepressants) and others for which suitable animal models have not yet been developed may be exceptions to this approach (SELLERS and SOBELL 1992). However, animal models of learned helplessness [e.g., inescapable shock, maternal deprivation (MCKINNEY 1988)] might be combined with operant drinking paradigms to permit preclinical investigation of the effect of anxiolytics and antidepressants in reducing both distress and alcohol consumption.

SELLERS and SOBELL (1992) argue that in addition to the use of random assignment, double-blind conditions, placebo controls, and parallel group designs, medication studies in alcoholics should include two or more dose levels of the active drug. Though simple random assignment is widely employed in alcoholism clinical trials, its utility requires that an adequate number of subjects be randomized to produce a balanced distribution of key demographic and clinical features between or among the treatment groups. Since exploratory clinical trials often involve relatively small sample sizes, it is not uncommon for group differences to arise on key pretreatment measures, creating the potential for difficulties in interpreting the results. A discussion of stratification and balancing procedures is provided by SPRIET and SIMON (1985). One useful alternative to simple randomization is urn randomization, a probabilistic balancing procedure that assigns patients to conditions so that groups are balanced on preselected key variables (WEI 1978). In factorial designs, which are often employed to test matching hypotheses, urn randomization can insure that there are adequate numbers of subjects in all cells and should result in groups that are comparable on a variety of key variables at the onset of treatment.

The integrity of the double-blind design can be questioned in cases where an active medication with discriminable effects is compared with an inert placebo as the only control treatment (FISHER and GREENBERG 1993; OXTOBY et al. 1989). Patients and/or investigators are often able to identify correctly the treatments received by the patients in controlled trials (FISHER and GREENBERG 1993; OXTOBY et al. 1989), which introduces an element of bias that double-blind designs are intended to avoid. Consequently, the use of an active placebo (i.e., one that produces side effects without the likelihood of producing a therapeutic effect) is recommended (FISHER and GREENBERG 1993). In addition, efforts should be made to evaluate the integrity of the double-blind. Unfortunately, the validity of the double-blind is often not evaluated in pharmacotherapy trials with alcoholics.

HUGHES and KRAHN (1985) have suggested an approach that can be employed to ascertain the degree to which differences between treatment groups are attributable to penetration of the double-blind by patients. However, their method depends on there being an overall sample size adequate to provide the statistical power necessary to identify differences in treatment response among up to six patient subgroups. FISHER and GREENBERG (1993) recommend that similar procedures be applied to assess the ability of the research personnel conducting the trial to penetrate the blind.

It cannot be assumed that medications developed for indications other than alcoholism will demonstrate the same dose-response relationship when used to treat alcoholism. SELLERS and SOBELL (1992) suggest that it may be preferable in alcoholism treatment studies to use a response-controlled design (i.e., one in which the dose is titrated to achieve maximal clinical response, taking adverse effects into account), rather than a dose-controlled design (i.e., a fixed dose) or a concentration-controlled design (i.e., dose titrated to achieve a particular concentration or concentration range). These authors also discuss issues such as the features of the patient population appropriate for initial clinical trials, key outcome variables, measurement of alcohol consumption, measurement of compliance, and the integration of pharmacotherapy with psychosocial treatment, most of which are discussed in detail below.

MEYER (1992) has also discussed a number of issues basic to the conduct of clinical trials in alcoholics, suggesting that studies of pharmacotherapy in alcoholics are more complex than many other areas of treatment research, including clinical trials in patients with other psychiatric disorders. He emphasizes the need to insure that active treatment and control groups receive comparable nonpharmacological treatment and that reliable and valid measures of treatment efficacy be employed.

I. Patient Assessment and Outcome Measurement: "For Whom Is One Drink Too Many?"

The technology for assessment of psychiatric disorders has grown rapidly (LEVINE and BAN 1987). The development of specific criteria for psychiatric diagnosis (e.g., DSM-IV: American Psychiatric Association 1994; ICD-10: World Health Organization 1992) has enhanced the acceptability of clinical diagnosis for the definition of clinical populations (OVERALL 1987). A number of structured and semistructured interview techniques (ROBINS et al. 1981, 1988; SPITZER et al. 1992, WING et al. 1990) have also been developed which increase the reliability and validity of psychiatric diagnosis.

BABOR and colleagues (in press) have reviewed the literature on measurement of drinking outcomes in alcoholism treatment research. They recommend that the evaluation of treatment outcome focus on specific indicators of drinking behavior, measures of life functioning and health, and global indicators of outcome. Instruments that have been tested parametrically for their psychometric features and that assess a variety of treatment-relevant domains are now available for use in clinical trials with alcoholics and other substance abuse patients (HESSELBROCK et al. 1983; LETTIERI et al. 1985; MCLELLAN et al. 1980, 1992; SKINNER and ALLEN 1982; SOBELL et al. 1980). When administered at specified intervals during the course of a clinical trial, these instruments permit the assessment of change in substance use and related behaviors. A promising area for assessment of alcohol consumption is the use of clinical markers, including those that can be used to validate self-reported alcohol consumption, a discussion of which is included in the chapter by Rommelspacher and Müller in this volume.

In the United States, successful alcoholism treatment outcome has traditionally been considered to be total and enduring abstinence, which is achieved by only a minority of patients (HELZER et al. 1985; POLICH et al. 1981). In other countries, particularly in Europe, reduction of harm through reduced alcohol consumption has been a desirable goal of intervention with heavy drinkers (LINDSTROM 1992). Greater tolerance of this approach has also developed more recently in the United States, though principally for heavy drinkers who are without substantial physical dependence.

In studies for which total abstinence is considered to be the only successful outcome, the measurement of outcome is relatively straightforward (i.e., abstinent vs drinking). In contrast, studies aimed at the reduction of alcohol consumption generally depend upon continuous outcome measures [e.g., drinking days (or conversely, days abstinent) and intensity of drinking]. A recurrent issue in relapse prevention trials that employ continuous outcome measures is how to define relapse. At the present time, despite the clinical and theoretical importance of this issue, there remains no consensus on its definition, which limits the ability to compare the efficacy of different treatments for relapse prevention.

The use of abstinence as the only measure of outcome in alcoholism treatment studies presents a number of clinical and research problems. From a clinical perspective abstinence may be an unrealistic expectation, insofar as alcohol dependence is a chronic, relapsing disorder. Similar expectations for outcomes following treatment for other chronic medical disorders such as hypertension, diabetes mellitus, or rheumatoid arthritis, would lead one to ask what proportion of patients treated for those disorders are discharged from treatment, free of all medications, and able to remain asymptomatic throughout a follow-up period. A more reasonable expectation for treatment of a chronic disorder is a reduction in primary symptoms and an increase in the functional status of patients, including a reduced need for hospitalization. By these standards, the treatment of alcoholism compares favorably with those of other medical disorders.

Since there is some evidence that abstinence may be important in the process of recovery from alcoholism (BABOR et al. 1988), this measure has some utility as a global indicator of outcome. However, some studies have shown that abstinence is not always associated with improvements in other areas of function (PATTISON 1969; SIMPSON and SAVAGE 1980). In addition, these studies include a significant minority of patients who, despite failing to become completely abstinent, nonetheless perform very well on all other measures of general functional status. Thus, from a clinical perspective, though abstinence may be a desirable goal of alcoholism treatment, it may be neither a reasonable standard of treatment effectiveness nor an adequate global measure of treatment outcome.

There are also statistical problems with the use of this dichotomous measure as the only outcome in a treatment study. Since a person who has drunk continuously every day since treatment began is counted the same as the person who has only drunk a few times, it is clear that the measure lacks sensitivity to change. In addition, there is substantial loss of statistical power when continuous measures of outcome such as frequency or intensity of drinking are converted into a single dichotomous measure. This lack of sensitivity translates into the requirement for substantially larger sample sizes than might otherwise be the case, in order to find a "true" difference in effectiveness.

Increased recognition of the multivariate nature of alcoholism (BABOR et al. 1988, 1992b) has also led investigators to examine a variety of outcomes in addition to alcohol consumption, including medical, social, family, and psychological functioning. The Addiction Severity Index (ASI, MCLELLAN et al. 1980, 1992) is a structured interview that is widely used to assess change in a variety of areas. The ASI can be scored in each of the problem areas to produce a quantitative problem severity profile that describes the patient's relative status in the following treatment-related problem areas: alcohol abuse, drug abuse, medical status, psychological adjustment, legal problems, family/social relations, and employment/financial support. Individual items, severity ratings, and composite scores can be used for the assessment of patients at intake to treatment and at various follow-up intervals. Treatment evaluation research findings (MCLELLAN et al. 1983, 1985) suggest that this information can be useful in assigning patients to appropriate treatment modalities and predicting treatment response. However, there remains a need for additional psychometrically sound instruments that are sensitive to both short-term and enduring change in the severity of a broad range of alcohol-related and other problems.

II. Treatment Retention and "How Much Additional Treatment Is Too Much Treatment?"

A thorny methodological problem in clinical trials with alcoholics is the high rate of attrition that is often observed. High rates of attrition seriously limit the strength of conclusions that can reasonably be drawn from such clinical trials in two ways. One major effect of attrition is the unavailability of key data on the effect of the study medication(s). This lack of data may require the elimination of subjects from statistical analyses and cause a loss of statistical power to detect treatment effects. One method that has been used to overcome the effects of loss of data is the technique of "endpoint analysis" (GOLDBERG 1987; OVERALL 1987; SPRIET and SIMON 1985). This technique involves carrying the last known data point, obtained from a subject who has left the study, forward to replace the unknown data points following the subject's termination. For example, if breathalyzer readings are collected weekly on two groups of subjects receiving either an experimental or a control treatment and a subject leaves treatment prematurely, the result of the last breathalyzer is carried forward to replace the missing values generated by the loss of the subject. This technique was originally developed for use in educational trials, where tests of acquired knowledge or skill (e.g., reading level) were accumulated over time. Under these conditions, it is conservative to estimate future acquired knowledge using a record of past acquired knowledge. However, in trials with alcohol- or drug-dependent patients it cannot be assumed that the level of performance seen during the earlier course of treatment is a good indication of the performance following dropout, which often reflects a return to alcohol or drug use. Consequently, endpoint analysis should probably be restricted to studies no longer than 6–12 weeks in duration (GOLDBERG 1987; KLEIN 1991).

A second problem with sample attrition is the *differential* loss of data that can occur when more subjects drop out of the test condition than from the control condition, or vice versa. In these cases there is not only the loss of statistical power to detect significant between-group effects, there is also the danger of misinterpreting the true effects of the treatment. For example, consider a situation in which a medication group has a 90% retention rate and a 60% rate of improvement on a target measure, compared with a placebo group in which there is a 50% retention rate and a 70% improvement rate on that measure. A direct comparison of the remaining subjects might lead to the erroneous conclusion that placebo was superior to the active medication. Under these circumstances it is likely that a lower rate of improvement was present among subjects who did not complete the study; consequently, if all subjects were included in the analysis a beneficial effect of the medication might become evident.

To minimize the effect of subject attrition in clinical trials with alcoholics, efforts should be made to reduce the loss of subjects from treatment and evaluation. Furthermore, appropriate statistical approaches should be employed to evaluate outcomes in subjects who are lost to attrition.

1. Preventing Attrition

Though it is not possible to prevent dropout entirely, there are methods that may serve to reduce the magnitude of the problem. SPRIET and SIMON (1985) discuss a number of causes of sample attrition and methods to prevent it (e.g., excluding uncooperative, unmotivated, or unstable patients – which, however, limits the ability to generalize the study's findings to the entire population of alcohol abusers).

Combining pharmacotherapy with psychotherapy has also been advocated for the prevention of attrition (KLEIN 1991). This may be particularly beneficial in reducing differential dropout among placebo-treated patients, for whom the absence of medication side effects may result in the correct identification of their treatment status, leading them to seek "active treatment" elsewhere. The provision of active, relevant treatment also has ethical advantages over an inactive placebo alone (KLEIN 1991). Combined use of both medication and psychotherapy in treatment trials for alcoholics also has "ecological validity." It is in the context of combined therapy that medication is most likely to play a role in alcoholism treatment (SELLERS et al. 1981), making that the most appropriate context in which to evaluate potential pharmacotherapies. There may be a concern, however, that the provision of psychotherapy will make it more difficult to detect a medication effect, since the between-group variability in outcome will be reduced by psychotherapeutic effects on subjects in both study groups (KLEIN 1991; Kranzler et al., to be published). A number of studies of the effects of medications on alcohol consumption, particularly in "heavy drinkers," have not included any psychosocial treatment, since the subjects were not seeking treatment (NARANJO et al. 1984, 1987, 1989, 1990; TONEATTO et al. 1991). Rather, these studies recruited actively drinking subjects who were not motivated to reduce their drinking. Based upon this, decreases in consumption that exceed those during the pretreatment baseline period or during placebo treatment were attributed to the pharmacologic effects of the medication being evaluated.

Studies have shown that the addition of psychosocial services and therapies can enhance the effects of a medication with well-demonstrated efficacy. A recent study by MCLELLAN and colleagues (1993) in methadone-maintained patients showed that the addition of drug counseling to methadone treatment resulted in reduced illicit drug use, as well as improvement on a variety of other outcome measures. The addition of other treatment services, including on-site medical, psychiatric, family, and employment counseling services, resulted in significantly greater improvement in outcomes. These data indicate that, despite receiving the same average dosage of methadone, the provision of even modest amounts of psychosocial services (i.e., drug counseling) significantly enhanced the effects of medication. Similarly, O'MALLEY and colleagues (1992) found that relapse prevention psychotherapy enhanced the efficacy of naltrexone in preventing alcoholic relapses. Thus, it is possible that the provision of medication without adequate concomitant psychosocial treatment might lead to a substantial underestimation of the effectiveness of the medication.

Comparative studies of medication and psychotherapy for major depression and panic disorder indicate that these treatments have different target symptoms (KLEIN 1991). Similar findings are beginning to emerge in studies of alcoholism treatment (O'MALLEY et al. 1992). Factorial study designs (KERLINGER 1986) make it possible to examine the effects of two or more independent variables and their interactive effects on a dependent variable (e.g., drinking behavior). Such designs may thereby yield findings of unique utility for alcoholism treatment through examination of the main and interactive effects of specific medications and specific psychotherapies with one another and with alcoholic subtypes.

2. Analysis of Outcomes When Attrition Occurs

Some of the problems associated with attrition can be avoided by using an "intention to treat" analysis plan (MEINERT 1986; SPRIET and SIMON 1985; SPILKER 1991), in which all subjects who enter a study are followed and their outcomes measured, regardless of whether they complete the planned treatment. This strategy is particularly useful when subjects can be located and evaluated at the time they were scheduled to complete treatment. The need to follow all patients, including those who do not complete the study, should be discussed with patients prior to initiating the study treatment (i.e., as part of the informed consent procedure).

Statistical methods for evaluating treatment outcome where subject attrition is a problem include life table (i.e., survival) analysis (MEYER 1992). This method, based on a definable event, makes it possible to compare the clinical course of subjects across treatment groups (LEE 1992). Though the parallel-groups design appears to lend itself well to repeated measures analysis of variance, missing data are particularly problematic for this approach (OVERALL 1987). Consequently, analysis of covariance (with pretreatment values entered as covariates) is the most widely used approach for testing the significance of treatment effects in clinical trials (OVERALL 1987). Though analysis of variance is the classical technique applied to experimental designs, statistical evaluation of treatment outcome may also be accomplished using multiple regression (COHEN and COHEN 1983; DERLINGER and PEDHAZUR 1973). In studies involving relatively small sample sizes (i.e., fewer than 50 subjects per treatment arm) and continuous independent variables (other than treatment condition), this approach provides a powerful method to evaluate the effect of interactions (including matching effects) on treatment outcome (AIKEN and WEST 1991; KADDEN et al. 1989; KRANZLER et al., in press). An alternative to regression analysis for evaluation of treatment matching, particularly in studies with larger sample sizes, is to block on a matching variable (i.e., create discrete subgroups based on that variable) and use it, along with treatment group (e.g., active medication vs control), as independent variables in a two-way analysis of covariance.

III. Maximizing Compliance and Determining "How Well You've Done"

Noncompliance with medication has been shown to range from about 15% to 93% (GREENBERG 1984). The rate of noncompliance varies with the illness being treated, the distress associated with its symptoms, and the adverse effects associated with the study medication. The potential for noncompliance is often greater when the risk from the illness is perceived by the patient to be small (SPRIET and SIMON 1985). It has been suggested that alcoholics are less likely than other patient populations to comply fully with medication regimens (MEYER 1992). Failure to ascertain the extent of noncompliance in an outpatient pharmacotherapy trial may lead to invalid conclusions (LASAGNA 1991), e.g., that the medication is efficacious (as a consequence of the fact that the placebo group was less compliant) or that the medication lacked efficacy (as a consequence of the fact that the active drug group was less compliant).

Efforts should be made to maximize compliance with the treatment regimen. One approach to maximizing compliance is to minimize side effects by using a low (i.e., minimum effective) dosage of the active medication. However, this may result in inadequate dosing and the mistaken conclusion that the medication lacks efficacy (KLEIN 1991). An alternative approach that should permit an adequate assessment of the medication without undue attrition or poor compliance is to determine the maximum tolerable dosage under carefully supervised conditions, prior to the initiation of a controlled clinical trial (KLEIN 1991). The use of a response-controlled design (SELLERS and SOBELL 1992), discussed above, may also enhance compliance.

Once the clinical trial has been initiated, careful monitoring of adverse effects and appropriate changes in medication dosage (as permitted in the study protocol) to minimize these effects should help to reduce attrition and maximize compliance. The less demanding the requirements of the study protocol (e.g., once-daily dosage of a medication), the greater the likelihood of compliance. As is true with efforts to reduce sample attrition, concomitant psychotherapy may enhance compliance with the study medication, through the therapist's active encouragement of the patient and by the development of individualized strategies to enhance compliance (RUSSELL 1984). Similarly, the attitudes of the treatment staff and the relationship that develops between staff and patients may have an impact on compliance. Attitudes toward alcoholism may be particularly important in this regard; a non-judgmental approach is an intrinic element in the psychotherapeutic relationship. Other methods of improving compliance are reviewed in detail by SPILKER (1991), including simplifying the demands of the protocol on patients, minimizing the number and duration of unpleasant or painful tests, maintaining relatively frequent contact with patients, allowing for flexible dosing regimens to deal with adverse reactions, and planning patient visits at a mutually convenient time. As can be seen from this list, methods to enhance compliance are largely commonsensical. Unfortunately, they have not been systematically evaluated (SPILKER 1991).

Compliance with both active and control medications should also be monitored to determine whether the study has provided an adequate evaluation of the active medication. A number of authors (GORDIS 1984; SPILKER 1991; SPRIET and SIMON 1985) discuss the relative merits of different approaches to evaluating compliance in clinical trials. Indirect methods depend on patient reports or on data that can be modified by the patient, while direct methods provide proof of the extent of compliance (SPILKER 1991). Questioning the patient is the simplest and most straightforward method, but this often results in an overestimate of compliance (SPRIET and SIMON 1985), a problem that theoretically would be greater among alcoholics. Counting tablets or capsules returned may also be useful, but tests of the validity of this method have yielded mixed results (PULLAR et al. 1989; ROTH et al. 1970; RUDD et al. 1989; YOUNG et al. 1984). Supervision of drug intake is feasible only with inpatient studies, making it unsuitable for the evaluation of medications for relapse prevention in alcoholics.

The detection in biological fluids of the drug being studied, its metabolites, or a tracer substance ingested at the same time as the drug, are direct methods that may be particularly suited to use in trials with alcoholics. Such methods do not depend on the accuracy of self-report. These techniques may be particularly useful in conjunction with behavioral counseling, in that they may permit the rapid identification of patients who are noncompliant with the dosing regimen, so that counseling to enhance compliance can be implemented (RUSSELL 1984). Plasma or urinary drug determinations can provide an accurate gauge of compliance with an active medication or an active control, but compliance with placebo treatment is not amenable to this approach. Furthermore, drug level determinations are often technically difficult and costly to perform and, in most cases, provide information only about recent ingestion of the drug (SPRIET and SIMON 1985). A tracer added to a medication must be nontoxic at the doses used, stable in biological fluids, easily and accurately detected, and biologically inert (SPILKER 1991). A variety of tracer substances, including methylene blue, phenol red, isoniazid, fluorescein, riboflavin, bromide, and phenobarbital, have been employed to measure compliance (KRAUS et al. 1987; PULLAR et al. 1989; ROTH et al. 1970; SPRIET and SIMON 1985; YOUNG et al. 1984). Depending on the rate of excretion of the tracer used, however, this method may also provide information only about recent ingestion of the drug. Furthermore, although a tracer substance may itself be biologically inert, its addition to an active drug preparation may result in an alteration of the drug's bioavailability.

LASAGNA (1991) has suggested that microelectronic monitoring techniques (e.g., a pill container that records the frequency with which it is opened) are more useful than pill counts or spot blood or urine levels (including, presumably, levels of compliance markers such as riboflavin). Such methods provide continuous assessments of the marked variability in medication compliance (GORDIS 1984; RUDD et al. 1989), which other methods do not adequately measure (CHEUNG et al. 1988; CRAMER et al. 1989; RUDD et al. 1981). Unfortunately, such technology is costly and offers no guarantee of accuracy. Subjects who turn the containers they receive to other purposes, such as the storage of drugs of abuse, will readily confound the assessment of compliance.

SULLIVAN et al. (1989) examined the impact of stringent selection criteria and compliance monitoring and reinforcement procedures in medication studies in heavy drinkers (NARANJO et al. 1984, 1987). These investigators used a variety of both direct and indirect methods to monitor compliance. They found that limiting study participation to those subjects who attended an initial assessment, met admission criteria, and successfully completed a baseline period resulted in good compliance with the study protocol. While this approach indicates that good compliance can be achieved in heavy drinkers, stringent selection criteria limit the extent to which the study results can be generalized to the population of alcoholics for whom medications may provide an important dimension of relapse prevention treatment.

IV. "How Long to Treat? Then What?"

Many psychiatric medications require substantial time for their specific beneficial effects to appear. The hypothesized mechanism by which a medication may exert its therapeutic effects will often set the lower bounds of study duration. Since alcoholism often has a chronic, fluctuating course, an adequate evaluation of treatment effects requires a considerable period of observation.

Medications with a short duration of effect may be useful in the initiation or early maintenance of abstinence or reduced alcohol consumption. In such cases, a trial as short as 3 weeks may be adequate. In contrast, medications may serve to enhance the acquisition of skills that enable patients to cope more effectively with precipitants to drinking, thereby preventing relapse or reducing its severity. Conversely, relapse prevention training (Marlatt and Gordon 1985) may serve to enhance the magnitude and duration of medication effects (Gallant 1993; Kranzler et al., in press). Under these circumstances a longer period of treatment (e.g., 12 or more weeks) may be required to evaluate the effects of the medication. In addition, a post-treatment follow-up period provides the opportunity to assess delayed and/or enduring treatment effects.

Though the combination of pharmacotherapy and psychotherapy might be expected to produce more persistent effects than either treatment modality alone, this has not yet been evaluated empirically. A similar commonsense assumption is that a longer duration of alcohol treatment results in better outcomes. Recent studies have not borne this out, however, and increasing the length of alcohol treatment does not result in better outcomes compared with briefer interventions (Institute of Medicine 1990). Unfortunately, the alcoholism treatment literature does not yet provide clear guidelines for determining the optimal duration of pharmacologic interventions. Both the replication of recent findings with medications such as buspirone and naltrexone and the identification of other medications that facilitate relapse prevention will make the optimal duration of treatment a more pressing question for investigation.

D. What May the Future Hold?

It appears that we may be in a watershed period for the development of pharmacotherapies for alcoholism. If the early studies with naltrexone (O'MALLEY et al. 1992; VOLPICELLI et al. 1992) provide an accurate estimate of the effectiveness of that medication for relapse prevention, the current situation may be similar to that which existed for the treatment of major depression in the mid-1950s. Although since that time numerous antidepressants have been developed and are now being prescribed regularly throughout the world, imipramine remains the standard against which they are measured. In addition, the treatment of comorbid psychiatric disorders in alcoholics appears to have benefited from the development of medications for treatment of psychiatric disorders in nonalcoholics. As unlikely as it may have seemed as recently as 15 years ago (MURRAY 1980), pharmaceutical companies have begun to take an interest in the development of medications for relapse prevention in alcoholics. Concerns that these companies have had over the potential for alcoholic subjects to experience adverse events that would adversely affect the acceptability of their medication in the treatment of "mainstream" psychiatric patients has given way to the recognition that alcoholics represent a large potential market. Furthermore, unlike the market for antidepressants and anxiolytics, the market for medications to treat alcoholism is not yet saturated.

As recently as 1980, however, Murray reported that the majority of alcoholics who sought treatment for their disorder received medications developed for use in other psychiatric disorders, rather than for primary use in the treatment of alcoholism. He argued that the reason for this was that psychiatrists felt so powerless to intervene in the process of alcohol dependence that whenever possible they redefined the disorder in terms of a psychiatric illness (e.g., major depression), which they believed they could treat. MURRAY (1980) attributed the historical lack of pharmaceutical industry support to a misconception by the industry that alcoholism is a moral, rather than a pharmacological, problem.

One would hope that the evidence adduced in the present volume concerning the pharmacology of alcohol abuse would persuade skeptics, within the pharmaceutical industry, government, and elsewhere, that medications are en route to becoming a basic element in the treatment armamentarium for alcoholism. The disorder is highly prevalent throughout much of the world (HELZER and CANINO 1992). A shift toward more realistic expectations about the potential role of medications in alcoholism treatment, combined with economic incentives to develop medications for the large population of potential recipients, should provide a much-needed impetus for pharmaceutical research in this area.

However, the responsibility for medications development cannot rest exclusively with the pharmaceutical industry. Fortunately, in recent years there has been a growing interest in the area by government, which has sponsored an increasing number of preclinical and clinical studies in the pharmacotherspy of alcoholism. A decade ago, JOHN LITTLETON (1984) wrote that the future for alcohol research, particularly in the area of medications development, could be bright. During the ensuing decade much promise has been realized. Continued efforts over the next decade can be expected to pay even greater dividends.

References

Aiken LS, West SG (1991) Multiple regression: testing and interpreting interactions. Sage Publications, Newbury Park CA

American Psychiatric Association (1994) Diagnostic and statistical manual of mental disorders, 4th edn. American Psychiatric Association, Washington DC

Babor TF, Lauerman RJ (1986) Classification and forms of inebriety: historical antecedents of alcoholic typologies. In: Galanter M (ed) Recent developments in alcoholism, vol 4. Plenum, New York, pp 113–144

Babor TF, Dolinsky Z, Rounsaville B, Jaffe J (1988) Unitary vs. multidimensional models of alcoholism treatment outcome: an empirical study. J Stud Alcohol 49:167–177

Babor TF, Dolinsky ZS, Meyer RE, Hesselbrock M, Hofmann M, Tennen H (1992a) Types of alcoholics: concurrent and predictive validity of some common classification schemes. Br J Addict 87:1415–1431

Babor TF, Hofmann M, DelBoca FK, Hesselbrock V, Meyer RE, Dolinsky ZS, Rounsaville B (1992b) Types of alcoholics: I. Evidence for an empirically-derived typology based on indicators of vulnerability and severity. Arch Gen Psychiatry 8:599–608

Babor TF, Longabaugh R, Zweben A, Fuller R, Stout R, Anton RF, Randall CL (in press) Issues in the definition and measurement of drinking outcomes in alcoholism treatment research. J Stud Alcohol

Ballenger JC, Goodwin FK, Major LF, Brown, GL (1979) Alcohol and central serotonin metabolism in man. Arch Gen Psychiatry 36:224–227

Balster RL (1990) Abuse potential of buspirone and related drugs. J Clin Psychopharmacol 10:31S–37S

Banki CJ (1981) Factors influencing monamine metabolites and tryptophan in patients with alcohol dependence. J Neural Trans 50:98–101

Benkelfat C, Murphy DL, Hill JL, George DT, Nutt D, Linnoila M (1991) Ethanol like properties of the serotonergic partial agonist *m*-chlorophenylpiperazine in chronic alcoholic patients. Arch Gen Psychiatry 48:383

Brown SA, Schuckit MA (1988) Changes in depression among abstinent alcoholics. J Stud Alcohol 49:412–417

Buydens-Branchey L, Branchey MH, Noumair D (1989a) Age of alcoholism onset. I. Relationship to psychopathology. Arch Gen Psychiatry 46:225–240

Buydens-Branchey L, Branchey MH, Noumair D, Lieber CS (1989b) Age of alcoholism onset. II. Relationship of susceptibility to serotonin precursor availability. Arch Gen Psychiatry 46:231–236

Cheung R, Dickins J, Nicholson PW, Thomas ASC, Smith HH, Larson HE, Deshmukh AA, Dobbs RJ, Dobbs SM (1988) Compliance with anti-tuberculous therapy: a field trial of a pill-box with a concealed electronic recording device. Eur J Clin Pharmacol 35:401–407

Ciraulo DA, Jaffe JH (1981) Tricyclic antidepressants in the treatment of depression associated with alcoholism. J Clin Psychopharmacol 1:146–150

Ciraulo DA, Alderson LM, Chapron DJ, Jaffe JH, Subbarao B, Kramer PA (1982) Imipramine disposition in alcoholics. J Clin Psychopharmacol 2:2–7

Ciraulo DA, Barnhill JG, Greenblatt DJ, Shader RI, Ciraulo AM, Tarmey MF, Molloy MA, Foti ME (1988a) Abuse liability and clinical pharmacokinetics of alprazolam in alcoholic men. J Clin Psychiatry 49:333–337

Ciraulo DA, Barnhill JG, Jaffe JH (1988b) Clinical pharmacokinetics of imipramine and desipramine in alcoholics and normal volunteers. Clin Pharmacol Ther 43:509–518

Cloninger CR (1987) Neurogenetic adaptive mechanisms in alcoholism. Science 236:410–416

Cloninger CR, Bohman M, Sigvardsson S (1981) Inheritance of alcohol abuse: cross-fostering analysis of adopted men. Arch Gen Psychiatry 38:861–868

Cloninger CR, Svrakic DM, Przybeck TR (1993) A psychobiological model of temperament and character. Arch Gen Psychiatry 50:975–990

Coccaro EF, Murphy DL (1990) (eds) Serotonin in major psychiatric disorders. American Psychiatric Press, Washington DC

Cohen J, Cohen P (1983) Applied multiple regression/correlation analyses for the behavioral sciences, 2nd edn. Lawrence Erlbaum, Hillsdale NJ

Cole JO, Orzack MH, Beake B, Bird M, Bar Tel Y (1982) Assessment of abuse liability of buspirone in recreational sedative users. J Clin Psychiatry 43:69–74

Cornelius JR, Fisher BW, Salloum IM, Cornelius MD, Ehler JG (1992) Fluoxetine trial in depressed alcoholics. Alcohol Clin Exp Res 16:362

Cramer JA, Mattson RH, Prevey ML, Scheyer RD, Ouellette VL (1989) How often is medication taken as prescribed? A novel assessment technique. JAMA 261: 3273–3277

Di Chiara G, Imperato A (1988) Drugs abused by humans preferentially increase synaptic dopamine concentrations in the mesolimbic system of freely moving rats. Proc Natl Acad Sci USA 85:5274–5278

Fisher S, Greenberg RP (1993) How sound is the double-blind design for evaluating psychotropic drugs? J Nerv Ment Dis 181:345–350

Gallant D (1993) Amethystic agents and adjunct behavioral therapy and psychotherapy. Alcohol Clin Exp Res 17:197–198

Gerra G, Caccavari R, Delsignore R, Bocchi R, Fertonani G, Passeri M (1992) Effects of fluoxetine and Ca-acetyl-homotaurinate on alcohol intake in familial and nonfamilial alcohol patients. Curr Ther Res 52:291–295

Glenn SW, Nixon SJ (1991) Applications of Cloninger's subtypes in a female alcoholic sample. Alcohol Clin Exp Res 15:851–857

Goldberg SC (1987) Persistent flaws in the design and analysis of psychopharmacology research. In: Meltzer HY (ed) Psychopharmacology: The third generation of progress. Raven, New York, pp 1005–1012

Gordis L (1984) General concepts for use of markers in clinical trials. Controlled Clin Trials 5:481–487

Greenberg RN (1984) Overview of patient compliance with medication dosing: a literature review. Clin Ther 6:592–599

Griffith JD, Jasinski DR, Casten GP, McKinney GR (1986) Investigation of the abuse liability of buspirone in alcohol-dependent patients. Am J Med 80 [Suppl 3B]:30–35

Havassy BE, Hall SM, Wasserman D (in press) Social supports and response to treatment for cocaine dependence. J Subst Abuse Treat

Helzer JE, Robins LN, Taylor JR, Carey K, Miller RH, Combs-Orme T, Farmer A (1985) The extent of long-term moderate drinking among alcoholics discharged from medical and psychiatric facilities. N Engl J Med 312:1678–1682

Helzer JE, Canino GJ (1992) Comparative analysis of alcoholism in ten cultural regions. In: Helzer JE, Canino GJ (eds) Alcoholism in North America, Europe, and Asia. Oxford University Press, New York, pp 289–308

Hesselbrock M, Babor TF, Hesselbrock V, Meyer RE, Workman K (1983) "Never believe an alcoholic?" On the validity of self-report measures of alcohol dependence and related constructs. Int J Addict 18:593–609

Hughes JR, Krahn D (1985) Blindness and the validity of the double-blind procedure. J Clin Psychopharmacol 5:138–142
Institute of Medicine (1990) Broadening the base of treatment for alcoholism. National Academy Press, Washington DC
Jacobson GR (1976) The alcoholisms: detection, diagnosis, and assessment. Human Sciences Press, New York
Kadden RM, Cooney NL, Getter H, Litt M (1989) Matching alcoholics to coping skills or interactional therapies: Posttreatment results. J Consult Clin Psychol 57:698–704
Kerlinger FN (1986) Foundations of behavioral research, 3rd edn. Holt, Rinehart and Winston, New York
Kerlinger FN, Pedhazur EJ (1973) Multiple regression in behavioral research. Holt, Rinehart and Winston, New York
Kissin B (1975) The use of psychoactive drugs in the long-term treatment of chronic alcoholics. Ann NY Acad Sci 252:385–395
Klein DF (1991) Improvement of phase III psychotropic drug trials by intensive phase II work. Neuropsychopharmacology 4:251–258
Kranzler HR, Liebowitz N (1988) Anxiety and depression in substance abuse: clinical implications. In: Frazier S (ed) Anxiety and depression, medical clinics of North America, vol 72. Williams and Wilkins, Philadelphia, pp 867–885
Kranzler HR, Burleson JA, Korner P, Del Boca FK, Bohn MJ, Brown J, Liebowitz N (submitted) Placebo-controlled trial of fluoxetine as an adjunct to relapse prevention in alcoholics.
Kranzler HR, Burleson JA, Del Boca FK, Babor TF, Korner PF, Brown JA, Bohn MJ (in press) Buspirone treatment of anxious alcoholics: A placebo-controlled trial. Arch Gen Psychiatry
Kranzler HR, Anton RF (in press) Implications of recent neuropsychopharmacologic research for understanding the etiology and development of alcoholism. J Consult Clin Psychol
Kraus RP, Grof P, Arana GW, Workman RJ, Harvey KJ, Hux M (1987) Methylene blue: a reliable and practical marker for validating compliance on the DST. J Clin Psychiatry 48:224–229
Krystal JH, Webb E, Kranzler HR, Cooney N, Heninger GR, Charney DS (1992) Yohimbine and MCPP effects in alcoholics and healthy controls. Alcohol Clin Exp Res 16:393
Lasagna L (1991) Commentary on "Improvement of phase III psychotropic drug trials by intensive phase II work". Neuropsychopharmacology 4:263–264
Lee MA, Meltzer HY (1991) Neuroendocrine responses to serotonergic agents in alcoholics. Biol Psychiatry 30:1017–1030
Lee ET (1992) Statistical methods for survival analysis. Wiley, New York
Lettieri DJ, Sayers MA, Nelson JE (eds) (1985) Summaries of alcoholism treatment assessment research. US Government Printing Office, Washington DC (DHHS pub. no. (ADM) 85-1379.)
Levine J, Ban TA (1987) Assessment methods in clinical trials. In: Meltzer HY (ed) Psychopharmacology: the third generation of progress. Raven, New York, pp 997–1003
Lindstrom L (1992) Managing alcoholism: matching clients to treatments. Oxford University Press, New York
Litt MD, Babor TF, DelBoca FK, Kadden RM, Conney N (1992) Types of alcoholics: II. Application of an empirically-derived typology to treatment matching. Arch Gen Psychiatry 8:609–614
Littleton JM (1984) The future could be bright. In: Edwards G, Littleton J (eds) Pharmacological treatments for alcoholism. Croom Helm, New York, pp 605–614
Marlatt GA, Gordon JR (eds) (1985) Relapse prevention: maintenance strategies in the treatment of addictive behaviors. Guilford, New York

Mason BJ, Kocsis JH (1991) Desipramine treatment of alcoholism. Psychopharmacol Bull 27:155–161
Mattila MJ, Aranko K, Seppala T (1982) Acute effects of buspirone and alcohol on psychomotor skills. J Clin Psychiatry 43:56–60
McKinney WT (1988) Models of mental disorders: a new comparative psychiatry. Plenum, New York
McLellan AT, Luborsky L, Woody GE, O'Brien, CP (1980) An improved diagnostic evaluation instrument for substance abuse patients: the addiction severity index. J Nerv Ment Dis 168:26–33
McLellan AT, Luborsky L, Woody GE, Druley KA, O'Brien CP (1983) Predicting response to alcohol and drug abuse treatments: role of psychiatric severity. Arch Gen Psychiatry 40:620–625
McLellan AT, Luborsky L, Cacciola J, Griffith JE (1985) New data from the Addiction Severity Index: reliability and validity in three centers. J Nerv Ment Dis 173:412–423
McLellan AT, Kushner H, Metzger D, Peters R, Smith I, Grissom G, Pettinati H, Argeriou M (1992) The fifth edition of the Addiction Severity Index. J Subst Abuse Treat 9:199–213
McLellan AT, Arndt LO, Woody GE, Metzger D (1993) Psychosocial services in substance abuse treatment?: a dose-ranging study of psychosocial services. JAMA 269:1953–1959
Meinert CL (1986) Controlled clinical trials. Oxford University Press, New York
Meyer RE (1986a) Anxiolytics and the alcoholic patient. J Stud Alcohol 47:269–273
Meyer RE (1986b) How to understand the relationship between psychopathology and addictive disorders: another example of the chicken and the egg. In: Meyer RE (ed) Psychopathology and addictive disorders. Guilford, New York
Meyer RE (1989) Prospects for a rational pharmacotherapy of alcoholism. J Clin Psychiatry 50:403–412
Meyer RE (1992) Some issues in the evaluation of a pharmacotherapy of alcoholism. In: Naranjo CA, Sellers EM (eds) Novel pharmacological interventions for alcoholism. Springer, New York, pp 40–55
Murray RM (1980) Why are the drug companies so disinterested in alcoholism? Br J Addict 75:113–115
Naranjo CA, Sellers EM, Roach CA, Woodley DV, Sanchez-Craig M, Sykora K (1984) Zimelidine-induced variations in alcohol intake by nondepressed heavy drinkers. Clin Pharmacol Ther 35:374–381
Naranjo CA, Sellers EM, Lawrin M (1986) Modulation of ethanol intake by serotonin uptake inhibitors. J Clin Psychiatry 47 [Suppl]:16–22
Naranjo CA, Sellers EM, Sullivan JT, Woodley DV, Kadlec K, Sykora K (1987) The serotonin uptake inhibitor citalopram attenuates ethanol intake. Clin Pharmacol Ther 41:266–274
Naranjo CA, Sullivan JT, Kadlec KE, Woodley-Remus DV, Kennedy G, Sellers EM (1989) Differential effects of viqualine on alcohol intake and other consummatory behaviors. Clin Pharmacol Ther 46:301–309
Naranjo CA, Kadlec KE, Sanhueza P, Woodley-Remus D, Sellers EM (1990) Fluoxetine differentially alters alcohol intake and other consummatory behaviors in problem drinkers. Clin Pharmacol Ther 47:490–498
Nunes EV, McGrath PJ, Quitkin FM, Stewart JP, Harrison W, Tricamo E, Ocepek-Welikson K (1993) Imipramine treatment of alcoholism with comorbid depression. Am J Psychiatry 6:963–965
O'Malley SS, Jaffe AJ, Chang G, Schottenfeld RS, Meyer RE, Rounsaville B (1992) Naltrexone and coping skills therapy for alcohol dependence: a controlled study. Arch Gen Psychiatry 49:894–898
Overall JE (1987) Introduction: methodology in psychopharmacology. In: Meltzer HY (ed) Psychopharmacology: the third generation of progress. Raven, New York, pp 995–996

Overall JE, Rhoades HM (1987) Adjusting *p* values for multiple tests of significance. In: Meltzer HY (ed) Psychopharmacology: the third generation of progress. Raven, New York, pp 1013–1018
Oxtoby A, Jones A, Robinson M (1989) Is your "double-blind" design truly double-blind? Br J Psychiatry 155:700–701
Pattison EM (1969) Evaluation of alcoholism treatment: a comparison of three facilities. Arch Gen Psychiatry 20:478–483
Pattison EM (1979) The selection of treatment modalities for the alcoholic patient. In: Mendelson JH, Mello NK (eds) The diagnosis and treatment of alcoholism. McGraw-Hill, New York, pp 229–255
Polich JM, Armor DJ, Braiker HB (1981) The course of alcoholism: four years after treatment. Wiley, New York
Prien R, Robinson D (eds) (1994) Clinical evaluation of psychotropic drugs: principles and guidelines. Raven, New York
Pullar T, Kumar S, Tindall H, Feely M (1989) Time to stop counting the tablets? Clin Pharmacol Ther 46:163–168
Robins LN, Helzer JE, Croughan H, Ratcliff KS (1981) National Institute of Mental Health Diagnostic Interview Schedule: its history, characteristics, and validity. Arch Gen Psychiatry 38:381–389
Robins LN, Wing N, Wittchen HU, Helzer JE, Babor TF, Burke J, Farmer A, Jablenski A, Pickens R, Regier DA, Sartorius N, Towle LH (1988) The Composite International Diagnostic Interview: an epidemiological instrument suitable for use in conjunction with different diagnostic systems and in different cultures. Arch Gen Psychiatry 45:1069–1077
Roth HP, Caron HS, Hsi BP (1970) Measuring intake of a prescribed medication: a bottle count and a tracer technique compared. Clin Pharmacol Ther 11:228–237
Rounsaville BJ, Kosten TR, Weissman MM, Kleber HD (1986) Prognostic significance of psychopathology in treated opiate addicts. Arch Gen Psychiatry 43: 739–745
Rounsaville BJ, Dolinsky ZS, Babor TF, Meyer RE (1987) Psychopathology as a predictor of treatment outcome in alcoholics. Arch Gen Psychiatry 44:505–513
Roy A, Virkkunen M, Linnoila M (1990) Serotonin in suicide, violence, and alcoholism. In: Coccaro EF, Murphy DL (eds) Serotonin in major psychiatric disorders. American Psychiatric Press, Washington DC
Rudd P, Marshall G, Taylor CB, Agras WS (1981) Medication monitor/dispenser for pharmaceutical and compliance research. Clin Pharmacol Ther 29:278
Rudd P, Byyny RL, Zachary V, LoVerde ME, Titus C, Mitchell WD, Marshall G (1989) The natural history of medication compliance in a drug trial: limitations of pill counts. Clin Pharmacol Ther 46:169–176
Russell ML (1984) Behavioral aspects of the use of medical markers in clinical trials. Controlled Clin Trials 5:526–534
Schuckit M (1979) Alcoholism and affective disorder: diagnostic confusion. In: Goodwin DW, Erickson C (eds) Alcoholism and affective disorders. Spectrum, New York, pp 9–19
Schuckit M (1983) Alcoholic patients with secondary depression. Am J Psychiatry 140:711–714
Schuckit MA, Irwin M, Mahler H (1990) Tridimensional Personality Questionnaire scores of sons of alcoholic and nonalcoholic fathers. Am J Psychiatry 147: 481–487
Schuckit MA (1994) The course of depression and anxiety symptoms in primary alcoholics during treatment. Paper presented at the annual meeting of the American Psychiatric Association, Philadelphia PA
Sellers EM, Sobell MB (1992) Medications for alcohol abuse and dependence: methodology for clinical studies. In: Naranjo CA, Sellers EM (eds) Novel pharmacological interventions for alcoholism. Springer, New York, pp 33–39

Sellers EM, Naranjo CA, Peachey JE (1981) Drugs to decrease alcohol consumption. N Engl J Med 305:1255–1262

Seppala T, Aranko K, Mattila MJ, Shrotriya RC (1982) Effects of alcohol on buspirone and lorazepam actions. Clin Pharmacol Ther 32:201–207

Simpson D, Savage L (1980) Drug abuse treatment readmissions and outcomes. Arch Gen Psychiatry 37:896–901

Skinner H, Allen BA (1982) Alcohol dependence syndrome: measurement and validation. J Abnorm Psychol 91:199–209

Sobell MB, Masito SA, Sobell LC, Cooper AM, Cooper TC, Sanders B (1980) Developing a prototype for evaluating alcohol treatment effectiveness. In: Sobell LC, Sobell MB, Ward E (eds) Evaluating alcohol and drug abuse treatment effectiveness: recent advances. Pergamon, New York

Spilker B (1991) Guide to clinical trials. Raven, New York

Spitzer RL, Williams JBW, Gibbon M, First MB (1992) The Structured Clinical Interview for DSM-III-R: I. History, rationale and description. Arch Gen Psychiatry 49:624–629

Spriet A, Simon P (1985) Methodology of clinical drug trials. Karger, Basel

Sullivan JT, Naranjo CA, Sellers EM (1989) Compliance among heavy alcohol users in clinical drug trials. J Subst Abuse 1:183–194

Taylor DP, Eison MS, Riblet LA, Vandermaelen CP (1985) Pharmacological and clinical effects of buspirone. Pharmacol Biochem Behav 23:687–694

Toneatto T, Romach MK, Sobell LC, Sobell MB, Somer GR, Sellers EM (1991) Ondansetron, a 5-HT_3 antagonist, reduces alcohol consumption in alcohol abusers. Alcohol Clin Exp Res 15:382

Volpicelli J, O'Brien C, Alterman A, Hayashida M (1992) Naltrexone in the treatment of alcohol dependence. Arch Gen Psychiatry 49:867–880

Wei LJ (1978) An application of an URN model to the design of sequential controlled clinical trials. J Am Stat Assoc 73:559–563

Wing JK, Babor T, Brugha T, Burke J, Cooper JE, Giel R, Jablenski A, Regier D, Sartorius N (1990) SCAN – Schedules for Clinical Assessment in Neuropsychiatry. Arch Gen Psychiatry 47:589–593

World Health Organization (1992) The ICD-10 classification of mental and behavioural disorders; clinical descriptions and diagnostic guidelines. World Health Organization, Geneva

Young LM, Haakenson CM, Lee KK, van Eeckout JP (1984) Riboflavin use as a drug marker in veterans administration cooperative studies. Controlled Clin Trials 5:497–504

Subject Index

Springer-Verlag and the Environment

We at Springer-Verlag firmly believe that an international science publisher has a special obligation to the environment, and our corporate policies consistently reflect this conviction.

We also expect our business partners – paper mills, printers, packaging manufacturers, etc. – to commit themselves to using environmentally friendly materials and production processes.

The paper in this book is made from low- or no-chlorine pulp and is acid free, in conformance with international standards for paper permanency.